Springer Series on Atomic, Optical, and Plasma Physics

Volume 104

The Springer Series on Atomic, Optical, and Plasma Physics covers in a comprehensive manner theory and experiment in the entire field of atoms and molecules and their interaction with electromagnetic radiation. Books in the series provide a rich source of new ideas and techniques with wide applications in fields such as chemistry, materials science, astrophysics, surface science, plasma technology, advanced optics, aeronomy, and engineering. Laser physics is a particular connecting theme that has provided much of the continuing impetus for new developments in the field, such as quantum computation and Bose-Einstein condensation. The purpose of the series is to cover the gap between standard undergraduate textbooks and the research literature with emphasis on the fundamental ideas, methods, techniques, and results in the field.

More information about this series at http://www.springer.com/series/411

Frank B. Rosmej · Valery A. Astapenko ·
Valery S. Lisitsa

Plasma Atomic Physics

Springer

Frank B. Rosmej
Faculty of Science and Engineering
Sorbonne University
Paris, France

Valery A. Astapenko
Moscow Institute of Physics and Technology
Moscow, Russia

Valery S. Lisitsa
National Research Center "Kurchatov Institute"
Moscow, Russia

ISSN 1615-5653 ISSN 2197-6791 (electronic)
Springer Series on Atomic, Optical, and Plasma Physics
ISBN 978-3-030-05966-8 ISBN 978-3-030-05968-2 (eBook)
https://doi.org/10.1007/978-3-030-05968-2

Library of Congress Control Number: 2018964256

This Springer imprint is published by the registered company Springer Nature Switzerland AG.
The registered company address is: Gewerbestrasse 11, 6330 Cham, Switzerland

Pour Geneviève

To the memory of
A. Ya. Faenov
and
L. A. Vainshtein

Preface

The emission of light is one of the most fascinating phenomena in nature. Everybody feels the beauty when looking at the colors appearing at sunset, when a bolt of lightning illuminates the night, or when the emission of the aurora moves like magic in the dark heaven. And since the discovery of the spectral analysis, no one doubted that the problems of describing atoms and matter would be solved once we had learned to understand the language of atomic spectra and the emission of light.

The book is devoted to the various aspects of light emission and the analysis of the radiative properties of matter and, in particular, the emission and absorption properties of atoms and ions in plasmas. It is based on lectures that we have given at the Sorbonne University in France and the Moscow Institute of Physics and Technology and the National Research Nuclear University in Russia.

The purpose of the presented material is to assist students and scientists investigating the complex atomic processes in different kinds of plasmas, by developing relatively simple but effective models. These models allow more generalized considerations and make it possible to extract also universal dependences (e.g., scaling laws) including both atomic and plasma parameters.

A characteristic feature of this book, therefore, is that, along with the presentation of strict quantum theories for the various electromagnetic and collisional processes, considerable attention is paid to a number of qualitative models that allow one to obtain an adequate comprehensive description, appealing more to physical intuition than to mathematical formalism. A distinctive feature of the approaches presented is the wide use of qualitative analogies, which makes it possible to transfer techniques and methods developed for particular processes to other phenomena that are important but have rarely been studied due to their complexity. There are a number of examples: the generalized quasi-classical Kramers approach for radiation transition probabilities, the Enrico Fermi method of equivalent photons as a unification of radiative and collisional processes, the local plasma frequency approximation (the so-called plasma atom) for multi-electron atomic processes, the Born–Compton model in the theory of ionization of an atom by electron impact, the quasi-classical methods for population kinetics of Rydberg atomic states and the very recent fascinating developments of quantum kinetics in dense plasmas. The advantage of these simplified models lies not only in the fact of a transparent presentation of the essential physical phenomena but also makes it possible to calculate atomic processes along with their necessary combination of a complex environment like plasma structure, transport, and turbulence.

The monograph also presents recent trends in atomic processes such as reduced population kinetics for the huge numbers of radiative–collisional transitions between autoionizing atomic states, the quantum mechanical interference effects in dense plasma atomic kinetics, hot electron effects, ionization potential depression in near-solid-density plasmas, description of exotic atomic states induced via the interaction of XFEL radiation with solid matter (warm dense matter, hollow ions, etc.) and the interaction of radiation with nanoparticles.

We believe that a description of both the general methods, the specific application and the presentation of a large number of experimental data, will be useful for wide readership including postgraduate, masters, and undergraduate students. The book will also be an important text and reference for teachers and scientists.

Paris, France
Moscow, Russia

Frank B. Rosmej
Valery A. Astapenko
Valery S. Lisitsa

Contents

About the Authors

Prof. Dr. Frank B. Rosmej is a Nationally Distinguished Professor at the Sorbonne University in Paris and is leading the research group "Atomic Physics in Dense Plasmas" at the LULI laboratory, Ecole Polytechnique, Palaiseau, France. He was born in 1962 in Germany; he obtained his Ph.D. 1991 and Habilitation thesis 1998 in Physics from the Ruhr University in Bochum, Germany. He received the Feodor Lynen award from the Alexander von Humboldt Foundation in Germany, the Eminent Scientist price from the RIKEN research center in Japan, the Kurchatov price from the National Research Center "Kurchatov Institute" in Russia, the Live Time Award in Atomic Physics from the "Int. Conf. on Atomic, Molecular, Nano and Optical Physics" and several pedagogic and research awards from the Sorbonne University; was invited Professor at the University of Tokyo, University of Nagoya, University of Osaka and the Research Center RIKEN in Japan, Scholar of the Stanford University and the University of Maryland in USA; had research stays at the Los Alamos National Laboratories, Lawrence Livermore Laboratories, Max Planck Institute for Quantum Optics. He was honored with professorships at the Excellence Universities "Moscow Institute of Physics and Technology MIPT" and the "National Research Nuclear University MEPhI" in Russia, was founding editor of "International Pedagogical Research Papers" and President of the National French Federation of "Fusion Science". His field of research covers experimental and theoretical atomic and plasmas

physics, quantum kinetics, non-equilibrium radiative properties, magnetic and inertial fusion science, the interaction of X-ray free electron lasers XFEL with dense matter, suprathermal electron generation in high energy density and ultrahigh-intensity lasers, plasma spectroscopy, and diagnostics. He is an author of 150 research papers and has held 70 invited talks at international conferences and workshops.
E-mail: frank.rosmej@sorbonne-universite.fr

Prof. Dr. Valery A. Astapenko is a Professor and principal researcher at the Moscow Institute of Physics and Technology MIPT in Dolgoprudny, Russia, and is a scientific expert of the Russian Ministry of Education Science. He was born in 1959 in Russia; he obtained his Ph.D. in 1985 and Habilitation thesis in 2000 from the Moscow Institute of Physics and Technology. He has been honored with the Kurchatov Prize, the Russian Federation Government Prize in the Field of Education and twice with the Moscow Government Prize for achievements in new technologies in education. He has been an invited Professor at the University Pierre and Marie Curie and Ecole Polytechnique in France. He is an author of 9 monographs and has published more than 170 research articles. His field of research is theoretical physics of photoprocesses in the field of ultrashort laser pulses and bichromatic radiation fields, interaction of radiation with matter, elementary processes in atoms and plasmas, coherent optical phenomena and polarization bremsstrahlung.
E-mail: astval@mail.ru

Prof. Dr. Valery S. Lisitsa heads the Laboratory of "Radiation Theory" at the National Research Center "Kurchatov Institute" in Moscow and is a Professor at the Moscow Institute of Physics and Technology MIPT in Dolgoprudny and the National Research Nuclear University "MEPhI" in Russia. He was born in 1945 in Russia; he obtained his Ph.D. in 1975 from the Moscow State University and his Habilitation thesis in 1979 from the Kurchatov Institute of Atomic Energy. He has been a member of the Scientific Council of Spectroscopy of the Russian Academy of Science and has been a leading research scientist of the Kurchatov Institute for more than 30 years. He has also been the deputy editor of the

International Journal of Experimental and Theoretical Physics JETP and a managing director of several research projects in the Kurchatov Institute. He was honored three times with the Kurchatov Prize and the best publication prize of the International Publishing Company "Nauka," and he has been an invited professor at the University of Bochum in Germany, the University Aix-Marseille, the University Pierre and Marie Curie, the Meudon Observatory in France and the National Research Center LHD in Japan. He is an author of 3 monographs and 20 scientific reviews and has published more than 200 research articles. His field of research is the theory of radiation of plasmas and gases including atomic spectral line broadening, interaction of intense laser radiation with matter, X-ray spectra from hot dense plasmas, elementary collisional–radiative processes in plasmas, radiation losses, nonlinear spectroscopy, and radiation transport by electromagnetic and plasma waves.

E-mail: vlisitsa@yandex.ru

Chapter 1
Introduction to Atomic Physics in Plasmas

Abstract This introductory chapter provides an overview on various elementary atomic physics processes and the general characteristics of radiation processes. The interrelation between radiation and matter is discussed in the framework of radiation trapping and the radiation transport theory. Simple, but efficient analytical approaches are presented to discuss the main effects of opacity and differential plasma motion on the spectral distribution. Particular attention is paid to autoionizing states together with corresponding Fano theory and dielectronic satellite emission. Novel properties like dielectronic satellite accumulation, transient three-body recombination, Stark broadening, nonlinear interference effects, and spatial properties are discussed along with many experimental data and spectroscopic applications. Finally, hollow ion X-ray emission of autoionizing states is presented that is still a mystery in high-current Z-pinch plasmas and dense laser-produced plasmas.

1.1 Atomic Physics and Plasma Physics

The plasma state (Chen 1984; Krall and Trivelpiece 1973; Schmidt 1979; Ichimaru 1973, 2004a, b; Zel'dovich and Raizer 2002) is characterized by a large degree of freedom for free and bound atomic particles. As a consequence, numerous atomic processes emerge (Griem 1964; Cowan 1981; Mihalas 1978; Sobelman et al. 1995; Sobelman 2006; Salzman 1998; Zel'dovich and Raizer 2002) that play an important role for the plasma evolution itself as well as for its diagnostics (Griem 1964, 1974, 1997; Fujimoto 2004; Kunze 2009; McWhirter 1965; Boiko et al. 1985; Michelis and Mattioli 1981; Mihalas 1978). Below, we will briefly summarize the main atomic processes in plasmas.

A characteristic feature of radiation processes is their interaction with matter and their eventual escape from the radiating matter, thereby contributing to an energy loss and the transmission of information (plasma diagnostics). These two features are of particular importance for astrophysical objects. The main mechanisms responsible for most of the radiation processes are related to the acceleration of charged particles (usually electrons) in external fields (electric and magnetic), whereas some of these

F. B. Rosmej et al., *Plasma Atomic Physics*, Springer Series on Atomic, Optical, and Plasma Physics 104, https://doi.org/10.1007/978-3-030-05968-2_1

processes are linked to the individual properties of the radiating particles itself. The various radiation–collisional processes can be summarized as follows:

(1) bremsstrahlung radiation (BrR) due to free–free radiative transitions of electrons in the fields of ions (or quadruple radiation in electron–electron collisions)

$$A^{+Z} + e \to A^{+Z} + e + \hbar\omega,$$

where A^{+Z} is the ion of type A with a charge Z;

(2) radiative (or photo) recombination (RR or PhR) due to free–bound transitions of electrons in the field of ions

$$A^{+Z} + e \to A^{+(Z-1)}(n) + \hbar\omega,$$

where $A^{+(Z-1)}(n)$ designates the captured electron into an atomic level with principal quantum number n;

(3) line radiation (LR) due to bound–bound transitions of electrons between upper and lower energy levels inside atoms or ions

$$A^*(n) \to A^*(n') + \hbar\omega.$$

*means an excited atomic state;

(4) polarization radiation (PR) due to free–free, free–bound, and bound–bound transitions of the electron with a virtual excitation of atomic cores

$$A^{+Z}(\gamma^*) + e \to A^{+Z}(\gamma) + e + \hbar\omega,$$
$$A^{+Z}(\gamma^*) + e \to A^{+(Z-1)}(\gamma, n) + \hbar\omega,$$
$$A^{+Z}(\gamma^*, m) + e \to A^{+Z}(\gamma, n) + \hbar\omega,$$

where γ^* designates the polarized core;

(5) cyclotron (or synchrotron) radiation (CR)

$$e + B \to e + B + N\hbar\omega,$$

where $N\hbar\omega$ denotes the emission of one or more quanta (N) of cyclotron radiation in a magnetic field B;

(6) radiation during dielectronic recombination (DR) that is a capture of an electron with simultaneous excitation of an atomic core leading to radiative stabilization of the core and the capture of the electron

$$A^{+Z}(\gamma) + e \to A^{+(Z-1)**}(\gamma', n) \to A^{+(Z-1)*}(\gamma, n) + \hbar\omega,$$

where $A^{+(Z-1)**}(\gamma', n)$ denotes a double-excited atomic state.

In plasmas, all the processes mentioned above are accompanied by a number of atomic–radiation interaction processes that are related to the radiation transport in a medium, namely:

(7) photoionization (PhI) and photoabsorption of radiation by atoms

$$A^{+Z} + \hbar\omega \to A^{+(Z+1)} + e,$$
$$A^{+Z}(n) + \hbar\omega \to A^{+Z}(m);$$

(8) scattering (both elastic and inelastic) of radiation on atomic systems

$$A^{+Z}(n) + \hbar\omega \to A^{+Z}(m) + \hbar\omega'.$$

Some of these processes are related to each other by the principle of detailed balance [where the direct and inverse flows of a system in thermodynamic equilibrium are exactly balanced (Reif 1965)]; e.g., photoabsorption is the inverse process of bound–bound spontaneous emission.

Together with the radiative processes, a broad variety of atomic collisional processes is involved in the calculation of the radiation emission:

(9) collisional ionization and excitation of atomic states

$$A^{+Z} + e \to A^{+(Z+1)} + 2e,$$
$$A^{*+Z}(n) + e \to A^{*+Z}(m) + e;$$

(10) charge exchange between neutral (or low ionized) atoms and highly charged ions

$$A + B^{+Z} \to A^{+} + B^{+Z-1}.$$

1.1.1 General Characteristics of Radiation Processes

Let us consider briefly a general picture for radiative–collisional processes in plasmas. An analysis of such processes must include three levels of description, that are connected with the three characteristics of radiation emission, namely: (1) radiation intensity driven by an elementary process; (2) spectral distribution of the emission, that is a distribution over frequencies ω of photons emitted in a unit volume of an optically thin plasma layer; and (3) the total radiation flow of a plasma system with accounting for reabsorption (numerous acts of emission and absorption) of the radiation in the volume (optically thick plasmas).

Mechanisms of plasma radiation emission are determined both by the individual properties of charged and neutral particles in the plasma itself and by its collective properties that are oscillation wave characteristics.

Plasma radiation connected with individual properties of particles includes the following types: spectral line radiation (LR) arising in electron radiative transitions in atoms or ions between two discrete energy levels (bound–bound transitions); radiative recombination (RR) radiation arising from radiative capture of a free electron on one atomic discrete energy level (free–bound transition); bremsstrahlung radiation (BrR) of an electron in an atomic potential (free–free transitions); cyclotron radiation (CR) of the electron during its rotation in a magnetic field with field strength B. All these types of radiation have the same classical roots, namely the acceleration $\left|\dot{\vec{\upsilon}}\right|^2 := w^2$ of the electron in external fields both electrical and magnetic ones. The total intensity of the radiation emission is determined by the value $I = 2e^2w^2/3c^3$ (e is the electron charge, and c is the velocity of light). A spectral distribution I_ω over frequencies ω is determined by the Fourier component $I_\omega = 2e^2w_\omega^2/3c^3$. A difference in the field type that is responsible for the electron acceleration results in sharp differences in both total intensity I and spectral intensity I_ω.

Thus for LR, $\omega_L = (E_2 - E_1)/\hbar$ (E_1, E_2 are level energies); for CR, $\omega_C = neB\sqrt{1 - \upsilon^2/c^2}/mc$ ($n = 1, 2, 3\ldots$, and m, υ are the electron mass and velocity); for BrR in the framework of classical electron motion, $\omega_{Br} \sim m\upsilon^3/Ze^2$ (Ze is the ion charge). If the rotation of the electron is periodical (as for LR and CR), the corresponding spectra are discrete; in the opposite case, they are continuous. The presence of jumps related to recombination on separate discrete atomic levels is typical for RR spectra. The discrete nature of spectra may be violated due to the broadening of separate spectral harmonics and lines leading to mutual overlapping. It is the case for highly excited (Rydberg) atomic spectral lines and high harmonics of CR ($n \gg 1$).

Plasma radiation emission related to collective motion is due to electron acceleration with phase correlated motion in the fields of plasma oscillations and the corresponding radiation may be coherent. Therefore, this radiation is connected with characteristics of plasma oscillations and can be considered as resonances in wave–particle, wave–wave, and wave–particle–wave interactions. This depends strongly on the equilibrium plasma conditions and its stability with respect to the excitation of specific waves. For stable plasmas close to thermodynamic equilibrium, the collective radiation is usually spontaneous and is determined by the plasma dielectric properties as well as boundary conditions. The main types of radiation under such conditions are:

(a) Cherenkov radiation of particles moving with a velocity $\vec{\upsilon}$ close to the phase velocity of electromagnetic waves (helicons); the resonance condition in the particle–wave system (Landau resonance) is written as $\omega = \vec{k} \cdot \vec{\upsilon}$ (ω is the frequency of electromagnetic wave, and k is its wave vector) and in a magnetic field as $\omega - l \cdot \omega_B = \vec{k} \cdot \vec{v}$ ($l = 0, 1, 2\ldots$, and $\omega_B^0 = eB/mc$ is the cyclotron frequency);

(b) Transition radiation is also determined by a particle–wave interaction and appears for transitions of a charged particle through boundaries with strong changes of electromagnetic wave properties (e.g., the plasma–vacuum boundary);

(c) Radiation determined by a transformation of longitudinal wave into transverse ones at a plasma boundary or inhomogeneities (linear wave–wave interaction). The radiation frequency in this case coincides with the frequency of the initial longitudinal wave (in the simplest case $\omega \approx \omega_{pe} = (4\pi n e^2/m)^{1/2}$, ω_{pe} is the plasma frequency and n the particle density);

(d) Radiation arising in nonlinear interaction between longitudinal and transverse waves; the conditions for such interaction are $\sum_i \omega_i = 0$, $\sum_i k_i = 0$, $i = 1, 2, 3\ldots$. For waves with relatively small amplitudes, the main effect is connected with the interaction of three waves. In the case of isotropic plasmas, the process results in radiation emission on frequencies $\omega \approx \omega_{pe}$ and $\omega = 2\omega_{pe}$. The emission at $\omega \approx \omega_{pe}$ is due to the "merging" of a Langmuir wave with a low-frequency plasma oscillation (e.g., ion sound wave), whereas the emission at the double frequency $\omega = 2\omega_{pe}$ is due to the merging of two Langmuir waves. The processes are due to a transformation of longitudinal waves into transverse ones via plasma fluctuations; electron bremsstrahlung takes place not in the field of a separate ion but in an electrical field of plasma density fluctuations (emission due to particle–wave–radiation interaction). The intensity of such transitions may increase by orders of magnitudes as compared with standard bremsstrahlung. This effect is responsible for the increased continuum radiation during solar flares. Intensities due to collective radiation emission mechanisms increase sharply in the presence of plasma instabilities. While usually induced radiation processes are observed, their corresponding intensities depend on the specific mechanisms of plasma instabilities.

1.1.2 Interrelation Between Radiation and Matter

The radiation–matter interrelation is typical for strongly radiating plasmas. In fact, from one side the radiation emission is due to particle accelerations and the spectrum is formed by their thermal motion, but from other side plasma radiative losses limit the matter temperature, that is in turn related to their motion. In a hot rarefied plasma the radiation processes influence strongly on the distribution of ions over ionization stages Z_i (ionization equilibrium) and for the fixed Z_i on the distribution over excited atomic states. These distributions together with the Maxwellian electron velocity distribution (which is easily established due to the fast collisions that dominate over the radiation) create the radiators responsible for LR, BrR, RR, and CR. Plasma particles in their turn influence on the shapes of radiating spectra (leading to spectral line broadening) as well as on the radiation transfer in media. The most important interrelation between plasma and radiation is for LR: here, the discrete nature of radiated spectra makes them sensitive to various broadening phenomena induced by electrons and ions, and the densities of radiating electrons on excited atomic energy levels are essentially determined by radiative processes of excitation and de-excitation.

The effect of the radiation emission on populations N_n of discrete atomic energy levels n is characterized by the parameter

$$\beta = N_e \langle \upsilon \sigma_{deex} \rangle \tau, \tag{1.1}$$

where σ_{deex} is the cross section of electron de-excitation, τ is the lifetime of the excited energy level due to the radiation emission, and $\langle \ldots \rangle$ denotes averaging over the electron velocity distribution. In the case where $\beta \gg 1$ (dense and cold plasmas), the de-excitation is dominated by collision processes driving the plasma to local thermodynamical equilibrium (LTE) where all level populations N_n are close to Boltzmann ones N^B. In the opposite case $\beta \ll 1$ (rarefied and hot plasmas), level populations are determined by radiation processes, so that every collisional excitation is accompanied by a radiative decay; this is the so-called coronal regime typical for the Sun's corona as well as for thermonuclear laboratory plasmas in the magnetic confinement scheme.

The main effects responsible for line broadening phenomena in plasmas are Doppler, Stark, and Zeeman effects. The thermal velocity distribution of the radiating particle results in a distribution of radiating frequencies of $\Delta\omega \sim \omega_0 \cdot \upsilon/c$ due to the Doppler effect.

A slowly varying electric field of ions E_i results in a static broadening where the spectral line shape is determined by the ion microfield distribution function $W(E_i)$ and the line width by the ion density N_i. The rapidly varying electron field results in an impact broadening where the spectral line shape has a dispersion (the Lorentz-type) shape $\Gamma_{im}/(\Delta\omega^2 + \Gamma_{im}^2)$ with a width Γ_{im} equal to the frequency of impact collisions. The non-homogeneous distribution of a magnetic field results also in broadening of CR lines which may exceed the Doppler broadening.

1.1.3 Radiative Emission and Volume Plasma Radiative Losses

The main characteristics of plasma radiation is the radiative emission $\eta(\omega)\, d\omega$ that is the energy radiated by unit volume of an optically thin (transparent) plasma in a unit time into a unit solid angle in a spectral frequency domain from ω to $\omega + d\omega$. The dependence of η on ω and temperature is rather specific for every mechanism of radiation emission, whereas the dependence on plasma density for a number of processes is of universal type; e.g., for CR, $\eta(\omega) \propto N_e$ (a radiation emission creates a continuous flow from every electron); for BrR and RR, $\eta(\omega) \propto N_e N_i$ (the radiation emission is a result of pair collisions of electrons and ions). For LR, the dependencies of $\eta(\omega)$ on N_e and N_i are incorporated as parameters into the expressions for the spectral line shapes. However, for the integral value $\int_0^\infty \eta(\omega)\,d\omega$ both dependencies mentioned may take place; namely in the limit of LTE ($\beta \gg 1$), the integral takes the form

$$\int \eta(\omega)d\omega \propto N^B/\tau, \tag{1.2}$$

whereas in the coronal limit ($\beta \ll 1$) one has

$$\int \eta(\omega)d\omega \propto N_e N_0, \tag{1.3}$$

where N_0 is the density of atoms (ions) in the ground state. For arbitrary β, we have

$$\int \eta(\omega)d\omega \propto N_e N_0/(1+\beta). \tag{1.4}$$

The dependence of radiative emission on other plasma parameters for BrR takes the form

$$\eta(\omega) \propto Z^2 T_e^{-1/2} \exp(-\hbar\omega/kT_e) g(T_e, \omega), \tag{1.5}$$

where Z is the nuclear charge, T_e is the electron temperature, k is the Boltzmann constant, and g is the Gaunt factor taking into account quantum effects [note that the Gaunt factor is defined as the ratio of a certain process – usually cross section – treated in the framework of the classical and quantum theory (Sobelman et al. 1995; Sobelman 2006; Griem 1964)], partially screenings of the nuclear charge by the electron core, etc., for CR for large values of T_e and n when the spectrum is already continuous,

$$\eta(\omega) \propto (B/T_e)^{1/4} \omega^{3/4} \exp\left[-\left(m^2 c^3 \omega/eBkT_e\right)^{1/2}\right], \tag{1.6}$$

where B is the magnetic field strengths. For LR $\eta(\omega) \propto \varphi(\omega)$, where the line shapes $\varphi(\omega)$ are determined by different broadening mechanisms.

Radiative losses (RL) of optically thin plasmas are determined by the value

$$Q = 4\pi \int_0^\infty \eta(\omega)\,d\omega. \tag{1.7}$$

For BrR, RR as well as for LR, the RL in the coronal regime (typical also for LR in optically transparent systems) are described by the universal formulae

$$Q = N_e N_i \langle \upsilon_e \sigma \hbar\omega \rangle \tag{1.8}$$

where σ is the cross section of the corresponding inelastic process (bremsstrahlung, radiative recombination, and collisional excitation), $\langle \ldots \rangle$ denotes averaging over velocities as well as the sum over ionization stages and energy levels. RL for these three mechanisms are usually expressed in the form of partial RL $q = Q/N_e N_Z$ (W cm^3), where N_Z is the density of the radiating highly charged ions.

Figure 1.1 shows the radiation loss of argon (Fournier et al. 1998) as a function of temperature in the low-density approximation (Corona model) from 2 eV until 20.000 eV. The figure shows the total radiation loss (solid black curve) as well as

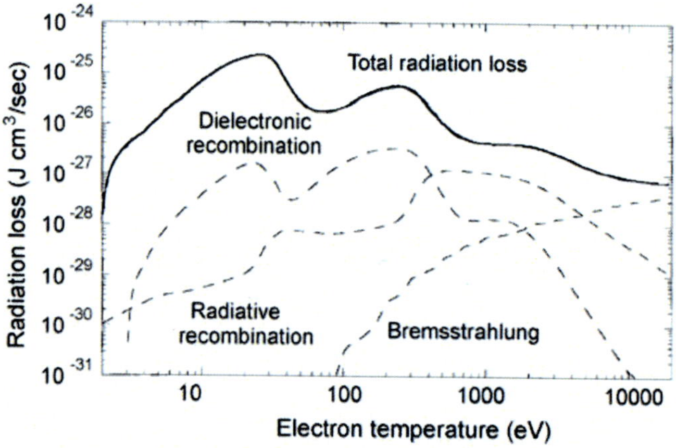

Fig. 1.1 Total and partial radiation losses of argon plasma in a low-density collisional–radiative approximation

Fig. 1.2 Xenon EUV emission from a hollow-cathode-triggered pinch plasma for applications in EUV lithography

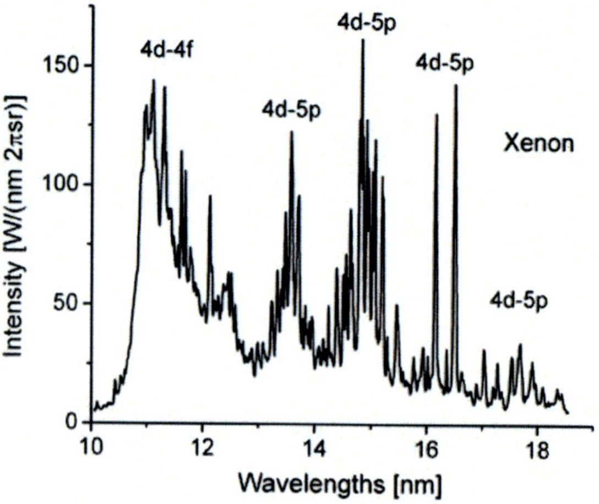

the contributions from dielectronic recombination (DR), radiative recombination (RR) and bremsstrahlung (BrR). It can clearly be seen that line radiation (LR) dominates in the whole temperature interval. Only for very large temperatures, the BrR dominates because the ionization is so high that bound states have negligible population. The peaks in total radiation losses near 20, 300, and 2000 eV correspond to the closed shell configuration of *M*-, *L*-, and *K*-shell.

Apart its fundamental interest in atomic physics, radiation emission of matter has multiple interests for applications: (a) the spectral distribution can be used for a unique characterization of the matter (spectroscopy); (b) it constitutes an important energy loss that in turn influences on the plasma evolution (hydrodynamics); and (c) radiation sources, e.g., for lithography (Krücken et al. 2004). Figure 1.2 shows the radiation emission of a Xenon plasma in the EVU range obtained from a hollow-cathode-triggered pinch plasma experiment designed to optimize the radiation emission for lithography (Vieker and Bergmann 2017). Temperatures are of the order of 50 eV, and electron densities are of the order of 10^{19} cm^{-3}. Relativistic Hartree–Fock calculations indicate that the bulk of emission is essentially due to *4d-4f* and *4d-5p* transitions (Fig. 1.2) of Xenon IX, X, XI, XII, and XIII.

1.1.4 Radiation Trapping and Plasma Radiation Losses in the General Case

To determine real intensities irradiated by different plasma objects, it is necessary to take into account possible absorption of the radiation inside the plasma itself resulting in the imprisonment of radiation when it does not escape from the whole plasma volume but rather from its external layers. Every radiating mechanism is accompanied by its inverse absorption mechanism characterized by an absorption coefficient $\kappa(\omega)$ at unity length. Under LTE conditions where the distribution of particles responsible for a specific mechanism of emission-absorption is at thermal equilibrium (it means for BrR and CR the Maxwellian distribution of electron velocities, for RR—the same plus distribution over ionization stages according to the Saha formula, for LR—Boltzmann distribution for population of excited atomic energy levels, i.e. $\beta \gg 1$) the coefficient is connected with the radiative emission $\eta(\omega)$ by the Kirchhoff law:

$$\eta(\omega)/\kappa(\omega) = B_{\rm P}(\omega), \tag{1.9}$$

where $B_{\rm P}(\omega)$ is the equilibrium radiation blackbody intensity [energy/time/area/solid angle/angular frequency] given by:

$$B_{\rm P}(\omega) = \frac{\hbar\omega^3}{4\pi^3 c^2} \frac{1}{\exp(\hbar\omega/kT) - 1}. \tag{1.10}$$

Correspondingly, the spectral intensity $I_\omega(L)$ of the radiation of an uniform plane layer with dimension L at thermal equilibrium is equal to

$$I_\omega(L) = B_P(\omega)\{1 - \exp[-\kappa(\omega)L]\}. \tag{1.11}$$

The integral intensity is given by

$$I(L) = \int_0^\infty I_\omega(L)\mathrm{d}\omega. \tag{1.12}$$

Let us define the frequency-dependent opacity according to

$$\tau_\omega(L) = \kappa(\omega) \cdot L \tag{1.13}$$

and the line center opacity according to

$$\tau_0(L) = \tau_{\omega=\omega_0}(L). \tag{1.14}$$

Equation (1.11) shows that if the opacity has reached values $\tau_\omega(L) > 5$, the observed spectral intensity approaches the Planck intensity. As the line center opacity (1.14) is always larger than the opacity for other frequencies, the approach to the Planck curve starts at frequencies $\omega = \omega_0$. This demonstrates that the realization of a Planck radiator requests not only LTE-atomic populations but also high opacity; i.e., if $\kappa(\omega)L \gg 1$ (the optically thick layer), one has

$$I_\omega(L) \approx B_P(\omega). \tag{1.15}$$

In this case, the plasma radiates as a blackbody from the surface. Surface radiation requests therefore a volume radiation under imprisonment. On the opposite hand, in spectral domains where $\kappa(\omega)L \ll 1$ (optically thin layer)

$$I_\omega(L) \approx \eta(\omega)L \tag{1.16}$$

that corresponds to the non-imprisonment volume radiation emission.

In the case of LR, a contribution of an imprisoned spectral line (reaching the universal Planck curve) with a central frequency ω_0 into the total radiation intensity $I(L)$ is given by

$$I(L) = B_P(\omega_0)\Delta\omega_{\mathrm{eq}}(L) \tag{1.17}$$

where $\Delta\omega_{\mathrm{eq}}$ is the so-called equivalent line width:

$$\Delta\omega_{\mathrm{eq}}(L) = \frac{\exp(\hbar\omega_0/kT)}{\omega_0^3} \cdot \int_0^\infty \omega^3 \cdot \frac{1 - \exp(-\tau_\omega(L))}{\exp(\hbar\omega/kT)} \cdot \mathrm{d}\omega. \tag{1.18}$$

For small opacities, the equivalent line width is smaller than the Full Width at Half Maximum (FWHM) of the original line shape because the emitted line intensity has not yet reached the Planck curve. In order to obtain the correct asymptotic limit also for the optically thin case, we obtain (V is the volume and F the surface)

$$L_{\text{limit}} = 4 \cdot \frac{V}{F}, \tag{1.19}$$

and i.e., for a cylindrical plasma with radius R, $L_{\text{limit}} = 2 \cdot R$. In the spectral domain where $\kappa(\omega = \omega_0)L \geq 10$, the intensity has dropped to the value α at frequency $\Delta\omega_{\text{eq}}/2$ from the central frequency, i.e.,

$$1 - \exp\left(-\kappa\left(\omega_{\text{eq}}/2\right)L\right) = \alpha. \tag{1.20}$$

If we define the equivalent width as the width where the Planck intensity has decreased by a factor of 2, i.e., $\alpha = 2^{-1}$ (according to the general convention in line shape theories to characterize the width of a spectral emission by the FWHM), we obtain for a Doppler spectral line profile

$$\varphi_{\text{D}}(\omega) = \frac{1}{\sqrt{\pi}\,\Gamma_{\text{D}}} \exp\left(-\frac{(\omega - \omega_0)^2}{(\Gamma_{\text{D}})^2}\right), \tag{1.21}$$

$$FWHM_{\text{D}} = 2\sqrt{\ln 2} \cdot \Gamma_{\text{D}}, \tag{1.22}$$

$$\Gamma_{\text{D}} = \omega_0 \sqrt{\frac{2\,k\,T_{\text{i}}}{M\,c^2}} \tag{1.23}$$

the following equivalent width (e = 2.71828…):

$$\Delta\omega_{\text{eq}} \approx 2 \cdot \sqrt{\ln 2} \cdot \Gamma_{\text{D}} \cdot \sqrt{\ln[\tau_0 + e]}. \tag{1.24}$$

For a Lorentz profile

$$\varphi_{\text{L}}(\omega) = \frac{\Gamma_{\text{L}}}{2\pi} \frac{1}{(\omega - \omega_0)^2 + (\Gamma_{\text{L}}/2)^2}, \tag{1.25}$$

$$FWHM_{\text{L}} = \Gamma_{\text{L}}, \tag{1.26}$$

$$\Gamma_{\text{L,ij}} = \sum_{\text{k}} W_{\text{ik}} + \sum_{\text{l}} W_{\text{jl}} \tag{1.27}$$

(where W_{ij} is the collisional–radiative population matrix) the equivalent line width is given by

$$\Delta\omega_{eq} \approx \sqrt{\ln^{-1}(2) - 1} \cdot \Gamma_{imp} \cdot \sqrt{\tau_0 + \ln(2) \cdot (1 - \ln^{-1}(2))}. \tag{1.28}$$

Γ_D, $\Gamma_L = \Gamma_{imp}$ are Doppler and impact widths, and $\kappa(\omega_0)L \gg 1$ is the optical thickness of the layer in the spectral line center. The equivalent line width can therefore be entirely expressed in terms of the line center optical thickness (1.14). The general expression for the equivalent line width (1.18) shows the following asymptotics for small optical thickness

$$\Delta\omega_{eq,D} \propto \tau_0, \quad \text{if } \tau_0 \ll 1, \tag{1.29}$$

$$\Delta\omega_{eq,imp} \propto \tau_0, \quad \text{if } \tau_0 \ll 1 \tag{1.30}$$

and the following asymptotes pour large optical thickness:

$$\Delta\omega_{eq,D} \propto \sqrt{\ln\,\tau_0}, \quad \text{if } \tau_0 \gg 1, \tag{1.31}$$

$$\Delta\omega_{eq,imp} \propto \sqrt{\tau_0}, \quad \text{if } \tau_0 \gg 1. \tag{1.32}$$

It is interesting to note that the asymptotes (1.29)–(1.32) as well as intermediate values deduced from the thermodynamical equilibrium situation are in good and partially excellent agreement with exact non-LTE radiation transport numerical calculations.

Due to these agreements, the concept of the equivalent line width permits likewise to define a useful escape factor for bound–bound transitions $i \to j$ according to

$$\Lambda_{ji} = \frac{\Delta\omega_{eq,ji} \cdot \varphi_{ji}(\omega_0)}{\tau_{0,ji}} \tag{1.33}$$

and an optically thick line shape according to

$$\Phi_{ji}(\omega) = \frac{\int_0^{\tau_\omega(z=L)} S_{\omega,ji} \exp(-\tau_{\omega,ij}) d\tau_{\omega,ij}}{\int_0^{\tau_\omega(z=L)} \int_0^\infty S_{\omega,ji} \exp(-\tau_{\omega,ij}) d\tau_{\omega,ij} d\omega}, \tag{1.34}$$

where the frequency-dependent opacity includes possible inhomogeneity

$$\tau_{\omega,ji}(z = L) = \int_0^L \kappa_{\omega,ji} dz = \int_0^L \frac{\pi^2 c^2}{\omega_{ji}^2} \frac{g_j}{g_i} A_{ji}\, n_i\, \varphi_{ji}(\omega) \left\{ 1 - \frac{n_j\, g_i}{n_i\, g_j} \right\} dz. \tag{1.35}$$

The expressions (1.33)–(1.35) are also very useful for non-LTE situations if single emission lines scan still be identified. Accordingly, the optically thick line emission [energy/time/area/solid angle/angular frequency] is given by

$$I(\omega) = \frac{1}{4\pi}\sum_{\mathrm{i,j}=1}^{N} \hbar\omega_{\mathrm{ji}}\, n_{\mathrm{j}}\, A_{\mathrm{ji}}\, \Lambda_{\mathrm{ji}}\, \Phi_{\mathrm{ji}}(\omega), \tag{1.36}$$

where ω_{ji} is the transition frequency of the bound–bound transition $j \rightarrow i$, $\Phi_{\mathrm{ji}}(\omega)$ the corresponding line profile, A_{ji} is the spontaneous transition rate, Λ_{ji} the corresponding escape factor, and n_{j} is the atomic population of the upper state j.

Equation (1.34) allows likewise to study the main effects on the line profile due to radiation transport effects, namely opacity broadening, non-homogenous density distribution, and differential plasma motion. Moreover, (1.34), (1.35) are extremely useful for rapid simulations when employing analytical expressions for Λ_{ji} (e.g., via the usual escape factor technique) and the spatial distribution $n_{\mathrm{j}}(z)$ because the asymptotic expressions and also intermediate values are in good and partially in excellent agreement with exact numerical calculation of the radiation transport equation. For example, with the help of the asymptotic expression for the equivalent line width at high opacities, we obtain the asymptotic expressions for the escape factor for $\tau_{0,\mathrm{ji}} \gg 1$:

$$\Lambda_{\mathrm{D,ji}} \propto \frac{\sqrt{\ln \tau_{0,\mathrm{ji}}}}{\tau_{0,\mathrm{ji}}}, \tag{1.37}$$

$$\Lambda_{\mathrm{L,ji}} \propto \frac{1}{\sqrt{\tau_{0,\mathrm{ji}}}}. \tag{1.38}$$

These asymptotic expressions are identical to the escape factors proposed by Holstein (1947), Irons (1979) that have been obtained from the numerical integration of the Biberman–Holstein transport equation and from the evaluation of the escape integral. For small optical thickness, i.e., $\tau_{0,\mathrm{ji}} \ll 1$ the escape factor is linear dependent on the line profile and given by

$$\Lambda_{\mathrm{D,ji}} \propto 1 - \frac{1}{2\sqrt{2}} \cdot \tau_{0,\mathrm{ji}}, \tag{1.39}$$

$$\Lambda_{\mathrm{L,ji}} \propto 1 - \frac{1}{4} \cdot \tau_{0,\mathrm{ji}}. \tag{1.40}$$

Of particular interest is a parabolic density distribution according to

$$n(z) = n_0 \left(1 - \left(\frac{L/2 - z}{L_{\mathrm{s}}}\right)^2\right) \tag{1.41}$$

with

$$s = \frac{1}{2}\frac{L}{L_s} \tag{1.42}$$

(where L_s is a scaling length and s is a dimensionless scaling length parameter) and a unique relative frequency shift according to

$$\Delta\omega = \omega_{ji}\frac{V_{rel}}{c} \tag{1.43}$$

of emission and absorption profiles (V_{rel} is the relative velocity of emitting and absorbing atoms, and c is the velocity of light), i.e.,

$$\varphi_{ij} = \varphi_{ij}\left(\omega_{ji} + \Delta\omega, \omega\right), \tag{1.44}$$

$$\varphi_{ji} = \varphi_{ji}\left(\omega_{ji}, \omega\right). \tag{1.45}$$

Equations (1.41)–(1.45) allow to study almost all principal radiation transport effects on the line profile and are particular well suited to provide fast simulations of optically thick radiation sources (e.g., Fig. 1.2). Moreover, when employing (1.41)–(1.45), the optically thick line profile (1.34) has an analytical solution that even can include line-overlapping effects:

$$\Phi_{ji}(\omega) = \frac{\tilde{\Phi}_{ji}(\omega)}{\int_0^\infty \tilde{\Phi}_{ji}(\omega)\mathrm{d}\omega}, \tag{1.46}$$

$$\tilde{\Phi}_{ji}(\omega) = \frac{\tilde{\tau}_{\omega,ji}}{\tau_\omega}\left\{1 - \exp(-\tau_\omega) + \left(\frac{L_{eff}}{L_s}\right)^2 K(\tau_\omega)\right\}, \tag{1.47}$$

$$K(\tau_\omega) = (1 - \exp(-\tau_\omega)) \cdot \left(\frac{1}{4} + \frac{1}{\tau_\omega^2}\right) + \frac{1}{\tau_\omega^2}(1 + \exp(-\tau_\omega)), \tag{1.48}$$

$$\tau_\omega = \sum_{ji} \kappa_{0,ij} L_{eff}\, \varphi_{ij}\left(\omega_{ij} + \Delta\omega, \omega, \alpha_{ij}\right), \tag{1.49}$$

$$\tilde{\tau}_{\omega,ji} = \kappa_{0,ij} L_{eff}\, \varphi_{ji}\left(\omega_{ji}, \omega, \alpha_{ji}\right), \tag{1.50}$$

$$\kappa_{0,ij} = \frac{\pi^2 c^2}{\omega_{ji}^2}\frac{g_j}{g_i} A_{ji}\, n_i \left\{1 - \frac{n_j}{n_i}\frac{g_i}{g_j}\right\}. \tag{1.51}$$

The function $K(\tau)$ describes inhomogeneity effects originating from the upper-level populations. As can be seen from (1.49), even line-overlapping effects can be included via the opacity τ_ω that originates from all possible line transitions.

Employing (1.46)–(1.48), even time-dependent simulations of the spectral distribution of differentially moving optically thick plasmas can be performed with reasonable numerical burden. We note that the remaining integral in (1.46) is just the sum for all frequency points of a line transition that has already been calculated for the spectral distribution of this transition.

Equation (1.47) shows that

$$\tilde{\Phi}_{\mathrm{ji}}(\omega) \rightarrow \frac{\varphi_{\mathrm{ji}}\left(\omega_{\mathrm{ji}}, \omega, \alpha_{\mathrm{ji}}\right)}{\varphi_{\mathrm{ij}}\left(\omega_{\mathrm{ji}} + \Delta\omega, \omega, \alpha_{\mathrm{ij}}\right)} \left\{1 - \exp\left(-\tau_{\omega,\mathrm{ji}}\right)\right\} \tag{1.52}$$

if the scaling lengths L_s is much larger than the effective photon path length L and line-overlapping effects are neglected. If also no differential plasma motion is encountered ($V_{\mathrm{rel}} = 0$, i.e., $\Delta\omega = 0$), we obtain the well-known result for a constant source function and complete frequency redistribution:

$$\tilde{\Phi}_{\mathrm{ji}}(\omega) \rightarrow \left\{1 - \exp\left(-\tau_{\omega,\mathrm{ji}}\right)\right\}. \tag{1.53}$$

Figure 1.3 shows the effects of non-homogenous plasma density employing the parabolic expression according to (1.41), (1.42) and the analytical expressions according to (1.46)–(1.48). For $s < 2$, strong self-reversal effects are seen.

The simulation of Fig. 1.3 has been carried out for the He-like resonance line of titanium at $\lambda_0 = 0.261$ nm, $kT_e = kT_i = 2000$ eV, $n_e = 10^{22}$ cm^{-3} and a line center

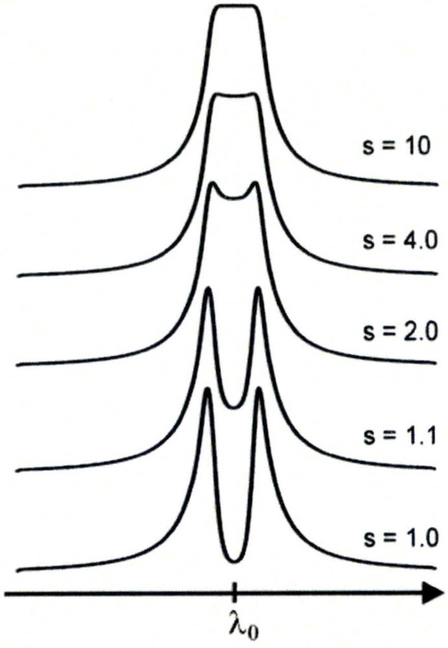

Fig. 1.3 Effects of an inhomogenous plasma on the optically thick line profile in dependence of the scaling parameters of (1.42)

optical thickness of $\tau_0 = 100$. A Voigt profile has been assumed as a local emission profile, and a convolution with an apparatus profile with $\lambda/\delta\lambda = 5000$ has been made. We note that a homogeneous plasma is represented by a scaling length parameter $s \to \infty$, maximal non-homogeneity, and strong self-reversal is obtained for $s = 1$.

Figure 1.4 shows the effects of differential plasma motion (other parameters are like in Fig. 1.3). The line profile is strongly asymmetric due to the relative frequency shift of emission and absorption profiles. An impressive impact of this differential plasma motion on the He-alpha resonance line of magnesium has been observed (Rosmej et al. 2000) in a high-contrast ultra-high-intensity laser-produced plasma experiment. A Nd-Glass laser with pulse duration of 400 fs, 1 J energy, wavelength 0.53 nm, focal spot diameter 10 μm, and intensity of 5×10^{18} W/cm^3 was brought to interaction with a solid magnesium target at normal incidence for different irradiation conditions. Figure 1.5 shows the spectral interval of He_α and satellites for a high contrast $1{:}10^{10}$–10^{11} experiment (solid blue curve in Fig. 1.5), and an experiment where a prepulse with energy of 0.03 J separated from the main pulse by 150 ps (solid black curve in Fig. 1.5) has been used. For the case of prepulse, the resonance line emission is dominating and the dielectronic satellite intensities are relatively small. With high contrast, however, the situation is dramatically different. A new type of spectra develops: Resonance lines seem to disappear and the dielectronic satellites become the most pronounced emission features in the spectrum. This phenomenon is known as dielectronic satellite accumulation DSA and will be discussed in detail below (Sect. 1.4.3). Also shown in Fig. 1.5, the simulation carried out with the MARIA code (Rosmej 1997, 2001,

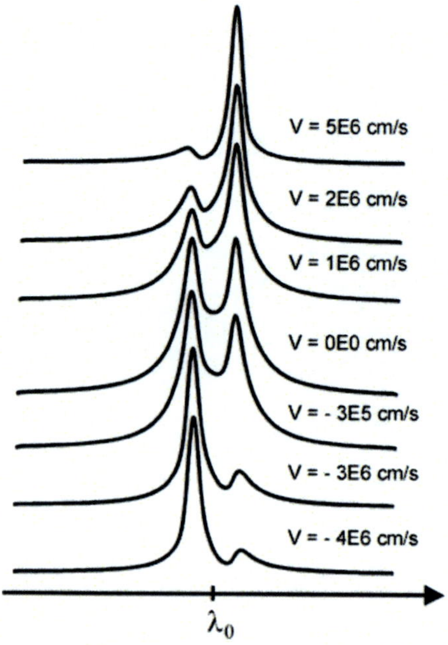

Fig. 1.4 Effects of differential plasma motion in inhomogeneous plasma on the optically thick line profile in dependence of differential plasma motion [velocities are defined according to (1.43)]

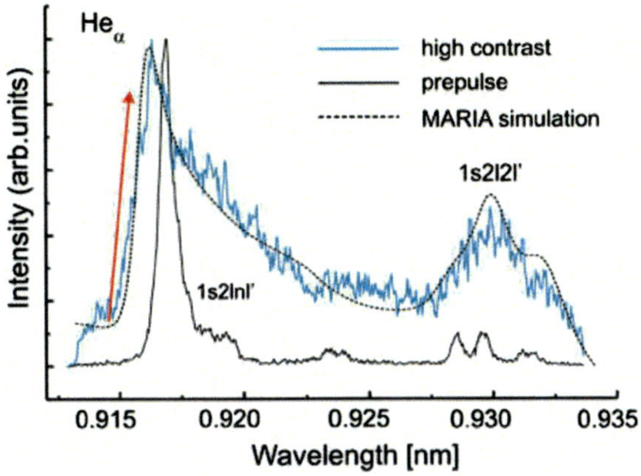

Fig. 1.5 *K*-shell X-ray emission of He_α and satellites of magnesium for different irradiation conditions: extremely high contrast 10^{10}–10^{11} and prepulse. The MARIA simulations indicate near-solid density $n_e = 4 \times 10^{23}$ cm^{-3}, $kT_e = 300$ eV and expansion velocities of the order of $V = 3 \times 10^7$ cm/s. The expansion velocity combined with radiation transport results in a characteristic steep rise of the He_α-intensity on the blue line wing (see red flash)

2006) indicates near-solid-density plasmas: electron density $n_e \approx 4 \times 10^{23}$ cm^{-3}, electron temperature $kT_e \approx 0.3$ keV, and expansion velocities of $V \approx 3 \times 10^7$ cm/s. The high density results in a high ion–ion coupling parameter ($\Gamma \approx 3$) and manifests itself in a strongly developed DSA, line broadening and line shift while the expansion velocity is responsible for the steep rise of He$_\alpha$ intensity on the blue wing (see red flash in Fig. 1.5) induced via asymmetric radiation transport in expanding plasmas as demonstrated in Fig. 1.4.

If two or more emission lines are very close to each other (means their spectral separation is less than about $\Delta\omega_{eq}$ from (1.18), the photon from one line might be strongly absorbed by another transition leading to "asymmetric re-pumping" and a corresponding strong distortion of line ratios (Rosmej et al. 1990). The inclusion of these effects in the escape factor approach is difficult. Figures 1.3, 1.4, and 1.5 demonstrate strong modifications of the line profiles in dense optically thick plasmas and subtle analysis of Stark broadening effects on the line profile appears to be difficult.

If free–free and bound–free absorption (the so-called continuum opacity) is important, (1.36) is no longer useful to approximate the spectral distribution because photons which are emitted from a spectral line may be redistributed with high probability elsewhere in the continuum. Therefore, photons from a well-defined bound–bound emission are lost for the particular line emission when they are reemitted far away from the line profile. Whether continuum absorption is important or not can be estimated from the following expressions:

$$\tau_{ff} \approx 2.4 \times 10^{-37} \bar{g}_{ff} \frac{n_e^2 Z_{eff}}{\sqrt{kT_e} E_\omega^3} (1 - \exp[-\hbar\omega/kT_e]) L_{eff}, \tag{1.54}$$

$$\tau_{fb} \approx 2.9 \times 10^{-17} \bar{g}_{fb} \frac{E_i^{5/2} n_i}{Z_{eff} E_\omega^3} (1 - \exp[-\hbar\omega/kT_e]) L_{eff}. \tag{1.55}$$

τ_{ff} is the free–free opacity, τ_{fb} is the bound–free opacity, E_ω is the photon energy in [eV], E_i is the ionization potential in [eV], kT_e is the electron temperature in [eV], n_e is the electron density in [cm^{-3}], n_i [cm^{-3}] is the population density from which photoionization proceeds, Z_{eff} is the effective charge of the plasma, L_{eff} is an effective photon path lengths in [cm], and $\bar{g}_{ff}$ and $\bar{g}_{fb}$ are the Gaunt factors for the free–free transitions (Bremsstrahlung) and the free–bound transitions (radiative recombination radiation). In order to operate in a meaningful manner with diagnostic line ratios, the continuum opacities τ_{ff} and τ_{fb} should be less than 1. Under quite many experimental conditions for dense laser-produced plasmas, the line center optical thickness of resonance lines is large, whereas the continuum opacities (1.54) and (1.55) are negligible. The line center opacity of a bound–bound transition can be estimated according to

$$\begin{aligned} \tau_{0,ij} &\approx \frac{2\pi^2 e^2 2\sqrt{\ln 2}}{m_e c \sqrt{\pi}\omega_{ji}} \frac{\omega_{ji}}{FWHM} f_{ij}\, n_i\, L_{eff} \left\{ 1 - e^{-\frac{\hbar\omega_{ji}}{kT_e}} \right\} \\ &= 8.32 \times 10^{-15} \left(\frac{\omega_{ji}}{FWHM} \right) f_{ij} \frac{n_i}{(m^{-3})} \frac{L_{eff}}{(m)} \left\{ 1 - e^{-\frac{\hbar\omega_{ji}}{kT_e}} \right\} \end{aligned} . \tag{1.56}$$

FWHM is the full width of the line profile at half maximum, ω_{ji} is the transition frequency, f_{ij} is the absorption oscillator strength, n_i is the density of the absorbing ground state "i" in [m^{-3}], and L_{eff} is a characteristic plasma dimension (often close to the value of L but not always) relevant for photoabsorption [m]. In the case of a Doppler line profile, the line center opacity is simply given by

$$\tau_{0,ij} \approx 1.08 \times^{-10} \frac{\lambda_{ji}}{(m)} \sqrt{\frac{M(amu)}{T_i(eV)}} f_{ij} \frac{n_i}{(m^{-3})} \frac{L_{eff}}{(m)} \left\{ 1 - e^{-\frac{\hbar\omega_{ji}}{kT_e}} \right\}. \tag{1.57}$$

τ_{ji} is the wavelength of the bound–bound transition [m], T_i is the ion temperature in [eV], and M is the atomic mass in [amu].

For the bremsstrahlung mechanism of emission–absorption, the typical imprisonment length L^* (that is a free path length averaged over frequencies) is equal to $L^* \approx 3 \times 10^{37} T^{7/2} \times (Z^2 N_i N_e)^{-1}$ (T is in eV, N_i and N_e are in cm^{-3}, and L^* is in cm).

The intensity of BrR escaping from a plasma layer in thermodynamic equilibrium with a thickness L is equal to

$$I(L) \propto Z^2 N_i N_e \sqrt{T} \cdot L \tag{1.58}$$

when $L \ll L^*$ (volume radiation); in the opposite case, $L \gg L^*$

$$I(L) \approx \sigma \cdot T^4 \tag{1.59}$$

(blackbody radiation, σ is the Stefan–Boltzmann constant). The last case is typical for astrophysical objects and especially stars. Here, the imprisoned radiation is responsible for the energy transfer from the hot star center to its more cold periphery. The value $T = T^*$ being a boundary between the volume and the surface RL is equal to $T^* \approx 2 \times 10^{-11} (Z^2 N_i N_e L)^{2/7}$. The dependence of T^* for star atmospheres $Z \approx 1$ is rather weak, so one may estimate approximately $T^*(\text{eV}) \approx 2 \times 10^{-11} (L N_i N_e)^{2/7}$.

The surface intensity of CR is determined by the Rayleigh–Jeans formula $T\omega^{*3}/12\pi^2 c^2$, where the value of effective frequency ω^* is determined by the optical thickness of the system $\kappa(\omega^*)L \approx 1$.

1.1.5 Excited Atoms Under Plasma Perturbations

An atom (or an ion) in a plasma is subject for various perturbations which lead to a definite distribution of the atom with regard to excited states as well as to changes of the atomic wave functions themselves. As a result of such perturbations, the atomic states turn out to be "mixed," strongly different from pure unperturbed atomic states. Consequently, the spectral characteristics of a radiation emission or absorption by the atom in media are also substantially different from the spectra of the unperturbed atom. The situation leads us to a conception of "a perturbed atom" that is an atom that is "dressed" by the media. The main consideration is devoted to an analysis of practically important situations when the atom is under simultaneous and combined action of perturbations of different natures: collisions, electrical and magnetic fields, oscillating and stochastic perturbation, blackbody radiation, etc. In this sense, our approach is quite different from the conventional consideration of atomic processes devoted to specific types of external influences on the atom with no regard to any other influence that could exist. The conception of the perturbed atom seems to be especially of interest for astrophysical and high-energy and high-intensity laser applications, where a variety of perturbations are present simultaneously.

The atomic state "mixing" mechanisms can easily be imagined considering a plasma as some external system of classically moving particles interacting with the atom (Griem 1964, 1974, 1997). If we consider plasmas from this point of view, the plasma action is composed of fast changing perturbations due to electrons on the one hand, and slow ones due to ions on the other hand. It is convenient to define these last perturbations by the electric microfields $F(t)$ caused by the ions. This defines a boundary for other types of electrical and magnetic fields in plasmas, such as plasma oscillation fields and fields of magnetic systems.

As far as the electron–atom interaction is concerned, it may be taken into account by the introduction of an additional atomic relaxation, described by the S-matrix. Thus, the plasma electrons induce non-elastic changes of an atomic electron from one sublevel to another. The action of an ion electric field F as well as other fields leads to periodic oscillations of the electron between different atomic states. Properly speaking, exactly these oscillations express the fact of atomic wave function "mixing" in an external field.

The usual approach to the atomic spectral description in plasmas is divided into two main problems (e.g., Griem 1974; Sobelman et al. 1995; Sobelman 2006): (1) the investigation of excited atomic state dynamics in external fields and (2) the investigation of atomic level population kinetics. For example, this approach is applied to the description of the Stark effect (Griem 1964, 1974; Sobelman et al. 1995; Sobelman 2006), where at first the true atomic states are defined (atomic dynamics), and then their populations and relaxation processes are analyzed. In reality, the values of the relaxation (kinetic) parameters have to be compared with the atomic interaction with external fields, and the whole character of the atomic state evolution is found to be more complicated. The situation is typical, for example, for highly charged ions in dense plasmas (Akhmedov et al. 1985; Anufrienko et al. 1993). Thus, it is more consistent to take into account both dynamical and kinetic parameters simultaneously.

This approach is used in laser physics (Rautian and Shalagin 1991; Klimontowitch 1983), where the atomic dynamics in external electromagnetic fields is described by kinetic equations, including relaxation parameters. Although we shall investigate dynamic and kinetic processes separately, it is necessary to consider a more general approach (than just used in laser physics (Rautian and Shalagin 1991)] for the description of atomic excited state evolution based on general kinetic equations (Blum 2012). The corresponding kinetic equation is written for the atomic density matrix: the diagonal elements determine the atomic state populations and the non-diagonal ones determine the polarization of the corresponding atomic transition (see also Chap. 7). Processes leading to atomic excitation as well as to atomic state mixing induce the polarization of media for the considered transition. The calculation of the polarization spectral distribution determines the characteristics of the radiation emitted or absorbed by atoms in plasmas. The main sources of atomic excitation in a plasma are electrons and light quanta. The electron action on the atom is characterized in the simplest cases by incoherent pumping of excited atomic states Q, being the rate of an atomic electron appearing at a given atomic level in a unit time and a unit volume. The excitation of atomic electrons by light quanta is of importance in optically dense plasmas (Burnett 1985) and requires more detailed information to investigate the spectral parameters. In fact, for excitation by incoherent pumping (electrons) the only spectral characteristics which define the media polarization is the line shape $I(\omega)$ of spontaneous radiation.

However, for excitation by a light quantum with frequency ω', the emitted radiation with frequency ω "remembers" the initial quantum ω' inducing a polarization for the considered transition. As a result, the polarization distribution induced by the quanta in the media is described by the more complicated frequency redistribution functions $R(\omega, \omega')$ determining the probability of the emission of a quantum with frequency ω when the absorption of the initial quantum with frequency ω' has taken place (Anufrienko et al. 1990; Demura et al. 1990; Burnett 1985; Talin and Klein 1982; Talin et al. 1997; Bulyshev et al. 1995; Mossé et al. 1999).

The calculation of spectral functions $I(\omega)$ and $R(\omega, \omega')$ in plasmas is a very complicated problem connected with both atomic state dynamics in external fields and the population kinetics of these very states.

To summarize, we may say that the atomic spectral formation in plasmas is a complex interference process in which act on a similar basis both relaxation and dynamical parameters of atomic interaction with a plasma. The state interference is essential for both the state population formation and the polarization of the medium. That is why the selection of purely dynamical effects in atomic spectral systems with subsequent relaxation parameters is a restrictive procedure.

The advantages of classical methods are connected with the universal nature of theoretical results based on them. It is just for astrophysical applications where it is desirable to have universal dependencies of atomic process probabilities on different parameters which make it possible to determine a correspondence with plasma parameters and to look for scaling parameters for the total problem.

Atomic states in plasmas are determined by a balance of radiative–collisional processes. One of them is determined by collisions of the atom with photons, another with electrons. However, the separation of these processes on pure collisional and radiative ones is also rather restrictive. From the point of view of classical mechanics, all these types of processes are determined by Fourier components of the electron motion in an atomic potential. The discrete or continuous nature of electron spectra is essential only under quantization conditions, determining an electron state density for both types of electron spectra. It will be shown below that for a broad domain of parameters, the dependencies of radiative–collisional processes on them are of pure classical nature and quantum parameters appear only after their substitution into classical formulae according to the correspondence principle. The most important parameter for radiative–collisional processes is the emitted frequency ω being a Fourier harmonic in a classical consideration and a discrete level energy difference in the quantum theory. Practically all universal dependencies of the transition probabilities are connected with the classical approach for a motion of external particles or atomic electrons in an atomic potential. We will demonstrate both these possibilities below.

1.2 Structure of Atoms and Ions

1.2.1 Symmetry Properties of the Coulomb Field

As is known (Huhges 1967; Landau and Lifschitz 2005), the Coulomb field ($U = -e/r$) possesses several symmetry properties. These properties involve together with the energy E and the orbital momentum $\vec{L}$ the presence of an additional integral of motion, the Runge–Lentz vector $\vec{A}$, which is equal to

$$\vec{A} = \vec{v} \times \vec{L} - e\vec{r}/r. \tag{1.60}$$

The physical meaning of vector $\vec{A}$ in classical mechanics follows from its connection with the coordinate $\langle \vec{r} \rangle$ averaged over the period of electron motion:

$$\vec{A} = -\frac{2e^2}{3a}\langle \vec{r} \rangle \quad (a = e^2/(2E)). \tag{1.61}$$

The quantum mechanical generalization of the Runge–Lentz vector is the operator

$$\hat{A} = \frac{1}{2m}(\hat{p} \times \hat{L} - \hat{L} \times \hat{p}) - e\vec{r}/r, \tag{1.62}$$

where $\hat{p}$, $\hat{L}$ and $\vec{r}$ are the operators of momentum, orbital momentum, and electron coordinates, respectively.

All components of the operator $\hat{A}$ commute with the Hamiltonian $\hat{H}$. The remainder of the commutation relations takes the form ($e_{\rm ijk}$ is an antisymmetrical tensor)

$$[L_{\rm i}, L_{\rm j}] = ie_{\rm ijk}L_{\rm k}; \quad [L_{\rm i}, A_{\rm j}] = ie_{\rm ijk}A_{\rm k}; \quad [A_{\rm i}, A_{\rm j}] = -2im^{-1}e_{\rm ijk} - L_{\rm k}\hat{H}. \tag{1.63}$$

Operators $\hat{L}$ and $\hat{A}$ are mutually perpendicular:

$$\hat{L}\hat{A} = \hat{A}\hat{L} = 0. \tag{1.64}$$

They satisfy the relation

$$\hat{A}^2 = \frac{2}{m}\hat{H}(\hat{L}^2 + 1) = e^2. \tag{1.65}$$

It is seen from the relations (1.63)–(1.65) that one may construct from operators $\hat{A}$ and $\hat{L}$ new vectors satisfying the commutation relations quite analogous for orbital momenta if one limits oneself to the subspace of atomic states belonging to a fixed value of the energy E (the isoenergetic surface $H = E$). Introducing new operators

$$\hat{J}_{1,2} = \frac{1}{2}\left[\hat{L} \pm \left(\frac{m}{a}\right)^{1/2}\hat{A}\right] \tag{1.66}$$

we can see that they satisfy the usual commutation relations for an orbital momentum:

$$[J_{1i}, J_{1j}] = ie_{ijk}J_{1k}; \quad [J_{2i}, J_{2j}] = ie_{ijk}J_{2k}. \tag{1.67}$$

Operators $\hat{J}_1$ and $\hat{J}_2$ commutate with each other. Thus, the subspace of atomic states belonging to the given isoenergetic surface $E = \text{const}$ possesses some transformation symmetry properties which may be described by the rotation of the operators $\hat{J}_{1,2}$ in two independent subspaces.

Equation (1.64) results in the equality of the squares of the vectors and their commutative properties which allow to achieve their simultaneous diagonalization:

$$\hat{J}_1^2 = \hat{J}_2^2 = j(j+1) \tag{1.68}$$

where the momentum magnitude of j defined from Landau and Lifschitz (2005) is equal to $(n-1)/2$. Correspondingly, the projections $m_{1,2}$ of the vectors vary within the limits:

$$m_{1,2} = -j, -j+1, \ldots, +j; j = \frac{n-1}{2}. \tag{1.69}$$

The presence of the two commutating operators $\hat{J}_1$ and $\hat{J}_2$ enables the construction of different states from wave functions with given n corresponding to definite magnitudes of the vector projections on different directions. The vectors $\hat{J}_1$ and $\hat{J}_2$ are independent and the directions of the vector quantization may not coincide. It is also possible, just as in angular momentum theory, to construct states corresponding to the squares of the vectors $\hat{J}_1$ and $\hat{J}_2$ and one of their projections. The convenience of using this or another representation becomes apparent in the presence of the diagonalization of perturbations possessing different symmetry. Essentially, the aforementioned method of states construction is an extension of the well-known method of the variable separation for a Coulomb field in spherical and parabolic coordinates.

Let us briefly dwell on the connection of parabolic and spherical quantization using the formalism mentioned above. As is well known (Landau and Lifschitz 2005; Bethe and Salpeter 1977), parabolic wave functions correspond to definite projections on one and the same extracted direction of the orbital momentum L defined by the quantum number m and the dipole momentum d, or the same Runge–Lentz vector A defined by the difference of the parabolic quantum numbers $n_2 - n_1$.

Taking into account the definition (1.66) of operators $\hat{J}_{1,2}$, it is easy to ascertain that the following relations hold true:

$$(J_1)_Z = m_1 = \frac{m + n_2 - n_1}{2}; \quad (J_2)_Z = m_2 = \frac{m + n_1 - n_2}{2};$$
$$L_Z = m; \quad (m/a)^{1/2} \cdot A_Z = n_2 - n_1. \tag{1.70}$$

Thus, the parabolic state corresponds to a definite z-projection of operators $\hat{J}_1$ and $\hat{J}_2$ (along with vectors $\hat{L}$ and $\hat{A}$) in a given direction.

It is also known that the spherical states ϕ_{nlm} correspond to definite magnitudes of $\hat{L}^2 = (\hat{J}_1 + \hat{J}_2)^2$, and the projection of L_Z equals m. Since the operators $\hat{J}_1$ and $\hat{J}_2$ obey the usual rules of the momentum composition, one can easily establish a connection between the wave function in parabolic $(\hat{J}_1^2, \hat{J}_{1Z}, \hat{J}_2^2, \hat{J}_{2Z})$ and spherical $(\hat{L}^2 = (\hat{J}_1 + \hat{J}_2)^2, \hat{J}_1^2, \hat{J}_2^2, \hat{L}_Z)$ coordinates, or bases. Using the standard results of the orbital momentum theory for the calculation of the transformation coefficient

$$\varphi_{\mathrm{n_1 n_2 m}} = \sum_{\mathrm{l,m'}} \langle n_1 n_2 m | n l m' \rangle \varphi_{\mathrm{nlm'}} \tag{1.71}$$

we arrive at the relation (Berestetskii et al. 2008; Sobelman 2006)

$$\langle n_1 n_2 m | n l m' \rangle = (-1)^{n-1+(2n_2 \pm |m| - m)/2} \times C\left[\frac{n-1}{2}, \frac{n-1}{2}, l; \frac{m + n_2 - n_1}{2}, \frac{m + n_1 - n_2}{2}, m'\right] \tag{1.72}$$

where $C[j_1, j_2, j; m_1, m_2, m]$ are the usual Clebsch–Gordan coefficients. The result (1.71), (1.72) is a specific case of the application of the Coulomb field symmetry. As has been discussed, it is possible to construct the wave functions of a more complicated symmetry, corresponding to the projections of the operators $\hat{J}_1$ and $\hat{J}_2$ on different directions.

The symmetry properties of the Coulomb field make it possible to construct coherent states for a hydrogen atom. These atomic states correspond to quantum states in the form of a wave packet with minimum uncertainty between canonically conjugate variables. Such wave packets make a transition to a classical motion along specific electron trajectories. To determine coherent states of an electron in a Coulomb field, one must determine the following variables, namely: the angular momentum of the electron motion, the eccentricity of the elliptic electron orbit, and its orientation in the space.

To do this, let us use the four-dimensional representation of the electron motion in terms of the operators $\hat{J}_{1,2}$ introduced above. According to the representation, the electron orbit orientation in the space can be determined by independent rotations of these vectors according to independent transformation rules for angular momenta. So one can introduce the following angular variables to determine the electron position in the space: θ is an angle of rotation along a classical trajectory with the standard classical angular velocity (or inverse rotation period), α and β are angles, determining the position of the operator $\hat{J}_1$ in the space, and γ and δ are similar angles for the operator $\hat{J}_2$. The eccentricity of the orbit is determined by the modulus of the Runge–Lentz vector being a combination of vectors $\hat{J}_{1,2}$, and the same is true for the electron angular momentum. The total energy E is determined by the action variable $I_1 \equiv R = n\hbar$, and the last is the well-known Bohr quantization condition.

Therefore, the evolution of an arbitrary atomic state can be constructed from the state $|jj\rangle$ which corresponds to the circular orbit by its rotation at the proper angles with the help of the operators $\hat{J}_{1,2}$. Accounting for the relation $j = (n-1)/2$ [see (1.69)], one can represent the total evolution as a sum of rotations with different values of j. As a result, one obtains:

$$|R, \alpha, \beta, \gamma, \delta, \theta\rangle = \sum_{\mathrm{j}} C_{\mathrm{j}}(R) \exp[-i\alpha J_1] \exp[-i\beta J_1] \times \exp[-i\gamma J_2] \exp[-i\delta J_2] \exp\left(\frac{-iR^3\theta}{2n^2\hbar^3}\right) |jj\rangle |jj\rangle \tag{1.73}$$

The constants $C_{\mathrm{j}}(R)$ are determined by normalization conditions and are equal to

$$C_{\mathrm{j}}(R) = \exp(-R/2\hbar) \frac{(R/\hbar)^j}{\sqrt{(2j)!}}. \tag{1.74}$$

Equations (1.73) and (1.74) determine a coherent state of hydrogen atoms in the $\hat{J}_{1,2}$ representation. Making a transformation to the spherical coordinate representation with the help of relations (1.66), (1.70)–(1.72), it is possible to obtain a coherent atomic state in the spherical basis.

The classical phase space measure in the variables (1.73) is $\mathrm{d}R\ \mathrm{d}\theta\ d((j+1/2)\hbar\cos\beta)\ \mathrm{d}\alpha\ d((j+1/2)\hbar\cos\delta)\ \mathrm{d}\gamma$. For large j, the state $|R, \alpha, \beta, \gamma, \delta, \theta\rangle$ gets its dominant contribution from $m_1 = (j+1/2)\cos\beta$ and $m_2 = (j+1/2)\cos\delta$. This corresponds just to the Bohr quantization of the action angles because $\cos\beta = 2J_{1z}/R$ and $\cos\delta = 2J_{2z}/R$.

The correspondence principle proposes that the expectation values of an operator X will be related to the corresponding classical variable $X(p,q)$ for a specific classical orbit.

1.2.2 Allowed and Forbidden Transitions

The main difference between the multielectron atomic spectra and the hydrogenic spectra is connected with the difference of the potential $V(r)$ of the atom from the Coulomb potential. As a consequence of this circumstance, "accidental" level degeneration is removed over orbital quantum numbers l, so that the transitions from sublevels corresponding to different values of l become spectroscopically distinguishable. The corresponding selection rules for individual transitions $l \longrightarrow l'$ become simple as compared with those for the hydrogen atom with two bases of quantization, namely parabolic and spherical (Bethe and Salpeter 1977; Landau and Lifschitz 2005). As expected from the selection rules, the transitions are divided into dipolar allowed and forbidden transitions. For example, $4d-2p$, $4f-2p$ transitions in the helium atom, the first one is allowed and the second one is forbidden.

It is known that non-hydrogenic atoms under the influence of an electrical field demonstrate an energy shift proportional to the square of the field strength (the quadratic Stark effect). The theory of the quadratic Stark effect is a typical example of the second-order perturbation theory (Bethe and Salpeter 1977). In fact, in the absence of accidental degeneration over the orbital quantum number l, the diagonal matrix elements of the perturbation $V = -\mathrm{d}F$ turn to zero and the energy shift ΔE arises only in the second-order perturbation theory:

$$\Delta E_{\mathrm{n}}^{(2)} = \sum \frac{V_{\mathrm{nm}} V_{\mathrm{mn}}}{\omega_{\mathrm{nm}}}. \tag{1.75}$$

The sum in (1.75) covers all intermediate states; however, the main contribution to the sum is made by nearby atomic levels. For example, in the helium atom these levels correspond to a change in the orbital momentum quantum number at a value of ± 1. The calculation of the matrix elements $V_{\mathrm{l}l'}$ between the states mentioned gives the following result (Bethe and Salpeter 1977):

$$\begin{aligned} \Delta E_{\mathrm{nlm}}^{(2)} &= C_4(nlm) F^2, C_4(nlm) = \frac{9}{4} n^2/(Z-1)^2(2l+1) \\ &\times \{[n^2-(l+1)^2][(l+1)^2-m^2]/(2l+3)(E_{\mathrm{nl}}-E_{\mathrm{nl+1}}) \\ &+ (n^2-l^2)(l^2-m^2)/(2l-1)(E_{\mathrm{nl}}-E_{\mathrm{nl-1}})\}. \end{aligned} \tag{1.76}$$

According to (1.76), the energy shift depends on "spherical" quantum numbers n, l, m, the dependency on n being very strong (which represents the increase of the polarizability of the excited states).

It should be noted that when applying a field F, there may be changes not only in the energy but also in the wave functions of the states. These changes are connected with the "mixing" of nearby atomic states that leads to the violation of the selection rules for dipolar radiative transitions. Let us estimate the ratio R of the intensity $I_{\mathrm{F}}(nl - n_0 l)$ of the forbidden atomic line to the intensity $I_{\mathrm{A}}(nl - n_0 l + 1)$ of the allowed atomic line in helium (Bethe and Salpeter 1977):

$$\begin{aligned} R(nlm) &= \frac{I_{\mathrm{F}}(nl \rightarrow n_0 l)}{I_{\mathrm{A}}(nl \rightarrow n_0 l+1)} = \frac{9}{4} F^2 [n^2-(l+1)^2] \\ &\times n^2[(l+1)^2-m^2]/[4(l+1)^2-1](E_{\mathrm{n,l+1}}-E_{\mathrm{nl}})^2. \end{aligned} \tag{1.77}$$

The spectral lines corresponding to the transitions between the excited helium levels lie in the visible spectral range, and therefore, they are convenient for electric field diagnostics in plasmas. As the electric field strength F increases (or with the transition to hydrogen-like highly excited atomic states), the quadratic Stark effect makes a transition to the linear effect. The transition is often observed in sufficiently dense plasmas. It is a typical situation when for the same transitions $n \rightarrow n_0$ some atomic states experience a linear Stark effect, whereas other states are subject to the quadratic Stark effect.

The energy spectrum of a multielectron atom is the spectrum of only one of its electrons. The rest of the electrons form an atomic (or ionic) core. The core is essentially an additional (third) body having influence on the interaction of an excited electron with the nuclear. The influence consists of the possibility of the core excitation leading to the appearance of a new type of transition, connected with the excitation of two (or more) atomic electrons. Examples of such transitions are considered below.

One of the important peculiarities of radiative transitions of an atomic electron in external shells of a neutral atom is the smallness of relativistic effects due to the smallness of the parameter $e^2/\hbar c \ll 1$. Therefore, the transitions of higher multipole order (e.g., quadrupole transitions as well as magnetic dipolar transitions) are strongly suppressed (forbidden) as compared to electric dipole transitions. This is the difference of the spectra mentioned with the spectra of highly charged ions. According to the order of magnitude estimate, the relative magnitudes of electric dipole ($E1$), magnetic dipole ($M1$), and magnetic quadrupole ($M2$) transition probabilities are given by Cowan (1981) $A_{E1}: A_{M1}: A_{E2} = 1: (\alpha Z)^2: (\alpha Z)^2$, where $\alpha = 1/137$ is the fine structure constant and Z is the effective nuclear charge.

1.2.3 Properties of Highly Charged Ion Spectra

A vast amount of the literature (Safronova and Senashenko 1983; Janev et al. 1985; Shevelko and Vainshtein 1993; Drake 2006) is devoted to calculations of highly charged ions (HCI) spectra. Here, we are interested in the general properties of these spectra for large nuclear charges $Z \gg 1$, which will be used later for the estimation of different effects. If the number of bound electrons is small compared with Z, then their motion is determined mostly by their interaction with the nucleus $V_{ie} \sim Ze^2/r$. Taking into account that $r \sim n^2/Z$ for a typical bound energy I_n of the electron on the atomic level n, one obtains the hydrogen-like formula: $I_n \sim Z^2/2n^2$. The difference of the transition energies $\Delta E_{nn'}$ is obviously given by the Rydberg formula [atomic units (a.u.)]:

$$\Delta E_{nn'} = \frac{Z^2}{2}\left(\frac{1}{n^2} - \frac{1}{n'^2}\right). \tag{1.78}$$

The interaction of electrons with each other $V_{ee} \sim r^{-1}$ leads to the splitting of energy sublevels $\sim Z$ for the given principal quantum number n (transitions with $\Delta n \equiv n - n' = 0$). The expansion may be continued and we obtain a series over degrees of Z^{-1} (Safronova and Senashenko 1983; Janev et al. 1985; Shevelko and Vainshtein 1993; Drake 2006). For example, for transitions in *Li*-like ions, the following relationship is obtained (Safronova and Senashenko 1983):

$$\Delta E_{2s-2p} = 0.0707Z - 0.120. \tag{1.79}$$

Analogous expansions may be obtained for transition oscillator strengths $f = 2\omega_{nn'}|d_{nn'}|^2$. For the same transitions with $\Delta n = 0$ in *Li*-like ions, we obtain (Safronova 1983) (with an account of $\omega \propto Z$, $d \propto Z^{-1}\ldots$):

$$f_{2s-2p} \simeq 1.35Z^{-1} + 2.20Z^{-2}. \tag{1.80}$$

The dependence on Z of the radiation transition width γ_r follows from the formula $\gamma_r = \frac{2}{3}\omega^3 d_{nn'}^2 c^{-3}$ that leads to

$$\gamma_r = Z^4 \gamma_{r_0}, \tag{1.80}$$

where γ_{r_0} is the radiation transition width in the hydrogen.

In the same manner, atomic level's fine structure splitting Δ_{fs} and the Lamb shift Δ_L increases. For hydrogenic ions, we have (Bethe and Salpeter 1977) (j being the total angular momentum):

$$\Delta_{fs} = -\frac{Z^4}{2n^3c^2} 1/(j+1/2), \quad \Delta_L(l=0) = \frac{4Z^4}{3\pi n^3 c^3} \ln\frac{1}{Z\alpha}. \tag{1.81}$$

Detailed calculations of the Lamb shift are performed by Erickson (1971), Drake (2006). The frequencies of the transitions between the fine structure sublevels extend into the visible and even the ultraviolet spectral domains for values $Z \geq 20$.

For non-hydrogenic ions, the spin–orbit interaction must be taken into account along with the interaction of the excited electron with a core, and the former may have the same order of magnitude as the latter one. For hydrogen-like ions with $Z \gg 1$, fine structure splitting leads to the partial removal of the degeneracy inside the given atomic level, splitting it into separate components with different values of total angular momentum j. The splitting may be large enough so that in case of an electric field interaction, some components experience a quadratic Stark effect, whereas the others follow the linear one. In principle, the calculations of the Stark effect for H-like ions do not differ in any way from corresponding calculations for neutral hydrogen (Zhidkoy et al. 1986; Lüders 1950; Ahmedov et al. 1985). However, the electric field strengths that cause the analogous splitting in a H-like ion are increased Z^4 times as compared with the neutral atom in accordance with the increase of fine structure splitting.

The energy-level behavior follows the corresponding results of Lüders (1950), Bethe and Salpeter (1977) for neutral hydrogen; however, the magnitudes of the field strengths are sharply increased. It is evident that for one and the same magnitude of field strength *F*, some components (e.g., $2p_{3/2}$ at the level $n = 2$) possess quadratic and others ($2p_{1/2}$, $2s_{1/2}$)-linear splitting over the field. In general, the case for intermediate values of *F*, the terms' behavior, depends on *F* and it is rather complicated to obtain analytical expressions.

The interaction *V* of the atomic dipolar momentum *d* with a field *F* in plasmas depends weakly on *Z*. In fact, the magnitude *d* decreases with increasing Z ($d \propto Z^{-1}$), whereas the field *F* created by the ion with charge *Z* at distance *R* increases with *Z* ($F \propto ZR^{-2}$). Altogether to observe a Stark structure of highly charged ions, it is necessary to have sufficiently high densities to provide large field strengths *F* at an average interparticle distance, to assure a splitting that may contain a fine structure.

The action of the electric field *F* on H-like ions leads, as in the case of helium, to a removal of interdictions for atomic transitions. In particular, the effects of atomic state "mixing" cause destruction of the metastable 2*s*-level in the field *F*. Indeed the squared amplitude due to changing of its lifetime has the form (Landau and Lifschitz 2005; Bethe and Salpeter 1977):

$$|a_s(t)|^2 \sim e^{-\mathrm{Im}(E_S(F)\cdot t)}, \tag{1.82a}$$

$$Im\{E_s(F)\} = \frac{\gamma}{2}\left(1 - \frac{\Delta E_{sp}}{\sqrt{\Delta E_{sp}^2 + 4d_{sp}^2 F^2}}\right), \tag{1.82b}$$

where γ is the decay rate of the $2p_{1/2}$ level and ΔE_{sp} is the Lamb shift dividing the metastable state ($2s_{1/2}$) and radiating state ($2p_{1/2}$). Given a large enough magnitude of the field, the lifetimes of the metastable and radiating levels become equal. The role of these effects in plasmas is examined in more detail below.

The calculations of the Stark spectral splitting of non-hydrogenic ions with regard to fine structure require simultaneous consideration of both the interelectron and spin–orbital interactions. The latter can be taken into account with the help of perturbation theory in the Pauli approximation up to ion charges $Z = 30$ (Klarsfeld 1969). The term structure is complex, and the same is true for the atomic state wave functions. Therefore, as a rule the spectral line intensity calculations are carried out with the help of numerical methods.

The increase of *Z* leads to a sharp increase of forbidden transitions. Thus, for example, the probabilities of quadrupole transitions *E*2 are $(Ze^2/\hbar c)^2$ times smaller than the probabilities of allowed dipole transitions, and as a consequence, they increase sharply with increasing *Z*. The width of the two-photon transition $2s-1s$ increases in a highly charged ion as (Klarsfeld 1969):

$$\gamma_{2s-1s} = (8.2283 \pm 0.0001)Z^6 \ (s^{-1}) \quad (1.83)$$

and for values as low as $Z = 20$ becomes of the same order of magnitude as the width of the allowed transition $2p-1s$ in the neutral hydrogen. The probabilities of magnetic dipole transitions ($M1$) increase analogously.

The transition probabilities $M1$ and $E2$ between the fine structure sublevels increase especially sharply. Indeed, for such a transition, the transition frequency ω itself increases proportionally to Z^4. Therefore, radiation widths proportional to ω^3 increase like Z^{12} or even more sharply. For example, for the transitions $2s_{1/2} - 1s_{1/2}$ in H-like ions we obtain:

$$\gamma_r(M1) = 2.46 \times 10^{-16} Z^{10} \ s^{-1}. \quad (1.84)$$

The sharp increase of transition probabilities of multipoles leads to the opportunity of observations even under the conditions of laboratory plasmas. An example is the $M1$ transition in the Fe XX ion with a wavelength of $\lambda = 2665$ Å observed in the plasma of the Princeton Tokamak PLT. The corresponding radiation width of the transition is equal to $570 \, s^{-1}$ (Sudkewer 1981).

From the above, it is evident that the main peculiarity of HCI spectra is a sharp increase of the transition probabilities as compared with the neutral atom. Under rarefied high-temperature plasma conditions of the Sun or of thermonuclear magnetically confined plasmas, the intensities of forbidden transitions are comparable with the intensities of the allowed transitions. This is due to the fact that for Z-values $Z \geq 20$ the radiation decay rates γ_r both for the allowed and forbidden transitions begin to dominate over the rates of the states' collisional excitation. Under such conditions of "coronal" equilibrium (e.g., taking place in the Sun's corona), the line intensities are obviously determined by atomic states excitation rates (since every act of an excitation is certainly accompanied by the radiation). The excitation rates for different sublevels, unlike radiation transition's rates, are of comparable order (Sudkewer 1981; Sobelman et al. 1995).

As an example, Fig. 1.6 shows the "Corona" X-ray spectrum of argon (right) observed in a tokamak. The left part of Fig. 1.6 schematically depicts the relevant atomic levels. Here, the resonance line corresponds to the dipole-allowed transition ($E1$)

$$W = 1s2p\,^1P_1 \rightarrow 1s^2\,^1S_0 + \hbar\omega_W$$

in He-like argon,

$$X = 1s2p\,^3P_2 \rightarrow 1s^2\,^1S_0 + \hbar\omega_X$$

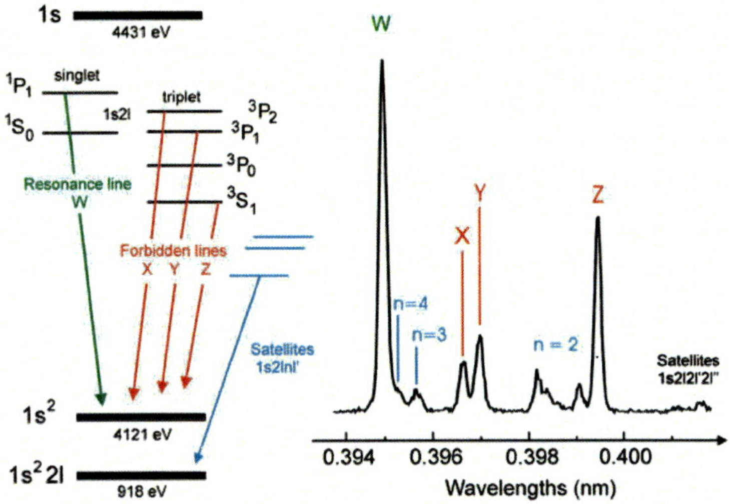

Fig. 1.6 High-resolution X-ray spectrum of argon observed in a tokamak

the magnetic dipole transition (*M1*),

$$Z = 1s2s\,^3S_1 \rightarrow 1s^2\,^1S_0 + \hbar\omega_Z$$

the magnetic quadrupole transition (*M2*), and

$$Y = 1s2p\,^3P_1 \rightarrow 1s^2\,^1S_0 + \hbar\omega_Y$$

the dipole-allowed intercombination transition (*E1*). It can clearly be seen that the line intensities of the *W*, *Y*, *X*, *Z* lines are of comparable order although their transition probabilities differ by many orders of magnitude (e.g., for argon: $A(W) = 1.07 \times 10^{14}\ \mathrm{s}^{-1}$, $A(Y) = 1.82 \times 10^{12}\ \mathrm{s}^{-1}$, $A(X) = 3.16 \times 10^{8}\ \mathrm{s}^{-1}$, $A(Z) = 4.80 \times 10^{6}\ \mathrm{s}^{-1}$). This is due to the strongly populated metastable states in low-density plasmas (see also Chap. 6). The spectrum of Fig. 1.6 also demonstrates a rather strong emission of Li-like satellites (designated as "$n = 2$", "$n = 3$", and "$n = 4$")

$$\text{Li-like satellites} = 1s2lnl' \rightarrow 1s^2nl' + \hbar\omega_{\text{satellite}}$$

and some emission of Be-like satellites

$$\text{Be-like satellites} = 1s2l2l'2l'' \rightarrow 1s^22l'2l'' + \hbar\omega_{\text{satellite}}.$$

The line emission of the He-like *W*-, *Y*-, *Z*-, *X*-lines as well as the Li-like satellites $1s2lnl'$ permits a unique characterization of the tokamak plasma, in particular with respect to electron temperature, ion temperature, charge exchange with the neutral background and neutral beam injection (Rosmej and Lisitsa 1998; Rosmej et al. 1999a, b, 2006; Bureyeva et al. 2003) as well as impurity diffusion.

1.3 Autoionizing Atomic States

1.3.1 Excitation of Core Hole States

The excitation of a number of electrons leads to the appearance of new autoionization lines in atomic or ionic spectra. The classic examples are the simplest autoionization states of a two-electron helium atom. The peculiarities of these states are connected with the fact that strictly speaking, they are not discrete but have a finite width Γ connected with the possibility of the atom's disintegration of the autoionization state into an ion and a free electron.

To discern the essence of the matter, let us consider an approximate model of the autoionization via the helium state $2s2p\,^1P$ emerging during the excitation of both of its electrons at the atomic level $n = 2$. The usual helium spectrum is the spectrum of excitation of one of its two electrons. This energy spectrum corresponds to the system of rigorously discrete levels, and the first excited state, the $n = 2$ level, lies approximately 20 eV above the ground state. These discrete levels condense towards the boundary of the discrete spectrum–continuum, situated approximately 25 eV above the ground state.

Let us now assume that the atom is divided into two parts, namely: The excited electron is in the state $n = 2\,(\Delta E_{12} \simeq 20\,\text{eV})$ and the remaining one-electron ion is called the core. The energy levels of such a core are hydrogen-like and they respond to the nuclear charge $Z = 2$, so that its excitation energy into the state $n = 2$ is approximately equal to $\Delta E_c = Z^2 Ry(1 - 1/4) \simeq 40\,\text{eV}$. Hence, the excitation energy of two electrons (the external and the core electrons) at a given time is approximately equal to $\Delta E^{(2)} = \Delta E_{12} + \Delta E_c \simeq 20\,\text{eV} + 40\,\text{eV} = 60\,\text{eV}$. Thus, the helium atom has a discrete state with the energy $\Delta E^{(2)} \simeq 60\,\text{eV}$ exceeding the atomic ionization potential $I \simeq 25\,\text{eV}$; that is, it is situated deeply in the continuum (or in the background of the continuum). However, this state is discrete only in the zeroth approximation when we have neglected the interaction of electrons between each other. Taking into account this electron interaction results in a decay of the autoionization state. One can imagine the mechanism of such decay in the following way: The inner core electron makes a transition to the ground state ($1s$) and the external electron is ejected into continuum. Such a process is named "core relaxation" and is accompanied by ejection of an external electron (Auger electron). Though the division of atomic electrons into the "core" and "external" parts is relative, it is at the same time rather visual and we shall use it in future discussions.

1.3.2 The Interaction of Discrete States with a Continuum: Fano resonances

As stated above, the autoionizing states are non-stationary and experience a decay described by the autoionization width Γ. This decay is due to the interaction of the discrete state with a continuous spectrum of free electron states. The description of the autoionization decay processes is in many respects similar to the description of radiative decays for which the role of the continuum is played by a number of electromagnetic field harmonics. A corresponding theory for autoionizing states may be developed by an analogy with the general theory of states damping. The initial research was performed by Fano (1961) applied to the photoeffect. The essence of Fano's approach involves the construction of correct stationary wave functions of a system of a "discrete level continuum," including the interaction between them. The problem is the generalization to the continuum spectrum from the known problem of the energy spectrum of two interacting discrete levels. The correctly constructed wave functions are then used later for the calculation of the transition probabilities between them and another discrete atomic level, or the atomic ground state for photoeffect problems, in particular.

Another approach was developed quite independently and somewhat later by Kompanets (1968), while somewhat later also a generalized formulation has been proposed (Lisitsa and Yakovlenko 1974). In contrast to the Fano method (Fano 1961), the Kompanets approach is based directly on time-dependent approaches being connected by a Fourier transform and they give identical results for the final transition probability. Nevertheless, the non-stationary approach is by far more easy and allows further advancement into the domain of strong interactions. Thus, we shall briefly consider both approaches.

The existence of autoionizing states leads to the appearance of sharp peaks for the probabilities of processes with the participation of a continuous spectrum, like photoionization, inelastic scattering, and so on.

To calculate effects of this kind, Fano (1961) suggested a method that is based on the determination of a mixed state of the system "discrete level continuum" by the diagonalization of their interaction. The essence of his methods encompasses the following. Consider the Hamiltonian

$$H = H_0 + V \tag{1.85}$$

having the states of discrete Ψ_{n} and continuous Ψ_ν spectra

$$\begin{aligned} \langle \Psi_{\mathrm{n}} | H | \Psi_{\mathrm{n}'} \rangle &= E_{\mathrm{n}} \delta_{\mathrm{nn}'} \\ \langle \Psi_\nu | H | \Psi_{\nu'} \rangle &= E_\nu \delta(E_\nu - E_{\nu'}). \end{aligned} \tag{1.86}$$

Here, E_{n} and E_ν are the energies of the discrete and continuous states. The magnitude V depicts the interaction between the discrete and continuous states:

$$\langle \Psi_\nu | H | \Psi_n \rangle = V_{\nu n}. \tag{1.87}$$

Let us find the eigenfunctions of the Hamiltonian (1.85). For this purpose, we expand the wave function Ψ into the states of discrete and continuous spectra

$$\Psi = a(E)\Psi_n + \int dE' \frac{d\nu}{dE'} b_{E'} \Psi_{\nu'}, \tag{1.88}$$

where the indices of the continuous states ν indicate the integration over the energy E by introducing the density of states $d\nu/dE$.

Fano determined the coefficients $a(E)$ and $b_{E'}$ from the wave function (1.88) of the discrete level against the background of the continuum. The result for coefficient $a(E)$ takes the form:

$$|a(E)|^2 = \frac{|V|^2}{(E - E_n - \Delta)^2 + \pi^2 |V|^4}, \tag{1.89}$$

where Δ represents a shift of the resonance transition with respect to E_n (Fano 1961). It is seen from (1.89) that the initial discrete state runs into the resonant curve with a half-width of $\pi|V|^2$ determined by the interaction with the continuum. This means that if at some moment in time the system has been in state Ψ_n, then its average lifetime determined with respect to the decay into the continuum is equal to $\hbar/2\pi|V|^2$. The further use of the Fano method consists in the calculation (with the help of perturbation theory) of the transition probability from the ground state Ψ_0 into a determined mixed state Ψ (1.88).

Kompanets (1968) suggested a non-stationary approach to the description of interactions with a continuum for resonance phenomena in the photoeffect. The basis of such a theory involves taking into account three types of atomic states, namely: the initial state with amplitude c_0, the discrete autoionizing state c_n, and the continuum state c_E. The matrix elements of the transitions from the initial state to the level n are designated by H_{0n}, one to the continuum by H_{0E}, and at least the interaction with the continuum characterizing the autoionization decay is designated by $V_{n,E}$. Then, for the amplitudes mentioned, we have the system of equations:

$$i\dot{c}_E = V_{E,n} e^{i\omega_{En} t} c_n + H_{0E} e^{i(\omega_e - \omega)t} c_0, \tag{1.90}$$

$$i\dot{c}_n = \sum_{E'} V_{n,E'} e^{i\omega_{E'n} t} c_{E'} + H_{0n} e^{i\Delta\omega_e t} c_0, \tag{1.91}$$

where $\omega_{En} \equiv \omega_e - \omega_n, \Delta\omega \equiv \omega - \omega_n$. The solution of the system is performed with the help of perturbation theory and assuming $c_0 = 1$. Writing down c_E from (1.90) in quadrature form

$$c_E = -ie^{(\omega_E - \omega)t} \cdot \int_0^t \left(H_{0E} + V_{E,n} \cdot c_n \cdot e^{i\Delta\omega t} \right) \cdot e^{(\omega_E - \omega)t'} \cdot dt' \tag{1.92}$$

let us substitute (1.92) into (1.91). Making the transition in (1.91) from the sum $\sum_E$ to the integral over energy $\rho(E)dE$ with the density of states $\rho(E)$, it is justified for $\omega t \to \infty$ to take out the slowly varying matrix elements at point $E' = \omega$. It results in the following relation:

$$\sum_{E'} V_{n,E'} c_{E'} e^{i(\omega_{E'} - \omega)t} = -\pi i [V_{n,\omega} H_{0\omega} \rho_\omega + |V_{n\omega}|^2 \rho_\omega c_n \cdot \exp(-i\Delta\omega t)]. \quad (1.93)$$

Substitution of (1.93) into (1.91) leads to an ordinary differential equation for c_0. Substituting its solution into (1.93), one finds the probability amplitude of the population of the continuum state c_E. Calculating the modulus of c_E to the second power, performing the integration over some continuous spectral interval E, and performing the limit $t \to \infty$, one obtains the final expression for the transition probability W:

$$W = 2\pi\rho(\omega) \frac{|H_{0\omega}(\omega_n - \omega) + V_{\omega,n} H_{0n}|^2}{(\omega_n - \omega)^2 + (\pi |V_{\omega,n}|^2 \rho_\omega)^2}. \quad (1.94)$$

Equation (1.94) coincides with the Fano formula (Fano 1961). Far from resonance, $\Delta\omega \gg \Gamma = \pi |V_{\omega,n}|^2 \rho_\omega$, the probability is reduced to the probability of the direct photoeffect:

$$W_{ph} = 2\pi\rho_\omega |H_{0\omega}|^2. \quad (1.95)$$

In the general case, it describes the resonance process of the transition into the continuum with regard to the interference of the direct and autoionization channels. The introduction of the dimensionless Fano parameters

$$x = \frac{\Delta\omega}{\Gamma}, \quad (1.96a)$$

$$q = \left| \frac{H_{n0}}{H_{0\omega} 2\pi \rho_\omega V_{n\omega}} \right| \quad (1.96b)$$

shows that the shape of the resonance curve is described by the function

$$f(x) = \frac{(x+q)^2}{x^2 + 1/4}, \quad (1.97)$$

having a typical asymmetrical form.

Let us discuss the physical sense of the Fano formula (1.94). First of all, it is evident that the photoionization process is of resonant type with a probability which increases sharply for $\Delta\omega \le \Gamma$. The first element in the numerator in (1.94) describes the decay of the initial state c_0 into the continuum with amplitude $(W_{ph})^{1/2}$ due to direct photoionization without any connection with state c_n. The second term describes the decay of the level c_0 into the continuum over the intermediate state c_n; namely, the decay amplitude is proportional to the matrix element H_{0n} of the transition from c_0 into c_n, multiplied by the amplitude $(\Gamma^{1/2})$ of the decay of state c_n. Thus, (1.94) allows for two opportunities for the state c_0 to decay into the continuum, namely directly (with probability W_{ph}) and over the state c_n. Both states c_0 and c_n "resound" against the background of the continuum in the interaction which leads to the decay of the states with probabilities W_{ph} and Γ as well as to the additional interference connection $(W_{ph}\Gamma)^{1/2}$ of these states. The connection is obviously realized by transitions from level c_0 into the continuum and from the continuum into level c_n.

1.4 Rydberg Atomic Autoionizing and Non-Autoionizing States in Plasmas

1.4.1 Rydberg Atomic States

We now pay attention to highly excited (Rydberg) states of atoms and ions. In recent years, the investigation of Rydberg states received keen interest (Stebbings and Dunning 1983). The great interest of Rydberg states for plasma applications is connected with the high multiplicity of atomic states from ionized atoms bringing their spectra to hydrogen-like ones, and with the effective population of highly excited atomic states in a plasma medium.

As a rule, the description of the Rydberg states is connected with quasi-classical or pure classical methods. A difficulty arises in the direct application of quantum mechanical methods for large quantum numbers that lead to great numerical difficulties caused by the sharp increase in the number of quantum states ($\propto n^2$). Therefore, the precision of quantum calculations usually diminishes with the increase of n. At the same time, it is clear that classical methods should work successfully for large n.

The usual channel of Rydberg state population for low-temperature plasma is three-body recombination and cascade excitation, and for high-temperature plasmas, these are photo and dielectronic recombinations. An important population channel for HCI excited states is their charge exchange with neutral atoms which exist already (neutral background) or are introduced into plasmas for heating (neutral beam injection or neutral beam heating). The typical values of the main quantum numbers n of the excited states are about $n \simeq 20$, for hydrogen-like ions and are $10 \leq n \leq 100$ for highly charged ions. The special cases are atoms in rarefied astrophysical plasmas that are excited as a rule by photoexcitation and leading to very large values of n up to $n = 400-600$ (Stebbings and Dunning 1983).

Also in dense plasmas, Rydberg transitions became of considerable interest. They can be used for diagnostic purposes, i.e., to determine the bulk electron temperature in non-Maxwellian plasmas (Rosmej 1995) and also for the determination of ionization potentials (Palchikov et al. 2002).

1.4.2 Autoionizing Rydberg Atomic States

Recently, also Rydberg satellite spectra from autoionizing states became of increasing interest for space-resolved X-ray spectroscopy. They have been identified in laser-produced plasmas with the help of atomic structure calculations (Rosmej et al. 2001a; Skobelev et al. 2002). Figure 1.7 shows a spectrum from a dense ns-laser-produced silicon plasma in the high-energy spectral range of the Li-like Rydberg satellites, i.e.,

$$1s2lnl' \rightarrow 1s^2 2l + \hbar\omega_{\text{sat}}.$$

The low-energy transitions

$$1s2lnl' \rightarrow 1s^2 nl' + \hbar\omega_{\text{sat}}$$

have serious impact on the interpretation of the resonance line emission. It has been demonstrated with the help of high-resolution X-ray spectroscopy and spectral simulations (Rosmej et al. 1998) that resonance line emission can effectively be replaced by an accumulation of Rydberg satellites in ns-laser-produced plasmas (Rosmej and Faenov 1997; Rosmej 1998; Rosmej et al. 1997a, b, 1998, 2000) as well as in fs-laser-produced plasmas.

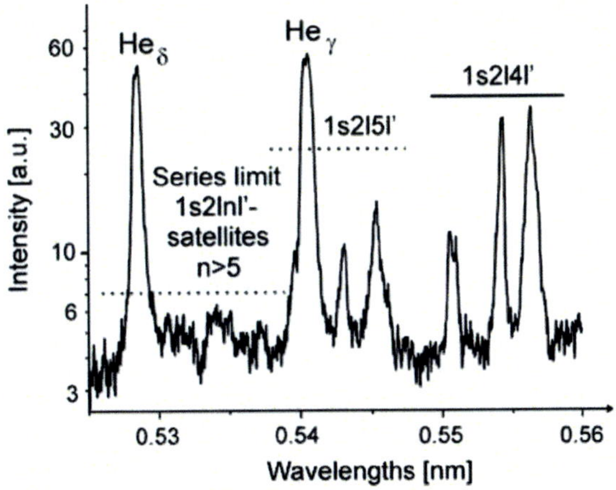

Fig. 1.7 Rydberg high-energy spectrum of hot dense silicon ns-laser-produced plasmas. For Li-like Rydberg $1s2lnl'$-satellites with $n > 3$, the spectral distribution splits essentially into three groups. For satellites with $n > 5$, the groups of different n-quantum number "mix" with each other and the series limit is located near the He-like He_δ transition

1.4.3 Dielectronic Satellite Accumulation in ns-Laser-Produced Plasmas

Figure 1.8 shows the space-resolved X-ray emission spectrum from a ns-laser irradiating a solid magnesium target. It can clearly be seen that far from the target surface, the emission is dominated by the He-like *W*- and *Y*-line. At closer distances to the target surface, intense satellite emission develops. The Li-like $1s2l2l'$-satellite emission is labeled according to the notation of Gabriel (1972), and the potential wavelengths interval is indicated by the dashed horizontal line (target distance about 90 μm).

At very small target distances (below 70 μm), the resonance line transforms into a very broad and complex structure and seems even to disappear very close to the target surface. This complex resonance line structure has been identified as a dielectronic satellite accumulation (Rosmej and Faenov 1997; Rosmej et al. 1997a, b, 1998), the so-called DSA model (Rosmej and Faenov 1997) in plasmas where the

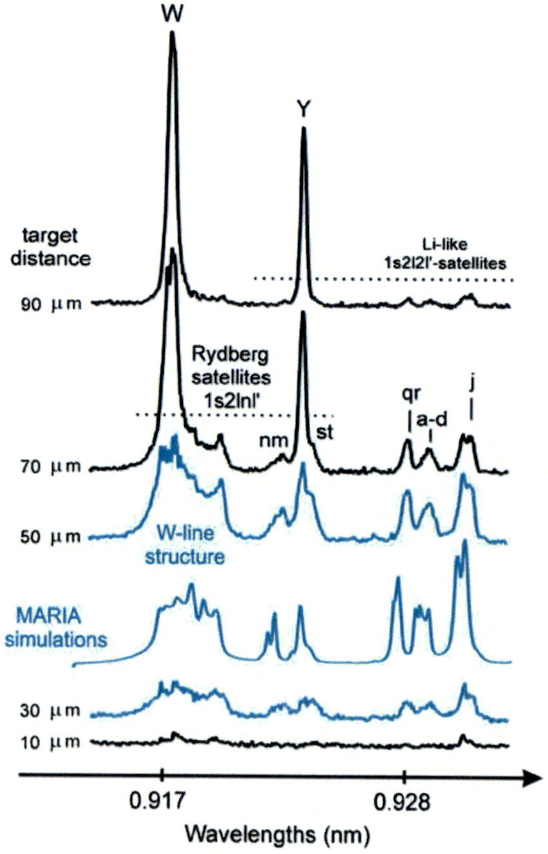

Fig. 1.8 Space-resolved X-ray emission spectrum from a ns-laser-produced Mg plasma. The MARIA simulations are carried out for a strongly coupled plasma with $n_e = 8 \times 10^{21}$ cm^{-3}, $kT_e = 100$ eV, $L = 15$ μm resulting in an ion–ion coupling parameter of $\Gamma_{ii} \approx 2$

ion–ion coupling parameter (Z is the ionic charge, n_i is the ionic density, and kT_e is the electron temperature)

$$\Gamma_{ii} = 2.32 \times 10^{-7} \cdot \frac{Z^2 \cdot n_i^{1/3}(\mathrm{cm})}{kT_e\ (\mathrm{eV})} \tag{1.98}$$

achieves values of the order of 1 (Rosmej et al. 1998, 2003). The basic mechanisms for the DSA in strongly coupled plasmas have been identified via spectral simulation with the MARIA suite of codes (Rosmej 1997, 1998, 2001, 2006): high line center opacity of the resonance line and low temperature. Figure 1.8 shows the spectral simulation carried for an optically thick magnesium plasma with electron density $n_e = 8 \times 10^{21}$ cm^{-3}, electron temperature $kT_e = 100$ eV, and effective plasma dimension of $L = 15$ μm. These parameters correspond to a strongly coupled plasma with an ion–ion coupling parameter (1.98) of $\Gamma_{ii} \approx 2$. The line center opacity of the He-like resonance line W is about $\tau_0\ (W) \approx 170$, and even the intercombination line Y and the Li-like j-satellite achieve values of optical thickness of $\tau_0\ (Y) \approx 0.3$ and $\tau_0\ (j) \approx 4$, respectively. Due to the large opacity of the resonance line and the Li-like n = 2 $1s2l2l'$-satellites, the Rydberg satellite emission starts to be dominating (negligible opacity) near the target surface. We note that Rydberg satellites accumulate not only near the He-like resonance line W but also near the He-like intercombination line Y (Rosmej et al. 1997a).

Similar observations have been made for the He-like $2l2l'$-satellites near Lyman α (Renner et al. 2001). The strongly coupled plasma regime can be likewise a favorable parameter regime for the creation of hollow ion configurations of the type $K^1L^0M^2$ via the so-called excited state coupling (Rosmej et al. 1998, 2003) that give rise to anomalous intense satellite emission near the He-like transition $1s3p\,^1P_1 \rightarrow 1s^2\,^1S_0 + \hbar\omega_{\mathrm{He}\beta}$.

1.4.4 Transient Three-body Recombination of Dielectronic Satellite Emission

In fs-laser-produced plasma, transient three-body recombination has been identified as an important mechanism for dielectronic satellite accumulation (Rosmej et al. 1997a), e.g., the population of autoionizing Rydberg levels via three-body recombination according to

$$1s2l + e + e \rightarrow 1s2lnl'$$

and subsequent radiation emission of Rydberg satellites in the low-energy region

$$1s2lnl' \rightarrow 1s^2nl' + \hbar\omega_{\mathrm{Ry,He}_\alpha}$$

and

$$1s2lnl' \rightarrow 1s^22l + \hbar\omega_{\mathrm{Ry,He}_n}$$

in the high-energy region.

Figure 1.9 shows the MARIA simulations of the transient spectral emission of a plasma produced by fs-laser irradiation of a solid Mg target. The simulations calculate the time-dependent atomic populations $n_{\mathrm{j}}(t)$ from the evolution of the temperature and density (including prepulse, rapidly heating, and the cooling phase) and include also hot electrons [characterized by the hot electron fraction $f_{\mathrm{hot}}(t)$ and hot electron temperature kT_{hot}; see Rosmej (1997)], opacity, and differential plasma motion:

$$\frac{\partial n_{\mathrm{j}}(t)}{\partial t} = -n_{\mathrm{j}}(t) \cdot \sum_{\mathrm{k}} W_{\mathrm{jk}}(t) + \sum_{\mathrm{l}} n_{\mathrm{l}}(t) \cdot W_{\mathrm{lj}}(t) \tag{1.99}$$

and

$$I_\omega(t) = \sum_{\mathrm{j}=1}^{N_{\max}} \sum_{\mathrm{i}=1}^{N_{\max}} \frac{\hbar\omega_{\mathrm{ji}}}{4\pi} \cdot n_{\mathrm{j}} \cdot A_{\mathrm{ji}} \cdot \Lambda_{\mathrm{ji}} \cdot \Phi_{\mathrm{ji}}, \tag{1.100}$$

where Λ_{ji} and Φ_{ji} are the time-dependent escape and optically thick line profiles [see, e.g., (1.33), (1.34)]. The matrix $W_{\mathrm{jk}}(t)$ includes the elementary atomic processes that are relevant for the population and depopulation of atomic levels (e.g., collisional excitation/de-excitation, ionization, three-body recombination, radiative recombination, dielectronic capture, autoionization, and radiative decay).

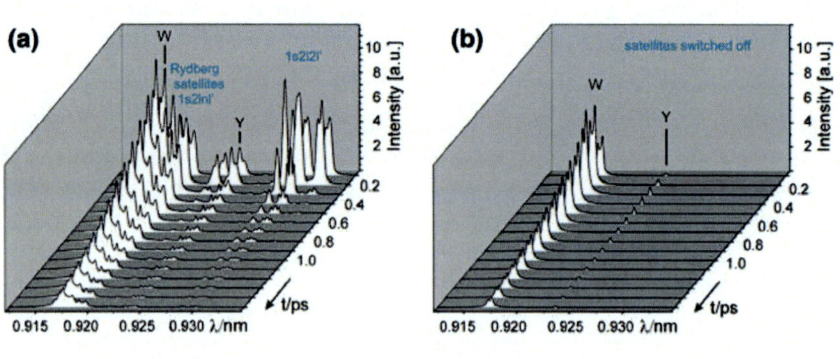

Fig. 1.9 MARIA simulations of the time-dependent evolution of the spectral distribution of a fs-laser-produced Mg plasma. **a** Simulations including Li-like $1s2l2l'$-satellites as well as Rydberg $1s2lnl'$-satellites, **b** same simulation as in (a), however, all satellite emission is artificially switched off

In order to compare the time-dependent spectral distribution of (1.100) with time integrated but spatially resolved measurements, (1.100) has to be integrated over time:

$$I_{\omega} = \int_{0}^{\infty} I_{\omega}(t)\mathrm{d}t. \tag{1.101}$$

Figure 1.10 compares the experimental data with the simulations according to (1.99)–(1.101). In order to obtain a best fit, time-dependent test functions of temperature, density, and hot electron fractions have been varied until a good agreement with the data was achieved.

Figure 1.10 demonstrates excellent agreement with the time-dependent non-LTE and non-Maxwellian MARIA simulations with the spectra at target distances between 45–100 μm. Apart the fundamental understanding of the complex line formation of resonance line emission and dielectronic satellite emission in strongly coupled plasmas, the analysis of the dielectronic satellite emission has outstanding advantages for diagnostic applications to provide a unique characterization of the plasma (see also Chap. 10).

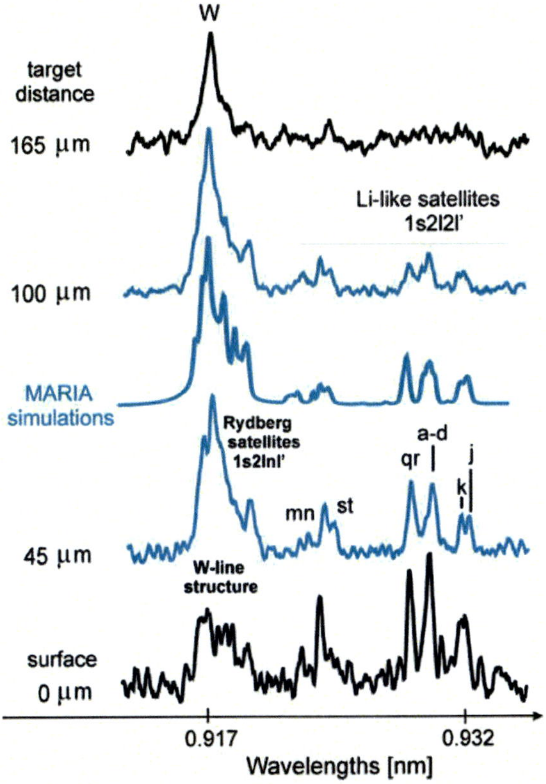

Fig. 1.10 MARIA simulations of the time-dependent evolution of the spectral distribution of a fs-laser-produced Mg plasma summed over the total emission to be compared with time-integrated but space-resolved measurements

1.5 Plasma Spectroscopy

The emission of light (Loudon 2000) is one of the most fascinating phenomena in nature, and we all use light to obtain information, to diagnose something, to control or optimize a process, or to understand what is true and what is right from spectroscopic diagnostics. Today plasma spectroscopy has proven to be one of the most powerful methods to understand various physical phenomena. It provides essential information about basic parameters (like temperature, density, chemical composition, and velocities) and relevant physical processes. The accessible parameter range of spectroscopy covers orders of magnitude in temperature and (especially) density, because practically all elements of particular, selected isoelectronic sequences can be used for diagnostic investigations. These elements can occur as intrinsic impurities or may be intentionally injected in small amounts (the so-called tracer elements). This makes plasma spectroscopy also a very interdisciplinary science, and several books and review articles concerning traditional plasma spectroscopic methods have been published (Griem 1964, 1974, 1997; Fujimoto 2004; Kunze 2009; McWhirter 1965; Boiko et al. 1985; Michelis and Mattioli 1981).

The rapid development of powerful laser installations (high-energy and high-intensity optical lasers, X-ray free electron lasers), intense heavy ion beams, and the fusion research (magnetic and inertial fusion) enables the creation of matter under extreme conditions never achieved in laboratories so far. An important feature of these extreme conditions is the non-equilibrium nature of the matter: e.g., a solid is heated by a fs-laser pulse and undergoes a transformation from a cold solid, warm dense matter, strongly coupled plasma to a highly ionized gas while time is elapsing. We might think about using time-dependent detectors to temporally resolve the light emission in the hope to have then resolved the problem. However, this is not so simple: There are serious technical obstacles and also basic physical principals to respect. A simple technical reason is that for 10 fs-laser radiation interacting with matter, we do not have any X-ray streak camera available (the current technical limit is about 0.5 ps). A principal reason is that the atomic system from which light originates has a characteristic time constant with might be much longer than the 10 fs (see also Chap. 6 for more details). The atomic system is therefore "shocked," and any light emission is highly out of equilibrium invalidating any standard methods for diagnostics (even if the experimental observation is time resolved).

More complex physical reasons are related to the fact that the term "non-equilibrium" concerns not only the time evolution but also the statistical properties of the system. One important example (known for laser-produced plasmas and inertial fusion research) is the creation of suprathermal electrons (or hot electrons). They seriously alter the light emission in a non-trivial manner (a trivial change is an enhanced ionization due to hot electrons, and a non-trivial is the qualitative distortion of ion charge stage distribution (Rosmej 1997) which in turn invalidates the application of any standard diagnostic methods). In consequence, traditional equilibrium methods are not anymore applicable. Moreover, theoretical analysis shows that often standard methods cannot be corrected or modified to be anymore useful like they have been at the beginning of their historical foundation.

In conclusion, an essential point in modern research is not only to understand the radiative properties by itself, but also to characterize an unknown phenomenon via spectroscopic and atomic physics methods that are essentially independent from other theories (plasma simulations, etc.).

1.5.1 Spatial Properties of Dielectronic Satellite Emission

Large efforts are made to create homogenous dense plasmas under extreme conditions to provide samples and emission properties, which can be directly compared with theory. Unfortunately, dense hot laboratory plasmas show almost always large variations of the plasma parameters over space and, in consequence, a large variation of the spectral emission. As spatial parameter grid reconstructions are difficult, X-ray spectroscopy with spatial resolution is frequently applied to obtain supplementary information. The spatial resolution can be realized either with a slit mounted at a suitable distance between the source and the X-ray crystal or by means of curved X-ray crystals. The most commonly used curved crystal arrangements are the Johann geometry, the Johannson geometry, and two-dimensional curved crystals; a review of these methods is given in Boiko et al. (1985).

It is worth emphasizing two particular methods that turned out to be extremely useful for dense plasma research:

(1) The vertical Johann geometry (Renner et al. 1997) which is extremely suitable for line profile investigations (spatial resolution of some μm with simultaneous extremely high spectral resolution of about $\lambda/\delta\lambda \approx 6000$, the spectral range, however, is rather small, permitting only to observe, e.g., the H-like aluminum Lyman$_\alpha$ line and corresponding satellite transitions (Renner et al. 2001). Due to the appearance of double-sided spectra, the geometry can provide line shift measurements without reference lines. Due to the high sensitivity of this method, even a spin-dependent line shift of He-like ions (W and W-lines) has recently been reported in dense laser-produced plasmas (Renner et al. 2006; Adamek et al. 2006). These results have initiated first attempts to investigate the density dependence of the exchange energy (Li et al. 2006; Rosmej et al. 2011),

(2) The spherical X-ray crystals (Faenov et al. 1994; Skobelev et al. 1995) which do provide simultaneously high spectral ($\lambda/\delta\lambda \approx 1000-6000$, dependent on the large varieties of possible geometries) and spatial resolution (about 10–30 μm), large spectral windows [permitting, e.g., to observe all the K_α-satellite series $K^1L^N \rightarrow K^2L^{N-1} + \hbar\omega_{K\alpha-sat}$ until the He-like resonance line W for aluminum (Rosmej et al. 2001a)] and large spatial intervals (up to cm with 10 μm resolution) and the possibility for X-ray microscopic applications (two-dimensional X-ray imaging).

Although the technology of space-resolved spectroscopy advances the analysis of the radiation emission considerably, a principle difficult remains: There is always a line of sight and a corresponding spatial integration of the radiation emission along this line of sight. Above-discussed spatial resolution can provide spatial resolution only for the coordinates perpendicular to the line of sight. Therefore, even when applying two-dimensional X-ray imaging methods the principle drawback of the line-of-sight integration remains.

It should be noted that Abel inversion methods which are widely used in plasma tube experiments, magnetic fusion research, etc., are difficult to implement for dense plasmas experiments: The emission from different coordinates corresponds often to different emission times and therefore quite different regimes are responsible for the line emission (with corresponding different excitation channels, e.g., heating or recombination regimes). An example is the radiation emission of the resonance line emission in dense laser-produced plasmas: During the heating process, collisional excitation from the ground state strongly populates the upper levels and drives correspondingly intensive line emission. When the laser is turned off, the plasma recombines and the upper levels are populated by recombination at much lower temperatures and often also at much lower (orders of magnitude) densities. Interesting high-density regimes are therefore masked by a low-density recombining regime.

The intensity driven by the low-density-recombining regime can drive even more intense line emission than the high-density plasma heating phase itself (a pitfall for Stark profile interpretations). This is easily seen in Fig. 1.11: A massive Mg target is irradiated with a ns-high-energy-density laser, and the line emission is observed with a spatial resolution in the direction of the expanding plasma. The almost horizontal emission feature corresponds to the bremsstrahlung emission of the target surface. Near the target surface (corresponding to the plasma heating regime), the intensities of the He-like intercombination line Y and the resonance line emission W are of similar order and strongly density dependent. At large target distances (z larger than about a few 100 μm corresponding to the recombining regime), the Y-line is as intense as the W-line and even dominates at very large distances (z larger than about 500 μm).

Proposing the use of a streak camera sounds good to distinguish the heating and recombination regime, but implies somehow to lose the resolution in space. We can continue with endless discussions of technical improvements of the spectroscopic equipment (note, e.g., there is limited hope for a 1 fs X-ray streak camera in nearest future), but it looks like that we somehow turn in a circle and do not reach our dream: a temporally resolved spectral photon distribution of useful atomic transitions (useful in the sense that they are sensitive either to temperature, density, hot electrons fractions, ion velocities, charge state distributions, chemical composition, …) originating from a single point entering the detector without interaction of the surrounding dense plasma. Although technical developments are extremely important, their developments alone are not sufficient to make our dream come true.

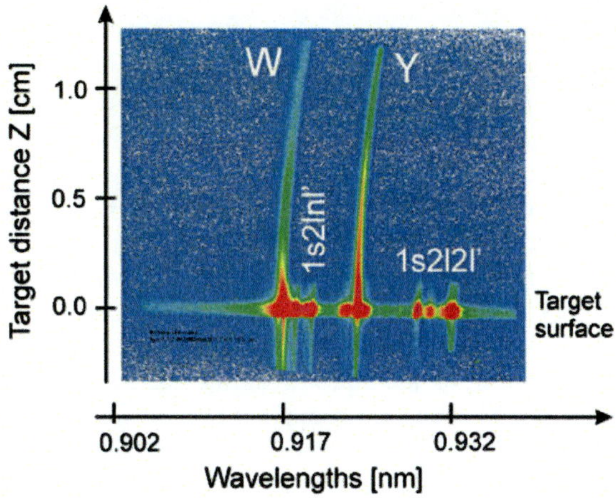

Fig. 1.11 Space-resolved X-ray emission spectrum from a ns-laser irradiating a solid Mg target. At large target distances ($z > 1$ cm), the intercombination line Y is even more intense than the resonance line emission W, whereas dielectronic $1s2l2l'$-,$1s2lnl'$-satellite emission is rather confined to the target surface

Let us return to Fig. 1.11: It looks like that there is another class of line emission that is important only at close target distances (note that the emission that seems to be below the target surface is an artifact due to the fact that the line of sight of the space-resolved spectrometer was not exactly parallel to the target surface but enclosed an angle of 20°). This emission is due to the so-called dielectronic satellite emission originating from autoionizing states of highly charged ions. Their primary excitation channels are dielectronic capture, inner-shell excitation, and collisional redistribution between the autoionizing levels. For atomic levels with large autoionizing rates, dielectronic capture is usually the most important excitation channel for the satellite intensity:

$$I^{\text{sat}}_{\text{k,ji}} = \alpha n_{\text{e}} n_{\text{k}} \frac{Q_{\text{k,ji}}}{g_{\text{k}}} \frac{\exp\left(-E_{\text{kj}}/kT_{\text{e}}\right)}{\left(kT_{\text{e}}\right)^{3/2}} \tag{1.102}$$

with

$$Q_{\text{k,ji}} = \frac{g_{\text{j}} A_{\text{ji}} \Gamma_{\text{jk}}}{\sum_{\text{l}} A_{\text{jl}} + \sum_{\text{m}} \Gamma_{\text{jm}}}. \tag{1.103}$$

$Q_{\text{k,ji}}$ is the so-called dielectronic satellite intensity factor for the satellite radiative transition from $j \rightarrow i$ after dielectronic capture from level k with population n_{k} into level j, n_{e} is the electron density, $\alpha = 1.656 \times 10^{-22}$ cm^3 s^{-1}, g_{j} and g_{k} are the statistical weights of the states j and k, Γ_{jk} is the autoionizing rate in [s^{-1}], E_{kj} is the dielectronic capture energy in [eV], and kT_{e} is the electron temperature [eV].

Inspection of the dielectronic capture channel and the correspondingly induced satellite line intensity (1.102) shows that the intensity is proportional to the square of the electron density (because the ground state n_{k} is proportional to the electron density):

$$I^{sat}(high\,\Gamma) \propto n_e^2. \qquad (1.104)$$

The emission is therefore confined to high-density and also high-temperature plasma areas because the capture energy E_{kj} is usually of the order of $Z_{eff}^2 Ry$ (where Z_{eff} is the effective charge of the autoionizing configuration seen by the radiative transition electron). This effect is clearly seen in Fig. 1.11: The $1s2l2l'$- and $1s2lnl'$-satellites are visible just around the target surface. Line-of-sight integration effects are therefore minimized, and (1.102)–(1.104) act like a "local emission source".

For satellite transitions whose dielectronic capture is coupled to excited states, the spatial confinement is even larger because the excited states are essentially proportional to the squared of the electron density (via dielectronic capture). Figure 1.12 shows an example for the Li-like $1s3l3l'$-satellites that are essentially coupled to the $1s2l$-excited states in high-density plasmas (Rosmej et al. 1998). Figure 1.12 shows a strong confinement of the satellite emission around the target surface, whereas the He_β-line expands far into space (up to several mm). The $1s3l3l'$-satellites emit almost only near the target surface and show negligible emission above the target surface because their intensity is almost proportional to the third power of the electron density. In low-density plasmas, the excited state population $1s2l$ is small and the $1s3l3l'$-satellites are rather weak because they are coupled essentially to the He-like ground state $1s^2\ ^1S_0$ via dielectronic capture (Rosmej 1998). We also note that autoionizing rates into the excited states $1s2l$ are usually orders of magnitude larger than corresponding ones into the He-like ground state $1s^2$ (Rosmej et al. 1998; Petitdemange and Rosmej 2013).

A further significant property of the satellite emission is that the radiation time of dielectronic satellite transitions is dominated by the large autoionizing rate rather than by radiative decay rate (as it is the case for usual resonance line emission), i.e.,

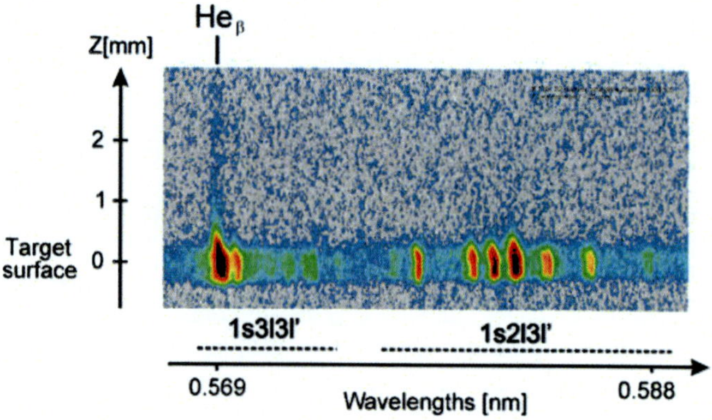

Fig. 1.12 Space-resolved X-ray emission spectrum of He_β and corresponding $1s2l3l'$- and $1s3l3l'$-satellite transitions from a high-density ns-laser-produced Si plasma. Satellite emission is strongly confined to the dense target surface plasma

$$\tau_{\text{sat}}(high\,\Gamma) \approx \frac{1}{A_{\text{rad}} + \Gamma_{\text{Auger}}} \approx \frac{1}{\Gamma_{\text{Auger}}}. \tag{1.105}$$

As characteristic autoionising rates are of the order of $10^{14}\ \text{s}^{-1}$, the radiative time scale of satellite transitions is typically of the order of some 10 fs only (see also Chap. 6).

The dielectronic satellite emission constitutes therefore a powerful class of X-ray transitions for spectroscopic diagnostics that are originating from multiple excited states of highly charged ions. Combined with modern spectroscopic techniques (high spectral resolution, high spatial resolution, high time resolution, high luminosity) unique and powerful characterizations of dense hot plasmas under extreme conditions are possible. Therefore, large theoretical and experimental efforts are currently devoted to the investigation of the radiation emission from multiple excited states of highly charged ions.

1.5.2 Stark Broadening Analysis of Rydberg Dielectronic Satellites

The analysis of space-resolved emission confined to the highest densities is of primary importance for Stark broadening analysis. Traditional applications (Griem 1964, 1974, 1997) are devoted to the analysis of non-autoionizing series lines that bare the experimental disadvantage of important spatial and temporal integration.

In order to profit from the particular properties of the dielectronic satellite transitions with respect to spatial and temporal integration, Stark broadening analysis has also been developed for Rydberg satellite transitions (Rosmej et al. 2003). These complex simulations became only recently numerically feasible with the development of powerful methods and computer codes (Talin and Klein 1982; Talin et al. 1995).

Figure 1.13 shows the Stark profile analysis of the Li-like Rydberg satellites $1s2l4l' \to 1s^2 2l + \hbar\omega_{\text{He}\gamma-\text{sat}}$ (Fig. 1.13a) and $1s2l5l' \to 1s^2 2l + \hbar\omega_{\text{He}\delta-\text{sat}}$ (Fig. 1.13b) carried out with the PPP code (Talin et al. 1995). Excellent agreement is obtained for the $1s2l4l'$-satellites. For the $1s2l5l'$-satellites, the emission group near 0.543 nm is likewise in excellent agreement, whereas the group near 0.545 nm shows discrepancies on the blue and red wing. These deviations from the data are due to the hollow ion Rydberg satellite transitions $K^1L^0M^1N^1 \to K^2L^0M^1 + \hbar\omega_{\text{Hollow}-\text{sat}}$ (Rosmej et al. 2003) that have not been included in the Stark broadening calculations.

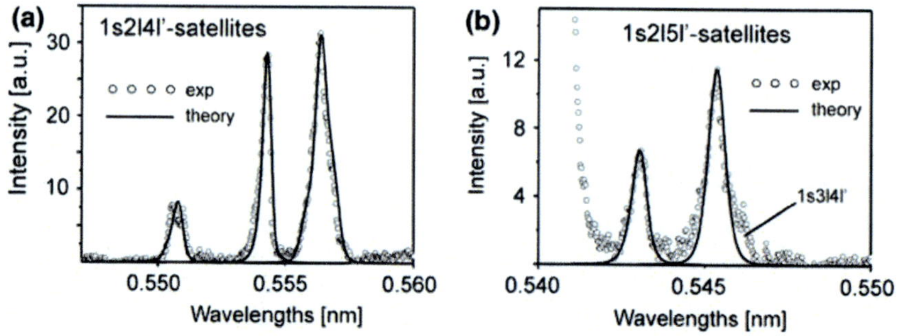

Fig. 1.13 Stark broadening analysis of Li-like Rydberg $1s2lnl'$-satellites of a dense ns-laser-produced silicon plasma. The simulations are carried out for $n_e = 4 \times 10^{20}$ cm^{-3}, $kT_e = kT_i = 350$ eV

1.5.3 Nonlinear Interference Effects in Stark Broadening of Multi-electron Configurations

The general theory of impact broadening is based on the density matrix and quantum kinetic approach (see Chap. 7) and considers the scattering amplitudes and phases thereby allowing to consider quantum mechanical interference effects. In the line broadening theory, interference effects arise due to transition frequencies that coincide or are so closely spaced that the corresponding spectral lines overlap. In some cases, the interference effects are so important that they alter the entire picture of the line broadening and it has been noted long time ago that interference effects may lead to a considerable line narrowing (Aleseyev and Sobelman 1969; Sobelman et al. 1995).

For satellites' transitions, or in general, transitions involving core hole states in upper and lower levels, interference effects result in important line narrowing or narrowing of the respective emission groups (Rosmej 2012a; Galtier et al. 2013) as is demonstrated in Fig. 1.14 via the Li-like dielectronic $1s2l2l'$- and $1s2l3l'$-satellites. It can clearly be seen that the omission of the interference effects considerably overestimates the width of the spectral distribution of the dielectronic satellite transitions (Rosmej 2012a). The situation is different for the high-energy channel $1s2l3l' \rightarrow 1s^2 2l + \hbar\omega$ of the same Li-like autoionizing configurations (satellites to $He_\beta = 1s3p\,^1P_1 \rightarrow 1s^2\,^1P + \hbar\omega$). Figure 1.15 demonstrates that the inclusion of interference effects practically do not change the overall shape of the satellite emission.

As Figs. 1.14 and 1.15 demonstrate, the high- and low-energy channel respond in a very different manner to quantum mechanical interference effects, that can schematically be summarized as follows

$$1s\,2l\,nl' \rightarrow \begin{cases} \text{low-energy channel} : 1s^2nl' & \text{strong interference effects} \\ \text{high-energy channel} : 1s^22l & \text{weak interference effects} . \end{cases} \quad (1.106)$$

The origin of this different behavior with respect to quantum interference effects is the different atomic structure of the $1s^22l$ levels with respect to the $1s^23l$ levels; see Table 1.1. The table demonstrates that for the low-energy channel, much more closely lying levels are encountered than for the high-energy channel. Correspondingly, the possibility for interference effects is larger for the low-energy channel.

We note that the fact that both channels originate from the same upper level makes studies of nonlinear interference effect more convenient, as different level populations for both channels (1.106) are not encountered.

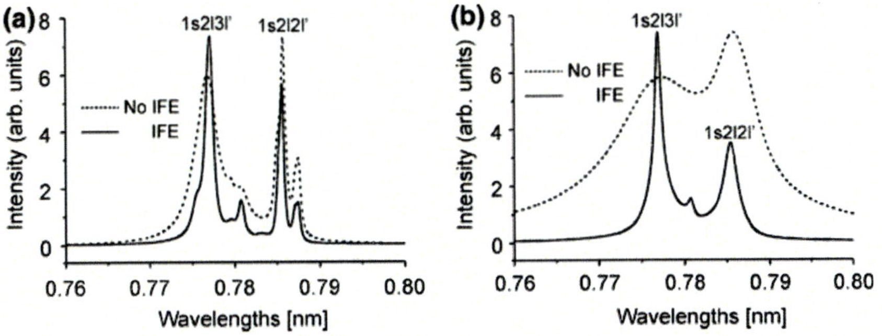

Fig. 1.14 Nonlinear interference effects in Stark broadening for Li-like satellite transitions $1s2l2l' \rightarrow 1s^22l + \hbar\omega$ and $1s2l3l' \rightarrow 1s^22l + \hbar\omega$ of a dense aluminum plasma at $kT_e = kT_i = 100$ eV, **a** $n_e = 1 \times 10^{23}$ cm^{-3} and **b** $n_e = 1 \times 10^{24}$ cm^{-3}

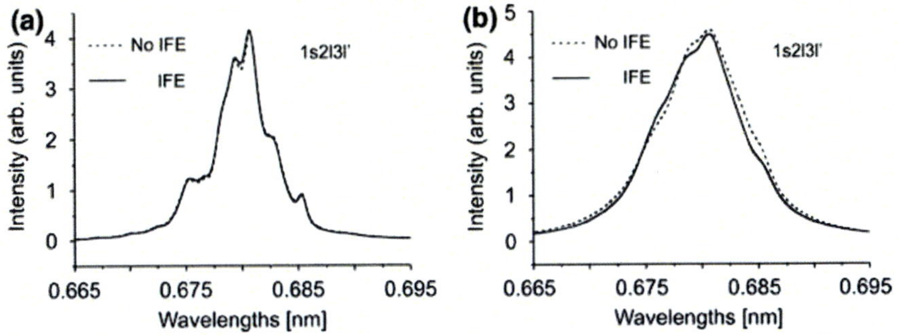

Fig. 1.15 Nonlinear interference effects in Stark broadening for Li-like satellite transitions $1s2l3l' \rightarrow 1s^22l + \hbar\omega$ of a dense aluminum plasma at $kT_e = kT_i = 100$ eV, **a** $n_e = 1 \times 10^{23}$ cm^{-3} and **b** $n_e = 3 \times 10^{23}$ cm^{-3}

Table 1.1 Energies of Li-like aluminum levels in [eV] counted from the ground state $1s^22s\ ^2S_{1/2}$

$1s^22s\ ^2S_{1/2}$	$1s^22p\ ^2P_{1/2}$	$1s^22p\ ^2P_{3/2}$	$1s^23s\ ^2S_{1/2}$	$1s^23p\ ^2P_{1/2}$	$1s^23p\ ^2P_{3/2}$	$1s^23d\ ^2D_{3/2}$	$1s^23d\ ^2D_{5/2}$
0.0	21.824	22.541	250.50	256.50	256.71	258.89	258.95

It is important to point out that the apparent width of a satellite group is not only a result of the Stark broadening but also due to the spectral distribution of the oscillator strengths (see more details in Chap. 6). Therefore, complex atomic structure calculations (involving intermediate coupling, multiconfigurations, and corresponding configuration interaction) are important to correctly predict the total width (Stark + oscillator strength distribution) of the satellite group emission. Detailed analysis of the precision of atomic structure calculations as well as different methods of calculations on Stark profiles has been studied in Rosmej et al. (2013).

Also interference effects in hollow ion X-ray emission have been considered (Rosmej 2012b). Hollow ion X-ray transitions are of great interest for the study of high-density matter due to their negligible bound–bound opacity: The lower state of the hollow ion X-ray transition is usually not an atomic ground state but an excited autoionizing state that is weakly populated even in very dense plasmas (Rosmej et al. 2015). For example, for the hollow ion configurations K^0L^N, the lower state of the X-ray transitions

$$K^0L^N \rightarrow K^1L^{N-1} + \hbar\omega_{\text{Hollow}} \tag{1.107}$$

is the K^1L^{N-1} which is an autoionizing state that is weakly populated. This is so, because the autoionizing rate is of the order of 10^{13}–10^{14} s^{-1} resulting in a rapid disintegration of the lower level population via the Auger effect:

$$K^1L^{N-1} \rightarrow K^2L^{N-3} + e_{\text{Auger}}. \tag{1.108}$$

Therefore, the bound–bound opacity [see (1.57)] for these hollow ion X-ray transitions

$$\tau_{\text{ij}}^{(\text{hollow})} \propto f_{\text{ij}} \cdot \lambda_{\text{ji}} \cdot L_{\text{eff}} \cdot n_{\text{i}}\left(K^1L^{N-1}\right) \tag{1.109}$$

is very low and these transitions can be observed in very high-density plasmas as their emission might readily leave the plasma while other transitions with highly populated lower states are strongly absorbed.

1.5.4 Hollow Ion X-Ray Emission in Dense Plasmas

A hollow ion is an ion, where one or more internal shells are entirely empty, whereas higher shells are filled with two or more electrons. Simple examples are the $K^0L^xL^y$ configurations for x = 2–8 and different numbers y of electrons in the M-shell: e.g., x = 1 and y = 0 corresponds therefore to the H-like Ly$_\alpha$ transitions, and x = 2 and y = 0 corresponds to the He-like $2l2l'$-satellites near Ly$_\alpha$.

Figure 1.16 shows the hollow ion X-ray emission of highly charged argon ions with two K-shell vacancies and different numbers of L-shell spectator electrons as observed with high-resolution X-ray spectroscopic methods in a dense Mega-Ampère Z-pinch experiment (Rosmej et al. 1993, 2001b), i.e.,

$$K^0L^N \to K^1L^{N-1} + \hbar\omega_{\mathrm{Hollow}}. \qquad (1.110)$$

Hollow ion transitions originating from the configurations K^0L^N of highly charged ions are of particular interest for dense plasma research. These transitions connect the most strongly bound states in atomic/ionic systems. Strongly coupled plasma effects can therefore be studied as perturbations to these stable transitions. Next, the response time of hollow ion emission is of the order of some 10 fs only and represents therefore a fast X-ray emission switch at times when the plasma density is highest (Rosmej et al. 2007). Likewise, as already discussed above, the intensity contributions from the low-density long-lasting recombining regime is negligible. Of particular importance for high-energy-density physics is the fact that the absorbing lower states are by themselves autoionizing states (namely K^1L^N) that are weakly populated even in dense plasmas. Radiation transport effects will therefore be small, e.g., even in ICF plasmas with well above solid density compression.

As concerns their line identification, the hollow ion X-ray transitions $K^0L^N \to K^1L^{N-1} + \hbar\omega_{\mathrm{Hollow}}$ are well separated from other transitions (Rosmej et al. 2015) and the identification of double K-hole configurations is therefore less complicated (for $N = 2$–5, the transitions are essentially located between the H-like Ly_α-line and the He-like He_α-line), as can be seen in Fig. 1.16.

Other well-observed examples of hollow ion configuration are of type $K^1L^0M^X$ for $x = 2$–18 (the L-shell does not contain any electrons, whereas the M-shell is filled with two or more electrons). The case $x = 2$ has been presented in Fig. 1.12 and is of particular importance for emission confined to the high-density plasma.

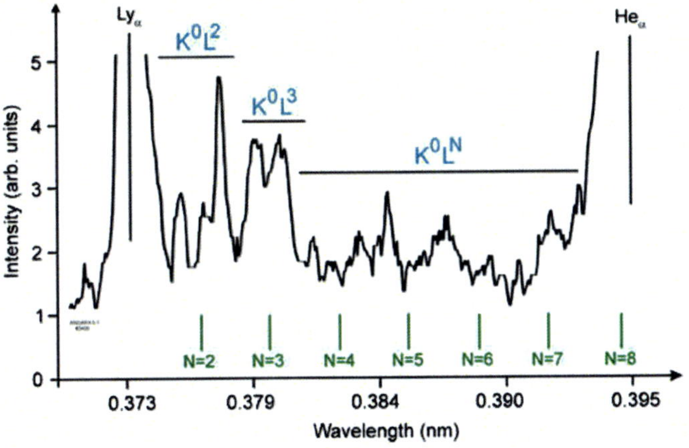

Fig. 1.16 Hollow ion X-ray emission from argon ions with two vacancies in the K-shell observed in a dense Mega-Ampère Z-pinch

Hollow ion configurations with no electrons in K- and L-shell have been considered in Rosmej et al. (1999b), Colgan et al. (2016).

Traditionally, hollow ion X-ray emission has been observed in ion beam surface interaction experiments (Armour et al. 1980; Briand et al. 1990) under low-density conditions: Highly charged ions are extracted from a heavy particle accelerator at definite energies and are then brought into interaction with matter. However, recently also in a few visible and IR laser-produced plasma experiments, observation of hollow ion emission has been claimed (Rosmej et al. 1999b, 2015; Faenov et al. 1999; Abdallah et al. 2000; Colgan et al. 2013), as well as in high current dense Z-pinch plasmas (Rosmej et al. 2001b). Although hollow ion emission has been identified on the basis of atomic structure calculations, the origin of their high intensities in dense laser-produced plasmas and Z-pinch plasmas is not yet well understood.

In Rosmej et al. (1999b, 2002, 2006, 2015), charge exchange between intermixing inhomogenous dense plasmas has been proposed to explain HI and dielectronic satellite emission. However, up to now no "ab initio" atomic kinetic models were developed to provide proof of any reasonable production mechanisms except transient three-body recombination (Rosmej et al. 1997b) and charge exchange (Rosmej et al. 2002). We note that usual collisional–radiative simulations indicate that HI emission in dense laser-produced plasmas should be quite small (Rosmej et al. 2015), i.e., below bremsstrahlung. Current achievements in the understanding of HI emission in dense plasmas concern essentially atomic structure calculations and corresponding identification of configurations. However, agreement with observed hollow ion emission (intensities) has only been achieved by means of numerous free intensity parameters (including LTE assumptions (Abdallah et al. 2000) to increase considerably (many orders of magnitude) the population of the hollow ion configurations with respect to atomic ground and singly excited states. The origin of these large parameters themselves remained not well justified.

It is therefore of fundamental interest to develop "ab initio" hollow ion population kinetics to shed more light into this mystery. The first attempt in this direction has been made by the development of ab initio HI population kinetics that is driven by intense radiation fields (Rosmej et al. 2007; Rosmej and Lee 2007). It was shown that effective hollow ion X-ray emission can be excited up to an observable level by the initiation of an effective photoionization chain reaction (Rosmej and Lee 2007):

$$K^2L^N + \hbar\omega_{\mathrm{XFEL}} \rightarrow K^1L^N + e, \tag{1.111a}$$

$$K^1L^N + \hbar\omega_{\mathrm{XFEL}} \rightarrow K^0L^N + e, \tag{1.111b}$$

$$K^0L^N \rightarrow K^1L^{N-1} + \hbar\omega_{\mathrm{Hollow}-K^0}. \tag{1.111c}$$

It is important to note that the fully time-dependent simulations carried out with the MARIA code (Rosmej 1997, 1998, 2001, 2006, 2012a, b) did not contain any free parameters. The excitation mechanism is therefore well identified: chain reaction-induced photoionization of the K-shell by intense radiation fields (for more details see also Chap. 10). The XFEL-driven hollow ion emission according to (1.111) predicted by the MARIA simulations for magnesium (Rosmej and Lee 2007) has finally later been observed at the X-ray free electron laser LCLS installation irradiating solid Mg samples (Ciricosta et al. 2016). It should be noted that also the predicted high intensity of the hollow ion emission (Rosmej and Lee 2007) (being of the same order like usual resonance line emission) corresponds to the observations made in (Ciricosta et al. 2016). Finally, we note that X-ray hollow ion emission has also been observed irradiating gases with intense XFEL (Young et al. 2010). It has also been claimed that observed intense hollow ion emission in petawatt laser-produced plasma experiments might be due to radiation fields driven by the laser–matter interaction (Colgan et al. 2013, 2016).

Figure 1.17 shows a high-quality high-resolution X-ray spectrum of aluminum (Colgan et al. 2013) where very intense and even dominating hollow ion emission from various charge stages are identified (indicated as *Hollow ions* $K^0L^XM^Y$ in Fig. 1.17). Likewise, Fig. 1.17 demonstrates, that even in a harsh radiation environment like petawatt installations, high-precision X-ray spectroscopy can be performed, thus allowing to apply X-ray spectroscopic methods and atomic physics for a unique characterization of matter under extreme conditions (Renner and Rosmej 2019). This is of great importance despite many controversially discussed phenomena like suprathermal electrons, radiation fields, opacity and near-solid-density atomic plasma physics. It is important to emphasize that the spectrum in Fig. 1.17 is in linear intensity scale and has been obtained from a direct scan of the exposed X-ray film. In order to avoid blacking of the X-ray film due to hot electrons, the spectrometer was protected with metal and plastic plates and the line of sight passed a strong static magnet (to avoid direct illumination of the crystal surface by hot electrons).

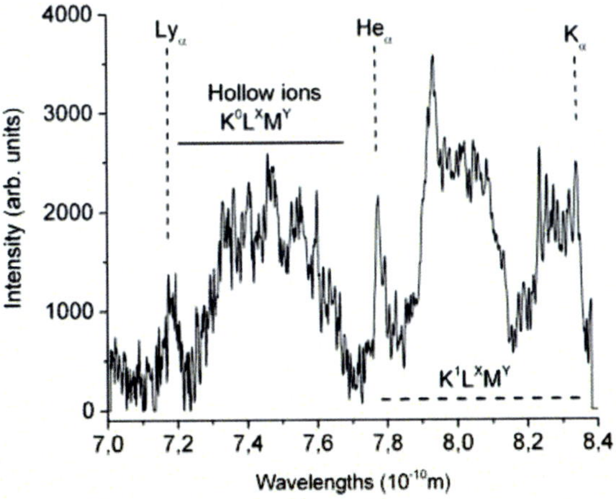

Fig. 1.17 Hollow ion aluminum X-ray emission induced by irradiating a 1.5 μm thin aluminum foil with a 160 J, 0.7 ps, 1.064 μm petawatt laser at $I = 3 \times 10^{20}$ W/cm^2

References

J. Abdallah Jr., I.Yu. Skobelev, A.Ya. Faenov, A.I. Magunov, T.A. Pikuz, F. Flora, S. Bollanti, P. Di Lazzaro, T. Letardi, E. Burattini, A. Grilli, A. Reale, L. Palladino, G. Tomassetti, A. Scafati, L. Reale, Spectra of multiply charged hollow ions in the plasma produced by a short-wavelength nanosecond laser. Quantum Electron. **30**, 694 (2000)

P. Adamek, O. Renner, L. Drska, F.B. Rosmej, J.-F. Wyart, Genetic algorithms in spectroscopic diagnostics of hot dense plasmas. Laser Part. Beams **24**, 511 (2006)

E.Kh. Akhmedov, A.L. Godunov, Yu.K. Zemtsov, V.A. Makhrov, A.N. Starostin, M.D. Thran, Profile of the Ly-alpha line of hydrogen-like ions in a dense plasma with allowance for the fine structure and the Lamb shift. JETP **62**, 266 (1985)

A.V. Anufrienko, A.L. Godunoy, A.V. Demura, Yu. K. Zemtzoy, V.S. Lisitsa, A.N. Starostin, M. D. Thran, V.A. Schipako, Nonlinear interference effects in Stark broadening of ion lines in a dense plasma. JETP **71**, 728 (1990)

A.V. Anufrienko, A.E. Bulyshev, A.L. Godunov, A.V. Demura, Yu.K. Zemtsov, V.S. Lisitsa, A. N. Starostin, Nonlinear interference effects and ion dynamics in the kinetic theory of Stark broadening of the spectral lines of multicharged ions in a dense plasma. JETP **76**, 219 (1993)

I.A. Armour, B.C. Fawcdtt, J.D. Silver, E. Träbert, X-ray spectra and satellite classification of foil-excited Mg and Al. J. Phys. B: At. Mol. Opt. Phys. **13**, 2701 (1980)

V.A. Aleseyev, I.I. Sobelman, Influence of collisions on stimulated random scattering in gases. JETP **28**, 991 (1969)

V.B. Berestetskii, E.M. Lifschitz, L.P. Pitaevskii, *Quantum Electrodynamics* (Elsevier Butterworth-Heinemann, Oxford, 2008)

H.A. Bethe, E.E. Salpeter, *Quantum Mechanics of One- and Two-electron Atoms* (Plenum Publishing, New York, 1977)

V.A. Boiko, A.V. Vinogradov, S.A. Pikuz, I.Yu. Skobelev, A.Ya. Faenov, X-ray spectroscopy of laser produced plasmas. J. Sov. Laser Research **6**, 82 (1985)

K. Blum, *Denstiy Matrix Theory and Applications* (Springer, New York, 2012)

A.E. Bulyshev, A.V. Demura, V.S. Lisitsa, A.N. Starostin, A.E. Suvorov, I.I. Yakunin, Redistribution function for resonance radiation in a hot dense plasma. JETP **81**, 113 (1995)

L.A. Bureyeva, V.S. Lisitsa, D.A. Petrov, D.A. Shuvaev, F.B. Rosmej, R. Stamm, The effect of charge exchange with neutrals on the saturation of the spectral line intensities of mulitcharged ions in plasmas. Plasma Phys. Rep. **29**, 835 (2003)

J.P. Briand, L. de Billy, P. Charles, S. Essabaa, P. Briand, R. Geller, J.P. Desclaux, S. Bliman, C. Ristori, Production of hollow atoms by the excitation of highly charged ions in interaction with a metallic surface. Phys. Rev. Lett. **65**, 159 (1990)

K. Burnett, Collisional redistriution of radiation. Phys. Rep. **118**, 339 (1985)

F.F. Chen, *Introduction to Plasma Physics and Controlled Fusion*, 2nd edn. (Plenum Press, New York, 1984)

O. Ciricosta, S.M. Vinko, B. Barbrel, D.S. Rackstraw, T.R. Preston, T. Burian, J. Chalupsky, B.I. Cho, H.-K. Chung, G.L. Dakovski, K. Engelhorn, V. Hájkova, P. Heimann, M. Holmes, L. Juha, J. Krzywinski, R.W. Lee, S. Toleikis, J.J. Turner, U. Zastrau, J.S. Wark, Measurements of continuum lowering in solid-density plasmas created from elements and compounds. Nat. Comm. **7**, 11713 (2016)

J. Colgan, J. Abdallah Jr., A.Ya. Faenov, S.A. Pikuz, E. Wagenaars, N. Booth, O. Culfa, R. J. Dance, R.G. Evans, R.J. Gray, T. Kaempfer, K.L. Lancaster, P. McKenna, A.L. Rossall, I. Yu. Skobelev, K.S. Schulze, I. Uschmann, A.G. Zhidkov, N.C. Woolsey, Exotic dense-matter states pumped by a relativistic laser plasma in the radiation-dominated regime. Phys. Rev. Lett. **110**, 125001 (2013)

J. Colgan, A.Ya. Faenov, S.A. Pikuz, E. Tubman, N.M.H. Butler, J. Abdallah Jr., R.J. Dance, T.A. Pikuz, I.Yu. Skobelev, M.A. Alkhimova, N. Booth, J. Green, C. Gregory, A. Andreev, R. Lötzsch, I. Uschmann, A. Zhidkov, R. Kodama, P. MicKenna, N. Woolsey, Evidence of high-n hollow-ion emission from Si ions pumped by ultraintense X-rays from relativistic laser plasma. Europhysics Lett. **114**, 35001 (2016)

R.D. Cowan, *The Theory of Atomic Structure and Spectra* (University of California Press, California, 1981)

A.V. Demura, A.V. Anufrienko, A.L. Godunov, Yu.K. Zemtsov, V.S. Lisitsa, A.N. Starostin, M.D. Taran, V.A. Schipakov, Nonlinear Interference Effects in Partial Redistribution of Radiation in Dense Plasmas. AIP Conf. Proc. **216**(1), 227 (1990)

G.W.F. Drake, *Handbook of Atomic, Molecular, and Optical Physics* (Springer, Berlin, 2006)

G.W. Erickson, Improved lamb-shift calculation for all values of Z. Phys. Rev. Lett. **27**, 780 (1971)

A.Ya. Faenov, S.A. Pikuz, A.I. Erko, B.A. Bryunetkin, V.M. Dyakin, G.V. Ivanenkov, A.R. Mingaleev, T.A. Pikuz, V.M. Romanova, T.A. Shelkovenko, High-performance X-ray spectroscopic devices for plasma microsources investigations. Phys. Scr. **50**, 333 (1994)

A.Ya. Faenov, A.I. Magunov, T.A. Pikuz, I.Yu. Skobelev, S.A. Pikuz, A.M. Urnov, J. Abdallah, R.E.H. Clark, J. Cohen, R.P. Johnson, G.A. Kyrala, M.D. Wilke, A. Maksimchuk, D. Umstadter, N. Nantel, R. Doron, E. Behar, P. Mandelbaum, J.J. Schwob, J. Dubau, F.B. Rosmej, A. Osterheld, High-resolved X-ray spectra of hollow atoms in a femtosecond laser-produced solid plasma. Phys. Scr. **T80**, 536 (1999)

U. Fano, Effects of configuration interaction on intensities and phase shifts. Phys. Rev. **124**, 1866 (1961)

K.B. Fournier, M. Cihen, M.J. May, W.H. Goldstein, Ionization state distribution and radiative cooling rate for argon in a low-density plasma. ADNDT **70**, 231 (1998)

T. Fujimoto, *Plasma Spectroscopy* (Clarendon Press, Oxford, 2004)

A.H. Gabriel, Dielectronic satellite spectra for highly-charged He-like ion lines. Mon. Not. R. Astr. Soc. **160**, 99 (1972)

E. Galtier, F.B. Rosmej, A. Calisti, B. Talin, C. Mossé, S. Ferri, V.S. Lisitsa, Interference effects and Stark broadening in XUV intra-shell transitions in aluminum under conditions of intense XUV free electron laser irradiation. Phys. Rev. A **87**, 033422 (2013)

H.R. Griem, *Plasma Spectroscopy* (McGraw-Hill Book Company, New York, 1964)

H.R. Griem, *Spectral Line Broadening by Plasmas* (Academic Press, New York and London, 1974)

H.R. Griem, *Principles of Plasma Spectroscopy* (Cambridge University Press, New York, 1997)

T. Holstein, Imprisonment of Resonance Radiation in Gases. Phys. Rev. **72**, 1212 (1947)

J.W.B. Hughes, Stark states and O(4) symmetry of hydrogenic atoms. Proc. Phys. Soc. **91**, 810 (1967)

S. Ichimaru, *Basic Principles of Plasma Physics: A Statistical Approach* (W.A. Benjamin Inc., London, 1973)

S. Ichimaru, *Statistical Plasma Physics I: Basic Principles* (Westview Press, 2004a)

S. Ichimaru, *Statistical Plasma Physics II: Condensed Plasmas* (Westview Press, 2004b)

F.E. Irons, The escape factor in plasma spectroscopy - I. The escape factor defined and evaluated. JQSRT **22**, 1 (1979)

R.K. Janev, L.P. Presnyakov, V.P. Shevelko, *Physics of Highly Charged Ions* (Springer, Berlin, 1985)

S. Klarsfeld, Radiative decay of metastable hydrogenic atoms. Phys. Lett. A **30**, 382 (1969)

Yu.A. Klimontowitch, *Kinetic Theory of Electromagnetic Processes* (Springer, Berlin, 1983)

A.K. Kompanets, Resonance processes in the photoeffect. JETP **27**, 519 (1968)

N.A. Krall, A.W. Trivelpiece, *Principles of Plasma Physics* (McGraw-Hill, New York, 1973)

T. Krücken, K. Bergmann, L. Juschkin, R. Lebert, Fundamentals and limits for the EUV emission of pinch plasma sources for EUV lithography. J. Phys. B: At. Mol. Opt. Phys. **37**, 3213 (2004)

H.-J. Kunze, *Introduction to Plasma Spectroscopy* (Springer, Berlin, 2009)

L.D. Landau, E.M. Lifschitz, *Quantum Mechanics: Non-relativistic theory* (Elsevier Butterworth-Heinemann, Oxford, 2005)

X. Li, Z. Xu, F.B. Rosmej, Exchange energy shifts in dense plasma conditions. J. Phys. B: At. Mol. Opt. Phys. **39**, 3373 (2006)

V.S. Lisitsa, S.I. Yakovlenko, Resonance of discrete states against the background of a continuous spectrum. JETP **39**(6), 975 (1974)

R. Loudon, *The Quantum Theory of Light*, 3rd ed. (Oxford Science Publications, Oxford, 2000)

G. Lüders, Der Starkeffect des Wasserstoffs bei kleinen Feldstärken. Ann. Phys. **443**, 301 (1950)

R.W.P. McWhirter, Spectral intensities, in *Plasma Diagnostic Techniques*, ed. by R.H. Huddelstone, S.L. Leonard (Academic Press, New York, 1965)

C. de Michelis, M. Mattioli, Soft X-ray spectroscoy of laboratory plasmas. Nucl. Fusion **21**, 677 (1981)

D. Mihalas, *Stellar Atmospheres*, 2nd edn. (W.H. Freeman & Co., San Francisco, 1978)

C. Mossé, A. Calisti, R. Stamm, B. Talin, R. Lee, L. Klein, Redistribution of resonance radiation in hot and dense plasmas. Phys. Rev. A **60**, 1005 (1999)

V.G. Palchikov, A.Ya. Faenov, I.Yu. Skobelev, A.I. Magunov, T.A. Pikuz, F.B. Rosmej, D.H.H. Hoffmann, W. Süß, M. Geißel, E. Biemont, P. Palmeri, P. Quinet, A. Sasaki, T. Utsumi, Measurements of the ground state ionization energy and wavelengths for the 1snp $^1P_1$ - 1 $s^2\,{}^1S_0$ (n = 4–9) lines of Mg XI and F VIII. J. Phys. B: At. Mol. Opt. Phys. **35**, 2741 (2002)

F. Petitdemange, F.B. Rosmej, Dielectronic satellites and Auger electron heating: irradiation of solids by intense XUV-Free Electron Laser radiation, in *New Trends in Atomic & Molecular Physics—Advanced Technological Applications*, vol. **76**, ed. by M. Mohan (Springer, Berlin, 2013), pp. 91–114. ISBN 978-3-642-38166-9

S.G. Rautian, A.M. Shalagin, *Kinetic Problems of Non-linear Spectroscopy* (North-Holland, Amsterdam, 1991)

F. Reif, *Fundamental of Statistical and Thermal Physics* (Mc-Graw Hill Book Company, New York, 1965)

O. Renner, F.B. Rosmej, Challenges in X-ray spectroscopy in investigations of matter under extreme conditions. Matter. Radiat. Under Extremes **4**, 024201 (2019)

O. Renner, T. Missalla, P. Sondhauss, E. Krousky, E. Förster, C. Chenais-Popovics, O. Rancu, Vertical dispersion Johann X-ray spectrometer with asymmetrically cut crystal. Rev. Sci. Instr. **68**, 2393 (1997)

O. Renner, E. Krouský, F.B. Rosmej, P. Sondhauss, M.P. Kalachnikov, P.V. Nickles, I. Uschmann, E. Förster, Observation of H-like Al Ly_α disappearance in dense cold laser produced plasmas. Appl. Phys. Lett. **79**, 177 (2001)

O. Renner, P. Adámek, P. Angelo, E. Dalimier, E. Förster, E. Krousky, F.B. Rosmej, R. Schott, Spectral line decomposition and frequency shifts in Al He_α group emission from laser produced plasma. JQSRT **99**, 523 (2006)

F.B. Rosmej, A spectroscopic method for the determination of the bulk-electron temperature in highly ionized plasmas containing Non-Maxwellian electrons. J. Phys. B Lett.: At. Mol. Opt. Phys. **28**, L747 (1995)

F.B. Rosmej, Hot electron X-ray diagnostics. J. Phys. B. Lett.: At. Mol. Opt. Phys. **30**, L819 (1997)

F.B. Rosmej, Kβ-line emission in fusion plasmas. Rapid Communication to Phys. Rev. E **58**, R32 (1998)

F.B. Rosmej, A new type of analytical model for complex radiation emission of hollow ions in fusion and laser produced plasmas. Europhys. Lett. **55**, 472 (2001)

F.B. Rosmej, An alternative method to determine atomic radiation. Europhys. Lett. **76**, 1081 (2006)

F.B. Rosmej, Exotic states of high density matter driven by intense XUV/X-ray Free Electron Lasers, in *Free Electron Laser, InTech*, , ed. by S. Varró (2012a), pp. 187–212. ISBN 978-953-51-0279-3. Free download: http://www.intechopen.com/books/free-electron-lasers/exotic-states-of-high-density-matter-driven-by-intense-xuv-x-ray-free-electron-lasers

F.B. Rosmej, X-ray emission spectroscopy and diagnostics of non-equilibrium fusion and laser produced plasmas, in *Highly Charged Ion Spectroscopic Research*, ed. by Y. Zou, R. Hutton (Taylor and Francis, New York, 2012b), pp. 267–341. ISBN: 978-1-4200-7904-3

F.B. Rosmej, A.Ya. Faenov, New Innershell Phenomena from Rydberg Series of Highly Charged Ions. Phys. Scr. **T73**, 106 (1997)

F.B. Rosmej, V.S. Lisitsa, A self-consistent method for the determination of neutral density from X-ray impurity spectra. Phys. Lett. A **244**, 401 (1998)

F.B. Rosmej, R.W. Lee, Hollow ion emission driven by pulsed X-ray radiation fields. Europhys. Lett. **77**, 24001 (2007)

F.B. Rosmej, A. Schulz, K.N. Koshelev, H.-J. Kunze, Asymmetric repumping of the Lyman-α components of hydrogen-like ions in a dense expanding plasma. JQSRT **44**, 559 (1990)

F.B. Rosmej, O.N. Rosmej, S.A. Komarov, V.O. Mishensky, J.G. Utjugov, Soft X-ray spectra analysis in a high current Z-pinch. AIP Conf. Proc. **299**, 552 (1993)

F.B. Rosmej, A.Ya. Faenov, T.A. Pikuz, F. Flora, P. Di Lazzaro, S. Bollanti, N. Lizi, T. Letardi, A. Reale, L. Palladino, O. Batani, S. Bossi, A. Bornardinello, A. Scafati, L. Reale, A. Zigler, M. Fraenkel, R.D. Cowan, Innershell satellite transitions in dense short pulse plasmas. JQSRT **58**, 859 (1997a)

F.B. Rosmej, A.Ya. Faenov, T.A. Pikuz, I.Yu. Skobelev, A.E. Stepanov, A.N. Starostin, B.S. Rerich, B.A. Makhrov, F. Flora, S. Bollanti, P. Di Lazzaro, T. Letardi, K. Wigli-Papadaki, A. Nottola, A. Grilli, L. Palladino, A. Reale, A. Scafati, L. Reale, Dominant role of dielectronic satellites in the radiation spectra of a laser plasma near the target surface. JETP Lett. **65**, 708 (1997b)

F.B. Rosmej, A.Ya. Faenov, T.A. Pikuz, F. Flora, P. Di Lazzaro, S. Bollanti, N. Lizi, T. Letardi, A. Reale, L. Palladino, O. Batani, S. Bossi, A. Bornardinello, A. Scafati, L. Reale, Line formation of high intensity He_{β}-Rydberg dielectronic satellites 1s3lnl′ in laser produced plasmas. J. Phys. B Lett.: At. Mol. Opt. Phys. **31**, L921 (1998)

F.B. Rosmej, V.S. Lisitsa, D. Reiter, M. Bitter, O. Herzog, G. Bertschinger, H.-J. Kunze, Influence of charge exchange processes on X-ray spectra in tokamak plasmas: experimental and theoretical investigation. Plasma Phys. Controlled Fusion **41**, 191 (1999a)

F.B. Rosmej, A.Ya. Faenov, T.A. Pikuz, A.I. Magunov, I.Yu. Skobelev, T. Auguste, P.D'Oliveira, S. Hulin, P. Monot, N.E. Andreev, M.V. Chegotov, M.E. Veisman, Charge exchange induced formation of hollow atoms in high intensity laser produced plasmas. J. Phys. B. Lett. : At. Mol. Opt. Phys. **32**, L107 (1999b)

F.B. Rosmej, U.N. Funk, M. Geißel, D.H.H. Hoffmann, A. Tauschwitz, A.Ya. Faenov, T.A. Pikuz, I.Yu. Skobelev, F. Flora, S. Bollanti, P.Di. Lazzaro, T. Letardi, A. Grilli, L. Palladino, A. Reale, A. Scafati, L. Reale, T. Auguste, P. D'Oliveira, S. Hulin, P. Monot, A. Maksimchuk, S. A. Pikuz, D. Umstadter, M. Nantel, R. Bock, M. Dornik, M. Stetter, S. Stöwe, V. Yakushev, M. Kulisch, N. Shilkin, X-ray radiation from ions with K-shell vacancies. JQSRT **65**, 477 (2000)

F.B. Rosmej, D.H.H. Hoffmann, M. Geißel, M. Roth, P. Pirzadeh, A.Ya. Faenov, T.A. Pikuz, I. Yu. Skobelev, A.I. Magunov, Space resolved observation of Si Li-like high energy Rydberg-transitions from autoionizing levels in dense laser produced plasmas. Phys. Rev. A **63**, 063409 (2001a)

F.B. Rosmej, D.H.H. Hoffmann, W. Süß, M. Geißel, O.N. Rosmej, A.Ya. Faenov, T.A. Pikuz, T. Auguste, P.D'Oliveira, S. Hulin, P. Monot, J.E. Hansen, G. Verbookhaven, High resolution X-ray imaging spectroscopy diagnostic of hollow ions in dense plasmas. Nuc. Instrum. Methods A **464**, 257 (2001b)

F.B. Rosmej, H.R. Griem, R.C. Elton, V.L. Jacobs, J.A. Cobble, A.Ya. Faenov, T.A. Pikuz, M. Geißel, D.H.H. Hoffmann, W. Süß, D.B. Uskov, V.P. Shevelko, R.C. Mancini, Investigation of charge exchange induced formation of two electron satellite transitions in dense laser produced plasmas. Phys. Rev. E **66**, 056402 (2002)

F.B. Rosmej, A. Calisti, R. Stamm, B. Talin, C. Mossé, S. Ferri, M. Geißel, D.H.H. Hoffmann, A. Ya. Faenov, T.A. Pikuz, Strongly coupled laser produced plasmas: investigation of hollow ion formation and line shape analysis. JQSRT **81**, 395 (2003)

F.B. Rosmej, V.S. Lisitsa, R. Schott, E. Dalimier, D. Riley, A. Delserieys, O. Renner, E. Krousky, Charge exchange driven X-ray emission from highly ionized plasma jets. Europhys. Lett. **76**, 815 (2006)

F.B. Rosmej, R.W. Lee, D.H.G. Schneider, Fast X-ray emission switches driven by intense X-ray free electron laser radiation. High Energy Density Phys. **3**, 218 (2007)

F.B. Rosmej, K. Bennadji, V.S. Lisitsa, Dense plasmas effects on exchange energy shifts in highly charged ions: an alternative approach for arbitrary perturbation potentials. Phys. Rev. A **84**, 032512 (2011)

F.B. Rosmej, B. Deschaud, K. Bennadji, P. Indelicato, J.P. Marquès, Study of electric dipole matrix elements of He-like ions for X-ray line shape calculations. Phys. Rev. A **87**, 022515 (2013)

F.B. Rosmej, R. Dachicourt, B. Deschaud, D. Khaghani, M. Dozières, M. Smîd, O. Renner, Exotic X-ray emission from dense plasmas. J. Phys. B: Review Special Topics **48**, 224005 (2015)

D. Salzmann, *Atomic Physics in Hot Plasmas* (Oxford University Press, New York, 1998)

U.I. Safronova, V.S. Senashenko, *Theory of Multicharged Ions Spectra* (Energoizdat, Moscow, 1983). (in Russian)

G. Schmidt, *Physics of High Temperature Plasmas*, 2nd edn. (Academic Press, New York, 1979)

V.P. Shevelko, L.A. Vainshtein, *Atomic Physics for Hot Plasmas* (Taylor & Francis, New York, 1993)

I.Yu. Skobelev, A.Ya. Faenov, B.A. Bryunetkin, V.M. Dyakin, Investigating the emission properties of plasma structure with X-ray imaging spectroscoy. JETP **81**, 692 (1995)

I.Yu. Skobelev, A.Ya. Faenov, T.A. Pikuz, A.I. Magunov, F. Flora, S. Bollanti, P. DiLazzaro, D. Murra, A. Reale, L. Reale, G. Tomassetti, A. Ritucci, G. Petrocelli, S. Martellucci, N. Lisi, F.B. Rosmej, Spectral transitions from the Rydberg autoionization states of a Li-like Mg X Ion. JETP **95**, 421 (2002)

I.I. Sobelman, *The Theory of Atomic Spectra* (Alpha Science, Oxford, 2006)

I.I. Sobelman, L.A. Vainshtein, E.A. Yukov, *Excitation of Atoms and Broadening of Spectral Lines* (Springer, Berlin, 1995)

R.F. Stebbings, F.B. Dunning, *Rydberg States of Atoms and Molecules* (Cambridge University Press, Cambridge, 1983)

S. Sudkewer, Spectroscopic diagnostics of tokamak plasmas. Phys. Scr. **23**, 72 (1981)

B. Talin, L. Klein, Model microfield theory of redistribution of laser radiation. Phys. Rev. A **26**, 2717 (1982)

B. Talin, A. Calisti, L. Godberg, R. Stamm, R.W. Lee, L. Klein, Frequency-fluctuation model for line-shape calculations in plasma spectroscopy. Phys. Rev. A **51**, 1981 (1995)

B. Talin, R. Stamm, V.P. Kaftandjian, L. Klein, Partial redistribution in high-density, highly ionized plasmas. Astr. J. **322**, 804 (1997)

J. Vieker, K. Bergmann, Influence of the electrode wear on the EUV generation of a discharge based extreme ultraviolet light source. J. Phys. D: Appl. Phys. **50**, 345601 (2017)

L. Young, E.P. Kanter, B. Krässig, Y. Li, A.M. March, S.T. Pratt, R. Santra, S.H. Southworth, N. Rohringer, L.F. DiMauro, G. Doumy, C.A. Roedig, N. Berrah, L. Fang, M. Hoener, P.H. Bucksbaum, J.P. Cryan, S. Ghimire, J.M. Glownia, D.A. Reis, J.D. Bozek, C. Bostedt, M. Messerschmidt, Femtosecond electronic response of atoms to ultra-intense X-rays. Nature **466**, 56 (2010)

Ya.B. Zel'dovich, Yu.P. Raizer, *Physics of Shock-Waves and High-Temperature Hydrodynamic Phenomena* (Dover Publications, New York, 2002)

A.G. Zhidkoy, A.N. Tkachev, S.I. Yakovlenko, Effect of fine structure on the line shape of hydrogenic ions. JETP **64**, 261 (1986)

Chapter 2
Radiative Characteristics of Polarized Atoms and Ions

Abstract The chapter is devoted to an introduction of the main radiative characteristics of atoms and ions to describe the light-matter interaction. The presentation is based on the correspondence principle between classical and quantum physics in order to derive the basic radiative properties, including the Einstein coefficients and the dynamic polarizability of the atom. Attention is also paid to the description of the dynamic polarizability of atoms and ions including the static, high-frequency and resonance case. The innovative concept of local plasma frequency is introduced for the atom that allows describing the polarizability of many-electron systems at a semi-quantitative level. Finally, the prospective interesting polarizability of metal nanoparticles placed in a solid-state matrix is considered within the framework of the Mie theory and the dipole approximation.

2.1 Oscillator Strengths

The Bohr theory is not only a theory applicable to the hydrogen atom, but also a theory of interaction of electromagnetic radiation with an atom since important features of this interaction are described by the second and third Bohr postulates. Moreover, the theory of interaction of radiation with atoms can be studied without recourse to the consistent quantum-electrodynamic formalism, but with the help of the so-called *correspondence principle* in the spirit of the semiclassical Bohr approach. A starting point of such a consideration is the expression for dipole radiation power known from classical electrodynamics (Amusia 1990):

$$Q(t) = \frac{2}{3c^3}\left|\ddot{\boldsymbol{d}}(t)\right|^2, \tag{2.1}$$

where

$$\boldsymbol{d}(t) = e\boldsymbol{r}(t) \tag{2.2}$$

F. B. Rosmej et al., *Plasma Atomic Physics*, Springer Series on Atomic, Optical, and Plasma Physics 104, https://doi.org/10.1007/978-3-030-05968-2_2

is the dipole moment of a particle with charge e (understood hereafter to be an electron). The dots above the dipole moment symbol in the formula (2.1) denote the second time derivative. The criterion of applicability of the dipole approximation [framework of formula (2.1)] can be formulated as

$$\lambda \gg a, \tag{2.3}$$

where a is the size of the spatial region for radiation, λ is the radiation wavelength. In case of an atom, when $a \approx 10^{-8}$ cm, the condition (2.3) covers a wide range of wavelengths down to X-rays.

The second time derivative of the dipole moment appearing in the right-hand side of (2.1) can be expressed in terms of the electron acceleration $\boldsymbol{w}$: $\ddot{\boldsymbol{d}} = e\boldsymbol{w}$ and formula (2.1) can be rewritten as

$$Q = \frac{2e^2w^2}{3c^3}. \tag{2.4}$$

Thus within the framework of classical physics, an accelerated charged particle will lose its energy by dipole radiation with a rate determined by the formula (2.4). It should be noted that energy loss by a charge in the Coulomb field results not in a decrease, but in an increase of its kinetic energy. Increased kinetic energy of a charge is accompanied by a twofold decrease in its potential energy, which is connected with decreasing distance to the center of the Coulomb field. As a result, the total electron energy decreases.

In case of periodic motion of a charge with a circular frequency ω_0, as it occurs with an atomic electron, the radiation power of interest is those averaged over the period of motion $T = 2\pi/\omega_0$. To perform this averaging in the formula (2.1), we will use the following equation that is valid for a real periodic function $f(t)$:

$$\left\langle f(t)^2 \right\rangle_T \equiv \frac{1}{T} \int\limits_{-T/2}^{T/2} f(t)^2 \mathrm{d}t = 2\sum_{n=1}^{\infty} |f_n|^2, \tag{2.5}$$

where

$$f_n = \frac{1}{T} \int\limits_{-T/2}^{T/2} f(t) \exp(-n\,\omega_0 t)\mathrm{d}t \tag{2.6}$$

is the nth harmonic of the Fourier expansion of the function $f(t)$. In derivation of the formula (2.5), it was assumed that the average over the period of the function under consideration is equal to zero, that is, $f_0 = 0$. It should be noted that the factor 2 in the right-hand side of the (2.5) is connected with taking into account the contribution of negative harmonics of the Fourier series ($n < 0$).

Using (2.5), in which it is assumed that $f(t) = \ddot{\boldsymbol{d}}(t)$, we obtain from (2.1) the following expression for dipole radiation power averaged over the period:

$$\langle Q(t)\rangle_{\mathrm{T}} = \sum_{\mathrm{n}=1}^{\infty} Q_{\mathrm{n}}, \tag{2.7}$$

where

$$Q_{\mathrm{n}} = \frac{4}{3c^3}\left|(\ddot{\boldsymbol{d}})_{\mathrm{n}}\right|^2. \tag{2.8}$$

In view of

$$(\ddot{\boldsymbol{d}})_{\mathrm{n}} = -(n\,\omega_0)^2(\boldsymbol{d})_{\mathrm{n}}, \tag{2.9}$$

we find with the help of (2.8)

$$Q_{\mathrm{n}} = \frac{4(n\,\omega_0)^4}{3c^3}\left|(\boldsymbol{d})_{\mathrm{n}}\right|^2. \tag{2.10}$$

The formula (2.10) describes the dipole radiation power at the nth harmonic frequency $\omega_{\mathrm{n}} = n\,\omega_0$. In particular, the radiation power at the frequency of periodic motion of an electron ω_0 (n = 1) is

$$Q(\omega_0) = \frac{4\omega_0^4}{3c^3}|\boldsymbol{d}(\omega_0)|^2. \tag{2.11}$$

Here, we re-designated the first Fourier harmonic of the dipole moment: $\boldsymbol{d}_1 = \boldsymbol{d}(\omega_0)$.

Now let us replace in the formula (2.11) the Fourier harmonic of the dipole moment with its matrix element calculated between the states $|m\rangle$ and $|n\rangle$ ($\Psi_{\mathrm{m,n}}(\boldsymbol{r})$ are the corresponding state wave functions):

$$\boldsymbol{d}(\omega_0) \to \boldsymbol{d}_{\mathrm{mn}} \equiv \langle m|\boldsymbol{d}|n\rangle = \int d\boldsymbol{r}\,\Psi_{\mathrm{m}}^*(\boldsymbol{r})\cdot\boldsymbol{d}\cdot\Psi_{\mathrm{n}}(\boldsymbol{r}), \tag{2.12}$$

where the frequency of the periodic motion ω_0 will be replaced by the frequency of the transition $|n\rangle \to |m\rangle$:

$$\omega_0 \to \omega_{\mathrm{mn}} = \frac{E_{\mathrm{n}} - E_{\mathrm{m}}}{\hbar}. \tag{2.13}$$

As a result, instead of the formula (2.11), we will obtain

$$Q_{\rm mn} = \frac{4\,\omega_{\rm mn}^4}{3\,c^3}\,|\boldsymbol{d}_{\rm mn}|^2. \tag{2.14}$$

The quantity (2.14) can be called electromagnetic radiation power in transition of an atomic electron from the stationary state $|n\rangle$ to the stationary state $|m\rangle$. It describes the intensity of radiation of different spectral series of a hydrogen atom: the Lyman series (m = 1), the Balmer series (m = 2), the Paschen series (m = 3), etc. However, it should be remembered that, in contrast to the classical radiation power (2.1), the quantity (2.14) should be understood statistically, that is, as a result of averaging over an ensemble of atoms.

If now the radiation power (2.14) is divided by the energy of the transition under consideration $\Delta E_{\rm mn} = \hbar\,\omega_{\rm mn}$, we obtain a quantity with dimension of reciprocal time coinciding with the Einstein coefficient for spontaneous radiation $A_{\rm mn}$:

$$\frac{Q_{\rm mn}}{\Delta E_{\rm mn}} = \frac{4\,\omega_{\rm mn}^3}{3\,\hbar\,c^3}\,|\boldsymbol{d}_{\rm mn}|^2 = A_{\rm mn} = \frac{1}{\tau_{\rm mn}}. \tag{2.15}$$

In the last equality of the formula (2.15), there is introduced the lifetime $\tau_{\rm mn}$ of the state $|n\rangle$ with respect to its spontaneous decay to the lower state $|m\rangle$. This time for the transition $|2\rangle \rightarrow |1\rangle$ in a hydrogen atom is $\tau_{12} \approx 1.6 \times 10^{-9}$ s.

Thus the use of the formula of classical electrodynamics (2.1) and replacements in (2.12)–(2.13) allowed to obtain a quantum result for the power of radiation of spectral lines (2.14) and the probability of spontaneous radiation (2.15). This reflects the *principle of correspondence* between classical and quantum physics. This principle can be formulated as follows: Quantum-mechanical expressions are obtained from classical expressions if in the latter the Fourier components of physical quantities are replaced by the corresponding matrix elements. In this case, the quantum transition frequency should coincide with the Fourier component frequency (Brandt and Lundqvist 1965).

Curiously, the existence of a finite lifetime of the excited state $|n\rangle$ can be interpreted in the spirit of the correspondence principle as "falling" of an electron into the nucleus due to photon radiation—which is just the process, against which the second Bohr postulate "struggles". This "falling" goes on until the electron reaches the ground state $m = 1$ with the lowest possible (from the point of view of quantum physics) energy.

To clear up the physical justification of the second Bohr postulate, we will introduce a classical period of electron revolution for the orbit with principal quantum number n, radius $r_{\rm n}$ and velocity $\upsilon_{\rm n}$:

$$T_{\rm n} = \frac{2\pi r_{\rm n}}{\upsilon_{\rm n}} = n^3\,2\pi\tau_{\rm a}. \tag{2.16}$$

Table 2.1 Oscillator strengths for a hydrogen atom

Initial state	1*s*	2*s*	2*p*		3*s*	3*p*		3*d*	
Final state	*np*	*np*	*ns*	*nd*	*np*	*ns*	*nd*	*np*	*nf*
$n = 1$	–	–	−0.139	–	–	−0.026	–	–	–
2	0.4162	–	–	–	−0.141	0.145	–	−0.417	–
3	0.0791	0.4349	0.014	0.696	–	–	–	–	–
4	0.0290	0.1028	0.0031	0.122	0.484	0.032	0.619	0.011	1.016
5	0.0139	0.0419	0.0012	0.044	0.121	0.007	0.139	0.0022	0.156
6	0.0078	0.0216	0.0006	0.022	0.052	0.003	0.056	0.0009	0.053
7	0.0048	0.0127	0.0003	0.012	0.027	0.002	0.028	0.0004	0.025
8	0.0032	0.0081	0.0002	0.008	0.016	0.001	0.017	0.0002	0.015
$\sum_{n=9}^{\infty} f_{n0}$	0.0109	0.0268	0.0007	0.023	0.048	0.002	0.045	0.0007	0.037
Asymptotic formula	$1.6n^{-3}$	$3.7n^{-3}$	$0.1n^{-3}$	$3.3n^{-3}$	$6.2n^{-3}$	$0.3n^{-3}$	$6.1n^{-3}$	$0.07n^{-3}$	$4.4n^{-3}$
Discrete spectrum	0.5650	0.6489	−0.119	0.928	0.707	−0.121	0.904	−0.402	1.302
Continuous spectrum	0.4350	0.3511	0.008	0.183	0.293	0.010	0.207	0.002	0.098
Total sum	1.000	1.000	−0.111	1.111	1.000	−0.111	1.111	−0.400	1.400

Now let us estimate the ratio of the period (2.16) to the lifetime $\tau_{\rm mn}$. Using (2.15), (2.16) and assuming that $r_{\rm mn} \approx a_{\rm B} R(m, n)$, where $R(m, n)$ is a function of the order of unity, we have approximately

$$\frac{T_{\rm n}}{\tau_{\rm mn}} \approx \frac{T_{\rm n}}{\tau_{\rm 1n}} \approx (\alpha n)^3 R^2(1,n) \approx 4 \times 10^{-7} n^3 R^2(1, n), \tag{2.17}$$

where $n^3 R^2(1,n) \to 1$ for large numbers n (see the asymptotic formula in Table 2.1). The second approximate equality in (2.17) reflects the fact that the lifetime of the excited state of a hydrogen atom is dominated by its transition to the ground state.

From the obtained relation (2.17), it follows that the period of electron revolution in the classical orbit is several orders of magnitude less than the lifetime in this state $|n\rangle$. Thus these states may be considered to a good degree of accuracy to be stationary according to the first two Bohr postulates. This stationary is a consequence of the small value of the fine structure constant α that is responsible for the electromagnetic interaction.

The principle of correspondence between classical and quantum physics concretized for radiative transitions in an atom is called the *spectroscopic principle of correspondence*. It can be formulated as follows: an atom in interaction with an electromagnetic field behaves as a set of classical oscillators with eigenfrequencies equal to frequencies of transitions between atomic energy levels. This means that to each transition between atomic states $|j\rangle$ and $|n\rangle$ is assigned an oscillator with an eigenfrequency $\omega_{\rm nj}$. Let us call these oscillators transition oscillators. The contribution of transition oscillators to the response of atoms to electromagnetic interaction is proportional to a dimensionless quantity called *oscillator strength*. The higher the value of the oscillator strength, the stronger is the corresponding transition. The oscillator strength for the transition between the discrete spectrum states

$|j\rangle \to |n\rangle, f_{nj}$ (in quantum mechanics, it is the practice to write the state indices from right to left—corresponding to absorption), is determined by the formula

$$f_{nj} = \frac{2\, m_e\, \omega_{nj} |\langle n|\boldsymbol{d}|j\rangle|^2}{3\, e^2\, \hbar\, g_j}, \tag{2.18}$$

where g_j is the statistical weight of the state j. From the formula (2.18), it follows the relation $g_j f_{nj} = -g_n f_{jn}$ since the oscillator strength for a transition with decreasing energy is negative. According to its physical meaning, the oscillator strength of a one-electron atom is always less than one.

The formulation of the correspondence principle via the oscillator strength in the form of (2.18) corresponds to the dipole approximation if the criterion of (2.3) is fulfilled. Otherwise, the definition (2.18) should be generalized to include also the non-dipole part of the interaction of electromagnetic radiation with atomic electrons. The non-dipole nature of interaction is found to be essential if the matrix element of the dipole moment in the formula (2.18) is equal to zero. Such transitions are called *dipole-forbidden* transitions in contrast to *dipole-allowed* transitions, when $\langle n|\boldsymbol{d}|j\rangle \neq 0$. Equality or inequality with respect to zero of the dipole moment of a transition can be predicted from an analysis of the symmetry of states involved in the transition. The relations between characteristics of atomic states allow predictions of a nonzero value of $\langle n|\boldsymbol{d}|j\rangle$ are called *selection rules for dipole radiation*. These rules have the simplest form for the hydrogen-like ion where the systematics of its electronic states (neglecting the spin–orbit interaction) is rather simple. An energy level with a principal quantum number n has a $2n^2$-fold degeneracy that occurs as follows. First, there is a degeneracy with respect to the orbital quantum number l that is specific for a hydrogen-like ion: states of an atomic electron with $l = 0, 1, \ldots n-1$ designated as $|n,\ l\rangle$ have energy E_n. We note that to numerical designations $l = 0,\ 1,\ 2,\ 3,\ 4,\ 5,\ 6\ldots$ there correspond letter designations: $s, p, d, f, g, h, i\ldots$ Then, each state $|n,\ l\rangle$ is degenerated with respect to the value of the projection of an electron orbital moment on a dedicated axis. This degeneracy is of general character and is connected with the spherical symmetry of the atomic potential. The quantum number of the projection of a moment of momentum m_l runs over $2l+1$ values: $m_l = -l, -l+1, \ldots, l-1, l$, to which the states $|n,\ l,\ m_l\rangle$ correspond. Finally, the electronic states $|n,\ l,\ m_l\rangle$ are doubly degenerated with respect to the electron spin projection, which results in a $2n^2$-fold degeneracy of an energy level of a hydrogen-like ion with a principal quantum number n. It should be noted that this classification is valid for discrete spectrum states. In the case of a continuous spectrum, there is an additional degeneracy of states resulting from different directions of electron momentum.

In terms of the given state classification, selection rules for dipole radiation of a hydrogen-like ion come into play as follows. Allowed transitions are transitions, for which an orbital quantum number changes by one: $l \to l \pm 1$. In this case, the magnetic quantum number changes by no more than one: $m_l \to m_l, m_l \pm 1$. In particular, if the magnetic quantum number does not change, linearly polarized radiation is emitted (absorbed), whereas in the opposite case circularly polarized radiation is emitted (absorbed). A particular case of the selection rules is the average dipole moment of an atom in the absence of external fields. It is equal to zero, i.e., $\langle \boldsymbol{d} \rangle = 0$ as a consequence of the spherical symmetry of an atom.

Besides electron transitions in a discrete spectrum (bound–bound transitions), there are also transitions from bound states to continuous ones (bound-free transitions), for which the concept of oscillator strength by a formula similar to (2.18) can also be introduced. Physically, to a bound-free transition, there corresponds atomic ionization. In contrast to the case of a bound–bound transition, the oscillator strength $f_{\varepsilon j}$ for a bound–free transition to a state with an energy ε is no longer a dimensionless quantity. The dimensionality of $f_{\varepsilon j}$ is equal to a reciprocal energy, which corresponds to the normalization of the wave function of the continuous spectrum to the energy delta function. Therefore, for a bound–free transition, instead of oscillator strength, the concept of oscillator strength density is used: $f_{\varepsilon j} \to \mathrm{d}f_j/\mathrm{d}\varepsilon$.

The oscillator strengths for bound–bound and bound–free transitions in an atom satisfy the so-called *golden sum rule*. For transitions from the ground state it is expressed by the equation

$$\sum_n f_{n0} + \int_{I_P}^{\infty} \frac{\mathrm{d}f_0}{\mathrm{d}\varepsilon} \mathrm{d}\varepsilon = N_e, \tag{2.19}$$

where I_p is the atomic ionization potential, N_e is the number of electrons of the atom.

The oscillator strengths for a number of electron transitions in a hydrogen atom are given in Table 2.1 taken from Bethe and Salpeter (1977).

From this table, we observe the following. First, for transitions with increasing energy, the oscillator strength is larger in the case of increasing orbital quantum number, i.e., the transition $n, l \to n', l+1$ is stronger than the transition $n, l \to n', l-1$ if $n < n'$. Secondly, the sum of oscillator strengths for transitions to the continuous spectrum decreases with increasing orbital quantum number of the initial state, that is, states with higher orbital moments are more difficult to ionize. Third, transitions to a state with a nearest principal quantum number have the highest oscillator strength. Fourthly, oscillator strengths for transitions from lower levels to states with large quantum numbers $n \geq 10$ decrease as n^{-3}. These regularities define the probabilities of corresponding radiative transitions in a hydrogen atom.

An important property of the oscillator strength for a hydrogen-like ion is its independence of the nucleus charge Z. This is easily seen from the definition (2.18). Here two values appear that depend on the nuclear charge: the transition frequency ω_{jn} and the matrix element of the dipole moment of a transition $\boldsymbol{d}_{nj}$. If it is remembered that $\omega_{jn} \propto Z^2$ and $\boldsymbol{d}_{nj} \propto Z^{-1}$, we immediately obtain $f_{nj} \propto Z^0$.

It is instructive to calculate oscillator strengths for transitions between energy levels of a quantum harmonic oscillator. In this case, the formula (2.18) can be rewritten as

$$f_{\mathrm{N'N}} = \frac{2\, m\, \omega_0 |\langle N'|\hat{z}|N\rangle|^2}{\hbar}, \tag{2.20}$$

where $\langle N'|\hat{z}|N\rangle$ is the matrix element to be determined of the harmonic oscillator with an eigenfrequency ω_0 and mass m. The oscillator coordinate operator is expressed in terms of the annihilation operator $\hat{a}$ and the creation operator $\hat{a}^+$ as follows: $\hat{z} = \sqrt{\hbar/2m\,\omega_0}(\hat{a} + \hat{a}^+)$. Using the definitions and the property of orthogonality of the state vectors $|N\rangle$, we obtain for the required matrix element:

$$\langle N'|\hat{z}|N\rangle = \sqrt{\frac{\hbar}{2m\,\omega_0}}\left[\sqrt{N+1}\,\delta(N', N+1) + \sqrt{N}\,\delta(N', N-1)\right], \tag{2.21}$$

where $\delta(m, n)$ is the Kronecker symbol. Substituting (2.21) into the definition (2.20), we find the simple relation

$$f_{\mathrm{N'N}} = (N+1) \cdot \delta(N', N+1) - N \cdot \delta(N', N-1). \tag{2.22}$$

Hence, it follows in particular that oscillator strengths are nonzero only for transitions to nearest energy states (the selection rule for a linear harmonic oscillator). From formula (2.22), it is likewise shown how the sum rule (2.19) is fulfilled (with $N_e = 1$). Finally, the considered case is an example of the situation when an oscillator strength can be an arbitrarily large quantity.

2.2 Classical and Quantum Expressions for Einstein Coefficients

Einstein coefficients were introduced phenomenologically for the description of the probability per unit time for three fundamental photoprocesses (photoabsorption, spontaneous radiation, and induced radiation) in interaction of thermal radiation with a two-level system. The consistent derivation of expressions for these coefficients is possible only within the framework of quantum electrodynamics, a complex physical discipline considering radiation and a substance on the quantum basis. Nevertheless, if the correspondence principle and the oscillator strength concept are used, corresponding formulas can be also obtained in a classical manner.

The spectroscopic principle of correspondence makes it possible to represent an atom and its interaction with radiation as a set of charged harmonic oscillators corresponding to transitions between atomic energy levels $n \rightarrow j$ ($E_j > E_n$). These oscillators describing the system response to electromagnetic disturbance are called transition oscillators. Their coordinate x_{nj} satisfies a damped oscillator equation with external force, in the right-hand side of which the oscillator strength $f_{jn} \neq 0$ (a dipole-allowed transition) is substituted as a factor:

$$\ddot{x}_{jn} + 2\,\delta_{jn}\,\dot{x}_{jn} + \omega_{jn}^2\,x_{jn} = f_{jn}\frac{e}{m}E(t). \tag{2.23}$$

Here $\omega_{jn} = (E_j - E_n)/\hbar$ is the transition frequency, δ_{nj} is the damping constant, e is the oscillator charge, dots above the coordinate symbol denote differentiation with respect to time, the electric field strength $E(t)$ is assumed to be independent of the coordinate [the dipole approximation (2.3)].

In the absence of a radiation field, the transition oscillator is at rest: $x_{jn} = 0$ and $\dot{x}_{jn} = 0$. An external field begins to "swing" the oscillator, imparting energy; forced oscillations of the transition oscillator occur, the time dependence of which, $x_{jn}(t)$, can be found from the (2.23).

We obtain the following expression for forced oscillations of an oscillator for the transition $j \rightarrow n$:

$$x_{jn}(t) = f_{jn}\frac{e}{m}\int\limits_{-\infty}^{\infty}\frac{E(\omega')\exp(-i\omega' t)}{\omega_{jn}^2 - \omega'^2 - 2i\omega'\delta_{jn}}\frac{d\omega'}{2\pi}, \tag{2.24}$$

where $E(\omega')$ is the Fourier transform of the electric field strength $E(t)$.

An oscillating charged oscillator radiates electromagnetic waves according to the formula for the radiation power of a dipole (2.1). In case of oscillations under the action of an external field, this radiation is induced radiation. If the pulse of the external field ceases, and the charged oscillator still oscillates, corresponding radiation is spontaneous. The probability per unit time for spontaneous radiation at the transition between atomic energy levels $j \rightarrow n$ is given by the Einstein coefficient A_{nj}. The explicit form of this coefficient can be obtained on the basis of classical consideration with the use of the spectroscopic principle of correspondence:

$$A_{nj} = \frac{2 f_{jn}\, e^2 \omega_{jn}^2}{3\, m\, c^3}. \tag{2.25}$$

The formula (2.25) gives the expression for the Einstein coefficient for spontaneous radiation in terms of the oscillator strength for a corresponding transition. It should be noted that the oscillator strength can be calculated theoretically or determined experimentally.

If now in the right-hand side of the (2.25) the expression for the oscillator strength (2.18) is substituted, we come to the following formula for the Einstein coefficient for spontaneous radiation:

$$A_{nj} = \frac{4\omega_{jn}^3 |\boldsymbol{d}_{nj}|^2}{3g_j \hbar c^3}, \tag{2.26}$$

where $\boldsymbol{d}_{nj}$ is the matrix element of the electric dipole moment (see also Chap. 7).

To derive the formula for the Einstein coefficient for absorption B_{jn} while remaining within the framework of classical physics, we proceed from the (2.24), from which the expressions for the transition oscillator rate can be obtained:

$$\dot{x}_{jn}(t) = -if_{jn}\frac{e}{m}\int\limits_{-\infty}^{\infty} \frac{\omega' E(\omega')\exp(-i\omega' t)}{\omega_{jn}^2 - \omega'^2 - 2i\omega'\delta_{jn}}\frac{d\omega'}{2\pi}. \tag{2.27}$$

From (2.27), we find for the period-averaged power absorbed by the transition under the action of radiation with a spectral energy density $\rho(\omega)$:

$$P_{jn} = f_{jn}\frac{2\pi^2 e^2}{3m}\rho(\omega_{jn}). \tag{2.28}$$

It will be remembered that in obtaining this relation it was assumed that the radiation spectrum width is much larger than the spectrum width for a transition in an atom which is true, for example, for thermal radiation.

By definition, the Einstein coefficient for absorption (in case of transition of an atom from the state $|n\rangle$ to the state $|j\rangle$) is

$$B_{jn} = \frac{w_{jn}}{\rho(\omega_{jn})}. \tag{2.29}$$

According to the physical picture of the process, the photoabsorption probability per unit time, w_{jn}, is equal to the ratio of absorbed power to the transition energy:

$$w_{nj} = \frac{P_{nj}}{\hbar\,\omega_{nj}}. \tag{2.30}$$

Gathering the formulas (2.28)–(2.30), we obtain:

$$B_{jn} = \frac{2\pi^2 e^2 f_{jn}}{m\,\hbar\omega_{jn}}. \tag{2.31}$$

In view of the explicit form of the oscillator strength (2.18), we find from (2.31) the expressions for the Einstein coefficient in terms of the matrix element of the transition dipole moment:

$$B_{\rm jn} = \frac{4\pi^2 |\boldsymbol{d}_{\rm jn}|^2}{3\, g_{\rm n} \hbar^2}. \tag{2.32}$$

The formula for the Einstein coefficient describing induced radiation follows from (2.32):

$$B_{\rm nj} = \frac{4\pi^2 |\boldsymbol{d}_{\rm nj}|^2}{3\, g_{\rm j}\, \hbar^2}. \tag{2.33}$$

It should be noted that the matrix element of the dipole moment can be considered to be symmetric with respect to its indices: $\boldsymbol{d}_{\rm nj} = \boldsymbol{d}_{\rm jn}$ (see also Chap. 7).

It is worth noting that the presented approach is valid if the external field is not too strong, i.e., when the amplitude of oscillations of a transition oscillator is linear with respect to the strength of the electric field in an electromagnetic wave. Otherwise, it is necessary to take into account nonlinear effects, and more complex considerations are required. It should be noted that the Einstein coefficient for spontaneous radiation can be represented as

$$A_{\rm nj} = B_{\rm nj}\, \rho_{\rm vac}(\omega_{\rm nj}), \tag{2.34}$$

where

$$\rho_{\rm vac}(\omega) = \frac{\hbar\omega^3}{\pi^2 c^3} \tag{2.35}$$

is a quantity that can be interpreted as the spectral density of energy of vacuum fluctuations of an electromagnetic field. Equations (2.34), (2.35) correspond to an interpretation of spontaneous radiation as radiation induced by vacuum fluctuations.

2.3 Dynamic Polarizability of Atoms

The expression for dynamic polarizability of an atom can likewise be obtained without recourse to quantum mechanics with the use of the spectroscopic principle of correspondence. Let us calculate the dipole moment of an atom $\boldsymbol{d}$ in the monochromatic field $\boldsymbol{E}(t) = 2Re\{\boldsymbol{E}_\omega \exp(-i\,\omega\, t)\}$ that is by definition equal to

$$\boldsymbol{d}(t) = 2Re\{\beta(\omega)\,\boldsymbol{E}_\omega \exp(-i\,\omega\, t)\}. \tag{2.36}$$

The Fourier component of the dipole moment is given by the expression

$$\boldsymbol{d}_{\omega} = \beta(\omega)\boldsymbol{E}_{\omega}. \quad (2.37)$$

In the formulas (2.36)–(2.37), $\boldsymbol{E}_{\omega}$ is the complex electric field vector for monochromatic radiation that is a Fourier component of $\boldsymbol{E}(t)$.

It will be remembered that the dipole moment of an atom in the absence of external fields is equal to zero due to spherical symmetry, so the value of an induced dipole moment can really be a measure of perturbation of an atom by external action. The linear dependence of $\boldsymbol{d}(t)$ on the electric field strength (2.36) is valid for small values of E (smallness of perturbation of an atomic electron state as a result of interaction with an electromagnetic field). Thus for sufficiently weak fields, the response of an atom to electromagnetic disturbance can be characterized by its polarizability $\beta(\omega)$.

According to the spectroscopic principle of correspondence, the change of an atomic state is made up of changes of motion of oscillators that correspond to transitions between atomic states (transition oscillators). Thus the deviations of transition oscillators from the equilibrium position under the action of the field $\boldsymbol{E}(t)$ can be considered to be small. For the nth oscillator, the equation of motion in the harmonic approximation is given by:

$$\ddot{\boldsymbol{r}}_{\mathrm{n}} + \delta_{0\mathrm{n}}\,\dot{\boldsymbol{r}}_{\mathrm{n}} + \omega_{0\mathrm{n}}^{2}\,\boldsymbol{r}_{\mathrm{n}} = \frac{e}{m} f_{0\mathrm{n}}\,\boldsymbol{E}(t), \quad (2.38)$$

where $\boldsymbol{r}_{\mathrm{n}}$ is the radius vector corresponding to the deviation of a transition oscillator from the equilibrium position, $\delta_{0\mathrm{n}}$, $\omega_{0\mathrm{n}}$, $f_{0\mathrm{n}}$ are the damping constant, the eigenfrequency, and the oscillator strength. For simplicity, we consider a one-electron atom in the ground state and its dipole moment $\boldsymbol{d} = e\boldsymbol{r}$. In case of a multielectron atom, the dipole moment is equal to the sum of dipole moments of atomic electrons. In view of the correspondence principle, an induced dipole moment of an atom is made up of the induced dipole moments $\boldsymbol{d}_{\mathrm{n}}$ of oscillators of transitions to the nth state: $\boldsymbol{d} = \sum_{\mathrm{n}} \boldsymbol{d}_{\mathrm{n}} = e \sum_{\mathrm{n}} \boldsymbol{r}_{\mathrm{n}}$. Going to Fourier components in this equation, we have

$$\boldsymbol{d}_{\omega} = e \sum_{\mathrm{n}} \boldsymbol{r}_{\mathrm{n}\omega}, \quad (2.39)$$

where $\boldsymbol{r}_{\mathrm{n}\omega}$ is the Fourier transform of the radius vector deviation (from the equilibrium position) of the transition oscillator. From the equation of motion (2.38), it follows:

$$\boldsymbol{r}_{\mathrm{n}\omega} = \frac{e}{m} \frac{f_{0\mathrm{n}}}{\omega_{0\mathrm{n}}^{2} - \omega^{2} - i\,\omega\,\delta_{0\mathrm{n}}} \boldsymbol{E}_{\omega}. \quad (2.40)$$

Substituting formula (2.40) into the (2.39) and using the definition of polarizability (2.37), we find

$$\beta(\omega) = \frac{e^2}{m} \sum_{n} \frac{f_{0n}}{\omega_{0n}^2 - \omega^2 - i\,\omega\,\delta_{0n}}. \tag{2.41}$$

Hence it follows that the dynamic polarizability of an atom, generally speaking, is a complex value with the dimensionality of a volume. The imaginary part of the polarizability is proportional to the damping constants of the transition oscillators. The sum in the right-hand side of the (2.41) includes both summations over the discrete energy spectrum and integration over the continuous energy spectrum. The imaginary part of the polarizability is responsible for absorption of radiation, and the real part defines the refraction of an electromagnetic wave in a medium. The expression (2.41) describes not only a one-electron atom, but also a multielectron atom. The multielectron nature of an atom is taken into account by the fact that in the definition of the oscillator strength (2.18) the dipole moment of an atom is equal to the sum of dipole moments of each of its electrons.

From the (2.41), several important limiting cases can be obtained. For example, if the frequency of an external field is equal to zero, the formula (2.41) gives the expression for the static polarizability of an atom:

$$\beta_0 \equiv \beta(\omega = 0) = \frac{e^2}{m} \sum_{n} \frac{f_{0n}}{\omega_{0n}^2}. \tag{2.42}$$

Hence it is seen that static polarizability is a real and positive value. It has large numerical values if in the spectrum transitions high oscillator strengths and low eigenfrequencies are encountered.

In the opposite (high-frequency) limit, when $\hbar\,\omega \gg I_P$, (I_P is the atomic ionization potential) and the eigenfrequencies in the denominators of (2.41) can be neglected, in view of the golden sum rule (2.19), we obtain from the formula (2.41):

$$\beta_\infty(\omega) = -\frac{e^2 N_e}{m\,\omega^2}. \tag{2.43}$$

The high-frequency polarizability of an atom (2.43) is a real and negative value.

If the frequency of an external field is close to one of the eigenfrequencies of the transition oscillators, so that the resonance condition

$$|\omega - \omega_{0n}| \leq \delta_{0n} \tag{2.44}$$

is fulfilled, and only one resonance term can be left in the sum (2.41), we obtain the following expression for the resonance polarizability:

$$\beta_{res}(\omega) = \left(\frac{e^2}{2\,m\,\omega_{0n}}\right) \cdot \frac{f_{0n}}{\omega_{0n} - \omega - i\,\delta_{0n}/2}. \tag{2.45}$$

In derivation of (2.45) from (2.41), in non-resonance combinations, the distinction of the external field frequency from the transition eigenfrequency was neglected. Resonance polarizability is a complex value, and the real part of which can be both positive and negative.

Equation (2.37) defines the dynamic polarizability. Taking the inverse Fourier transform, we obtain

$$\boldsymbol{d}(t) = \int_{-\infty}^{\infty} \beta(\tau)\,\boldsymbol{E}(t-\tau)\,\mathrm{d}\tau, \tag{2.46}$$

where $\beta(\tau)$ is a real function of time where its Fourier transform is equal to the dynamic polarizability $\beta(\omega)$. The simplest expression for $\beta(\tau)$ follows from the formula (2.45):

$$\beta_{\mathrm{res}}(\tau) = -i\frac{e^2 f_{0\mathrm{n}}}{2m\,\omega_{0\mathrm{n}}} \cdot \theta(\tau) \cdot \exp(-i\,\omega_{0\mathrm{n}}\,\tau - \delta_{0\mathrm{n}}\,\tau/2), \tag{2.47}$$

where $\theta(\tau)$ is the Heaviside step function. The time dependence of the induced dipole moment $\boldsymbol{d}(t)$ coincides with the time dependence of the right-hand side of (2.47) for a delta pulse of the field: $\boldsymbol{E}(t) = \boldsymbol{E}_0\,\delta(t)$, where $\delta(t)$ is the Dirac delta function. In the general case, the expression for $\beta(\tau)$ can be obtained by replacement of the frequency $\omega_{0\mathrm{n}} \to \sqrt{\omega_{0\mathrm{n}}^2 - (\delta_{0\mathrm{n}}/2)^2}$ and summation over all transition oscillators. It should be noted that a decrease in the oscillation eigenfrequency in view of damping that follows from this replacement is quite natural since friction (the analog of damping) decreases the rate of motion.

2.4 General Relations of Atomic Polarizability

From the formula (2.47), it follows in particular that the function $\beta(\tau)$ is zero for times $\tau < 0$, which is a reflection of the *causality* principle. Really, as seen from (2.46), in order for the effect to appear after its cause, the fulfillment of the condition $\beta(\tau < 0) = 0$ is necessary. The causality principle imposes certain restrictions on the form of the function $\beta(\omega)$, from which the *Kramers–Kronig* relations follow that connect the real and imaginary parts of dynamic polarizability:

$$Re\{\beta(\omega)\} = \frac{1}{\pi}\,V.P. \int_{-\infty}^{\infty} \frac{Im\{\beta(\omega')\}}{\omega' - \omega}\,\mathrm{d}\omega', \tag{2.48}$$

$$Im\{\beta(\omega)\} = \frac{1}{\pi} V.P. \int_{-\infty}^{\infty} \frac{Re\{\beta(\omega')\}}{\omega - \omega'} d\omega', \tag{2.49}$$

where the principal-value integral is given by:

$$V.P. \int_{-\infty}^{+\infty} \frac{f(x)}{x-a} dx = \lim_{\Delta \to 0} \left\{ \int_{-\infty}^{a-\Delta} \frac{f(x)}{x-a} dx + \int_{a+\Delta}^{+\infty} \frac{f(x)}{x-a} dx \right\}. \tag{2.50}$$

With the use of (2.48), (2.49), it is possible to obtain the real part of the polarizability via the imaginary part and vice versa. For practical purposes, it is convenient to integrate (2.48) over positive frequencies and to represent the principal value of the integral as a "punctured" integral:

$$Re(\beta(\omega)) = \frac{2}{\pi} \int_{0}^{\infty} \frac{\omega' Im(\beta(\omega')) - \omega\, Im(\beta(\omega))}{\omega'^2 - \omega^2} d\omega'. \tag{2.51}$$

In derivation of (2.51), it was assumed that the imaginary part of the polarizability is an odd function of frequency.

There is an important relation connecting the imaginary part of the dynamic polarizability and the photoabsorption cross section $\sigma_{ph}(\omega)$. This relation is called the *optical theorem* and looks like

$$Im(\beta(\omega)) = \frac{c}{4\pi\omega} \sigma_{ph}(\omega). \tag{2.52}$$

With the use of the optical theorem we find with (2.51)

$$Re(\beta(\omega)) = \frac{c}{2\pi^2} \int_{0}^{\infty} \frac{\sigma_{ph}(\omega') - \sigma_{ph}(\omega)}{\omega'^2 - \omega^2} d\omega'. \tag{2.53}$$

Equation (2.53) expresses the real part of polarizability in terms of the photoabsorption cross section. In particular, for static polarizability, we have:

$$\beta(0) = \frac{c}{2\pi^2} \int_{0}^{\infty} \frac{\sigma_{ph}(\omega)}{\omega^2} d\omega. \tag{2.54}$$

From (2.53), it is possible to obtain the sum rule for the photoabsorption cross section when going to the limit $\omega \to \infty$ and using the expression (2.43) for the high-frequency polarizability:

$$\frac{mc}{2\pi^2 e^2}\int\limits_0^\infty \sigma_{\text{ph}}(\omega)\,\text{d}\omega = N; \quad \frac{2m}{\pi e^2}\int\limits_0^\infty \omega\, Im\{\beta(\omega)\}\,\text{d}\omega = N. \tag{2.55}$$

The second equality in (2.55) follows from the first equality in view of the optical theorem (2.52).

As it was noted, the above formulas for the polarizability are valid for a spherically symmetric atomic state. In the general case, the polarizability β_{ij} is a tensor. For an atom being in the state $|n\,l\,m\rangle$, where n, l, m are the principal, orbital, and magnetic quantum numbers, the tensor can be represented as (Amusia 1990)

$$\beta_{\text{ij}} = \begin{bmatrix} \beta^s_{\text{nl}} - \frac{1}{2}P_2\,\beta^a_{\text{nl}} & \frac{i}{2}P_1\,\beta^a_{\text{nl}} & 0 \\ -\frac{i}{2}P_1\,\beta^a_{\text{nl}} & \beta^s_{\text{nl}} - \frac{1}{2}P_2\,\beta^a_{\text{nl}} & 0 \\ 0 & 0 & \beta^s_{\text{nl}} + P_2\,\beta^t_{\text{nl}} \end{bmatrix}. \tag{2.56}$$

Here β^s_{nl}, β^a_{nl}, β^t_{nl} are the scalar, antisymmetric, and tensor components of the polarizability. The functions of magnetic and orbital quantum numbers $P_{1,\,2}(m)$ for $l \neq 0$ are

$$P_1(m) = \frac{m}{l}; \quad P_2(m) = \frac{3\,m^2 - l(l+1)}{l(2l-1)}. \tag{2.57}$$

For a spherically symmetric state, when $l = m = 0$, $P_1 = P_2 = 0$, formula (2.56) shows that the atomic polarizability becomes a scalar.

With the use of (2.56), it is possible to write the shift and splitting of magnetic sublevels of the state $|n\,l\,m\rangle$ in the ac field $\boldsymbol{E}(t) = E\,Re\{\boldsymbol{e}\,\exp(-i\,\omega\,t)\}$ (in the general case, $\boldsymbol{e}$ is a complex vector, and $|\boldsymbol{e}| = 1$) far from the resonance as

$$\Delta E_{\text{nlm}} = -\frac{1}{4}E^2\left[\beta^s_{\text{nl}}(\omega) + P_2(m)\,\beta^t_{\text{nl}}(\omega)\right] \tag{2.58}$$

for linear polarization of the electric field and

$$\Delta E_{\text{nlm}} = -\frac{1}{4}E^2\left[\beta^s_{\text{nl}}(\omega) \pm P_1(m)\,\beta^a_{\text{nl}}(\omega) - \frac{1}{2}P_2(m)\,\beta^t_{\text{nl}}(\omega)\right] \tag{2.59}$$

for right-hand $(+)$ and left-hand $(-)$ circular polarizations of the unit vector $\boldsymbol{e}$. The components of the polarizability tensor $\beta^s_{\text{nl}}, \beta^a_{\text{nl}}, \beta^t_{\text{nl}}$ are

$$\beta^s_{\text{nl}}(\omega) = \frac{1}{3\,(2l+1)}\left[l\,\sigma_{\text{l}-1}(\omega) + (l+1)\,\sigma_{\text{l}+1}(\omega)\right], \tag{2.60}$$

$$\beta^a_{\text{nl}}(\omega) = \frac{1}{2\,l+1}\left[\sigma^-_{\text{l}-1}(\omega) - \sigma^-_{\text{l}+1}(\omega)\right], \tag{2.61}$$

$$\beta^{t}_{\rm nl}(\omega) = -\frac{1}{3\,(2l+1)}\cdot\left[\sigma_{l-1}(\omega) + \frac{2l-1}{2l+3}\,\sigma_{l+1}(\omega)\right], \tag{2.62}$$

where

$$\sigma_{l'}(\omega) = \sum_{n'} \frac{f^{n'l'}_{\rm nl}}{\omega^2_{\rm n'n} - \omega^2 - i\omega\delta_{\rm n'n}}, \tag{2.63}$$

$$\sigma^{-}_{l'}(\omega) = \omega \sum_{n'} \frac{f^{n'l'}_{\rm nl}}{\omega_{\rm n'n}\left(\omega^2_{\rm n'n} - \omega^2 - i\omega\delta_{\rm n'n}\right)} \tag{2.64}$$

are the corresponding spectral sums. Thus the formulas (2.56)–(2.64) generalize the concept of dynamic (dipole) polarizability of an atom to the general non-spherical case of the state $|n\,l\,m\rangle$ $(l \neq 0)$ when polarizability is a tensor.

In the foregoing, we were dealing with dipole polarizability that describes the response of the atom to a spatially homogeneous electric field. If the characteristic dimension of the spatial homogeneity of a field is less than the size of the atom, the dipole polarizability should be replaced by the generalized polarizability of an atom $\beta(\omega,\, q)$ that depends also on the impulse $\hbar\boldsymbol{q}$ transmitted to the atom as a result of the atom–field interaction. With the use of $\beta(\omega,\; q)$, the dipole polarizability of the atomic core due to an external electric field is given by

$$\boldsymbol{D}(\omega) = \int \beta(\omega,\, q)\,\boldsymbol{E}(\omega, \boldsymbol{q})\frac{d\boldsymbol{q}}{(2\,\pi)^3}, \tag{2.65}$$

where $\boldsymbol{E}(\omega,\, \boldsymbol{q})$ is the space-time Fourier transform of the electric field. For the spatially homogeneous field $\boldsymbol{E}(\omega,\, \boldsymbol{q}) = \boldsymbol{E}(\omega)\,\delta(\boldsymbol{q})$ and $\beta(\omega) = \beta(\omega,\, \boldsymbol{q} = 0)$ (2.65) turns (in case of a spherically symmetric atomic state) into an entirely local approximation $D(\omega) = \boldsymbol{E}(\omega)\cdot\beta(\omega)$.

2.5 Static Polarizability of Atoms and Ions

The static polarizability β_0 of a hydrogen-like ion in the ground 1s-state is

$$\beta_0 = \frac{9}{2}\frac{a_{\rm B}^3}{Z^4}, \tag{2.66}$$

where $a_{\rm B}$ is the Bohr radius. The value of the static polarizability β_0 rapidly decreases with increasing nuclear charge and increases sharply with principal quantum number n. These regularities are easily understood at the qualitative level if it is remembered that static polarizability is proportional to the volume of an atom that decreases with increasing Z and increases strongly with n.

Table 2.2 Static dipole polarizability: The value of static polarizability is given in atomic units

Ion	Ca^{+10}	Ti^{+12}	Fe^{+16}	Co^{+17}	Ni^{+18}
I_n, eV	558.2	737.8	1168	1293	1419
Z_{eff}	12.8	14.73	18.54	19.5	20.43
β_0^{exp}	1.74×10^{-2}	1.04×10^{-2}	4.44×10^{-3}	3.69×10^{-3}	3.08×10^{-3}
β_0	1.89×10^{-2}	1.09×10^{-2}	4.33×10^{-3}	3.53×10^{-3}	2.94×10^{-3}

An atomic unit of polarizability is equal to 0.149 Å^3 (1 Å = 10^{-8} cm)

For multiply charged ions ($N \ll Z$, N is the number of electrons in the ion core, Z is the nuclear charge), the following approximate expression for the static polarizability can be obtained:

$$\beta_0 = \frac{63}{16}\frac{N^3}{Z^4}a_B^3. \tag{2.67}$$

The dependence of the static polarizability of a multiply charged ion on the nuclear charge is the same as for a hydrogen-like ion (2.66). This follows from quantum considerations for ions with filled shells. In this case, if $Z \gg N$, the minimum frequency of a virtual transition is proportional to the squared nuclear charge. Then from the general quantum-mechanical expression for the polarizability (2.41), it follows the above Z-dependence.

For ions with partially filled electron shells, the main contribution to static polarizability is made by a virtual transition with no change in principal quantum number $\Delta n = 0$. In this case, $\beta_0 \propto Z^{-3}$.

For ions with a filled outer shell, rather good agreement with experimental data is provided by the simple empirical formula for the static dipole polarizability of an outer shell with principal quantum number n:

$$\beta_0 = N_n \frac{n^6}{Z_{eff}^4} a_B^3, \tag{2.68}$$

where N_n and Z_{eff} are, respectively the number of electrons in the outer electron shell and the effective nuclear charge. The latter value can be determined from $Z_{eff} = n\sqrt{I_n/Ry}$, where I_n is the ionization potential of the outer shell. The formula (2.68) provides particular good results for neon-like ions ($n = 2,\ N_n = 8$). This can be seen from Table 2.2.

2.6 Local Plasma Frequency Model of Polarizability of Many Electron Systems

The local plasma model was proposed by Brandt and Lundqvist in the mid-1960s (Brandt and Lundqvist 1965) for the description of the photoabsorption by multi-electron atoms in the spectral range $\omega \sim Z$ a.u. (here 1 a.u. = 27.2 eV is the atomic unit of energy). In this case, in contrast to ranges of high ($\omega \sim Z^2$ a.u.) and low ($\omega \sim 1$ a.u.) frequencies, photon absorption is defined more by collective effects rather than by one-particle interaction. Based on these qualitative considerations, the electron core of an atom is approximated by the inhomogeneous distribution of the charge and the interaction which an electromagnetic field is defined by the plasma resonance condition:

$$\omega = \omega_{\mathrm{p}}(r) = \sqrt{\frac{4\,\pi\, n(r)\, e^2}{m}}, \tag{2.69}$$

where $n(r)$ is the local electron density and $\omega_{\mathrm{p}}(r)$ is its associated local plasma frequency. It can be shown that to the condition (2.69) there corresponds the following expression for the dipole dynamic polarizability that satisfies the Kramers–Kronig relations (2.48), (2.49) and the sum rule (2.55):

$$\beta^{\mathrm{BL}}(\omega) = \int_0^{R_0} \frac{\omega_{\mathrm{p}}^2(r) r^2\, \mathrm{d}r}{\omega_{\mathrm{p}}^2(r) - \omega^2 - i \cdot 0} = \int \beta^{\mathrm{BL}}(r, \omega)\, \mathbf{dr}. \tag{2.70}$$

Here, the value $\beta^{\mathrm{BL}}(r, \omega)$ is introduced that corresponds to a so-called spatial polarizability density in the Brandt–Lundqvist approximation, R_0 is the size of the atom (ion). In the denominator of the integrand of (2.70) an infinitesimal imaginary additive (designated as "$i \cdot 0$") is introduced that indicates the rule of pole bypass in calculation of the integral.

The expression (2.70) has correct high-frequency asymptotics (2.43). In the low-frequency limit, it gives

$$\beta^{\mathrm{BL}}(\omega \to 0) \to R_0^3/3, \tag{2.71}$$

that is, static polarizability is found to be proportional to the volume of an atom.

Despite of its apparent simplicity, the formula (2.71) well describes available experimental data. First of all, this concerns multielectron atoms with filled shells since in this case the main contribution to polarizability is made by the continuous energy spectrum of an atom, and the local plasma frequency approximation (2.70) is the most adequate. This fact is demonstrated by Table 2.3, where the values

Table 2.3 Static polarizability

Atom (ion)	Kr I	Xe I	KI I	Rb II	Cs II	Sr III	Ba III
β_0^{exp}	17	27	7.5	12	16.3	6.6	11.4
β_0^{var}	26.8	30.9	9.1	14.3	17.8	8.7	11.4
β_0^{USh}	21.1	25.5	6.6	11.9	15.3	7.5	9.7
β_0^{StZ}	17.2	27.3	5.25	8.5	14.6		
β_0^{BL}	24	27	8.6	11.6	13.5	7	8.4

(in atomic units) of the static polarizability of atoms and ions with filled electron shells are calculated within the framework of different models as well as experimental data. For the calculations of the static polarizability in the framework of the Brandt–Lundqvist model, the used atomic/ionic radius was calculated from the Thomas–Fermi–Dirac model including correlation allowance.

$\beta_0^{\rm var}$ is the calculation from the variational method, $\beta_0^{\rm USh}$ is the calculation of Shevelko and Ulanzev (1994), $\beta_0^{\rm StZ}$ is the calculation of Stott and Zaremba (1980) within the framework of the electron density formalism, $\beta_0^{\rm BL} = R^3/3$ is the calculation in the framework of the Brandt-Lundqvist model.

With the use of the formula (2.70) and the statistical model of an atom, the following expression for the dynamic polarizability can be obtained:

$$\beta(\omega, Z) = r_{\rm TF}^3\, \tilde{\beta}\left(\frac{\omega}{Z}\right) = \frac{b^3\, a_{\rm B}^3}{Z}\, \tilde{\beta}\left(\frac{\omega}{Z\, v_{\rm a}}\right),$$
$$\tilde{\beta}(v) = \int_0^{x_0} \frac{4\pi f(x)\, x^2\, {\rm d}x}{4\pi f(x) - v^2 - i\cdot 0}, \tag{2.72}$$

where $r_{\rm TF} = b\, a_{\rm B}/Z^{1/3}$ is the Thomas–Fermi radius, Z is the atomic nucleus charge, $a_{\rm B}$ is the Bohr radius, $b \cong 0.8853$, $\tilde{\beta}(v)$ is the dimensionless polarizability as a function of the reduced frequency $v = \omega/Z\, v_{\rm a}$, $x_0 = R_0/r_{\rm TF}$ is the reduced atomic radius, $f(x)$ is the universal function describing the distribution of the electron density $n(r)$ in an atom according to the formula $n(r) = Z^2 f(r/r_{\rm TF})$, $v_{\rm a} \cong 4.13 \times 10^{16}\,{\rm s}^{-1}$ is the atomic unit of frequency.

It should be emphasized that the dimensionless polarizability $\tilde{\beta}(v)$ does not depend on the atomic nuclear charge. Thus, the representation of the dynamic polarizability of a statistical atom (2.72) reveals a scaling law with respect to the parameter $v = \omega/Z\, v_{\rm a}$.

The results of calculation of the real and imaginary parts of the dipole dynamic polarizability of a krypton atom within the framework of the local plasma frequency method by the formula (2.70) using the Slater and Lenz–Jensen electron densities are presented in Fig. 2.1. Given in the same figure (curve 3) are the results of calculation for the corresponding values in the random phase exchange approximation that is today the most consistent quantum-mechanical method of description of the electronic structure of atoms. It is seen that the dynamic polarizability of a krypton atom calculated in the local plasma frequency model for the Lenz–Jensen electron density renders in a smoothed manner the quantum-mechanical features of the frequency behavior of dynamic polarizability. These are most pronounced in the vicinity of potentials of ionization of electron subshells. The use of Slater wave functions allows to some extent to resolve the spectral fluctuations of the polarizability in the vicinity of the ionization potentials of the electron subshells. However, in this case, we loose the universality of the description like in the statistical model.

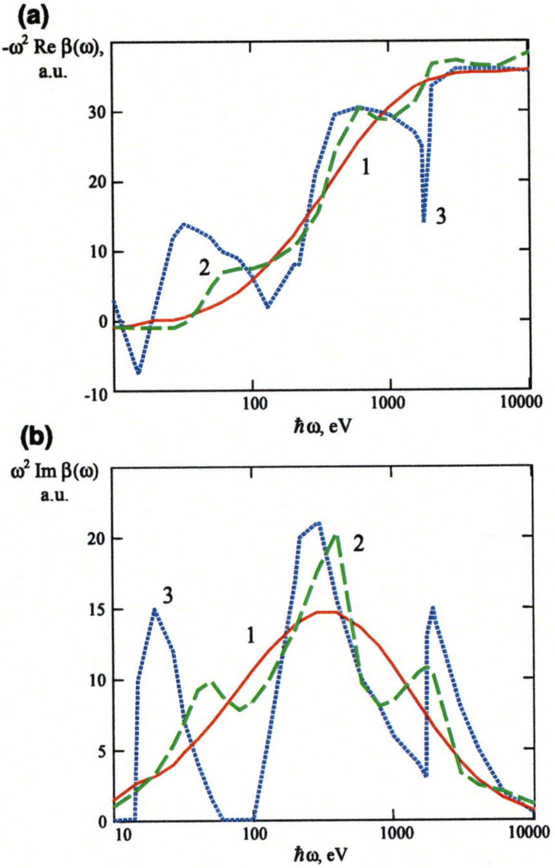

Fig. 2.1 Frequency dependences of the real (**a**) and imaginary (**b**) parts of the polarizability of a krypton atom calculated in different approximations: in the local plasma frequency approximation for the Lenz–Jensen electron density (1), in the local plasma frequency approximation for the Slater electron density (2), in the random phase exchange approximation (3) (Korol et al. 1998)

Let us present the high-frequency asymptotics of dimensionless polarizability that follows from the formula (2.72) in view of the explicit form of the function $f(x)$ for distribution of the Thomas–Fermi and Lenz–Jensen electron densities. For the imaginary part $\tilde{\beta}(\nu)$, we have

$$Im\left\{\tilde{\beta}^{T-F}(\nu \to \infty)\right\} \to \frac{4.35}{\nu^4}, \tag{2.73a}$$

$$Im\left\{\tilde{\beta}^{L-J}(\nu \to \infty)\right\} \to \frac{4.615}{\nu^4}. \tag{2.73b}$$

From the formulas (2.73), it is seen that the above statistical models give a similar result for the imaginary part of polarizability. It should be noted that in the hydrogen-like approximation (for a spherically symmetric atomic state) the imaginary part of polarizability decreases as $\nu^{-4.5}$. The high-frequency asymptotics of the real part of the dimensionless polarizability $\tilde{\beta}(\nu)$ in both models of atomic core electron density looks like

$$Re\left\{\tilde{\beta}(\nu \to \infty)\right\} \to -\frac{b^{-3}}{\nu^2}, \qquad (2.74)$$

which is in qualitative agreement with the general formula (2.43). From comparison of the expressions (2.73) and (2.74), it follows in particular that at high frequencies the imaginary part of polarizability decreases much more rapidly than its real part.

2.7 Dynamic Polarizability of Nanoparticles

Let us calculate the dynamic polarizability of a sphere placed in a dielectric medium. We assume that the sphere radius is much larger than the distance between atoms in the substance. In this case, for the description of the interaction of the sphere with an electromagnetic field the concept of the dielectric permittivity can be used.

Using the Mie theory for calculation of the cross section of radiation scattering by a spherical particle of radius r_s, it is possible to obtain the following expression for the polarizability of a spherical particle in terms of the Mie coefficients:

$$\beta_s = \frac{3}{2}\varepsilon_m \left(\frac{r_s}{x}\right)^3 \cdot [-i a_1(x, mx, m) + b_1(x, mx, m)], \qquad (2.75)$$

where

$$x = k r_s = \sqrt{\varepsilon_m}\frac{\omega}{c} r_s, \qquad (2.76)$$

$$m = \sqrt{\frac{\varepsilon_s(\omega)}{\varepsilon_m}}. \qquad (2.77)$$

m is the relative refractive index; ε_m, $\varepsilon_s(\omega)$ are the dielectric permittivities of the matrix and the material of the sphere; a_n and b_n are the Mie coefficients in the Legendre polynomial and spherical Hankel function expansions of a scattered wave outside the sphere.

Hereafter, we assume that the dependence of the value ε_m on the radiation frequency ω can be neglected. With the use of boundary conditions (for a more detailed description, see Sect. 4.5.1), the following formulas for the Mie coefficients can be obtained (Van der Hulst 1981):

$$a_n(x,y,m) = \frac{\psi_n'(y)\,\psi_n(x) - m\,\psi_n'(x)\,\psi_n(y)}{\psi_n'(y)\,\zeta_n(x) - m\,\zeta_n'(x)\,\psi_n(y)}, \tag{2.78}$$

$$b_n(x,y,m) = \frac{m\,\psi_n'(y)\,\psi_n(x) - \psi_n'(y)\,\psi_n'(x)}{m\psi_n'(y)\,\zeta_n(x) - \zeta_n'(x)\,\psi_n(y)}, \tag{2.79}$$

where

$$\psi_n(z) = z j_n(z) = \sqrt{\frac{\pi z}{2}} J_{n+1/2}(z), \tag{2.80}$$

$$\zeta_n(z) = z h_n^{(1)}(z) = \sqrt{\frac{\pi z}{2}} H_{n+1/2}^{(1)}(z) \tag{2.81}$$

are the functions introduced by Debye in 1909; $j_n(z)$ is the spherical Bessel function, $J_{n+1/2}(z)$ and $H_{n+1/2}^{(1)}(z)$ are the Bessel and Hankel functions of half-integer order. For $n = 1$, we have

$$j_1(z) = \frac{\sin(z)}{z^2} - \frac{\cos(z)}{z}, \tag{2.82}$$

$$h_1^{(1)}(z) = \frac{\sin(z)}{z^2} - \frac{\cos(z)}{z} - i\left[\frac{\sin(z)}{z} + \frac{\cos(z)}{z^2}\right]. \tag{2.83}$$

In the limit of small parameters $x = k\,r_s$, the following expansions for the Mie coefficients can be obtained from the formulas (2.78)–(2.83) (Van der Hulst 1981):

$$a_1 \cong i s x^3 \left(1 + t x^2 - i s x^3\right), \quad b_1 = i s u x^5, \tag{2.84}$$

where

$$s = \frac{2}{3}\frac{m^2 - 1}{m^2 + 2}, \quad t = \frac{3}{5}\frac{m^2 - 2}{m^2 + 2}, \quad u = \frac{1}{30}\left(m^2 + 2\right) \tag{2.85}$$

are the auxiliary functions of the relative refractive index. In view of the formulas (2.84), (2.85), in the limit $x = k\,r_s \ll 1$ we obtain

$$\beta_s(\omega, r_s) = r_s^3\, \varepsilon_m \frac{\varepsilon_s(\omega) - \varepsilon_m}{\varepsilon_s(\omega) + 2\,\varepsilon_m}, \tag{2.86}$$

the so-called Lorentz formula for the polarizability of a small spherical particle of radius $r_s \ll \lambda/2\,\pi\,\sqrt{\varepsilon_m}$ and dielectric permittivity $\varepsilon_s(\omega)$ that is placed in a matrix with a dielectric permittivity ε_m.

The expression (2.86) is valid for a sufficiently small nanosphere radius $r_s \ll 2\pi\sqrt{\varepsilon_m}\,\lambda$, where λ is the wavelength of a scattered photon. The analysis shows that for a metal sphere in glass the formula (2.86) for the optical spectral and adjacent spectral ranges works well for nanosphere radii less than 20 nm.

It should be noted that the expression (2.86) can be obtained to an accuracy of the factor ε_m with the use of the Clausius–Mossotti formula $\frac{\varepsilon(\omega)-1}{\varepsilon(\omega)+2} = \frac{4}{3}\pi N_a \beta(\omega)$ if in the latter it is assumed that $\varepsilon(\omega) = \varepsilon_s(\omega)/\varepsilon_m$ and $N_a = 1/V_s$, where V_s is the sphere volume. From the formula (2.86), it follows that the dynamic polarizability of a sphere with small radius has a maximum for the frequency ω_r defined by setting the real part of the denominator in the right-hand side of the (2.86) to zero:

$$Re\{\varepsilon_s(\omega_r) + 2\,\varepsilon_m\} = 0. \qquad (2.87)$$

The resonance (2.87) is connected with excitation of plasmons at the surface of the sphere.

The dielectric permittivity $\varepsilon_s(\omega)$ appearing in (2.86) can be expressed in terms of the real n_s and imaginary κ_s parts of the refractive index of the sphere material by the known formula:

$$\varepsilon_s(\omega) = \varepsilon_1(\omega) + i\,\varepsilon_2(\omega) = [n_s(\omega)]^2 - [\kappa_s(\omega)]^2 + 2\,i n_s(\omega)\,\kappa_s(\omega). \qquad (2.88)$$

The spectral dependences of the refractive index components $n_s(\omega)$ and $\kappa_s(\omega)$ for a number of metals are determined experimentally in the work of Johnson and Christy (1972) in a range of photon energies $\hbar\,\omega$ from 0.64 to 6.6 eV. Experiments were carried with thin films with thickness' ranging from 18.5 to 50 nm. In doing so, it was found that the complex refractive index does not depend on the film thickness in the range from 25 to 50 nm. Corresponding plots for silver and gold are given in Fig. 2.2.

From the plots in Fig. 2.2, it follows that practically throughout the presented range of photon energies the condition $n_s(\omega) < \kappa_s(\omega)$ is fulfilled, which corresponds to a negative value of the real part of dielectric permittivity. Negative dielectric permittivity means the impossibility of propagation of an electromagnetic wave in a volumetric sample as well as reflection of radiation from such a medium.

The results of calculation from the formula (2.86) of the real and imaginary parts of the polarizability of a silver nanosphere placed in glass $\varepsilon_m = 2.25$ with the use of the data on the refractive index of silver of Fig. 2.2a are presented in Fig. 2.3. From the figure, there follows the presence of a polarizability resonance at a photon energy of about 3.1 eV. Within the framework of the simplest model, when the dielectric permittivity of a metal is described by the plasma formula $\varepsilon_s(\omega) = 1 - \omega_p^2/\omega^2$ (ω_p is the plasma frequency), the position of the maximum of the imaginary part of the metal sphere polarizability is given by the equation following from the resonance condition (2.87):

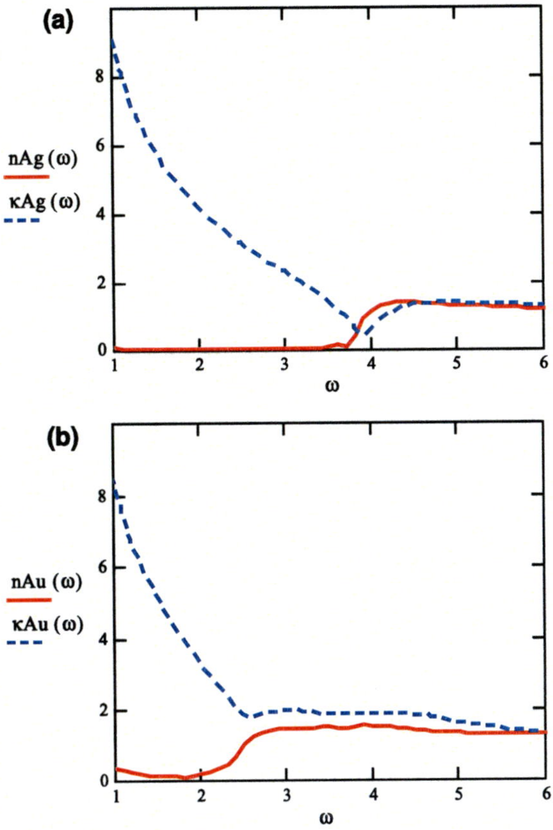

Fig. 2.2 Experimental dependences of the real (solid curve) and imaginary (dotted curve) parts of the refractive index of silver (**a**) and gold (**b**) films as functions of the photon energy (Johnson and Christy 1972); the photon energy in electron–volts is plotted on the abscissa

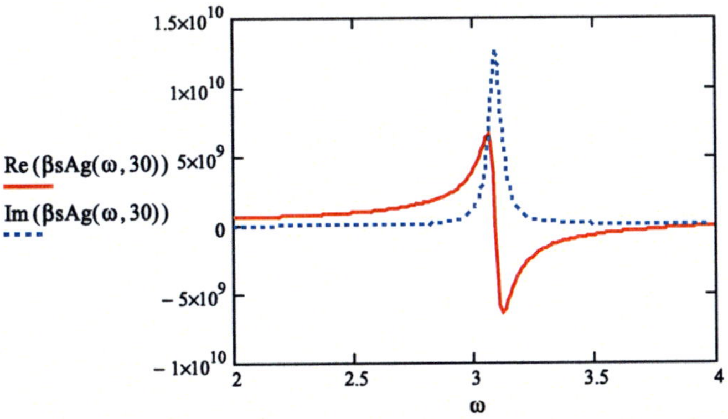

Fig. 2.3 The real and imaginary parts of the polarizability of a silver sphere with radius of 30 nm in a glass matrix; the abscissa is plotted in electron–volts, the ordinate is plotted in atomic units

$$\omega_{res} = \frac{\omega_p}{\sqrt{1 + 2\,\varepsilon_m}}. \tag{2.89}$$

In derivation of (2.89), it was assumed that the dielectric permittivity of a matrix in the frequency range under consideration does not depend on frequency. It should be noted that in vacuum the resonance frequency of a plasmon at the sphere surface is $\omega_{res} = \omega_p/\sqrt{3}$.

For silver $\hbar\,\omega_p \approx 9$ eV, so in case of a glass matrix, it follows from (2.89) that $\hbar\,\omega_{res} \approx 3.8$ eV. The difference of this value from the data of Fig. 2.3 is connected with the fact that the plasma formula for the dielectric permittivity used in derivation of (2.89) has an appreciable error due to neglect of the contribution of bound electrons.

A resonance increase in the polarizability of a metal nanosphere results in a number of important optical effects caused by increasing the electric field strength of the electromagnetic wave near the nanosphere if the condition (2.87) is fulfilled. It should be noted that one of such effects is an increase (by more than 10 orders of magnitude) in the probability of spontaneous Raman scattering of light by molecules attached to metal nanoparticles. This effect was observed experimentally now several times, and it has great prospects for applications.

References

M.Y. Amusia, *Atomic Photoeffect* (Springer, 1990)

H.A. Bethe, E.E. Salpeter, *Quantum Mechanics of One- and Two-Electron Atoms* (Springer, 1977)

W. Brandt, S. Lundqvist, Atomic oscillations in the statistical approximation. Phys. Rev. **139**, A612–A617 (1965)

H.C. Van der Hulst, *Light Scattering by Small Particles* (Dover Publications, NY, 1981)

P.B. Johnson, R.W. Christy, Optical constants of the noble metal. Phys. Rev. B **6**, 4370 (1972)

A.V. Korol, A.G. Lyalin, O.I. Obolensky, A.V. Solovyov, Investigation of the role of the polarization mechanism of emission by atoms in a wide range of photon frequencies. JETP **87**, 251 (1998)

V.P. Shevelko, A.D. Ulanzev, Static multipole polarizability of atoms and ions in the Thomas-Fermi model. J. Russ. Laser Res. **15**, 529 (1994)

M.J. Stott, E. Zaremba, Linear-response theory within the density-functional formalism: application to atomic polarizabilities. Phys. Rev. A. **21**, 12 (1980)

Chapter 3
Probabilities of Radiative Transitions

Abstract In this chapter, we consider the probability of photoprocesses including bound–bound, bound–free, and free–free electronic transitions. This concerns atomic radiation transitions in the discrete energy spectrum, radiative recombination, Bremsstrahlung including polarization channel, photoionization, photodetachment of negative ions, and phase control of photoprocesses by ultrashort laser pulses. Considerable attention has been paid to various types of broadening of the spectral lines of atomic radiative transitions, including plasma broadening mechanisms. The rotational approximation of the Kramers electrodynamics is presented which is suitable for describing both free–free and free–bound electronic transitions in the high frequency limit. The photoionization of atoms is described both within the framework of a rigorous quantum mechanical approach and with the help of a number of approximate methods. Analytical generalized photoionization cross section formulas from K-, L-, M-, N-, and O-shell that include also possible inner-shell photoionization are presented. Finally, generalized scaled formulas for radiation recombination rates into all states with principal quantum numbers n = 1–9 and orbital quantum numbers l = 0–8 are given that can be applied for a large variety of practical cases.

3.1 Radiative Transition Cross Sections

To calculate the probability per unit time (rate) for a phototransition between atomic energy levels, together with the Einstein coefficient for an induced process (2.29), a radiative transition cross section is widely used. In a monochromatic field of frequency ω it is determined by the equation

$$\sigma_{\mathrm{jn}}(\omega) = \frac{w_{\mathrm{jn}}}{j(\omega)}, \tag{3.1}$$

F. B. Rosmej et al., *Plasma Atomic Physics*, Springer Series on Atomic, Optical, and Plasma Physics 104, https://doi.org/10.1007/978-3-030-05968-2_3

where

$$j(\omega) = \frac{cE_0^2}{8\pi\hbar\omega} \tag{3.2}$$

is the photon flux density at a specified frequency and E_0 is the amplitude of the strength of the electric field of monochromatic radiation. Since the dimensionality of the probability per unit time w_{jn} is [s^{-1}] and the dimensionality of the photon flux density is [$s^{-1}m^{-2}$], it follows from (3.1) that the cross section has the dimensionality [m^2], i.e., the dimensionality of an area. Thus, we can say that the cross section describes some effective area of a hard sphere: If the projectile trajectory touches the hard sphere, a transition takes place, and if the trajectory lies outside, no transition occurs.

Using the formulas (2.28), (2.30), we find from (3.1) to (3.2) the following expression for the cross section of the transition $n \to j$:

$$\sigma_{jn}(\omega) = \frac{2\pi^2 e^2}{mc} f_{jn} G_{jn}^{(h)}(\omega), \tag{3.3}$$

where $G_{jn}^{(h)}(\omega)$ is the spectral form of a line of the homogeneously broadened transition, and f_{jn} is the transition oscillator strength (2.18). From the expression (3.3), it follows in particular that a large oscillator strength implies a large cross section. For dipole-forbidden transitions, when $f_{jn} = 0$, the cross section is equal to zero.

Substituting the expression for the oscillator strength (2.18) into (3.3), we obtain the expression for the phototransition cross section in terms of the matrix element of the dipole moment $\boldsymbol{d}_{jn}$:

$$\sigma_{jn}(\omega) = \frac{4\pi^2 \omega_{jn}}{3\hbar c g_n} |\langle n|\boldsymbol{d}|j\rangle|^2 G_{jn}^{(h)}(\omega). \tag{3.4}$$

g_n is the statistical weight (the degeneracy factor) of a nth atomic energy level. Because $G_{jn}^{(h)}(\omega)$ has it maximum value at $\omega = \omega_{jn}$, the maximum value of the cross section corresponds to the frequency $\omega = \omega_{jn}$ (the resonance condition):

$$\sigma_{jn}^{(max)} = \sigma_{jn}(\omega = \omega_{jn}) = \frac{8\pi}{3\hbar c g_n} |\langle n|\boldsymbol{d}|j\rangle|^2 \frac{\omega_{jn}}{\Delta\omega_{jn}}, \tag{3.5}$$

that is, it is proportional to the ratio of the eigenfrequency to the spectral characteristic width of an atomic transition line. This ratio is similar to the resonator Q-factor and is equal to the number of free oscillations of the oscillator until their total damping. It should be noted that in the early works of M. Planck and A. Einstein on the quantum theory of electromagnetic interaction atoms and molecules of a substance were called resonators.

In case of natural broadening, when $\Delta\omega_{jn} = A_{nj}$, (3.5) is transformed with (2.26) to the form:

$$\sigma_{jn}^{(max)} = \frac{g_j}{2\pi g_n}\lambda_{jn}^2. \tag{3.6}$$

Hence, it follows that the resonance cross section of a phototransition in case of natural broadening of a line is proportional to the squared of the resonant wavelength $\lambda_{jn} = 2\pi c/\omega_{jn}$; that is, in a wide spectral range, it exceeds by many orders of magnitude the geometrical cross section of an atom (equal to its area). It should be noted that the maximum cross section does not depend on the matrix element of the dipole moment.

For an atom under the action of radiation with a finite spectral width $\Delta\omega \neq 0$, the probability per unit time for the excitation of an atomic transition $n \to j$ is given by the integral

$$w_{jn} = \int \sigma_{jn}(\omega')j(\omega')d\omega'. \tag{3.7}$$

If the radiation spectrum considerably exceeds the spectral width of the transition line, i.e. $\Delta\omega \gg \Delta\omega_{jn}$, as it is the case, for example, for the thermal radiation, the spectral form of a transition line in the definition of the cross section can be replaced by the delta function $G_{jn}^{(h)}(\omega') \to \delta(\omega' - \omega_{jn})$, and the integral (3.7) gives:

$$w_{jn} = \frac{4\pi^2\omega_{jn}}{3\hbar c g_n}|\langle n|\mathbf{d}|j\rangle|^2 j(\omega_{jn}) = B_{jn}\rho(\omega_{jn}) \tag{3.8}$$

since $j(\omega) = c\rho(\omega)/\hbar\omega$, $\rho(\omega)$ is the spectral radiation density. The second equality in formula (3.8) coincides with the relation that was used by A. Einstein to describe the interaction of thermal radiation with a two-level atom.

In the general case of an arbitrary relation between the widths of radiation spectra and the atomic transition in an atom, the expression (3.7) should be used for the calculation of the photoexcitation rate. It should be noted that for the calculation of the probability per unit time for a phototransition, instead of a cross section, the concept of a spectral Einstein coefficient

$$D_{jn}(\omega) = B_{jn}G_{jn}(\omega) \tag{3.9}$$

is sometimes used. The formula (3.7) can be rewritten in terms of the spectral Einstein coefficient (3.9) with replacement of the photon flux density $j(\omega)$ by the spectral radiation energy density $\rho(\omega)$.

The atomic transition $n \to j$ induced by the action of radiation corresponds to photon absorption, if $E_n < E_j$. The reverse transition $j \to n$ corresponds to photon radiation that can be both induced (under the action of an external field) and spontaneous. The cross section of induced radiation $\sigma_{nj}(\omega)$ is described by formulas similar to (3.3)–(3.4) because the matrix element of the dipole moment can be made symmetric by selection of the wave functions, i.e. $\langle n|\boldsymbol{d}|j\rangle = \langle j|\boldsymbol{d}|n\rangle$ and $G_{\rm jn}^{(h)}(\omega) = G_{\rm nj}^{(h)}(\omega)$ and the difference in expressions for the cross sections of induced radiation and absorption consists only in different statistical weights of states. Therefore, the expression for the cross section of induced radiation $\sigma_{\rm nj}(\omega)$ for the transition $j \to n$ is obtained from (3.4) by the replacement $g_{\rm n} \to g_{\rm j}$.

It is easy to express the absorption coefficient $k_{\rm jn}$ in terms of the absorption cross section in order to describe electromagnetic wave damping in the propagation in a medium with resonant atoms. The damping is a result of a transition in an atom from a lower to a higher energy state, i.e. $n \to j$ $(E_{\rm n} < E_{\rm j})$: $k_{\rm jn} = N_{\rm n}\sigma_{\rm jn}$, where $N_{\rm n}$ is the concentration of atoms in the state n. It is obvious that the absorption coefficient has the dimensionality of reciprocal length [m^{-1}]. By analogy with the absorption coefficient, it is possible to introduce an amplification coefficient for the reverse transition $j \to n$ corresponding to induced radiation of a photon and an increase in radiation intensity: $k_{\rm nj} = N_{\rm j}\sigma_{\rm nj}$. Therefore, the effective amplification coefficient taking into account induced radiation and absorption can be expressed as follows:

$$k_{\rm nj}^{(\rm tot)} = k_{\rm nj} - k_{\rm jn} = g_{\rm j}\sigma_{\rm nj}(\omega)\left\{\frac{N_{\rm j}}{g_{\rm j}} - \frac{N_{\rm n}}{g_{\rm n}}\right\}. \tag{3.10}$$

If $k_{\rm nj}^{(\rm tot)} > 0$ (induced radiation dominates), radiation is amplified by the medium of resonant atoms. Otherwise radiation is attenuated since photon absorption dominates over induced radiation. Since the atomic transition cross section is a positive number, then, as follows from the formula (3.10), amplification of radiation is given by the condition:

$$\frac{N_{\rm j}}{g_{\rm j}} - \frac{N_{\rm n}}{g_{\rm n}} > 0, \quad E_{\rm n} < E_{\rm j}. \tag{3.11}$$

The relation (3.11) is called the *population inversion* condition. It is widely used in quantum electronics, laser physics, and analogous to the study of possible radiation amplification. In thermodynamic equilibrium, when the Boltzmann formula for energy level populations holds true, an inequality reverse to (3.11) is fulfilled. Thus, to obtain population inversion, it is necessary to disturb a substance considerably from its thermal equilibrium state. This is usually achieved by the action of external photons on a medium being called pumping.

The concept of a cross section is used not only for the description of absorption and induced radiation, but it also characterizes other photoinduced processes such as photon scattering, photorecombination, and bremsstrahlung. The cross section concept is likewise used in calculations of the interaction of other elementary particles (electrons, protons, neutrons) with atoms and molecules. In all cases, the cross section is determined by a formula similar to (3.1), with replacement of the photon flux density by the flux density of particles that induce the process under consideration.

It should be noted that the concept of probability per unit time for a photoinduced process loses its physical meaning for ultrashort electromagnetic pulses where the duration of radiation is of the order of the period of the electric field oscillations at the carrier frequency. Ultrashort pulses can be produced with the use of special methods of time compression of laser radiation (CPA: chirped pulse amplification). To ultrashort pulses in the optical range there corresponds a duration of several femtoseconds. At present, in the visible range of laser wavelengths of well-controlled shapes with durations of only 1.5 periods of the optical frequency have been produced. In the UV range, pulse durations down to several hundreds of attoseconds have been achieved. To describe the interaction of ultrashort radiation pulses with a substance, it is more adequate to use the concept of the total probability for a process (i.e., during the total action of a pulse) that can be expressed in terms of a cross section (Rosmej et al. 2014, 2016, 2021).

3.2 Spectral Line Shapes of Atomic Radiative Transitions

The relations (2.28)–(2.29) were obtained in the limit of radiation with a broad spectrum (the so-called broadband illumination). In the general case, using the same approach as for the derivation of (2.37), it is possible [using (2.27)] to find the following expression for absorbed power by the transition $n \to j$:

$$P_{\mathrm{jn}} = \frac{2\pi^2 e^2}{3m} f_{\mathrm{jn}} \int_0^\infty G_{\mathrm{jn}}^{(h)}(\omega')\rho(\omega')\mathrm{d}\omega', \tag{3.12}$$

where

$$G_{\mathrm{jn}}^{(h)}(\omega') = \frac{(\delta_{\mathrm{jn}}/\pi)}{(\omega_{\mathrm{jn}} - \omega')^2 + (\delta_{\mathrm{jn}})^2} \tag{3.13}$$

is the spectral form (profile) of a line transition in case of homogeneous broadening. The function (3.13) is called the Lorentz function or Lorentzian. From (3.12) to (3.13), it follows that external field frequencies are most effectively absorbed in the vicinity of the transition eigenfrequency ω_{jn}. The characteristic frequency interval, in which the interaction between radiation and the atom is most intensive, is given by the damping constant: $|\omega_{\mathrm{jn}} - \omega'| \leq \delta_{\mathrm{jn}}$. Hence, it follows that a characteristic width of an atomic transition line (in view of both detuning signs) is given by $\Delta\omega_{\mathrm{jn}}^{(h)} = 2\delta_{\mathrm{jn}}$. Substituting the damping constant from the last equation into (3.13), we obtain the form of a line of a homogeneously broadened transition in an atom or in any other quantum system.

Instead of the characteristic width $\Delta\omega_{\text{jn}}^{(h)}$, a characteristic time can be introduced according to

$$T_2 = \frac{2}{\Delta\omega_{\text{jn}}^{(h)}}. \tag{3.14}$$

The parameter T_2 is called the transverse relaxation time or the phase relaxation time. It is related to the damping constant of a transition oscillator according to $T_2 = 1/\delta_{\text{jn}}$ and therefore, as it follows from (3.14), defines the lifetime of a transition oscillator in the mode of free oscillation. The time T_2 (as will be shown below) is called the *irreversible* phase relaxation time.

Thus, from the point of view of the spectroscopic principle of correspondence, homogeneous broadening of a spectral line is defined by the damping of a transition oscillator occurring in free oscillation without external field.

An important particular case of homogeneous broadening is *natural broadening* of a line due to spontaneous radiation: $\Delta\omega_{\text{jn}}^{(h)} = A_{\text{nj}}$. In this case, transition oscillator damping is caused by the interaction of an atomic electron with the vacuum fluctuations of an electromagnetic field. Natural broadening for an atom in free space is the minimum possible, and it defines the degree of radiation monochromaticity that is in principle achievable.

Another type of broadening occurs in interaction of radiation with an ensemble of atoms when the transition eigenfrequency is spread over $\Delta\omega_{\text{jn}}^{(\text{inh})}$ for different atoms. This means that every atom has a specific frequency shift and the observed frequency spread belongs to the ensemble of atoms. This spread defines *inhomogeneous broadening* of a line. The spectral form (shape) is determined by the distribution function of the frequency shifts. In the case of a Gaussian distribution function, one has (Gaussian):

$$G_{\text{jn}}^{(\text{inh})}(\omega') = \frac{1}{\sqrt{2\pi}\Delta\omega_{\text{jn}}^{(\text{inh})}} \exp\left\{ -\frac{\left(\omega' - \omega_{\text{jn}}^{(c)}\right)^2}{2\left(\Delta\omega_{\text{jn}}^{(\text{inh})}\right)^2} \right\}, \tag{3.15}$$

where $\omega_{\text{jn}}^{(c)}$ is the central frequency of a transition in an ensemble of atoms.

The spectral forms of lines for homogeneous (Lorentzian) and inhomogeneous (Gaussian) broadening are presented in Fig. 3.1 for $\omega_{\text{jn}}^{(c)} = \omega_{\text{jn}}$ and $\Delta\omega_{\text{jn}}^{(\text{inh})} = \Delta\omega_{\text{jn}}^{(h)}$; the coordinate axes are plotted in relative units, i.e. $(\omega' - \omega_{\text{jn}}^{(c)})/\Delta\omega_{\text{jn}}^{(\text{inh})}$.

It is seen that the spectral form of a homogeneous broadened line has more extended "wings" compared to inhomogeneous Gaussian broadening. The functions (3.13), (3.15) have two common properties: (1) normalization to unity and (2) both functions tend to the Dirac delta function in the limit of zero width.

Different causes can produce inhomogeneous broadening of a line depending on the concrete realization of the radiative transition. For atoms in a gas and ions in

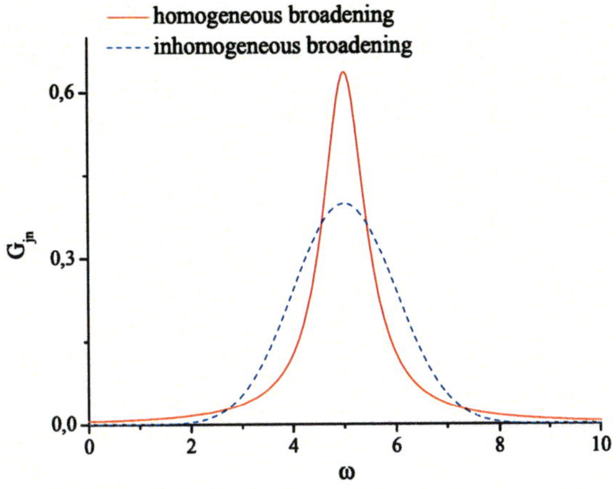

Fig. 3.1 Homogeneous and inhomogeneous broadening of a spectral line

a plasma, an important mechanism of inhomogeneous broadening is the Doppler effect. According to this effect, the radiation frequency from the intrinsic frame of reference is connected with an atom in the laboratory frame of reference by the formula

$$\omega = \gamma\omega_0\left(1 - \frac{\upsilon}{c}\cos\theta\right) \equiv \gamma\omega_0\left(1 - \frac{\upsilon_k}{c}\right), \tag{3.16}$$

where ω_0 is the frequency in the frame of reference of the radiating atom, $\gamma = \left(1 - (\upsilon/c)^2\right)^{-1/2}$ is the Lorentz factor, θ is the angle between the atomic velocity vector and the photon wave vector (the angle of radiation), and υ_k is the projection of the atomic (ionic) velocity on the direction of photon radiation. In gases and low-temperature plasmas $\upsilon \ll c$ and $\gamma \cong 1$. Since the velocity of atoms/ions in thermal equilibrium has a distribution, the width that is defined by the thermal velocity $\upsilon_T = \sqrt{2T/M}$ (M is the mass of an atom/ion) and the radiation frequency in the laboratory frame of reference [see (3.16)] will also have a distribution. The Doppler profile can be represented in terms of the integral

$$G_{ki}^{(D)}(\omega) = \int_{-\infty}^{\infty} \delta\left(\omega - \omega_{ki}\left(1 - \frac{\upsilon_k}{c}\right)\right) f(\upsilon_k)d\upsilon_k, \tag{3.17}$$

where $f(\upsilon_k)$ is the function of distribution of atomic (ionic) velocity projections on a specified direction. The delta function in the right-hand side of (3.17) marks out those values of velocity projections, at which, according to the Doppler effect (3.16), the frequency in the laboratory frame of reference is equal to a specific value ω. The formula (3.17) corresponds to an interpretation of inhomogeneous broadening as a distribution function of atomic eigenfrequencies. We can say that the Doppler effect maps the velocity distribution of atoms onto their eigenfrequency distribution. It should be noted that using the second equality in the right-hand side of (3.16) considerably simplifies the derivation of the expression for the Doppler profile.

In case of a Maxwellian velocity distribution, i.e.

$$f(\upsilon_{\mathrm{k}}) = \frac{1}{\sqrt{\pi}\upsilon_{\mathrm{T}}} \exp\left(-\left(\frac{\upsilon_{\mathrm{k}}}{\upsilon_{\mathrm{T}}}\right)^2\right), \tag{3.18}$$

the calculation of the integral (3.17) gives

$$G_{\mathrm{ki}}^{(D)}(\omega) = \frac{1}{\sqrt{2\pi}\Delta\omega_{\mathrm{ki}}^{(D)}} \exp\left[-\frac{(\omega - \omega_{\mathrm{ki}})^2}{2\left(\Delta\omega_{\mathrm{ki}}^{(D)}\right)^2}\right], \tag{3.19}$$

where

$$\Delta\omega_{\mathrm{ki}}^{(D)} = \frac{1}{\sqrt{2}} \frac{\upsilon_{\mathrm{T}}}{c} \omega_{\mathrm{ki}} \equiv \sqrt{\frac{T}{M}} \frac{\omega_{\mathrm{ki}}}{c} \tag{3.20}$$

is the Doppler line width (note that the FWHM is given by $2\sqrt{\ln 2} \cdot \Delta\omega_{\mathrm{ki}}^{(D)}$; see also (1.22)).

The right-hand side of (3.20) includes a temperature defining the translational velocity of atoms (ions) in the plasma and is called the ion temperature. As can be seen from (3.20), the measurement of the Doppler line width for radiation with a specified central frequency allows determining the ion temperature of the plasma if the mass of the radiating atom (ion) is known. According to the formula (3.20), the Doppler line width is proportional to the central radiation frequency, so the role of the Doppler effect increases with energy of an atomic transition. For the optical range and room temperatures, Doppler broadening is of the order of 10 GHz.

A Doppler profile being of the form (3.19) assumes an unchanged velocity of the atom during radiation emission. This condition is realized if the free path of an atom/ion in a gas or a plasma is larger than the wavelength of the radiation. However, it can be violated for sufficiently dense and hot plasmas such as laser-produced plasmas. Then, the line profile is close to be a Lorentz profile (3.13), with a line width that is inversely proportional to the atomic collision frequency. In rarefied low-temperature plasmas, the line profile is described by the formula (3.19).

Another mechanism of inhomogeneous line broadening characteristic to atoms/ions in plasma is connected with the Stark effect (Griem 1974, 1997; Sobelman et al. 1995). The Stark effect represents a shift (and, generally speaking, splitting) of the spectral line radiation of an atom under the action of an external electric field. Inhomogeneous broadening under the action of the Stark effect occurs in case of a static (or sufficiently slowly varying with time) electric field. Such fields are produced by plasma ions due to relatively low (in comparison with electrons) velocity (because the mass of ions is much larger than the electron mass). The expression for the Stark effect that is linear with respect to the electric field looks like

$$\omega - \omega_{ik} = C_{ik}F, \tag{3.21}$$

where F is the magnitude of the electric field strength at the location of an atom (ion) and C_{ik} is the Stark constant for the atomic transition. The linear Stark effect is characteristic for hydrogen-like ions where we encounter degeneracy in orbital quantum number.

Let us consider the influence of the Stark effect on the form of a spectral line in the static case when the electric field that acts on the radiating atom is a result of a large number of plasma ions. Then for the strength of the total electric field, we have

$$\boldsymbol{F} = \sum_{j=1}^{N_0} \boldsymbol{F}_j = \sum_{j=1}^{N_0} e\frac{\boldsymbol{r}_j}{r_j^3}, \tag{3.22}$$

where the summation is carried out over all plasma ions, with N_0 being the number of ions. For simplicity, we assumed $Z = 1$ in (3.22). The relation (3.21) establishes a univocal correspondence between the electric field and the frequency shift. Therefore, if the distribution function for the electric field strength $W(\boldsymbol{F})$ is known, it is easy to find with the help of the formula (3.21) the frequency distribution function that is a line profile for inhomogeneous broadening. This function looks like

$$G_{ik}^{(St)}(\omega) = \frac{1}{C_{ik}} W\left(\frac{\omega - \omega_{ik}}{C_{ik}}\right). \tag{3.23}$$

The factor in front of the ion field distribution function results from the relation $dF/d\omega = 1/C_{ik}$ being the frequency transformation of the field distribution.

Thus, a key problem is the determination of the ion field distribution function $W(\boldsymbol{F})$. Following Holtsmark, simple expressions can be obtained in the approximation of negligible correlation between the plasma ions. Then, the probability for ions having a radius vector in the interval $(\boldsymbol{r}_j, \boldsymbol{r}_j + d\boldsymbol{r}_j)$ is proportional to the product $V^{-N_0} \prod_{j=1}^{N_0} d\boldsymbol{r}_j$, where V is the plasma volume. The probability $W(\boldsymbol{F})$ is proportional to that part of the N_0-dimensional space in variables $\boldsymbol{r}_j$, in which (3.22) is fulfilled. According to this, the formula for the ion field distribution function is given by:

$$W(\boldsymbol{F}) = \left\langle \delta\left(\boldsymbol{F} - \sum_{j=1}^{N_0} e\frac{\boldsymbol{r}_j}{r_j^3}\right)\right\rangle, \tag{3.24}$$

where averaging is done over the above distribution of ion radii. The explicit form of the function $W(\boldsymbol{F})$ can be obtained when using the integral representation of the delta function and going to the limit N_0, $V \to \infty$ at constant concentration of the ions $N = N_0/V$. As a result, the following expression is obtained (Unsöld 1955):

$$W(\boldsymbol{F})\mathrm{d}\boldsymbol{F} = W(F)\mathrm{d}F = H\left(\frac{F}{F_0}\right)\frac{\mathrm{d}F}{F_0}, \tag{3.25}$$

where

$$H(\beta) = \frac{2}{\pi}\beta\int_0^\infty x\sin(\beta x)\exp\left(-x^{3/2}\right)\mathrm{d}x \tag{3.26}$$

is the so-called Holtsmark distribution, $\beta = F/F_0$, $F_0 = \alpha e N^{2/3}$ is a scaling factor of the electric field strength and $\alpha = 2\pi(4/15)^{2/3} \cong 2.603$. From (3.25), it follows that the ion field distribution depends only on the field strength. The plot of the Holtsmark function is presented in Fig. 3.2. The maximum of this function corresponds to a value $\beta = 1.607$.

The Holtsmark distribution function is normalized according to $\int_0^\infty H(\beta)\mathrm{d}\beta = 1$. In the vicinity of zero, $H(\beta)$ increases as β^2, and at $\beta \gg 1$ it decreases as $\beta^{-5/2}$. The Holtsmark distribution differs essentially from the Gaussian distribution (3.15) that describes spectral line broadening as a result of the Doppler effect. It is close to the Gaussian velocity distribution for low $\beta \ll 1$, when the contribution to the field originates from a large number of ions. As is known, statistical regularities in ensembles of particles are described just by the Gaussian distribution. For strong fields, the Holtsmark distribution coincides with the distribution of a field from one nearest particle (the binary approximation). Thus, the Holtsmark function describes the transition from the Gaussian distribution for weak fields to the binary

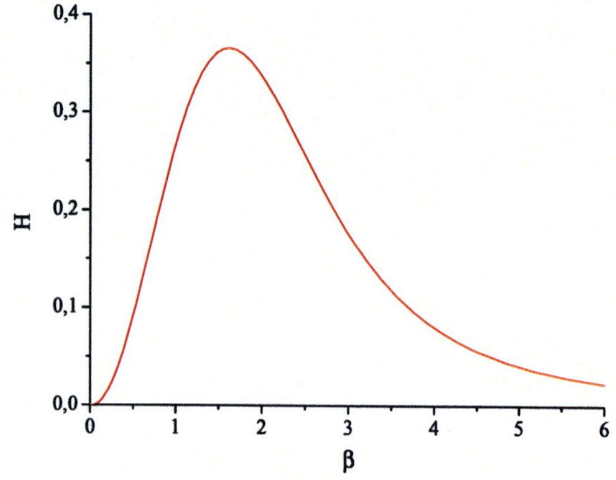

Fig. 3.2 Holtsmark function according to (3.26)

distribution for strong fields. Curiously, the scaled field strength of the Holtsmark distribution $F_0 = 2.603\, eN^{2/3}$ differs from the field at an average interionic distance $\bar{F} = e(4\pi N/3)^{2/3} \approx 2.61\, eN^{2/3}$ by less than one percent.

In plasma, a change of the phase of a transition oscillator (an equivalent oscillator) leading to homogeneous broadening can be a result from a collision between an atom and a charged particle. Let us consider this process for the case of interaction of electrons with a hydrogen-like atom. Then, an instantaneous change of the oscillator frequency $\omega(t)$ under the action of the electric field $F(t)$ of a plasma electron can be represented as

$$\omega(t) = \frac{C}{e} F(t), \tag{3.27}$$

where C is the Stark constant. The change of the phase of an equivalent oscillator during the time of electron transit close to the atom (ion) is given by

$$\Delta\phi = \int_{-\infty}^{\infty} \omega(t) \mathrm{d}t = \int_{-\infty}^{\infty} \frac{C}{\rho^2 + \upsilon^2 t^2} \mathrm{d}t = \pi \frac{C}{\rho \upsilon}, \tag{3.28}$$

ρ is the impact parameter, υ is the velocity of the perturbing particle assumed to be moving uniformly on a straight trajectory. From (3.28), it follows that a phase change $\Delta\varphi = \pi$ is achieved at

$$\rho_{\mathrm{W}} = \frac{C}{\upsilon}. \tag{3.29}$$

This value is called the Weisskopf radius; it defines the region of impact parameters corresponding to an essential change of the equivalent oscillator phase. To the radius ρ_{W}, there is related the Weisskopf cross section

$$\sigma_{\mathrm{W}} = \pi \rho_{\mathrm{W}}^2 = \pi \frac{C^2}{\upsilon^2} \tag{3.30}$$

and the Weisskopf frequency

$$\Omega = \frac{\upsilon}{\rho_{\mathrm{W}}} = \frac{\upsilon^2}{C}. \tag{3.31}$$

The parameter ρ_{W} defines the effective volume of interaction of an atom with perturbing particles. If there is one particle in this volume, the binary interaction takes place. Otherwise many particles interact simultaneously with an atom (non-binarity).

The Weisskopf frequency (3.31) separates impact and static limits for the spectral line shape. The profile of a line broadened by collisions with electrons can be represented in the form

$$G(\Delta\omega) = \frac{1}{2\pi} \frac{\Delta\omega(\omega - \omega_{\text{ik}})}{(\omega - \omega_{\text{ik}})^2 + (\Delta\omega(\omega = \omega_{\text{ik}})/2)^2} \tag{3.32}$$

that would coincide with the Lorentz profile (3.13) if the line width $\Delta\omega$ in the numerator of the right-hand side of the (3.32) would be a constant value (as in the denominator). In fact, the line width in the numerator depends on frequency detuning. If this detuning is much less (in magnitude) than the Weisskopf frequency (the impact limit), it can be considered that $\Delta\omega(\omega - \omega_{\text{ik}}) = \Delta\omega(\omega = \omega_{\text{ik}})$, and the Lorentz profile of a line (3.13) is realized. In the impact case, interaction of a quantum system with a perturbing particle is of instantaneous nature. As a result, the phase of oscillations of a transition oscillator "jumps". A typical value of impact broadening in gas at normal pressure and room temperature is about 100 GHz, so in the optical range it far exceeds natural broadening of a line, that has characteristic values of 100 MHz.

In the opposite case $|\omega - \omega_{\text{ik}}| \gg \Omega$ (the far wing of a line), it is necessary to take into account the frequency dependence of the line width in formula (3.32). Then, static broadening is realized, and the line profile (in the binary approximation) looks like

$$G_{\text{ik}}^{(\text{st})}(\omega) = \frac{2\pi N C^{3/2}}{|\omega - \omega_{\text{ik}}|^{5/2}}. \tag{3.33}$$

The dependence on frequency detuning in the right-hand side of (3.33) coincides with the result of the Holtsmark approximation in the strong field limit, when $H(\beta \gg 1) \propto \beta^{-5/2}$. As was already noted, this coincidence is connected with the fact that in the strong field limit the static Holtsmark distribution coincides with the distribution of a field from one particle nearest to an atom; that is, the binary approximation is valid.

The condition $|\omega - \omega_{\text{ik}}| \gg \Omega$ means that in the static limit the time of interaction (ρ_{W}/v) should be much larger than the time of spectral line formation $(\approx 1/|\omega - \omega_{\text{ik}}|)$. In the impact limit, on the contrary, the time of formation of a line is long, and perturbation can be considered to be instantaneous.

The expression for a line broadening in the static limit and the binary approximation (3.33) was obtained for an interaction charge-dipole. In the general case, when the energy of interaction of "broadening particles" with an atom (ion) looks like $U(R) = C_n/R^n$ (R is the distance between particles), the line profile in the static limit is described by the expression

$$G_{\rm ik}^{(\rm st)}(\omega) = \frac{4\pi N(C_{\rm n}/\hbar)^{3/n}}{n|\omega - \omega_{\rm ik}|^{(n+3)/n}}. \tag{3.34}$$

The obtained expression is valid in the spectral range $|\omega - \omega_{\rm ik}| \gg \Omega$. From the formula (3.34), it follows that the frequency dependence in the static wing of a line coincides with the Lorentz dependence (3.13) only for $n = 3$, that is, for dipole–dipole interaction.

3.3 Quasi-classical and Quantum Radiative Transition Probabilities

3.3.1 *Kramers Electrodynamics*

The foundations of radiation theory for a classically moving particle (electron) in a given potential $U(r)$ are described in numerous books on classical electrodynamics (Jackson 2007; Landau and Lifschitz 2003). In accordance with Gervids and Kogan (1975, 1991), Kogan and Kukushkin (1984), Kogan et al. (1992) we shall dwell on a number of classical spectral peculiarities connected with the attractive potential $U(r) = -|U(r)|$ playing an important role in the applicability of classical mechanics to atomic physics. The essence of the problem involves the situation when an emitting electron in an attractive field experiences an acceleration and may obtain a kinetic energy $W = E + |U(r)|$, that considerably exceeds its initial energy E at infinity. In this case, the classical nature of the electron motion is even preserved when the quantum energy $\hbar\omega$ emitted by the electron exceeds its initial energy E. This circumstance essentially expands the domain of applicability of the classical methods to atomic processes, including the inelastic domain $\hbar\omega \geq E$. Below, we will focus on the Coulomb field case playing an important role for the atomic processes in plasmas. Atomic potentials of more general type are investigated in (Kogan and Kukushkin 1984). The results of the following considerations will be used later in the quasi-classical approximation for radiation transitions probabilities.

Classical electrodynamics (CED) operates with an effective spectral radiation yield $dk(\omega)$, cross section $d\sigma(\omega)$ and energy $dE(\rho, \omega)$ emitted during the time of the collision with an impact parameter ρ in a frequency domain $d\omega$. These values are connected by the relation

$$\frac{dk(\omega)}{d\omega} = \hbar\omega\frac{d\sigma}{d\omega} = \int_0^\infty 2\pi\rho d\rho [dE(\rho, \omega)/d\omega]. \tag{3.35}$$

The spectral distribution of the emitted energy is defined by the Fourier coefficients $\dot{x}(\rho, \omega)$ and $\dot{y}(\rho, \omega)$ of the electron velocity components $\dot{x}(\rho, t)$ and $\dot{y}(\rho, t)$:

$$\frac{dE(\rho,\omega)}{d\omega} = \frac{e^2}{3\pi c^3}\omega^2\left[|\dot{x}(\rho,\omega)+i\dot{y}(\rho,\omega)|^2 + |\dot{x}(\rho,\omega)-i\dot{y}(\rho,\omega)|^2\right]. \quad (3.36)$$

These Fourier components for a motion in a Coulomb field are expressed in terms of the Hankel functions $H_{iv}^{(1)}(iv\varepsilon)$ and their derivatives $H_{iv}^{(1)'}(iv\varepsilon)$:

$$\frac{dE(\rho,\omega)}{d\omega} = \frac{2\pi Z^2 e^6 \omega^2}{3m^2 \upsilon^4 c^3} f(v,\varepsilon), \quad (3.37)$$

$$f(v,\varepsilon) = \left\{\left[H_{iv}^{(1)'}(iv\varepsilon)\right]^2 - \left(1-\frac{1}{\varepsilon^2}\right)\left[H_{iv}^{(1)}(iv\varepsilon)\right]^2\right\}. \quad (3.38)$$

Here, $v = \omega/\tilde{\omega}$ is a dimensionless frequency in units of the "classical" Coulomb frequency, $\tilde{\omega} = \upsilon/a = m\upsilon^3/Ze^2$ ($a = Ze^2/m\upsilon^2$ is the Coulomb length), $\varepsilon^2 = 1+\rho^2/a^2$ is the eccentricity of the hyperbolic trajectory (ρ is the impact parameter): parameters Ze, m, υ, and c are the standard designations for nuclear charge, mass, electron velocity, and the speed of light, respectively.

The function $f(v,\varepsilon)$ is well known (Landau and Lifschitz 2003) being the complete derivative of the function

$$g(v) = \frac{\pi\sqrt{3}}{4} iv H_{iv}^{(1)}(iv\varepsilon) H_{iv}^{(1)'}(iv\varepsilon), \quad (3.39)$$

that makes it possible to perform the integration over $d\rho$ in (3.35) and to obtain the cross section

$$d\sigma(\omega) = \frac{16\pi e^2 \upsilon^2}{3\sqrt{3}\hbar c^3} a^2 \frac{dv}{v} g(v). \quad (3.40)$$

The function $g(v)$ is named the Gaunt factor. For large radiation frequencies $\omega/\tilde{\omega} \gg 1$, the factor $g(v)$ approaches unity, and the numerical factor before $g(v)$ in (3.40) is the so-called Kramers bremsstrahlung cross section.

The total (integral over ω) effective radiation k is expressed in terms of the total radiation energy loss $\Delta E(\rho)$ during the collision:

$$k = \int \hbar\omega \frac{d\sigma}{d\omega} d\omega = \int_0^\infty 2\pi\rho d\rho \Delta E(\rho). \quad (3.41)$$

The magnitude $\Delta E(\rho)$ may be expressed, in turn, with the help of a time integral from the square of electron acceleration $w(\rho,t)$:

$$\Delta E(\rho) = \frac{2e^2}{3c^3} \int_{-\infty}^{\infty} [w(\rho, t)]^2 \mathrm{d}t. \tag{3.42}$$

For the central field, (3.42) is rewritten in the form

$$\Delta E(\rho) = \frac{4e^2}{3m^2c^3} \int_{r_0(\rho)}^{\infty} \left(\frac{\mathrm{d}U}{\mathrm{d}r}\right)^2 \frac{\mathrm{d}r}{\upsilon_r}, \tag{3.43}$$

$\upsilon_r(\rho)$ is the radial velocity, and $r_0(\rho)$ is the classical turning point defined from the relation

$$1 = \frac{\rho^2}{r_0^2} - \frac{|U(r_0)|}{E}. \tag{3.44}$$

Let us write down the spectral distributions of the emitted energy in the domain of large and small frequencies. Following (Kogan and Kukushkin 1984; Gervids and Kogan 1991), let us use the normalized spectral functions being the ratio of the spectral distribution $\mathrm{d}E(\rho, \omega)$ to the total radiation $\Delta E(\rho)$

$v \ll 1, \rho \gg a$:

$$\frac{\mathrm{d}E(\rho, \omega)}{\Delta E(\rho)} = \frac{8}{\pi^2}\left\{[sK_0(s)]^2 + [sK_1(s)]^2\right\}\mathrm{d}s, \tag{3.45}$$

$$s = M\omega/2E;$$

$v \gg 1, \rho \ll a$:

$$\frac{\mathrm{d}E(\rho, \omega)}{\Delta E(\rho)} = \frac{12}{\pi^2} G(u) u \mathrm{d}u, \tag{3.46}$$

$$u = M^3\omega/3Z^2me^4,\ G(u) = u\left[K_{1/3}^2(u) + K_{2/3}^2(u)\right], \tag{3.47}$$

where $M = m\upsilon\rho$ is the electron orbital momentum, and $K_\nu(x)$ are the Mcdonald functions.

Let us analyze in more detail the high-frequency case (3.46), (3.47). First, it is obvious that the spectral distribution described by the variable u does not depend on the initial electron energy E. This suppression of the energy integral is due to the aforementioned electron acceleration in an attractive potential $U(r) = -Ze^2/r$. As a matter of fact, the radiation of large frequencies $\omega \gg \tilde{\omega} = \upsilon/a$ originates from the part of sharp curvature of the impact electron trajectory where its acceleration is maximum. It is obvious that the largest acceleration is observed near the trajectory turning point r_0 (3.43), (3.44). In this domain, the potential energy $U(r_0)$ is much larger compared with the initial energy ($|U(r_0)| \gg E$) and that is the reason why the latter does not influence the spectral distribution of the emitted energy. This domain for the Coulomb field was first indicated by Kramers (1923). The non-Coulomb generalization of the approximation to suppress the energy integral forms the basis of the so-called Kramers electrodynamics introduced by Kogan et al. (1992).

According to (3.36), the energy spectral distribution consists of the two polarizations corresponding to the rotation directions along and against the electron trajectory. In accordance with the total intensity (3.46, 3.47), the sum of two contributions from two polarizations mentioned is

$$\begin{aligned}\frac{\mathrm{d}E(\rho,\omega)}{\Delta E(\rho)} &= \frac{6}{\pi^2}u^2\left\{\left[K_{1/3}(u)+K_{2/3}(u)\right]^2+\left[K_{1/3}(u)-K_{2/3}(u)\right]^2\right\}\mathrm{d}u \\ &\propto \left[F_-^2(u)+F_+^2(u)\right].\end{aligned} \tag{3.48}$$

It is easy to see that in almost all domains, the change of the function $F_-(u)$ (corresponding to the sum of functions $K_{1/3}$ and $K_{2/3}$) substantially exceeds the function $F_+(u)$ (corresponding to their difference). This particular circumstance is caused mathematically by the compensation of the functions $K_{1/3}$ and $K_{2/3}$ and reflects an important feature of radiation formation in the high-frequency domain; namely, the radiation is basically caused by the electron rotation near the turning point r_0 of the trajectory. The angular velocity of such a rotation $\omega_\mathrm{R}(r_0)$ is defined by the relation:

$$\omega_\mathrm{R}(r_0) = M/mr_0^2 = \sqrt{2(E+|U(r_0)|)/mr_0^2} = \frac{\upsilon_{\max}}{r_0}, \tag{3.49}$$

where $\upsilon_{\max}$ is the maximum electron velocity.

The aforementioned nature of the spectral formation becomes apparent if one writes down the functions $F_\pm(\omega,E,M)$ for an arbitrary potential $U(r)$ in the form of the Fourier components of the electron trajectory (Kogan and Kukushkin 1984; Kogan et al. 1992):

$$F_\pm(\omega,E,M) = \int\limits_{r_0(E,M)}^{\infty}\frac{\cos\left(\int\limits_{r_0(E,M)}^{\infty}[\omega\pm\omega_\mathrm{R}(r')]\left\{\frac{2}{m}\left[E+|U(r)|-\frac{M\omega_\mathrm{R}(r')}{2}\right]\right\}^{-1/2}\mathrm{d}r'\right)}{\sqrt{E+|U(r)|-\frac{M\omega_\mathrm{R}(r)}{2}}}r\mathrm{d}r. \tag{3.50}$$

For large frequencies $\omega \gg \tilde{\omega}$ the integrated expressions in (3.50) promptly oscillate everywhere, excluding the points of oscillation compensation $\omega \approx \omega_\mathrm{R}(r_\omega)$. The compensation takes place only for the function F_- (which explains the definition of its index) but not for the function F_+. Therefore, the more differ the F_- and F_+ contributions to the intensity $I(\omega)$, the higher the frequency ω becomes. This circumstance follows from pure classical mechanics and also manifests itself in quantum calculations of transitions probabilities, known as the Bethe rule.

The above analysis of the radiation mechanism in the high-frequency domain reveals a means for an universal spectra description (Kogan et al. 1992). The description is reached by the replacement of the real electron motion by its rotation along a circle with angular velocity $\omega_R(r)$. This approximation is obtained by the introduction of the delta function $\delta[\omega - \omega_R(r)]$ into (3.43), leading to the following spectral distribution (Kogan and Kukushkin 1984; Gervids and Kogan 1991):

$$\left(\frac{dk}{d\omega}\right)_R \simeq \frac{8\pi e^2}{3c^3 m^2 \upsilon} \int_0^\infty \left(\frac{\partial U}{\partial r}\right)^2 \sqrt{1 - \frac{U(r)}{E}} r^2 \delta[\omega - \omega_R(r)] dr. \tag{3.51}$$

Calculation of the integral in (3.51) leads to the following expression for the Gaunt factor (3.40) of the bremsstrahlung (Kogan and Kukushkin 1984; Kogan et al. 1992; Gervids and Kogan 1991)

$$g_{rot}(\omega) = \frac{6}{Z^2 e^4} \frac{D_\omega^2}{2 + D_\omega} \frac{[E + |U(r_\omega)|]^3}{m\omega^2}, \tag{3.52a}$$

$$D_\omega = -d \ln[E + |U(r_\omega)|]/d \ln r_\omega. \tag{3.52b}$$

In correspondence with the ideas presented above, the radiation radius r_ω is defined by the relation

$$\omega_R(r_\omega) = \omega \tag{3.53a}$$

or

$$\frac{E + |U(r_\omega)|}{r_\omega^2} = \frac{m\omega^2}{2}. \tag{3.53b}$$

The rotational approximation (3.52), (3.53) is of high precision. For example, for a Coulomb potential the error of the approximation does not exceed 5% even for a frequency as low as $\omega = \tilde{\omega}/2$. The detailed analysis of the rotational approximation results in a more general class of atomic potentials (Kogan and Kukushkin 1984; Kogan et al. 1992; Gervids and Kogan 1991).

The supression of the energy integral in the Kramers electrodynamics and the peculiarities of the radiation spectra connected with it work well in the high-frequency domain $\omega \gg \tilde{\omega}$. This domain makes the main contribution to the total bremsstrahlung intensity. As far as the low-frequency domain $\omega \ll \tilde{\omega}$ is concerned, (3.45) shows that there is no compensation of K_0 and K_1 and consequently there is no domination of the spectral function F_- with respect to F_+.

The independence of the radiation characteristics on the energy indicates the universal nature of the radiation spectral dependence on the frequency, not only for the infinite motion ($E > 0$) considered above but also for the finite motion ($E<0$) of the electron along an elliptical trajectory, as well. This is easy to verify by analyzing the finite motion intensity distribution $I_n(\omega)$, being the sum of harmonics $n = \omega/\omega_0$, where

$$\omega_0 = (2|E|)^{3/2}/Ze^2\sqrt{m} \tag{3.54}$$

is the typical frequency of the finite motion (the analog of the frequency $\tilde{\omega}$ in a continuum spectrum). The intensity I_n of a given harmonic is equal to (Gervids and Kogan 1991):

$$\begin{aligned} I_n &\propto n^2E^4\left(1-\tilde{\varepsilon}^2\right)^2 \\ &\times\left\{K_{1/3}^2\left[\frac{n}{3}\left(1-\tilde{\varepsilon}^2\right)^{3/2}\right]+K_{2/3}^2\left[\frac{n}{3}\left(1-\tilde{\varepsilon}^2\right)^{3/2}\right]\right\}, \end{aligned} \tag{3.55}$$

where $\tilde{\varepsilon} = \left(1 - 2|E|M^2/Z^2me^4\right)^{1/2}$ is the eccentricity of elliptical trajectory. It is simple to ensure that the argument of the K-functions in (3.55) is reduced, as in the continuum spectrum case, to the universal variable $u \sim M^3\omega/Z^2$. Independence of the spectrum on the energy is realized for the radiation intensity of the classical motion averaged over the period $T = 2\pi/\omega_0$, namely for the quantity (Gervids and Kogan 1991):

$$T\mathrm{d}I = TI_n\mathrm{d}n = \frac{2\pi}{\omega_0}I_n\frac{\mathrm{d}\omega}{\omega_0} \propto \frac{I_n}{\omega_0^2}\mathrm{d}\omega. \tag{3.56}$$

One can see that the quantity (3.56) becomes independent of the electron energy after substitution of (3.55). To summarize, it should be noted that Kramers high-frequency spectral domain possesses a universal intensity distribution for transitions in both continuum and discrete spectra. The universality is connected with the supression of the energy integral for attractive atomic potentials.

3.3.2 *Discrete Energy Spectrum*

Let us consider the results of Kramers electrodynamics (KrED) in the application to transitions in the discrete energy spectrum. As it follows from the general properties of the KrED, the dependencies of spectral characteristics of the radiation remain the same as in the case of a continuous energy spectrum. The only additional fact to be taken into account is the discrete nature of the energy spectrum which corresponds to the following relation between the emitted photon frequency ω and the difference between the initial E_{nl} and final $E_{n'l'}$ atomic state energies:

$$\omega = (E_{\rm nl} - E_{\rm n'l'})/\hbar. \tag{3.57}$$

The values for the energies $E_{\rm nl}$ should be taken from the results of quantum mechanical calculations or from corresponding experimental data.

Equation (3.57) leads to the relationship between the spectral interval $d\omega$ of the emitted photon frequencies and the density dn' of final states:

$$d\omega/dn' = 2\pi/T_{\rm n'l'}. \tag{3.58}$$

Here, $T_{\rm n'l'}$ is the period of classical motion (in the general case being only the radial period) of the electron with the energy equal to the energy $E_{\rm n'l'}$ of the final state. The value $T_{\rm nl}$ is determined by the conventional formulae of classical mechanics (Landau and Lifschitz 2005; Naccache 1972; Kogan and Kukushkin 1984; Kogan et al. 1992) for the case of the central potential $U(r)$.

The general expression for the probability of the transition $\Gamma \to \Gamma'$ ($\Gamma \equiv \{nl\}$) may be obtained from the classical spectral distribution for the emitted energy whose terms are to be separated with respect to the increase and decrease of the electron angular momentum:

$$\Delta E_\omega(\rho)_\pm = \frac{2e^2 m\omega^4}{3\pi c^3}[F_\pm]^2, \tag{3.59}$$

where the functions $F_\pm$ defined by (3.50) correspond to the radiation emission processes with the increase (F_+) and decrease (F_-) of the electron angular momentum.

In order to obtain the probability $W(\Gamma \to \Gamma')$ of a radiative transition per unit time, we divide the quantity (3.59) by the energy $\hbar\omega$ of an emitted photon and by the period $T_{\rm nl}$ of the classical motion with given initial energy and then multiply the result by the final state density (3.58). Thus, we obtain

$$W_{\Gamma\to\Gamma'} = \frac{\Delta E_\omega(\rho)}{\hbar\omega}\left|\frac{d\omega}{dn'}\right|\frac{1}{T_{\rm nl}} = \frac{2\pi}{T_{\rm nl}T_{\rm n'l'}}\frac{\Delta E_\omega(\rho)}{\hbar\omega}. \tag{3.60}$$

Equation (3.60), with account of (3.59) and relation $M = \hbar(l + 1/2) = mv\rho$, takes the form

$$W(nl \to n'l \pm 1) = \frac{4}{3}\left(\frac{\omega}{c}\right)^3 \frac{me^2}{\hbar}\frac{1}{T_{\rm nl}T_{\rm n'l'}}F_\pm^2(\omega, l). \tag{3.61}$$

This result coincides with the result of the corresponding quasi-classical calculation (Gantsev et al. 1985; Kogan et al. 1992), in the limit of $n \gg 1$, $l \gg 1$. The periods $T_{\rm nl}$ and $T_{\rm n'l'}$ in the latter calculation originate from the normalization constants of quasi-classical wave functions (for the relations between the functions $F_\pm$ and quasi-classical matrix elements, see Goreslavski et al. (1982), Gantsev et al. (1985), Landau and Lifschitz (1977), Naccache (1972), Kogan and Kukushkin (1984), Kogan et al. (1992)).

The KrED result for a bound–bound transition corresponds to the high-frequency domain where the emitted frequency ω largely exceeds the frequency of the electron revolution around the field center on its classical trajectory:

$$\omega_{\mathrm{nn'}}T_{\mathrm{nl}} \gg 1. \tag{3.62}$$

The KrED method for the description of an electron bound–bound radiative transition in a central potential $U(r)$ may be considered as some alternative to the well-known quantum defect method (Bates and Damgaard 1949; Davydkin and Zon 1981; Sobelman 1972, 2006; Cowan 1981). The latter is based on the following relation for the final state density

$$\partial\omega/\partial n' = Z_{\mathrm{i}}^2(\nu_{\mathrm{n'l'}})^{-3}, \quad \nu_{\mathrm{n'l'}} = n' - \mu_{\mathrm{l'}}, \tag{3.63}$$

where $\mu_{\mathrm{l'}}$ is the quantum defect value. Equation (3.63) originates from the corresponding dependence in the Coulomb potential generalized onto the case of non-integer quantum number ν_{nl}. The essential feature of the quantum defect method is the use of the Coulomb results for the spectral distribution of the transition probabilities with the subsequent replacement of the originally integer quantum number n by the non-integral quantity ν_{nl}. This approach may be ultimately interpreted as a Coulomb-type approximation for the potential $U(r)$.

It should be noted that the KrED approach does not require such an approximation for the potential. Thus, for free–free radiative transitions (bremsstrahlung) in the field of a many-electron atom, the use of the Thomas–Fermi (TF) potential in (3.39) leads to a successful description of the radiation spectral distribution. The validity of the TF model for the description of bound–bound transitions and the comparison of corresponding results of the KrED with the quantum defect method (Bates and Damgaard 1949) [and its classical analog (Davydkin and Zon 1981)] for the case of an arbitrary deviation of the potential from the Coulomb-type are to be investigated in future.

The most detailed comparison of quasi-classical results for bound–bound transitions with the corresponding quantum numerical calculations has been carried out for the case of the Coulomb field (Gantsev et al. 1985):

$$\sum_{\mathrm{l'}=l\pm1} W(nl \to n'l') = 2(l+1/2)G_0\left[\omega(l+1/2)^3/3Z^2\right]/3\pi^2c^3(nn')^3, \tag{3.64}$$

where

$$\begin{aligned} G_0(x) &\equiv x\left[K_{1/3}^2(x) + K_{2/3}^2(x)\right] \\ &= \frac{x}{2}\left\{\left[K_{1/3}(x) + K_{2/3}(x)\right]^2 + \left[K_{1/3}(x) - K_{2/3}(x)\right]\right\}^2 \propto x\left[F_-^2(x) + F_+^2(x)\right]. \end{aligned} \tag{3.65}$$

The integration of the spectral probability (3.65) over frequencies gives the total probability (per unit time) of the radiative decay of the state $\{nl\}$:

$$W^{\rm tot}(nl) = 4Z^4/\pi\sqrt{3}c^3n^3l^2. \tag{3.66}$$

The quantity inverse to (3.66) determines the mean lifetime of this state. Equation (3.66) weighted by the factor $(2l+1)$ and averaged over the values of angular momentum l gives the probability

$$W(n) = \frac{8Z^4\ln(n)}{\pi\sqrt{3}c^3n^5} \tag{3.67}$$

which is close to the results of quantum numerical calculations (Goreslavsky et al. 1982; Gantsev et al. 1985; Bethe and Salpeter 1977) that give (for $n > 20$): $W(n) \approx 7.89\times 10^9 \cdot \frac{Z^4}{n^5}\cdot \{3\ln(n) - 0.247\ [{\rm s}^{-1}]\}\cdot$

Using explicit expressions for the functions F_- and F_+ in the case of the Coulomb potential, it appears possible to trace the origin of the success of the KrED approach. These functions determine the probabilities of transitions with the decrease and increase of the electron angular momentum, respectively. This fact can be proven in the framework of classical radiation theory by calculating the rate of angular momentum loss $dM/d\omega dt$ caused by the classical emission of radiation with frequency ω [e.g., (Landau and Lifschitz 2003; Jackson 2007)]. Though the net rate of the angular momentum change is negative, as it should be, the term containing the function F_+ is positive and therefore corresponds to the increase of the electron angular momentum. Note that the relation between the functions F_+ and F_- indicated is especially transparent within the framework of the quasi-classical approach. In this case, these functions correspond to the transitions with a positive and negative change of the electron orbital quantum number ($\Delta l = \pm 1$), respectively (3.64), (3.65).

In the Coulomb case, the values of the function $F_-^2(x)$ largely exceed the values of the function $F_+^2(x)$ in a wide range, $x \geq 10^{-2}$. The predominance of the transition with $\Delta l = -1$ over a transition with $\Delta l = +1$ and the growth of this predominance with the growth of the transition frequency ($x \propto \omega M^3$) constitutes the essence of the well-known Bethe empirical rule (Bethe and Salpeter 1977) derived originally from the results of quantum numerical calculations in the Coulomb case. However, it follows from our consideration (Kogan et al. 1992) that the physical nature of this rule is purely classical. Indeed, this phenomenon can entirely be interpreted classically in terms of the correlation between the angular momentum and polarization of classical radiation. The qualitative explanation can be based on the fact that the intensity of the emission of the radiation, circularly polarized along the direction of the radiating electron rotation, largely exceeds the intensity corresponding to the case of opposite directions of the electron and radiated electric field rotation (the situation is similar, e.g., to the cyclotron radiation emission). The degree of predominance discussed evidently predetermines the accuracy of the "rotational approximation" (RA). For a Coulomb field, a quantitative estimate of the accuracy of the RA can be found from a comparison of the corresponding contributions of the $\Delta l = \pm 1$ transitions. Their ratio is equal to (Gantsev et al. 1985):

$$\int_0^\infty xF_-^2(x)\mathrm{d}x \Big/ \int_0^\infty xF_+^2(x)\mathrm{d}x = \frac{1+\pi\sqrt{3}/6}{1-\pi\sqrt{3}/6} \approx 20.5\,. \qquad (3.68)$$

Thus, the accuray (integral in ω) of the RA is of the order of 5% that agrees with the results of the classical calculations. A detailed numerical comparison of the results of the quantum and quasi-classical calculations for the transition probabilities has been carried out (Gantsev et al. 1985). The result of this calculations of the quantum corrections to the classical limit of the transition probabilities in the Kramers domain is presented in Kogan and Kukushkin (1984).

The degree of deviation of the transition probability from the Bethe rule can be clearly characterized by the function

$$D(x) \equiv xF_+^2(x) = x\left[K_{2/3}(x) - K_{1/3}(x)\right]^2. \qquad (3.69)$$

It is appropriate to designate this quantity as the *Bethe rule defect*. A useful analytic approximation of this function is presented in Kogan et al. (1992). The aforementioned predominance of $F_-(x)$ over $F_+(x)$ can be written according to (3.69) in the form of the ratio $D(x)/G_2(x)$ (for G_2 see (3.74) below), which is small in the domain of the applicability of the Bethe rule.

Thus, the KrED method provides the clues for an universal description of the transitions between those discrete spectral energy states which dominantly contribute to the total integral of radiation emission rates. Even for the Coulomb case, in spite of its detailed investigation in the literature (Landau and Lifschitz 1977; Bethe and Salpeter 1977), the KrED approach yields simple analytic results. Thus, the replacement of the Gordon formulae (Bethe and Salpeter 1977) for the transition probabilities by the corresponding KrED formulae appreciably reduces the number of variables since these classical formulae contain a smaller number of independent variables. An application of the KrED method to the non-Coulomb case and a comparison with already existing methods (e.g., the quantum defect method) are subjects for research.

3.4 Radiative Recombination

3.4.1 Kramers Photorecombination Cross Section

The suppression of the energy integral that is fundamental for the KrED approach manifests itself most strongly in the process of photorecombination radiation emission. Indeed, this process is strongly inelastic since the energy $\hbar\omega$ of an emitted photon is in any case larger than the initial energy E of the recombining electron:

$$\hbar\omega = E + |E_{\mathrm{nl}}|, \tag{3.70}$$

where E_{nl} is the energy of electron bound (final) state.

The possibility to suppress the energy integral (see Sect. 3.3) permits the use of the classical approach even for the description of such strongly inelastic process. We shall use the universal classical formula (3.59) for the spectral distribution of the emitted energy to describe the photorecombination cross section. It may be achieved by means of the continuation of the corresponding results for the bremsstrahlung radiation (BR) cross section onto the domain of negative values of the final electron energy with account of its quantization law (3.58).

For the PhR, in contrast to the BR, not only the cross section of the integral over the orbital momentum l is of essential interest, but the cross section $\sigma_{\mathrm{PhR}}(E \to n'l')$, differential with respect to l, as well. In order to obtain $\sigma_{\mathrm{PhR}}(E \to n'l')$, one has to replace the integration over the impact parameters ρ in the BR formulae by a summation over the final state orbital momentum $l' = l \pm 1$:

$$\sigma_{\mathrm{PhR}}\left(E \to n'l'\right) = \frac{\hbar^2 \pi l'}{mE} \frac{\Delta E_\omega(l')}{\hbar\omega} \frac{2\pi}{T_{\mathrm{n'l'}}}, \tag{3.71}$$

where the relation $\hbar(l' + 1/2) = \rho m v$ is to be used in $\Delta E_\omega(\rho)$. The cross section (3.71) is a functional of the atomic potential $U(r)$ which enters to the spectral functions $F_-(\omega)$ and $F_+(\omega)$ and for the period $T_{\mathrm{n'l'}}$. A detailed comparison of the quasi-classical result (3.71) with exact quantum computations for the non-Coulomb potentials has not yet been performed. The Coulomb potential (3.71) has been investigated by Kukushkin and Lisitsa (1985), Kim and Pratt (1983). In this case, the spectral dependence of the PhR cross section is described in terms of the universal spectral function $G_0(x)$ (3.65):

$$\sigma_{\mathrm{PhR}}(E \to nl) = 8Z^2 (l + 1/2)^2 G_0\left[\omega(l + 1/2)^3/3Z^2\right]/3c^3 n^3 E. \tag{3.72}$$

Note that (3.72) shows a universal dependence of the cross section on the classical parameter in the argument of the function G_0 similar to the case of the BR.

The total (integral over ρ or l) PhR cross section $\sigma_{\mathrm{PhR}}(E \to n)$ obtained by integration of (3.72) for the Coulomb case is given by

$$\sigma_{\mathrm{PhR}}(E \to n) = \frac{8\pi}{3\sqrt{3}} \frac{Z^4}{c^3 n^3} \frac{1}{E\omega}, \tag{3.73}$$

where $\omega = E + Z^2/2n^2$ (in atomic units). Equation (3.73) is known as Kramers formula.

3.4.2 Radiative Recombination Rates

The analytic result (3.72) allows a derivation of a simple formula for the photorecombination rate q_{nl} into the state with given quantum numbers n and l for a Maxwellian energy distribution with temperature T. Multiplying (3.72) by the electron velocity and then averaging over the Maxwellian velocity distribution, we obtain (Kukushkin and Lisitsa 1985) (in atomic units):

$$q_{\mathrm{PhR}}^{\mathrm{nl}} = 4\left[\frac{2}{2+x_{\mathrm{T}}} G_2(x_{\mathrm{m}}) + \frac{x_{\mathrm{T}}}{2+x_{\mathrm{T}}} \Psi(x_{\mathrm{m}}, x_{\mathrm{T}}) e^{x_{\mathrm{m}} x_{\mathrm{T}}}\right] / \pi^2 n^3 c^3 l^2. \tag{3.74a}$$

Here, the universal dimensionless parameters are introduced:

$$x_{\mathrm{m}} = E_{\mathrm{n}} M^3/3 = (l+1/2)^3/6n^2, \quad x_{\mathrm{T}} = 3/TM^3, \tag{3.74b}$$

which determine the dependencies of the rate q on the level with energy $E = 1/2n^2$ and angular momentum $M = \hbar(l+1/2)$. The function G_2 is related to the universal function G_0 by the equation

$$G_2 = \int_{x}^{\infty} G_0(x')\mathrm{d}x' = xK_{1/3}(x)K_{2/3}(x) \tag{3.74c}$$

and the function Ψ is expressed in terms of the above-defined "Bethe rule defect" $D(x)$ (3.69):

$$\Psi(x_{\mathrm{m}}, x_{\mathrm{T}}) = \int_{x_{\mathrm{m}}}^{\infty} D(y) \exp(-x_{\mathrm{T}} y)\mathrm{d}y. \tag{3.74d}$$

It follows from (3.74d) that the photorecombination rate is determined by two different terms. In the first term described by the G_2 function, the Bethe rule defect is neglected, whereas the second term is caused exclusively by the Bethe rule defect and becomes appreciable for small x_{m} and large x_{T}.

Thus in the KrED, the PhR rate into the energy state with given n and l quantum numbers is described by an universal function of two parameters. This universal dependence of the PhR rate in the Kramers domain is in agreement with the exact quantum numerical calculations (as discussed below).

The accuracy of the quasi-classical calculations of the PhR rate turns out to be fairly good (within $\sim 20\%$, according to the results obtained from exact numerical calculations). Detailed tables for the PhR rate in the Coulomb case, obtained in quasi-classical approximation, are presented in Gantsev et al. (1985).

The applicability of KrED analytical results for the Coulomb field case to the description of PhR cross sections for an ion with a core was investigated in detail (Kim and Pratt 1983). The authors use the approximation for the potential of such an ion in the form of a modified Coulomb potential with an effective charge Z_{eff}. It has been shown (Kim and Pratt 1983) that in a wide range of electron energies and ion charges, this Coulomb-type approximation of the potential yields a satisfactory description of the PhR cross sections provided the value Z_{eff} is taken equal to the mean value of the charges of the nuclei Z and of the ion Z_i, $Z_{eff} = (Z+Z_i)/2$.

The cross section for the recombination as a function of the Kramers parameter $\omega l^3/Z_{eff}^2$ for ions with various values of atomic number Z, ion charge Z_i and electron energy $E(keV)$ and different sets of parameters Z, Z_i, E prove to be satisfactorily described by an universal classical formula with the aforementioned value of Z_{eff}. The same agreement between classical and quantum results occurs also for the dependencies of the cross section on $n(\propto n^{-3})$ which follows from (3.73). The substitution of the Z_{eff} value in this equation gives the following simple analytic approximation for the PhR cross section summed over l (Kim and Pratt 1983):

$$\sigma_{\mathrm{PhR}}^{\mathrm{eff}}(n) = \frac{8\pi}{3\sqrt{3}} Z_{\mathrm{eff}}^4 / c^3 n^3 E\left(E + Z_{\mathrm{eff}}^2/2n^2\right). \tag{3.75}$$

The total PhR cross section is obtained by the summation of (3.75) over all allowed (non-occupied) quantum states according to the following equation:

$$\sigma_{\mathrm{PhR}}^{\mathrm{eff}} = W_{\mathrm{n}_0}\sigma_{\mathrm{n}_0} + \sum_{n \geq n_0+1} \sigma_{\mathrm{n}}. \tag{3.76}$$

Here, n_0 is the value of the principal quantum number of the filled atomic shell, and W_{n_0} is the statistical weight determined by the ratio of the number of free places in this shell to their total number. The results for the total PhR cross section (3.76) are in good agreement with the results of quantum numerical calculations.

The agreement between the quasi-classical and quantum results may be improved by means of a proper choice of the lower limit of summation $n_{0\mathrm{eff}}$ in (3.76) which depends on the effects of screening and correlations of the electrons in the filled atomic shell. In this case, the value $n_{0\mathrm{eff}}$ will be an universal parameter for a given isoelectronic sequence. Modifying the Kramers formula (3.75), it is easy to obtain an analytical approximation for the total PhR cross section by replacing the summation over n by an integration Kim and Pratt (1983):

$$\sigma^{\rm tot} \approx \int_{n_{0\rm eff}}^{\infty} \sigma_{\rm n} {\rm d}n = \frac{8\pi Z_{\rm eff}^2}{3\sqrt{3}c^3 E} \ln\left(1 + Z_{\rm eff}^2/2En_{0\rm eff}^2\right). \tag{3.77}$$

The values of the cutoff parameter $s = 1/2n_{0\rm eff}^2$, obtained from the comparison of (3.77) with the results of quantum calculations, are as follows: $s = 1.1$ for fully stripped ions (bare nuclei), $s = 0.065$ for *Ne*-like shells, $s = 0.045$ for *Ar*-like shells. The numerical quantum results are described satisfactorily by a linear dependence of the logarithm argument on s, namely $(1 + sZ_{\rm eff}^2/E)$ for $Z_{\rm eff} = (Z + Z_{\rm i})/2$.

The simple approximations (3.75), (3.77) for the partial and total PhR cross sections lead to a simple and reliable analytical result (Kim and Pratt 1983) for some important characteristics of a plasma. The rate of photorecombination $\alpha = <\langle v\sigma^{\rm tot}\rangle>$ is given by

$$\alpha = \frac{16}{3}\sqrt{\frac{2\pi}{3}} \frac{Z^2 c^{-3}}{\sqrt{T}} [\exp(b)E_1(b) + C + \ln(b)], \quad b = \frac{sZ^2}{T}. \tag{3.78}$$

3.4.3 Radiative Losses

The rate of electron energy loss $\beta = \langle Ev\sigma^{\rm tot}\rangle$ can be approximated by

$$\beta = \frac{T}{c^2}\left[1 - \frac{16}{3}\sqrt{\frac{2\pi}{3}} \frac{Z^4}{cT^{3/2}} \exp(b)E_1(b)\right]. \tag{3.79}$$

The total rate of radiation losses of plasma due to photorecombination $\gamma = \sum_{\rm n} \gamma_{\rm n} = \sum \langle \omega_{\rm n} v\sigma_{\rm n}^{\rm tot}\rangle$ requires a computation of the sum over n with the aid of approximations for partial cross sections:

$$\gamma = \frac{16}{3}\sqrt{\frac{2\pi}{3}} s \frac{Z^4}{c^3 T^{1/2}}. \tag{3.80}$$

The brackets denote the averaging over the Maxwellian velocity distribution function, $C = 0.577$ is the Euler constant, $E_1(x)$ is the integral exponential function. Comparison with numerical calculations show (Kim and Pratt 1983) that the precision of (3.78, 3.79, 3.80) are considerably improved is one sets $Z = Z_{\rm eff}$.

3.4.4 Generalized Scaled Empirical Formulas for Radiative Recombination Rates

Quantum mechanical numerical calculations have been performed for the photoionization cross sections. In order to obtain radiative recombination rates, these photoionization cross sections have been transformed with the help of the Milne relation to radiative recombination cross sections:

$$g_i \sigma_{ij}^{(\mathrm{phi})}(\omega) = \frac{2m_e c^2 E}{\hbar^2 \omega^2} g_j \sigma_{ji}^{(\mathrm{RR})}(E), \tag{3.81a}$$

$$\hbar\omega = \hbar\omega_0 + E = E_i + E, \tag{3.81b}$$

where E is the energy of the photoelectron. Note that the energy level i corresponds to charge state Z, whereas the level j belongs to charge state $Z + 1$. The rate coefficients of the spontaneous radiative recombination are then given by:

$$R_{ji}^{\mathrm{spon}} = n_e \int_0^\infty \mathrm{d}E \sigma_{ji}^{(\mathrm{RR})}(E) \upsilon(E) F(E), \tag{3.82a}$$

where $F(E)$ is the electron energy distribution function of the continuum electrons, n_e is the electron density, and $\upsilon(E)$ is the electron velocity given by

$$\upsilon(E) = \sqrt{\frac{2E}{m_e}}. \tag{3.82b}$$

Numerical results have been scaled with respect to charge Z and temperature parameter β and fitted to the following analytical formula for convenient application (P_1, P_2, and P_3 are fitting parameters) (Rosmej et al. 2022):

$$\left\langle \upsilon \cdot \sigma^{(\mathrm{RR})} \right\rangle = 10^{-8} \cdot Z_{\mathrm{eff}} \cdot Q \cdot P_1 \cdot \sqrt{\beta} \cdot \frac{\beta + P_2}{\beta + P_3} \quad \left[\mathrm{cm}^3\,\mathrm{s}^{-1}\right], \tag{3.83a}$$

$$\beta = \frac{Z_{\mathrm{eff}}^2 \cdot Ry}{kT_e}. \tag{3.83b}$$

$Ry = 13.606$ eV, kT_e is the electron temperature in [eV], Z_{eff} is the effective charge determined by $Z_{\mathrm{eff}} = n(E_n(\mathrm{eV})/Ry)^{1/2}$ where E_n is the ionization potential of the state n of the ion before recombination and Q is a factor depending on the quantum numbers of angular momentum for the considered transition:

$$Q\left(n_0 l_0^{m-1} SL \rightarrow n_0 l_0^m S_0 L_0\right) = m \cdot \left|G_{\mathrm{SL}}^{S_0 L_0}\right|^2 \cdot \frac{(2S_0+1)(2L_0+1)}{2 \cdot (2l_0+1)(2S+1)(2L+1)}. \quad (3.83c)$$

m is the number of equivalent electrons, $G_{\mathrm{SL}}^{S_0 L_0}$ is the fractional parentage coefficient. For example, for radiative recombination into the 4*d*-shell from the bare nuclei, we have $n_0 = 4$, $l_0 = 2$, $S_0 = 0.5$, $L_0 = 2$, $m = 1$, $S = 0$, $L = 0$, $G_{\mathrm{SL}}^{S_0 L_0} = 1$ from which it follows $Q(bare\,nuc \rightarrow 4d\,^2D) = 1$. For the radiative recombination into the He-like ground state $1s\,^2S \rightarrow 1s^2\,^1S$, $n_0 = 1$, $l_0 = 0$, $S_0 = 0$, $L_0 = 0$, $m = 2$, $S = 0.5$, $L = 0$, $G_{\mathrm{SL}}^{S_0 L_0} = 1$, it follows $Q(1s\,^2S \rightarrow 1s^2\,^1S) = 0.5$, and for the radiative recombination into the triplet $n = 2$ *P*-state, i.e. the transition $1s\,^2S \rightarrow 1s2p\,^3P$, $n_0 = 2$, $l_0 = 1$, $S_0 = 1$, $L_0 = 1$, $m = 1$, $S = 0.5$, $L = 0$, $G_{\mathrm{SL}}^{S_0 L_0} = 1$, it follows $Q(1s\,^2S \rightarrow 1s2p\,^3P) = 0.75$.

Table 3.1 shows the numerical result for hydrogen for all *nl*-states from n = 1–9 and l = 0–8. The before last line provides the sum of the recombination rates over all states with n = 1–9 and l = 0–8 obtained from detailed numerical quantum calculations, whereas the last line provides the sum of the numerical results for n = 1–9 and l = 0–8 and the Kramers approximation for $n > 9$.

Table 3.1 Numerical calculation of the radiative recombination into hydrogen, $Q = 1$, $Z_{\mathrm{eff}} = 1$ in (3.83)

nl	P_1	P_2	P_3
1*s*	4.07×10^{-6}	0.05	0.516
2*s*	6.03×10^{-7}	0.04	0.530
2*p*	1.57×10^{-6}	0.04	2.59
3*s*	2.03×10^{-7}	0.06	0.666
3*p*	5.94×10^{-7}	0.03	2.56
3*d*	6.52×10^{-7}	0.01	7.39
4*s*	9.31×10^{-8}	0.06	0.727
4*p*	2.86×10^{-7}	0.04	2.79
4*d*	4.10×10^{-7}	0.01	7.49
4*f*	3.00×10^{-7}	0.01	17.5
5*s*	5.10×10^{-8}	0.07	0.822
5*p*	1.57×10^{-7}	0.03	2.78
5*d*	2.51×10^{-7}	0.01	7.63
5*f*	2.64×10^{-7}	0.01	17.6
5*g*	1.39×10^{-7}	0.00	32.4
6*s*	3.09×10^{-8}	0.07	0.869
6*p*	9.66×10^{-8}	0.03	2.89
6*d*	1.61×10^{-7}	0.01	7.79
6*f*	1.97×10^{-7}	0.01	17.8

(continued)

Table 3.1 (continued)

nl	P_1	P_2	P_3
6*g*	1.58×10^{-7}	0.00	32.5
6*h*	6.74×10^{-8}	0.00	55.8
7*s*	2.03×10^{-8}	0.08	0.954
7*p*	6.37×10^{-8}	0.03	2.98
7*d*	1.09×10^{-7}	0.01	7.94
7*f*	1.44×10^{-7}	0.01	17.9
7*g*	1.39×10^{-7}	0.00	32.6
7*h*	9.44×10^{-8}	0.00	55.7
7*i*	3.33×10^{-8}	0.00	88.5
8*s*	1.41×10^{-8}	0.08	0.989
8*p*	4.43×10^{-8}	0.03	3.06
8*d*	7.71×10^{-8}	0.01	8.09
8*f*	1.06×10^{-7}	0.01	18.1
8*g*	1.14×10^{-7}	0.00	32.8
8*h*	9.58×10^{-8}	0.00	55.9
8*i*	5.53×10^{-8}	0.00	88.1
8*k*	1.48×10^{-8}	0.00	115
9*s*	1.01×10^{-8}	0.08	1.02
9*p*	3.20×10^{-8}	0.03	3.14
9*d*	5.64×10^{-8}	0.01	8.21
9*f*	8.02×10^{-8}	0.01	18.3
9*g*	9.16×10^{-8}	0.00	32.9
9*h*	8.70×10^{-8}	0.00	56.1
9*i*	6.32×10^{-8}	0.00	87.7
9*k*	3.18×10^{-8}	0.00	131
9*l*	5.85×10^{-9}	0.00	127
Total (1*s*...9 *l*)	1.09×10^{-5}	0.16	1.69
Total (1*s*...∞)	1.28×10^{-5}	0.18	2.14

Fitting parameters approximate the numerical results typically accurate better than 10% in the large temperature range $1/8 < \beta < 64$

Table 3.2 shows the numerical result for H-like molybdenum for all *nl*-states from $n = 1$–9 and $l = 0$–8. The before last line provides the sum of the recombination rates over all states with $n = 1$–9 and $l = 0$–8 obtained from detailed quantum mechanical calculations, whereas the last line provides the sum of the numerical results for $n = 1$–9 and $l = 0$–8 and the Kramers approximation for $n > 9$.

The numerical data have been scaled with respect to Z and β and finally fitted with three parameters. The fitting parameters of Table 3.2 can be used for all ions with $Z > 1$ due to the scaled representation of numerical results. Let us demonstrate an example for application of (3.83), namely radiative recombination into the 6g-state from fully stripped carbon at $kT_e = 15.3$ eV: $n_0 = 6$, $l_0 = 4$, $S_0 = 0.5$,

$L_0 = 4$, $m = 1$, $S = 0$, $L = 0$, $G_{SL}^{S_0L_0} = 1$ from which it follows $Q(bare\,nuc \to 6g\,^2G) = 1$, $Z = Z_{eff} = 6$, $\beta = 32$ and (from Table 3.2) $P_1 = 1.58 \times 10^{-7}$, $P_2 = 0.00$, $P_3 = 32.5$, from which it follows from (3.83a) $\langle \upsilon \cdot \sigma^{(RR)}(6\,g) \rangle = 2.66 \times 10^{-14}\,\text{cm}^3\,\text{s}^{-1}$. The exact numerical quantum mechanical result calculated specifically for carbon provides $\langle \upsilon \cdot \sigma^{(RR)}(6\,g) \rangle = 2.65 \times 10^{-14}\,\text{cm}^3\text{s}^{-1}$. This example demonstrates the high precision of the fitting formulas (3.83) and the advantageous representation of numerical results in Z- and β-scaled representation.

Let us now consider the application of Table 3.2 to estimate the radiative recombination rates for non-hydrogen-like ions with the help of an effective charge Z_{eff}. One of the most difficult tests is the radiative recombination into the ground state of neutral helium, i.e. the transition $1s\,^2S \to 1s^2\,^1S$. The ionization potential of the helium ground state is $E_i(1s^2\,^1S) = 24.587\,\text{eV}$ from which it follows an effective charge $Z_{eff} = n_0 \cdot \sqrt{E_i/Ry} = 1 \cdot \sqrt{24.587/13.606} = 1.3443$, $Q = 0.5$ and (from Table 3.2) $P_1 = 4.37 \times 10^{-6}$, $P_2 = 0.06$, $P_3 = 0.574$. Let us consider radiative recombination at $kT_e = 0.425$ eV from which it follows (3.83b) $\beta = Z_{eff}^2 Ry/kT_e = 57.85$ and from (3.83a) $\langle \upsilon \cdot \sigma^{(RR)}(1s^2\,^1S) \rangle = 2.21 \times 10^{-13}\,\text{cm}^3\,\text{s}^{-1}$. The exact numerical quantum mechanical result calculated specifically for the Helium ground state $\langle V \cdot \sigma^{(RR)}(1s^2\,^1S) \rangle = 2.53 \times 10^{-13}\,\text{cm}^3\,\text{s}^{-1}$.

Let us finish with a consideration of the recombination into the triplet $n = 2$ S-state of He I, i.e. the transition $1s\,^2S \to 1s2s\,^3S$: $n_0 = 2$, $l_0 = 0$, $S_0 = 1$, $L_0 = 0$, $m = 1$, $S = 0.5$, $L = 0$, $G_{SL}^{S_0L_0} = 1$ it follows $Q(1s\,^2S \to 1s2s\,^3S) = 0.75$, $Z_{eff} = n_0 \cdot \sqrt{E_i/Ry} = 2 \cdot \sqrt{4.7677/13.606} = 1.1839$, $Q = 0.75$ and (from Table 3.2) $P_1 = 6.53 \times 10^{-7}$, $P_2 = 0.06$, $P_3 = 0.642$. Let us consider radiative recombination at $kT_e = 3.4$ eV from which it follows (3.83b) $\beta = Z_{eff}^2 Ry/kT_e = 4.738$ and from (3.83a) $\langle \upsilon \cdot \sigma^{(RR)}(1s^2\,^1S) \rangle = 0.951 \times 10^{-14}\,\text{cm}^3\,\text{s}^{-1}$. The exact numerical quantum mechanical result calculated specifically for the Helium triplet $1s2s\;^3S$-state provides $\langle \upsilon \cdot \sigma^{(RR)}(1s2s\,^3S) \rangle = 1.03 \times 10^{-14}\,\text{cm}^3\,\text{s}^{-1}$. These examples demonstrate that the use of the generalized scaled fitting parameters of

Table 3.2 Numerical calculation of the radiative recombination into H-like ions, $Q = 1$, $Z_{eff} = Z_n$

nl	P_1	P_2	P_3
1*s*	4.37×10^{-6}	0.06	0.574
2*s*	6.53×10^{-7}	0.06	0.642
2*p*	1.63×10^{-6}	0.04	2.65
3*s*	2.13×10^{-7}	0.06	0.704
3*p*	6.19×10^{-7}	0.04	2.74
3*d*	6.61×10^{-7}	0.01	7.44
4*s*	9.69×10^{-8}	0.07	0.798

(continued)

Table 3.2 (continued)

nl	P_1	P_2	P_3
4*p*	2.97×10^{-7}	0.04	2.85
4*d*	4.15×10^{-7}	0.01	7.54
4*f*	3.01×10^{-7}	0.01	17.5
5*s*	5.23×10^{-8}	0.07	0.848
5*p*	1.62×10^{-7}	0.04	2.96
5*d*	2.53×10^{-7}	0.01	7.68
5*f*	2.66×10^{-7}	0.01	17.7
5*g*	1.39×10^{-7}	0.00	32.4
6*s*	3.16×10^{-8}	0.07	0.891
6*p*	9.81×10^{-8}	0.03	2.93
6*d*	1.63×10^{-7}	0.01	7.83
6*f*	1.98×10^{-7}	0.01	17.8
6*g*	1.59×10^{-7}	0.00	32.5
6*h*	6.76×10^{-8}	0.00	55.8
7*s*	2.07×10^{-8}	0.08	0.973
7*p*	6.45×10^{-8}	0.03	3.01
7*d*	1.10×10^{-7}	0.01	7.98
7*f*	1.45×10^{-7}	0.01	18.0
7*g*	1.40×10^{-7}	0.00	32.7
7*h*	9.47×10^{-8}	0.00	55.7
7*i*	3.33×10^{-8}	0.00	88.6
8*s*	1.42×10^{-8}	0.08	1.01
8*p*	4.47×10^{-8}	0.03	3.09
8*d*	7.76×10^{-8}	0.01	8.12
8*f*	1.07×10^{-7}	0.01	18.2
8*g*	1.15×10^{-7}	0.00	32.8
8*h*	9.61×10^{-8}	0.00	56.0
8*i*	5.54×10^{-8}	0.00	88.2
8*k*	1.48×10^{-8}	0.00	115
9*s*	1.02×10^{-8}	0.08	1.03
9*p*	3.23×10^{-8}	0.03	3.16
9*d*	5.67×10^{-8}	0.01	8.25
9*f*	8.05×10^{-8}	0.01	18.3
9*g*	9.19×10^{-8}	0.00	33.0
9*h*	8.72×10^{-8}	0.00	56.1
9*i*	6.33×10^{-8}	0.00	87.7
9*k*	3.18×10^{-8}	0.00	131
9*l*	5.86×10^{-9}	0.00	127
Total (1*s*…9*l*)	1.09×10^{-5}	0.16	1.69
Total (1*s*…∞)	1.28×10^{-5}	0.18	2.14

Fitting parameters can also be used for any non-H-like ion with $Z_{\mathrm{eff}} > 1$ in the large temperature range $1/8 < \beta < 64$ because the numerical results have been scaled not only with respect to Z but at the same time with respect to β too. Typical accuracy of fitted rates is 10%

Table 3.2 provide likewise a good accuracy for non-hydrogen-like ions (if the H-like approximation holds true reasonably well) if the charge Z is replaced by the effective charge Z_{eff}. Note, that detailed radiative recombination rates for H I, He I and He II are presented in Annex A.2 and A.3.

3.4.5 Enhanced Radiative Recombination in Storage Rings

Several observations in storage rings have identified enhanced radiative recombination by about a factor of 10 at very low energies (Gao et al. 1995, 1997). The essence of the effect is an anomalous increase of recombination rates when the relative energy of the electron–ion collision becomes of comparable value with the transverse electron beam temperature that is of the order of 0.1 meV. The first observations have been made with multicharged ions with a core, and it was suggested that dielectronic recombination might contribute. However, also measurements with bare nuclei indicated enhanced radiative recombination rates by a factor of 4 (Gao et al. 1995). It was found that the excess rates defined as a difference between the measured and standard ones increase sharply as a function of an ion charge (as $Z^{2.8}$) and fall with the increase of the electron energy (Gao et al. 1997). However, for very highly charged ions, the $Z^{2.8}$-scaling could not be confirmed (Hoffknecht et al. 2001).

It has been discussed (Heerlein et al. 2002, 2004a) that three-body recombination can be excluded as an explanation of enhanced rates, as the typical density dependence was not observed. Also Zeeman and Stark effects, stimulated emission, multiphoton effects and QED effects could finally not be made responsible for the enhanced rates, and it was proposed that the observations are driven by an electron distribution function that includes electrons with negative energy when the ions merge the electron beam (Heerlein et al. 2002). However, the modification of the electron distribution function has been controversely discussed (Hörndl et al. 2004; Heerlein et al. 2004b).

In further investigations (Hörndl et al. 2005), dense plasma screening effects as well as B-field effects on the cross sections have been excluded as an explanation for the enhanced rates. In fact, the B-field cross section calculations did not reproduce the observed $B^{0.5}$-field dependence. Finally, transient electric-field-induced enhanced recombination (Hörndl et al. 2005) has been proposed. In this scenario, radiative decay of transiently formed Rydberg states inside the solenoid can stabilize a fraction of these bound electrons by preventing field ionization in the toroidal demerging section. Thus, sufficiently deeply bound electrons with principal quantum numbers contribute, in addition to the RR channel, to the observed electron ion recombination rate. This model is also geometry dependent, but at present, respective observations have not been possible as all experimental setups are very similar and therefore enhanced radiative recombination rates seem to request further investigations.

3.5 Two-Channel Bremsstrahlung in Electron–Atom Collisions

Photon radiation in scattering of a charged particle by an atom (ion, molecule, cluster, etc.) is called *bremsstrahlung*. The initial and final states of a radiating particle in this process belong to the continuous spectrum, and radiation energy originates from its kinetic energy. Let us consider at first a simple case when a nonrelativistic electron is scattered by a "bare" nucleus (that is, a nucleus without bound electrons) with a charge number Z. We use the classical expression for the dipole radiation power Q in terms of acceleration $\mathbf{w}$ of a scattered electron (an acceleration of a nucleus can be neglected because of its heavy mass):

$$Q(t) = \frac{2e^2}{3c^3} w^2(t). \tag{3.84}$$

The total energy of the bremsstrahlung is

$$E = \frac{4e^2}{3c^3} \int_0^\infty |\mathbf{w}(\omega)|^2 \frac{d\omega}{2\pi}. \tag{3.85}$$

In derivation of (3.85), the following relation was used:

$$\int_{-\infty}^{\infty} f^2(t) dt = 2 \int_0^\infty |f(\omega)|^2 \frac{d\omega}{2\pi}, \tag{3.86}$$

where $f(t)$ is a real function of time and $f(\omega)$ is its Fourier component. To calculate the Fourier component of the acceleration $\mathbf{w}(\omega)$, it is necessary to concretize the character of motion of the particle. It is well known that in the case of a central force field the momentum of an electron $M = m\,\upsilon\,\rho$ (note that $|\upsilon| = \upsilon$) is conserved, where υ is the electron velocity (at an infinite distance from a nucleus) and, ρ is the impact parameter (see Fig. 3.3).

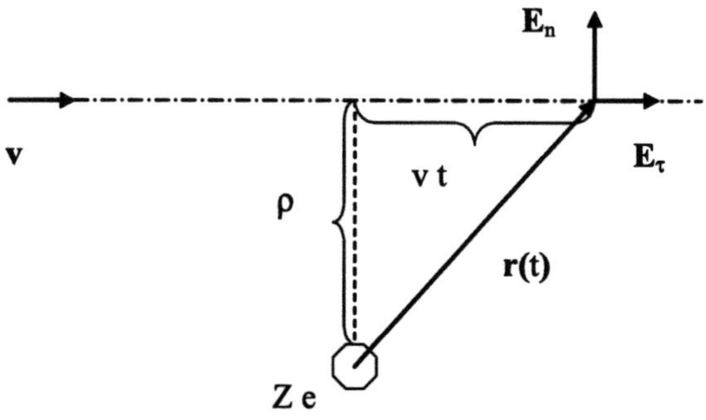

Fig. 3.3 Diagram of electron scattering by a nucleus in the approximation of straight trajectories, ρ is the impact parameter

Thus, the motion of a particle in the potential $U(r = |\boldsymbol{r}|)$ is characterized by two values: the initial velocity and the impact parameter, so the Fourier component of acceleration depends also on ρ: $\boldsymbol{w}(\omega) \to \boldsymbol{w}_\rho(\omega)$. For the last value, we have:

$$\boldsymbol{w}_\rho(\omega) = \frac{e}{m}\boldsymbol{E}(\omega, \rho), \tag{3.87}$$

where $\boldsymbol{E}(\omega, \rho)$ is the Fourier component of the strength of the electric field of a nucleus acting on a scattered electron with a specified impact parameter. Let us calculate $\boldsymbol{E}(\omega, \rho)$ in the *approximation of straight trajectories* of electron motion. This approximation is valid for "distant" collisions, when $\rho > a_C$ ($a_C = Ze^2/mv^2$ is the Coulomb length). It should be noted that this approach was used by E. Fermi for the calculation of excitation of atoms by charged particles. Using elementary electrodynamic formulas, we find for the Fourier component of the strength of the electric field of a nucleus:

$$\boldsymbol{E}(\omega, \rho) = \frac{2Ze}{\rho v}\left\{F\left(\frac{\omega\rho}{v}\right)\boldsymbol{e}_\mathrm{n} - iF'\left(\frac{\omega\rho}{v}\right)\boldsymbol{e}_\tau\right\}, \tag{3.88}$$

where $\boldsymbol{e}_{\mathrm{n},\tau}$ are the normal and tangent (with respect to the velocity vector $\boldsymbol{v}$) unit vectors (see Fig. 3.3) and

$$F(\zeta) = \int_0^\infty \frac{\cos(\zeta x)}{(1 + x^2)^{3/2}}\mathrm{d}x, \tag{3.89}$$

where the prime ($F'(x)$) denotes differentiation with respect to the argument.

In view of (3.87), it follows from (3.85) an expression for the bremsstrahlung energy differential with respect to photon frequency:

$$\frac{\mathrm{d}E_\rho}{\mathrm{d}\omega} = \frac{2e^4}{3\pi m^2 c^3}|\boldsymbol{E}(\omega, \rho)|^2. \tag{3.90}$$

The probability of bremsstrahlung in a scattering process of an electron with specified impact parameter and frequency is related to the energy of (3.90) by the relation:

$$\frac{\mathrm{d}W_\rho}{\mathrm{d}\omega} = \frac{1}{\hbar\omega}\frac{\mathrm{d}E_\rho}{\mathrm{d}\omega}, \tag{3.91}$$

and the differential (with respect to frequency) cross section is therefore

$$\frac{\mathrm{d}\sigma}{\mathrm{d}\omega} = 2\pi \int_{\rho_{\min}}^{\rho_{\max}} \frac{\mathrm{d}W_\rho}{\mathrm{d}\omega}\rho \mathrm{d}\rho. \tag{3.92}$$

ρ_{min} and ρ_{max} are the minimum and maximum impact parameters. Gathering the formulas (3.90)–(3.92) together, we obtain:

$$\frac{d\sigma}{d\omega} = \frac{4e^4}{3m^2c^3\hbar\omega} \int\limits_{\rho_{min}}^{\rho_{max}} |\boldsymbol{E}(\omega,\rho)|^2 \rho d\rho. \tag{3.93}$$

Hence, in the approximation of straight trajectories, we have obtained for the spectrally resolved cross section for the electron bremsstrahlung in the field of a "bare" nucleus:

$$\frac{d\sigma}{d\omega} = \frac{16Z^2e^6}{3m^2\upsilon^2c^3\hbar\omega} \int\limits_{\rho_{min}}^{\rho_{max}} \frac{d\rho}{\rho} \left\{ F^2\left(\frac{\omega\rho}{\upsilon}\right) + F'^2\left(\frac{\omega\rho}{\upsilon}\right) \right\}, \tag{3.94}$$

where the function $F(\zeta)$ is given by the formula (3.89).

Classical consideration is found to be insufficient for the determination of the limits of integration ρ_{min}, ρ_{max} in (3.94). For this purpose, it is necessary to invoke quantum considerations. For example, the minimum value ρ_{min} is defined by the de Broglie wavelength of a scattered electron:

$$\rho_{min} \approx \lambda_{DB}/2\pi = \frac{\hbar}{m\upsilon}. \tag{3.95}$$

The relation (3.95) reflects the fact that the location of a quantum particle cannot be determined more exactly than spatial "diffusiveness" of its wave function that is characterized by the de Broglie wavelength. To determine the maximum impact parameter ρ_{max}, it is necessary to use the law of conservation of energy and the relation between the change of momentum Δp of an incident electron and the impact parameter ρ: $\Delta p \approx \hbar/\rho$. Then, it can be obtained:

$$\rho_{max} \approx \frac{\upsilon}{\omega}. \tag{3.96}$$

In derivation of (3.96), the energy conservation law was used in the form $\hbar\omega = \upsilon\,\Delta\boldsymbol{p}$ that is true for small changes of an electron momentum $|\Delta\boldsymbol{p}| \ll p$, which corresponds to the approximation of straight trajectories. This approximation realized in case of distant collisions $\rho > a_C$ implies the weakness of interaction of a projectile with the target (nucleus). Naturally, in weak interaction mainly low-frequency photons will be emitted. It can be shown that a corresponding condition looks like: $\omega < \omega_C$, where $\omega_C = \upsilon/a_C$ is the Coulomb frequency. In the low-frequency region, the argument of the function $F(\zeta)$ and of its derivative $F'(\zeta)$ is less than one: $\zeta = \omega\rho/\upsilon < 1$, so, as it follows from the definition (3.89), the following approximate equalities can be used: $F(\zeta) \approx 1$ and $F'(\zeta) \approx 0$. As a result, instead of (3.94) we have:

$$\frac{\mathrm{d}\sigma}{\mathrm{d}\omega}=\frac{16Z^2e^6}{3m^2\upsilon^2c^3\hbar\omega}\ln\left(\frac{\rho_{\max}}{\rho_{\min}}\right). \quad (3.97)$$

It is easy to generalize the obtained expression to an arbitrary scattered particle, for which the used approximations are fulfilled. For this purpose, it is necessary to make replacements in the formulas (3.84) and (3.87) according to $e \to e_\mathrm{p}$, $m \to m_\mathrm{p}$, where e_p and m_p are the charge and the mass of the projectile. Then, in view of (3.95) and (3.96) we obtain from (3.97) the expression for the spectral bremsstrahlung of a nonrelativistic charged particle on a "bare" nucleus in the low-frequency approximation ($\hbar\omega \ll m_\mathrm{p}\upsilon^2/2$):

$$\frac{\mathrm{d}\sigma}{\mathrm{d}\omega}=\frac{16Z^2e^2e_\mathrm{p}^4}{3m_\mathrm{p}^2\upsilon^2c^3\hbar\omega}\ln\left(\frac{m_\mathrm{p}\upsilon^2}{\hbar\omega}\right). \quad (3.98)$$

From the obtained equation, it follows that the bremsstrahlung cross section is *inversely proportional to the squared mass of the projectile*. Thus, in going from light charged particles (electron, positron) to heavy particles (proton, alpha particle, etc.), the cross section of the process under consideration decreases more than million times. This conclusion led to the well-known statement that heavy charged particles do not emit bremsstrahlung photons. As it will be clear from the following, this statement needs considerable correction.

The spectral intensity of radiation is equal to the process cross section multiplied by the projectile flux and the energy of an emitted photon, so (3.98) gives:

$$\frac{\mathrm{d}I}{\mathrm{d}\omega}=\frac{16Z^2e^2e_\mathrm{p}^4}{3m_\mathrm{p}^2\upsilon c^3}\ln\left(\frac{m_\mathrm{p}\upsilon^2}{\hbar\omega}\right). \quad (3.99)$$

As discussed above, the formulas (3.98), (3.99) were obtained in the approximation of distant collisions corresponding to emission of low-frequency photons. The contribution of high-frequency photons $\omega > \omega_\mathrm{C}$ to the bremsstrahlung is made by close collisions $\rho < a_\mathrm{C}$ corresponding to strongly curved trajectories. The spectral cross section and intensity of the electron bremsstrahlung in this case are described by the Kramers formulas:

$$\frac{\mathrm{d}\sigma^{(\mathrm{Kram})}}{\mathrm{d}\omega}=\frac{16\pi Z^2e^6}{3\sqrt{3}m^2\upsilon^2c^3\hbar\omega}, \quad (3.100)$$

$$\frac{\mathrm{d}I^{(\mathrm{Kram})}}{\mathrm{d}\omega}=\frac{16\pi Z^2e^6}{3\sqrt{3}m^2\upsilon c^3}. \quad (3.101)$$

The right-hand side of the (3.101) does not include the Planck constant, which is indicative of a purely classical nature of this expression.

It is interesting to note that the Kramers formulas describe not only bremsstrahlung, but also photorecombination, when the final state of a radiating electron belongs to the discrete ion spectrum. This circumstance follows from the fact that radiation in the high-frequency limit $\omega > \omega_C$ is "gathered" from a section of the trajectory of closest approach to the nucleus, so a radiating electron "does not know" where it is scattered after emission of a photon.

The expressions (3.98)–(3.99) were obtained within the framework of classical consideration with quantum "insertions" (3.95), (3.96). It is clear that such an approach is not consistent, but its important advantage is physical transparency and mathematical simplicity. It is pertinent to note here that the use of the quantum mechanical formalism within the framework of the *Born approximation* results in the same formulas for the cross section and intensity of bremsstrahlung of low-frequency photons as (3.98), (3.99).

The criterion of the Born approximation is given by the relation:

$$\frac{Z|ee_p|}{\hbar \upsilon} \ll 1, \qquad (3.102)$$

Relation (3.102) corresponds to sufficiently fast projectiles. The condition (3.102) allows calculation of the scattering cross section according to the perturbation theory; the ratio $Z|ee_p|/\hbar\upsilon$ serves as a small parameter in the theory. The possibility of classical consideration is given by an inequality that is reverse to (3.102), so the above coincidence of results is to a certain extent by accident. A similar accidental coincidence of classical and quantum results holds true for the Rutherford cross section of electron scattering by a nucleus.

3.6 Bremsstrahlung in Many Electron–Atom Collisions and Mass-Independent Radiation

When going to bremsstrahlung on an atom, it is necessary to take into account the screening effect of bound electrons, resulting in the replacement

$$\rho_{max} \to \min(\upsilon/\omega, r_a), \qquad (3.103)$$

(r_a is the atomic radius) in the expressions for the cross section and intensity. In fact, for impact parameters $\rho > r_a$ the atomic field is equal to zero, so the acceleration of a projectile vanishes, and, according to (3.84), the bremsstrahlung vanishes too. It is obvious that screening is essential for sufficiently low frequencies $\omega < \upsilon/r_a$; otherwise, a projectile should pass sufficiently close to the nucleus to emit a photon at specified frequency.

In case of bremsstrahlung on multielectron atoms, when the Thomas–Fermi model is valid, the Thomas–Fermi radius can be used as an atomic radius:

$r_a \approx r_{TF} = a_B b/\sqrt[3]{Z}$, where $a_B \approx 0.53$ Å is the Bohr radius, Z is the charge number of the atomic nucleus, and $b \cong 0.8553$ is a constant.

The replacement (3.103) corresponds to the *screening approximation* in the bremsstrahlung theory used by H. Bethe and W. Heitler to generalize the cross section formulas to the atomic case. Physically, the screening approximation means the replacement of atomic electrons by a nucleus with effective charge. Thus, bound electrons are excluded from consideration as a dynamical degree of freedom that might occur during bremsstrahlung. In fact, during emission of high-energy photons an energy–momentum excess can be transferred to atomic electrons, resulting in their excitation and ionization.

Besides a real excitation, atomic electrons can be excited *virtually* in a collision of an atom with a charged particle. Virtual excitation corresponds to the occurrence of a variable dipole moment in an atom that, according to the fundamentals of electrodynamics, should radiate electromagnetic waves. Such a process is called *polarization bremsstrahlung* since it is connected with the dynamic polarizability of an atom defining a radiating dipole moment.

Polarization bremsstrahlung can be interpreted also as a process of scattering of the eigenfield of a projectile (a virtual photon) contributing to the radiation field (a real photon) of atomic electrons. Polarization bremsstrahlung is therefore an additional radiation channel in charge scattering by a target with a system of bound electrons. We will call a bremsstrahlung that exists also on a "bare" nucleus ordinary or static bremsstrahlung. The latter term implies that this channel is the only channel in the model of static distribution of the electron charge of bound electrons.

Let us derive formulas for the polarization bremsstrahlung of a fast charged particle on an atom, considering the atom to be an elementary dipole with polarizability $\beta(\omega)$. For the description of the projectile motion, we use, as above, the classical approach and the approximation of straight trajectories. We proceed from the formula for dipole radiation power, but this time we will formulate it in terms of the dipole moment of the radiating system:

$$Q(t) = \frac{2}{3c^3}\left|\ddot{\boldsymbol{d}}(t)\right|^2. \tag{3.104}$$

Here, two dots denote the second time derivative. Integrating the (3.104) with respect to time and using the formula (3.87) for the squared second derivative of the dipole moment, we come to the expression for the total energy of polarization bremsstrahlung for the time of a collision corresponding to the impact parameter ρ:

$$E = \frac{4e^2}{3c^3}\int\limits_0^\infty \omega^4\left|\beta(\omega)\boldsymbol{E}^{(p)}(\omega,\rho)\right|^2\frac{d\omega}{2\pi}, \tag{3.105}$$

where $\boldsymbol{E}^{(p)}(\omega, \rho)$ is the Fourier component of the strength of the electric field of a charged projectile at the location of an atom. In derivation of this formula, the relation $\ddot{\boldsymbol{d}}(\omega) = -\omega^2 \boldsymbol{d}(\omega)$ was used that follows from the definition of the Fourier components.

Going from the total radiated energy to the spectral cross section as was done in derivation of the formula (3.93), we obtain the following expression for the polarization bremsstrahlung:

$$\frac{d\sigma^{PB}}{d\omega} = \frac{4\omega^3 |\beta(\omega)|^2}{3c^3\hbar} \int\limits_{\tilde{\rho}_{min}}^{\tilde{\rho}_{max}} \left|\boldsymbol{E}^{(p)}(\omega,\rho)\right|^2 \rho \, d\rho. \tag{3.106}$$

The upper limit of integration in this formula follows from the energy conservation law (3.96), being of the same value as for static bremsstrahlung. But the lower limit of integration is essentially different. In the elementary dipole approximation under consideration, it is defined by the size of an atom:

$$\tilde{\rho}_{min} = r_a. \tag{3.107}$$

As the analysis shows, scattering with small impact parameters $\rho < r_a$ makes small contributions to the polarization bremsstrahlung cross section since in this case the coherence in re-emission of the projectile eigenfield by atomic electrons to a real photon is lost.

From Fig. 3.3, it is easy to see that the Fourier component of the strength of the electric field of a projectile in the approximation of straight trajectories can be calculated by a formula similar to (3.88), with replacement of the nuclear charge by the projectile charge. As a result, for the strength $\boldsymbol{E}^{(p)}(\omega,\rho)$ we have:

$$\boldsymbol{E}^{(p)}(\omega,\rho) = \frac{2e_p}{\rho \upsilon}\left\{-F\left(\frac{\omega\rho}{\upsilon}\right)\boldsymbol{e}_n + iF'\left(\frac{\omega\rho}{\upsilon}\right)\boldsymbol{e}_\tau\right\}, \tag{3.108}$$

where $\boldsymbol{e}_n$ and $\boldsymbol{e}_\tau$ are the normal and tangent unit vectors, and the function $F(\zeta)$ is given by the (3.89). Substituting (3.108) in (3.106), we obtain the spectral cross section of polarization bremsstrahlung:

$$\frac{d\sigma^{PB}}{d\omega} = \frac{16 e_p^2 \omega^3 |\beta(\omega)|^2}{3\upsilon^2 c^3 \hbar} \int\limits_{r_a}^{\upsilon/\omega} \frac{d\rho}{\rho}\left\{F^2\left(\frac{\omega\rho}{\upsilon}\right) + F'^2\left(\frac{\omega\rho}{\upsilon}\right)\right\} d\rho. \tag{3.109}$$

Hence, we find for the intensity:

$$\frac{dI^{PB}}{d\omega} = \frac{16 e_p^2 \omega^4 |\beta(\omega)|^2}{3\upsilon c^3} \int\limits_{r_a}^{\upsilon/\omega} \frac{d\rho}{\rho}\left\{F^2\left(\frac{\omega\rho}{\upsilon}\right) + F'^2\left(\frac{\omega\rho}{\upsilon}\right)\right\} d\rho. \tag{3.110}$$

It should be noted that the formula (3.110) does not contain the Planck constant, which is indicative of the classical nature of the polarization bremsstrahlung.

In the low-frequency limit, when $F(\zeta) \approx 1$ and $F'(\zeta) \approx 0$, the formula (3.109) gives:

$$\frac{d\sigma^{PB}}{d\omega} = \frac{16 e_p^2 \omega^3 |\beta(\omega)|^2}{3 \upsilon^2 c^3 \hbar} \ln\left(\frac{\upsilon}{\omega r_a}\right). \tag{3.111}$$

This expression is valid for frequencies $\omega < \upsilon / r_a$; otherwise, it is necessary to employ formula (3.109). However, the calculation shows that in the frequency range $\omega > \upsilon / r_a$ polarization bremsstrahlung is low.

The cross section (3.111) can be obtained within the framework of the quantum approach in the domain of validity of the Born approximation (3.92), that means for fast (but nonrelativistic) projectiles.

It should be emphasized that the polarization bremsstrahlung cross sections (3.109), (3.111) do not depend on the projectile mass in contrast to the static bremsstrahlung cross section (3.99). Thus, the long existing statement in physics that heavy charged particles do not emit bremsstrahlung photons does not extend to the polarization channel. This circumstance is connected with the fact that the static bremsstrahlung cross section is proportional to the squared acceleration of a projectile, while the polarization cross section does not depend on this acceleration.

The polarization bremsstrahlung cross section (3.111) can be obtained from the static cross section (3.98) via the replacements $m_p \to m$, $e_p^4 \to e^2 e_p^2$, $\rho_{min} \to \tilde{\rho}_{min}$ and

$$Z \to Z_{pol}(\omega), \tag{3.112}$$

where

$$Z_{pol}(\omega) = \frac{m\omega^2}{e^2} |\beta(\omega)|^2 \tag{3.113}$$

is the effective polarization atomic charge (in units of the electron charge e) (Rosmej et al. 2017). A polarization charge characterizes the ability of the electron core of an atom to emit a photon under the action of an ac field. In contrast to an ordinary charge, a polarization charge depends on radiation frequency. The frequency dependence of polarization charges of silver and krypton atoms is presented in Fig. 3.4.

From Fig. 3.4, it is seen that in the high-frequency range the polarization charge is equal to the number of bound electrons of an atom (or the charge number of its nucleus). This fact follows from the definition (3.113) and the formula for high-frequency polarizability (2.43). In the low-frequency region $\omega \to 0$ a polarization charge, according to (3.113), decreases quadratically since in this case atomic polarizability is equal to its static value (2.42) that does not depend on frequency. Finally, in the intermediate spectral range the polarization charge is a non-monotonic

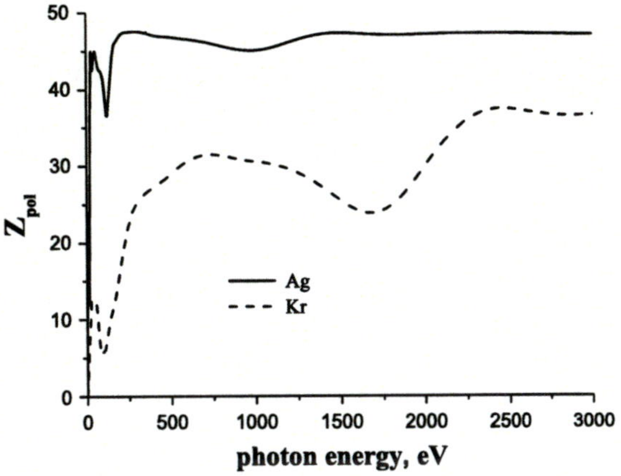

Fig. 3.4 Spectral dependence of the polarization charges of silver and krypton atoms

function that reflects the features of the energy spectrum of an atom. For example, a wide "dip" in the dashed curve of Fig. 3.4 in the range of 1600–1750 eV corresponds to the binding energy for $2p$-electrons in the krypton atom. A minimum in the low-frequency region corresponds to virtual excitation of atomic subshells with principal quantum number $n = 3$. Thus, the spectral cross section of polarization bremsstrahlung reflects the atomic core dynamics as a function of frequency.

In the high-frequency limit, when $\omega \gg \omega_a$(ω_a is the characteristic frequency of excitation of an atom in the discrete spectrum), but still $\omega < \upsilon / r_a$, $\beta(\omega) \approx -Ze^2/m\omega^2$ ($Z_{pol}(\omega) = Z$) the formula (3.111) gives:

$$\frac{d\sigma^{PB}}{d\omega} = \frac{16 Z^2 e^4 e_p^2}{3 m^2 \upsilon^2 c^3 \hbar \omega} \ln\left(\frac{\upsilon}{\omega r_a}\right). \tag{3.114}$$

Curiously, in case of an incident electron (positron), the obtained expression differs from the formula for the static bremsstrahlung cross section only by the logarithmic factor.

Let us now consider the resonance case, when the bremsstrahlung frequency is close to one of the eigenfrequencies of an atom, i.e. $\omega \approx \omega_0$. The dynamic polarizability looks like

$$\beta(\omega \approx \omega_0) \cong \frac{e^2}{m} \frac{f_0}{\omega_0^2 - \omega^2 - 2i\omega\gamma_0}. \tag{3.115}$$

This expression for the resonance polarizability follows from the general formula (2.41) if only the resonance summand is considered for which $\omega_{nm} \equiv \omega_0, f_{nm} \equiv f_0$, and $\gamma_{nm} \equiv \gamma_0$. Substituting the formula (3.115) in (3.111), we obtain:

$$\frac{d\sigma^{\mathrm{res}}}{d\omega} = \frac{4}{3}\frac{e_{\mathrm{p}}^2}{\hbar c}\left(\frac{c}{\upsilon}\right)^2 \frac{r_{\mathrm{e}}^2 f_0^2 \omega_0}{(\omega_0 - \omega)^2 + \gamma_0^2} \ln\left(\frac{\upsilon}{\omega r_{\mathrm{a}}}\right), \tag{3.116}$$

where $r_{\mathrm{e}} = e^2/mc^2 \approx 2.8 \times 10^{-13}$ cm is the classical radius of an electron.

From the expression (3.116), it is seen that resonance polarization bremsstrahlung has a sharp maximum at the frequency $\omega = \omega_0$ if $\gamma_0 \ll \omega_0$. The last inequality is satisfied in case of excitation of electrons of the outer atomic shell into the discrete spectrum. For neutral atoms, energies of the resonant photons are about 10 eV. In case of multiply charged ions with a system of bound electrons (an electron core), these energies can be much higher and reach values of the order of several keV. However, in this case the transition damping constant being equal to the Einstein coefficient A_{mn} is also large, and therefore, the resonance is not sharp anymore. At frequencies corresponding to virtual excitation of inner atomic shells, the resonance structure in the spectral dependence of the dynamic polarizability $\beta(\omega)$ disappears. In the spectral curves, "dips" appear that correspond to the beginning of the photoionization of an atomic subshell (see Fig. 3.4).

Thus, resonance effects in the above-considered *spontaneous* polarization bremsstrahlung are essential only in a narrow frequency interval in the vicinity of a resonance and are ill-defined in the integral characteristics of radiation. The situation changes when going to the *induced* bremsstrahlung (also called the *induced bremsstrahlung effect*).

3.7 Photoionization

3.7.1 General Relations

Let us consider at first a bound–free transition of a quantum system with photon absorption under the action of electromagnetic radiation of moderate intensity, when the condition of applicability of the perturbation theory is satisfied. Let the atom be excited as a result of absorption of a photon of an external field. Photoabsorption is characterized by the spectral cross section that is connected with the probability per unit time for excitation of a bound electron under the action of electromagnetic radiation with a specified frequency ω. For the photoabsorption cross section $\sigma(\omega)$, there exist a number of general relations that are used for the construction of approximate models to quantitatively describe the photoeffect. For example, it is convenient to express the value $\sigma(\omega)$ in terms of the *spectral function of dipole excitations* $g(\omega)$ according to the formula (Amusia 1990):

$$\sigma(\omega) = \frac{2\pi^2}{137} a_{\mathrm{B}} \upsilon_{\mathrm{a}} g(\omega), \tag{3.117}$$

where a_B is the Bohr radius, and $\upsilon_a \cong 2.18 \times 10^8$ cm/s is the velocity of an electron in the first Bohr orbit in a hydrogen atom (an atomic unit of velocity). Hereafter, the number 137 resulted from writing the velocity of light in atomic units: $c/\upsilon_a \cong 137$. The function $g(\omega)$ is very convenient because it satisfies the sum rule:

$$\int g(\omega)\mathrm{d}\omega = N_n, \tag{3.118}$$

where N_n is the total number of electrons in an atomic shell n. Besides, the spectral function $g(\omega)$ satisfies also the equation:

$$g(\omega) = \sum_j f_{ij}\delta(\omega - \omega_{ij}), \tag{3.119}$$

where f_{ij} is the strength of an oscillator for the transition $i \rightarrow j$ and ω_{ij} is the eigenfrequency of this transition. It should be noted that if we are dealing with the photoionization of an electron in subshell nl with specified principal n and orbital l quantum numbers, the expressions (3.117)–(3.119) should be related to this subshell and designated with corresponding indices: X_{nl}.

The above formulas (3.117)–(3.119) concern not only photoionization, but also photoabsorption that is accompanied by electron transitions in the discrete spectrum, i.e. photoexcitation. In case of a photoionization, the summation in (3.119) is replaced by an integration over states of the continuous spectrum, the integrand being a differential oscillator strength for transition to the continuum $\mathrm{d}f/\mathrm{d}\varepsilon$, where ε is the energy of a state of the continuous spectrum of an electron. The differential oscillator strength is expressed in terms of the matrix element $d_{i\varepsilon}$ of a transition dipole moment operator for transitions to the continuum in the same manner as for transitions to the discrete spectrum:

$$\frac{\mathrm{d}f}{\mathrm{d}\varepsilon} = \frac{2\omega|d_{i\varepsilon}|^2}{3e^2a_B^2}, \tag{3.120}$$

where e is the elementary charge.

It is useful to introduce the concept of an oscillator strength density for transitions into the discrete spectrum too. The oscillator strength of such transitions has to be divided by the energy interval from the given level to the nearest energy level. It can be shown that in this case the following relation is valid (Cowan 1981; Sobelman 1972, 2006):

$$\lim_{n'\to\infty} \frac{4\pi^2 Rya_B^2}{137}\frac{f_{nl,n'l'}}{E_{n'+1l'} - E_{n'l'}} = \sigma_{nl,\varepsilon l'}(I_{nl}). \tag{3.121}$$

This means, the normalized oscillator strength density for infinitely large principal quantum numbers goes over to the threshold value of the partial (corresponding to a given value of orbital quantum number l') cross section of photoionization of an electron subshell nl. The limiting transition (3.121) is a demonstration of a smooth conjugation of optical characteristics of discrete and continuous spectra.

The most general expression for the cross section of photoionization of an electron subshell in the one-electron approximation (that is, with neglect of inter-electron correlations) looks like

$$\sigma_{\mathrm{nl}}(\omega) = \frac{4\pi^2 N_{\mathrm{nl}} \upsilon_a}{3e^2 a_{\mathrm{B}} \omega 137(2l+1)} \left[\left| d_{\mathrm{nl},\varepsilon(l+1)} \right|^2 + \left| d_{\mathrm{nl},\varepsilon(l-1)} \right|^2 \right], \tag{3.122}$$

where N_{nl} is the number of equivalent electrons, that is, electrons with the same values of principal and orbital quantum numbers. Here, there are introduced the matrix elements of a dipole moment operator for transitions to states of the continuous spectrum with orbital quantum numbers allowed by selection rules. These matrix elements can be expressed in terms of radial wave functions of the initial ($R_{\mathrm{nl}}(r)$) and final ($R_{\varepsilon l'}$) states as follows:

$$d^r_{\mathrm{nl},\varepsilon l'} = \frac{e\omega \upsilon_{\mathrm{a}}}{a_{\mathrm{B}}} \sqrt{(2l+1)(2l'+1)} \begin{pmatrix} l & 1 & l' \\ 0 & 0 & 0 \end{pmatrix} \int\limits_0^\infty R_{\mathrm{nl}}(r) r R_{\varepsilon l'}(r) r^2 \mathrm{d}r, \tag{3.123}$$

where $\begin{pmatrix} l & 1 & l' \\ 0 & 0 & 0 \end{pmatrix}$ is the so-called 3*j*-symbol. It results from integration with respect to angular variables in the definition of the matrix element of the dipole moment. The $3j$ symbol describes the *selection rules for dipole radiation*, according to which $l' = l \pm 1$. Naturally, in the case $l = 0$ there is one allowed value of a quantum number of an orbital moment in the final state: $l' = 1$. As a rule, the main contribution to the photoionization cross section is made by a transition with increasing quantum number of an orbital moment $l \to l+1$. Exceptions to this rule occur if for some specific reasons the matrix element $d_{\mathrm{nl,n'}l+1}$ is small or goes to zero. On the other hand for the angular distribution of ionized electrons (that we do not consider here), the transition $l \to l-1$ can play an important role.

3.7.2 *Hydrogen-like Approximation*

As was shown for the first time by Sommerfeld (1978), the total (integrated with respect to the electron escape angle) photoionization cross section for the ground 1*s*-state of a hydrogen-like ion is

$$\sigma^{\mathrm{H-like}}_{\mathrm{ph}\ 1s}(\omega) = \frac{2^9 \pi^2}{3Z^2 137} \left(\frac{I_{1\mathrm{s}}}{\hbar\omega} \right)^4 a_{\mathrm{B}}^2 \frac{\exp(-4\zeta \mathit{arcctg}\zeta)}{1 - \exp(-2\pi\zeta)}, \tag{3.124}$$

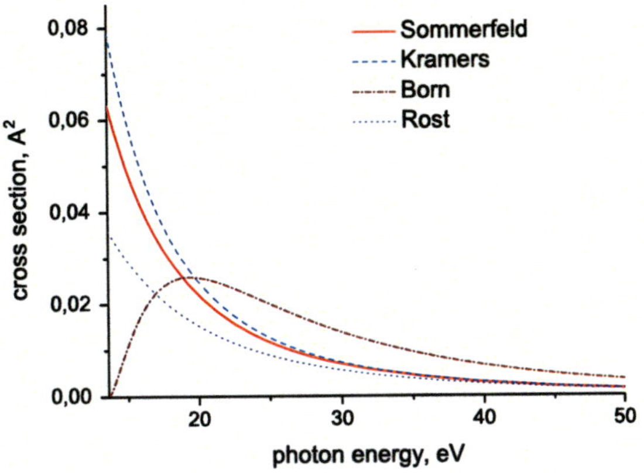

Fig. 3.5 Sommerfeld, Kramers, and Born cross sections of photoionization of the ground state of a hydrogen atom and the cross section in the Rost approximation

where ω is the ionizing radiation frequency, Z is the nuclear charge, $I_{1s} = Z^2Ry$ is the ionization potential of the 1*s*-state (*Ry*=13.6 eV), $p = \sqrt{2m(\hbar\omega - I_{1s})}$ is the momentum of the ionized electron, and $\zeta = Zme^2/p\hbar$ is the so-called *Born parameter*. The Born parameter (see also (3.102)) is a dimensionless quantity characterizing the force of interaction between an electron and a charged particle. This parameter is introduced in the electron scattering theory and usually written in terms of the electron velocity: $\zeta = Ze^2/\hbar v$. The dependence of the Sommerfeld photoeffect cross section (3.124) on the photon energy is presented in Fig. 3.5 (solid curve).

It should be noted that photoionization is a process of the first order with a smallness parameter being the electromagnetic interaction constant ($e^2/\hbar c \cong 1/137$ in ordinary units). This manifests itself in the presence of the velocity of light (the number 137) to the first power in the denominator of the formula (3.124).

In the vicinity of the photoionization threshold, when $p \to 0, \quad \zeta \to \infty$, we obtain from the formula (3.124) the following approximate expression for the photoeffect cross section:

$$\sigma_{1s}(\omega) \approx \frac{2^9\pi^2 a_B^2}{3e^4 Z^2 137}\left(1 - \frac{8}{3}\frac{(\hbar\omega - I_{1s})}{I_{1s}}\right) \approx \frac{0.23 a_B^2}{Z^2}\left(1 - \frac{8}{3}\frac{(\hbar\omega - I_{1s})}{I_{1s}}\right), \quad (3.125)$$

where e is the base of the natural logarithm (not to be confused with an elementary charge). Thus, the photoeffect cross section for a hydrogen atom ($Z = 1$) at the threshold ($\hbar\omega = Ry$) is equal to 0.063 Å^2 or 6.3 Mb. It should be noted that the cross section of photoionization of atoms is often given in megabarns: 1 Mb = 10^{-18} cm^2.

An important feature of the photoeffect of hydrogen-like atoms follows from formulas (3.124)–(3.125): The maximum of the cross section value is achieved at threshold, that is, at the minimum radiation frequency, at which photoionization is still possible. For higher frequencies, the cross section monotonically decreases. This property is caused by the fact that an ionized electron experiences Coulomb attraction of a nucleus that increases the cross section.

From the formula (3.125), it follows that the cross section of photoionization of the ground state of a hydrogen-like ion decreases at the threshold inverse proportionally to the squared nuclear charge. Such a behavior of the cross section has a simple qualitative interpretation: With increasing nuclear charge the radius of the ground state of a hydrogen-like ion decreases $r_{1s} \propto Z^{-1}$, whence (on the assumption that $\sigma_{1s} \propto r_{1s}^2$) there follows the threshold dependence of the photoionization cross section that can also be represented as $\sigma_{1s}^{\text{thres}} \propto 1/I_{1s}$. Hence, it follows that the threshold value of the photoeffect cross section for *ns*-states (with another principal quantum number *n*) can be represented as

$$\sigma_{\text{ns}}^{\text{thres}} = (I_{1s}/I_{\text{ns}})\sigma_{1s}^{\text{thres}}. \tag{3.126}$$

Thus, the threshold value of the photoionization cross section increases with principal quantum number. Curiously, this relation is validated by experimental cross sections even in the case of non-hydrogen-like atoms. For example, for an argon atom we have $I_{1s} : I_{2s} : I_{3s} \approx 150 : 10 : 1$, while the ratio of experimental threshold cross sections for these shells is 300 : 30 : 1.

In the high-frequency mode $\hbar\omega \gg I_{1s}$ we obtain from (3.124)

$$\sigma_{1s}(\omega) \approx \frac{2^8\pi}{3}\frac{a_B^2}{Z^2 137}\left(\frac{I_{1s}}{\hbar\omega}\right)^{7/2}\left[1 - \pi\sqrt{\frac{I_{1s}}{\hbar\omega}}\right]. \tag{3.127}$$

The formula (3.127) reflects the well-known asymptotic decrease in the hydrogen-like photoeffect cross section with increasing frequency: $\omega^{-7/2}$. It makes sense to emphasize that the photoionization cross section (3.124) goes to its asymptotic behavior (3.127) only at rather high values of frequency, i.e. about $\omega > 40 I_{1s}/\hbar$ since the expansion parameter $(-2\pi\zeta)$ of the exponent in (3.124) becomes much less than unity only at such high frequencies.

For the photoionization of *nl*-subshells (with $l \neq 0$), the photoeffect cross section decreases also monotonically with increasing frequency, and for $\omega \gg I_{\text{nl}}/\hbar$ we have

$$\sigma_{\text{nl}}(\omega) \propto 1/\omega^{l+7/2}, \tag{3.128}$$

that is, the cross section decrease is more rapid.

As discussed above in relation with the Sommerfeld formula (3.124), there follow characteristic features for the cross sections of photoionization of a hydrogen-like ion, i.e. a maximum at threshold, a monotonic decrease with increasing frequency. These characteristic features, when going to multielectron atoms, are, generally speaking, violated. Nevertheless, the hydrogen-like formula for the photoionization cross section is a starting point for construction of an approximate method of description by an order of magnitude. For example, if the high-frequency dependence (3.128) is employed from the threshold and combined with the sum rule $\frac{137}{2\pi^2 a_B v_a}\int_{I_{nl}}^{\infty}\sigma_{nl}(\omega)d\omega = N_{nl}$, we obtain the following photoionization cross section in the hydrogen-like approximation:

$$\sigma_{nl}(\omega) = \frac{4\pi^2 a_B^2}{137} N_{nl}\left(\frac{5}{2}+l\right)\frac{I_{nl}^{5/2+l} Ry}{(\hbar\omega)^{7/2+l}}. \tag{3.129}$$

The cross section (3.129) applied for the 1 s-electron gives a 3.2-fold excess over the exact value near the threshold, and far from the threshold an underestimation of 2.7 times. Therefore, (3.129) defines the cross section within an order of magnitude (in the hydrogen-like approximation).

For semiquantitative characterization of radiative phenomena, simple formulas obtained by Kramers within the framework of classical physics are often used. They describe cross sections for radiative processes in electron scattering in the field of a point charge. These formulas are valid for non-small values of the Born parameter $\zeta = Ze^2/\hbar v \geq 1$, that is, for large charge numbers and low electron velocities. In this case, the electron motion is quasi-classical and can be described to a good degree of accuracy as a motion along a classical trajectory.

Within the framework of the Kramers approach for the cross section of photoionization of an atomic subshell with quantum numbers *nl*, the following expression can be obtained (see Sect. 3.3):

$$\sigma_{nl}^{(Kr)}(\omega) = \frac{64\pi}{3\sqrt{3}} N_{nl}\frac{a_B^2}{137 Z^2}\sqrt{\frac{Ry}{I_{nl}}}\left(\frac{I_{nl}}{\hbar\omega}\right)^3. \tag{3.130}$$

Hence, the formula (3.130) corresponds to the cross section of photoionization of a hydrogen atom in the ground state if it is assumed that $Z = N_{nl} = 1$ and $I_{nl} = Ry$. Figure 3.5 presents these results (dashed line) that demonstrate that, despite its simplicity, the Kramers formula adequately describes the cross section of photoionization of a hydrogen atom. The most distinction from the exact cross section is at threshold. The Kramers formula overestimates the Sommerfeld threshold value of the cross section by about 30%. In the high-frequency mode, the expression (3.130) gives another asymptotics than the Sommerfeld formula (3.124): ω^{-3} instead of $\omega^{-3.5}$. However, since the cross section goes to the high-frequency asymptotics only very far from the threshold (more than ten times), this distinction has little effect in the actual region of photon energies where the cross section is high.

3.7.3 Photoeffect Cross Section in the Born Approximation

In the mode of small values of the Born parameter $\zeta = Ze^2/\hbar v \ll 1$, the influence of an atomic core on the motion of an ionized electron can be considered to be a small perturbation. This is the case for high velocities and low nuclear charges. In this case, in calculation of the matrix elements $d_{\mathrm{nl},\varepsilon l+1}$ appearing in the general formula for the photoeffect cross section (3.122), plane waves corresponding to free motion can be used as a wave function of an ionized electron. Then for the cross section of photoionization of an atomic subshell, the following expression can be obtained:

$$\sigma_{\mathrm{nl}}(\omega) = \frac{8\pi^2}{3\cdot 137} N_{\mathrm{nl}} a_{\mathrm{B}}^2 \frac{Ry}{\hbar\omega} \left(\frac{p(\omega)}{\hbar}\right)^3 \left| g_{\mathrm{nl}}\left(\frac{p(\omega)}{\hbar}\right)\right|^2, \tag{3.131}$$

where $p(\omega) = \sqrt{2m(\hbar\omega - I_{\mathrm{nl}})}$ is the momentum of the ionized electron, I_{nl} is the potential of ionization of an electron subshell, $g_{\mathrm{nl}}(k) = \sqrt{\frac{2}{\pi}} \int_0^\infty j_l(kr) R_{\mathrm{nl}}(r) r^2 \mathrm{d}r$ is the Fourier transform of the radial wave function of the initial state of an atom, $R_{\mathrm{nl}}(r)$ is the radial wave function of the initial state of an atomic electron normalized according to $\int_0^\infty |R_{\mathrm{nl}}(r)|^2 r^2 \mathrm{d}r = 1$, and $j_l(kr)$ is the spherical Bessel function of the lth order.

Let us give for reference several spherical Bessel functions: $j_0(x) = \sin x/x$, $j_1(x) = \sin x/x^2 - \cos x/x$, $j_2(x) = (3x^{-3} - x^{-1})\sin x - 3\cos x/x^2$. Spherical Bessel functions describe the radial dependence of a spherical wave with a specified value of an orbital quantum number l.

In case of photoionization of the ground state of a hydrogen atom, we have: $R_{10}(r) = \left(2/\sqrt{a_{\mathrm{B}}^3}\right)\exp(-r/a_{\mathrm{B}})$ and $g_{10}(k) = \sqrt{\frac{2}{\pi}} \frac{4a_{\mathrm{B}}^{3/2}}{(1+k^2a_{\mathrm{B}}^2)^2}$, $N_{\mathrm{nl}} = 1$, $I_{\mathrm{nl}} = Ry$. Substituting these equations in the formula (3.131), we find the following expression for the cross section of photoionization of a hydrogen atom in the Born approximation:

$$\sigma_{1\mathrm{s}}^{(B)}(\omega) = \frac{2^8\pi}{3\cdot 137} a_{\mathrm{B}}^2 \frac{Ry}{\hbar\omega} \frac{(p(\omega)a_{\mathrm{B}}/\hbar)^3}{\left[1 + (p(\omega)a_{\mathrm{B}}/\hbar)^2\right]^4}. \tag{3.132}$$

The plot of the function $\sigma_{1\mathrm{s}}^{(B)}(\omega)$ is presented in Fig. 3.5 as a dashed-dotted line.

From Fig. 3.5 a characteristic feature of the Born cross section is seen: It goes to zero at the threshold in contrast to the exact Sommerfeld cross section and the approximate Kramers cross sections that have a maximum at the threshold. This is connected with the fact that the Born approximation does not take into account nuclear attraction that increases the cross section value. At the same time, the function (3.132) has correct high-frequency asymptotics $\sigma_{1\mathrm{s}}^{(B)}(\hbar\omega \gg Ry) \propto \omega^{-7/2}$

since in the mode of high photon energies an ionized electron can be considered to be free, which corresponds to the condition of applicability of the Born approximation. Nevertheless, the ratio of the Born cross section to the exact cross section for $\hbar\omega = 100$ eV is 2.1, at $\hbar\omega = 1$ keV it is 1.38, and only for $\hbar\omega = 10$ keV, this ratio is equal to 1.12; that is, the convergence is rather slow.

Thus for not too high photon energies the Kramers photoeffect cross section for a hydrogen atom describes the real situation better than the Born cross section.

3.7.4 Local Plasma Frequency Model

So far the photoionization cross section was calculated with neglect of interelectron interaction; that is, it was assumed that photon absorption occurs as a result of interaction of an electromagnetic field with individual electrons, the contributions of which are additively summed, giving the total cross section. There is a rather simple alternative approach to the description of an atomic photoeffect based on purely classical considerations. It is the *local plasma frequency model* or the Brandt–Lundqvist approximation (Brandt and Lundqvist 1965) that was considered in Sect. 2.6. Within the framework of this approach, an atom is approximated by an inhomogeneous distribution of electron density with concentration $n(r)$ (plus nucleus) [Rosmej et al. 2020]. Each spatial point corresponds to its own local plasma frequency $\omega_p(r) = \sqrt{4\pi e^2 n(r)/m}$, and interaction of an external electromagnetic field of frequency ω with atomic electrons is defined by the plasma resonance condition

$$\omega = \omega_p(r) = \sqrt{\frac{4\pi e^2 n(r)}{m}}. \tag{3.133}$$

From this equation, it follows that absorption of electromagnetic field energy by atomic electrons occurs at those distances from a nucleus where the local plasma frequency coincides with the ionizing radiation frequency. This model results in the following simple expression for the spectral function:

$$g(\omega) = \int d^3r\, n(r)\, \delta\big(\omega - \omega_p(r)\big). \tag{3.134}$$

It is easy to see that the spectral function (3.134) satisfies the sum rule (3.118). For the photoionization cross section, according to (3.117), we have

$$\sigma(\omega) = \frac{2\pi^2}{137} a_B v_a \int d^3r\, n(r)\, \delta\big(\omega - \omega_p(r)\big). \tag{3.135}$$

The presence of the delta function in (3.135) allows easy integration with respect to spatial variables. As a result, we obtain the so-called Brandt–Lundqvist approximation for the photoionization cross section:

$$\sigma_{\rm ph}^{\rm B-L}(\omega) = \frac{4\pi^2\omega}{137 v_a} r_\omega^2 \frac{n(r_\omega)}{|n'(r_\omega)|}, \tag{3.136}$$

where r_ω is the solution of (3.133). This value corresponds to the radial distance (from the nucleus) of the plasma resonance, and the prime denotes differentiation with respect to the radius. Thus, within the framework of the model, the photoeffect cross section is defined only by the distribution of the electron density $n(r)$. For the last value, it is convenient to use the statistical model of an atom, in which $n(r) = (Z^2/a_{\rm B}^3) f(x = r/r_{\rm TF})$, where $f(x)$ is the universal function of the reduced distance $x = r/r_{\rm TF}$, Z is the nuclear charge, $r_{\rm TF} = b a_{\rm B}/Z^{1/3}$ is the Thomas–Fermi radius, and $b \cong 0.8853$. Substituting the above expression for electron density in the formula (3.136), we find

$$\sigma_{\rm ph}^{\rm B-L}(\omega) = s\left(\nu = \frac{\hbar\omega}{2Z\,Ry}\right) = \frac{9\pi^4\nu}{32\cdot 137} x_\nu^2 \frac{f(x_\nu)}{|f'(x_\nu)|} a_{\rm B}^2, \tag{3.137}$$

here the reduced frequency $\nu = \hbar\omega/(2Z\,Ry)$ is introduced, and x_ν is the solution of the equation $\nu = \sqrt{4\pi f(x)}$ being a result of (3.133).

As seen from (3.137), the photoionization cross section in the Brandt–Lundqvist approximation is found to be a universal function, that is, independent of nuclear charge but a function of the reduced frequency: $s(\nu)$. The formula (3.137) reveals a corresponding scaling law for the cross section with respect to the variable ν. The universal function $s(\nu)$ is defined by the type of the statistical model of atom, that is, by the dependence of $f(x)$.

Figure 3.6 shows the calculation of the photoionization cross section of a krypton atom carried out within the framework of two alternative approaches: the quantum hydrogen-like approximation (3.129) (solid curve), and the classical local plasma model (3.137) that employs the Thomas–Fermi electron density (dotted curve).

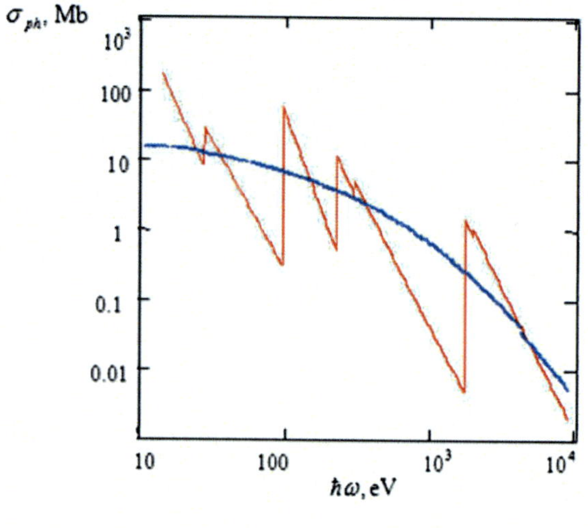

Fig. 3.6 Cross section of photoionization of a krypton atom: solid red curve—hydrogen-like approximation (3.129); dotted blue curve—local plasma model (3.137) with electron density according to the Thomas–Fermi model

It is seen that the first dependence is a saw-toothed curve with jumps at frequencies corresponding to the ionization potentials of electron subshells. The value of a jump decreases with increasing potential of subshell ionization according to the formula (3.126). The cross section of photoionization of an atom in the local plasma model (for the Thomas–Fermi electron density) is a smooth monotonically decreasing curve that describes in a smooth manner the quantum jumps of the hydrogen-like approximation.

The main advantages of the Brandt–Lundqvist approximation are simplicity, clearness, and universality. It gives the worst description of the process in spectral intervals in the vicinity of thresholds of ionization of electron subshells as it is seen from Fig. 3.6. In the original work of Brandt and Lundqvist (1965), it was noted that the local plasma model is adequate to physics of electromagnetic field photoabsorption by an atom not throughout the frequency range, but at frequencies $\omega \approx Z\,Ry/\hbar$ ($Ry = 13.6\,\text{eV}$), when collective interactions dominate over one-particle interactions. For such frequencies, the distance to a nucleus, at which the plasma resonance condition (3.133) is satisfied (in the Thomas–Fermi model), coincides with the Thomas–Fermi radius, that is equal to the distance where the electron density is maximum. Therefore, the assumption of domination of collective phenomena in the photoeffect at frequencies $\omega \approx Z\,Ry/\hbar$ seems to be reasonable, at least at a qualitative level.

The use of the exponential screening model for the normalized function of the electron density $f(x = r/r_{\text{TF}})$, i.e.

$$f_{\text{exp}}(x) = \frac{128}{9\pi^3} e^{-2x} \tag{3.138}$$

allows obtaining a simple analytical expression for the photoeffect cross section. In this case, then the transcendental (3.133) is easily solved, and we obtain with the use of (3.137)

$$\sigma_{\text{ph}}^{\text{B-L(exp)}}(\omega = (2Z\,Ry/\hbar)\nu) = \frac{9\pi^4 a_{\text{B}}^2 \nu}{64 \cdot 137} \ln^2\left(\frac{16\sqrt{2}}{3\pi\nu}\right); \quad \nu \leq \frac{16\sqrt{2}}{3\pi} \cong 2.4. \tag{3.139}$$

A characteristic feature of the cross section (3.139) is the existence of a "cutoff frequency", which is connected with limited radial electron density near a nucleus in the model (3.138). Therefore, there exists a radiation frequency, for which the plasma resonance condition is not satisfied. Another characteristic feature of the photoeffect cross section calculated with the function (3.138) is the presence of a pronounced maximum at $\hbar\omega_{\text{max}}^{(\text{exp})} \cong 8.8Z$ eV.

The atomic photoeffect cross section calculated within the framework of the Brandt–Lundqvist approximation (3.137) while employing different statistical atomic models is presented in Fig. 3.7. We note, that the comparison of the local plasma frequency approach employing various atomic models shows often surprisingly good agreement with experimental photoionization cross section data (Rosmej et al. 2020).

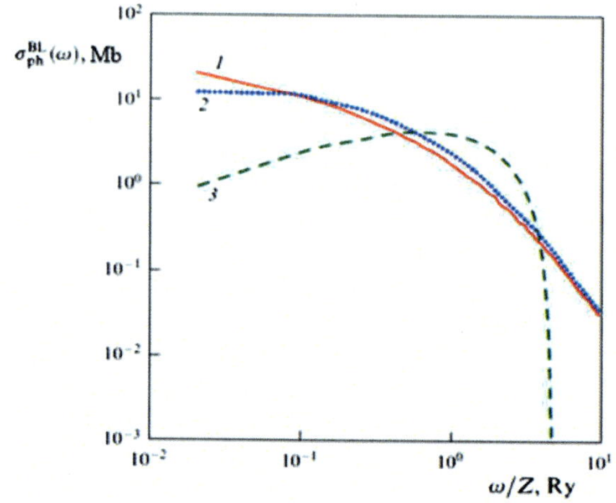

Fig. 3.7 Photoeffect cross section in the Brandt–Lundqvist approximation employing different statistical atomic models: 1—Thomas–Fermi, 2—Lenz-Jensen, 3—exponential screening

3.7.5 *Approximate Quantum Methods of Calculation of Photoabsorption Cross Sections*

Along with the above classical method of consideration of interparticle correlations in photoabsorption, there are approximate quantum methods taking into account multiparticle effects, in which the photoionization cross section is calculated with the use of somewhat more simplified approaches in comparison with consistent quantum mechanical consideration, such as the *random phase exchange approximation* (RPEA).

One of such methods is based on the *local electron density functional* (DFT) formalism. The simplification of calculation is achieved due to introduction of a local effective potential for the determination of one-particle wave functions of the ground state of a system. For this purpose, a *non-local* exchange-correlation energy is calculated in the local density approximation according to the equations

$$V_{xc}(r) = -\frac{0.611e^2}{r_s(r)} - \frac{0.1e^2}{3a_B}\ln\left(1 + \frac{11.4a_B}{r_s(r)}\right); \quad \frac{4}{3}\pi r_s^3(r) = n^{-1}(r). \qquad (3.140)$$

Equation (3.140) is the so-called exchange-correlation potential. As a result, the solutions of the corresponding equations are found to be no more difficult than the solution of the Hartree differential equations. The effects of interelectron interaction are taken into account with the use of the introduction of a self-consistent field representing the sum of external and induced fields and being a solution of an integral equation.

The results of such calculations are in excellent agreement with available experimental data. Besides, they are indicative of an important role of multiparticle effects in photoionization of atoms with filled electron shells. These effects result (except for the case of neon) in a considerable shift of the photoionization cross section maximum to the region of higher frequencies in comparison with the independent electron approximation, when the position of a maximum practically coincides with the threshold energy of a photon. For example, the maximum of the cross section of photoionization of a xenon atom in the vicinity of the 4*d*-threshold is shifted by about 2.5 *Ry* in the direction of high frequencies. In this case, there is no strong resonance connected (within the framework of one-particle consideration) with transition from the 4*d*-subshell to the virtual *f*-state located in the continuous spectrum.

It is interesting to note that the local DFT method predicts a lower (by several electron-volts) value of the photoeffect threshold in comparison with its observed value. At the same time, this method does not describe highly excited states of the discrete spectrum of an atom. It should be emphasized that in this case the sum rule for the photoabsorption cross section is fulfilled because the "non-physical" contribution of the continuous spectrum to the cross section is compensated by the contribution of the discrete spectrum adjacent to the photoionization threshold that is not taken into account. This fact is present in a much more pronounced form in the above versions of classical description of the atomic photoeffect. As can be seen from Fig. 3.6, the Thomas–Fermi model for the atomic electron density gives a photoionization cross section that is strongly pulled into the low-frequency region, though the sum rule for corresponding cross sections is fulfilled. Within the framework of statistical models, naturally, there is no discrete energy spectrum of an atom at all, so the "non-physical" region of the continuous spectrum below photoionization threshold simulates to a certain extent the contribution of bound states not taken into account.

3.7.6 Rost Hybrid Method

Let us consider a simple model of an atomic photoeffect that admits the analytical representation of the process cross section, known as the Rost hybrid method (Rost 1995). From the formal point of view, this approach is based on the approximate operator equation:

$$\exp\left\{-\frac{i(\hat{H}_0+\Delta_1)t}{\hbar}\right\}\exp\left\{\frac{i\hat{H}_0 t}{\hbar}\right\}\approx\exp\left(-\frac{i\Delta_1 t}{\hbar}\right),\quad \Delta_1=\frac{e^2 a_B}{r^2},\tag{3.141}$$

where H_0 is the unperturbed Hamiltonian of the atom. Hence, the expression for the cross section is given by

$$\sigma_{ph}(\omega)\approx\frac{2\pi Z^2 v_a^2}{3\cdot 137\omega}\int_{-\infty}^{+\infty} dt\langle\psi|\exp(-i\Delta_1 t/\hbar)|\psi\rangle e^{i\omega t}\tag{3.142}$$

The representation of the cross section via (3.142) is called the "hybrid" approximation: It is quantum mechanically due to the general operator approach and at the same time has classical features since the approximate commutation of operator exponents (3.141) is used. It should be noted that the formula (3.142) can be rewritten in terms of the electron density if the following replacement is made:

$$|\psi(\mathbf{r})|^2 \rightarrow 4\pi r^2 n(r). \tag{3.143}$$

After integration with respect to time, the remaining integral (due to the presence of the delta function) can be represented as

$$\sigma_{\mathrm{ph}}(\omega) = \frac{8\pi^3 Z^2}{3\cdot 137} a_{\mathrm{B}}^5 \left(\frac{2Ry}{\hbar\omega}\right)^{7/2} n\left(r = \sqrt{\frac{a_{\mathrm{B}} v_{\mathrm{a}}}{\omega}}\right). \tag{3.144}$$

In particular, from (3.144) it follows the hydrogen-like high-frequency asymptotics of the photoionization cross section if $n(r \rightarrow 0) \rightarrow const$. The dependence (3.144) is presented in Fig. 3.5 as a dotted curve.

Thus, as in the Brandt–Lundqvist approximation, the photoeffect cross section in the Rost hybrid approximation is found to be an electron density functional. But in this case the characteristic distance of the radiative process r_ω is not defined by the plasma resonance condition (3.133), but by the difference of the atomic Hamiltonians H_l with orbital quantum numbers differing (according to the dipole selection rules) by one:

$$\hbar\omega = H_1(r) - H_0(r). \tag{3.145}$$

Equation (3.145) immediately follows from (3.141) in view of the energy conservation law. Based on (3.145), it is possible to give a physical interpretation of the Rost approximation. From this equation, it follows that photon absorption occurs with a fixed electron coordinate as in the Born–Oppenheimer approximation, where the values of coordinates of molecular nuclei do not change during an electron transition. It should be noted that the formula (3.141) is just a mathematical expression of this fact. So the Rost hybrid approximation can be considered as a generalization of the adiabatic principle to the case of electron transitions in atoms.

It should be emphasized that the Rost model does not fulfill the sum rule for the photoabsorption cross section (3.118) in contrast to the Brandt–Lundqvist approximation. This hints to the inconsistency of the hybrid approach used in the derivation of the expression for the photoeffect cross section within the framework of this model.

3.7.7 Generalized Scaled Empirical Photoionization Cross Sections from K-, L-, M-, N- and O-Shell

Quantum mechanical numerical calculations for the photoionization cross sections of different subshells have been performed in a Z- and energy threshold-scaled representation that allow to establish a generalised scaled photoionization model GSPM (Rosmej et al. 2020):

$$\sigma^{(\text{phi})} = \frac{\pi a_0^2}{Z_{\text{eff}}^2} \cdot \frac{m}{2l_0 + 1} \cdot P_1 \cdot \frac{u + P_2}{u + P_3} \cdot \frac{1}{(u + P_4)^{7/2 + l_0}}, \tag{3.146a}$$

$$u = \frac{E - E_{n_0 l_0}}{\tilde{Z}^2 \cdot Ry}, \tag{3.146b}$$

$$Z_{\text{eff}} = n_0 \sqrt{\frac{E_{n_0 l_0}}{Ry}}, \tag{3.146c}$$

$$\tilde{Z} = Z_{\text{eff}} + (Z_{\text{eff}} - z) \quad \text{for single electrons in outer shell } n_0 l_0, \tag{3.146d}$$

$$\tilde{Z} = Z_n - N_{\text{bound}} + N_{\text{nl} \geq n_0 l_0} \quad \text{for inner-shell ionization.} \tag{3.146e}$$

a_0 is the Bohr radius ($\pi a_0^2 = 8.79 \times 10^{-17} \text{cm}^2$), m is the number of equivalent electrons in the subshell $n_0 l_0$, n_0 and l_0 are principal and orbital quantum number, respectively, $Ry = 13.606$ eV, Z_n is the nuclear charge, $E_{n_0 l_0}$ is the ionization potential, N_{bound} is the number of bound electrons, $z = Z_n - N_{\text{bound}} + 1$ is the spectroscopic symbol, $N_{\text{nl} \geq n_0 l_0}$ is the number of electrons in subshells higher or equal than

Table 3.3 Numerical quantum mechanical calculations of the photoionization cross section from H-like ions from the $n_0 l_0$-subshells

$n_0 l_0$	P_1	P_2	P_3	P_4
1*s*	4.667×10^{-1}	2.724×10^{0}	9.458×10^{0}	1.189×10^{0}
2*s*	5.711×10^{-2}	6.861×10^{-1}	7.768×10^{0}	3.644×10^{-1}
2*p*	8.261×10^{-2}	1.843×10^{-1}	7.340×10^{0}	2.580×10^{-1}
3*s*	1.682×10^{-2}	1.436×10^{-1}	7.356×10^{0}	1.436×10^{-1}
3*p*	2.751×10^{-2}	1.742×10^{-1}	7.162×10^{0}	1.742×10^{-1}
3*d*	3.788×10^{-3}	1.566×10^{-1}	7.880×10^{0}	1.566×10^{-1}
4*s*	7.096×10^{-3}	8.799×10^{-2}	7.308×10^{0}	8.799×10^{-2}
4*p*	1.493×10^{-2}	1.197×10^{-1}	1.027×10^{1}	1.197×10^{-1}
4*d*	1.769×10^{-3}	1.205×10^{-1}	6.346×10^{0}	1.205×10^{-1}
4*f*	1.092×10^{-4}	1.055×10^{-1}	9.231×10^{0}	1.055×10^{-1}
5*s*	3.956×10^{-3}	5.846×10^{-2}	8.651×10^{0}	5.846×10^{-2}

For H-like ions, $\tilde{Z} = Z_{\text{eff}} = Z_n$. Fitting parameters are generally accurate within 20% in the large energy range from $10^{-3} < u < 32$(i.e., from threshold to about 30 times threshold)

the subshell n_0l_0, and P_1, P_2, P_3, P_4 are fitting parameters that are given in Table 3.3. The scaled formula (3.146) provides a precision of about 20% of the photoionization cross sections of H-like ions when employing the parameters given in Table 3.3. The particular advantage of the developed formula (3.146) is that it shows the right high-energy and low-energy asymptotics that have been discussed above.

Let us first consider the application of formulas (3.146) to the threshold value of hydrogen discussed above (Sect. 3.7.2). From (3.146) and the parameters for the 1*s* state in Table 3.1, it follows with $l_0 = 0$, $m = 1$, $Z_{eff} = 1$, $\sigma^{(phi)}(1s) \approx 6.4 \times 10^{-20}\text{cm}^2$. This is in excellent agreement with the exact value of the Sommerfeld formula (3.124) that gives $\sigma^{(phi)}(1s) = 6.3 \times 10^{-20}\text{cm}^2$.

Now, we consider photoionization from the 2*p*-shell of H-like helium, i.e. the transition $2p + \hbar\omega \rightarrow \text{nuc} + e$ at a photon energy of 122 eV: $Z_{eff} = 2$, $u = 2$, $l_0 = 1$, $m = 1$ and the fitt-parameters for the 2*p* state in Table 3.1 it follows $\sigma^{(phi)}(2p) \approx 3.7 \times 10^{-21}\text{cm}^2$ whereas the exact quantum mechanical numerical result is $\sigma^{(phi)}(2p) = 3.7 \times 10^{-21}\text{cm}^2$.

The fitt-parameters might also be used to estimate the photoionization cross section for non-H-like ions in the framework of the H-like approximation with effective charges. Let us consider for demonstration of the application of (3.146) the photoionization from Li-like aluminum: (a) transition $1s^22s + \hbar\omega \rightarrow 1s^2 + e$ at a photon energy of 7020 eV: $E_{2s} \approx 442\,\text{eV}$ form which it follows $Z_{eff} \approx 11.4$. Because the considered 2*s*-electron corresponds to the photoionization of a single outer electron $\tilde{Z} = 11.8$ and $u = 3.47$. With $l_0 = 0$ and $m = 1$ and the fitt-parameters for the 2*s* state in Table 3.1 it follows $\sigma_{fit}^{(phi)}(2s) \approx 1.3 \times 10^{-22}\text{cm}^2$. The quantum mechanical numerical result is $\sigma^{(phi)}(2s) = 1.3 \times 10^{-22}\text{cm}^2$, (b) transition $1s^23d + \hbar\omega \rightarrow 1s^2 + e$ at a photon energy of 1010 eV: $E_{3d} \approx 183\,\text{eV}$ form which it follows $Z_{eff} \approx 11.0$. Because the considered "d-electron" corresponds to the photoionization of a single outer electron $\tilde{Z} = Z_{eff}$ and $u = 0.5$. With $l_0 = 2$ and $m = 1$ and the fitt-parameters for the "d-state" in Table 3.1, it follows $\sigma_{fit}^{(phi)}(3d) \approx 4.3 \times 10^{-22}\text{cm}^2$. The quantum mechanical numerical result is $\sigma^{(phi)}(3d) = 4.4 \times 10^{-22}\text{cm}^2$, (c) let us consider a more complicated ground state, the transition $1s^22s^22p^2 + \hbar\omega \rightarrow 1s^22s^22p^1 + e$ in B-like neon at a photon energy of 2117 eV: $E_{2p} \approx 158\text{eV}$ from which it follows $Z_{eff} \approx 6.81$, $\tilde{Z} = 7.62$ and $u \approx 2.48$. With $l_0 = 1$ and $m = 2$ and the fitt-parameters for the 2p state in Table 3.1, it follows $\sigma^{(phi)}(2p) \approx 1.5 \times 10^{-22}\text{cm}^2$. The quantum mechanical numerical result is $\sigma^{(phi)}(2p) = 2.3 \times 10^{-22}\text{cm}^2$. These examples have general character: The H-like approximation is valid for excited states, whereas for ground and close to ground state excited states, specific numerical calculations are requested and in these cases (3.146) might be used only for an order of magnitude estimate.

We now apply the parameters of Table 3.1 to non-hydrogen-like ions and inner-shell photoionization (of primary interest for X-ray Free Electron Laser applications) with the help of the rescaling relation (3.146b, e). We consider the photoionization of the 1*s*- and 2*s*-shells of B-like Neon at a photon energy of

2000 eV: (a) the ionization potential for the transition $1s^2 2s^2 2p^1 + \hbar\omega \rightarrow 1s^1 2s^2 2p^1 + e$ is $E_{1s} \approx 1050\,\mathrm{eV}$ from which it follows $Z_{\mathrm{eff}} \approx 8.78$. Because the ionization of the $1s$-electron corresponds to inner-shell ionization, $Z_n = 10$, $N_{\mathrm{bound}} = 5$, $N_{\mathrm{nl} \geq n_0 l_0}(1s) = 5$ and $\tilde{Z} = 10 - 5 + 5 = 10$ from which it follows $u = 0.70$. With $l_0 = 0$ and $m = 2$ and the fitt-parameters for the $1s$ state in Table 3.1, it follows $\sigma^{(\mathrm{phi})}(1s) \approx 3.8 \times 10^{-20}\mathrm{cm}^2$. The quantum mechanical numerical result is $\sigma^{(\mathrm{phi})}(1s) = 3.9 \times 10^{-20}\mathrm{cm}^2$; (b) we now consider the transition $1s^2 2s^2 2p^1 + \hbar\omega \rightarrow 1s^2 2s^1 2p^1 + e$ for a photon energy of 2132 eV: $E_{2s} \approx 173\,\mathrm{eV}$ from which it follows $Z_{\mathrm{eff}} \approx 7.13, N_{\mathrm{nl}} > n_0 l_0(2s) = 3$, $\tilde{Z} = 10 - 5 + 3 = 8$ and $u \approx 2.10$. With $l_0 = 0$, $m = 2$ and the fitt-parameters for the $2s$ state in Table 3.1 it follows $\sigma^{(\mathrm{phi})}(2s) \approx 2.0 \times 10^{-21}\mathrm{cm}^2$. The quantum mechanical numerical result is $\sigma^{(\mathrm{phi})}(2s) = 2.3 \times 10^{-21}\mathrm{cm}^2$. The general precision for inner-shell photoionization is difficult to estimate, but (3.146) might estimate inner-shell photoionization cross sections within a factor of 2 or so but might be only an order of magnitude estimate in more complex cases (Rosmej et al. 2020).

3.8 Photodetachment from Negative Ions

Several atoms and molecules, having captured an electron, can form negative ions. As a result of such a capture, energy is released that is called electron affinity energy. In Table 3.4, the electron affinity energy ε_a is given for a number of atoms and molecules.

An affinity energy means that energy is necessary to move an outer electron from the negative ion to infinity. This energy can be transferred to an ion as a result of absorption of a photon of sufficiently high frequency $\omega > \varepsilon_a/\hbar$. The process of detachment of an outer electron of a negative ion under the action of an electromagnetic field is called *photodestruction*. In case of photodestruction of a negative ion, a detached electron is in a neutral atom field that is much more weak than the long-range Coulomb field of an ion. So in the first approximation, it is possible to neglect the influence of an atom on a detached electron, assuming it to be free, and to use a plane wave for its wave function. In other words, for description of photodestruction of a negative ion the Born approximation is adequate, so the formula (3.131) should be valid. The specific photodestruction cross section is then defined by the form of the functions $g_{\mathrm{nl}}(k)$ representing the Fourier transform of a wave function of an outer electron of a negative ion. The simplest model that can be used for the radial wave function $R_{\mathrm{nl}}(r)$ of a negative ion is called the zero radius

Table 3.4 Electron affinity energies

Ion	H	Li	O	F	S	Cl	O_2	H_2O
ε_a, eV	0.75	0.58	1.47	3.45	2.08	3.6	0.45	0.9

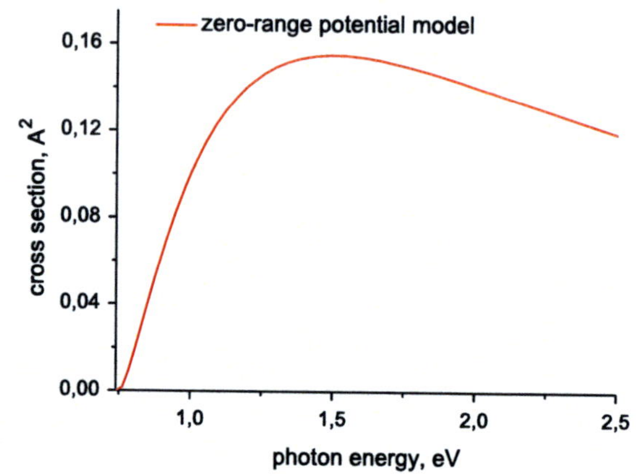

Fig. 3.8 Cross section of photodestruction of a negative ion of a hydrogen atom calculated in the zero radius potential approximation

potential approximation. In this approximation, we have: $R_{10}^{(0)}(r) = \sqrt{2\gamma}\exp(-\gamma r)/r$, where $\gamma = \sqrt{2m\varepsilon_a}/\hbar$ is the parameter of the reciprocal characteristic length of the potential. Using this wave function, we obtain from (3.131):

$$\sigma_{\mathrm{ph}}^{(0)}(\omega) = \frac{4\pi}{3 \cdot 137} a_{\mathrm{B}}^2 \sqrt{\frac{\varepsilon_{\mathrm{a}}}{Ry}} \left(\frac{\sqrt{2m(\hbar\omega - \varepsilon_{\mathrm{a}})}}{m a_{\mathrm{B}} \omega} \right)^3. \tag{3.147}$$

The spectral dependence of the cross section (3.147) is shown in Fig. 3.8.

It is seen that at threshold the cross section goes to zero as it should be in the Born approximation. The maximum of the spectral dependence is reached at a photon energy approximately equal to the double affinity energy of a hydrogen atom (0.75 eV). The value of the cross section itself at maximum is 2.6 times higher than the maximum cross section of photoionization of the neutral hydrogen.

From the formula (3.147), there follows the high-frequency asymptotics of the cross section of photodestruction of a negative hydrogen ion in the zero radius potential approximation: $\sigma_{\mathrm{ph}}^{(0)}(\omega) \propto \omega^{-1.5}$, that is, with increasing frequency the cross section decreases much more slowly than in case of photoionization of neutral hydrogen.

The zero radius potential approximation is favorably distinguished by its simplicity but gives results differing essentially from those of more accurate models. Besides, in case of negative ions of atoms with high polarizability, in calculation of the photodestruction cross section it is necessary to take into account variable polarization of the atomic core that appreciably changes the cross section.

The polarization of a core can also play an essential role in the process of ionization of neutral multielectron atoms. This multiparticle effect at a quantum level is taken into account in the random phase exchange approximation, and within the framework of the classical picture, it is considered in the local plasma frequency approximation.

3.9 Phase Control of Photoprocesses by Ultrashort Laser Pulses

For ultrashort laser pulses, we have to consider the total probability W for the elementary atomic physics processes instead of the probability per unit time (Rosmej et al. 2014, 2016, 2021):

$$W = \frac{c}{4\pi^2}\int_0^\infty \sigma(\omega')\frac{|E(\omega',\tau)|^2}{\hbar\omega'}\mathrm{d}\omega', \tag{3.148}$$

where c – velocity of light, $E(\omega')$ – Fourier transform of electric field in the pulse, τ – pulse duration, and $\sigma(\omega')$ – cross section of the elementary process under consideration. Let us apply the so-called corrected Gaussian pulse to obtain explicit expressions for the probabilities for the photoexcitation. The Fourier transform of this pulse has the form (Rosmej et al. 2014):

$$E_{\mathrm{CGP}}(\omega',\omega,\tau,\varphi) = iE_0\tau\sqrt{\frac{\pi}{2}}\frac{\omega'^2\tau^2}{1+\omega^2\tau^2}\left\{e^{-i\varphi-(\omega-\omega')^2\tau^2/2} - e^{i\varphi-(\omega+\omega')^2\tau^2/2}\right\}, \tag{3.149}$$

where E_0, ω, and τ are the pulse amplitude, carrier frequency, and duration, respectively, and φ is the initial phase. An important feature of (3.149) is the absence of an electric field component at zero current frequency in contrast with the widely used expression of the standard Gaussian shape. Let us apply the formula (3.148) for calculation of the photoexcitation of a multielectron atom by an ultrashort Gaussian pulse (3.149) in the local plasma frequency model. Within the framework of this model, the expression for the photoabsorption cross section looks like

$$\sigma_{\mathrm{ph}}^{(\mathrm{BL})}(\omega') = \frac{2\pi^2 e^2}{mc}\int n(r)\delta(\omega' - \omega_{\mathrm{pl}}(r))\mathrm{d}\boldsymbol{r}, \tag{3.150}$$

where $\omega_{\mathrm{pl}}(r) = \sqrt{4\pi e^2 n(r)/m}$ is the local plasma frequency, and $n(r)$ is the spatial distribution of electron density in an atom. Substituting (3.150) in (3.148), we find

$$W_{\mathrm{tot}}^{(\mathrm{ph})} = \frac{\sqrt{\pi}e}{\sqrt{m}\hbar}\int_0^\infty \left|E(\omega_{\mathrm{pl}}(r),\varphi)\right|^2\sqrt{n(r)}r^2\mathrm{d}r, \tag{3.151}$$

where $\left|E(\omega_{\mathrm{pl}}(r),\varphi)\right|^2$ is the squared absolute value of the Fourier transform of the electric field calculated at the local plasma frequency, in which the carrier envelope phase dependence is clearly indicated. To analyze phase effects in the total

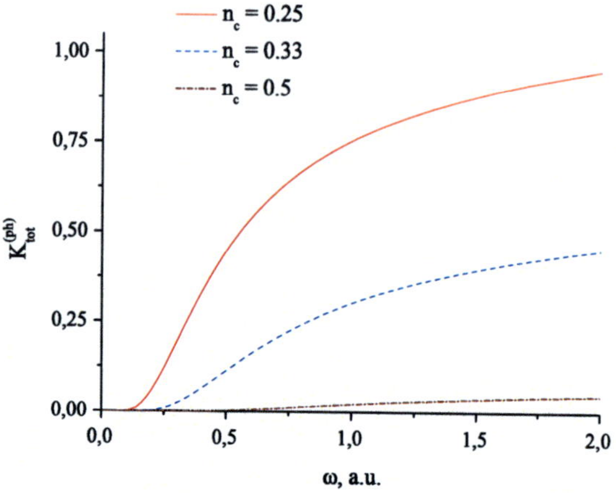

Fig. 3.9 Phase modulation factor for the total probability of photoabsorption of an ultrashort pulse by an atom as a function of carrier frequency

probability of photoexcitation by ultrashort laser pulses, we will introduce a phase modulation factor according to

$$K_{\text{tot}}^{(\text{ph})} = 2\frac{W_{\text{tot}}^{(\text{ph})}(\varphi = 0) - W_{\text{tot}}^{(\text{ph})}(\varphi = \pi/2)}{W_{\text{tot}}^{(\text{ph})}(\varphi = 0) + W_{\text{tot}}^{(\text{ph})}(\varphi = \pi/2)}. \tag{3.152}$$

The phase modulation factor for the total probability of photoabsorption by an atom with the charge $Z = 30$ calculated within the framework of the statistical model for electron density is presented in Fig. 3.9 for three pulse durations as a function of carrier frequency. The dimensionless parameter n_c is the number of periods in the radiation pulse at given carrier frequency.

It is seen that an appreciable dependence of the photoabsorption probability on the carrier envelope phase exists only for $n_c < 0.5$. The phase modulation factor for the fixed parameter n_c increases with carrier frequency. It should be noted that the photoabsorption probability at the high-frequency boundary of Fig. 3.9 is 15% of its maximum value (that in this model corresponds to the frequency $\omega_{\max} = 0.4$ a.u.).

The expression for the total photoabsorption probability (3.148) can be used to study the interaction of an ultrashort pulse with a *metal nanosphere* in a dielectric medium. If the radiation wavelength far exceeds the nanoparticle radius r_s, the dynamic polarizability of a nanoparticle can be described by the Lorentz formula:

$$\beta_s(\omega) = \frac{\varepsilon_s(\omega) - \varepsilon_m}{\varepsilon_s(\omega) + 2\varepsilon_m} r_s^3, \tag{3.153}$$

$\varepsilon_s(\omega)$ is the dielectric permittivity of a nanoparticle metal, and ε_m is the dielectric permittivity of the matrix. Hence, with the use of the optical theorem (2.52), it is possible to find the photoabsorption cross section in the dipole approximation and

Fig. 3.10 Total probability of photoabsorption of an ultrashort pulse ($n_c = 0.25$) on a silver sphere ($r_s = 5.3$ nm) as a function of carrier frequency for two values of the carrier envelope phase

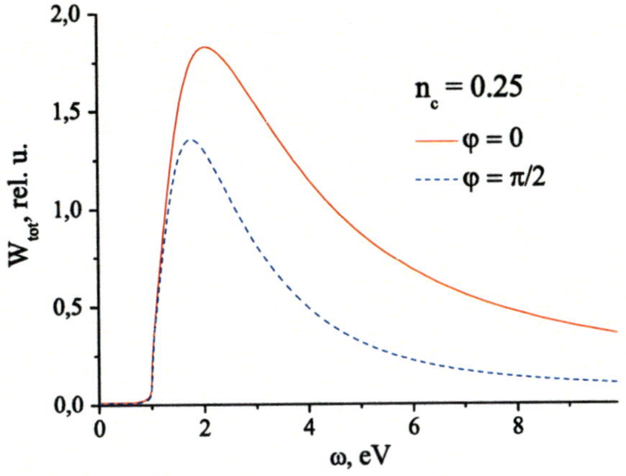

with the help of the formula (3.148) the total photoabsorption probability during the action of the pulse.

The photoabsorption probabilities of an ultrashort pulse by a silver nanoparticle in a glass matrix are given in Fig. 3.10 for two values of carrier envelope phase. The frequency dependence of the dielectric permittivity of silver is restored with the use of data on the real and imaginary parts of the refractive index.

It is seen that for the present case ($n_c = 0.25$) there is an essential dependence of photoabsorption on the carrier envelope phase, especially for photon energies at the carrier frequency exceeding the energy at the maximum. With increasing radiation pulse duration, the dependence of the probability on the carrier envelope phase becomes less appreciable and for $n_c > 0.5$ practically disappears (see Fig. 3.9).

In a number of cases, for excitation of a quantum system a sequence of identical pulses separated by a time interval T (not to be confused with the oscillation period designation) is used. It is not difficult to obtain the Fourier transform of the electric field strength for such a sequence consisting of N identical pulses in terms of the Fourier transform of a single pulse $E(\omega)$:

$$E_{\mathrm{N}}(\omega) = \frac{\sin(\omega TN/2)}{\sin(\omega T/2)} \exp\left[i\frac{(N-1)\omega T}{2}\right] E(\omega). \tag{3.154}$$

Substituting (3.154) in the right-hand side of the (3.148), we find the probability of photoexcitation of a quantum transition under the action of N identical pulses:

$$W_{21}(N) = \frac{c}{4\pi^2\hbar} \int \frac{\sigma_{21}(\omega)}{\omega} \left[\frac{\sin(\omega TN/2)}{\sin(\omega T/2)}\right]^2 |E(\omega)|^2 d\omega. \tag{3.155}$$

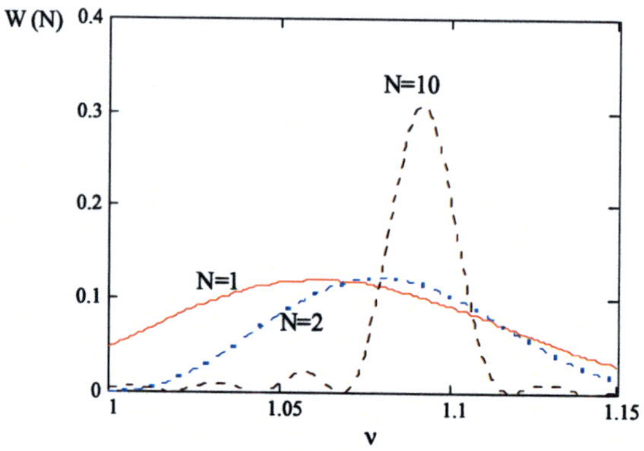

Fig. 3.11 Probability of H-atom photoionization by the action of a train of N two-cycle long laser pulses

Let us use these expressions for the description of the photoionization of a hydrogen atom under the action of a train of short pulses. In this case, the process cross section $\sigma_{21}(\omega)$ is given by the Sommerfeld formula that can be written as [see (3.124)]:

$$\sigma_{\rm H}\left(\nu = \frac{\hbar\omega}{Ry}\right) = \frac{2^9\pi^2 a_{\rm B}^2}{3\cdot 137\cdot \nu^4}\frac{\exp\left(-\dfrac{4arctg\sqrt{\nu-1}}{\sqrt{\nu-1}}\right)}{1-\exp\left(-2\pi/\sqrt{\nu-1}\right)}, \tag{3.156}$$

where $a_{\rm B} \cong 0.53$ Å is the Bohr radius, and $Ry \cong 13.6$ eV is the Rydberg constant.

The photoionization probabilities of a hydrogen atom due to the action of N laser pulses with duration equal to two cycles in dependence of the carrier frequency are presented in Fig. 3.11 [using formulas (3.155)–(3.156)]. The abscissa is expressed in terms of the dimensionless parameter $\nu = \varepsilon/Ry + 1$, where $\varepsilon = \hbar\omega - Ry$ is the energy of ionized electron.

One can see from Fig. 3.11 that the spectral dependence of the photoionization probability shows a narrowing that increases with the number N of laser pulses. The value of the parameter ν at maximum is determined by the equation $\omega T = 2\pi k$ (here k is a natural number). Since the energy of the ionized electron is equal to $Ry(\nu - 1)$ and the number of these electrons is proportional to the probability $W(N)$, one can conclude that it is possible to manipulate considerably the energy spectrum of the photoelectron by changing the laser pulse parameters.

References

M.Y. Amusia, *Atomic Photoeffect* (Springer, Berlin, 1990)

D.R. Bates, A. Damgaard, The calculation of the absolute strengths of spectral lines. Phil. Trans. **242**, 101 (1949)

H.A. Bethe, E.E. Salpeter, *Quantum Mechanics of One- and Two Electron Atoms* (Springer, Berlin, 1977)

W. Brandt, S. Lundqvist, Atomic oscillations in the statistical approximation. Phys. Rev. **139**, A612 (1965)

R.D. Cowan, *The Theory of Atomic Structure and Spectra* (University of California Press, Berkeley, 1981)

V.A. Davydkin, B.A. Zon, Radiation and polarization characteristics of Rydberg atomic states. Opt. Spectrosk. **51**, 13 (1981)

R.A. Gantsev, N.F. Kazakova, V.P. Krainov, Radiation transition rates in hydrogen-like plasmas, in *Plasma Chemistry*, vol. 12, ed. by B.M. Smirnov (Energoatomizdat, Moscow, 1985), p. 96

H. Gao, D.R. DeWitt, R. Schuch, W. Zong, S. Asp, M. Pajek, Observation of enhanced electron-ion recombination rates at very low energies. Phys. Rev. Lett. **75**, 4381 (1995)

H. Gao, R. Schuch, W. Zong, E. Justinianoz, D.R. DeWitt, H. Lebius, W. Spies, Energy and charge dependence of the rate of electron–ion recombination in cold magnetized plasmas. J. Phys. B **30**, L499–L506 (1997)

V.I. Gervids, V.I. Kogan, Penetration and screening effects in electron bremsstrahlung on ions. JETP Lett. **22**, 142 (1975)

V.I. Gervids, V.I. Kogan, Electron bremsstrahlung in a static potential, in *Polarization Bremsstrahlung of Particles and Atoms*, eds. by V.N. Tsytovich, I.M. Oiringel (Plenum, New York, 1991)

S.P. Goreslavsky, N.B. Delone, V.P. Krainov, Probabilities of radiative transitions between highly excited atomic states. JETP **55**, 1032 (1982)

H.R. Griem, *Spectral Line Broadening by Plasmas* (Academic Press, New York, 1974)

H.R. Griem, *Principles of Plasma Spectroscopy* (Cambridge University Press, New York, 1997)

C. Heerlein, G. Zwicknagel, C. Toepffer, Radiative recombination enhancement of bare ions in storage rings with electron cooling. Phys. Rev. Lett. **89**, 83202 (2002) and erratum Phys. Rev. Lett. **93**, 209901(E) (2004a)

C. Heerlein, G. Zwicknagel, C. Toepffer, Reply to Hörndl et al. PRL **93**, 209301 (2004). Phys. Rev. Lett. **93**, 209302 (2004b)

A. Hoffknecht, C. Brandau, T. Bartsch, C. Böhme, H. Knopp, S. Schippers, A. Müller, C. Kozhuharov, K. Beckert, F. Bosch, B. Franzke, A. Krämer, P.H. Mokler, F. Nolden, M. Steck, Th Stöhlker, Z. Stachura, Recombination of bare Bi83+ ions with electron. Phys. Rev. A **63**, 012702 (2001)

M. Hörndl, S. Yoshida, K. Tokési, J. Burgdörfer, Comment on radiative recombination enhancement of bare ions in storage rings with electron cooling. Phys. Rev. Lett. **93**, 209301 (2004)

M. Hörndl, S. Yoshida, A. Wolf, G. Gwinner, J. Burgdörfer, Enhancement of low energy electron-ion recombination in a magnetic field: Influence of transient field effects. Phys. Rev. Lett. **95**, 243201 (2005)

J. Jackson, *Classical Electrodynamics*, 3rd edn. (Wiley, New York, 2007)

L. Kim, R.H. Pratt, Direct radiative recombination of electrons with atomic ions: cross sections and rate coefficients, Phys. Rev. A **27**, 2913 (1983)

V.I. Kogan, A.B. Kukushkin, Radiation emission by quasiclassical electrons in an atomic potential. JETP **60**, 665 (1984)

V.I. Kogan, A.B. Kukushkin, V.S. Lisitsa, Kramers electrodynamics and electron-atomic radiative-collisional processes. Phys. Rep. **213**, 1 (1992)

H.A. Kramers, On the theory of X-ray absorption and of the continuous X-ray spectrum. Phil. Mag. **46**, 836 (1923)

A.B. Kukushkin, V.S. Lisitsa, Radiative cascades between Rydberg atomic states. JETP **61**, 937 (1985)

L.D. Landau, E.M. Lifschitz, *Quantum Mechanics: Nonrelativistic Theory* (Pergamon, Oxford, 1977)

L.D. Landau, E.M. Lifschitz, *The Classical Theory of Fields* (Pergamon, Oxford, 2003)

L.D. Landau, E.M. Lifschitz, *Mechanics* (Pergamon, Oxford, 2005)

P.F. Naccache, Matrix elements and correspondence principles. J. Phys. B: At. Mol. Opt. Phys. **5**, 1308 (1972)

F.B. Rosmej, V.A. Astapenko, V.S. Lisitsa, Effects of ultrashort laser pulse durations on Fano resonances in atomic spectra. Phys. Rev. A **90**, 043421 (2014)

F.B. Rosmej, V.A. Astapenko, V.S. Lisitsa, Scaling laws for ionization of atomic states by ultra-short electromagnetic pulses. J. Phys. B **49**, 025602 (2016)

F.B. Rosmej, V.A. Astapenko, V.S. Lisitsa, XUV and X-ray elastic scattering of attosecond electromagnetic pulses on atoms. J. Phys. B: At. Mol. Opt. Phys. **50**, 235601 (2017)

F.B. Rosmej, V.A. Vainshtein, V.A. Astapenko, V.S. Lisitsa, *Statistical and quantum photoionization cross sections in plasmas: analytical approaches for any configurations including inner shells*, Matter and Radiation at Extremes (Review) **5**, 064202 (2020). Open access: https://aip.scitation.org/doi/10.1063/5.0022751

F.B. Rosmej, V.A. Astapenko, E. Khramov, *XFEL and HHG interaction with matter: effects of ultrashort pulses and random spikes*, Letter to Matter and Radiation at Extremes **6**, 034001 (2021).Open access: https://doi.org/10.1063/5.0046040

F.B. Rosmej, V.A. Vainshtein, V.A. Astapenko, V.S. Lisitsa, *Semi-empirical analytical radiative recombination rates into nl = 1s−9l states for hydrogen and non H-like ions*, in preparation for Matter and Radiation at Extremes (2022)

J.M. Rost, Analytical total photo cross section for atoms. J. Phys. B. **28**, L601 (1995)

I.I. Sobelman, *Introduction to the Theory of Atomic Spectra* (Pergamon, Oxford, 1972)

I.I. Sobelman, *Theory of Atomic Spectra* (Alpha Science Int. Limited, Oxford, UK, 2006)

I.I. Sobelman, L.A. Vainshtein, E.A. Yukov, *Excitation of Atoms and Broadening of Spectral Lines*, 2nd edn. (Springer, Berlin, 1995)

A. Sommerfeld, *Atombau and Spektrallinien: Band I und II* (Harri Deutsch, Frankfurt, 1978)

A. Unsöld, *Physik der Sternatmosphären* (Springer, Berlin, 1955)

Chapter 4
Radiation Scattering on Atoms, Plasmas, and Nanoparticles

Abstract Radiation scattering on free electrons, atoms, plasmas, and nanoparticles is considered using various approaches both quantum and classical ones. Scattering on atoms is described in dipole and non-dipole approximations while the high-frequency limit is applied for elastic (Rayleigh) and Compton scattering. The high-frequency limit is treated using the dynamical form factor (DFF), the Compton profile, and the impulse approximation. The DFF of the plasma component is also used for the description of the radiation scattering in plasmas, namely Compton scattering, transient scattering, and scattering with plasmon generation and absorption. Radiation scattering and absorption on nanoparticles placed in homogeneous media are presented within the framework of the Mie theory. Numerical examples are given for radiation scattering and absorption on silver nanospheres with different radii in glass in the vicinity of surface plasmon resonances.

4.1 Photon Scattering by a Free Electron

Scattering of a photon can be interpreted as virtual absorption of an incident photon and simultaneous emission of a scattered photon. At first, let us consider the simplest case, when a photon is scattered by a free electron. Strictly speaking, the model of a free (not interacting with the environment) electron is always approximate. Nevertheless, in a number of cases, this approximation is well satisfied. Moreover, as we will see further, even a bound electron under certain parameter conditions of scattering can be considered to be free.

Scattering of a photon with a frequency ω and a wave vector $\boldsymbol{k}$ by a free electron is accompanied by a change in frequency $\omega \rightarrow \omega'$ and in wave vector $\boldsymbol{k} \rightarrow \boldsymbol{k}'$. If the angle of photon scattering θ is introduced (Fig. 4.1), then due to the conservation of energy–momentum it is possible to obtain the following relation between the frequency change and the angle of photon scattering:

$$\frac{1}{\omega'} - \frac{1}{\omega} = \frac{\hbar}{m c^2}(1 - \cos\theta), \tag{4.1}$$

F. B. Rosmej et al., *Plasma Atomic Physics*, Springer Series on Atomic, Optical, and Plasma Physics 104, https://doi.org/10.1007/978-3-030-05968-2_4

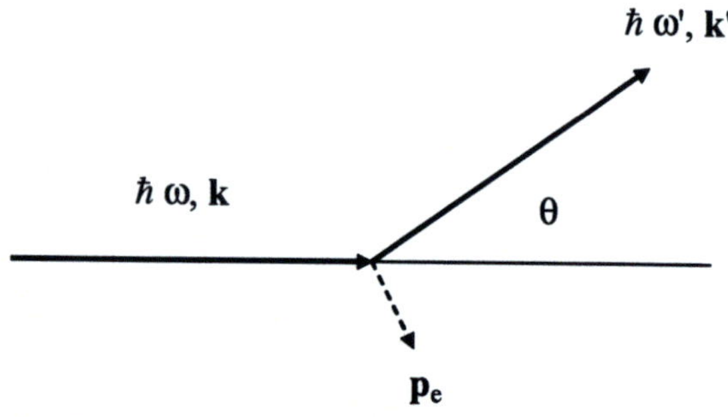

Fig. 4.1 Scattering of a photon by an electron at rest, p_e is the recoil momentum of the electron

where m is the electron mass and c is the velocity of light. The factor in front of the parentheses in the right-hand side of (4.1) can be rewritten as $\lambda_C/2\pi c$, where $\lambda_C = h/mc \cong 2.42 \times 10^{-10}$ cm is the Compton wavelength of an electron.

It is easily seen from the expression (4.1) that at zero-scattering angle $\theta = 0$ the change in photon frequency is equal to zero: $\omega = \omega'$. If $\theta \neq 0$, the frequency of a scattered photon is found to be less than its initial frequency $\omega' < \omega$ since part of the photon energy is transferred to a scattered electron (the "recoil" energy). A decrease in frequency corresponds to an increase in radiation wavelength, so photon scattering through a nonzero angle is accompanied by an increase in wavelength.

With the use of standard rules of quantum electrodynamics (Berestetskii et al. 1982), the following expression (the Klein-Nishina-Tamm formula) can be obtained for the cross section of photon scattering by a free electron in the laboratory frame of reference (connected with an electron):

$$d\sigma = \frac{r_e^2}{2}\left(\frac{\omega'}{\omega}\right)^2\left(\frac{\omega}{\omega'} + \frac{\omega'}{\omega} - \sin^2\theta\right)d\Omega', \tag{4.2}$$

where $r_e = e^2/mc^2 \cong 2.82 \times 10^{-13}$ cm is the so-called classical electron radius, $d\Omega'$ is an element of the solid angle in the direction of the wave vector of a scattered photon. If $\hbar\omega \ll mc^2$, the frequency change is small (see (4.1)) in comparison with the frequency itself: $|\Delta\omega| \ll \omega$, and in (4.2) it can be assumed $\omega \approx \omega'$. Then from the expression (4.2), the well-known nonrelativistic *Thomson formula* is obtained for the cross section of photon scattering by a free electron at rest that is differential with respect to the angle:

$$d\sigma^{\mathrm{Th}} = \frac{1}{2} r_e^2 (1 + \cos^2\theta)\left(\frac{\omega'}{\omega}\right)^2 d\Omega'. \tag{4.3}$$

It is seen that in this case the value of the cross section is defined by the squared of the classical electron radius, that is, a very small quantity.

The cross section integrated with respect to the angle of scattering in the case of an arbitrary photon energy can be obtained from (4.2):

$$\sigma(x) = 2\pi r_e^2 \frac{1}{x}\left\{\left(1 - \frac{4}{x} - \frac{8}{x^2}\right)\ln(1+x) + \frac{1}{2} + \frac{8}{x} - \frac{1}{2(1+x)^2}\right\}, \tag{4.4}$$

where $x = 2\hbar\omega/mc^2$. The scattering cross section (4.4) normalized to the squared classical electron radius is given in Fig. 4.2.

In the nonrelativistic limit of low photon energy, when $x \ll 1$, the first terms of the expansion of the right-hand side of the (4.4) give

$$\sigma = \sigma^{\mathrm{Th}}\left(1 - \frac{2\hbar\omega}{mc^2}\right), \quad \hbar\omega \ll mc^2, \tag{4.5}$$

here, the Thomson cross section $\sigma^{\mathrm{Th}} = 8\pi r_e^2/3$ of photon scattering by an electron integrated with respect to the angle is introduced. This fact is illustrated by Fig. 4.2, in which it is seen that the normalized scattering cross section in the limit of low photon energy is $8\pi/3 \cong 8.378$. The inflection point of the curve shown in Fig. 4.2 corresponds to the approximate equality of the photon energy and the electron rest energy, when $x = 1$.

In the ultrarelativistic case $x \gg 1$, when the photon energy is much higher than the electron rest energy $\hbar\omega \gg mc^2$, we obtain from (4.4):

$$\sigma = \pi r_e^2 \frac{mc^2}{\hbar\omega}\left(\ln\left(\frac{2\hbar\omega}{mc^2}\right) + \frac{1}{2}\right), \tag{4.6}$$

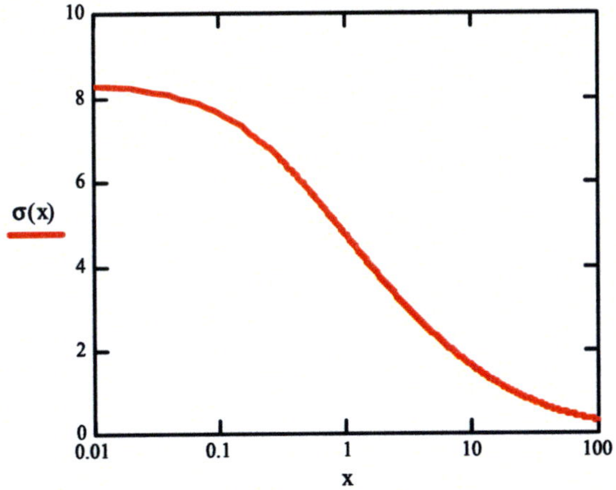

Fig. 4.2 The normalized cross section of photon scattering by a free electron in a wide range of photon energies, $x = 2\hbar\omega/mc^2$

that is, the scattering cross section decreases in inverse proportion to the photon energy.

It should be noted that in the ultrarelativistic limit the differential scattering cross section in the laboratory frame of reference has a sharp maximum in the direction of the initial propagation of the photon. On the other hand in the nonrelativistic limit, the distribution of scattered radiation is of dipole character [as can be seen from the formula (4.3)].

4.2 Radiation Scattering on Atoms

Within the classical framework, when electromagnetic radiation acts on an atom, bound electrons begin oscillation, which, according to known electrodynamic laws, results in the emission of a secondary or scattered electromagnetic wave. In quantum terms, this process represents a scattering of the photon by an atomic electron: the atomic electron is virtually excited to some intermediate atomic state and finally reaches the steady state. We note that by virtual excitation it is meant the transition of an electron to a state with another energy: this transition is then "instantly" followed by the reverse transition to a steady state with the initial or another energy. If this final steady state coincides with the initial state, we encounter the so-called *Rayleigh scattering*, and in the other case, there is *Raman scattering*.

Quite another situation is possible in X-ray scattering, namely when an atomic electron in the course of photon scattering is ionized. This process is referred to as *Compton scattering* of a photon by an atomic electron.

4.2.1 Classical Description

We will dwell first on Rayleigh scattering of radiation by an atom. In order not to resort to the quantum mechanical formalism, we employ the spectroscopic principle of correspondence for the description of this process. According to this principle, an atom behaves as a set of oscillators with frequencies equal to eigenfrequencies of atomic electrons when interacting with radiation. These oscillators are called transition oscillators since each of them corresponds to some transition between two steady states of an electron in an atom.

To calculate the Rayleigh scattering cross section, we proceed from the expression for the power of dipole radiation induced by the action of an electromagnetic wave on an atom (Landau and Lifschitz 1975):

$$Q(t) = \frac{2}{3c^3}\left|\ddot{\mathbf{d}}(t)\right|^2, \tag{4.7}$$

where c is the speed of light, two dots above the dipole moment vector denote differentiation with respect to time. The criterion of applicability of the dipole approximation (formula (4.7)) can be formulated by the following inequality:

$$\lambda \gg a, \tag{4.8}$$

where a is the size of the region of space responsible for radiation, λ is the radiation wavelength. In case of an atom, when $a \approx 10^{-8}$ cm, the condition (4.8) covers a wide range of wavelengths down to soft X-rays. We note that for modern X-ray Free Electron Laser installations, where photon energies up to about 20 keV are reached, the dipole condition (4.8) is not anymore valid.

The expression (4.7) describes the instantaneous power of radiation at a given instant of time t. In experiments, the power averaged over the period of oscillation of the field in an electromagnetic wave causing dipole moment oscillations $T = 2\pi/\omega$ is measured. In case of monochromatic radiation, the formula for the average radiation power follows from (4.7), i.e.,

$$\langle Q(t)\rangle_{\mathrm{T}} = \sum_{n=1}^{\infty} Q_{\mathrm{n}}, \tag{4.9}$$

$$Q_{\mathrm{n}} = \frac{4}{3c^3}\left|(\ddot{\boldsymbol{d}})_{\mathrm{n}}\right|^2, \tag{4.10}$$

where $(\ddot{\boldsymbol{d}})_{\mathrm{n}}$ is the nth Fourier harmonic of the second derivative of the dipole moment. Using the periodicity of dipole moment oscillations, with the use of the Fourier transform, it is possible to obtain the radiation power at a frequency of periodic motion of an electron $\omega(n = 1)$ as

$$Q \equiv Q_{n=1} = \frac{4\omega^4}{3c^3}|\boldsymbol{d}_{\omega}|^2. \tag{4.11}$$

Substituting (2.37) in (4.11) and using the formula for the intensity of monochromatic radiation in terms of the Fourier harmonic of the electric field strength, i.e.,

$$I = \frac{c}{2\pi}|\boldsymbol{E}_{\omega}|^2, \tag{4.12}$$

we obtain for the cross section of the Rayleigh scattering of radiation by an atom in an ith state

$$\sigma_{\mathrm{i}}^{(\mathrm{Rsc})}(\omega) = \frac{8\pi}{3}\left|\frac{\omega^2}{c^2}\beta_{\mathrm{i}}(\omega)\right|^2, \tag{4.13}$$

where $\beta_i(\omega)$ is the dynamic polarizability of an atom in an ith state. The detailed consideration of the dynamic polarizability was presented in Chap. 2.

Let us consider for comparison the classical Thomson formula for the cross section of radiation scattering by a free electron

$$\sigma_{sc}^{(Th)}(\omega) = \frac{8\pi}{3}\left(\frac{e^2}{mc^2}\right)^2. \tag{4.14}$$

In order to compare (4.13) and (4.14), we consider different limiting cases of the expression for the cross section of radiation scattering by an atom (4.13). The first case corresponds to the scattering of low-frequency radiation, when $\omega \ll \omega_{ni}$: the radiation frequency is much less than the frequencies of the transition of an atom from the initial state to excited states. In this limit, the scattering cross section is given by a formula similar to (4.13), in which the frequency-independent static polarizability of an atom $\beta_0 = \beta(\omega = 0)$ is dominating. Thus from the expression (4.13), it follows that in the low-frequency limit the scattering cross section increases as the fourth power of frequency, which, in particular, defines the blue color of the sky (solar radiation scattered in the air).

Resonant scattering of radiation by an atom occurs if the radiation frequency is close to one of the atomic eigenfrequencies. Then the previous formulas give

$$\sigma_i^{(ressc)}(\omega \approx \omega_{ni}) = \frac{2\pi}{3} f_{ni}^2 r_e^2 \frac{\omega_{ni}^2}{(\omega - \omega_{ni})^2 + (\delta_{ni}/2)^2}, \tag{4.15}$$

where $r_e \approx 2.8 \times 10^{-13}$ cm is the electron classical radius. In case of an exact resonance with natural broadening of a transition, when $(\delta_{ni})_{nat} = A_{ni}$ (A_{ni} is the Einstein coefficient for spontaneous radiation), it follows from (4.15) that

$$\sigma_i^{(ressc)}(\omega = \omega_{ni}) \approx \lambda_{ni}^2, \tag{4.16}$$

i.e., the resonant scattering cross section is proportional to the squared of the radiation wavelength, which in the optical range exceeds the geometrical size of an atom by several orders of magnitude.

Finally, in the high-frequency limit, when the eigenfrequencies of the atom can be neglected in comparison with the radiation frequency, we obtain

$$\sigma_i^{(Rsc)}(\omega \gg \omega_{ni}) = \frac{8\pi}{3} r_e^2 N_a^2, \tag{4.17}$$

where N_a is the number of atomic electrons. In derivation of (4.17), the sum rule for oscillator strengths, i.e.,

$$\sum_n f_{ni} = N_a \tag{4.18}$$

was used. The condition of the high-frequency approximation $\omega \gg \omega_{ni}$ can be rewritten as $\omega \gg I_P/\hbar$, where I_P is the ionization potential of the atomic shell that mostly contributes to the scattering cross section.

The Rayleigh scattering cross section in the high-frequency limit (4.17) obtained in the dipole approximation is proportional to the squared number of atomic electrons. This reflects the fact that if the dipole condition (4.8) is valid the electronic charge of an atom oscillates as a single entity under the action of an electromagnetic wave. As a result, secondary radiation leading to scattering is coherent with respect to the contributions of all atomic electrons. It is the presence of the factor N_a^2 in the cross section that manifests this coherence.

For atoms with only one electron, (4.17) coincides with the Thomson formula (4.14), i.e., with the cross section of radiation scattering by a free electron. This coincidence is not accidental since neglecting the binding energy of the atomic electron corresponds to the condition $\omega \gg \omega_{ni}$.

The above cross sections describe photon scattering in total solid angle. These are the so-called integrated cross sections. To obtain a differential cross section describing photon scattering by a spherically symmetric system into a specified solid angle $\Omega' + d\Omega'$, one has to multiply the integrated cross sections by the factor of the angular dependence of dipole radiation, $3(1+\cos^2\theta)/16\pi$, where θ is the angle between the wave vector of incident radiation $\boldsymbol{k}$ and the wave vector of scattered radiation $\boldsymbol{k}'$. Then, instead of formula (4.17), we have

$$\frac{d\sigma_i^{(Rsc)}(\omega \gg \omega_{ni})}{d\Omega'} = \frac{1+\cos^2\theta}{2} r_e^2 N_a^2. \tag{4.19}$$

Hence, it is seen that the cross section maximum corresponds to forward ($\theta = 0$) and backward ($\theta = 180°$) scattering of photons.

The expression (4.19) is valid for spherically symmetric systems in the dipole approximation, when the change in the wave vector of radiation as a result of scattering $\Delta\boldsymbol{k} = \boldsymbol{k}' - \boldsymbol{k}$ is small in comparison with the atomic momentum, i.e., $|\boldsymbol{r}_j||\Delta\boldsymbol{k}| \ll 1$. It can be generalized to the non-dipole case via the replacement $N_a \to n(\Delta\boldsymbol{k})$, where $n(\Delta\boldsymbol{k})$ is the spatial Fourier transform of the atomic electron density. The latter statement means that radiation scattering "tests" the distribution of the electron density inside an atom. Within the framework of the quantum mechanical formalism, the Fourier transform $n(\Delta\boldsymbol{k})$ is equal to the atomic form factor in a specified electronic state:

$$n(\Delta\boldsymbol{k}) \to n_{ii}(\Delta\boldsymbol{k}) \equiv F_i(\Delta\boldsymbol{k}) = \langle i|\sum_j \exp(i\,\Delta\boldsymbol{k}\,\boldsymbol{r}_j)|i\rangle, \tag{4.20}$$

where the sum is calculated over all electrons of an atom, and the symbol $\langle i|\ldots|i\rangle$ is the Dirac notation for the matrix element (the integral of an operator .... calculated with atomic wave functions $\psi_i(\boldsymbol{r}_j)$). It immediately follows from (4.20) that in the framework of the dipole approximation [see relation (4.8)] $F_i(\Delta\boldsymbol{k}) \cong N_a$ since in this case the exponents in the sum (4.20) are small.

In the exponential screening approximation, the electron density decreases with increasing distance to the nucleus on a characteristic scale length R_a (the average atomic radius). The atomic form factor and accordingly the Fourier transform of the electron density is given by

$$F_i(\Delta \boldsymbol{k}) = \frac{N_a}{1 + \Delta k^2 R_a^2}. \tag{4.21}$$

The radius R_a for multielectron atoms can be assumed to be equal to the Thomas–Fermi radius $R_a \cong r_{TF} = b\hbar^2/\sqrt[3]{Z} m_e e^2$ ($b \cong 0.8853$, Z is the nuclear charge).

From the given expression and the formulas (4.17), (4.19), it follows that if a condition opposite to the dipole condition is fulfilled, i.e., $|\Delta \boldsymbol{k}| > R_a^{-1}$, the cross section of the Rayleigh scattering of radiation by an atom starts to decrease because of the loss of coherence between the contributions of different atomic electrons. Thus in the general case, (4.19) is replaced by the expression for the differential cross section of Rayleigh scattering in the high-frequency limit:

$$\frac{d\sigma_i^{(Rsc)}(\omega \gg \omega_{ni})}{d\Omega'} = \frac{1 + \cos^2\theta}{2} r_e^2 |n_{ii}(\boldsymbol{k}' - \boldsymbol{k})|^2. \tag{4.22}$$

As follows from (4.20)–(4.22), the dipole expression for the cross section (4.19) works reasonably well for small scattering angles even if the dipole condition (4.8) is violated.

Formula (4.22) is valid for radiation scattering by nonrelativistic electrons. It should be noted that in Rayleigh scattering the radiation frequency remains the same despite of the fact that a considerable momentum $\Delta \boldsymbol{p} = \hbar \Delta \boldsymbol{k}$ is transferred from the photon to the atom. This circumstance is connected with the fact that the momentum excess $\Delta \boldsymbol{p}$ is finally absorbed not by an atomic electron, but by the nucleus where the recoil energy is low (due to the heavy mass).

In the case of Raman scattering, when the state of an atom changes, the cross section can be expressed in terms of the scattering tensor, a quantity being the generalization of the dynamic polarizability. On the other hand, the radiation frequency changes by an amount equal to the change in energy of the atom.

4.2.2 *Quantum Description*

The quantum description of radiation scattering by an atom is based on the formula for the cross section of electromagnetic field scattering by a quantum system (Berestetskii et al. 1982):

$$\frac{d\sigma_{\rm fi}^{\rm (scat)}}{d\Omega'} = \frac{\omega(\omega')^3}{\hbar^2 c^4 g_{\rm i}} \left| \sum_{\rm n} \left[\frac{(\boldsymbol{e}'^* \boldsymbol{d}_{\rm fn})(\boldsymbol{e}\boldsymbol{d}_{\rm ni})}{\omega_{\rm ni} - \omega - i\gamma_{\rm ni}} + \frac{(\boldsymbol{e}\boldsymbol{d}_{\rm fn})(\boldsymbol{e}'^* \boldsymbol{d}_{\rm ni})}{\omega_{\rm ni} + \omega' - i\gamma_{\rm ni}} \right] \right|^2, \tag{4.23}$$

where $\omega' = \omega - \omega_{\rm fi}$, $\boldsymbol{e}$ and $\boldsymbol{e}'$ are unit polarization vectors of incident and scattered radiation, respectively. The formula (4.23) describes radiation scattering that can be accompanied by simultaneous excitation/de-excitation of a quantum system. Scattering in the case $\omega_{\rm fi} \neq 0$ is called *Raman scattering*, and if $\omega_{\rm fi} = 0$, there is *Rayleigh scattering*. It should be noted that the formula (4.23) was obtained by Kramers and Heisenberg before the advent of quantum mechanics. It is convenient to rewrite the expression (4.23) as

$$\frac{d\sigma_{\rm fi}^{\rm (scat)}}{d\Omega'} = \frac{\omega(\omega')^3}{c^4 g_{\rm i}} \left| e^l (e'^*)^s c_{\rm fi}^{\rm ls}(\omega', \omega) \right|^2, \tag{4.24}$$

where

$$c_{\rm fi}^{\rm ls}(\omega', \omega) = \frac{1}{\hbar} \sum_{\rm n} \left[\frac{(\boldsymbol{d}_{\rm fn})^l (\boldsymbol{d}_{\rm ni})^s}{\omega_{\rm ni} - \omega - i\gamma_{\rm ni}} + \frac{(\boldsymbol{d}_{\rm fn})^s (\boldsymbol{d}_{\rm ni})^l}{\omega_{\rm ni} + \omega' - i\gamma_{\rm ni}} \right] \tag{4.25}$$

is the *electromagnetic field scattering tensor* of the quantum system; l, s are the three-dimensional vector indices, over which summation in the formula (4.24) is assumed.

In the case of Rayleigh scattering (when $\omega_{\rm fi} = 0$), the scattering tensor changes to a *polarizability tensor*:

$$c_{\rm fi}^{\rm ls} \to \beta_{\rm i}^{\rm ls}(\omega) = \frac{1}{\hbar} \sum_{\rm n} \left[\frac{d_{\rm in}^l d_{\rm ni}^s}{\omega_{\rm ni} - \omega - i\gamma_{\rm ni}} + \frac{d_{\rm in}^s d_{\rm ni}^l}{\omega_{\rm ni} + \omega - i\gamma_{\rm ni}} \right]. \tag{4.26}$$

Note that the polarizability tensor relates the electric dipole moment vector to the strength of the electric field of frequency ω:

$$d_{\rm i}^l(\omega) = \beta_{\rm i}^{\rm ls}(\omega) E^s(\omega). \tag{4.27}$$

Here, summation over the index s is implied.

In the case of a spherically symmetric state of a quantum system, the polarizability tensor changes to a scalar: $\beta_{\rm i}^{\rm ls}(\omega) = \beta_{\rm i}(\omega)\delta^{\rm ls}$. Then it can be shown that the expression (4.26) coincides with (2.41) if the explicit form for the oscillator strength is used (2.18) and the relationship between the damping constants is taken into account ($\delta_{\rm in} = 2\gamma_{\rm in}$). The differential cross section of Rayleigh scattering by a quantum system in the spherically symmetric state is then given by

$$\frac{d\sigma_{\rm ii}^{(\rm scat)}}{d\Omega'} = \frac{\omega^4}{c^4}|\beta_{\rm i}(\omega)|^2|(\boldsymbol{e}'^*\boldsymbol{e})|^2. \tag{4.28}$$

We will sum this expression over the polarization directions of a scattered photon with the use of the formula

$$2\left\langle |(\boldsymbol{e}'^*\boldsymbol{e})|^2\right\rangle = \sin^2\theta. \tag{4.29}$$

In (4.29), averaging is carried out over all possible directions of the vector $\boldsymbol{e}'$ at a specified angle θ between the vectors $\boldsymbol{e}$ and $\boldsymbol{n}'$ ($\boldsymbol{n}'$ is the unit vector in the direction of the scattered photon). As a result, we have

$$\frac{d\sigma_{\rm ii}^{(\rm scat)}}{d\Omega'} = \frac{\omega^4}{c^4}|\beta_{\rm i}(\omega)|^2\left(1-(\boldsymbol{n}'\boldsymbol{e})^2\right). \tag{4.30}$$

For non-polarized incident radiation, formula (4.30) should be averaged over the direction of the vector $\boldsymbol{e}$, and we obtain

$$\frac{d\sigma_{\rm ii}^{(\rm scat)}}{d\Omega'} = \frac{\omega^4}{c^4}|\beta_{\rm i}(\omega)|^2\,\frac{1+(\boldsymbol{n}'\boldsymbol{n})^2}{2}, \tag{4.31}$$

where $\boldsymbol{n}$ is the unit vector in the direction of an incident photon. In derivation of (4.31), the averaging rule, i.e., $\langle e^l e^s\rangle = (\delta^{ls} - n^l n^s)/2$ was used.

The expression (4.31) describes the cross section of Rayleigh scattering of radiation by a spherically symmetric system as a function of frequency ω and scattering angles $\theta = \arccos(\boldsymbol{n}'\boldsymbol{n})$. At high frequencies $\omega \gg \omega_{\rm a}$ ($\omega_{\rm a}$ is the characteristic frequency of transitions in a discrete spectrum), we obtain with (2.43) and (4.31)

$$\frac{d\sigma_{\rm ii}^{(\rm scat)}}{d\Omega'} = N_{\rm e}^2 r_{\rm e}^2\frac{1+\cos^2\theta}{2}, \tag{4.32}$$

where θ is the angle of scattering, $N_{\rm e}$ is the number of electrons in the quantum system, $r_{\rm e} = e^2/(m_{\rm e}c^2)$ is the electron classical radius ($r_{\rm e} = 2.8\times10^{-13}$ cm). It should be noted that the expression (4.32) was obtained on the basis of the quantum approach that is identical to the result of classical scattering in the high-frequency approximation (4.19).

For $N_{\rm e} = 1$, the expression (4.32) coincides with the Thomson formula for the cross section of radiation scattering by a free electron. This coincidence is not accidental since there are no bound states for a free electron, i.e., $\omega_{\rm a} = 0$, and the condition of the high-frequency approximation is fulfilled automatically. The squared number $N_{\rm e}^2$ appearing in (4.32) corresponds to the coherence of photon scattering by all electrons of the quantum system. Coherence of scattering is a consequence of the dipole approximation $\lambda \gg a$ used in derivation of all formulas of this chapter. In the opposite limit, i.e., $\lambda \ll a$, coherence is disturbed, and N_e^2 in (4.32) is replaced by N_e.

In the opposite limit (low frequencies), when the dynamic polarizability can be replaced by its static value, the Rayleigh scattering cross section increases with the fourth power of frequency as it follows from formulas (4.28)–(4.31).

Let the frequency of scattered radiation be close to one of the eigenfrequencies of a transition in the quantum system, where the following relation holds true:

$$|\omega - \omega_{\mathrm{ni}}| \leq \gamma_{\mathrm{ni}}. \tag{4.33}$$

In this case, the sum over the intermediate states in (4.23) can be replaced by one resonance summand:

$$\frac{d\sigma_{\mathrm{fi}}^{(\mathrm{scat})}}{d\Omega'} = \frac{\omega(\omega')^3}{\hbar^2 c^4 g_{\mathrm{i}}} \frac{\left|\sum_{M_{\mathrm{n}}} (\boldsymbol{e}'^* \boldsymbol{d}_{\mathrm{fn}})(\boldsymbol{e}\boldsymbol{d}_{\mathrm{ni}})\right|^2}{(\omega - \omega_{\mathrm{ni}})^2 + \gamma_{\mathrm{ni}}^2}. \tag{4.34}$$

Here, only the summation over the degenerate states of the resonance level is left. The phenomenon described by the cross section (4.34) is called *resonance fluorescence*. Let us consider a Rayleigh case, when $\omega_{\mathrm{fi}} = 0$ and accordingly the frequencies of incident and scattered photons coincide: $\omega = \omega'$. We will assume for simplicity that the initial state is non-degenerate: $g_{\mathrm{i}} = 1$. Then (4.34) takes the following form

$$\frac{d\sigma_{\mathrm{i}}^{(\mathrm{RF})}}{d\Omega'} = \frac{1}{16}\left(\frac{c}{\omega_{\mathrm{ni}}}\right)^2 \frac{A_{\mathrm{ni}}^2 |(\boldsymbol{e}'^* \boldsymbol{e})|^2}{(\omega - \omega_{\mathrm{ni}})^2 + \gamma_{\mathrm{ni}}^2}, \tag{4.35}$$

where A_{ni} is the Einstein coefficient of spontaneous radiation.

In the case of natural line broadening, when the half-width is defined by spontaneous radiation, i.e., $\gamma_{\mathrm{ni}} = A_{\mathrm{ni}}/2$ and of an exact resonance $\omega = \omega_{\mathrm{ni}}$, we obtain from (4.35)

$$\frac{d\sigma_{\mathrm{i}}^{(\mathrm{RF})}}{d\Omega'} \propto \lambda^2. \tag{4.36}$$

Hence it follows that the cross section of resonance fluorescence in the optical range in the case of natural line broadening exceeds by many orders of magnitude the scattering cross section far from the resonance. It should be noted that relation (4.36) has already be obtained (see (4.19)) on the basis of a classical consideration.

4.3 High-Frequency Radiation Scattering on Atoms

For scattering of electromagnetic radiation in the X-ray range $\lambda \leq 1$ Å ($k = 2\pi/\lambda \geq 10$ Å^{-1}), the dipole approximation (4.8) is violated, and instead (4.23) has to be used to take into account the wave vectors of incident and scattered photons. This is of particular interest for modern X-ray Free Electron Laser (i.e., LCLS in USA, EU-XFEL in Germany, SACLA in Japan) driven scattering experiments where photon energies larger than 10 keV can be generated.

4.3.1 Non-dipole Character of Scattering

The non-dipole scattering tensor defining the scattering cross section with the use of (4.24) is given by

$$c_{\text{fi}}^{\text{ls}}(\boldsymbol{k}',\boldsymbol{k}) = \frac{e^2}{m\omega'\omega}\left\{\frac{m}{\hbar}\sum_{\text{n}}\left[\frac{j_{\text{fn}}^{l}(\boldsymbol{k})j_{\text{ni}}^{s}(\boldsymbol{k}')}{\omega_{\text{fn}}+\omega-i0}+\frac{j_{\text{fn}}^{s}(\boldsymbol{k}')j_{\text{ni}}^{l}(\boldsymbol{k})}{\omega_{\text{in}}-\omega-i0}\right]-\delta^{\text{ls}}n_{\text{fi}}(\boldsymbol{q})\right\}, \quad (4.37)$$

where $\boldsymbol{q} = \boldsymbol{k}' - \boldsymbol{k}$ is the change in the photon wave vector, δ^{lk} is the Kronecker symbol, $\hat{j}^l(\boldsymbol{k}) = \frac{1}{2m}\sum_{j=1}^{N}\left\{\hat{p}_{\text{j}}^{l}\exp(-i\boldsymbol{k}\boldsymbol{r}_{\text{j}}) + \exp(-i\boldsymbol{k}\boldsymbol{r}_{\text{j}})\hat{p}_{\text{j}}^{l}\right\}$ and $\hat{n}(\boldsymbol{q}) = \sum_{j=1}^{N}\exp(-i\boldsymbol{q}\boldsymbol{r}_{\text{j}})$ are the Fourier transforms of the atomic electron current density and electron density operators, $|i\rangle, |f\rangle$ are the initial and final atomic states.

It should be noted that in the high-frequency limit that can be determined by the inequality $\hbar\omega \gg I$ (I is the characteristic atomic ionization potential) the main contribution to the scattering tensor is made by the second summand in braces of (4.37). The sum in square brackets has an order of magnitude of $(I/\omega)^2$. In the high-frequency domain, the scattering tensor $c_{\text{fi}}^{\text{ls}}(\boldsymbol{k}',\boldsymbol{k})$ is found to be a scalar equal to

$$c_{\text{fi}}^{\text{hf}}(\boldsymbol{k}',\boldsymbol{k}) = -\frac{e^2}{m\omega'\omega}n_{\text{fi}}(\boldsymbol{q} = \boldsymbol{k}' - \boldsymbol{k}). \quad (4.38)$$

It should be emphasized that the condition of the high-frequency limit $\hbar\omega \gg I$ is fulfilled for the majority of atomic electrons in the X-ray range of photon energies, i.e. when $\hbar\omega \geq 10$ keV. Exceptions are electrons of inner shells (in particular *K*- and *L*-shells) of heavy elements with a nuclear charge $Z \geq 60$.

Substituting the expression for the scattering tensor in the high-frequency limit (4.38) into the formula for the scattering cross section (4.24) and averaging over photon polarizations, we find

$$d\sigma^{\text{hf}} = \frac{1}{2}r_{\text{e}}^2\left(1+\cos^2\vartheta\right)\left(\frac{\omega'}{\omega}\right)|n_{\text{fi}}(\boldsymbol{q})|^2 d\Omega'. \quad (4.39)$$

The obtained expression differs from the classical Thomson formula (4.3) by the presence of an additional factor that is proportional to the squared absolute value of the matrix element of the Fourier transform of the atomic electron density operator calculated for the wave vector $\boldsymbol{q} = \boldsymbol{k}' - \boldsymbol{k}$.

4.3.2 Dynamic Form Factor of an Atom

The expression (4.39) should be supplemented with (4.24) expressing the law of conservation of energy in scattering. It is convenient to write this equation in terms of the delta function $\delta(\omega - \omega' + (E_i - E_f)/\hbar)$. Using its integral representation $\delta(x) = (1/2\pi)\int \exp(ixt)dt$ and the Heisenberg operator $\hat{q}(t) = \exp(i\hat{H}t/\hbar)\hat{q}\exp(-i\hat{H}t/\hbar)$, it is possible to obtain the following frequency-angular cross section of photon scattering by an atom in the initial state $\langle i\rangle$ in the *high-frequency approximation*:

$$\frac{d\sigma_i}{d\Omega' d(\Delta\omega)} = \left(\frac{d\sigma}{d\Omega'}\right)^{Th}\left(\frac{\omega}{\omega'}\right)S_i(\Delta\omega, \boldsymbol{q}), \tag{4.40}$$

where $(\frac{d\sigma}{d\Omega'})^{Th}$ is given by the formula (4.3). $S_i(\Delta\omega, \boldsymbol{q})$ is the dynamic form factor (DFF) of an atom (Platzman and Wolf 1973), i.e.,

$$S_i(\Delta\omega, \boldsymbol{q}) = \frac{1}{2\pi}\int_{-\infty}^{\infty} dte^{-i\Delta\omega t}\langle i|\hat{n}(\boldsymbol{q}, t)\hat{n}(-\boldsymbol{q})|i\rangle. \tag{4.41}$$

As can be seen from (4.41), the DFF is a space-time Fourier transform of the density–density correlation function for atomic electrons. The dynamic form factor satisfies a number of sum rules. In the case of a hydrogen-like atom we have the simple relations

$$\int_{-\infty}^{\infty} S(\omega, \boldsymbol{k})d\omega = 1, \qquad \int_{-\infty}^{\infty} S(\omega, \boldsymbol{k})\omega d\omega = \hbar k^2/2m. \tag{4.42}$$

Thus the frequency-angular distribution of photons scattered by an atom in the high-frequency approximation $\hbar\omega \gg I$ is determined by the formula (4.40). If $\hbar\omega < I$, the expression (4.40) becomes, generally speaking, invalid. This is particularly the case, when the frequency of the scattered radiation approaches one of the atomic eigenfrequencies (then, so-called resonant scattering arises and the cross section has a pronounced maximum).

In case of Rayleigh scattering ($|i\rangle = |f\rangle$) by a ground state of a hydrogen-like atom, the diagonal matrix element $n_{ii}(\boldsymbol{q})$ included in the expression for the cross section (4.39) is easily calculated analytically, and (4.39) is transformed to the form

$$\frac{d\sigma_{ii}}{d\Omega'} = \left(\frac{d\sigma}{d\Omega'}\right)^{Th} \frac{1}{[1 + a_H^2 q^2/4]^4}, \tag{4.43}$$

where $a_H = \hbar^2/Z m e^2$ is the characteristic radius of a hydrogen-like ion. From the given expression, it follows that the Rayleigh scattering cross section sharply decreases ($\propto (1/a_H q)^8$) if $q > a_H^{-1}$.

Thus for high recoil momenta (in comparison with the characteristic atomic momentum) scattering proceeds with a change of the atomic state. In this case, if an atom is ionized, there is Compton scattering; if it is excited to the state of a discrete spectrum, scattering is combinational (Raman). It should be noted that the Raman scattering cross section increases at small q from zero, reaches its maximum at $q \approx 1/a$, and then decreases. The calculation for the ground state of a hydrogen-like atom shows that the value of this cross section is always less than the value of its Rayleigh analog. At maximum, it reaches a value about 20% of the Thomson cross section.

In the opposite case ($q < a_H^{-1}$), photon scattering proceeds with no changes in the atomic states. This can be ascertained with the use of the formulas (4.40)–(4.41). Really, in case of small changes in the wave vector of the scattered photon, the Fourier transform of the electron density operator is equal to the number of atomic electrons $\hat{n}(q < a_H^{-1}) \cong N$. Then from (4.41), it follows that the DFF of an atom looks like $S_i = N^2 \delta(\omega - \omega')$, that is, $\omega = \omega'$ and accordingly $E_i = E_f$. Hence, in the case of small transfered momenta, the scattering is of coherent character with respect to the contribution of atomic electrons and the cross section is proportional to the squared number of electrons. On the contrary, in the case of high transferred momenta $q > 1/a$, scattering is incoherent, and its cross section is proportional to N. In view of these dependencies, the following approximate expression for the DFF of an atom can be proposed:

$$S(\Delta\omega, q) \approx \theta(1/a - q) N^2 \delta(\Delta\omega) + \theta(q - 1/a) N \delta\left(\Delta\omega + \hbar q^2/2m\right). \tag{4.44}$$

The first summand in the right-hand side of this equation describes coherent processes occurring in the case of small recoil. The second summand in the right-hand side of (4.44) relates to incoherent phenomena, when an energy–momentum excess (that appears in the course of scattering) is carried away by an ionized electron.

4.3.3 *Impulse Approximation in the Theory of Compton Scattering*

It was indicated above that (leaving out the contribution of inner K-shells) for the description of the X-ray scattering by an atom, the high-frequency approximation

(4.40) can be used, where the dynamic properties of an atom are essential for photon scattering (defined by its dynamic form factor (4.41)). As a rule, the DFF of an atom can not be calculated in the general form. The qualitative formula for the DFF (4.44) does not take into account the details of the electronic structure of an atom and is not precise enough for comparison with experimental data. However, there exist an approximate method for the calculation of the cross section of X-ray scattering by an atom that was used at the very early days for the analysis of experimental data. This is the so-called impulse approximation (IA) assuming that atomic electrons in the course of X-ray scattering behave as free electrons to a greater extent than as bound electrons. The exact definition of the IA will be given below.

The impulse approximation for the description of the Compton scattering of X-rays by an atom, as was shown by Platzman and Wolf (1973), can be obtained from the first principles. For this purpose, it is necessary to use the expression for the DFF (4.41) and the explicit form of the spatial Fourier transform of the electron density operator $\hat{n}(\boldsymbol{q}) = \sum_{j=1}^{N} \exp(-i\boldsymbol{q}\boldsymbol{r}_j)$. In the expression for the Heisenberg operator $\hat{n}(\boldsymbol{q}, t) = \exp(i\hat{H}t/\hbar)\hat{n}(\boldsymbol{q}) \exp(-i\hat{H}t/\hbar)$, the complete Hamiltonian of the system $\hat{H}$ is the sum of the operators of kinetic $\hat{T}$ and potential $\hat{V}$ energies: $\hat{H} = \hat{T} + \hat{V}$ (Note that the potential energy operator commutes with the electron density operator, whereas the kinetic energy operator does not commute). Then we will write down the known expansion of the operator exponent included in the determination of the DFF:

$$\exp\left(\frac{i\hat{H}t}{\hbar}\right) = \exp\left(\frac{i\hat{T}t}{\hbar}\right) \exp\left(\frac{i\hat{V}t}{\hbar}\right) \exp\left(\frac{-i\left[\hat{T},\hat{V}\right]}{\hbar^2}\frac{t^2}{2}\right)\ldots, \tag{4.45}$$

where $\left[\hat{A},\hat{B}\right]$ is the commutator of the operators $\hat{A}$ and $\hat{B}$, the dots at the end of (4.45) denote exponents with multiple commutators containing higher powers of the time parameter t. In fact, (4.45) is a power expansion of the time interval t. The value of this interval is given by the value $1/\Delta\omega$ as it follows from (4.41). Really, the contribution to the time integral in the determination of the DFF for large values of the variable t is small due to strong oscillations of the exponent. The IA condition is given by the equation

$$\exp\left(-i\left[\hat{T},\hat{V}\right]t^2/2\hbar^2\right) = 1, \tag{4.46}$$

according to which the non-commutativity of the operators $\hat{T}$ and $\hat{V}$ can be neglected *for times responsible for the process*. Physically, this means that during photon scattering, the potential in which an atomic electron moves practically does not change. If (4.46) is valid, the contribution of exponents with multiple commutators to the expansion (4.45) can be neglected and the expression for the DFF in the impulse approximation looks simply like

$$S_{\mathrm{i}}^{\mathrm{IA}}(\Delta\omega, \boldsymbol{q}) = \frac{1}{2\pi} \int\limits_{-\infty}^{\infty} \mathrm{d}t e^{-i\Delta\omega t} \langle i| e^{i\hat{T}t/\hbar} \hat{n}(\boldsymbol{q}) e^{-i\hat{T}t/\hbar} \hat{n}(-\boldsymbol{q}) |i\rangle, \tag{4.47}$$

where $\boldsymbol{q} = \boldsymbol{k}' - \boldsymbol{k}$. From this equation, it follows that in the framework of the IA the potential $\hat{V}$ vanishes from the determination of the DFF. However, it should be emphasized that this does not mean to neglect the binding of atomic electrons to a nucleus. As was already noted, in the impulse approximation, the potential $\hat{V}$ is assumed to be constant, and therefore it cancels out in the expressions for the energy of initial and final atomic states. In other words, the energy of the initial and final states of an ionized electron within the framework of the IA is measured with respect to a constant instantaneous value of the potential $\hat{V}$. The expression (4.47) can be evaluated approximately replacing the sum over the complete set of wave functions by plane waves. Using the fact that a plane wave is an eigenfunction of the operator $\exp(-i\hat{T}t/\hbar)$ we then obtain for the DFF of a (nl)th electron subshell of an atom

$$S_{\mathrm{nl}}^{\mathrm{IA}}(\Delta\omega, \boldsymbol{q}) = \int \frac{\mathrm{d}\boldsymbol{p}}{4\pi\hbar^3} \delta\left(\Delta\omega + \frac{(\boldsymbol{p} - \hbar\boldsymbol{q})^2}{2m\hbar} - \frac{\varepsilon_{\mathrm{nl}}}{\hbar}\right) |R_{\mathrm{nl}}(p)|^2, \tag{4.48}$$

$\varepsilon_{\mathrm{nl}}$ is the subshell binding energy, $R_{\mathrm{nl}}(p)$ is the radial wave function of the electron subshell in the momentum representation that is determined by the expression

$$R_{\mathrm{nl}}(p) = \sqrt{\frac{2}{\pi}} \int\limits_{0}^{\infty} R_{\mathrm{nl}}(r) j_{\mathrm{l}}(pr) r^2 \mathrm{d}r, \tag{4.49}$$

here $j_{\mathrm{l}}(p\,r)$ is the spherical Bessel function of the first kind, $R_{\mathrm{nl}}(r)$ is the normalized radial wave function of the electron subshell. If the relation between a momentum and the energy of atomic electrons in the quasi-free approximation is used, i.e.,

$$p^2/2m = \varepsilon_{\mathrm{nl}}, \tag{4.50}$$

we have instead of (4.48)

$$S_{\mathrm{nl}}^{\mathrm{IA}}(\Delta\omega, \boldsymbol{q}) = \int \frac{\mathrm{d}\boldsymbol{p}}{4\pi\hbar^3} \delta\left(\Delta\omega + \frac{\hbar q^2}{2m} - \frac{\boldsymbol{pq}}{m}\right) |R_{\mathrm{nl}}(p)|^2. \tag{4.51}$$

As can be seen from this formula, the frequency shift in Compton scattering within the framework of the impulse approximation is related to the Doppler shift in photon scattering due to a moving atomic electron (the summand $\boldsymbol{p}\boldsymbol{q}$ under the sign of the delta function in (4.51)). In this case, the spectrum of scattered photons is defined by the distribution of atomic electrons by momenta given by the function $|R_{\mathrm{nl}}(p)|^2$.

The cross section of Compton scattering of X-rays by an atom is usually expressed in terms of the Compton profile (CP) of an electron subshell. The CP is determined by the formula

$$J_{\mathrm{nl}}(Q) = \frac{1}{2\hbar^3} \int_Q^\infty |R_{\mathrm{nl}}(p)|^2 p\mathrm{d}p. \tag{4.52}$$

In view of this determination, the dynamic form factor of an electron subshell within the framework of the impulse approximation (4.51) can be expressed in terms of the CP as follows:

$$S_{\mathrm{nl}}^{\mathrm{IA}}(q) = \frac{m}{|\boldsymbol{q}|} J_{\mathrm{nl}}\left(Q = \frac{m|\Delta\omega| - \hbar q^2/2}{|\boldsymbol{q}|}\right). \tag{4.53}$$

It should be noted that the CP satisfies the normalizing condition:

$$2\int_0^\infty J_{\mathrm{nl}}(Q)\mathrm{d}Q = 1. \tag{4.54}$$

Since Compton scattering is an incoherent process, the atomic CP is equal to the sum of CPs of all electron subshells.

Using the relation (4.40) and the formula (4.53), we find for the cross section of Compton scattering of X-rays by an atom:

$$\frac{\mathrm{d}\sigma^{\mathrm{IA}}}{\mathrm{d}\Omega' d(\Delta\omega)} = \left(\frac{\mathrm{d}\sigma}{\mathrm{d}\Omega}\right)^{\mathrm{Th}} \left(\frac{\omega}{\omega'}\right) \frac{m}{q} J(Q), \tag{4.55}$$

where $Q = \dfrac{m|\Delta\omega| - \hbar q^2/2}{|\boldsymbol{q}|}$. Thus the expression (4.55), together with the formula (4.3) for $(\mathrm{d}\sigma/\mathrm{d}\Omega')^{\mathrm{Th}}$, gives the frequency-angular distribution of X-rays scattered by an atom within the framework of the high-frequency and impulse approximations. It is seen that the frequency distribution for a specified angle of scattering is defined by the Compton profile of an atom (4.52) that in turn depends on the wave functions $R_{\mathrm{nl}}(r)$ of the atomic electron. The corresponding frequency dependence is a bell-shaped curve. Its width is defined by the width of the distribution of the atomic momenta wave functions. Let us give the expression for the CP of the ground state of a hydrogen-like ion:

$$J_{1s}^{H}(Q) = \frac{8p_{\mathrm{H}}^5}{3\pi(p_{\mathrm{H}}^2 + Q^2)^3}, \tag{4.56}$$

where $p_H = Zme^2/\hbar$ is the characteristic momentum of a hydrogen-like ion. Substituting (4.56) in (4.55), we find the frequency function of the distribution of X-rays within the framework of the IA for a fixed angle of scattering by a hydrogen-like ion in the 1s-state:

$$\frac{d\sigma_{1s}^{H}}{d\Delta\omega} = \sigma_{1s}(q)\frac{q^6 p_H^6}{\left[q^2 p_H^2 + (m|\Delta\omega| - \hbar q^2/2)^2\right]^3}, \tag{4.57}$$

where $q = |\boldsymbol{k}' - \boldsymbol{k}|$ is the magnitude of the photon wave vector change in scattering. From this formula, it follows that the central frequency of scattered radiation is determined by the equation $m(\omega - \omega') = \hbar q^2/2$. This equation represents the law of conservation of energy–momentum for the process of radiation scattering by a free electron at rest. The width of the frequency distribution (4.57) is proportional to the parameter $q p_H/m$, which corresponds to a Doppler broadening of the spectrum of radiation scattered by an electron moving with the velocity $\upsilon_H = p_H/m = Ze^2/\hbar$.

Presented in Fig. 4.3 are the spectral cross sections of Compton scattering (in relative units) by a hydrogen-like ion in the 1s-state calculated in the high-frequency approximation with the use of the exact wave functions (curves 1, 3) and within the framework of the impulse approximation (curves 2, 4) according to formulas (4.55)–(4.56). Two values of ionic charges are considered: $Z = 1$ (curves 1, 2) and $Z = 2$ (curves 3, 4). The incident photon energy is $\hbar\omega = 17.4$ keV (639.7 a.u.) (which corresponds to the line K_{α_1} in a molybdenum atom), the angle of radiation scattering is $\vartheta = 133.75°$. As can be seen from Fig. 4.3, the

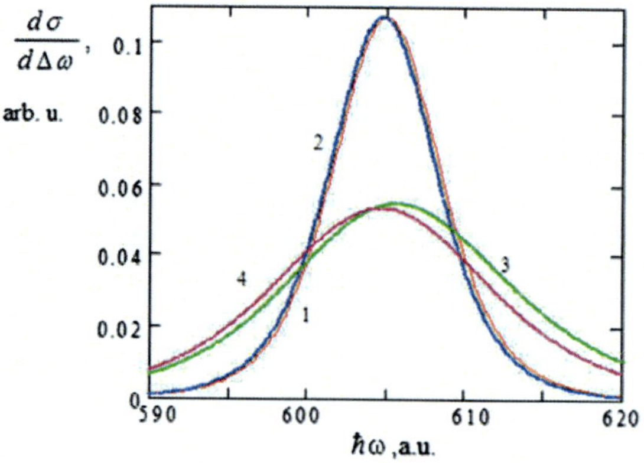

Fig. 4.3 The spectral cross section of Compton scattering of X-rays ($\hbar\omega = 17.4$ keV, the angle of scattering is $\vartheta = 133.75°$) by a ground state hydrogen-like ion calculated in the high-frequency approximation with the use of the exact functions (curves 1, 3) and within the framework of the impulse approximation (curves 2, 4) for different ionic charges: curves 1, 2—$Z = 1$, curves 3, 4—$Z = 2$

maximum of the spectral cross section corresponds to a scattered photon energy of $\hbar\omega = 605$ a.u. (16.456 keV) and a recoil energy of $E_R = \hbar|\Delta\omega| = 944$ eV.

It should be noted that the recoil energy in scattering by a free electron at rest for otherwise same parameters is 949 eV, being in good agreement with the above value E_R at the maximum of the spectral cross section. It should be emphasized that for the case of a free electron there is a univocal correspondence between the scattering angle and the frequency of the scattered photon and the spectral cross section represents a delta function of the frequency detuning $\omega - \omega' - E_R/\hbar$. Binding of an electron to a nucleus results in a finite spectral width $\delta\omega$ of the frequency distribution of the scattered photons at a specified angle of scattering θ. Naturally, the stronger the binding to a nucleus is, the higher the value of the parameter $\delta\omega$ is. This also follows from Fig. 4.3: for larger charge values, the frequency dependence is broader. Figure 4.3 also demonstrates that the accuracy of the IA decreases with increasing binding energy E_B ($E_B = Z^2Ry$ for a hydrogen-like ions) at fixed recoil energy E_R. Thus the value of the ratio $\xi = E_B/E_R$ can serve as a criterion for the applicability of the IA; with decreasing parameter ξ the IA accuracy increases. Figure 4.3 shows also an asymmetry of the frequency distribution of the scattered photons when the exact wave functions are employed (curves 1, 3). In the high-frequency wing of the line, the intensity of the scattered radiation decreases more slowly than in the low-frequency wing. This asymmetry increases with the binding energy E_B. At the same time, the calculation within the framework of the IA gives a symmetric (for the present case "Lorentz") profile. It should be noted that the asymmetry could be to some extent taken into account within the framework of the impulse approximation if in the right-hand side of (4.45) succeeding expansion terms are taken into account.

4.4 Scattering on Plasmas

We now consider the scattering of electromagnetic radiation by electrons in plasmas. The cross section of radiation scattering by ions is negligible because of the heavy mass of an ion. In contrast to atomic electrons, plasma electrons execute an infinite motion, that is, they are delocalized throughout the plasma volume. Therefore, strictly speaking, the condition for the application of the dipole approximation (4.8) is not fulfilled for plasma electrons because the radius of the area of their localization is very large. Since plasma electrons are quasi-free, the potential of their ionization I_P is equal to zero, the condition of the high-frequency approximation is naturally fulfilled, and it is possible to use the expression for the scattering cross section according to formula (4.22).

The adjective "quasi-free" is used in reference to plasma electrons is not accidentally. The prefix "quasi" reflects the fact that plasma electrons interact with other particles and collective plasma excitations. This interaction manifests itself in the scattering of electromagnetic waves by plasma electrons in the vicinity of frequencies of collective plasma excitations. For example, a momentum excess transferred from a photon in scattering can be absorbed by a plasma electron itself but can be as well transferred to another particle or a quasi-particle. Depending on these possibilities, the frequency of scattered radiation differs. Therefore, by recording the spectrum of the scattered photons, it is possible to obtain information on the plasma properties.

4.4.1 General Expression for the Cross Section of Radiation Scattering in Plasmas

Let us generalize the scattering cross section (4.22) taking into account a possible change in the state of a plasma electron in the course of scattering:

$$\frac{d\sigma_{\rm fi}^{(\rm plas)}(\boldsymbol{k},\boldsymbol{k}')}{d\Omega' d\omega'} = \frac{1+\cos^2\theta}{2}\delta(\Delta\omega + \omega_{\rm fi}) r_{\rm e}^2 |n_{\rm fi}(\boldsymbol{k}'-\boldsymbol{k})|^2 dV, \tag{4.58}$$

where $\Delta\omega = \omega' - \omega$ is the change in radiation frequency, dV is the element of the volume in which the scattering plasma electrons are located (the volume of interaction). The given expression describes radiation scattering with the transition of a plasma electron from the state $|i\rangle$ to the state $|f\rangle$ (here Dirac ket vectors are used to designate electronic states). Expression (4.58) gives the cross section of photon scattering in the frequency interval $\omega' + d\omega'$ and is therefore differential not only with respect to the angle, but also with respect to the frequency of the scattered radiation. The delta function in the right-hand side of (4.58) expresses the conservation law of energy in the scattering process.

Since in the experiment the initial and final states of a plasma electron are not fixed, formula (4.58) should be summed over the final states and averaged over the initial ones. As a result, we obtain

$$\frac{d\sigma_{\Sigma}^{(\rm plas)}(\boldsymbol{k},\boldsymbol{k}')}{d\Omega' d\omega'} = \frac{1+\cos^2\theta}{2} r_{\rm e}^2 S(\Delta\boldsymbol{k}, \Delta\omega)) dV, \tag{4.59}$$

where the function $S(\Delta\boldsymbol{k}, \Delta\omega)$ is the so-called electron *dynamic form factor* (DFF) of the plasma or the spectral density function. The DFF reflects the influence of plasma characteristics on the radiation scattering cross section. In the general case, the determination of the DFF in terms of the Fourier component of the plasma density time-domain correlator looks like (Platzman and Wolf 1973).

$$S(\Delta\boldsymbol{k}, \Delta\omega) = \frac{1}{2\pi}\int_{-\infty}^{\infty} dt\, e^{i\Delta\omega t}\langle \hat{n}(\Delta\boldsymbol{k}, t)\hat{n}(-\Delta\boldsymbol{k})\rangle, \tag{4.60}$$

where $\hat{n}$ is the electron density operator, the angle brackets include quantum mechanical and statistical averaging. Equation (4.58) can be obtained from the formula

$$S(\Delta \mathbf{k}, \Delta\omega) = \sum_{f,i} w(i)\delta(\Delta\omega + \omega_{fi})|n_{fi}(\Delta \mathbf{k})|^2, \tag{4.61}$$

in which averaging over initial states and summation over final states of plasma electrons are carried out explicitly ($w(i)$ is the probability of a plasma electron being in the ith state).

Physically, the electron DFF defines the probability of absorption of a four-dimensional energy–momentum wave vector $\Delta k = (\Delta \mathbf{k}, \Delta\omega)$ by a plasma in terms of the action of external disturbance on an electronic component. It should be emphasized that in case of a homogeneous distribution of charges in plasmas this probability would be equal to zero since then the Fourier transform of the electron density is given by the delta function $n(\Delta \mathbf{k}) \to \delta(\Delta \mathbf{k})$, that is, $\Delta \mathbf{k} = 0$ and $\mathbf{k} = \mathbf{k}'$. Radiation scattering is therefore a result of plasma density fluctuations. These fluctuations can be due to various reasons. A typical cause of electronic charge fluctuations is Debye screening of ions by plasma electrons, when the electron density is increased in the vicinity of a positive ion. Another type of fluctuations is connected with collective excitations in plasmas, for example, plasmons. Then, electronic charge fluctuations are of nonstationary character: they oscillate at a plasma frequency $\omega_{pe} = \sqrt{4\pi e^2 n_e / m_e}$ (n_e is the density of the plasma electrons). Scattering by nonstationary fluctuations changes the radiation spectrum. Thus, from the observed scattered radiation spectrum the fluctuation spectrum of the electron density in a plasma that is given by the DFF (4.60)–(4.61) can be deduced.

To calculate the DFF, it is possible to use the formula relating the DFF of a plasma component to the function of plasma response to external electromagnetic disturbance (a fictitious external potential). This relation is known as the fluctuation–dissipative theorem (Platzman and Wolf 1973):

$$S(\Delta \mathbf{k}, \Delta\omega) = \frac{Im\{F_{ee}(\Delta \mathbf{k}, \Delta\omega)\}}{\pi e^2[\exp(-\hbar\Delta\omega/T) - 1]}, \tag{4.62}$$

where $F_{ee}(\Delta \mathbf{k}, \Delta\omega)$ is a linear function of the response of the electronic component to a fictitious external potential that acts on plasma electrons, T is the plasma temperature (in energy units). The imaginary part of the response function appearing in (4.62) describes energy dissipation in plasmas. For this reason, the theorem is called the fluctuation–dissipation theorem. The response function is expressed in terms of the dielectric permittivity of the plasma and describes the propagation of various electromagnetic disturbances. After a number of mathematical transformations, the following result can be obtained from (4.62) for the dynamic form factor of an electronic component in a plasma:

$$S(\Delta k) = \left|\frac{\varepsilon^{l(i)}}{\varepsilon^l}\right|^2 |\delta n_e|^2 + Z_i \left|\frac{1 - \varepsilon^{l(e)}}{\varepsilon^l}\right|^2 |\delta n_i|^2, \tag{4.63}$$

where

$$\left|\delta n_{\mathrm{e,i}}\right|^2 = \frac{n_{\mathrm{e,i}}}{\sqrt{2\pi}\upsilon_{\mathrm{Te}}|\Delta \boldsymbol{k}|} \exp\left(-\frac{\Delta\omega^2}{2\Delta \boldsymbol{k}^2 \upsilon_{\mathrm{T,ie}}^2}\right) \tag{4.64}$$

are the Fourier transforms of the squared thermal fluctuations of the electronic and ionic components of the plasma, $\varepsilon^l = \varepsilon^l(\Delta k)$ is the longitudinal part of the plasma dielectric permittivity that describes the propagation of longitudinal plasma waves (corresponding to the density fluctuations), υ_{T} is the thermal velocity of the plasma particles. The indices (i) and (e) designate for the dielectric permittivity and the thermal velocity that these quantities belong to an ionic or electronic plasma component.

Hereafter, the expressions for the longitudinal part of the electronic and ionic dielectric plasma permittivities appearing in the right-hand side of the (4.63) are needed. Their explicit form depends on the relationship between the spatial and time components of the four-vector $\Delta k = (\Delta \boldsymbol{k}, \Delta\omega)$, that is, between the spatial and time characteristics of the electronic and ionic components of the plasma. Thus, the electronic part of the function $\varepsilon^{l(e)}$, for which the relation $|\Delta\omega| < \upsilon_{\mathrm{Te}}|\Delta \boldsymbol{k}|$ holds true, describes the effect of charge screening in a plasma. The corresponding expression looks like

$$\varepsilon^{l(e)}(\Delta \boldsymbol{k}) \cong 1 + \frac{1}{d_{\mathrm{e}}^2|\Delta \boldsymbol{k}|^2}, \tag{4.65}$$

where d_{e} is the Debye screening radius of the electronic component. On the contrary, the reverse inequality $|\Delta\omega| > \upsilon_{\mathrm{Ti}}|\Delta \boldsymbol{k}|$ is characteristic for the ionic component of the longitudinal dielectric permittivity (because of the low velocity of ions). This results in the frequency-dependent relation

$$\varepsilon^{l(i)}(\Delta\omega) \cong 1 - \frac{\omega_{\mathrm{pi}}^2}{\Delta\omega^2}, \tag{4.66}$$

where $\omega_{\mathrm{pi}} = \sqrt{4\pi e^2 n_{\mathrm{i}}/m_{\mathrm{i}}}$ is the ion plasma frequency.

4.4.2 Radiation Scattering by Plasma Electrons

Substituting the DFF of (4.63)–(4.64) into the expression for the cross section of radiation scattering in a plasma (4.59), it is possible to obtain a cross section being a sum of two terms. The first of them is related to the first summand in braces of (4.63). It looks like

$$\frac{d\sigma_e^{(plas)}(\boldsymbol{k},\boldsymbol{k}')}{d\Omega' d\omega'} \cong \frac{\exp\left(-\frac{\Delta\omega^2}{2\Delta k^2 v_{Te}^2}\right)}{\sqrt{2\pi} v_{Te}|\Delta \boldsymbol{k}|}\left[\frac{\Delta k^2 d_e^2}{1+\Delta k^2 d_e^2}\right]^2 \frac{1+\cos^2\theta}{2} r_e^2 n_e dV. \tag{4.67}$$

The above cross section is connected with radiation scattering by an electronic charge that screens the electron density fluctuation. The electron density fluctuation is described by the factor in the first row of (4.67) containing the thermal velocity of plasma electrons. The factor in square brackets is responsible for screening. It is small in the case of small transferred wave vectors $|\Delta \boldsymbol{k}| < d_e^{-1}$ and tends to one in the opposite limit. Thus scattering by electron density fluctuations is suppressed for small angles of scattering $\theta < 2\arcsin(\lambda/4\pi d_e)$ (this inequality is valid for sufficiently short wavelengths $\lambda < 4\pi d_e$).

As can be seen from the frequency dependence in the formula (4.67), the width of the radiation spectrum (scattered by the electron density fluctuations) is defined by the thermal velocity of plasma electrons that in turn is a measure of plasma temperature. The higher the temperature, the wider is the spectrum of scattered photons. The value of the cross section is proportional to the concentration of plasma electrons. Therefore, measuring the intensity and the spectrum of scattered radiation, it is possible to judge such important characteristics of plasma as the temperature and the electron concentration.

The process described by the cross section (4.67) is similar to Compton scattering since a momentum excess is transferred to a plasma electron that carries away considerable recoil energy as in case of the Compton effect with an atom. If the change in wave vector is larger than the reciprocal of the Debye radius, a photon is scattered by each plasma electron "separately", and collective plasma effects can be neglected. Compton scattering of radiation in a plasma is shown in Fig. 4.4, in which the recoil momentum of a plasma electron $\boldsymbol{p}_e = \hbar \cdot \Delta \boldsymbol{k}$ is presented.

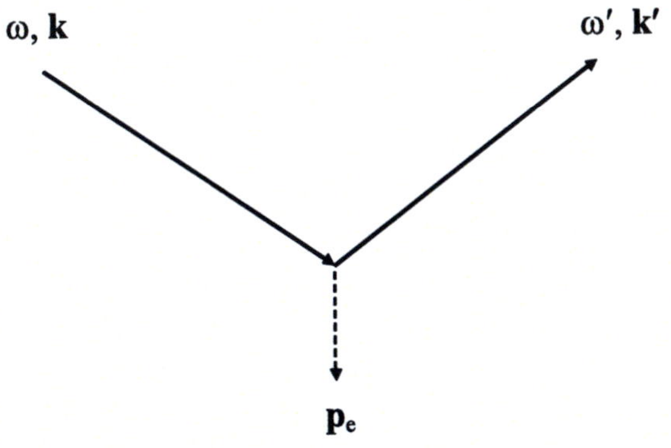

Fig. 4.4 Radiation scattering by a plasma electron

The value of the radiation scattering cross section for one electron is proportional to the squared electron classical radius $r_e^2 \cong 7.84 \times 10^{-25}$ cm^2, i.e. a very small quantity. Note that an atomic unit of cross section that defines, for example, the cross section of electron scattering by atoms is equal to 2.8×10^{-17} cm^2 . Therefore, the possibility to study plasma characteristics with the use of radiation scattering appeared only after the development of laser radiation sources of high spectral radiance (Kunze 1968; Sheffield 1975).

4.4.3 Transient Scattering of Radiation in Plasmas

The cross section of radiation scattering by an electronic charge screening ion density fluctuations looks like

$$\frac{d\sigma_i^{(plas)}(\boldsymbol{k},\boldsymbol{k}')}{d\Omega' d\omega'} \cong \frac{\exp\left(-\dfrac{\Delta\omega^2}{2\Delta\boldsymbol{k}^2 \upsilon_{Ti}^2}\right)}{\sqrt{2\pi}\upsilon_{Ti}|\Delta\boldsymbol{k}|}\left[\frac{Z_i}{1+\Delta\boldsymbol{k}^2 d_e^2}\right]^2 \frac{1+\cos^2\theta}{2} r_e^2 n_i dV, \qquad (4.68)$$

Z_i is the charge of a plasma ion, $n_i = n_e/Z_i$ is the concentration of ions in an electrically neutral plasma. Here the first factor describes ion fluctuations, and the expression in square brackets represents the electronic charge of a Debye sphere that scatters radiation with a specified value $\Delta\boldsymbol{k}$. In the case $|\Delta\boldsymbol{k}| < d_e^{-1}$, the scattering cross section is proportional to the squared charge of an ion, in which the coherent character of the process with respect to the contribution of electrons inside the Debye sphere manifests itself. In case of fulfillment of the reverse inequality, coherence is disturbed, and the scattering cross section decreases as $\Delta\boldsymbol{k}^{-4} d_e^{-4}$. For sufficiently long wavelengths of scattered radiation $\lambda > d_e$, coherence takes place for all angles of scattering. In the opposite limit of short-wavelength radiation, the scattering cross section is maximum in the range $\theta < 2\arcsin(\lambda/4\pi d_e)$ and sharply decreases with increasing angle of scattering.

Radiation scattering by a Debye sphere (4.68) is called *transient scattering* (Ginzburg and Tsytovich 1990). This process is represented schematically in Fig. 4.5, in which the recoil momentum of an ion $\boldsymbol{p}_i = \hbar|\Delta\boldsymbol{k}|$ arises in the course of scattering.

Transient scattering is accompanied by the transfer of a momentum excess to a heavy ion, so the change in photon energy is insignificant. The last fact can be demonstrated with the use of the limiting transition

$$\frac{\exp\left(-\dfrac{\Delta\omega^2}{2\Delta\boldsymbol{k}^2 \upsilon_{Ti}^2}\right)}{\sqrt{2\pi}\upsilon_{Ti}|\Delta\boldsymbol{k}|} \to \delta(\Delta\omega) \qquad (4.69)$$

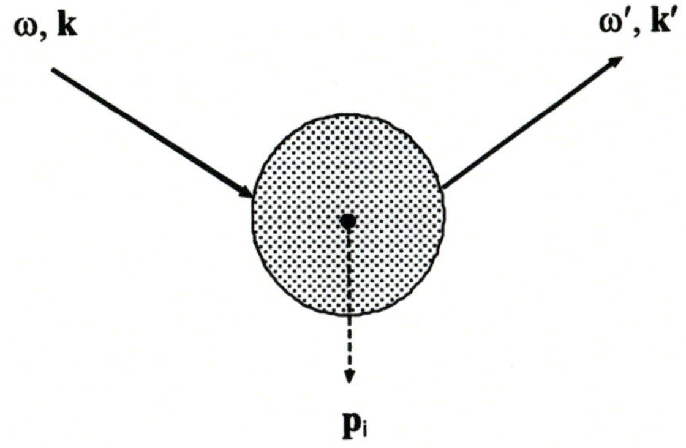

Fig. 4.5 Transient scattering of radiation by a Debye sphere in plasma

that is valid if $|\Delta\omega| \gg \upsilon_{Ti}|\Delta \boldsymbol{k}|$. From the relation (4.69), it follows that in this case scattering proceeds practically with no change in radiation frequency: $\omega' \cong \omega$. Thus radiation scattering by a Debye sphere is an analog of Rayleigh scattering of radiation by an atom, when the radiation frequency does not change.

4.4.4 Radiation Scattering by a Plasmon

So far, we have considered a photon momentum excess $\Delta \boldsymbol{p}_{ph} = \hbar \Delta \boldsymbol{k}$ that was transferred to individual plasma excitations: electrons and ions. A process involving collective plasma excitation is also possible, when a momentum excess is transferred to a *plasmon* representing a coupled oscillation of an electronic charge and a longitudinal electric field. This process is represented schematically in Fig. 4.6.

A plasmon is characterized by a corresponding dispersion law. It will be remembered that the dispersion law describes the relation between a frequency and a wave vector. For collective excitations (quasi-particles) in plasmas, the dispersion can be obtained from the condition that the dielectric permittivity of a medium is zero, that is, from the condition of the plasma wave propagation. In the case of longitudinal electric field oscillations (a plasmon), the dispersion law is determined by the equation

$$\varepsilon^{(l)}(\boldsymbol{k}, \omega) = 0, \tag{4.70}$$

where $\varepsilon^{(l)}(\boldsymbol{k}, \omega)$ is the longitudinal part of the dielectric permittivity of the medium. In view of the explicit form of the dielectric permittivity of a plasma for $|\boldsymbol{k}| < d_e^{-1}$ according (4.65), we obtain from (4.70):

$$\omega_{pl}(\boldsymbol{k}) \cong \omega_{pe}\sqrt{1 + 3\boldsymbol{k}^2 d_e^2} \approx \omega_{pe}. \tag{4.71}$$

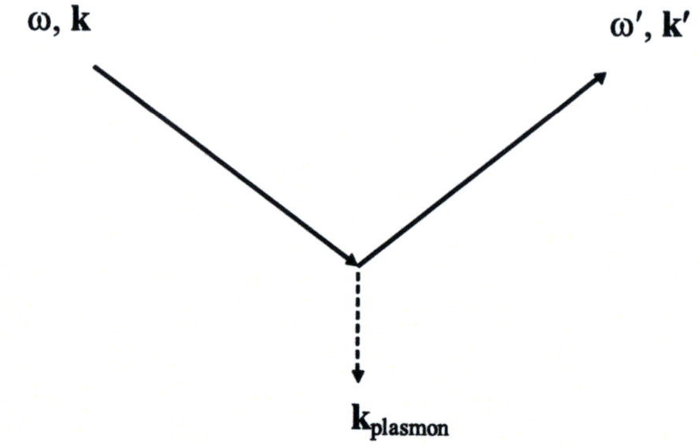

Fig. 4.6 Radiation scattering by a plasmon

Equation (4.71) represents the dispersion law for a plasmon. It corresponds to the propagation of plasmons at the plasma frequency ω_{pe}. For sufficiently large wave vectors $|\boldsymbol{k}| > d_e^{-1}$, a plasmon is a not well-defined excitation since it disappears during times of the order of the plasmon oscillation period.

In case of fulfillment of the plasmon resonance condition (4.70), a singularity appears in the expression for the electron DFF (4.63) that corresponds to emission/absorption of a plasmon. This singularity is caused by the presence of the function $\varepsilon^{(l)}(\Delta\boldsymbol{k}, \Delta\omega)$ in the denominator of the formula (4.63). Due to the presence of the imaginary part in the dielectric permittivity, it approaches a delta function of the form $\delta(\Delta\omega \pm \omega_{pl}(\Delta\boldsymbol{k}))$ describing the energy conservation law (the plus sign corresponds to emission of a plasmon, the minus sign corresponds to absorption). Omitting mathematical details, we will write down the expression for the electron DFF with the transfer of an energy–momentum to an emitted plasmon:

$$S_{pl}(\Delta\boldsymbol{k}, \Delta\omega) = \left(N_{pl}(\Delta\boldsymbol{k}) + 1\right)\frac{\hbar\omega_{pl}(\Delta\boldsymbol{k})\Delta\boldsymbol{k}^2}{8\pi e^2}\delta\left(\Delta\omega + \omega_{pl}(\Delta\boldsymbol{k})\right)\Theta(1 - |\Delta\boldsymbol{k}|d_e), \tag{4.72}$$

where $\Theta(x)$ is the Heaviside step theta function and

$$N_{pl}(\Delta\boldsymbol{k}) = \frac{1}{\exp\left(\omega_{pl}(\Delta\boldsymbol{k})/T\right) - 1} \tag{4.73}$$

is the number of plasmons at a specified temperature in thermodynamic equilibrium. Substituting (4.72) into the formula (4.59), after integration with respect to the scattered photon frequency, we obtain the cross section of radiation scattering with generation of a plasmon:

$$\frac{d\sigma_{\mathrm{pl}}^{(\mathrm{plas})}}{d\Omega'} = \frac{1+\cos^2\theta}{2} r_{\mathrm{e}}^2 \Theta(1-|\Delta k|d_{\mathrm{e}})\left(N_{\mathrm{pl}}(\Delta k)+1\right) \frac{\hbar\omega_{\mathrm{pl}}(\Delta k)\Delta k^2}{8\pi e^2} \mathrm{d}V. \quad (4.74)$$

The frequency of the scattered radiation is $\omega' = \omega - \omega_{\mathrm{pl}}(\Delta \boldsymbol{k})$.

For the scattering cross section with plasmon absorption, the expression is similar to (4.74) and the replacement $N_{\mathrm{pl}}(\Delta \boldsymbol{k})+1 \rightarrow N_{\mathrm{pl}}(\Delta \boldsymbol{k})$ and $\omega' = \omega + \omega_{\mathrm{pl}}(\Delta \boldsymbol{k})$ is quite accurate. From the calculations it follows that by recording the dependence of the scattered photon frequency on the angle of scattering it is possible to determine experimentally the dispersion law for plasmons.

Similar expressions are valid for cross sections of radiation scattering at collective plasma excitations of other types, such as ion-sound waves.

By recording the radiation spectrum (scattered by the plasma), it is possible to study various quasi-particles in plasma. The width of the radiation spectrum scattered by some collective excitation defines the damping constant and accordingly the lifetime of the quasi-particle. Scattering by collective excitations in plasmas accompanied by a change in radiation frequency is an analog of Raman scattering of radiation by atoms and molecules.

Scattering in the X-ray spectral range became an important diagnostic tool for high-energy density science (Glenzer and Redmer 2009; Sheffield et al. 2010) and attracted in particular attention to study the so-called Warm Dense Matter regime (Lee et al. 2003) produced by various means (Kozyreva et al. 2003; Soho et al. 2008; Tauschwitz et al. 2007; Sheffield et al. 2010). The small scattering cross section, however, makes it rather difficult to obtain good signal-to-noise ratios for spectrally resolved analysis unless X-ray scattering sources are not driven by multikilojoule lasers (Glenzer and Redmer 2009; Gamboa et al. 2012).

The advent of the X-ray Free Electron Lasers installations, however, has entirely changed the experimental landscape: dense plasmas are produced with powerful optical lasers (that are built near the experimental XFEL facilities itself) and are then diagnosed with the XFEL (Fletcher et al. 2013; MacDonald et al. 2016). The first experiment where a dense plasma was created with an optical laser and then probed by XFEL has been carried out in 2011 at LCLS (Seely et al. 2011): here, X-ray pumping of inner-shell transitions in dense aluminum plasma has been demonstrated for various charge states (Rosmej et al. 2016).

It should be noted, however, that the high intensity of the XFEL beam perturbs considerably the material to be studied itself, in particular, via strong heating that proceeds from inner-shell ionization followed by subsequent equilibration of photoelectrons (kinetic energy being equal to the access energy of the ionization potential) and Auger electrons in the valence band (so-called Auger electron heating) (Galtier et al. 2011; Rosmej et al. 2012). Therefore, high-resolution X-ray spectroscopy remains a primary diagnostic tool (Renner and Rosmej 2019) although many important parameters can be accessed with scattering diagnostics (e.g., temperature, density, average charge number, collective plasma effects). The access to atomic structure in dense plasmas via X-ray scattering diagnostics is only indirect (via the dynamic form factor) and also therefore high-resolution X-ray spectroscopy appears to be an important complementary diagnostic.

4.5 Scattering on Nanoparticles

The color of the strongest colorants (organic dyes) is defined, as a rule, by absorption of radiation by dye molecules. In interaction of radiation with metal nanoparticles, we encounter a different situation. For small nanosphere radii $r_s < 15 \div 35$ nm absorption dominates, for large radii radiation scattering begins to play a decisive role. The combined effect of absorption and scattering on the passage of a light beam through a substance is described by the extinction cross section $\sigma_{ext} = \sigma_{scat} + \sigma_{abs}$. The quantity $\sigma_{ext} I_0$ represents the power removed from a light beam of intensity I_0 due to absorption and scattering. The ratio of the scattering cross section to the extinction cross section is called *quantum efficiency* $\eta = \sigma_{scat}/\sigma_{ext}$. The quantum efficiency characterizes the relative value of power removed from a light beam that goes into radiation scattering, that is, can be recorded by a photodetector. From the aforesaid, it follows that the quantum efficiency increases with nanoparticle size, but, as it will be seen from further consideration, another important characteristic come into play, namely the quality factor of resonant scattering that decreases with increasing radius.

An important specific feature of photoprocesses with metal nanoparticles is the absence of luminescence under the action of radiation (characteristic of targets with strong optical transitions in the discrete spectrum). As a result, radiation of metal nanoparticles is completely driven by the scattering of the electromagnetic waves. This results, in particular, in a narrow radiation spectrum, when a metal nanosphere is excited by a laser in narrow spectral range.

Spherical nanoparticles of noble metals (gold and silver) find wide application in various fields of research and technologies. In particular, gold nanospheres are used as active nanomarkers in medicine and biology. Besides, the use of metal nanoparticles in various sensors shows promising results.

From the fundamental point of view, the advantage of the spherical form of a nanoparticle consists first of all in the fact that there is no dependence of its optical properties (absorption and scattering cross sections) on the form, i.e. the optical properties depend essentially only on size. This makes it possible to control the characteristics of the electromagnetic response of a nanosphere by changing its radius. Besides, cross sections of radiation scattering and absorption by metal spheres can be described within the framework of a relatively simple analytical approach.

4.5.1 Mie Theory of Radiation Scattering and Absorption

The Mie theory of radiation scattering by a metal sphere in a homogeneous medium is based on the expansion of an electromagnetic field in terms of cylindrical harmonics (due to the cylindrical symmetry of the problem) and "joining" of tangential components of the strength of electric and magnetic fields at the boundary of the sphere (Jackson 2007). To satisfy these boundary conditions, it is necessary to take into account the field irradiating the sphere. A field inside the sphere and a secondary wave field representing the scattered wave. The derivation of corresponding formulas is rather cumbersome, but they are available in classical monographs on optics (Van de Hulst 1981; Born and Wolf 1999), and we will give here only the final result. The cross section of radiation scattering by a metal sphere according to the Mie theory is

$$\sigma_{\mathrm{scat}}^{(\mathrm{Mie})} = \frac{2\pi c^2}{\varepsilon_{\mathrm{m}}\omega^2}\sum_{n=1}^{\infty}(2n+1)\left\{|a_{\mathrm{n}}(x,mx,m)|^2 + |b_{\mathrm{n}}(x,mx,m)|^2\right\}, \tag{4.75}$$

where a_{n}, b_{n} are the Mie coefficients that are given by the formulas (2.78)–(2.81), the parameters x and m are given by:

$$x = kr_{\mathrm{s}} = \sqrt{\varepsilon_{\mathrm{m}}}\frac{\omega}{c}r_{\mathrm{s}}. \tag{4.76}$$

The variable x from (4.76) is the product of the wave vector of radiation in the matrix with the nanosphere radius and

$$m = \sqrt{\varepsilon_{\mathrm{s}}(\omega)/\varepsilon_{\mathrm{m}}} \tag{4.77}$$

is the ratio of the refractive indices of the nanosphere and the matrix material.

Equation (4.75), in contrast to (4.74), takes into account not only the dipole summand ($n = 1$), but also the terms corresponding to the contribution of higher order multipoles ($n > 1$). This is essential for short wavelengths, when the condition of the dipole approximation ($\lambda \gg r_{\mathrm{s}}$) is not fulfilled.

The results of calculation of the cross section of radiation scattering by silver spheres of different sizes within the framework of the Mie theory are presented in Fig. 4.7 in the spectral range close to the plasmon resonance.

From Fig. 4.7, it follows that with increasing size of the metal nanosphere the position of the scattering cross section maximum is shifted to the long-wavelength region, and the width of the spectral maximum increases. For radii of 30 and 40 nm, a second peak appears due to a quadrupole resonance in excitation of a surface plasmon. The quadrupole term corresponds to $n = 2$ in the sum (4.75).

A decrease in frequency and an increase in the spectral width of the plasmon resonance with increasing nanosphere size result in a decrease of the quality factor Q of plasmon excitation that by definition is $Q = \omega_{\mathrm{res}}/\Delta\omega$. For a gold nanosphere, the value of the quality factor decreases from 8 to 2.5 as the radius changes from 30 to 80 nm. In this case, the resonance energy decreases from 2.2 to 1.7 eV. In case of a similar change in the radius of a silver nanosphere, the quality factor of a surface plasmon decreases from 11 to 4, and the resonance energy decreases from 2.7 to 2.5 eV.

The quality factor of a plasmon resonance is very important since it is equal to the coefficient of amplification of a local field, i.e., it defines the value of the

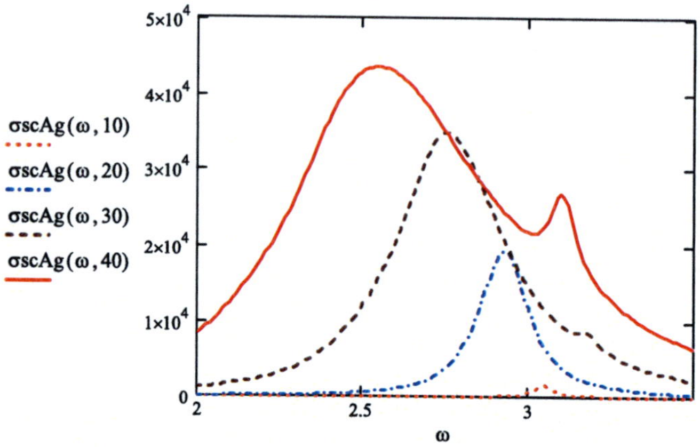

Fig. 4.7 Cross sections of radiation scattering by silver nanospheres of different radii (10, 20, 30, 40 nm) in a medium with $\varepsilon_m = 2.25$ and in the vicinity of a plasmon resonance; the abscissa is the photon energy in eV, the ordinate is the cross section in nm^2

strength of an electric field connected with a surface plasmon (this field is acting on particles in the vicinity of the surface). As a result of this amplification, it is possible to observe a whole class of nonlinear optical phenomena, such as Raman scattering that are otherwise practically not observable (or would require otherwise super-strong electromagnetic fields for their recording).

For the cross section of radiation extinction by a metal sphere, the Mie theory gives (Van de Hulst 1981)

$$\sigma_{\mathrm{ext}}^{(\mathrm{Mie})} = \frac{2\pi c^2}{\varepsilon_m \omega^2} \sum_{n=1}^{\infty} (2n+1) Re\{a_n(x, mx, m) + b_n(x, mx, m)\}. \tag{4.78}$$

The expressions for the expansion coefficients a_n, b_n are given in Chap. 2 [formulas (2.78)–(2.81)], the parameters x and m are given by (4.76)–(4.77). The extinction describes the radiation intensity as a result of absorption and scattering by the substance particles.

As was already mentioned, the cross section of radiation absorption is equal to the difference of the cross sections of extinction and scattering $\sigma_{\mathrm{abs}} = \sigma_{\mathrm{ext}} - \sigma_{\mathrm{scat}}$. With the use of this relation and the expressions (4.75), (4.78), it is possible to calculate the cross section of photoabsorption by a metal sphere. The calculations for silver spheres in a medium with $\varepsilon_m = 2.25$ (glass) of different radii are presented in Fig. 4.8. It is seen that in contrast to the scattering cross section the absorption cross section has a considerably more strong quadrupole resonance that even dominates over the dipole resonance even at a sphere radius of 30 nm.

The numerical analysis within the framework of the Mie theory shows that in the case of silver nanospheres the cross sections maxima of absorption and scattering of radiation at a plasmon become equal for a radius of 15 nm; for larger radii, scattering dominates over absorption. In the case of gold nanospheres, scattering

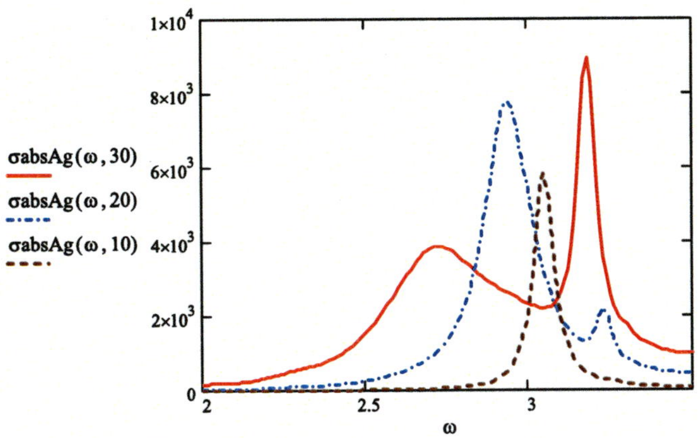

Fig. 4.8 The cross section of radiation absorption by silver spheres of different radii (10, 20, 30 nm) calculated within the framework of the Mie theory in the region of a plasmon resonance. The abscissa is the photon energy in eV, the ordinate is the cross section in nm^2

becomes equal to absorption for a radius of 37 nm and the maximum of the spectral cross section of scattering is shifted (with respect to the absorption maximum) to the region of lower photon energies by about 0.1 eV. The ratio of scattering and absorption cross sections is related to the quantum efficiency of scattered radiation that is an important parameter for the practical application of nanoparticles.

It should be emphasized that the strong dependence of spectra of radiation scattering and absorption in the vicinity of a plasmon resonance on the sphere radius results in inhomogeneous broadening for an ensemble of nanoparticles with an appreciable spread of radii. This should be taken into account in the analysis of corresponding experimental data.

Naturally, the Mie theory has a certain range of applicability. This is connected first of all with the phenomenological description of a nanoparticle material by the use of the dielectric permittivity. Such an approach is valid for sufficiently large nanoparticles and radiation wavelengths, when the substance can be considered as a continuum. In the case of small-sized nanoparticles, local effects become important that are not taken into account when using the bulk dielectric permittivity of a metal. Finally, the Mie theory assumes the presence of a sharp boundary of a metal sphere, which is also an approximation.

References

V.B. Berestetskii, L.P. Pitaevskii, E.M. Lifshitz, *Quantum Electrodynamics* (Elsevier, Oxford, 1982)

M. Born, E. Wolf, *Principles of Optics*, 7th ed. (Cambridge University Press, 1999)

L.B. Fletcher, E. Galtier, P. Heimann, H.J. Lee, B. Nagler, J. Welch, U. Zastrau, J.B. Hastings, S. H. Glenzer, Plasmon measurements with a seeded x-ray laser. J. Instrum. **8**, C11014 (2013)

E. Galtier, F.B. Rosmej, D. Riley, T. Dzelzainis, F.Y. Khattak, P. Heimann, R.W. Lee, S.M. Vinko, T. Whitcher, B. Nagler, A. Nelson, J.S. Wark, T. Tschentscher, S. Toleikis, R. Fäustlin, R. Sobierajski, M. Jurek, L. Juha, J. Chalupsky, V. Hajkova, M. Kozlova, J. Krzywinski, Decay of crystaline order and equilibration during solid-to-plasma transition induced by 20-fs microfocused 92 eV free electron laser pulses. Phys. Rev. Lett. **106**, 164801 (2011)

E.J. Gamboa, C.M. Huntington, M.R. Trantham, P.A. Keiter, R.P. Drake, D.S. Montgomery, J.F. Benage, S.A. Letzring, Imaging x-ray Thomson scattering spectrometer design and demonstration. Rev. Sci. Instrum. **83**, 10E108 (2012)

V.L. Ginzburg, V.N. Tsytovich, *Transition Radiation and Transition Scattering* (CRC Press, 1990)

S.H. Glenzer, R. Redmer, X-ray Thomson scattering in high energy density plasmas. Rev. Mod. Phys. **81**, 1625 (2009)

J. Jackson, *Classical Electrodynamics*, 3rd ed. (Wiley, New York, 2007)

A. Kozyreva, M. Basko, F.B. Rosmej, T. Schlegel, A. Tauschwitz, D.H.H. Hoffmann, Dynamic confinement of targets heated quasi-isochorically with heavy ion beam. Phys. Rev. E **68**, 056406 (2003)

H.J. Kunze, in *The laser as a tool for plasma diagnostics*, Plasma Diagnostics ed. by W. Lochte-Holtgreven) (North-Holland Publishing Company, Amsterdam, 1968)

L.D. Landau, E.M. Lifschitz, *The Classical Theory of Fields*, 4th edn. (Pergamon, New York, 1975)

R.W. Lee, S.J. Moon, H.-K. Chung, W. Rozmus, H.A. Baldis, G. Gregori, R.C. Cuable, O.L. Landen, J. Wark, A. Ng, S.J. Rose, C.L. Lewis, D. Riley, J.-C. Gauthier, P. Audebert, Finite temperature dense matter studies on next-generation light sources. J. Opt. Soc. Am B **20**, 770 (2003)

M.J. MacDonald, T. Gorkhover, B. Bachmann, M. Bucher, S. Carron, R.N. Coffee, R.P. Drake, K. R. Ferguson, L.B. Fletcher, E.J. Gamboa, S.H. Glenzer, S. Göde, S.P. Hau-Riege, D. Kraus, J. Krzywinski, A.L. Levitan, K.-H. Meiwes-Broer, C.P. O'Grady, T. Osipov, T. Pardini, C. Peltz, S. Skruszewicz, M. Swiggers, C. Bostedt, T. Fennel, T. Döppner, Measurement of high-dynamic range x-ray Thomson scattering spectra for the characterization of nano-plasmas at LCLS. Rev. Sci. Instrum. **87**, 11E709 (2016)

P.M. Platzman, P.A. Wolf, *Waves and Interactions in Solid State Plasmas* (Academic Press, 1973)

O. Renner, F.B. Rosmej, Challenges of X-ray spectroscopy in investigations of matter under extreme conditions. Matter Radiat. Extrem. Rev. **4**, 024201 (2019)

F.B. Rosmej, Exotic states of high density matter driven by intense XUV/X-ray Free Electron Lasers, in *Free Electron Laser*, ed. by S. Varró (Tech 2012), pp. 187–212, ISBN 978-953-51-0279-3. Free download: http://www.intechopen.com/books/free-electron-lasers/exotic-states-of-high-density-matter-driven-by-intense-xuv-x-ray-free-electron-lasers

F.B. Rosmej, A. Moinard, O. Renner, E. Galtier, H.J. Lee, B. Nagler, P.A. Heimann, W. Schlotter, J.J. Turner, R.W. Lee, M. Makita, D. Riley, J. Seely, XFEL resonant photo-pumping of dense plasmas and dynamic evolution of autoionizing core hole states, in Proceedings of the 12th International Conference on Fusion Science and Applications—IFSA-2013, Nara, Japan. J. Phys.: Conf. Ser. **688**, 012093 (2016)

S. Sahoo, G.F. Gribakin, G. Shabbir Naz, J. Kohanoff, D. Riley, Compton scatter profiles for warm dense matter, Phys. Rev. E **77**, 046402 (2008)

J. Seely, F.B. Rosmej, R. Shepherd, D. Riley, R.W. Lee, Proposal to Perform the 1st High Energy Density Plasma Spectroscopic Pump/Probe Experiment, approved LCLS proposal n° L332 (carried out in 2011)

J. Sheffield, *Plasma Scattering of Electromagnetic Radiation*, 1st edn. (Academic Press, 1975)

J. Sheffield, D. Froula, S.H. Glenzer, N.C. Luhmann, *Plasma Scattering of Electromagnetic Radiation*, 2nd edn. (Academic Press, 2010)

A. Tauschwitz, J.A. Maruhn, D. Riley, G. Shabbir Naz, F.B. Rosmej, S. Borneis, A. Tauschwitz, Quasi-isochoric ion beam heating using dynamic confinement in spherical geometry for X-ray scattering experiments in WDM regime. High Energy Density Phys. **3**, 371 (2007)

H.C. Van de Hulst, *Light Scattering by Small Particles* (Dover Publications, New York, 1981)

Chapter 5
Electron–Atom Collisions

Abstract The theory of electron–atom collisions including excitation, ionization, and recombination is presented in the framework of Fermi's equivalent photon method, the similarity function approach, and semi-empirical analytical formulas. Collisional excitation is described via a quasi-classical consideration. Dipole-allowed, dipole-forbidden, and intercombination electron transitions are considered including intermediate coupling effects. Comparisons between different theoretical approaches and experimental data for excitation cross-sections are provided for various atoms and type of electronic transitions. Semi-empirical analytical formulas for excitation, de-exciation, ionization, and three-body recombination are given. Complex dielectronic recombination rates in dense plasmas are presented with account for density and electric field effects. Extensive numerical data for dielectronic recombination into H-, He-, and Li-like ions taking into account multi-channel Auger and radiative decay are given for all elements with nuclear charge $Z_n = 2–42$ together with easy to use scaled semi-empirical formulas. The theory of excited states coupling and collisional redistribution for dielectronic recombination is developed.

5.1 Fermi Equivalent Photon Method

The radiative–collisional processes appear to be a wide domain for the application of Kramers electrodynamics (KrED). These are processes in which the electron participates while moving along a classical highly curved quasi-parabolic orbit (Kogan et al. 1992). The most natural domain for the application of the KrED is the physics of multicharged ions (MCI).

According to the Fermi concept (Fermi 1924) of equivalent photons (EPh), the electromagnetic field produced by an external particle (e.g., an electron) in the vicinity of a MCI location may be interpreted as a flux of equivalent photons incident on the MCI. It can be shown that this description is applicable provided the dipole approximation for describing the interaction between the bound electron of the MCI and the incident electron of the plasma holds true. In this case, the dipole approximation describes in an universal manner all the processes of energy loss

F. B. Rosmej et al., *Plasma Atomic Physics*, Springer Series on Atomic, Optical, and Plasma Physics 104, https://doi.org/10.1007/978-3-030-05968-2_5

induced by the incident electron (either due to radiation emission during a collision with an ion or due to an inelastic non-radiative collision with an ion) as the processes of the emission of real or equivalent photons. The probability of both processes is determined by the dipole matrix element for the corresponding inelastic (radiative or non-radiative) transition of the incident electron.

The spectral intensity distribution of the EPh may be described on the basis of the classical radiation theory. In this case, the intensity of the EPh flux is simply determined by the Fourier transforms of the electron coordinates determined in turn by its classical trajectory. This approach makes it possible to treat several important radiative–collisional processes:

(a) excitation of an ion by electron impact treated as an absorption of the EPh by this ion,
(b) the same excitation as in a) but with subsequent re-emission of a real photon as a resonance fluorescence of the EPh,
(c) dielectronic recombination as a resonance fluorescence of the EPh, which results in a recombination of the incident electron.

An essential advantage of the EPh method is related to applications of purely radiative processes to the description of non-radiative processes (both, collisional and radiative–collisional). Several processes are of resonant character with respect to the absorption of the EPh by the ion and of non-resonant character, for which the intermediate state of a two-step "absorption–re-emission" process is not real and consequently is not obeying the energy conservation law (this state is formed by the process of virtual excitation with the energy $E \neq \hbar\omega_0$, where ω_0 is the frequency of a resonant transition). These non-resonant processes are known as polarization radiation (Thytovich and Oiringel 1991) and can be treated as the non-resonant scattering of the EPh by the ion. The polarization radiation is determined by the dynamical polarizability of ion in the domain of non-resonant frequencies.

For the application of the Fermi method, it is necessary that effective distances r_{eff} which are responsible for the main contribution to the inelastic collision cross-section are much greater than the characteristic size of the bound electron orbit. This requirement is especially well fulfilled for MCI. Let us illustrate this for the excitation of a $\Delta n = 0$ transition. The electron orbit size is of the order $1/Z$ (in atomic units), transition energies ΔE for $\Delta n = 0$ transitions in MCI are typically of the order of Z, and the values of r_{eff} for the corresponding cross-section can be estimated as

$$r_{\text{eff}} \sim r_{\omega=\Delta E/\hbar} \sim (Z/\Delta E^2)^{1/3} \sim Z^{-1/3} \gg Z^{-1}. \tag{5.1}$$

This inequality justifies the use of the dipole approximation for the potential V of interaction between bound and incident electrons (with space coordinate vectors $\mathbf{r}_i$ and $\mathbf{r}_e$, respectively), $V = e^2\mathbf{r}_i \cdot \mathbf{r}_e/\mathbf{r}_e^3$. In this framework, the static Coulomb interaction between the bound and incident electrons transforms to the processes of emission and absorption of the EPh by electrons, and the corresponding probabilities are determined by the conventional dipole matrix elements.

The electric field produced by the incident electron at the location $r = 0$ of the ion is equal to

$$\mathbf{F}(0,t) = -e\mathbf{r}_e(t)/\mathbf{r}_e^3(t), \tag{5.2}$$

where the dependence $\mathbf{r}_e(t)$ describes the classical trajectory of the incident electron. Using the equation of the motion of the incident electron in the field of the MCI $m\ddot{\mathbf{r}}_e = -Ze^2\mathbf{r}_e/\mathbf{r}_e^3$, it is convenient to transform (5.2) into the form

$$F(t) = \ddot{\mathbf{r}}_e(t)m/Ze. \tag{5.3}$$

The spectral distribution for the EPh flux I_ω of the electric field of the incident electron can be expressed in terms of the Fourier transforms:

$$I_\omega = \frac{c}{4\pi^2}\frac{1}{\omega}\left\{|F_{x,\omega}|^2 + |F_{y,\omega}|^2\right\} = \frac{c\omega^3}{4\pi^2 Z^2}\left\{|x_\omega|^2 + |y_\omega|^2\right\}, \tag{5.4}$$

where x and y are the coordinates of the incident electron in the plane of its motion. The Fourier transforms of the electron space coordinates in the Coulomb field are well known (Berestetskii et al. 1982; Landau and Lifschitz 2003; Jackson 1998). Thus, we obtain

$$I_\omega = \frac{c\omega^2}{4\upsilon^4}\left\{\left[H_{i\nu}^{(1)\prime}(i\nu\varepsilon)\right]^2 - \frac{\varepsilon^2 - 1}{\varepsilon^2}\left[H_{i\nu}^{(1)}(i\nu\varepsilon)\right]^2\right\}, \tag{5.5}$$

where υ is the electron initial velocity, $H_{i\nu}^{(1)}$ is the Hankel function, ε is the orbital eccentricity;

$$\varepsilon = 1 + 2EM^2/Z^2; \quad \nu = \omega Z/\upsilon^3; \; E = m\upsilon^2/2; \tag{5.6}$$

E and M are the energy and angular momentum of the incident electron, respectively. In the limit of low EPh frequencies, $\nu \ll 1$, the main contribution to the spectral distribution of the EPh flux integrated over the electron impact parameters ρ is due to the distance from the field center trajectories, ($\rho \gg a \equiv Z/2E$), which are nearly rectilinear, with eccentricity $\varepsilon \gg 1$. In this case, (5.5) is transformed to

$$I_\omega = \left(c\omega/2\pi^2\upsilon^4\right)\left\{K_0^2(\omega\rho/\upsilon) + K_1^2(\omega\rho/\upsilon)\right\}, \tag{5.7}$$

where $K_0(x)$ and $K_1(x)$ are the Macdonald functions. Fermi (1924) used (5.7) to describe atomic excitation by a rectilinearly moving particle.

For the description of the processes resulting in a loss of a considerable part of the incident electron energy, it is necessary to consider the EPh with high frequencies, namely $\nu \gg 1$. The main contribution to the emission of such EPh comes from the strongly curved electron trajectories, $\varepsilon - 1 \ll 1$, which are close to the field center, $\rho \ll a$. In this Kramers domain, we arrive at the result (see Chap. 3)

$$I_{\omega} = \pi^{-2} Z^{-2} c M G_0(\omega M^3/3Z^2) \tag{5.8a}$$

with [see also (3.47)]

$$G_0(x) = x\left[K_{1/3}^2(x) + K_{2/3}^2(x)\right]. \tag{5.8b}$$

5.1.1 *Excitation by Electron Impact as Absorption of Equivalent Photons by an Ion*

The equivalent photons method makes it possible to obtain a simple analytical description of the collisional processes and treat them as purely radiative. Within this framework, the excitation of multicharged ions (MCI) by electron impact may be clearly considered as absorption of the equivalent photons (EPh) with a resonant frequency $\omega_0 = \Delta E_{\mathrm{if}}/\hbar$. The relationship between the collisional cross-section σ_{exc} and the cross-section σ_{abs} for the absorption of the EPh can be obtained by means of equating the number of excitation events, during the time interval $\mathrm{d}t$ (caused by the collisions of the MCI with the electron flux) with a space density n_{e} and a particle velocity υ_{e} to the corresponding number of transitions caused by the absorption of the EPh produced by a single electron, i.e., $\mathrm{d}N_{\mathrm{exc}} = n_{\mathrm{e}}\upsilon_{\mathrm{e}}\sigma_{\mathrm{exc}}\mathrm{d}t$. This is multiplied by the total number of electrons in the volume $\mathrm{d}V$ corresponding to the time interval $\mathrm{d}t$, $\mathrm{d}V = 2\pi\rho\mathrm{d}\rho\upsilon_{\mathrm{e}}\mathrm{d}t$,

$$\mathrm{d}N_{\mathrm{abs}} = \int 2\pi\rho\mathrm{d}\rho n_{\mathrm{e}}\upsilon_{\mathrm{e}}\mathrm{d}t \int \mathrm{d}\omega\left(cE_{\omega}^2/4\pi^2\hbar\omega\right)\sigma_{\mathrm{abs}}(\omega), \tag{5.9}$$

where the expression in brackets corresponds to the spectral distribution of the EPh flux (5.4) produced by a single electron with a fixed value of the impact parameter ρ. Assuming $\sigma_{\mathrm{exc}} = \int \sigma_{\mathrm{exc}}^l \mathrm{d}l$ for the relation of the total and partial (with respect to the orbital quantum number l) cross-sections, we arrive at the result

$$\sigma_{\mathrm{exc}}^l = 2\pi(\hbar/m\upsilon_{\mathrm{e}})^2(l+1/2)\int \sigma_{\mathrm{abs}}(\omega)\left(cE_{\omega}^2/4\pi^2\hbar\omega\right)\mathrm{d}\omega. \tag{5.10}$$

Furthermore, the expression for the EPh flux can be taken out of the integral at the frequency ω_0 of the radiative transition in the MCI core because of its weak frequency dependence in comparison with the absorption cross-section. The resulting integral over ω gives the well-known expression

$$\int \sigma_{abs}(\omega)d\omega = \pi^2(c/\omega)^2 g_f 4\omega_0^2 |d_{if}|^2/3\hbar c^3, \quad (5.11)$$

where d_{if} is the dipole moment matrix element of the transition and g_f is the statistical weight of the upper level.

Substituting the spectral distribution (5.5) into (5.10) for the EPh flux, produced by the electron in the Coulomb field of the MCI, we finally obtain

$$\sigma^l_{exc} = \frac{8\pi^3}{3}(\hbar/mv_e)^2\omega_0^2|d_{if}|^2 g_f v_e^{-4}(1+1/2) \cdot \left\{\left[H^{(1)'}_{i\nu}(i\nu\varepsilon)\right]^2 - (\varepsilon^2-1)\varepsilon^{-2}\left[H^{(1)}_{i\nu}(i\nu\varepsilon)\right]^2\right\}. \quad (5.12)$$

The transition in (5.12) to the Kramers electrodynamics (KrED) domain ($\nu \gg 1$) corresponds to the transition from (5.5) to (5.8). Thus, we obtain the result in the KrED domain:

$$\sigma^1_{exc} = (8\pi/3)(\hbar/mv_e)^2(g_f/g_i)f_{if}Z^{-2}(l+1/2)^2 G_0[\omega(l+1/2)^3/3Z^2], \quad (5.13)$$

where f_{if} is the oscillator strength for the transition considered and g_i is the statistical weight of the lower level.

Equation (5.13) manifests explicitly the interrelation between the independence of the radiation characteristics on the energy (see also Chap. 3) and the well-known fact of the finiteness of the excitation cross-section at threshold. Thus, we face once more the phenomenon, inherent to the KrED, of the independence of the spectral distribution on the energy, which leads to a smooth transition between the discrete and continuous energy spectra for the processes with both real (from the BR to the PR) and equivalent photons (the transition from the Born approximation domain for the excitation to its threshold and further to the DR).

The total excitation cross-section is obtained by summing up the partial cross-section (5.10) over l, yielding an expression in terms of the well-known spectral distribution for the Coulomb bremsstrahlung Gaunt-factor $g(\nu)$ (Kogan et al. 1992)

$$\sigma^{if}_{exc} = \frac{8\pi^2}{\sqrt{3}}|d_{if}|^2 g_i^{-1} v_e^{-2} g[Z\omega_0/(2E)^{3/2}]. \quad (5.14)$$

The function $g(\nu)$ has a simple analytic approximation:

$$g(\nu) \approx \frac{\sqrt{6}}{\pi}\cdot \ln\left[\left(\frac{2}{\gamma\cdot\nu}\right)^{1/\sqrt{2}} + \exp\left(\frac{\pi}{\sqrt{6}}\right)\right]. \quad (5.15)$$

The result (5.14) was derived earlier (Bazylev and Chibisov 1981) in a somewhat different way. It should be noted that (5.14) is valid up to the excitation threshold where Kramers EPh spectrum (5.8) does not depend on the incident electron energy at all. In the opposite limit of a fast-incident particle, the cross-section (5.14) exhibits a logarithmic (Born-type) structure. It is this result that was derived by Fermi for atomic excitation and ionization by fast particles. Equation (5.14) is in good agreement with quantum numerical calculations as well as with experimental data (Bazylev and Chibisov 1981; Gau and Henry 1977).

It should be also noted that for the first time, the interrelation between the excitation cross-section for allowed dipole transitions and the Gaunt factor for bremsstrahlung in a Coulomb field for the general quantum case (Sommerfeld formula) was investigated by Gailitis (1963). So far we have restricted ourselves to the case of quasi-classical incident electron motion; however, the applicability of the KrED approach to the calculation of excitation cross-sections remains valid also for an arbitrary (not necessarily purely Coulombic) ionic potential (e.g., of a Thomas–Fermi ion). The description of these excitation cross-sections may be achieved by the replacement of the Coulomb EPh intensity by the corresponding EPh spectral intensity in (5.5).

5.1.2 Autoionization Decay and Dielectronic Capture

Let us recall the essence of the dielectronic recombination (DR) process. An incident electron with the energy $E_{\rm i}$ excites an ion core with an excitation energy $\Delta E = \hbar\omega_0$. In this case, if the energy $E_{\rm i}$ is smaller than ΔE, the electron is finally captured by the ion into a state with the energy $E_{\rm f} = -{\rm Ry}/n_{\rm f}^2$ obeying the condition

$$E_{\rm i} - E_{\rm f} = E_{\rm i} + 1/2n_{\rm f}^2 = \Delta E = \hbar\omega_0. \qquad (5.16)$$

This capture results in a double excited state of the ion, namely the ion core electron is excited with energy ΔE while the captured electron occupies a highly excited level of the ion. This state of the ion can decay in two possible ways:

(i) by relaxation of the ion core electron into the initial ground state with the simultaneous ejection of the captured electron from the ion: This process is known as autoionization;
(ii) by radiative decay of the ion core electron, resulting in its return to the initial state after the emission of a photon of energy $\hbar\omega \simeq \hbar\omega_0 = \Delta E$, whereas the captured electron remains in the ion.

For illustration, Fig. 5.1 shows the relevant energy level diagram for the He-like $2l2l'$-satellites close to the H-like Lyman-alpha transition (so-called Ly$_\alpha$-satellites). The energy of the $2l2l'$-satellites is approximately (in the H-like approximation) $E_{\rm sat} \approx 2 \cdot Z^2{\rm Ry}/4 = Z^2{\rm Ry}/2$ which is half of the ionization potential of the H-like ground state (Z^2Ry). The series limit of the autoionizing levels $2lnl'$ is the first excited state $2l$. Radiative decay from the $2l2l'$-levels populates the single excited levels $1s2l\ ^{1,3}$L, from which radiative decays (W- and Y-line) populate finally the

Fig. 5.1 Energy level diagram relevant for the He-like autoionizing levels $2l2l'$, so-called Ly_α-satellites

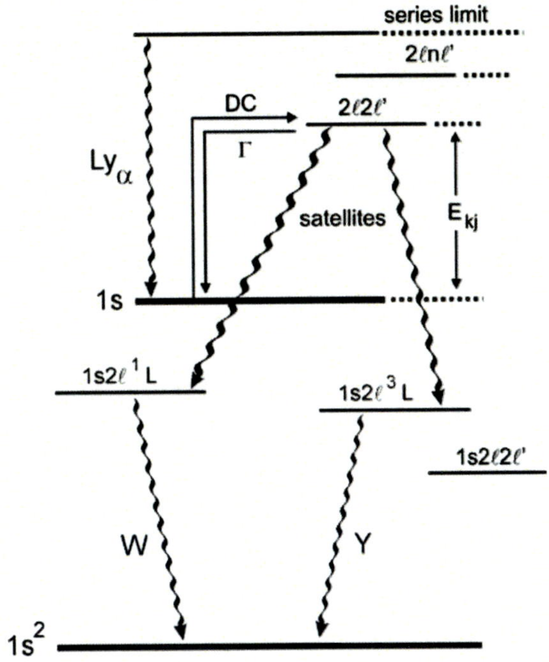

ground state $1s^2$. The chain of processes, dielectronic capture ($1s + e - 2l2l'$), radiative decay to single excited levels ($2l2l' - 1s2l + h\nu$) and radiative decay to the ground state ($1s2l\ ^{1,3}\mathrm{L} - 1s^2 + h\nu'$), is called dielectronic recombination (DR channel) because an effective recombination has taken place (from the H-like ground state $1s$ to the He-like ground state $1s^2$).

Thus, the DR process as well as the photorecombination (PhR) process result in the capture of an incident electron and its simultaneous photon emission. The difference is that the photon is emitted by the ion core electron in the DR process rather than by the incident electron as in the PhR process. The relationship between the PhR and the DR is analogous to the interrelation between conventional and polarization bremsstrahlung (Astapenko 2013).

As a rule, the DR rate is large for ions with a complex core which possesses transitions between the levels with the same quantum number n (the transitions with $\Delta n = 0$, e.g., $2s \to 2p$ transitions in lithium-like and more complex ions). The transition energy $\Delta E = \hbar\omega_0$ for $\Delta n = 0$ and $Z \gg 1$ is of the order of $Z \cdot \mathrm{Ry}$, while the ionization energy is of the order of $Z^2 \cdot \mathrm{Ry} \gg \Delta E$. Since the energy E of the incident recombining electron is in any case smaller than the excitation energy, this implies the following inequality

$$(Z^2\mathrm{Ry}/E)^{1/2} \sim Ze^2/\hbar\upsilon \gg 1, \tag{5.17}$$

which justifies the application of a quasi-classical approach to the description of dielectronic recombination process.

An application of the proposed approach to the DR implies the treatment of a DR process as a resonance fluorescence with a complicated intermediate state which appears after the capture of the incident electron by an ion and possesses an additional channel of decay via the autoionization. The resonance fluorescence thus involves three types of quantum states:

- an initial state with total energy E_1 of a non-excited ion and an initial spectral distribution I_0 of equivalent photons;
- an intermediate state with total energy E_2 of an excited ion with a captured incident electron on a highly excited ionic level (double excited ion with an ion charge reduced by unity) and an EPh distribution I_0 reduced by one EPh of energy ω_{eq};
- a final state with total energy E_3 of a single excited ion with a charge reduced by unity, an EPh of energy ω and a EPh distribution I_0. The state energies are connected by conservation laws:

$$E_3 - E_1 = \omega - \omega_{eq}, \quad E_3 - E_2 = \omega - \omega_0. \tag{5.18}$$

The resonance fluorescence probability has the form (Heitler 1984)

$$w_{RF} = \frac{|V_{21}|^2 |V_{32}|^2}{[(\omega - \omega_{eq})^2 + \Gamma^2/4][(\omega - \omega_0)^2 + \gamma^2/4]}, \tag{5.19}$$

where V_{21} and V_{32} are the matrix elements that correspond to the absorption of an EPh at frequency ω_{eq} and the emission of a real photon at frequency ω, respectively; γ and Γ are the total probabilities (per unit time) of photon absorption and emission, i.e.,

$$\gamma(E) = 2\pi \sum_{\mathbf{k}} |V_{32}|^2 \delta(E - E_3), \tag{5.20}$$

$$\Gamma(E) = \gamma \sum_{k_{eq}} |V_{21}|^2 \frac{1}{[(E - E_2)^2 + \gamma^2/4]}. \tag{5.21}$$

The quantities $\Gamma(E)$ and $\gamma(E)$ in (5.21) should be taken at the energy $E = E_3$, but in fact they depend only weakly on energy.

It is noteworthy to recall an implication of (5.19): For the elementary process of absorption–emission, the energies of the absorbed and emitted photons are equal (within the very small width Γ). This "memory" about the absorbed photon by the ion (atom) manifests itself in the probability (5.19) which does not reduce to the product of absorption and emission probabilities. Indeed, the first factor in the denominator of (5.19) does not bind (approximately via the corresponding δ function) the energies of the initial (ω_{eq}) and exactly resonant (ω_0) photons, but only the energies of the initial and final (ω) photons. It is the continuity of the incident photon spectrum that reduces the resonance fluorescence to the two independent processes of absorption and subsequent emission. Thus, the DR width Γ_{DR} is formed by the width γ (5.20) and by the probability (per unit time) of the autoionization process of the intermediate state:

$$\Gamma_{DR} = \gamma + \Gamma_A . \tag{5.22}$$

This relation implies the possibility of the return of the recombined electron to the continuum energy state with the simultaneous equivalent photon re-emission by the ion core.

The matrix element V_{21} is determined by the oscillator strength of the radiative transition in an ion ($V_{21} \sim d_{21}$, where d_{21} is the matrix element of the dipole moment of a bound radiating electron in an ion) and is proportional to the flux density of the EPh incident on an ion. For the continuous EPh spectrum, the total probability of absorption Γ_{RF} is expressed in terms of I_0:

$$\Gamma_{RF} = 2\pi I_{0,\omega} \overline{|V_{21}|^2}, \tag{5.23}$$

where the bar ($\overline{|V_{21}|^2}$) denotes the averaging over the angles of the absorbed photon wave vector. As applied to the specific conditions of the model for the DR considered, the summation over k_{eq} in (5.21) should be supplemented by a summation over the final states of the captured electron. This procedure combined with the conservation law for the incident electron energy leads to the result:

$$\Gamma_{DR} = \Gamma_{RF} Z^2/n^3. \tag{5.24}$$

Using then the expression for the DR total probability summed over the EPh frequencies

$$\sum_{\omega} w_{DR} = \frac{\gamma \Gamma_A}{\gamma + \Gamma_A} \tag{5.25}$$

we obtain the rate of the autoionization process

$$\Gamma_A = \frac{f_{12}}{\pi n^3} l G_0\left(\frac{\omega_0 M^3}{3Z^2}\right), \tag{5.26}$$

where f_{12} is the oscillator strength of the excited radiative transition in the ion core, and the function $G_0(x)$ is given by (5.8b). Equation (5.26) coincides with the result which may be obtained from the exact quantum calculation (Beigman et al. 1981; Sobelman and Vainshtein 2006) of autoionization rate in Kramers domain (quasi-classical motion of the incident electron along a quasi-parabolic orbit, $Ze^2/\hbar v \gg 1$, $\rho \ll a$). The total DR rate corresponding to the capture of an electron into an ionic nl state is then given by the expression

$$\alpha_{\mathrm{DR}} = \left(\frac{2\pi}{T}\right)^{3/2} \frac{g(2)}{g(1)} \frac{(2l+1)\gamma\,\Gamma_{\mathrm{A}}}{(\gamma + \Gamma_{\mathrm{A}})} \exp\left[-\frac{\omega}{T} + \frac{Z^2}{2n^2T}\right], \tag{5.27}$$

where T is the electron temperature, and $g(1)$ and $g(2)$ are the statistical weights for the ground and excited ion levels, respectively.

The result (5.26) may also be derived on the basis of the relation between the autoionization probability and the cross-section for the ion excitation by electron impact near the ionization threshold. This relation follows from the detailed balance equation for the mutually inverse processes of autoionization and electron capture into an ion nl-level with the excitation of the $1 \to 2$ transition in an ion core, i.e.,

$$(2l+1)g_2\Gamma_{\mathrm{A}}(nl) = Z^2 n^{-3} \omega g_1 \sigma_{\mathrm{exc}}(l)/\pi^2 a_0^2. \tag{5.28}$$

Substituting the KrED result (5.13) for the excitation cross-section σ_{exc}, we obtain (5.26).

All the methods for the derivation of autoionization, dielectronic capture, and dielectronic recombination are equivalent in the sense that they are based on the dipole approximation for the interaction between an incident and a bound electron. It is exactly this approximation that allows us to treat all of the processes related to an energy loss of the incident electron as processes of effective radiation of either real (bremsstrahlung and photorecombination radiation) or equivalent (excitation, dielectronic recombination, polarization bremsstrahlung, and polarization recombination) photons.

5.2 Ionization by Electron Impact

5.2.1 Thomson Formula

Ionization of atoms by electron impact is one of the most important elementary processes defining the characteristics both of laboratory and astrophysical plasmas. A consistent description of this phenomenon involves the quantum mechanical approach; however, its main qualitative features can be determined also within the framework of classical mechanics. A corresponding formula for the cross-section was first proposed by J. Thomson in 1912, even before the development of the quantum theory. The classical consideration of ionization of an atom by electron impact

$$e + A \rightarrow A^{+} + 2e \tag{5.29}$$

carried out by Thomson is based on the assumption of elastic scattering of the projectile by bound electrons of the target. In this case, ionization results from the transfer of energy to an atom, that is, higher than the ionization potential of the electron subshell under consideration (as a result of projectile scattering). The applicability of the classical method to estimate the atomic ionization cross-section is based on the exact coincidence of quantum mechanical and classical cross-section of elastic electron–electron scattering.

Without considering the binding of atomic electrons to the nucleus, the expression for the integrated (with respect to the angle of scattering) cross-section of collisional ionization can be represented as (Astapenko and Lisitsa 2007)

$$\sigma_{\mathrm{i}} = \int\limits_{\Delta E > I} \mathrm{d}\sigma, \tag{5.30}$$

where $\mathrm{d}\sigma$ is the differential cross-section of electron–electron scattering, ΔE is the energy transferred from a projectile, I is the atomic ionization potential. Neglecting the change of the energy of the incident electron in comparison to its initial energy, we can derive the cross-section $\mathrm{d}\sigma$ with the help of the Rutherford formula (that describes the cross-section of elastic scattering of a charged particle in the Coulomb potential):

$$\frac{\mathrm{d}\sigma^{(R)}}{\mathrm{d}\Omega} = \left(\frac{Z\,e^2}{2\,m\upsilon^2\,\sin^2(\theta/2)}\right)^2. \tag{5.31}$$

$\mathrm{d}\Omega$ is an element of the solid angle, into which a projectile is scattered, θ is the angle of scattering, and υ is the velocity of the incident electron. For electron–electron scattering, we have $Z = 1$. Assuming that the value of energy transferred to an atomic electron ΔE is equal to the recoil energy, it is easy to find that

$$\Delta E = 4\,E\,\sin^2(\theta/2). \tag{5.32}$$

For the derivation of (5.32), the relation for a transmitted pulse that neglects the binding of an atomic electron to the nucleus was used, i.e.,

$$\Delta p = 2\,m\upsilon\,\sin(\theta/2), \tag{5.33}$$

that follows from elementary consideration of elastic scattering (Landau and Lifschitz 2005). With the use of (5.32), it is possible to express the angle of scattering in terms of energy transferred to an atomic electron ΔE. As a result, we obtain from (5.31) for a one-electron atom

$$d\sigma = \frac{\pi e^4 \, d\Delta E}{E(\Delta E)^2}, \tag{5.34}$$

where $E = mv^2/2$ is the energy of the incident electron. Substituting the expression (5.34) in (5.30) and integrating over the possible values of the transferred energy ΔE, we obtain the so-called *Thomson formula* for the cross-section of collisional ionization of a one-electron atom:

$$\sigma_i = \int_I^E d\sigma = \frac{\pi e^4}{E}\left(\frac{1}{I} - \frac{1}{E}\right), \quad E > I. \tag{5.35}$$

If the dimensionless parameter $x = E/I$ is introduced, the right-hand side of (5.35) can be rewritten as

$$\sigma_i^{(\mathrm{Th})}(E) = \frac{\pi e^4}{I^2} \frac{x-1}{x^2}, \quad x \geq 1. \tag{5.36}$$

It is clear that the value $x_{\mathrm{th}} = 1$ ($E_{\mathrm{th}} = I$) is the threshold value: At $x < 1$, the process cross-section is equal to zero since the energy of an incident electron is insufficient for ionization of an atom.

5.2.2 *Similarity Function Method for the Ionization Cross-Section*

Qualitatively, the Thomson formula (5.36) renders properly the features of the collisional ionization of a one-electron atom. However, from the quantitative point of view the formula is not very precise. To obtain a realistic description, it is advisable to represent the expression (5.36) as follows (Astapenko and Lisitsa 2007):

$$\sigma_i^{(\mathrm{Th})}(E) = \pi a_I^2 f^{(\mathrm{Th})}(E/I). \tag{5.37}$$

Here the dimensionless Thomson *similarity function* $f^{(\mathrm{Th})}(x) = (x-1)/x^2$ describing the dependence of the collisional ionization cross-section on a projectile energy and the "ionization radius" are introduced:

$$a_I = \frac{e^2}{I}. \tag{5.38}$$

As one can easily see from (5.38), the ionization radius is equal to the distance between incident and atomic electrons, at which the energy of the Coulomb interaction is equal to the atomic ionization potential. The radius a_{I} defines the cross-section $\sigma_{\mathrm{I}} = \pi a_{\mathrm{I}}^2$ that characterizes the cross-section of collisional ionization of a one-electron atom by an order of magnitude.

From formula (5.37), it follows that the projectile energy enters in the expression for the collisional ionization cross-section only in terms of the ratio E/I. This circumstance forms the basis for the similarity function method that assumes that the ratio of the ionization cross-section to the value $\sigma_{\mathrm{I}} = \pi a_{\mathrm{I}}^2$ is a universal function of the dimensionless variable $x = E/I$. Thus, the formula (5.36) is generalized by (5.37): The similarity function $f(x)$ can be determined either within the framework of a chosen theoretical or empirical approach.

The Thomson similarity function gives a value for the cross-section maximum of $1/4$ at $x_{\mathrm{max}} = 2$, i.e., at $E_{\mathrm{max}} = 2I$. The comparison with experimental data shows that this value x_{max} is too low. In other words, the Thomson formula shifts the true position of the cross-section maximum closer to the threshold value $E_{\mathrm{th}} = I$. In fact, the maximum of the collisional ionization cross-section is found from $E_{\mathrm{max}} = 3I$ to $E_{\mathrm{max}} = 4I$ (see the next section).

Curiously, the Thomson formula gives a value for the cross-section maximum of collisional ionization of a hydrogen atom of $\sigma_{\mathrm{i}}^{(H)}(E_{\mathrm{max}}) = \pi a_{\mathrm{B}}^2 \approx 0.88\ \text{Å}^2$ (a_{B} is the Bohr radius) that is equal to the area of the first Bohr orbit. According to (5.36), the energy at cross-section maximum is $E_{\mathrm{max}} = 2I_{1s}^{(H)} \cong 27.2\,\mathrm{eV}$, which coincides with the atomic unit of energy.

For more realistic descriptions of the cross-section of atomic ionization by electron impact, it is necessary to develop other similarity functions. For ionization potentials $I > 10$ eV, good agreement with experimental data is obtained by the Gryzinski similarity function (Gryzinski 1959, 1965a, b, c)

$$f^{(\mathrm{Gryz})}(x) = \frac{1}{x}\left(\frac{x-1}{x+1}\right)^{3/2}\left[1 + \frac{2}{3}\left(1 - \frac{1}{2x}\right)\ln\left(2.7 + \sqrt{x-1}\right)\right] \tag{5.39}$$

derived within the framework of classical consideration but taking into account the velocity distribution function of the bound electrons. For completeness, we mention also the Eletskii–Smirnov similarity function

$$f^{(\mathrm{ES})}(x) = \frac{10\,(x-1)}{\pi x\,(x+8)}, \tag{5.40}$$

obtained empirically (on the basis of comparison with experimental data). It should be noted that the functions (5.39) and (5.40) are similar to each other.

The formulas (5.36), (5.37) were obtained for a one-electron atom. For the calculation of collisional ionization of multielectron atoms, the expression (5.37) should be generalized so that the contribution of different subshells of an atom (in particular the number of electrons in each subshell) to the total cross-section is taken into account. The resultant formula looks like

$$\sigma_{\rm i}(E) = \sum_{\rm nl} N_{\rm nl}\, \pi\, a_{I_{\rm nl}}^2 f(E/I_{\rm nl})\; \theta(E - I_{\rm nl}), \tag{5.41}$$

where $N_{\rm nl}$ is the number of equivalent electrons, $I_{\rm nl}$ the ionization potential of a *nl*-subshell (*n*, *l* are the principal and orbital quantum numbers), and $\theta(x)$ is the Heaviside step function that describes the "inclusion" of inner atomic subshells in the process at $E > I_{\rm nl}$.

The comparison with experimental data shows that the similarity functions (5.39), (5.40) give the best result for atoms with ionization potentials $I > 10$ eV (remember that $I = \min\{I_{\rm nl}\}$). In case of multielectron atoms with a moderate value of I (10 eV $> I >$ 6 eV), the Born–Compton similarity function is more adequate for the description of collisional ionization (Astapenko 2001):

$$f^{\rm (BC)}(x) = \frac{2.5}{\pi}\frac{1}{x}\int\limits_1^{y_{\rm m}} {\rm d}y \int\limits_{\sqrt{x}-\sqrt{x-y}}^{\sqrt{x}+\sqrt{x-y}} \frac{{\rm d}t}{t^2\left[t^2 + (y - t^2)^2/0.64\right]}, \tag{5.42}$$

where $y_{\rm m} = (x + 1)/2$. The similarity function (5.42) is obtained in the first Born approximation for the interaction of a projectile with atomic electrons. For its derivation, the analogy between collisional ionization and Compton scattering of the projectile eigenfield by an atom was used, and the electron shell of the atom was described within the framework of the Thomas–Fermi model.

In the case that a significant contribution to the ionization process is made by atomic subshells with large orbital quantum numbers ($l = 2$, 3), the use of the similarity functions (5.42) may be insufficient. In this case, the similarity function of the so-called Binary Encounter Bethe (BEB) approximation can be used (Kim and Rudd 1994):

$$f^{\rm (BEB)}(x, u) = \frac{1}{1 + x + u}\left[\frac{\ln(x)}{2}\left(1 - \frac{1}{x^2}\right) + 1 - \frac{1}{x} - \frac{\ln(x)}{1 + x}\right]. \tag{5.43}$$

In (5.43), the additional parameter u is introduced that represents the ratio of the average kinetic energy of a subshell to its ionization potential. It should be noted that for a hydrogen atom (in view of the virial theorem) $u = 1$. The parameter u takes into account the decrease of the cross-section of collisional ionization of atomic subshells with high orbital momenta. This decrease is connected with the fact that at equal ionization potentials, a subshell with a higher orbital moment has a smaller radius, which results in a decrease of process cross-section.

Figure 5.2 shows the similarity functions (5.39), (5.40) as well as (5.42), (5.43). A common feature of the similarity functions presented in Fig. 5.2 is their identical near-threshold dependence: $f(x) \propto x - 1$. This dependence is a consequence of the common quantum mechanical regularity connected with the behavior of the statistical weight of the final state of an ionized electron as a function of the projectile energy. On the other hand, the asymptotic behavior of the similarity functions at

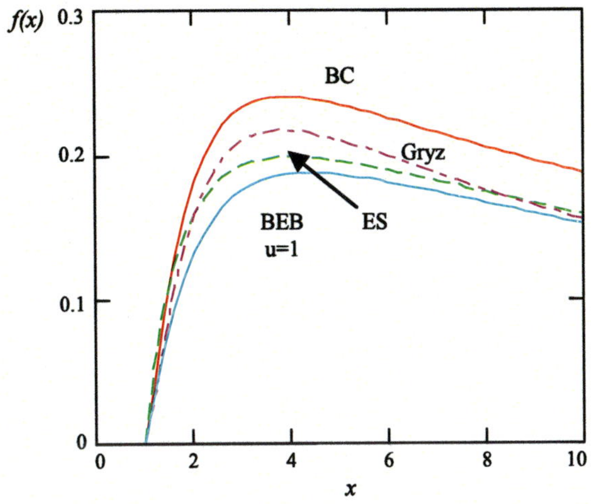

Fig. 5.2 Comparison of different similarity functions for the collisional ionization cross-section of an atom: Gryz—Gryzinski; ES—Eletskii–Smirnov; BC—Born–Compton; BEB—BEB approximation

high-incident electron energies $E \gg I$ ($x \gg 1$) is different. From formulas (5.39), (5.40) and (5.42), (5.43) and $x \gg 1$, we obtain: $f^{(\mathrm{Gryz}),(\mathrm{BEB})} \propto \ln(x)/x$, $f^{(\mathrm{ES}),(\mathrm{BC})} \propto 1/x$. The first asymptote coincides with the high-energy limit of the inelastic cross-section of the Bethe quantum mechanical theory, the second asymptote deviates from it, but provides better agreement with experimental data at not too high projectile energies.

The similarity function method favorably differs from other methods of calculation of the collisional ionization cross-section by its simplicity and reliability; it is often used for fast evaluation of the cross-sections (in particular for bulky complex configurations).

5.2.3 *Comparison with Experimental Data*

At present days, electron impact ionization cross-sections of neutral atoms have been measured for the majority of elements from the periodic table. The experimental measurements are carried out with the help of the so-called crossed-beam technique that consists of several steps:

(a) the beam of fast neutrals is produced by neutralization of fast (with an energy of several keV) ions in a chamber with low-pressure gas (about 10^{-4} torr),
(b) the ion beam is then pre-extracted from a DC gas discharge, focused, and passed through a special filter that sorts ions with respect to their velocities,
(c) the resultant beam of fast neutrals with specified energy retains the collimation of the initial ion beam, which is mandatory for high-precision measurements,
(d) the absolute neutral atom flux is measured with the help of a calibrated detector,

(e) the measurement of the collisional ionization cross-section is then carried out intersecting the neutral beam of atoms with the electron beam,
(f) ions resulting from electron–atom collisions are focused on an electrostatic analyzer extracting ions with a given charge number that are then recorded by an electron multiplier,
(g) based on the measured data, the experimental value of the ionization cross-section is finally calculated by the formula

$$\sigma_i^{(\exp)}(E) = \frac{J_i(E)\, \upsilon_e\, \upsilon_n}{J_e(E)\, R F \sqrt{\upsilon_e^2 + \upsilon_n^2}}, \tag{5.44}$$

where $J_{i,e}$ are the ion and electron currents, $\upsilon_{n,e}$ are the velocities of neutrals and electrons, R is the neutral flux, F is the value characterizing the degree of intersection of the neutral beam and the electron beam.

Thus, the measurement of the absolute value of the collisional ionization cross-section consists in the measurement of each value appearing in the right-hand side of (5.44).

Figure 5.3 shows the comparison of the experimental cross-section of collisional ionization of a hydrogen atom with the results of calculations by the similarity function method.

It is seen that in this case, the Eletskii–Smirnov similarity functions and the BEB approximation provide better agreement with the experiment than the Born–Compton method that considerably overestimates the cross-section. This circumstance is connected with the fact that the formulas for the Born–Compton method were obtained in the Thomas–Fermi approximation for an electron shell of an

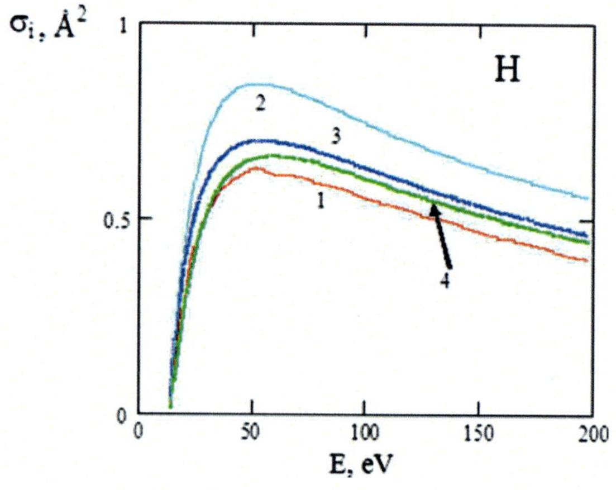

Fig. 5.3 Ionization cross-sections of a hydrogen atom by electron impact: 1—experiment, 2—Born–Compton method, 3—Eletskii–Smirnov formula, 4—BEB approximation for u = 1

ionized atom that is only valid for multielectron atoms. Moreover, in this case the ionization potential of a hydrogen atom (13.6 eV) is overestimated and the use of the Born–Compton similarity function is beyond its range of applicability.

From Fig. 5.3, it follows that the maximum of the ionization cross-section of a hydrogen atom by electron impact is reached at an incident electron energy $E_{max} \approx$ 50 eV in contrast to the value $E_{max} \approx 27.2$ eV following from the Thomson formula (5.36). The value of the cross-section at the maximum is about 0.6 Å^2, which is somewhat less than the prediction of the Thomson theory (0.88 Å^2). Thus in the case of a hydrogen atom, the classical approach of J. Thomson strongly shifts the position of a maximum to a too low energy range, but gives a satisfactory value of the maximum cross-section.

It is of interest to compare the cross-section of photoionization of a hydrogen atom $\sigma_{ph}(\omega)$ with the cross-section of ionization by electron impact. The cross-section of photoionization of a hydrogen atom reaches its maximum at threshold, i.e., at a photon energy of 13.6 eV, whereas the maximum of the photoionization cross-section (0.064 Å^2) is about an order of magnitude less. Different asymptotic behavior of the cross-sections can also be noted: $\sigma_{ph}(\omega) \propto \omega^{-7/2}$, while $\sigma_i(E) \propto \ln(E)/E$, i.e., the photoionization cross-section decreases much more rapidly.

As an example we compare experimental data with different theories of collisional ionization of a multielectron atom. Figure 5.4 shows an example for the tellurium atom (atomic number Z = 52): The configurations of outer electron shells are $(4d)^{10}(5s)^2(5p)^4$, and their ionization potentials are $I_{4d} = 47$ eV, $I_{5s} = 18$ eV, $I_{5p} \cong 9$ eV. As follows from these energies and formulas (5.37)–(5.38), the cross-section maximum of collisional ionization of the 5*p*-subshell is about an order of magnitude higher than the corresponding value for the 5*s*-subshell, indicating

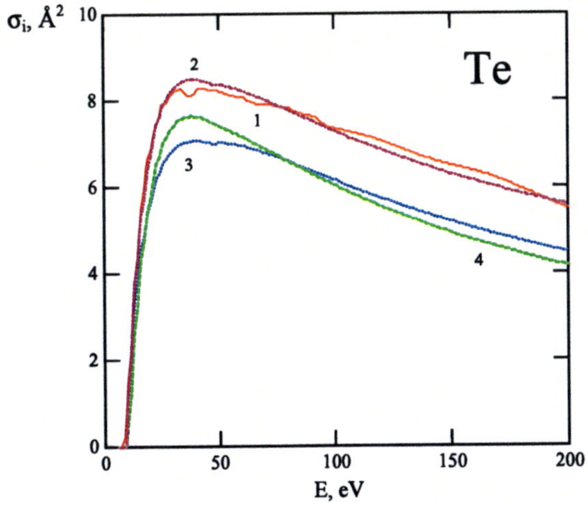

Fig. 5.4 Cross-section of ionization of a tellurium atom by electron impact: 1—experiment, 2—Born–Compton method, 3—Eletskii–Smirnov formula, 4—Gryzinsky formula

that the main contribution to the ionization process stems from the outer atomic subshell. This situation is typical: As a rule, the cross-section of collisional ionization of an atom is defined by its outer subshell since it has the smallest ionization potential.

From Fig. 5.4, it is seen that in this case the best agreement with experimental data is achieved by the Born–Compton similarity function (5.42). The Eletskii–Smirnov and Gryzinsky formulas somewhat underestimate the cross-section.

The value of the cross-section of collisional ionization of neutral atoms in the ground state varies in a relatively narrow range: from 0.5 Å^2 (for helium) to about 10 Å^2 for heavy atoms such as tellurium. Atoms in excited states with low ionization potential have large collisional ionization cross-sections that at the maximum can make up several hundreds of squared angstroms. For multiply charged positive ions with a high ionization potential, the cross-section can be rather small because $\sigma \propto 1/I \propto 1/Z^2$.

The cross-section of ionization of atoms by electron impact defines an important parameter for atomic population kinetic equations that is called the rate coefficient [typical employed units are (cm^3 s^{-1})] that is determined by the expression

$$k_{\rm i}(T_{\rm e}) = \int \sigma_{\rm i}(E)\, \upsilon_{\rm e}(E)\, F_{\rm e}(E, T_{\rm e})\, {\rm d}E, \tag{5.45}$$

where $F_{\rm e}(E, T_{\rm e})$ is the energy distribution function of plasma electrons at a given temperature $T_{\rm e}$ of an electron subsystem, $\upsilon_{\rm e}(E) = \sqrt{2E/m_{\rm e}}$ is the velocity (non-relativistic) of an electron.

With the use of the universal expression (5.41) for the atomic ionization cross-section, it is easy to estimate the rate coefficient for a Maxwellian electron energy distribution function:

$$F_{\rm e}(E, T_{\rm e}) = \frac{2}{\sqrt{\pi}} \frac{\sqrt{E}}{(T_{\rm e})^{3/2}} \exp\left[-\frac{E}{T_{\rm e}}\right], \tag{5.46}$$

where $T_{\rm e}$ is expressed in energy units. Substituting the formulas (5.41) and (5.46) in the expression (5.45), we obtain

$$k_{\rm i}(T_{\rm e}) = \frac{2}{\pi} \sum_{\rm nl} \sqrt{2\, m_{\rm e} I_{\rm nl}} \left(\frac{e^2}{I_{\rm nl}}\right)^2 H(I_{\rm nl}/T_{\rm e}), \tag{5.47}$$

where electron temperature is included in the dimensionless function

$$H(y) = y^{3/2} \int_1^\infty f(x)\, {\rm e}^{-xy}\, x\, {\rm d}x, \tag{5.48}$$

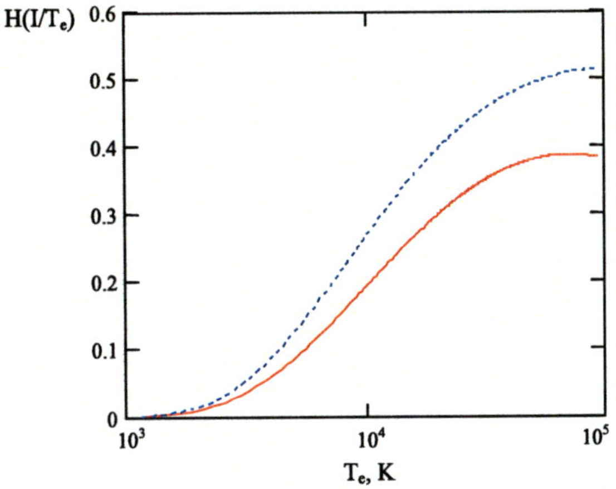

Fig. 5.5 Dependence of the function $H(I/T_e)$ [the formulas (5.47), (5.48)] on electron temperature calculated with the Eletskii–Smirnov (solid curve) and Born–Compton (dotted curve) similarity functions for a ionization potential of I = 1Ry (ionization potential of a hydrogen atom)

that depends itself on the similarity function $f(x)$. $H(y)$ reaches a maximum at about values y = 0.06–0.08, i.e., at a temperature of $T_{\max} \approx (12 - 15)\,I$. It should be noted that at such temperatures, an atom is already ionized. The functions $H(I/T)$ calculated with the Eletskii–Smirnov and Born–Compton similarity functions for the ionization potential of a hydrogen atom are shown in Fig. 5.5.

5.3 Analytical Empirical Formulas for Ionization, Single, and Total Recombination Rates

5.3.1 *Ionization*

Among a vast amount of fitting formulas and numerical calculations (Sobelman and Vainshtein 2006; Voronov 1997; Lotz 1970; Kato et al. 1991), we point out here one of the most used formulas from Lotz providing a modified explicit analytic expression for the ionization from shell "n" of an ion with charge state "Z" (i.e., $X_Z(n) + e \rightarrow X_{Z+1}(m) + e + e$) averaged over a Maxwellian electron energy distribution function:

$$I_{Z,Z+1}(n,m) \approx 6 \times 10^{-8}\, P_n \left(\frac{\text{Ry}}{E_{Z,Z+1}(n,m)}\right)^{3/2} \sqrt{\beta_{nm}}\, e^{-\beta_{nm}}\, \alpha(\beta_{nm})\, [\text{cm}^3\,\text{s}^{-1}], \tag{5.49a}$$

$$\alpha(\beta_{nm}) \approx \ln\left[1 + \frac{0.562 + 1.4\,\beta_{nm}}{\beta_{nm}\,(1 + 1.4\,\beta_{nm})}\right], \tag{5.49b}$$

$$\beta_{\mathrm{nm}} = \frac{E_{Z,Z+1}(n,m)}{kT_{\mathrm{e}}}. \tag{5.49c}$$

kT_e is the electron temperature in [eV], $E_{Z,Z+1}(n,m)$ is the ionization energy in [eV] from state "*n*" of ion "*Z*" to state "*m*" of ion "Z + 1", and P_n is the number of equivalent electrons in the state "*n*". Note, that detailed ionization rate coefficients for H I, He I and He II are presented in Annex 2 and 3.

5.3.2 *Three-Body Recombination in Dense Plasmas*

Three-body recombination (i.e., $X_{Z+1}(m) + e + e \rightarrow X_Z(n) + e$) is the inverse process of ionization and can be approximated by the following analytical expression:

$$T_{Z+1,Z}(m,n) \approx 2 \times 10^{-31} \left(\frac{\mathrm{Ry}}{E_{Z,Z+1}(n,m)} \right)^3 \frac{P_{\mathrm{n}}\, g_Z(n)}{g_{Z+1}(m)}\, \beta_{\mathrm{nm}}^2\, \alpha(\beta_{\mathrm{nm}})\ [\mathrm{cm}^6\, \mathrm{s}^{-1}], \tag{5.50}$$

with $\alpha(\beta_{\mathrm{nm}})$ and β_{nm} given by (5.49b, c). $g_{Z+1}(m)$ is the statistical weight of the state before recombination (usually the strongly populated ground state) and $g_Z(n)$ is the statistical weight of the recombined state. Note, that detailed three-body recombination rate coefficients for H I, He I and He II are presented in Annex 2 and 3.

Of particular interest for the calculation of the ionic fraction and radiation losses is the total three-body recombination rate, i.e., the summation over principal quantum number "*n*" until $N_{\max}$ in (5.50):

$$T_{Z+1,Z} = \sum_{n=1}^{N_{\max}} T_{Z+1,Z}(n). \tag{5.51}$$

The summation over the principal quantum number "*n*" in (5.51) has to be taken out with care. In fact, (5.51) assumes that all recombination into excited states finally populate the ground state via radiative cascades. At large quantum numbers, however, collisional processes become so important that that the recombination flow to an excited state might even be transferred back before it can decay to the ground state by radiative cascades. As collisional rates strongly increase with principal quantum number "*n*" but radiative decay rates decrease with principal quantum number "*n*", there exist a critical electron density $n_{\mathrm{e,crit}}$ where collisional processes are equally important as radiative decay for a given principal quantum number $n = n_{\mathrm{crit}}$. A rough guideline for the selection of the maximum principal quantum number in (5.51) is therefore

$$N_{\max} \approx n_{\text{crit}}. \tag{5.52}$$

If the atomic structure is such that the atomic ground state has the principal quantum number $n = 1$, critical electron density and critical principal quantum number are related by:

$$n_{\text{e,crit}} \geq 6 \times 10^{19} Z^7 \frac{(n_{\text{crit}} - 1)^{2n_{\text{crit}}-2}}{n_{\text{crit}}^3 (n_{\text{crit}} + 1)^{2n_{\text{crit}}+2}} \left(\frac{kT_{\text{e}}(\text{eV})}{Z_{\text{eff}}^2 \,\text{Ry}} \right)^{1/2} \left[\text{cm}^{-3}\right]. \tag{5.53}$$

kT_{e} is the electron temperature in [eV], Z_{eff} is the effective ionic charge and Ry = 13.6 eV. With the help of (5.53), for each given electron density $n_{\text{e,crit}}$, the critical principal quantum number n_{crit} can be calculated. Equation (5.53) has a well-defined asymptote for large quantum numbers:

$$\lim_{n_{\text{crit}} \to \infty} \left\{ \frac{(n_{\text{crit}} - 1)^{2n_{\text{crit}}-2}}{n_{\text{crit}}^3 (n_{\text{crit}} + 1)^{2n_{\text{crit}}+2}} \right\} = \lim_{n \to \infty} \left\{ \frac{1}{n_{\text{crit}}^3 (n_{\text{crit}} + 1)^4} \left(\frac{n_{\text{crit}} - 1}{n_{\text{crit}} + 1} \right)^{2n_{\text{crit}}-2} \right\}$$
$$\approx \frac{0.0183}{n_{\text{crit}}^7} \tag{5.54}$$

because

$$\lim_{n_{\text{crit}} \to \infty} \left\{ \left(\frac{n_{\text{crit}} - 1}{n_{\text{crit}} + 1} \right)^{2n_{\text{crit}}-2} \right\} \approx \frac{1}{54.6}. \tag{5.55}$$

Therefore, we can write

$$n_{\text{e,crit}} \approx 6 \times 10^{19} Z_{\text{eff}}^7 \frac{1}{n_{\text{crit}}^3} \frac{0.0183}{n_{\text{crit}}^4} \left(\frac{kT_{\text{e}}(\text{eV})}{Z_{\text{eff}}^2 \text{Ry}} \right)^{1/2} \approx 10^{18} \frac{Z_{\text{eff}}^7}{n_{\text{crit}}^7} \left(\frac{kT_{\text{e}}(\text{eV})}{Z_{\text{eff}}^2 \text{Ry}} \right)^{1/2} \left[\text{cm}^{-3}\right]. \tag{5.56}$$

Equation (5.56) shows that the critical electron density scales with the seventh power of the principal quantum number and with the seventh power of the effective charge. Equation (5.52) can therefore be estimated as follows:

$$n_{\text{crit}} \approx 373 \frac{Z_{\text{eff}}}{n_{\text{e,crit}}^{1/7}} \left(\frac{kT_{\text{e}}(\text{eV})}{Z_{\text{eff}}^2 \text{Ry}} \right)^{1/14}. \tag{5.57}$$

The maximum principal quantum number is not a very critical issue for radiative and dielectronic recombination, as both processes decrease rapidly with the principal quantum number itself. For the three-body recombination, however, the recombination rates increase strongly with principal quantum number and $N_{\max}$ has to be chosen with care. Monte Carlo simulations (Mansbach and Keck 1969) that take into account the complex movement of the electron in the excitation–ionization among the numerous excited states indicate the following:

$$T_{Z+1,Z} \approx 2 \times 10^{-27} \zeta \frac{Z_{\mathrm{eff}}^3}{(kT_{\mathrm{e}})^{9/2}} \left[\mathrm{cm}^6\,\mathrm{s}^{-1}\right] \tag{5.58}$$

with $\zeta = 1$. Most of the results obtained with different methods propose expressions similar to (5.58) but differ by the numerical coefficient $\zeta = 0.1$–10 (Hahn 1997; Mayorov et al. 1994). Note that if all three-body recombination rates are summed up and the upper limit is identified with the collisional ionization limit, $\zeta = 3.1$ (Hahn and Li 1996) (note that the ionization limit employed in (Hahn and Li 1996) does not depend on density).

In dense and cold plasmas, the classical three-body recombination rate is diverging because the Maxwell electron distribution function becomes very narrow. This is unphysical, because it can be shown that this violates the Pauli principle. A recent investigation based on a consistent use of the Fermi–Dirac distribution function and Pauli-blocking factors has shown (Deschaud et al. 2014, 2015) that the three-body recombination rate is then well defined for all transitions from the hot dense plasma to the warm dense matter (WDM), to the hot solid, and to the cold solid.

5.3.3 Radiative Recombination in Dense Plasmas

In a similar manner, the total radiative recombination is the sum of all radiative recombination into the ground and excited states (N_{max} is the largest principal quantum number to be taken into account):

$$R_{Z+1,Z} = \sum_{n=1}^{N_{\mathrm{max}}} \sum_{l=0}^{n-1} R_{Z+1,Z}(nl). \tag{5.59}$$

In the optical electron model (hydrogenic approximation), the radiative recombination can be directly represented by a sum over the orbital l-quantum numbers (Baker and Menzel 1938; Sobelman and Vainshtein 2006):

$$R_{Z+1,Z}(n) = \sum_{l=0}^{n-1} R_{Z+1,Z}(nl), \tag{5.60}$$

$$R(n) \approx 5.2 \times 10^{-14} Q_{\mathrm{n}} Z_{\mathrm{eff}} \beta_{\mathrm{n}}^{3/2} \gamma(\beta_{\mathrm{n}}) \left[\mathrm{cm}^3\,\mathrm{s}^{-1}\right], \tag{5.61a}$$

$$Z_{\mathrm{eff}} = n_{\mathrm{gr}} \cdot \sqrt{\frac{E_{Z,Z+1}(n_{\mathrm{gr}})}{\mathrm{Ry}}}, \tag{5.61b}$$

$$\gamma(\beta_n) \approx \ln\left[1 + \frac{0.562 + 1.4\,\beta_n}{\beta_n(1 + 1.4\,\beta_n)}\right], \tag{5.61c}$$

$$\beta_n = \frac{E_{Z,Z+1}(n)}{kT_e}, \tag{5.61d}$$

$$Q_n \approx 1 - \frac{N}{2\,n^2}. \tag{5.61e}$$

kT_e is the electron temperature in [eV], $E_{Z,Z+1}(n)$ is the ionization energy of the state "n" of ions "Z" into state "m" of ion "Z + 1" in [eV], Z_{eff} is the effective charge of the ion before recombination, n_{gr} is the principal quantum number of the ground state, Ry = 13.6 eV and Q_n is an angular factor which takes into account the Pauli principle (means the reduced probability to be captured into a certain level that is already partially occupied with electrons), (5.61e) is the corresponding hydrogenic approximation. For example, for radiative recombination into He-like neon (i.e., $Ne^{9+}(1s) + e \rightarrow Ne^{8+}(1s^2)$), we have $Q_n = 1 - 1/(2 \cdot 1^2) = 0.5$ and $Z_{eff} = 1 \cdot \sqrt{1195.81\,\text{eV}/\text{Ry}} = 9.4$. Note, that detailed radiative recombination rate coefficients for H I, He I and He II are presented in Annex 2 and 3.

Using (5.61) and Q_n = 1, the sum of (5.60) over the n-quantum numbers can be approximated by the following analytic expression:

$$R^{tot}(n \geq n_1) = 2.6 \times 10^{-14}\, Z_{eff}\, n_1 \beta_1^{1/2} \cdot \{\ln(1.78\beta_1) + g(\beta_1)(1 + \beta_1/n_1)\}\left[\text{cm}^3\,\text{s}^{-1}\right], \tag{5.62a}$$

$$\beta_1 = \frac{Z_{eff}^2\,\text{Ry}}{n_1^2 kT_e}, \tag{5.62b}$$

$$g(\beta_1) \approx \ln\left[1 + \frac{0.562 + 1.4\,\beta_1}{\beta_1\,(1 + 1.4\,\beta_1)}\right]. \tag{5.62c}$$

n_1 is the principal quantum number from which the sum is taken (usually over all higher lying excited states with $n > n_{gr}$). In practice, the calculations of the total radiative recombination rate employ detailed calculations for the recombination into the ground state and states that have the same principal quantum number as the ground state, i.e., $R(n = n_{gr})$ (either via (5.61) or more advanced detailed quantum mechanical calculations) and employ (5.62) for the excited states with an effective charge given by (5.61b). In this case, $n_1 = n_{gr} + 1$. In other words, one employs detailed calculations for the states with the same principal quantum number as the ground state and the hydrogenic approximation with effective charge for the excited states.

In the framework of the hydrogenic approximation for the contribution of the excited states, dense plasma effects can be estimated with the help of (5.52) and (5.62) assuming that all recombination is suppressed for quantum numbers larger than the critical quantum number, i.e., $n > n_{crit}$. In this case, $n_1 = n_{crit}$ in (5.62) and the total radiative recombination in dense plasmas is given by

$$R_{dense}^{tot} \approx R^{tot}(n \leq n_{crit}) \approx R(n = n_{gr}) + \left\{R^{tot}(n \geq n_{gr} + 1) - R^{tot}(n \geq n_{crit} + 1)\right\}. \tag{5.62d}$$

5.4 Classical Consideration of Collisional Excitation of an Atom

5.4.1 *Fermi Photon Equivalent Method and Oscillator Strength Method*

Excitation of atoms in collisions with electrons is another example of an inelastic collisional process that plays an important role in various fields of physics and technology. In contrast to collisional ionization, when an atomic electron is excited to the continuous energy spectrum (corresponding to infinite motion), excitation of an atomic electron goes to the discrete spectrum, that is (within the framework of the classical picture), to another atomic orbit with higher energy. This phenomenon is responsible for emission of photons in plasmas resulting from a transition of an atomic excited state to the ground state. It is also one of the mechanisms to achieve population inversion in gas lasers (so-called electron beam pumping). Also the population inversion of the soft X-ray Ne-like and Ni-like lasers is based on collisional excitation in plasmas (Sobelman and Vinogradov 1985; Elton 1990). Note that X-ray laser schemes without population inversion have also been proposed (Braunstein and Shuker 2003) that is based on a complex interplay of the atomic master equations that include the atomic coherences (Loudon 2000).

Electron collisional excitation is schematically described as

$$e + A \rightarrow A^* + e, \tag{5.63}$$

where the symbol A^* denotes an atom in the excited state of a discrete spectrum. For calculation of the reaction cross-section (5.63), we will use the spectroscopic principle of correspondence between quantum physics and classical physics. According to this principle, an atom in interaction with an electromagnetic field behaves as a set of oscillators that are assigned to a pair of energy levels E_i and E_j of the atomic spectrum. Let us assume that $E_i < E_j$. The eigenfrequencies of these oscillators are equal to the eigenfrequency of the transition $i \rightarrow j$: $\omega_{ji} = \left(E_j - E_i\right)/\hbar$, and the efficiency of their interaction with an electromagnetic field is defined by the *oscillator strength*:

$$f_{ji} = \frac{2\,m\omega_{ji}\left|\mathbf{d}_{ji}\right|^2}{3\,\hbar\,e^2\,g_i}, \tag{5.64}$$

where g_i is the statistical weight of the initial state. In the quantum mechanical description of the dipole moment of a transition oscillator, $\mathbf{d}_{ji}$ is a matrix element of an electric dipole moment operator calculated between states $|i\rangle$ and $|j\rangle$. In the case of excitation of an atom, $\omega_{ji} > 0$ and $f_{ji} > 0$; for an electron transition with decreasing energy ($\omega_{ij} < 0$), $f_{ij} < 0$. Since $\left|\mathbf{d}_{ji}\right| = \left|\mathbf{d}_{ij}\right|$, we obtain from (5.64) $g_i f_{ji} = -g_j f_{ij}$.

Atomic excitation $i \to j$ via collisions with an electron corresponds therefore to the interaction between the electric field of the scattered electron and the transition oscillator. Assuming a homogenous incident electron field close to the atom, it is possible to write the following equation for the radius vector of the oscillator $\mathbf{r}_{ji}$:

$$\ddot{\mathbf{r}}_{ji} + \gamma_{ji}\,\dot{\mathbf{r}}_{ji} + \omega_{ji}^2\,\mathbf{r}_{ji} = f_{ji}\frac{e}{m}\mathbf{E}(t,\,\rho), \tag{5.65}$$

where γ_{ij} is the damping constant, $\mathbf{E}(t,\,\rho)$ is the strength of the electric field that is produced by an incident electron moving along a trajectory with an impact parameter ρ relative to the atom.

Let us assume $f_{ji} \neq 0$. Corresponding transitions are called dipole (or optically)-allowed transitions. Otherwise, the transitions are called dipole or optically forbidden transitions.

The Fourier transform of (5.65) is given by:

$$\mathbf{v}_{ji}(\omega) = f_{ji}\cdot\frac{e}{m}\cdot\frac{(-i\,\omega)\,\mathbf{E}(\omega,\,\rho)}{\omega_{ji}^2 - \omega^2 - i\,\gamma_{ji}\,\omega}, \tag{5.66}$$

where $\mathbf{E}(\omega,\,\rho)$ is the Fourier component of the electric field strength of a scattered electron. In order to determine the excitation cross-section of an atom for the transition $i \to j$, we calculate the work done on a transition oscillator by an incident electron during the duration of the collision:

$$A_{ji}(\rho) = \int_{-\infty}^{\infty} e\,\mathbf{v}_{ji}(t)\,\mathbf{E}(t,\,\rho)\,\mathrm{d}t = \frac{1}{2\,\pi}\int_{-\infty}^{\infty} e\,\mathbf{v}_{ji}(\omega)\,\mathbf{E}^*(\omega,\,\rho)\,\mathrm{d}\omega. \tag{5.67}$$

The second equality of (5.67) is a Fourier representation using the relation $\mathbf{E}(-\omega,\,\rho) = \mathbf{E}^*(\omega,\,\rho)$ and the integral representation of the delta function, i.e.,

$$\int_{-\infty}^{\infty} \exp(i\,\omega\,t)\,\mathrm{d}t = 2\,\pi\,\delta(\omega). \tag{5.68}$$

It should be emphasized that the expression (5.67) is valid only for sufficiently high-impact parameters $\rho > a$, where a is the "cutoff" length. If $\rho > a$, penetration of an incident electron into an atomic core can be neglected. The analysis shows that essentially distant collisions contribute to the excitation cross-section of a dipole-allowed transition. Hereafter, we assume $A_{\mathrm{ji}}(\rho < a) = 0$. The cutoff parameter a is of the order of several Bohr radii; its exact value depends on the atom and the specific transition.

Substituting (5.66) in (5.67) and transforming the integration over positive frequencies only, we obtain

$$A_{\mathrm{ji}}(\rho) = f_{\mathrm{ji}} \frac{e^2}{2m} \int\limits_0^\infty G_{\mathrm{ji}}^{(h)}(\omega - \omega_{\mathrm{ji}}) \left|\mathbf{E}(\omega, \rho)\right|^2 \mathrm{d}\omega. \tag{5.69}$$

$$G_{\mathrm{ji}}^{(h)}(\Delta\omega) = \frac{\gamma_{\mathrm{ji}}/2\pi}{\Delta\omega^2 + \left(\gamma_{\mathrm{ji}}/2\right)^2} \tag{5.70}$$

is the spectral line shape of a transition for homogeneous broadening. Equation (5.70) shows that the damping constant γ_{ji} defines the spectral width of a line. The function (5.70) satisfies the asymptotic relation

$$G_{\mathrm{ji}}^{(h)}\left(\Delta\omega, \gamma_{\mathrm{ji}} \to 0\right) \to \delta(\Delta\omega). \tag{5.71}$$

The spectral width of the function $\left|\mathbf{E}(\omega, \rho)\right|^2$ in (5.69) is defined by the ratio υ/ρ. This value is much larger than the width of the spectral line of an atomic transition γ_{ji} for $\rho > a$ and $\upsilon > \sqrt{2\hbar\,\omega_{\mathrm{ji}}/m}$. Therefore, the asymptotic formula (5.71) can be used.

The probability of excitation of the transition $i \to j$ is equal to the ratio

$$W_{\mathrm{ji}}(\rho) = \frac{A_{\mathrm{ji}}(\rho)}{\hbar\,\omega_{\mathrm{ji}}}, \tag{5.72}$$

where $\hbar\,\omega_{\mathrm{ji}} = \Delta E_{\mathrm{ji}}$ is the atomic excitation energy. The analysis shows that $W_{\mathrm{ji}} < 1$ for the considered range of impact parameters ρ and impact velocities υ as it should be according to the physical meaning of the probability.

The cross-section integrated with respect to the impact parameter is given by

$$\sigma_{\mathrm{ji}} = 2\pi \int\limits_a^\infty W_{\mathrm{ji}}(\rho)\,\rho\,\mathrm{d}\rho. \tag{5.73}$$

Here the upper limit of integration with respect to the impact parameter is assumed to be equal to infinity according to the classical picture. Substituting the formulas (5.72) and (5.69) in (5.73), we find in view of (5.71):

$$\sigma_{\mathrm{ji}} = \pi f_{\mathrm{ji}} \frac{e^2}{m\,\Delta E_{\mathrm{ji}}} \int_a^{\infty} \left|\mathbf{E}(\omega_{\mathrm{ji}}, \rho)\right|^2 \rho\, \mathrm{d}\rho. \tag{5.74}$$

To proceed further, let us consider the *approximation of straight trajectories*. In this case, it is easy to obtain an expression for the Fourier component of an incident electron field:

$$\mathbf{E}(\omega, \rho) = \frac{2e}{\rho\upsilon}\left\{F\left(\frac{\omega\rho}{\upsilon}\right)\mathbf{e}_{\mathrm{n}} - iF'\left(\frac{\omega\rho}{\upsilon}\right)\mathbf{e}_{\tau}\right\}, \tag{5.75}$$

where $\mathbf{e}_{\mathrm{n},\tau}$ are the normal and tangent (with respect to the velocity vector $\mathbf{v} = \mathrm{const}$) unit vectors and

$$F(\zeta) = \int_0^{\infty} \frac{\cos(\zeta x)}{(1+x^2)^{3/2}}\,\mathrm{d}x \tag{5.76}$$

(the prime in (5.75) denotes differentiation with respect to the argument).

Substituting (5.75)–(5.76) in (5.74), we obtain an expression for the collisional excitation cross-section as a function of the incident electron energy $E = m\upsilon^2/2$:

$$\sigma_{\mathrm{ji}}(E) = 2\pi f_{\mathrm{ji}} \left(\frac{e^2}{\Delta E_{\mathrm{ji}}}\right)^2 \frac{\Delta E_{\mathrm{ji}}}{E} \int_1^{\infty} H\left(\frac{\omega_{\mathrm{ji}}\, a\,\tilde{\rho}}{\sqrt{2E/m}}\right) \frac{\mathrm{d}\tilde{\rho}}{\tilde{\rho}}. \tag{5.77}$$

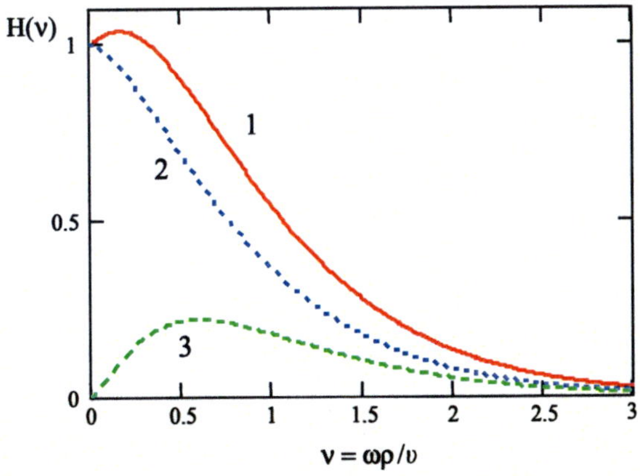

Fig. 5.6 Spectrum of the electric field of an incident electron (5.78) as a function of the dimensionless frequency: 1—total, 2—normal component of the field, 3—tangential component of the field

$$H(v) = F^2(v) + F'^2(v) \tag{5.78}$$

is the spectral function of the electric field strength of an incident electron (see Fig. 5.6). In (5.77), integration with respect to the dimensionless variable $\tilde{\rho} = \rho/a$ is introduced. From (5.77) to (5.78), it follows that the spectrum of the electric field of the scattered electron depends only on the dimensionless parameter $v = \omega\,\rho/\upsilon$.

Figure 5.6 shows that the main contribution to the spectral function $H(v)$ near its maximum is essentially only due to the normal component of the electric field of the electron. In this parameter region, the spectrum width is of the order of magnitude of the ratio υ/ρ.

It is convenient to rewrite formula (5.77) for the collisional excitation cross-section of an atom according

$$\sigma_{\rm ji}(E) = 2\,\pi f_{\rm ji}\left(\frac{e^2}{\Delta E_{\rm ji}}\right)^2 \phi\left(\frac{E}{\Delta E_{\rm ji}}, \eta\right), \tag{5.79}$$

$$\phi(x, \eta) = \frac{1}{x}\int\limits_1^\infty H\left(\frac{\eta\,\tilde{\rho}}{\sqrt{x}}\right)\frac{{\rm d}\tilde{\rho}}{\tilde{\rho}} \tag{5.80}$$

is a dimensionless function that depends on the ratio $E/\Delta E_{\rm ji}$ and on the dimensionless parameter

$$\eta = \frac{1}{\sqrt{2}}\frac{a}{a_{\rm B}}\sqrt{\frac{\Delta E_{\rm ji}}{2{\rm Ry}}}, \tag{5.81}$$

where ${\rm Ry} = 13.6$ eV, $a_{\rm B} = 0.53 \times 10^{-8}$ cm is the Bohr radius. The numerical value η depends on the value of the cutoff length a.

The expression (5.72) for the excitation probability of an atom can be rewritten as

$$W_{\rm ji}(\rho) = \int\limits_0^\infty \sigma_{\rm ji}^{\rm (ph)}(\omega)\,\frac{{\rm d}N(\omega, \rho)}{{\rm d}S\,{\rm d}\omega}\,{\rm d}\omega, \tag{5.82}$$

where

$$\sigma_{\rm ji}^{\rm (ph)}(\omega) = f_{\rm ji}\frac{2\,\pi^2\,e^2}{m\,c}G_{\rm ji}^{(h)}(\omega) \tag{5.83}$$

is the cross-section of photoabsorption of an atom for the transition $i \rightarrow j$. The expression

$$\frac{dN(\omega,\rho)}{dS\,d\omega}=\frac{c}{(2\pi)^2}\frac{|\mathbf{E}(\omega,\rho)|^2}{\hbar\,\omega} \tag{5.84}$$

can be interpreted as the number of photons (per unit area in a unit frequency interval) contained in the electric field of an incident electron during the collision time. Based on the formulas (5.82)–(5.84), the process of atomic collisional excitation can be represented as an atomic absorption of photons forming the eigenfield of a scattered charged particle. Such photons are called equivalent photons. The knowledge of the photoabsorption cross-section (e.g., from experimental data) and the number of equivalent photons allow to obtain the transition probability according (5.82). This approach was used by E. Fermi in 1924 (even before the development of quantum mechanics) for the calculation of the atomic excitation induced by fast-charged particles (Fermi 1924). This theory is called the *Fermi equivalent photon method* (discussed in Sect. 5.1).

5.4.2 Similarity Function Method for Collisional Excitation of an Atom

The formulas (5.79)–(5.81) for the cross-section of collisional excitation of a dipole-allowed transition in an atom in the approximation of straight trajectories are valid for sufficiently high-incident electron energies $E \gg \Delta E_{\rm ji}$. In the vicinity of the excitation threshold, i.e., $E \approx \Delta E_{\rm ji}$, an electron loses practically its whole of kinetic energy and the approximation $\mathbf{v} = \text{const}$ becomes obviously incorrect. The analysis shows that the expression (5.79) can be extended to the whole range of incident electron energies if the function $\varphi\,(E/\Delta E_{\rm ji})$ is properly chosen. This choice can be made either empirically on the basis of comparison with experimental data or on general theoretical considerations. The basis of this approach is the assumption that the ratio

$$\varphi\left(E/\Delta E_{\rm ji}\equiv x\right)=\frac{\sigma_{\rm ji}(E)}{2\,\pi f_{\rm ji}\left(e^2/\Delta E_{\rm ji}\right)^2} \tag{5.85}$$

is a universal function of the dimensionless variable $x = E/\Delta E_{\rm ji}$ only. Equation (5.85) expresses the essence of the *similarity function method* for the calculation of the cross-section of collisional excitation of an atom. Quantum mechanical consideration shows that the similarity function $\varphi(x)$ should satisfy two asymptotic relations:

$$\varphi(x\to 1)\propto\sqrt{x-1}\ \text{and}\ \varphi(x\gg 1)\propto\frac{\ln(x)}{x}. \tag{5.86}$$

In view of (5.86), $\varphi(x)$ can be approximated by Astapenko et al. (2000)

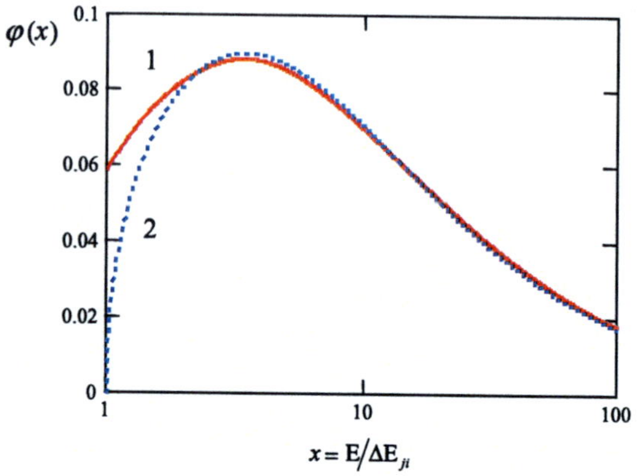

Fig. 5.7 Empirical similarity function (1) and the similarity function calculated in the approximation of straight trajectories (2) for the cross-section of collisional excitation of a dipole-allowed transition in an atom

$$\varphi(x) = \frac{\ln\left(1 + a\sqrt{x-1}\right)}{x+b}. \qquad (5.87)$$

The values of the parameters a and b can be obtained from experimental data: $a \approx 0.5$, $b \approx 3$. Let us note the difference of the near-threshold behavior of the similarity function for the excitation of an atom, i.e., $\varphi(x \approx 1) \propto \sqrt{x-1}$ from the corresponding dependence for ionization of an atom by electron impact, i.e., $f(x \approx 1) \propto x - 1$.

It is of interest to compare the similarity function (5.87) with the expression obtained in the approximation of straight trajectories (5.80), (5.78). This comparison for $\eta = 1.7$ is given in Fig. 5.7. From this figure, it follows that for $x > 3$ both functions practically coincide. In particular, the position of maxima $x_{max} \cong 3.45$ and the maximum values $\varphi_{max} \cong 0.09$ coincide. A noticeable difference exists only in the near-threshold region $1 < x < 2$, where the approximation of straight trajectories is inadequate.

Therefore, the cross-section of collisional excitation of a dipole-allowed transition in an atom for arbitrary energies of an incident electron, including the near-threshold region, can be represented as

$$\sigma_{\mathrm{ji}}(E) = 2\pi\,\sigma_{\mathrm{a}} f_{\mathrm{ji}} \left(\frac{2\,\mathrm{Ry}}{\Delta E_{\mathrm{ji}}}\right)^2 \varphi\left(\frac{E}{\Delta E_{\mathrm{ji}}}\right), \qquad (5.88)$$

where the similarity function $\varphi(x)$ is given by the formula (5.87). Note that (5.88) is expressed in atomic units ($e = 1$, $E_{\mathrm{a}} = 2\,\mathrm{Ry} \cong 27.2$ eV, and $\sigma_{\mathrm{a}} = a_{\mathrm{B}}^2 \cong 2.8 \times 10^{-17}$ cm^2). In view of the above values for x_{max}, φ_{max}, the formula (5.88) gives for the maximum cross-section

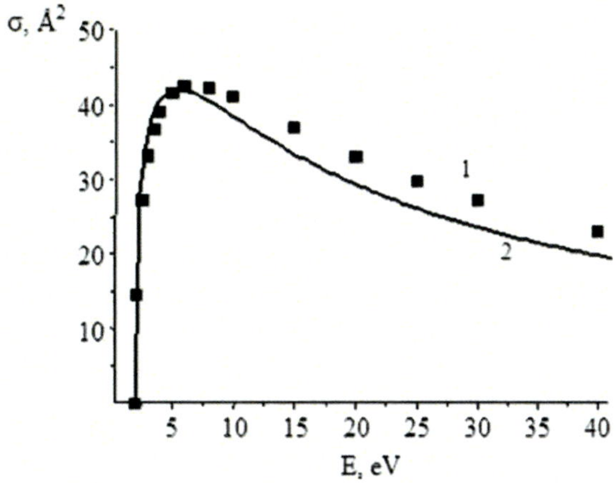

Fig. 5.8 Cross-section of excitation of a lithium atom by electron impact for the transition $2s \rightarrow 2p$: 1—experimental data of the American National Standards Institute (NIST 2019), 2—calculation by the similarity function method according (5.88)

$$\sigma_{ji}\left(E_{max} = 3.45\,\Delta E_{ji}\right) \cong 0.63 \times 10^{-16} f_{ji}\left(\frac{Ry}{\Delta E_{ji}}\right)^2 \,[cm^2]. \qquad (5.89)$$

Thus, the cross-section of collisional excitation of a dipole-allowed transition is directly proportional to the transition oscillator strength and inversely proportional to the squared excitation energy. A characteristic value for the transition energies in a neutral atom is 1–10 eV. The oscillator strengths vary in a more wide range: from 10^{-6} to 2. At $f_{ji} < 10^{-6}$, an electron transition in an atom can be considered to be forbidden. The maximum value $f_{ji} \cong 2$ is reached for transitions with no change in a principal quantum number in atoms of alkaline-earth elements.

Figure 5.8 shows the experimental (curve 1) and calculated (curve 2) excitation cross-sections for the transitions $2s \rightarrow 2p$ in the lithium atom. Because there is no change in principal quantum number, the oscillator strength is rather high ($f_{2p \rightarrow 2s} = 0.75$) and the excitation energy rather low ($\Delta E_{2p \rightarrow 2s} \approx 1.85$ eV). Therefore, the maximum value of the cross-section is rather large: $\sigma_{max} \cong 4.3 \times 10^{-15}$ cm^2. The position of the maximum corresponds to an energy of about 7.5 eV. Figure 5.8 demonstrates rather good agreement between theory and experiment, especially in the vicinity of threshold.

5.4.3 *Analytical Empirical Formulas for Excitation and De-excitation Rates*

5.4.3.1 Dipole Excitation and De-excitation of Ions

For dipole-allowed transitions (i.e., $\Delta l = \pm 1$), one of the most used general formulas has been proposed by Van Regemorter (1962). The corresponding excitation rates (integration of the cross-section over a Maxwellian energy distribution function) can be cast in the following analytical form:

$$\langle v\sigma(\alpha \to \alpha')\rangle := C_{\alpha\alpha'} = 3.15 \times 10^{-7} f_{\alpha\alpha'} \left(\frac{\mathrm{Ry}}{\Delta E_{\alpha\alpha'}}\right)^{3/2} \cdot \sqrt{\beta} \cdot \mathrm{e}^{-\beta} \cdot p^{(Z>0)}(\beta), \quad (5.90a)$$

$$\Delta E_{\alpha\alpha'} = E_{\alpha} - E_{\alpha'}, \quad (5.90b)$$

$$\beta = \frac{\Delta E_{\alpha\alpha'}}{T_{\mathrm{e}}}. \quad (5.90c)$$

$f_{\alpha\alpha'}$ is the oscillator strength for the dipole transition from state α to state α', and $p^{(Z>0)}(\beta)$ is an effective Gaunt factor. The required oscillator strength $f_{\alpha\alpha'}$ is easily obtained from the spontaneous transition probability $A_{\alpha'\alpha}$

$$f_{\alpha\alpha'} = \frac{1}{4.339 \times 10^{7}\,(E_{\alpha} - E_{\alpha'})^{2}} \frac{g_{\alpha'}}{g_{\alpha}} A_{\alpha'\alpha} \quad (5.91)$$

with E_{α} and E_{α} expressed in [eV], g_{α} and $g_{\alpha'}$ are the statistical weights of the lower and upper states, respectively. Note that (5.91) expresses the absorption oscillator strength in terms of the spontaneous emission coefficient. The effective Gaunt-factor $p(\beta)$ can be approximated by an analytical expression:

$$p^{(Z>0)}(\beta) = 0.2757\,\mathrm{e}^{-1.3\beta}\left(\beta - \frac{\beta^{2}}{4} - \ln(\beta) - 0.5772\right) + 0.2\left(1 - \mathrm{e}^{-4.5\beta}\right). \quad (5.92)$$

This formula provides the correct asymptotic behavior for low and high energies and an accuracy better than 5% for all values of β.

De-excitation rates are obtained from the principle of detailed balance, i.e.,

$$C_{\alpha'\alpha} = C_{\alpha\alpha'} \frac{g_{\alpha}}{g_{\alpha'}} \exp(\beta) \quad (5.93)$$

providing

$$C_{\alpha'\alpha} = 3.15 \times 10^{-7} f_{\alpha\alpha'} \frac{g_\alpha}{g_{\alpha'}} \left(\frac{\text{Ry}}{\Delta E_{\alpha\alpha'}} \right)^{3/2} \cdot \sqrt{\beta} \cdot p(\beta). \tag{5.94}$$

Equation (5.94) demonstrates that de-excitation rates do not contain the exponential factor from the Maxwellian average. For low temperatures, i.e., for parameters $\beta > 10$, the effective Gaunt-factor approaches $p^{(Z>0)}(\beta) \to 0.2$ (approaching a finite value for the Gaunt factor is due to the attraction of the electron via the Coulomb potential of the atoms, i.e., the electron practically falls into the potential of the ion and excites the ion) and we find formally the following asymptotic expressions:

$$C_{\alpha\alpha'} \to 6.3 \times 10^{-8} f_{\alpha\alpha'} \left(\frac{\text{Ry}}{\Delta E_{\alpha\alpha'}} \right)^{3/2} \sqrt{\beta} \cdot \exp(-\beta) \propto \frac{e^{-\Delta E_{\alpha\alpha'}/T_e}}{\sqrt{T_e}}, \tag{5.95}$$

$$C_{\alpha'\alpha} \to 6.3 \times 10^{-8} f_{\alpha\alpha'} \frac{g_\alpha}{g_{\alpha'}} \left(\frac{\text{Ry}}{\Delta E_{\alpha\alpha'}} \right)^{3/2} \sqrt{\beta} \propto \frac{1}{\sqrt{T_e}}. \tag{5.96}$$

It should be noted that (5.95), (5.96) provide only approximate values because for low temperatures, the Born approximation is not valid anymore and normalization of the transition probability becomes of importance. Equations (5.95) and (5.96) demonstrate that for small temperatures, the excitation rate vanishes whereas the de-excitation rate is rather large. Therefore, in low-temperature recombining plasmas, the collisional processes are dominated by the de-excitation of the populated levels. For high temperatures, i.e., $\beta < 10$, we find the following asymptotic expressions:

$$C_{\alpha\alpha'} \to 8.7 \times 10^{-8} f_{\alpha\alpha'} \left(\frac{\text{Ry}}{\Delta E_{\alpha\alpha'}} \right)^{3/2} \sqrt{\beta} \cdot \ln(1/\beta) \propto \frac{\ln T_e}{\sqrt{T_e}}, \tag{5.97}$$

$$C_{\alpha'\alpha} \to 8.7 \times 10^{-8} f_{\alpha\alpha'} \frac{g_\alpha}{g_{\alpha'}} \left(\frac{\text{Ry}}{\Delta E_{\alpha\alpha'}} \right)^{3/2} \sqrt{\beta} \cdot \ln(1/\beta) \propto \frac{\ln T_e}{\sqrt{T_e}}. \tag{5.98}$$

Therefore, for high temperatures (i.e., hot electrons), excitation and de-excitation rates have the same asymptotic behavior and are identical except the ratio of the statistical weights.

5.4.3.2 Dipole Excitation and De-excitation of Neutral Atoms

Dipole excitation and de-excitation by collisions between electrons and neutral atoms can likewise be described by the formulas (5.90), (5.91), (5.93), however, with a modified effective Gaunt factor. The following analytical formula is proposed:

$$p^{(Z=0)}(\beta) = \left\{ \begin{array}{ll} \text{if } \beta \leq 0.4: & 0.27566 \cdot \left(\beta - \dfrac{\beta^2}{4} + \dfrac{\beta^3}{12} - \ln(\beta) - 0.57722 \right) \\ \text{else} & 0.066 \dfrac{\sqrt{\beta + 2}}{\beta + 0.127} \end{array} \right\}. \tag{5.99}$$

These formulas provide the correct asymptotic behavior for low and high energies and an accuracy better than 3% for all values of β.

For low temperature, we find formally the following asymptotic expressions:

$$C_{\alpha\alpha'} \to 2.1 \times 10^{-8} f_{\alpha\alpha'} \frac{g_\alpha}{g_{\alpha'}} \left(\frac{\text{Ry}}{\Delta E_{\alpha\alpha'}} \right)^{3/2} \frac{\exp(-\beta)}{\sqrt{\beta}} \propto \sqrt{T_e} \cdot \exp(-\Delta E_{\alpha\alpha'}/T_e), \tag{5.100}$$

$$C_{\alpha'\alpha} \to 2.1 \times 10^{-8} f_{\alpha\alpha'} \frac{g_\alpha}{g_{\alpha'}} \left(\frac{\text{Ry}}{\Delta E_{\alpha\alpha'}} \right)^{3/2} \frac{1}{\sqrt{\beta}} \propto \sqrt{T_e}. \tag{5.101}$$

It should likewise be noted here that (5.100), (5.101) provide only very approximate values because for low temperatures, the Born approximation is not valid, and normalization of the transition probability becomes of importance. Equations (5.100), (5.101) indicate that, unlike for the case of electron excitation of ions, the excitation and de-excitation rates vanish both for low temperatures. For high temperatures, the following asymptotes are obtained:

$$C_{\alpha\alpha'} \to 8.7 \times 10^{-8} f_{\alpha\alpha'} \left(\frac{\text{Ry}}{\Delta E_{\alpha\alpha'}} \right)^{3/2} \sqrt{\beta} \cdot \ln(1/\beta) \propto \frac{\ln T_e}{\sqrt{T_e}}, \tag{5.102}$$

$$C_{\alpha'\alpha} \to 8.7 \times 10^{-8} f_{\alpha\alpha'} \frac{g_\alpha}{g_{\alpha'}} \left(\frac{\text{Ry}}{\Delta E_{\alpha\alpha'}} \right)^{3/2} \sqrt{\beta} \cdot \ln(1/\beta) \propto \frac{\ln T_e}{\sqrt{T_e}}. \tag{5.103}$$

Comparing (5.102), (5.103) with (5.97), (5.98), we find that the high-temperature asymptotes of neutral atoms and ions are identical.

Finally we note that numerous variants of the effective Gaunt factors $p^{(Z>0)}(\beta)$ and $p^{(Z=0)}(\beta)$ are proposed in the literature, see, e.g., (Fischer et al. 1996). They practically do all not differ very much.

5.5 Excitation of Dipole-Forbidden Transitions in Atoms

5.5.1 *Intercombination Transitions*

The previous section considered collisional excitation of dipole-allowed transitions in atoms that can be described classically with the use of the concept of a transition oscillator. For dipole-forbidden transitions, this approach is not applicable because $f_{ji} = 0$. In this case, the interaction between an incident electron and an atom/ion is of *non-dipole* nature.

Dipole-forbidden transitions can be of two types: (1) no change in an atomic spin and (2) with a change in an atomic spin. In the first case, there is no dipole moment of a transition because of non-fulfillment of selection rules (in pure *LS*-coupling) for the orbital quantum number *L*: $|L_j - L_i| > 1$ or $L_i = L_j = 0$. Excitation for these atomic transitions is due to *direct* Coulomb interaction of an incident electron with quadrupole or other more higher multipole moments.

The dependences of the cross-section of collisional excitation of transitions of the first type on the incident electron energy E in the near-threshold region ($x = E/\Delta E_{ji} \approx 1$) and for high energies ($x \gg 1$) are the same as for dipole-allowed transitions [see (5.86)]. In view of this fact, the most simple approximation of the excitation cross-section of a dipole-forbidden transition *with no change in spin* can be represented as

$$\sigma_{ji}\left(x = E/\Delta E_{ji}\right) = c\frac{\sqrt{x-1}}{a + x^{3/2}}, \tag{5.104}$$

where a, c are the parameters that define the incident electron energy at the cross-section maximum ($E_{max} = \Delta E_{ji}\, x_{max}$) and the value of the cross-section maximum σ_{max} itself. It should be noted that formula (5.104) can be used for an approximate description of the cross-section of a dipole-allowed transition. As a rule, the maximum cross-section σ_{max} for dipole-forbidden transitions with no change in atomic spin is much less than a corresponding dipole-allowed transition. The maximum of the cross-section of dipole-forbidden transitions is shifted to the region of lower energies in comparison with dipole-allowed transitions (in the majority of cases $1.5 < x_{max} < 2$, whereas with increasing excitation energy x_{max} decreases).

Now let us consider collisional excitation of dipole-forbidden transitions with a change in atomic spin, the so-called *intercombination* transitions (see also Sect. 1.2.2). In this case, excitation of an atom occurs due to *exchange* interaction between incident and atomic electrons. Exchange interaction is essentially of quantum mechanical nature. At a qualitative level, the process can be described as follows. An incident electron transfers a considerable part of its energy ΔE to an atomic electron that is in a state with energy E_i. As a result, the incident electron is captured into an atomic orbit with an energy $E_j > E_i$, and the atomic electron is ionized. Thus, the incident and atomic electrons seem to exchange their roles that are based on the indistinguishability of electrons.

The energy transfer, at which exchange excitation of an atom at the transition $i \to j$ occurs, is determined by the inequality

$$E + |E_{\rm j}| \le \Delta E \le E + |E_{\rm i}|. \tag{5.105}$$

We note that bound states of atomic electrons correspond to negative energies $E_{\rm i,j} < 0$. The simple classical consideration is valid far from threshold $x \gg 1$ when [according to (5.105)] the energy transferred from an incident electron to an atomic electron is much larger than the excitation energy: $\Delta E \gg \Delta E_{\rm ji}$. Therefore, exchange interaction is more strong than direct interaction if $\Delta E = \Delta E_{\rm ji}$. It occurs at small distances from the atomic nucleus (of the order of the size of an excited electron orbit), in contrast to the electron–dipole interaction for the excitation of dipole-allowed transitions that occur at long distances from an atom.

According to the above-developed physical picture of cross-section calculation, it is possible to use the expression (5.34) for the cross-section of electron energy transfer in Rutherford scattering. Integrating this formula within the limits determined by the relation (5.105), we find

$$\sigma_{\rm ji}^{(\rm inter)}\left(E \gg \Delta E_{\rm ji}\right) = \frac{\pi e^4}{E} \frac{\Delta E_{\rm ji}}{\left(E + |E_{\rm j}|\right)\left(E + |E_{\rm i}|\right)}. \tag{5.106}$$

It is convenient to rewrite this equation in terms of the dimensionless variable $x = E/\Delta E_{\rm ji}$. From (5.106), it follows then the asymptotic expression for the intercombination transition cross-section in the high-energy domain, i.e., $E \gg |E_{\rm i}|$:

$$\sigma_{\rm ji}^{(\rm inter)}(E \gg |E_{\rm i}|) = \frac{\pi e^4}{\left(\Delta E_{\rm ji}\right)^2} \frac{1}{x^3}. \tag{5.107}$$

Therefore at high-incident electron energies, the excitation cross-section of an intercombination transition decreases more rapidly than the cross-section with no change in spin (5.104). This fact is connected with the necessity of transfer of a large quantity of energy in exchange interaction [see the relations (5.105)] resulting in intercombination excitation.

In the near-threshold region of energies, i.e. $x = E/\Delta E_{\rm ji} \approx 1$, the same asymptotic is valid for the intercombination cross-section as for the dipole case [$\sigma \propto \sqrt{x-1}$, see (5.86)]. Combining the limiting cases [see analogy for the derivation of (5.104)], the following simple approximation can be obtained for the excitation cross-section of an intercombination transition in an atom by electron impact (Astapenko et al. 2000):

$$\sigma_{\rm ji}^{(\rm inter)}\left(x = \frac{E}{\Delta E_{\rm ji}}\right) = c\frac{\sqrt{x-1}}{a + x^{7/2}}, \tag{5.108}$$

where a and c are the parameters characterizing the transition under consideration. The obtained expression is valid in a wide range of projectile energies up to relativistic values. Figure 5.9 shows the cross-section of collisional excitation of the intercombination transition $3s\,{}^1S \to 3p\,{}^3P$ in a magnesium atom calculated by (5.108) for $c = 2 \times 10^{-14}$ cm^2 and $a = 10$ together with experimental data.

In the initial state, there are two valence electrons with antiparallel spins in the 3*s*-subshell of a magnesium atom, i.e., the total spin is zero. In a collision with an incident electron, one of the 3*s-valence* electrons is excited to the 3*p*-subshell due to exchange interaction with a spin flip, resulting in a total atomic spin of one. From Fig. 5.9, it is seen that the cross-section is rather large and that $x_{max} = E_{max}/\Delta E_{ji} \approx 1.85$, which is considerably less than for the case of a dipole-allowed transition, when $x_{max} \cong 3.45$. It follows from it that the large value of the cross-section maximum is caused by a relatively low value of the excitation threshold ($\Delta E_{ji} \cong 2.7$ eV).

5.5.2 *Intermediate Coupling Effects*

The asymptotic behavior of the intercombination cross-sections for high-energy is strictly valid only in the *LS*-coupling scheme. In general, however, intermediate coupling admixes spin-allowed cross-sections to the exchange cross-section thereby changing entirely the asymptotic behavior. As has been shown by Vainshtein (Sobelman and Vainshtein 2006), the excitation cross-section can be expressed in terms of products of radial cross-sections and angular factors. In the intermediate coupling scheme, this can be formulated for the mixed states a_0 and a_1 in the following way:

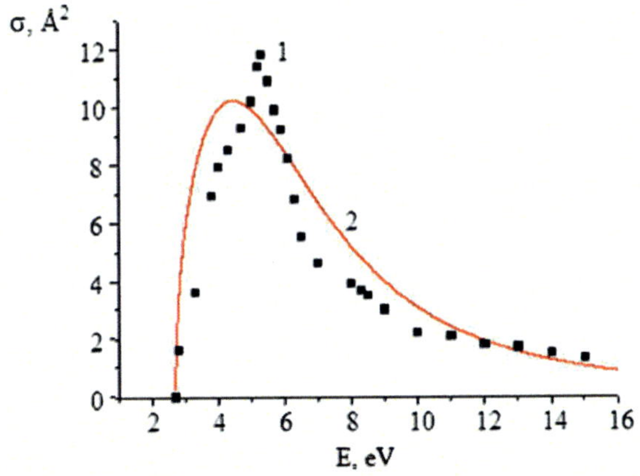

Fig. 5.9 Excitation cross-section by electron impact of the intercombination transition $3s\,{}^1S \to 3p\,{}^3P$ in a magnesium atom: 1—experimental data (NIST 2019), 2—model cross-section (5.108)

$$\begin{aligned}\sigma_{a_0a_1} = \sum_{\kappa} \{ & Q_{\kappa}^{d}(a_0,a_1)\sigma_{\kappa}^{d}(n_0l_0,n_1l_1) + Q_{\kappa}^{e}(a_0,a_1)\sigma_{\kappa}^{e}(n_0l_0,n_1l_1) \\ & + Q_{\kappa+1}^{e}(a_0,a_1)\sigma_{\kappa+1}^{e}(n_0l_0,n_1l_1)\} \end{aligned} . \quad (5.109)$$

$\sigma_{\kappa}^{d}(n_0l_0,n_1l_1)$, $\sigma_{\kappa}^{e}(n_0l_0,n_1l_1)$, and $\sigma_{\kappa+1}^{e}(n_0l_0,n_1l_1)$ are the radial parts of the one-electron cross-sections of direct and exchange contributions of multiplicity κ (the multiplicity can vary in the interval $\kappa = \kappa_{\min}, \kappa_{\min}+2, \ldots, \kappa_{\max}$ with $\kappa_{\min} = |l_0 - l_1|$ and $\kappa_{\max} = l_0 + l_1$), $Q_{\kappa}^{d}(a_0,a_1)$, $Q_{\kappa}^{e}(a_0,a_1)$, and $Q_{\kappa+1}^{e}(a_0,a_1)$ are the corresponding angular factors of direct and exchange terms. If the mixed state is represented by

$$\Psi(a) = \sum_{\mathrm{LS}} \langle a \mid a_{\mathrm{LS}} \rangle \Psi(a_{\mathrm{LS}}), \quad (5.110)$$

where $\langle a \mid a_{\mathrm{LS}} \rangle$ are the mixing coefficients, the angular factors are given by

$$Q_{\kappa}^{d} = \frac{2l_0+1}{2J_0+1} \cdot b_{\kappa}^{2}(a_0,a_1), \quad (5.111)$$

$$Q_{\kappa}^{e} = \frac{2l_0+1}{2J_0+1} \cdot \left(\frac{b_{\kappa}^{2}(a_0,a_1)}{4} + \sum_{x} b_{\kappa,x}^{2}(a_0,a_1) \right). \quad (5.112)$$

The respective amplitudes in intermediate coupling can be expressed in terms of the known mixing coefficients and amplitudes in *LS*-coupling:

$$b_{\kappa,x}(a_0,a_1) = \sum_{L_0S_0L_1S_1} \langle a_0 \mid a_{0,L_0S_0} \rangle b_{\kappa,x}^{(LS)}(a_{0,L_0S_0}, a_{1,L_1S_1}) \langle a_{1,L_1S_1} \mid a_1 \rangle, \quad (5.113)$$

$$b_{\kappa}(a_0,a_1) = \sum_{L_0S_0L_1S_1} \langle a_0 \mid a_{0,L_0S_0} \rangle b_{\kappa}^{(LS)}(a_{0,L_0S_0}, a_{1,L_1S_1}) \langle a_{1,L_1S_1} \mid a_1 \rangle. \quad (5.114)$$

$b_{\kappa}^{(LS)}(a_{0,L_0S_0}, a_{1,L_1S_1})$ and $b_{\kappa,x}^{(LS)}(a_{0,L_0S_0}, a_{1,L_1S_1})$ are the amplitudes in *LS*-coupling that have analytical solution in terms of the $3j$ and $6j$ symbols, the quantum numbers of the atomic core L_PS_P and of the optical electron l_0 and l_1, the number of equivalent electrons m, and the fractional parentage coefficient $G_{L_PS_P}^{L_0S_0}$:

$$b_{\kappa}^{(LS)}(a_{0,L_0S_0}, a_{1,L_1S_1}) = \delta_{S_0S_1} \cdot (-1)^{J_1-S_0} \cdot [J_0J_1] \cdot \begin{Bmatrix} \kappa & J_0 & J_1 \\ S_0 & L_1 & L_0 \end{Bmatrix} \cdot \tilde{b}_{\kappa}^{(LS)}(L_0L_1), \quad (5.115)$$

$$b_{\kappa,x}^{(LS)}(a_{0,L_0S_0},a_{1,L_1S_1}) = \sqrt{3/2}\cdot(-1)^{S_P-S_1+1/2+L_1}\cdot[J_0J_1S_0S_1x]\cdot \begin{Bmatrix} \kappa & J_0 & x \\ S_0 & L_1 & L_0 \end{Bmatrix}\begin{Bmatrix} x & J_1 & 1 \\ S_1 & S_0 & L_1 \end{Bmatrix}\begin{Bmatrix} 1 & S_0 & S_1 \\ S_P & 1/2 & 1/2 \end{Bmatrix}\cdot\tilde{b}_{\kappa}^{(LS)}(L_0L_1), \quad (5.116)$$

$$\tilde{b}_{\kappa}^{(LS)}(L_0L_1) = \sqrt{m}\cdot(-1)^{L_P}\cdot[L_0L_1]\cdot\begin{Bmatrix} \kappa & L_0 & L_1 \\ L_P & l_1 & l_0 \end{Bmatrix}\cdot G_{L_PS_P}^{L_0S_0}, \quad (5.117)$$

$$[J_0J_1\ldots] = [(2J_0+1)\cdot(2J_1+1)\cdot\ldots]^{1/2}. \quad (5.118)$$

Table 5.1 Intermediate coupling angular coefficients and fitting parameters for rate coefficients

Transition	Q^d	Q^e	A^d/A^e	χ^d/χ^e	D^d/D^e
$1s^2\ ^1S_0 - 1s2s\ ^1S_0$					
LS-coupling	2	0.5			
$Z_n = 9$	2	0.5	3.31/2.09	0.782/0.608	0.350/0.00
$Z_n = 18$	2	0.5	3.40/2.18	0.641/0.665	0.30/0.00
$Z_n = 42$	2	0.5	3.33/2.48	0.933/1.18	1.00/0.05
$1s^2\ ^1S_0 - 1s2s\ ^3S_1$					
LS-coupling	0	1.5			
$Z_n = 9$	0	1.5	2.13/1.83	0.0651/0.587	-0.80/0.00
$Z_n = 18$	0	1.5	2.83/2.07	0.287/0.657	-0.40/0.00
$Z_n = 42$	0	1.5	3.33/2.48	0.933/1.18	1.00/0.05
$1s^2\ ^1S_0 - 1s2p\ ^1P_1$					
LS-coupling	2	0.5			
$Z_n = 9$	2.00	0.5	5.90/12.7	0.378/1.08	4.50/0.00
$Z_n = 18$	1.97	0.5	8.91/12.6	0.217/1.07	2.10/0.00
$Z_n = 42$	1.52	0.5	9.35/14.3	0.392/1.69	3.25/0.05
$1s^2\ ^1S_0 - 1s2p\ ^3P_0$					
LS-coupling	0	1.67×10^{-1}			
$Z_n = 9$	0	1.67×10^{-1}	6.19/13.5	0.386/1.07	4.50/0.00
$Z_n = 18$	0	1.67×10^{-1}	9.03/13.0	0.222/1.06	2.15/0.00
$Z_n = 42$	0	1.67×10^{-1}	9.35/14.3	0.392/1.69	3.25/0.05
$1s^2\ ^1S_0 - 1s2p\ ^3P_1$					
LS-coupling	**0**	0.5			
$Z_n = 9$	$\mathbf{6.27\times10^{-4}}$	0.5	6.19/13.5	0.386/1.07	4.50/0.00
$Z_n = 18$	$\mathbf{3.21\times10^{-2}}$	0.5	9.02/13.0	0.221/1.07	2.15/0.00
$Z_n = 42$	$\mathbf{4.85\times10^{-1}}$	0.5	9.35/14.3	0.392/1.69	3.25/0.05
$1s^2\ ^1S_0 - 1s2p\ ^3P_2$					
LS-coupling	0	8.33×10^{-1}			
$Z_n = 9$	0	8.33×10^{-1}	6.19/13.5	0.386/1.07	4.50/0.00
$Z_n = 18$	0	8.33×10^{-1}	9.06/1.29	0.218/1.06	2.10/0.00
$Z_n = 42$	0	8.33×10^{-1}	9.35/14.3	0.392/1.69	3.25/0.05

The fitting coefficients for $Z_n = 42$ can be used for any ion with $Z_n > 10$. The precision of the fitting coefficients β, χ, D is typically better than 10% in a large temperature interval of $1/8 < \beta < 64$

It is difficult to obtain general expressions for the cross-sections in intermediate coupling because the calculation of the mixing coefficients $\langle a \mid a_{LS}\rangle$ requests numerical calculations of the atomic structure and the sums in (5.109), (5.111)–(5.114) are rather cumbersome.

In order to provide some insight into the intermediate coupling effects on excitation cross-sections, let us consider the excitation of the He-like excited levels $a_1 = 1s2l\,LSJ$ from the ground state $a_0 = 1s^2\ ^1S_0$. In this case, formulas (5.109)–(5.114) can be considerably simplified ($n_0 = 1;\ l_0 = 0;\ n_1 = 2;\ l_1 = 0, 1;\ \kappa = 1$):

$$\sigma_{a_0a_1} = Q_\kappa^d \sigma_\kappa^d(n_0l_0, n_1l_1) + Q_\kappa^e \sigma_\kappa^e(n_0l_0, n_1l_1). \tag{5.119}$$

Table 5.1 shows the angular Q-factors for LS-coupling and intermediate coupling for various elements. The angular factors for the intercombination transition $1s^2\ ^1S_0 - 1s2p\ ^3P_1$ show that in LS-coupling the contribution of the direct cross-section is zero ($Q^d = 0$) and gradually increases with increasing nuclear charge (see bold values in Table 5.1). For molybdenum ($Z_n = 42$), the intermediate coupling effect is so strong that the angular factor reaches a value of $Q^d = 0.485$ which is about one-third of the angular factor for the resonance transition $1s^2\ ^1S_0 - 1s2p\ ^1P_1$. Therefore, the high-energy asymptote is entirely dominated by the direct cross-section rather than by the exchange cross-section, i.e.,

$$\sigma(1s^2\ ^1S_0 - 1s2p^3P_1) = Q_1^d\sigma_1^d(1,0;2,1) + Q_1^e\sigma_1^e(1,0;2,1) \underset{E\gg\Delta E}{\rightarrow} Q_1^d\sigma_1^d(1,0;2,1). \tag{5.120}$$

For completeness, Table 5.1 provides also the fitting parameters for the explicit calculation of the corresponding rate coefficients averaged over a Maxwellian distribution function.

The adopted fitting formulas are as follows:

$$\frac{C_{\rm ij}}{({\rm cm^3\,s^{-1}})} = \frac{10^{-8}}{Z^3}\left(\frac{E_{\rm j}}{E_{\rm i}}\right)^{3/2}\sqrt{\beta}\cdot\exp\left(-\frac{\Delta E_{\rm ij}}{T_{\rm e}}\right) \cdot\left\{\frac{Q^d\cdot A^d\cdot(\beta+1+D^d)}{\beta+\chi^d} + \frac{Q^e\cdot A^e\cdot(\beta+D^e)}{\beta+\chi^e}\right\}, \tag{5.121}$$

$$\beta = \frac{Z^2{\rm Ry}}{T_{\rm e}}, \tag{5.122}$$

$$\Delta E_{\rm ij} = E_{\rm i} - E_{\rm j}, \tag{5.123}$$

where Z is the spectroscopic symbol and Ry = 13,606 eV. Let us consider an example for molybdenum and the intercombination transition $1s^2\ {}^1S_0 - 1s2p\ {}^3P_1$ for an electron temperature of $T_e = 20{,}000$ eV: $Z = 42 - 2 + 1 = 41$, $E_{i=1s^2\ {}^1S_0} = 23{,}810.6\,\text{eV}$, $E_{j=1s2p\ {}^3P_1} = 5903.7\,\text{eV}$, $\Delta E_{ij} = 17{,}906.9\,\text{eV}$, $\beta = 1.277$: $C_{ij} \approx 1.5 \times 10^{-13}\,\text{cm}^3\,\text{s}^{-1}$. This is the intermediate coupling cross-section whereas in pure LS-coupling $C_{ij}^{LS} \approx 2.6 \times 10^{-14}\,\text{cm}^3\,\text{s}^{-1}$. This example demonstrates not only the importance of intermediate coupling on the high-energy asymptotes of cross-sections and rates but also its impact on the total rate coefficient for rather moderate temperatures (β—values of the order of one). We note that the fitting coefficients in Table 5.1 do not include resonance contributions. These contributions are most pronounced for the excitation of the $1s2s\ {}^3S_1$ state (some 10%) and the $1s2p\ {}^3P_2$ state (about 10%). For applications in plasma atomic physics, resonance contributions can be rather well included in atomic kinetics via explicit inclusion of multiple excited autoionizing states as suggested by Cowan (1980, 1983).

It should be noted that the decreasing of the cross-section σ_κ with decreasing κ is not connected with any small parameter. This differs radically from the interaction of an atom with an electromagnetic field where higher multipoles contain the factor $(Z \cdot \alpha)^{2\kappa+1}$ making each successive term smaller by a factor of about $5 \times 10^{-5} \cdot Z^2$. In the case of electron–atom collisions, such small parameter does not exist. Numerical calculations show that the multipole cross-section $\sigma_{\kappa+2}$ is usually about ten times smaller than σ_κ (Sobelman and Vainshtein 2006) but might be in some cases have even larger contributions than the lowest one (Rosmej 2000). Finally we note that unlike radiative transitions, non-dipole transitions (e.g., monopole, quadrupole) can have rather large cross-section values and are not at all negligible for high-precision calculations.

For rapid calculations of large transition arrays, the Regemorter approach provides a reasonable estimate of the collisional cross-sections for the total emission group. For similar purposes, the plane wave Born PWB approximation attracts interest up to present days, because this approach can be easily incorporated in atomic structure codes and allows to estimate also monopole and quadrupole transitions and shows a correct high-energy behavior (which is difficult to obtain in more complex numerical methods like the R-matrix and the convergent-close-coupling method). The pathological behavior at threshold of the PWB approximation can be removed by the empirical Robb–Cowan approach (Cowan 1981) that has recently been improved by the so-called Elwert–Sommerfeld correction factor (Kilcrease and Brookes 2013).

Other important corrections to the first-order cross-sections have been proposed by Vainshtein (Sobelman and Vainshtein 2006): one-channel normalization, K-matrix, inclusion of exchange and intermediate coupling effects (note that the above fitting parameters of Table 5.1 are based on numerical calculations of the one-channel normalized Coulomb–Born approximation including exchange and intermediate coupling effects). For example, it has been demonstrated for neutral helium HeI (Beigman et al. 2000) that the application of the K-matrix method to pure Born cross-sections results in rather good agreement with the numerically very complex convergent-close-coupling calculations CCC. Detailed electron collisional excitation and deexcitation rate coefficients (including intercombination transitions) for H I, He I and He II are presented in Annex 2 and 3.

Finally we note that the application of PWB collisions strengths to line polarization provides reasonable numerical values as essentially ratios of cross-sections enter to the line polarization formulas (Cowan 1981; Percival and Seaton 1958).

5.6 Analytical Empirical Formulas for Dielectronic Recombination in Dense Plasmas

Dielectronic recombination is a combination of dielectronic capture and subsequent radiative stabilization competing with multi-channel autoionization. In non-LTE plasmas, excited states coupling, angular momentum changing collisions, collisional depopulation and ionization potential depression strongly alter the stabilization processes. A consistent description of non-LTE dielectronic recombination involves therefore atomic kinetics and electric field perturbations of the atomic structure and related matrix elements (Rosmej et al. 2020).

5.6.1 *Autoionization, Dielectronic Capture, and Dielectronic Recombination*

Dielectronic recombination can easily be calculated from the autoionizing rate of a certain atomic state with the help of the principle of detailed balance. The first step is the application of the principle of detailed balance to dielectronic capture, i.e.,

$$n_{\mathrm{j}}^{Z} \cdot \Gamma_{\mathrm{jk}}^{Z,Z+1} = n_{\mathrm{k}}^{Z+1} \cdot n_{\mathrm{e}} \cdot \langle \mathrm{DC} \rangle_{\mathrm{kj}}, \tag{5.124}$$

where n_{j}^{Z} is the atomic population of the upper state, $\Gamma_{\mathrm{jk}}^{Z,Z+1}$ is the autoionizing rate from the upper state to a state k with population n_{k}^{Z+1}, and $\langle \mathrm{DC} \rangle_{\mathrm{kj}}$ is the dielectronic capture rate from state k to the upper state j. In thermodynamic equilibrium, the populations n_{j}^{Z} and n_{k}^{Z+1} are linked via the Saha–Boltzmann equation because the states j and k belong to different ionic states, Z and $Z+1$ respectively, i.e.,

$$\frac{n_{\mathrm{j}}^{Z}}{n_{\mathrm{k}}^{Z+1}} = n_{\mathrm{e}} \cdot \frac{g_{\mathrm{j}}^{Z}}{2g_{\mathrm{k}}^{Z+1}} \cdot \left(\frac{2\pi\hbar^2}{m_{\mathrm{e}} k T_{\mathrm{e}}} \right)^{3/2} \cdot \exp\left(\frac{\Delta E_{\mathrm{k,j}}^{Z+1,Z}}{kT_{\mathrm{e}}} \right). \tag{5.125}$$

g_j^Z and g_k^{Z+1} are the statistical weights of the states j and k, n_e is the electron density, m_e the electron mass, and T_e the electron temperature. The energy difference $\Delta E_{k,j}^{Z+1,Z}$ is related to the so-called dielectronic capture energy E_{kj}^{DC} by

$$\Delta E_{k,j}^{Z+1,Z} = -E_{kj}^{DC}. \tag{5.126}$$

E_{kj}^{DC} is the energy of the Auger electron, if the autoionizing state j decays via autoionization to state k. Combining (5.124)–(5.126), we find the general expression for the dielectronic capture rate:

$$\langle DC\rangle_{kj} = \frac{g_j^Z}{2g_k^{Z+1}} \cdot \left(\frac{2\pi\hbar^2}{m_e}\right)^{3/2} \cdot \Gamma_{jk}^{Z,Z+1} \cdot \frac{\exp(-E_{kj}^{DC}/kT_e)}{(kT_e)^{3/2}} \tag{5.127a}$$

or, in convenient units

$$\langle DC\rangle_{kj} = 1.656 \times 10^{-22} \cdot \frac{g_j^Z}{g_k^{Z+1}} \cdot \Gamma_{jk}^{Z,Z+1} \cdot \frac{\exp\left(-E_{kj}^{DC}(\mathrm{eV})/T(\mathrm{eV})\right)}{(T_e(\mathrm{eV}))^{3/2}} \left[\frac{\mathrm{cm}^3}{\mathrm{s}}\right]. \tag{5.127b}$$

If $P_{j,gr}^Z$ is the probability that the autoionizing state j of charge state Z decays to the ground state gr of the same charge state, the quantity $P_{j,gr}^Z \cdot \langle DC\rangle_{kj}$ is called the dielectronic recombination rate coefficient [$\mathrm{cm}^3\ \mathrm{s}^{-1}$] into state k via the intermediate state j:

$$\langle DR\rangle_{kj}^{Z+1,Z} = P_{j,gr}^Z \cdot \langle DC\rangle_{kj}^{Z+1,Z}. \tag{5.128}$$

In general, the probability $P_{j,gr}^Z$ is a function of density and temperature, i.e.,

$$P_{j,gr}^Z = P_{j,gr}^Z(n_e, T_e). \tag{5.129}$$

The probability function (5.129) has to be determined from numerical calculations of a multilevel, multicharge state atomic population kinetics that explicitly involves all necessary autoionizing states as "active levels" (means the populations of the autoionizing levels are calculated on the same footing as ground and single excited states). If collisions are negligible compared to spontaneous radiative decay rates as well as autoionizing rates, the probability $P_{j,gr}^Z$ can be approximated by the so-called satellite branching factor

$$P_{j,gr}^Z \to \sum_i B_{ji}^Z = \sum_i \left\{\frac{A_{ji}^Z}{\sum_l A_{jl}^Z + \sum_k \Gamma_{jk}^{Z,Z+1}}\right\}. \tag{5.130}$$

Let us illuminate the situation with the help of the most simple autoionizing states $2l2l'$, in particular the state $j = 2p^2\ {}^1D_2$ of He-like argon. In this case, $Z = 16$,

$k = 1s\ ^2S_{1/2}$, and $i = 1s2p\ ^1P_1, i' = 1s2p\ ^3P_1,\ i'' = 1s2p\ ^3P_2$: $\Gamma_{\mathrm{kj}}^{Z,Z+1} = 3.09 \times 10^{14}\,\mathrm{s}^{-1}$, $A_{\mathrm{ji}}^{Z} = 1.22 \times 10^{14}\,\mathrm{s}^{-1}$, $A_{\mathrm{ji'}}^{Z} = 1.92 \times 10^{9}\,\mathrm{s}^{-1}$, $A_{\mathrm{ji''}}^{Z} = 6.21 \times 10^{12}\,\mathrm{s}^{-1}$. Therefore, $P_{\mathrm{j,gr}}^{Z=16} \approx \sum_{\mathrm{i}} \mathrm{B}_{\mathrm{ji}}^{Z=16} = 2.78 \times 10^{-1} + 4.38 \times 10^{-6} + 1.42 \times 10^{-2} = 2.92 \times 10^{-1}$. The approximation $P_{\mathrm{j,gr}}^{Z} \approx \sum_{\mathrm{i}} \mathrm{B}_{\mathrm{ji}}^{Z}$ assumes that all single excited states, namely $i = 1s2p\ ^1P_1,\ \ i' = 1s2p\ ^3P_1,\ i'' = 1s2p\ ^3P_2$, decay entirely to the ground state $gr = 1s^2\ ^1S_0$ via the radiative transitions $1s2p\ ^1P_1 \rightarrow gr$, $1s2p\ ^3P_1 \rightarrow gr$, and $1s2p\ ^3P_2 \rightarrow gr$. Because $A(1s2p\ ^1P_1 \rightarrow gr) = 1.07 \times 10^{14}\,\mathrm{s}^{-1}$, $A(1s2p\ ^3P_1 \rightarrow gr) = 1.82 \times 10^{12}\,\mathrm{s}^{-1}$, $A(1s2p\ ^3P_2 \rightarrow gr) = 3.16 \times 10^{8}\,\mathrm{s}^{-1}$, the assumption that the excited state j decays to the ground state via the intermediate states i, i', i'' is a good assumption because the sum in (5.130) is dominated by the strongest transition $j \rightarrow i$ where collisional "competition" starts to be important only for near-solid density plasmas (and contributions $j \rightarrow i'$ and $j \rightarrow i''$ are small). Therefore, the dielectronic recombination rate can be approximated by the following expression:

$$\langle \mathrm{DR} \rangle_{\mathrm{kj}}^{Z+1,Z} \approx \sum_{\mathrm{i}} \left\{ \frac{A_{\mathrm{ji}}^{Z}}{\sum_{\mathrm{l}} A_{\mathrm{jl}}^{Z} + \sum_{\mathrm{k}} \Gamma_{\mathrm{jk}}^{Z,Z+1}} \cdot \langle \mathrm{DC} \rangle_{\mathrm{kj}}^{Z+1,Z} \right\}. \tag{5.131}$$

With the help of (5.125), (5.131) can be written as follows:

$$\langle \mathrm{DR} \rangle_{\mathrm{kj}}^{Z+1,Z} \approx \frac{1}{2g_{\mathrm{k}}^{Z+1}} \cdot \left(\frac{2\pi\hbar^2}{m_{\mathrm{e}}} \right)^{3/2} \cdot \frac{\exp(-E_{\mathrm{kj}}^{\mathrm{DC}}/kT_{\mathrm{e}})}{(kT_{\mathrm{e}})^{3/2}} \cdot \sum_{\mathrm{i}} \left\{ \frac{g_{\mathrm{j}}^{Z} \cdot \Gamma_{\mathrm{jk}}^{Z,Z+1} \cdot A_{\mathrm{ji}}^{Z}}{\sum_{\mathrm{l}} A_{\mathrm{jl}}^{Z} + \sum_{\mathrm{k}} \Gamma_{\mathrm{jk}}^{Z,Z+1}} \right\}. \tag{5.132}$$

The term in parenthesis is the so-called dielectronic satellite intensity factor

$$Q_{\mathrm{k,ji}}^{Z+1,Z} = \frac{g_{\mathrm{j}}^{Z} \cdot \Gamma_{\mathrm{jk}}^{Z,Z+1} \cdot A_{\mathrm{ji}}^{Z}}{\sum_{\mathrm{l}} A_{\mathrm{jl}}^{Z} + \sum_{\mathrm{k}} \Gamma_{\mathrm{jk}}^{Z,Z+1}}. \tag{5.133}$$

Therefore, under the assumptions made in (5.130), the dielectronic recombination due to the autoionizing states $2l2l'$ is given by the sum of the dielectronic satellite intensity factors, for the present example of Ar, numerical calculations (including intermediate coupling and configuration interaction) provide $\sum_{\mathrm{j,i}} Q_{\mathrm{k,ji}} = 8.29 \times 10^{14}\,\mathrm{s}^{-1}$ and $E_{\mathrm{k,ji}=2l2l'}^{\mathrm{DC}} \approx 2.302 \times 10^{3}\,\mathrm{eV}$. For the autoionizing states $2l3l'$, we obtain $\sum_{\mathrm{j,i}} Q_{\mathrm{k,ji}} = 7.95 \times 10^{14}\,\mathrm{s}^{-1}$ and $E_{\mathrm{k,ji}=2l3l'}^{\mathrm{DC}} \approx 2.875 \times 10^{3}\,\mathrm{eV}$, for the $2l4l'$ states $\sum_{\mathrm{j,i}} Q_{\mathrm{k,ji}} = 4.77 \times 10^{14}\,\mathrm{s}^{-1}$ and $E_{\mathrm{k,ji}=2l4l'}^{\mathrm{DC}} \approx 3.072 \times 10^{3}\,\mathrm{eV}$, for the $2l5l'$ states $\sum_{\mathrm{j,i}} Q_{\mathrm{k,ji}} = 2.96 \times 10^{14}\,\mathrm{s}^{-1}$ and $E_{\mathrm{k,ji}=2l5l'}^{\mathrm{DC}} \approx 3.16 \times 10^{3}\,\mathrm{eV}$, for the $2l6l'$ states $\sum_{\mathrm{j,i}} Q_{\mathrm{k,ji}} = 1.89 \times 10^{14}\,\mathrm{s}^{-1}$ and $E_{\mathrm{k,ji}=2l6l'}^{\mathrm{DC}} \approx 3.21 \times 10^{3}\,\mathrm{eV}$, for the $2l7l'$ states $\sum_{\mathrm{j,i}} Q_{\mathrm{k,ji}} = 1.29 \times 10^{14}\,\mathrm{s}^{-1}$ and $E_{\mathrm{k,ji}=2l7l'}^{\mathrm{DC}} \approx 3.24 \times 10^{3}\,\mathrm{eV}$. One can see that the convergence for high n-spectator electrons is not quite rapid and follows approximately the scaling law $\sum_{\mathrm{j}} Q_{\mathrm{k,ji}=2\mathrm{lnl'}}^{\mathrm{DC}} \propto 1/n^3$.

Despite of the most simple configurations $2lnl'$ for the dielectronic recombination, one can see that the numerical calculations are rather cumbersome: Very large quantum numbers nl' have to be involved in the numerical calculations to reach convergence. For large quantum numbers, however, convergence is difficult to achieve in purely quantum numerical calculations and quasi-classical approaches are mandatory to practically solve the problem.

Next, to obtain the total dielectronic recombination rate from H-like to He-like ions, one needs to invoke all possible intermediate states $j = 3lnl'; 4lnl'; 5lnl', \ldots$ One can easily understand that for more complex configurations, the number of autoionizing states to be involved becomes rapidly numerically prohibitive for purely quantum mechanical numerical calculations.

Moreover, dielectronic recombination is not only related to corresponding atomic structure calculations, but also to the collisional radiative interplay for the calculation of the probability $P_{\mathrm{j,gr}}$. In general, one needs to include explicitly all relevant autoionizing states in a collisional radiative model in order to correctly predict the ionic fractions for given temperature and density. In this case, however, the atomic state population kinetics is entirely dominated by the number of autoionizing high-n-states and numerically prohibitive. It likewise looks rather strange, to dominate an atomic state population kinetics by autoionizing states just for the purpose to calculate one recombination coefficient. It is essentially for these reasons that numerical calculations of ionic fractions are still under controversial discussion up to present days (and in particular for high-Z elements) (NIST 2019; Rubiano et al. 2007; Chung et al. 2013; Colgan et al. 2015).

Moreover, dielectronic recombination is therefore not only a theoretical subject in atomic physics, but it has important impact to the radiative properties of atoms and ions, plasma spectroscopy, and technical applications (e.g., radiation sources). Note also that historically, dielectronic recombination has been invented to understand the order of magnitude discrepancies in the ionic abundance between calculations and spectroscopic observation from the solar corona emission (Burgess 1964).

5.6.2 *Total Rates of Dielectronic Recombination and Multichannel Approach*

In order to obtain the total dielectronic recombination rate $\langle \mathrm{DR} \rangle_{\mathrm{tot}}^{Z+1,Z}$, all dielectronic recombination rates $\langle \mathrm{DR} \rangle_{\mathrm{kj}}^{Z+1,Z} = P_{\mathrm{j,gr}}^{Z} \cdot \langle \mathrm{DC} \rangle_{\mathrm{kj}}^{Z+1,Z}$ have to be summed with respect to the initial state k and also with respect to the intermediate states j, i.e.,

$$\langle \mathrm{DR} \rangle_{\mathrm{tot}}^{Z+1,Z} = \sum_{\mathrm{k}} \sum_{\mathrm{j}} \langle \mathrm{DR} \rangle_{\mathrm{kj}}^{Z+1,Z} = \sum_{\mathrm{k}} \sum_{\mathrm{j}} P_{\mathrm{j,gr}}^{Z} \cdot \langle \mathrm{DC} \rangle_{\mathrm{kj}}^{Z+1,Z}. \tag{5.134}$$

Because the probability $P_{\mathrm{j,gr}}^{Z}$ is a function of density and temperature [see discussion related to (5.129)], it is difficult to obtain general and closed formulas for the dielectronic recombination rate coefficient. Only in the low-density approximation, where relation (5.130) approximately holds true, general formulas in terms of dielectronic recombination rate coefficients that depend only on temperature can be obtained.

5.6.2.1 Burgess Formulas

One of the most used general approximate empirical formula in this framework is the so-called Burgess formula (Burgess 1964) that assumes that the *nl*-spectator electron which is not interacting with the core is treated in the hydrogenic approximation and that the capture cross-section can be expressed in terms of the excitation cross-section for transitions $\alpha_0 \to \alpha$ using the principle of correspondence discussed above:

$$\langle \mathrm{DR} \rangle_{\mathrm{kj}}^{Z+1,Z} := D^{Z+1,Z}(\alpha_0 \to \alpha, nl). \tag{5.135}$$

For the total dielectronic recombination rate, we have

$$\langle \mathrm{DR} \rangle_{\mathrm{tot}}^{Z+1,Z} := D^{Z+1,Z} = \sum_{\alpha_0} \sum_{\alpha} \sum_{\mathrm{n}} \sum_{l=0}^{n-1} D^{Z+1,Z}(\alpha_0 \to \alpha, nl). \tag{5.136}$$

For the simplest example of autoionizing states $2l2l'$ outlined above, $\alpha_0 = 1s$ and $\alpha_1 = 2p$, i.e., the transition $\alpha_0 \to \alpha$ corresponds to the Ly-alpha transition in H-like ions. For these configurations, dielectronic recombination into the ground state is the most important one, i.e., there exists a single state $k = \alpha_0 = 1s$. α_0 coincides therefore with the atomic ground state and the sum over α_0 can be suppressed, i.e.,

$$D^{Z+1,Z} \approx \sum_{\alpha} \sum_{\mathrm{n}} \sum_{l=0}^{n-1} D^{Z+1,Z}(\alpha_0 \to \alpha, nl). \tag{5.137}$$

The dielectronic recombination rate coefficient $D^{Z+1,Z}(\alpha_0 \to \alpha, nl)$ can then be expressed via the following analytical empirical expression (Burgess 1964)

$$D^{Z+1,Z}(\alpha_0 \to \alpha, nl) = 4.8 \times 10^{-11} f_{\alpha_0 \alpha} B_{\mathrm{d}} \beta^{3/2} \mathrm{e}^{-\beta \chi_{\mathrm{d}}} \left[\mathrm{cm}^3\,\mathrm{s}^{-1}\right], \tag{5.138}$$

where

$$\beta = \frac{(z+1)^2 \, \mathrm{Ry}}{kT_{\mathrm{e}}}, \tag{5.139}$$

$$\chi_{\mathrm{d}} = \frac{\chi}{1 + 0.015 \dfrac{z^3}{(z+1)^2}}, \tag{5.140}$$

$$\chi = \frac{\Delta E(\alpha_0 \to \alpha)}{(z+1)^2 \, \mathrm{Ry}}. \tag{5.141}$$

z is the so-called spectroscopic symbol of the double excited ion after recombination. The spectroscopic symbol is given by $z = Z_n - N_{bound} + 1$, where N_{bound} is the number of bound electrons. For example, for neutral helium the spectroscopic symbol is $z = 1$ (He I), singly ionized helium has $z = 2$ (He II).

If the first resonance transition is a $\Delta n = 0$ transition, the branching factor B_d is given by Cowan (1981):

$$B_d = \left(\frac{z\chi}{z^2 + 13.4}\right)^{1/2} \frac{1}{1 + 0.105(z+1)\chi + 0.015(z+1)^2\chi^2}. \tag{5.142}$$

For $\Delta n \neq 0$, the branching factor is approximated by Cowan (1981):

$$B_d = \left(\frac{z\chi}{z^2 + 13.4}\right)^{1/2} \frac{0.5}{1 + 0.210(z+1)\chi + 0.030(z+1)^2\chi^2}. \tag{5.143}$$

The branching factor B_d has the following meaning: After dielectronic capture, a double excited state is formed that can decay via autoionization or radiative decay. For the dielectronic recombination, only the radiative decays contribute finally to recombination as autoionization returns the autoionizing state to the original state.

According to (5.138), α_0 is the ground state and therefore $f_{\alpha_0\alpha}$ is the dipole oscillator strength for the resonance transition $\alpha_0 \to \alpha$ with transition energy $\Delta E(\alpha_0 \to \alpha)$ in [eV]. As the oscillator strength drops rapidly with principal quantum number, it is usually sufficient to consider only the first two α-terms in the sum of (5.137) and we finally obtain the desired expression for the total dielectronic recombination $[D^{Z+1,Z}(\alpha_0 \to \alpha) := \sum_n \sum_{l=0}^{n-1} D^{Z+1,Z}(\alpha_0 \to \alpha, nl)]$:

$$D^{Z+1,Z} \approx D^{Z+1,Z}(\alpha_0 \to \alpha_1) + D^{Z+1,Z}(\alpha_0 \to \alpha_2). \tag{5.144}$$

Let us consider the dielectronic recombination into neutral helium as an example:

$$\mathrm{He}^{1+}(1s) + e \to \mathrm{He}^{0+**}(nln'l') \to \mathrm{He}^{0+}(1s^2). \tag{5.145}$$

For this example, $\alpha_0 = 1s$, $\alpha_1 = 2p$, $\alpha_2 = 3p$, …. Therefore, the oscillator strength $f_{\alpha_0\alpha_1}$ corresponds to the oscillator strength of the resonance line, namely the H-like Lyman-alpha line of singly ionized helium, $f_{\alpha_0\alpha_2}$ corresponds to the Lyman-beta line. The oscillator strengths are $f_{1s\to 2p} = 0.4164$ and $f_{1s\to 3p} = 0.07914$, respectively, and their transition energies are $\Delta E(1s \to 2p) = 40.81\,\mathrm{eV}$ and $\Delta E(1s \to 3p) = 48.37\,\mathrm{eV}$, respectively. The spectroscopic symbol is

$z = 1$ and $\Delta n \neq 0$ (therefore (5.143) applies). As one can see, higher n oscillator strengths provide almost negligible contribution to the total dielectronic recombination rate. From (5.139)–(5.143), we obtain $\beta = 4\mathrm{Ry}/kT_e$, $B_d(1s \to 2p) = 0.0825$, $B_d(1s \to 3p) = 0.0846$, $\chi_d(1s \to 2p) = 0.747$, and $\chi_d(1s \to 3p) = 0.886$. More precise quantum mechanical calculations (Sobelman and Vainshtein 2006; Wang et al. 1999) provide $B_{d,ref}(1s \to 2p) = 0.155$, $B_{d,ref}(1s \to 3p) = 0.0144$, $\chi_{d,ref}(1s \to 2p) = 0.744$, and $\chi_{d,ref}(1s \to 3p) = 0.888$. For the rate coefficient at $kT_e = \mathrm{Ry}$ $(\beta = 4)$, we obtain: $D^{He1+,He0+}(1s \to 2p) = 1.65 \times 10^{-12} [\mathrm{cm^3\,s^{-1}}]$, $D^{\mathrm{He1+,He0+}}(1s \to 3p) = 3.2 \times 10^{-13}\,[\mathrm{cm^3\,s^{-1}}]$ and for the more precise quantum mechanical calculations, $D_{\mathrm{ref}}^{\mathrm{He1+,He^{0+}}}(1s \to 2p) = 3.10 \times 10^{-12}\,[\mathrm{cm^3\,s^{-1}}]$, $D_{\mathrm{ref}}^{\mathrm{He1+,He0+}}(1s \to 3p) = 5.46 \times 10^{-14}\,[\mathrm{cm^3\,s^{-1}}]$. This confirms that the leading terms for the dielectronic recombination are indeed given by (5.144).

5.6.2.2 Multichannel Approach

The comparison of the results from formulas (5.138)–(5.143) with more precise calculations (Sobelman and Vainshtein 2006; Wang et al. 1999; Kato and Asano 1999) shows that the χ_d-values are in quite good agreement, whereas the B_d values differ strongly. For the resonance transition $1s \to 2p$, the B_d-value obtained from (5.143) is about a factor of 2 smaller than more precise values. This is a general observation: The precision of formulas (5.142), (5.143) is about a factor of 2 for the resonance transition.

For the transition $1s \to 3p$, the B_d-value obtained from (5.143) is about a factor of 6 larger than the more precise values. This large overestimation is also a general observation and related to the fact that formulas (5.142), (5.143) take into account only one autoionizing channel. For the $3lnl'$-configurations (that are related to the transition $\alpha_0 \to \alpha_2 = 1s \to 3p$), however, autoionization decays not only to the ground state but to excited states too:

$$3lnl' \to \left\{ \begin{array}{c} 1s + e_{\mathrm{Auger}} \\ 2l + e_{\mathrm{Auger}} \end{array} \right\}. \tag{5.146}$$

Numerical calculations show (Rosmej et al. 1998; Petitdemange and Rosmej 2013) that the autoionizing rates to the excited states "$2l$" are even more important than to the ground state "$1s$". This reduces considerably the branching factor for the dielectronic recombination [the B_d-factor in (5.143)]. As the more precise calculations take into account many channels of Auger decay, the branching factor is therefore systematically smaller that those of (5.142), (5.143). In fact, as one can see from (5.143) very similar branching factors are provided for the transitions $\alpha_0 \to \alpha_1 = 1s \to 2p$ and $\alpha_0 \to \alpha_2 = 1s \to 3p$ due to the consideration of one autoionizing channel only.

Table 5.2 Fitting coefficients according (5.147) for the dielectronic recombination into H-like ions originating from the $2lnl'$- and $3lnl'$-autoionizing levels, $Z = Z_n$, $m = 1$, $l_0 = 0$ in (5.147)

	$2lnl'$: $\alpha_0 = 1s \rightarrow \alpha = 2p$		$3lnl'$: $\alpha_0 = 1s \rightarrow \alpha = 3p$	
Element	B_d	χ_d	B_d	χ_d
He	3.12×10^{-4}	0.744	5.48×10^{-6}	0.888
Li	3.72×10^{-4}	0.736	6.41×10^{-6}	0.887
Be	3.67×10^{-4}	0.727	6.53×10^{-6}	0.885
B	3.42×10^{-4}	0.718	6.47×10^{-6}	0.883
C	3.13×10^{-4}	0.709	6.32×10^{-6}	0.881
N	2.85×10^{-4}	0.700	6.31×10^{-6}	0.879
O	2.58×10^{-4}	0.691	5.92×10^{-6}	0.877
F	2.33×10^{-4}	0.682	5.70×10^{-6}	0.874
Ne	2.11×10^{-4}	0.673	5.48×10^{-6}	0.872
Na	1.90×10^{-4}	0.665	5.26×10^{-6}	0.870
Mg	1.72×10^{-4}	0.657	5.04×10^{-6}	0.868
Al	1.56×10^{-4}	0.649	4.84×10^{-6}	0.866
Si	1.41×10^{-4}	0.642	4.63×10^{-6}	0.863
P	1.27×10^{-4}	0.636	4.43×10^{-6}	0.861
S	1.15×10^{-4}	0.630	4.24×10^{-6}	0.859
Cl	1.05×10^{-4}	0.624	4.05×10^{-6}	0.857
Ar	9.50×10^{-5}	0.620	3.87×10^{-6}	0.856
K	8.61×10^{-5}	0.616	3.69×10^{-6}	0.854
C	7.82×10^{-5}	0.612	3.52×10^{-6}	0.852
Sc	7.09×10^{-5}	0.609	3.35×10^{-6}	0.851
Ti	6.45×10^{-5}	0.606	3.19×10^{-6}	0.849
V	5.85×10^{-5}	0.604	3.04×10^{-6}	0.848
Cr	5.33×10^{-5}	0.602	2.89×10^{-6}	0.847
Mn	4.85×10^{-5}	0.601	2.74×10^{-6}	0.846
Fe	4.42×10^{-5}	0.599	2.60×10^{-6}	0.845
Co	4.03×10^{-5}	0.598	2.47×10^{-6}	0.844
Ni	3.68×10^{-5}	0.598	2.34×10^{-6}	0.843
Cu	3.37×10^{-5}	0.597	2.22×10^{-6}	0.842
Zn	3.08×10^{-5}	0.597	2.10×10^{-6}	0.842
Ga	2.83×10^{-5}	0.596	1.99×10^{-6}	0.842
Ge	2.60×10^{-5}	0.596	1.88×10^{-6}	0.841
As	2.39×10^{-5}	0.596	1.78×10^{-6}	0.841
Se	2.20×10^{-5}	0.596	1.68×10^{-6}	0.841
Br	2.03×10^{-5}	0.596	1.59×10^{-6}	0.841
Kr	1.88×10^{-5}	0.596	1.50×10^{-6}	0.841
Rb	1.74×10^{-5}	0.597	1.42×10^{-6}	0.841
Sr	1.61×10^{-5}	0.597	1.34×10^{-6}	0.842
Y	1.50×10^{-5}	0.597	1.27×10^{-6}	0.842

(continued)

Table 5.2 (continued)

	$2lnl'$: $\alpha_0 = 1s \to \alpha = 2p$		$3lnl'$: $\alpha_0 = 1s \to \alpha = 3p$	
Element	B_d	χ_d	B_d	χ_d
Zr	1.39×10^{-5}	0.598	1.20×10^{-6}	0.842
Nb	1.30×10^{-5}	0.599	1.13×10^{-6}	0.843
Mo	1.21×10^{-5}	0.599	1.07×10^{-6}	0.843

The numerical data include corrections for multidecay channels (two channels for $2l2l'$ and four channels for $3lnl'$)

Due to the existence of multichannel autoionization decay, the multichannel radiative decay and the complexity of configurations involved, quantum numerical calculations of the dielectronic recombination are very complex and the precision of the Burgess formula is difficult to determine. This is another reason why the recourse to quasi-classical methods appears to be mandatory for a comprehensive description of the dielectronic recombination phenomenon.

Below, we provide numerical data for dielectronic recombination into H-, He-, and Li-like ions taking into account multichannels for Auger and radiative decay (Beigman 1981; Shevelko and Vainshtein 1993; Vainshtein and Shevelko 1996). The numerical results have been fitted to a simple analytical expression in order to facilitate the application of these complex calculations:

$$D^{Z+1,Z}(\alpha_0 \to \alpha, nl) = 10^{-8} \cdot \frac{m}{2l_0+1} \cdot B_d \cdot \beta^{3/2} \cdot e^{-\beta\chi_d}\,[\mathrm{cm}^3\,\mathrm{s}^{-1}], \tag{5.147a}$$

$$\beta = \frac{Z^2 \cdot \mathrm{Ry}}{kT_e}, \tag{5.147b}$$

where Ry = 13,606 eV, kT_e is the electron temperature in [eV], m is the number of equivalent electrons of state α_0, Z is the charge of the ion where the core transition $\alpha_0 \to \alpha$ takes place (e.g., for the $2lnl'$-autoionizing states of He-like argon the core transition is the $1s \to 2p$ transition in H-like argon, $Z = 18$), l_0 is the corresponding orbital momentum of state α_0. The physical meaning of the parameter χ_d is related to the fact that all contributions of the configuration with different spectator electrons nl have to be summed up for the total dielectronic recombination rate with different energies [see (5.16)]. The parameter χ_d provides a fit to the numerical results to replace the sum of different energies in a best manner by an average energy $\chi_d \cdot \beta$. Finally, the total sum is replaced by an average amplitude B_d to provide a simple analytical expression without any summation.

Table 5.2 presents the numerical calculation of the total dielectronic recombination rate into H-like ions for the core transitions $1s$–$2p$ and $1s$–$3p$ for all elements from He ($Z = 2$) (see also Annex 3) until Mo ($Z = 42$) and the corresponding fitting parameters according (5.147). It can be seen that for low-Z elements, the dielectronic recombination related to the core transition $1s$–$2p$ is dominating, for large Z-values, the relative contribution of the dielectronic recombination with the core

Table 5.3 Fitting coefficients according to (5.147) for the dielectronic recombination into He-like ions originating from the $1s2lnl'$- and $1s3lnl'$-autoionizing levels, $Z = Z_n - 1$, $m = 2$, $l_0 = 0$ in (5.147)

	$1s2lnl'$: $\alpha_0 = 1s^2 \to \alpha = 1s2p$		$1s3lnl'$: $\alpha_0 = 1s^2 \to \alpha = 1s3p$	
Element	B_d	χ_d	B_d	χ_d
Li	3.39×10^{-5}	1.11	1.57×10^{-6}	1.27
Be	9.94×10^{-5}	0.961	2.12×10^{-6}	1.14
B	1.53×10^{-4}	0.891	2.51×10^{-6}	1.07
C	1.93×10^{-4}	0.848	2.98×10^{-6}	1.03
N	2.17×10^{-4}	0.818	3.40×10^{-6}	1.00
O	2.34×10^{-4}	0.795	3.92×10^{-6}	0.983
F	2.17×10^{-4}	0.775	4.23×10^{-6}	0.967
Ne	2.05×10^{-4}	0.757	4.50×10^{-6}	0.956
Na	1.88×10^{-4}	0.740	4.56×10^{-6}	0.945
Mg	1.72×10^{-4}	0.726	4.54×10^{-6}	0.937
Al	1.57×10^{-4}	0.713	4.47×10^{-6}	0.929
Si	1.43×10^{-4}	0.701	4.36×10^{-6}	0.922
P	1.30×10^{-4}	0.690	4.22×10^{-6}	0.916
S	1.18×10^{-4}	0.681	4.07×10^{-6}	0.910
Cl	1.07×10^{-4}	0.672	3.92×10^{-6}	0.905
Ar	9.72×10^{-5}	0.664	3.76×10^{-6}	0.901
K	8.83×10^{-5}	0.658	3.61×10^{-6}	0.897
C	8.02×10^{-5}	0.652	3.45×10^{-6}	0.893
Sc	7.28×10^{-5}	0.647	3.30×10^{-6}	0.889
Ti	6.62×10^{-5}	0.642	3.15×10^{-6}	0.886
V	6.02×10^{-5}	0.638	3.01×10^{-6}	0.883
Cr	5.47×10^{-5}	0.635	2.87×10^{-6}	0.880
Mn	4.98×10^{-5}	0.632	2.73×10^{-6}	0.877
Fe	4.54×10^{-5}	0.629	2.60×10^{-6}	0.875
Co	4.14×10^{-5}	0.627	2.47×10^{-6}	0.873
Ni	3.78×10^{-5}	0.625	2.35×10^{-6}	0.871
Cu	3.46×10^{-5}	0.623	2.23×10^{-6}	0.869
Zn	3.16×10^{-5}	0.622	2.11×10^{-6}	0.868
Ga	2.90×10^{-5}	0.620	2.00×10^{-6}	0.867
Ge	2.67×10^{-5}	0.619	1.90×10^{-6}	0.865
As	2.45×10^{-5}	0.619	1.80×10^{-6}	0.864
Se	2.26×10^{-5}	0.618	1.70×10^{-6}	0.864
Br	2.08×10^{-5}	0.617	1.61×10^{-6}	0.863
Kr	1.93×10^{-5}	0.617	1.52×10^{-6}	0.862
Rb	1.78×10^{-5}	0.616	1.44×10^{-6}	0.862
Sr	1.65×10^{-5}	0.616	1.36×10^{-6}	0.861

(continued)

Table 5.3 (continued)

	$1s2lnl'$: $\alpha_0 = 1s^2 \rightarrow \alpha = 1s2p$		$1s3lnl'$: $\alpha_0 = 1s^2 \rightarrow \alpha = 1s3p$	
Element	B_d	χ_d	B_d	χ_d
Y	1.53×10^{-5}	0.616	1.29×10^{-6}	0.861
Zr	1.43×10^{-5}	0.616	1.22×10^{-6}	0.861
Nb	1.33×10^{-5}	0.616	1.15×10^{-6}	0.861
Mo	1.24×10^{-5}	0.616	1.09×10^{-6}	0.861

The numerical data include corrections for multidecay channels (two channels for $1s2lnl'$ and four channels for $1s3lnl'$)

transition $1s$–$3p$ increases. The Burgess formula provides amplitudes B_d that are about a factor of 3 smaller than the present numerical calculations. For the $3lnl'$-states, the Burgess formula considerably overestimates the dielectronic recombination rate because it does not take into account the multichannel radiative and Auger decay. This is of particular importance for low-Z elements. The one channel approximation, e.g., for C provides $B_d = 6.75 \times 10^{-5}$, whereas the four-channel approximation provides $B_d = 6.32 \times 10^{-6}$, i.e., a reduction by a factor of 10. The multichannel decay is much less important for higher Z-values, e.g., for Fe $B_d = 5.13 \times 10^{-6}$ whereas the four-channel approximation provides $B_d = 2.60 \times 10^{-6}$.

Table 5.3 presents the numerical calculation of the total dielectronic recombination rate into He-like ions for the core transitions $1s$-$2p$ and $1s$-$3p$ for all elements from He (Z = 2) until Mo (Z = 42) and the corresponding fitting parameters according (5.147). It can be seen that for low-Z elements, the dielectronic recombination related to the core transition $1s$-$2p$ is dominating, for large Z-values, the relative contribution of the dielectronic recombination with the core transition $1s$-$3p$ increases. The Burgess formula provides amplitudes B_d that are about a factor of 3 smaller than the present numerical calculations. For the $1s3lnl'$-states, the Burgess formula considerably overestimates the dielectronic recombination rate because it does not take into account the multichannel radiative and Auger decay. This is of particular importance for low-Z elements. The one channel approximation, e.g., for C provides $B_d = 6.76 \times 10^{-5}$, whereas the four-channel approximation provides $B_d = 2.98 \times 10^{-6}$, i.e., a reduction by a factor of 20. The multichannel decay is much less important for higher Z-values, e.g., for Fe $B_d = 5.34 \times 10^{-6}$ whereas the four-channel approximation provides $B_d = 2.60 \times 10^{-6}$.

Table 5.4 provides the numerical results of dielectronic recombination into Li-like ions related to a core transition $2s$-$2p$, i.e., the core transition is a $\Delta n = 0$ transition. Therefore, the fitting parameter χ_d is rather small and the associated exponential factor for the dielectronic recombination does not vary much. In addition, the configurations $1s^2 2lnl'$ are only autoionizing for rather high principal quantum numbers. This is quite different for the dielectronic recombination related to the core transition $2s$–$3p$: The states are autoionizing for rather low quantum numbers nl and the temperature dependence is much different due to an order of

Table 5.4 Fitting coefficients according to (5.147) for the dielectronic recombination into Li-like ions originating from the $1s^22lnl'$- and $1s^23lnl'$-autoionizing levels, $Z = Z_n - 2$, $m = 1$, $l_0 = 0$ in (5.147)

	$1s^22lnl'$: $\alpha_0 = 1s^22s \to \alpha = 1s^22p$		$1s^23lnl'$: $\alpha_0 = 1s^22s \to \alpha = 1s^23p$	
Element	B_d	χ_d	B_d	χ_d
Be	8.09×10^{-5}	0.0571	1.97×10^{-6}	0.197
B	6.86×10^{-5}	0.0400	2.85×10^{-6}	0.173
C	5.18×10^{-5}	0.0306	6.61×10^{-6}	0.161
N	3.95×10^{-5}	0.0248	1.06×10^{-5}	0.153
O	3.09×10^{-5}	0.0207	1.47×10^{-5}	0.149
F	2.47×10^{-5}	0.0179	1.85×10^{-5}	0.145
Ne	2.02×10^{-5}	0.0156	2.17×10^{-5}	0.142
Na	1.69×10^{-5}	0.0139	2.41×10^{-5}	0.140
Mg	1.43×10^{-5}	0.0126	2.57×10^{-5}	0.138
Al	1.23×10^{-5}	0.0115	2.67×10^{-5}	0.136
Si	1.07×10^{-5}	0.0105	2.71×10^{-5}	0.135
P	9.43×10^{-6}	0.00981	2.69×10^{-5}	0.133
S	8.41×10^{-6}	0.00914	2.60×10^{-5}	0.131
Cl	7.57×10^{-6}	0.00858	2.53×10^{-5}	0.130
Ar	6.87×10^{-6}	0.00809	2.42×10^{-5}	0.128
K	6.25×10^{-6}	0.00772	2.31×10^{-5}	0.127
C	5.76×10^{-6}	0.00736	2.19×10^{-5}	0.126
Sc	5.35×10^{-6}	0.00704	2.09×10^{-5}	0.124
Ti	5.00×10^{-6}	0.00677	1.97×10^{-5}	0.123
V	4.67×10^{-6}	0.00658	1.86×10^{-5}	0.122
Cr	4.42×10^{-6}	0.00637	1.76×10^{-5}	0.120
Mn	4.20×10^{-6}	0.00620	1.66×10^{-5}	0.119
Fe	4.02×10^{-6}	0.00605	1.57×10^{-5}	0.118
Co	3.86×10^{-6}	0.00592	1.48×10^{-5}	0.117
Ni	3.72×10^{-6}	0.00581	1.40×10^{-5}	0.116
Cu	3.61×10^{-6}	0.00571	1.32×10^{-5}	0.115
Zn	3.51×10^{-6}	0.00564	1.25×10^{-5}	0.114
Ga	3.42×10^{-6}	0.00558	1.18×10^{-5}	0.113
Ge	3.35×10^{-6}	0.00553	1.11×10^{-5}	0.112
As	3.25×10^{-6}	0.00556	1.05×10^{-5}	0.111
Se	3.20×10^{-6}	0.00554	9.96×10^{-6}	0.110
Br	3.20×10^{-6}	0.00546	9.43×10^{-6}	0.109
Kr	3.17×10^{-6}	0.00546	8.92×10^{-6}	0.108
Rb	3.15×10^{-6}	0.00547	8.45×10^{-6}	0.107
Sr	3.13×10^{-6}	0.00548	8.01×10^{-6}	0.106
Y	3.12×10^{-6}	0.00551	7.59×10^{-6}	0.105
Zr	3.11×10^{-6}	0.00554	7.20×10^{-6}	0.105
Nb	3.11×10^{-6}	0.00558	6.83×10^{-6}	0.104
Mo	3.11×10^{-6}	0.00563	6.48×10^{-6}	0.103

The numerical data include corrections for multidecay channels (one channel for $1s^22lnl'$ and four channels for $1s^23lnl'$)

Table 5.5 B_d-factors according to (5.147) for the dielectronic recombination into Li-like ions originating from the $1s^2nln'l'$-autoionizing levels, $Z = Z_n - 2$, $m = 1$, $l_0 = 0$ in (5.147)

Element	$1s^22lnl'$: $\alpha_0 = 1s^22s \rightarrow \alpha = 1s^22p$		
	B_d (one channel)	B_d (multichannel)	B_d (Burgess)
Be	8.09×10^{-5}	–	1.34×10^{-4}
C	5.18×10^{-5}	–	7.99×10^{-5}
Mg	1.34×10^{-5}	–	1.94×10^{-5}
Ar	6.87×10^{-6}	–	8.65×10^{-6}
Fe	4.02×10^{-6}	–	4.88×10^{-6}
Mo	3.11×10^{-6}	–	3.87×10^{-6}
	$1s^23lnl'$: $\alpha_0 = 1s^22s \rightarrow \alpha = 1s^23p$		
Be	3.44×10^{-5}	1.97×10^{-6}	2.88×10^{-5}
C	6.45×10^{-5}	6.61×10^{-6}	6.98×10^{-5}
Mg	6.43×10^{-5}	2.57×10^{-5}	6.96×10^{-5}
Ar	4.55×10^{-5}	2.42×10^{-5}	5.15×10^{-5}
Fe	2.61×10^{-5}	1.57×10^{-5}	3.54×10^{-5}
Mo	8.61×10^{-6}	6.48×10^{-6}	1.89×10^{-5}
	$1s^24lnl'$: $\alpha_0 = 1s^22s \rightarrow \alpha = 1s^24p$		
Be	1.60×10^{-5}	3.47×10^{-7}	1.10×10^{-5}
C	2.52×10^{-5}	3.39×10^{-7}	2.23×10^{-5}
Mg	2.06×10^{-5}	1.30×10^{-6}	1.87×10^{-5}
Ar	1.29×10^{-5}	2.05×10^{-6}	1.27×10^{-5}
Fe	6.54×10^{-6}	2.00×10^{-6}	8.01×10^{-6}
Mo	1.87×10^{-6}	1.17×10^{-6}	3.82×10^{-6}

The numerical data show single- and multiple-channel approximation as well as corresponding factors according the theory of Burgess (note that the different numerical coefficients in (5.138) compared to (5.147a) have been included in the value for B_d-Burgess for comparison of the different methods)

magnitude different χ_d-factor. Unlike the dielectronic recombination into H- and He-like ions (Tables 5.2 and 5.3), the dielectronic recombination related to the $n = 3$ core transition is very important compared to the $2s$–$2p$ related recombination. Due to this reason, the temperature dependence of the total recombination rate (being the sum of the $2s$–$2p$, $2s$–$3p$,… transitions) is complex and differs qualitatively from the dielectronic recombination into H- and He-like ions that are dominated by a single exponential factor.

The influence of multichannel Auger and radiative decay on the dielectronic recombination related to the $2s$–$3p$ core transition is likewise important for low-Z elements. For Be atoms, the multichannel decay reduces the B_d—factor by more than a factor of 10, whereas for Ar the multichannel decay decreases the dielectronic recombination only by a factor of 2.

It is interesting to discuss the influence of the various mechanisms related to the dielectronic recombination with the core hole transition $2s$–$4p$. The single channel approximation leads to a wrong estimation of the importance of high-order

Table 5.6 Fitting coefficients according to (5.147) for the dielectronic recombination into excited states of Li-like ions originating from the $1s^23lnl'$- and $1s^24lnl'$-autoionizing levels, $Z = Z_n - 2$, $m = 1$, $l_0 = 1$ in (5.147)

	$1s^23lnl'$: $\alpha_0 = 1s^22p \to \alpha = 1s^23d$		$1s^24lnl'$: $\alpha_0 = 1s^22p \to \alpha = 1s^24d$	
Element	B_d	χ_d	B_d	χ_d
Be	1.78×10^{-4}	0.140	1.88×10^{-5}	0.190
B	2.99×10^{-4}	0.137	2.01×10^{-5}	0.189
C	3.74×10^{-4}	0.135	2.04×10^{-5}	0.188
N	4.44×10^{-4}	0.133	2.18×10^{-5}	0.187
O	5.15×10^{-4}	0.131	2.35×10^{-5}	0.187
F	5.52×10^{-4}	0.130	2.53×10^{-5}	0.186
Ne	5.65×10^{-4}	0.128	2.67×10^{-5}	0.185
Na	5.76×10^{-4}	0.127	2.88×10^{-5}	0.181
Mg	5.73×10^{-4}	0.125	3.28×10^{-5}	0.174
Al	5.61×10^{-4}	0.124	3.32×10^{-5}	0.172
Si	5.39×10^{-4}	0.122	3.33×10^{-5}	0.171
P	5.19×10^{-4}	0.120	3.48×10^{-5}	0.167
S	4.96×10^{-4}	0.119	3.46×10^{-5}	0.165
Cl	4.71×10^{-4}	0.117	3.44×10^{-5}	0.164
Ar	4.48×10^{-4}	0.115	3.41×10^{-5}	0.163
K	4.25×10^{-4}	0.114	3.38×10^{-5}	0.161
C	4.04×10^{-4}	0.112	3.34×10^{-5}	0.160
Sc	3.83×10^{-4}	0.110	3.30×10^{-5}	0.159
Ti	3.64×10^{-4}	0.109	3.25×10^{-5}	0.158
V	3.45×10^{-4}	0.107	3.20×10^{-5}	0.157
Cr	3.27×10^{-4}	0.105	3.14×10^{-5}	0.156
Mn	3.11×10^{-4}	0.104	3.08×10^{-5}	0.156
Fe	2.95×10^{-4}	0.102	3.02×10^{-5}	0.155
Co	2.80×10^{-4}	0.101	2.95×10^{-5}	0.154
Ni	2.66×10^{-4}	0.0992	2.88×10^{-5}	0.154
Cu	2.53×10^{-4}	0.0978	2.80×10^{-5}	0.153
Zn	2.40×10^{-4}	0.0964	2.72×10^{-5}	0.153
Ga	2.28×10^{-4}	0.0951	2.64×10^{-5}	0.153
Ge	2.17×10^{-4}	0.0939	2.56×10^{-5}	0.152
As	2.06×10^{-4}	0.0927	2.47×10^{-5}	0.152
Se	1.96×10^{-4}	0.0916	2.39×10^{-5}	0.152
Br	1.86×10^{-4}	0.0905	2.30×10^{-5}	0.152
Kr	1.77×10^{-4}	0.0895	2.22×10^{-5}	0.152
Rb	1.68×10^{-4}	0.0885	2.14×10^{-5}	0.152
Sr	1.60×10^{-4}	0.0876	2.05×10^{-5}	0.152
Y	1.52×10^{-4}	0.0867	1.97×10^{-5}	0.152
Zr	1.45×10^{-4}	0.0859	1.89×10^{-5}	0.152
Nb	1.38×10^{-4}	0.0851	1.82×10^{-5}	0.152
Mo	1.31×10^{-4}	0.0844	1.74×10^{-5}	0.152

The numerical data include corrections for multidecay channels (three channels for $1s^23lnl'$ and six channels for $1s^24lnl'$)

dielectronic recombination rates. For example, the numerical calculations in the single channel approximation for Be atoms give $B_{\mathrm{d}}^{(1\,\mathrm{Channel})}(2s-3p)=3.4\times10^{-5}$, whereas $B_{\mathrm{d}}^{(1\,\mathrm{Channel})}(2s-4p)=1.6\times10^{-5}$, i.e., $2s$-$4p$ transitions are only reduced by a factor of about 2 compared to the $2s$-$3p$ transitions. Numerical calculation including the multichannel decay provide an entirely different picture: $B_{\mathrm{d}}^{(6\,\mathrm{Channel})}(2s-3p)=2.0\times10^{-6}$ but $B_{\mathrm{d}}^{(6\,\mathrm{Channel})}(2s-4p)=3.5\times10^{-7}$, i.e., the numerical calculations including the multichannel decay indicate that higher-order dielectronic recombination rates are strongly suppressed.

This is a general observation that multichannel decay reduces the dielectronic recombination. Table 5.5 shows the numerical calculation for the dielectronic recombination B_{d}-factors for single- and multiple- channel decay into Li-like ions for different orders and elements in comparison to the standard Burgess formula. One observes that the Burgess formula is in reasonable agreement with the numerical results for the one-channel decay although it might differ up to a factor of 3 in some cases. Comparing, however, the numerical calculations for the multi-channel decay (which is the most correct approach as discussed above) discovers extremely large overestimations of the dielectronic recombination by the Burgess formula. In particular for light elements, the overestimation might be as large as 1–2 orders of magnitude, e.g., for the dielectronic recombination related to the autoionizing states $1s^2 4lnl'$ of Be, we have $B_{\mathrm{d}}^{(\mathrm{multi-channel})}(2s-4p)=3.47\times10^{-7}$, whereas $B_{\mathrm{d}}^{(\mathrm{Burgess})}(2s-4s)=1.10\times10^{-5}$, i.e., an overestimation by more than factor of 30 compared to the Burgess formula. It is therefore *not* recommended (Rosmej et al. 2020) to calculate higher-order contributions to the dielectronic recombination via the Burgess approach.

5.6.2.3 Excited State-Driven Dielectronic Recombination

Table 5.6 shows the dielectronic recombination rates related to the excited states $1s^2 2p$ of Li-like ions. It can be seen from a comparison of the numerical data from Tables 5.4 and 5.6 that the excited state contribution is even more important than the ground state contribution. For example, for Be, $B_{\mathrm{d}}(2s-3p)=1.97\times10^{-6}$, whereas $B_{\mathrm{d}}(2p-3d)=1.78\times10^{-4}$ and $B_{\mathrm{d}}(2p-4d)=1.88\times10^{-5}$. This means the excited state contribution is up to 2 orders of magnitude more important than the ground state contribution. Therefore, even for rather moderate densities with small population of the excited states, their contribution to dielectronic recombination can be important.

Particular important cases are encountered if the first excited states are related to $\Delta n=0$ radiative transitions. Because these transition probabilities are by orders of magnitude lower than those for $\Delta n>0$ transitions, Boltzmann populations with respect to the ground state are already achieved for rather low electron densities. For example, for Be, at densities of about $10^{15}\ \mathrm{cm}^{-3}$, the population of the excited states $1s^2 2p$ is more important than those of the ground state $1s^2 2s$ (Rosmej 1994).

Therefore, all excited state contributions of Beryllium, e.g., for tokamaks at typical divertor densities have larger contribution than the ground state.

The excited state contribution could be even important for very low densities if the excited states are metastable states. Therefore, in heavy ions, where we met excited states that are close to ground states, either related by a dipole-allowed radiative transition or by multipole transitions, the dielectronic recombination is extremely complex even for rather low densities. This is the main reason that up to present days, ionic balance calculations of heavy elements differ strongly from one method to another and that dielectronic recombination remains an active field of research and of considerable interest for applications (nuclear fusion, astrophysics, radiation sources, and spectroscopic diagnostics).

In conclusion, the excited state contribution is driven by atomic kinetics that can have much more important impact, than any other sophisticated effects related to ground state contributions. This points again to the great practical importance of quasi-classical methods (albeit of limited precision) that provide the possibility to obtain numerical data even for large quantum numbers that are important for the dielectronic recombination. It is important to emphasize that the inclusion of excited state contributions for the dielectronic recombination up to high quantum numbers for the corresponding core transitions may exceed ground state contributions by many orders of magnitude and it is for this reason that it is more precise to include as much as possible excited state contributions with the limited precision rather than improve via sophisticated atomic structure calculations the simplest core-transition-related dielectronic recombination but ignoring higher-order and excited states contributions.

5.6.3 *Dense Plasma and Electric Field Effects on Dielectronic Recombination*

Dense plasma effects are of multiple origins. The first one (as discussed above in relation with Table 5.6) is related to the atomic population kinetics where highly populated excited states directly contribute to the recombination process (Rosmej 1994). The second one is related to the shift of bound states into the continuum, and the third one concerns the change of atomic processes itself (cross-sections) due to the plasma electric microfield.

5.6.3.1 Atomic Population Kinetics

Let us begin with the atomic population kinetics and the existence of a critical principal quantum number, where collisional processes start to dominate over radiative ones [see discussion related to (5.52)–(5.57)]. The physical meaning of this critical quantum number N_{max} (5.52) is that above this principal quantum number, the population kinetics is essentially governed by the Saha–Boltzmann relation between highly excited states and the next ionization stage and any single recombination processes (whatever its magnitude is) are rapidly assimilated via collisions that effectively suppress the recombination processes. In fact, the analytical formulas for total dielectronic recombination and also for total radiative recombination assume that finally all excited states effectively recombine (via radiative cascading to the ground state). As has been demonstrated (Rosmej et al. 2006; Rosmej 2012), even the large rates of charge exchange recombination processes from excited states are effectively assimilated by collisions [although the collision limit itself changes (Rosmej 2012)] so that there exists a critical quantum number where the Saha–Boltzmann relations hold true and where recombination into higher states does not effectively contribute to recombination (see also Sect. 10.3.1).

5.6.3.2 Limitation of Bound States

The second phenomenon is related to the fact that the plasma electric microfield limits the number of bound states. Electric field ionization starts at the critical field strength F_{crit} that is given by Bethe and Salpeter (1977)

$$F_{\text{crit}} = 6.8 \times 10^8 \frac{\text{V}}{\text{cm}} \cdot \frac{Z_{\text{eff}}^3}{n_{\text{F}}^4}, \tag{5.148}$$

where Z_{eff} is the effective ion charge and n_{F} the principal quantum number from which on field ionization starts. In order to estimate the limited number of quantum states that take effectively part in the recombination process, we identify the critical field strength F_{crit} with the Holtsmark field

$$F = 1.3 \times 10^{-6} \cdot Z_{\text{i}} \cdot N_{\text{i}}^{2/3}(\text{cm}^{-3}) \frac{\text{V}}{\text{cm}}, \tag{5.149}$$

and the principal quantum number n_{F} with the maximum quantum number, i.e.,

$$n_{\max} \approx 4.8 \times 10^3 \cdot \frac{Z_{\text{eff}}^{3/4}}{Z_{\text{i}}^{1/4} \cdot N_{\text{i}}^{1/6}(\text{cm}^{-3})}. \tag{5.150}$$

Let us compare the maximum quantum number from (5.150) with the critical quantum number of (5.57) assuming that $n_{\text{e,crit}} = N_{\text{i}} \cdot Z_{\text{i}}$ and $kT_{\text{e}}(\text{eV}) = \alpha \cdot Z_{\text{eff}}^2 \text{Ry}$, where α is a constant (of the order of $\alpha \approx 0.1 - 1$):

$$n_{\text{crit}} \approx 373 \cdot \alpha^{1/14} \cdot \frac{Z_{\text{eff}}}{Z_{\text{i}}^{1/7} \cdot N_{\text{i}}^{1/7}}. \tag{5.151}$$

With the help of (5.149) and (5.150), we find

$$\frac{n_{\max}}{n_{\text{crit}}} \approx \frac{13}{\alpha^{1/14}} \cdot \frac{1}{Z_{\text{eff}}^{1/4} \cdot Z_{\text{i}}^{3/28} \cdot N_{\text{i}}^{1/42}(\text{cm}^{-3})}. \tag{5.152}$$

Let us illuminate the relation (5.152) with a typical example: near-solid density high-temperature H-like aluminum, i.e., $Z_{\text{eff}} = 13$, $Z_{\text{i}} = 12$, $\alpha = 0.3$, $N_{\text{i}} = 10^{22}\,\text{cm}^{-3}$ providing $n_{\max}/n_{\text{crit}} \approx 1.7$. Therefore, for almost all practical applications, we encounter the relation

$$\frac{n_{\max}}{n_{\text{crit}}} > 1. \tag{5.153}$$

The physical interpretation of relation (5.153) means that the collisional thermalization and the associated effective suppression of single recombination rates into excited states are therefore related to a large ensemble of high-n quantum numbers below those that merge into the continuum. For this reason, a detailed quantum mechanical treatment of high-n-states and their merging into the continuum appears to be not critical and quasi-classical estimates seem to be well adapted to the problem.

5.6.3.3 Effects of Angular Momentum Changing Collisions

It is evident that a strict consideration of angular momentum changing collisions requests a very extended atomic level system that includes all details of the autoionizing states in order to treat properly the collisional population redistribution effects. We restrict here ourself to a principle discussion with the help of the most frequently employed formula for dielectronic recombination proceeding from dielectronic capture from channel k and with radiative transition $j \to i$ [see also (5.131)–(5.133)]:

$$\langle \text{DR} \rangle_{\text{k,ji}}^{Z+1,Z} \approx \frac{1}{2g_{\text{k}}^{Z+1}} \cdot \left(\frac{2\pi\hbar^2}{m_{\text{e}}}\right)^{3/2} \cdot \frac{g_{\text{j}}^{Z} \cdot \Gamma_{\text{jk}}^{Z,Z+1} \cdot A_{\text{ji}}^{Z}}{\sum_{\text{l}} A_{\text{jl}}^{Z} + \sum_{\text{k}} \Gamma_{\text{jk}}^{Z,Z+1}} \cdot \frac{\exp(-E_{\text{kj}}^{\text{DC}}/kT_{\text{e}})}{(kT_{\text{e}})^{3/2}}. \tag{5.154}$$

Let us now consider a simple illustrative example, namely the Ly-alpha dielectronic $2l2l'$-satellites of He-like ions and depict two levels, one that has very large autoionizing rate and one that has a negligible ones. For the first case, we consider the level $j' = 2p^2\,{}^1D_2$, $k = 1s\,{}^2S_{1/2}$ and the radiative transition $j' = 2p^2\,{}^1D_2 \to i' = 1s2p\,{}^1P_1$. Atomic structure calculations for carbon ($Z_{\text{n}} = 6$) deliver: $\Gamma_{\text{j'k}}^{Z,Z+1} = 2.5 \times 10^{14}\,\text{s}^{-1}$, $A_{\text{j'i'}}^{Z} = 1.4 \times 10^{12}\,\text{s}^{-1}$, $\sum_{\text{l}} A_{\text{jl}}^{Z} = 1.4 \times 10^{12}\,\text{s}^{-1}$, $\sum_{\text{k}} \Gamma_{\text{j'k}}^{Z,Z+1} = 2.5 \times 10^{14}\,\text{s}^{-1}$. For the second case, we consider the autoionizing configuration $j = 2p^2\ ^3P_1$, $k = 1s\ ^2S_{1/2}$ and the radiative transition $j = 2p^2\ ^3P_1 \to i = 1s2p\ ^3P_2$. Atomic structure calculations ($Z_n = 6$) provide: $\Gamma_{jk}^{Z,Z+1} = 0$, $A_{ji}^Z = 6.0 \times 10^{11}\ \mathrm{s}^{-1}$, $\sum_l A_{jl}^Z = 1.4 \times 10^{12}\ \mathrm{s}^{-1}$, $\sum_k \Gamma_{jk}^{Z,Z+1} = 0$ from which it follows $Q_{k,ji}^{Z+1,Z} = 0$.

Assuming a two-level system where only dielectronic capture and angular momentum changing collisions (characterized by the rate coefficient $C_{j'j}$) contribute, the atomic populations n_j^Z and $n_{j'}^Z$ are given by

$$n_{j'}^Z\left(\sum_l A_{j'l}^Z + \sum_k \Gamma_{j'k}^{Z,Z+1} + n_e C_{j'j}\right) = n_k^{Z+1} \cdot n_e \cdot \langle \mathrm{DC}\rangle_{k,ji'}^{Z+1,Z} + n_e \cdot n_j^Z \cdot C_{jj'}, \tag{5.155}$$

$$n_j^Z\left(\sum_l A_{jl}^Z + \sum_k \Gamma_{jk}^{Z,Z+1} + n_e C_{jj'}\right) = n_k^{Z+1} \cdot n_e \cdot \langle \mathrm{DC}\rangle_{k,ji}^{Z+1,Z} + n_e \cdot n_{j'}^Z \cdot C_{j'j}, \tag{5.156}$$

where

$$\langle \mathrm{DC}\rangle_{k,q}^{Z+1,Z} = \frac{1}{2g_k^{Z+1}} \cdot \left(\frac{2\pi\hbar^2}{m_e}\right)^{3/2} \cdot g_q^Z \cdot \Gamma_{qk}^{Z,Z+1} \cdot \frac{\exp\left(-E_{kq}^{DC}/kT_e\right)}{(kT_e)^{3/2}} \tag{5.157}$$

with $q = j, j'$. In the absence of collisions, (5.154)–(5.156) transform to

$$n_q^{(0),Z}\left(\sum_l A_{ql}^Z + \sum_k \Gamma_{qk}^{Z,Z+1}\right) = n_k^{Z+1} \cdot n_e \cdot \langle \mathrm{DC}\rangle_{q,k}^{Z+1,Z}, \tag{5.158}$$

providing

$$n_q^{(0),Z} = n_k^{Z+1} \cdot n_e \cdot \frac{1}{2g_k^{Z+1}} \cdot \left(\frac{2\pi\hbar^2}{m_e}\right)^{3/2} \cdot \frac{g_q^Z \cdot \Gamma_{qk}^{Z,Z+1}}{\sum_l A_{ql}^Z + \sum_k \Gamma_{qk}^{Z,Z+1}} \cdot \frac{\exp\left(-E_{kq}^{DC}/kT_e\right)}{(kT_e)^{3/2}}, \tag{5.159}$$

where the index "(0)" indicates the low-density case.

In order to understand the effect of angular momentum changing collisions on the total dielectronic recombination rate, we need to consider the sum for the two levels, i.e.,

$$\langle \mathrm{DR_{coll}} \rangle_{\mathrm{tot}}^{Z+1,Z} = \langle \mathrm{DR_{coll}} \rangle_{\mathrm{k,ji}}^{Z+1,Z} + \langle \mathrm{DR_{coll}} \rangle_{\mathrm{k,j'i'}}^{Z+1,Z}, \tag{5.160}$$

where the index "coll" for the single dielectronic recombination rates $\langle \mathrm{DR_{coll}} \rangle_{\mathrm{k,ji}}^{Z+1,Z}$ and $\langle \mathrm{DR_{coll}} \rangle_{\mathrm{k,j'i'}}^{Z+1,Z}$ indicates that these rates include the collisional processes. This has to be distinguished from (5.154) which is a low-density approximation. It is of principal interest to understand the change of the dielectronic recombination due to collisions with reference to the low-density case, i.e., we consider the ratio

$$\frac{\langle \mathrm{DR_{coll}} \rangle_{\mathrm{tot}}^{Z+1,Z}}{\langle \mathrm{DR} \rangle_{\mathrm{tot}}^{Z+1,Z}} = \frac{\langle \mathrm{DR_{coll}} \rangle_{\mathrm{k,ji}}^{Z+1,Z} + \langle \mathrm{DR_{coll}} \rangle_{\mathrm{k,j'i'}}^{Z+1,Z}}{\langle \mathrm{DR} \rangle_{\mathrm{k,ji}}^{Z+1,Z} + \langle \mathrm{DR} \rangle_{\mathrm{k,j'i'}}^{Z+1,Z}}. \tag{5.161}$$

The collisional dielectronic recombination rates cannot be determined from relations like (5.154) but need to be determined directly from the populations, i.e.,

$$\langle \mathrm{DR_{coll}} \rangle_{\mathrm{k,ji}}^{Z+1,Z} \propto n_{\mathrm{j}}^{Z} \cdot A_{\mathrm{ji}}^{Z}, \tag{5.162}$$

because the product of the level population with the radiative decay is the rate at which the excited state decays to the ground state which is equivalent to dielectronic recombination (note that the usual branching ratios that appear in formulas like those of (5.154) are already included via the equilibrium population) if the right-hand side of (5.155), (5.156) is driven by dielectronic capture and angular momentum changing collisions between the autoionizing levels under consideration. Combining the relations (5.161) and (5.162), we obtain

$$\frac{\langle \mathrm{DR_{coll}} \rangle_{\mathrm{tot}}^{Z+1,Z}}{\langle \mathrm{DR} \rangle_{\mathrm{tot}}^{Z+1,Z}} = \frac{n_{\mathrm{j}}^{Z} \cdot A_{\mathrm{ji}}^{Z} + n_{\mathrm{j'}}^{Z} \cdot A_{\mathrm{j'i'}}^{Z}}{n_{\mathrm{j}}^{(0),Z} \cdot A_{\mathrm{ji}}^{Z} + n_{\mathrm{j'}}^{(0),Z} \cdot A_{\mathrm{j'i'}}^{Z}}, \tag{5.163}$$

i.e.,

$$\frac{\langle \mathrm{DR_{coll}} \rangle_{\mathrm{tot}}^{Z+1,Z}}{\langle \mathrm{DR} \rangle_{\mathrm{tot}}^{Z+1,Z}} = \frac{\dfrac{n_{\mathrm{j'}}^{Z}}{n_{\mathrm{j'}}^{(0),Z}} + \dfrac{n_{\mathrm{j}}^{Z} \cdot A_{\mathrm{ji}}^{Z}}{n_{\mathrm{j'}}^{(0),Z} \cdot A_{\mathrm{j'i'}}^{Z}}}{\dfrac{n_{\mathrm{j}}^{(0),Z} \cdot A_{\mathrm{ji}}^{Z}}{n_{\mathrm{j'}}^{(0),Z} \cdot A_{\mathrm{j'i'}}^{Z}} + 1}. \tag{5.164}$$

Because $E_{\mathrm{kj}}^{DC} \approx E_{\mathrm{kj'}}^{DC}$, we have for the population ratio in the low-density case (for the example given above)

$$\frac{n_{\mathrm{j}}^{(0),Z}}{n_{\mathrm{j'}}^{(0),Z}} \approx \frac{g_{\mathrm{j}}^{Z} \cdot \Gamma_{\mathrm{jk}}^{Z,Z+1}}{g_{\mathrm{j'}}^{Z} \cdot \Gamma_{\mathrm{j'k}}^{Z,Z+1}} \cdot \frac{\sum_{\mathrm{l}} A_{\mathrm{j'l}}^{Z} + \sum_{\mathrm{k}} \Gamma_{\mathrm{j'k}}^{Z,Z+1}}{\sum_{\mathrm{l}} A_{\mathrm{jl}}^{Z} + \sum_{\mathrm{k}} \Gamma_{\mathrm{jk}}^{Z,Z+1}} \approx 0, \tag{5.165}$$

because $\Gamma_{\mathrm{jk}}^{Z,Z+1} \ll \Gamma_{\mathrm{j'k}}^{Z,Z+1}$. Therefore, population is essentially transferred by angular momentum changing collision form level j' to level j but not vice versa. Let us now specify the above example for the populations of (5.155), (5.156):

$$n_{\mathrm{j}}^{Z}\left(\sum_{\mathrm{l}} A_{\mathrm{jl}}^{Z} + n_{\mathrm{e}} C_{\mathrm{jj'}}\right) \approx n_{\mathrm{e}} \cdot n_{\mathrm{j'}}^{Z} \cdot C_{\mathrm{j'j}}, \tag{5.166}$$

and

$$n_{\mathrm{j'}}^{Z}\left(\sum_{\mathrm{l}} A_{\mathrm{j'l}}^{Z} + \sum_{\mathrm{k}} \Gamma_{\mathrm{j'k}}^{Z,Z+1}\right) \approx n_{\mathrm{k}}^{Z+1} \cdot n_{\mathrm{e}} \cdot \langle \mathrm{DC} \rangle_{\mathrm{k,j'}}^{Z+1,Z}, \tag{5.167}$$

Equations (5.155), (5.156), (5.158) indicate that for autoionizing levels with very large autoionizing rates, the populations are close to the low-density case. Equation (5.167) corresponds therefore to the case of low density (5.158), i.e.,

$$n_{\mathrm{j'}}^{Z} \approx n_{\mathrm{j'}}^{(0),Z}. \tag{5.168}$$

Injecting relations (5.165), (5.166), (5.168) into (5.164), we obtain

$$\frac{\langle \mathrm{DR}_{\mathrm{coll}} \rangle_{\mathrm{tot}}^{Z+1,Z}}{\langle \mathrm{DR} \rangle_{\mathrm{tot}}^{Z+1,Z}} \approx 1 + \frac{n_{\mathrm{e}} \cdot C_{\mathrm{j'j}}}{\sum_{\mathrm{l}} A_{\mathrm{jl}}^{Z} + n_{\mathrm{e}} C_{\mathrm{jj'}}} \cdot \frac{A_{\mathrm{ji}}^{Z}}{A_{\mathrm{j'i'}}^{Z}}. \tag{5.169}$$

Because $g_{\mathrm{j'}} \cdot C_{\mathrm{j'j}} \approx g_{\mathrm{j}} \cdot C_{\mathrm{jj'}}$ for closely spaced levels, relation (5.169) takes the form

$$\frac{\langle \mathrm{DR}_{\mathrm{coll}} \rangle_{\mathrm{tot}}^{Z+1,Z}}{\langle \mathrm{DR} \rangle_{\mathrm{tot}}^{Z+1,Z}} \approx 1 + \frac{g_{\mathrm{j'}}^{Z}}{g_{\mathrm{j}}^{Z}} \cdot \frac{A_{\mathrm{ji}}^{Z}}{A_{\mathrm{j'i'}}^{Z}} \cdot \left(\frac{1}{1 + \sum_{\mathrm{l}} A_{\mathrm{jl}}^{Z} / n_{\mathrm{e}} C_{\mathrm{jj'}}}\right). \tag{5.170}$$

If $\sum_{\mathrm{l}} A_{\mathrm{jl}}^{Z} \approx n_{\mathrm{e}} C_{\mathrm{jj'}}$, the term in parenthesis of relation (5.170) is about one half and the relation indicates that the total dielectronic recombination rate is enhanced (i.e., $\langle \mathrm{DR}_{\mathrm{coll}} \rangle_{\mathrm{tot}}^{Z+1,Z} / \langle \mathrm{DR} \rangle_{\mathrm{tot}}^{Z+1,Z} > 1$) due to angular momentum changing collisions. This can be understood in a transparent qualitative picture: For the level j' with high autoionization rate, the dielectronic capture is high and due to the large autoionizing rate, the branching factor for radiative de-excitation is small. If, however, a certain fraction of population is collisionally transferred to another level before autoionization and radiative decay disintegrate the upper level j', the level j is

Table 5.7 Field-free autoionization decay rates in [s^{-1}] including intermediate coupling, configuration, and magnetic interaction

State	$Z_n = 3$	$Z_n = 6$	$Z_n = 13$	$Z_n = 18$	$Z_n = 26$	$Z_n = 42$
$2p^2\ {}^1S_0$	8.4×10^{10}	5.1×10^{12}	1.3×10^{13}	1.9×10^{13}	3.4×10^{13}	7.0×10^{13}
$2p^2\ {}^1D_2$	1.5×10^{14}	2.5×10^{14}	3.1×10^{14}	3.1×10^{14}	2.3×10^{14}	2.1×10^{14}
$2p^2\ {}^3P_0$	2.9×10^{7}	2.3×10^{9}	2.3×10^{11}	1.2×10^{12}	3.7×10^{12}	2.8×10^{12}
$2p^2\ {}^3P_1$	0	0	0	0	0	0
Breit interaction	2.6×10^{7}	6.8×10^{8}	1.9×10^{10}	7.2×10^{10}	3.2×10^{11}	2.2×10^{12}
$2p^2\ {}^3P_2$	1.1×10^{9}	3.1×10^{10}	3.0×10^{12}	2.1×10^{13}	1.1×10^{14}	1.5×10^{14}

effectively populated by collisions from $j' \to j$ (because the population of the level j is small as dielectronic capture is small due to small autoionization rate). The transferred population, however, has a very favorable branching factor for the level j (e.g., in the above example $A^Z_{ji}/\left(\sum_l A^Z_{jl} + \sum_k \Gamma^{Z,Z+1}_{jk}\right) = 6.0\times 10^{11}/1.4\times 10^{12} = 0.43$) compared to the level j' ($A^Z_{j'i'}/\left(\sum_l A^Z_{j'l} + \sum_k \Gamma^{Z,Z+1}_{j'k}\right) = 1.4\times 10^{12}/2.6\times 10^{14} = 0.0088$) and the transferred population is more effectively transferred to the ground state to finally contribute to the dielectronic recombination.

5.6.3.4 Electric Field Effects on Cross-Sections

The influence of the electric field on autoionization and corresponding dielectronic recombination rates was studied by (Davis and Jacobs 1975; Jacobs et al. 1976; Jacobs and Davis 1979) with the simplest atomic system of He-like autoionizing states $2l2l'$. It was realized that forbidden autoionizing processes (forbidden in LS-coupling scheme) become allowed by electric field mixing of autoionizing bound state wave functions. The allowed autoionization width is given by the first-order transition rate

$$\Gamma(d \to c) = \frac{2\pi}{\hbar} \cdot |\langle d|V|c\rangle|^2 \delta(E_d - E_c), \tag{5.171}$$

where V is the electrostatic interaction. Because V is a scalar operator, the autoionization vanishes unless there are available adjacent continuum states c with the same angular momentum and parity as the discrete levels d (Cowan 1981). Because of the absence of even parity P states below the second ionization threshold, the $2p^2\ {}^3P$-state of He-like ions is metastable against autoionization decay. In the presence of perturbing electric fields, however, autoionization of the state $a = 2p^2\ {}^3P$ may occur by a second-order process involving the field-induced transition to the nearby autoionizing state $d = 2s2p\ {}^3P$. In a quasi-static ion field, the field-induced autoionization rate is given by

$$\Gamma(a \to c) = \frac{2\pi}{\hbar} \cdot \left| \sum_{\mathrm{d}} \frac{\langle a|\vec{Q} \cdot \vec{E}|d\rangle \langle d|V|c\rangle}{(E_{\mathrm{a}} - E_{\mathrm{d}}) + i\hbar(\Gamma_{\mathrm{d}} + A_{\mathrm{d}})/2} \right|^2 \delta(E_{\mathrm{a}} - E_{\mathrm{c}}), \tag{5.172}$$

where $\vec{Q}$ is the electric dipole moment operator, Γ_{d} and A_{d} are the autoionization and radiative width of the state d, respectively. Therefore, the first-order contribution from the field-induced transition decays directly into the non-resonant continuum $c = 1s\epsilon p\ ^3P$.

It should be noted that for practical applications, not only field-induced transitions have to be considered, but intermediate coupling, configuration, and magnetic interactions too. In particular for highly charged ions, these "non-electric field effects" may have likewise a considerable contribution to the forbidden autoionization width, as is demonstrated in the following Table 5.7. In addition, the Breit interaction induces an autoionization rate for the $2p^2\ ^3P_1$—state (see second line for the state $2p^2\ ^3P_1$ in Table 5.7).

Table 5.7 shows also the general effect that if the nuclear charge increases, the autoionizing widths are more and more distributed over the levels. Therefore, electric field effects are best studied for low-Z elements.

From the relationship between the corresponding capture and autoionizing rates, it follows that the electric field can induce dielectronic recombination through normally inaccessible high angular momentum states which have large statistical weights (Jacobs et al. 1976). In fact, in a plasma, the angular momentum l is no longer a good quantum number, because the presence of an electric field destroys the spherical symmetry. However, the projection m which is defined with respect to the direction of the electric field remains a good quantum number. For nonzero quantum numbers m, this results in a twofold degeneracy of the outer electron in addition to the twofold degeneracy due to the spin. The appropriate transformation of the field-free substates l has the form

$$|n\lambda m\rangle = \sum_{l=|m|}^{n-1} |nlm\rangle\langle nlm \mid n\lambda m\rangle, \tag{5.173}$$

where the quantum number λ, which replaces l in the presence of the electric field, can have integer values from $\lambda = 0\ldots(n - |m| - 1)$. The calculations demonstrate (Jacobs et al. 1976; Jacobs and Davis 1979; Bureyeva et al. 2001, 2002) that the dependence of the autoionization rates on the quantum number λ is rather smooth in contrast to the field-free case where the autoionization rates decrease rapidly with quantum number l. Due to this reason, dielectronic capture in the presence of electric field increases because it is proportional to the autoionizing rate and the statistical weight, i.e., $\langle \mathrm{DC}\rangle_{\mathrm{k,j}}^{Z+1,Z} \propto g_{\mathrm{j}}^{Z} \cdot \Gamma_{\mathrm{jk}}^{Z,Z+1}$. Because the dielectronic recombination is proportional to the dielectronic capture [see (5.154)], this results in a considerable increase of the total dielectronic recombination rate. For example, for the autoionizing states $1s^2 2pnl$ in Be-like Fe^{22+} an about threefold increase of the dielectronic recombination rate was found even for densities as low as 10^{14} cm^{-3} (Jacobs et al. 1976). This dramatic increase for rather low densities is particularly connected with the fact that for the $1s^2 2pnl$-configuration, the resonance spontaneous transition probability $2p$–$2s$ is not very large and high-n-states have autoionizing rates larger than radiative decay rates for n-quantum numbers up to about 100. In consequence, high-n-states contribute considerably to the dielectronic recombination rate. As high-n-states are likewise strongly affected by rather small electric fields, a considerable impact on the total recombination rate is encountered even for rather low plasma densities (being of importance for typical densities of solar corona or magnetic fusion plasmas).

The interaction with an electric field makes atomic structure calculations extremely complex, and it is difficult to derive general conclusions. It has been, however, demonstrated (Bureyeva et al. 2001, 2002) that the quasi-classical approach combined with a transformation to parabolic quantum numbers (5.173) provides results that are in surprisingly good agreement with extremely complex numerical calculations (Robicheaux and Pindzola 1997). Moreover, the quasi-classical approach combined with the transformation to parabolic quantum numbers allowed deriving a closed expression for the autoionization rate in an electric field:

$$\Gamma(n,\lambda,m) = \int_{l_{\min}}^{l_{\max}} P(nl;\lambda m)\cdot\Gamma(nl)\cdot \mathrm{d}l, \tag{5.174}$$

with

$$l_{\min}^2 = \frac{1}{2}\left\{\left[(n-1)^2+m^2-\lambda^2\right] - \sqrt{\left[(n-1)^2+m^2-\lambda^2\right]^2 - 4(n-1)^2 m^2}\right\} \tag{5.175}$$

and

$$l_{\max}^2 = \frac{1}{2}\left\{\left[(n-1)^2+m^2-\lambda^2\right] + \sqrt{\left[(n-1)^2+m^2-\lambda^2\right]^2 - 4(n-1)^2 m^2}\right\}. \tag{5.176}$$

$\Gamma(nl)$ is the standard autoionizing rate in spherical coordinates and $P(nl;\lambda m)$ is a joint probability (with normalization equal to one) for the appearance of spherical (nl) and parabolic (λm) quantum numbers that can be expressed in terms of Clebsch–Gordan coefficients. For large quantum numbers and the condition $m < l \ll n$ (quasi-classical limit of Clebsch–Gordan coefficients that is of practical interest), the joint probability can be approximated by Bureyeva et al. (2002)

$$P(nl;\lambda m) \approx \frac{1}{\pi} \cdot \left\{ \frac{2l}{\sqrt{(l^2 - l^2_{\min}) \cdot (l^2_{\max} - l^2)}} \right\}. \quad (5.177)$$

Substituting quasi-classical values for the autoionization rate $\Gamma(nl)$ into (5.174) and using (5.173), we obtain an autoionizing rate in parabolic quantum numbers expressed in terms of universal functions ($t = l/l_{\text{eff}}$, $l_{\text{eff}} = (3Z^2/\omega)^{1/3}$):

$$\Gamma(n,\lambda,m) = \frac{f_{\text{ij}}}{\pi \cdot n^3} \cdot I(t_{\min}, t_{\max}), \quad (5.178)$$

$$I(t_{\min}, t_{\max}) \approx \frac{2}{l_{\max}} \cdot \left(\frac{3Z^2}{\omega}\right)^{2/3} \cdot Y\left(l_{\min} \cdot (\omega/3Z^2)^{1/3}\right), \quad (5.179)$$

$$Y(x) \approx 0.284 \cdot \exp(-2x^3). \quad (5.180)$$

f_{ij} is the oscillator strength of the core transition with charge Z (e.g., the oscillator strength corresponding to the transition $1s - 2p$ in H-like Al for the He-like $2lnl'$-satellites, $Z = 13$). The formulas (5.174)–(5.180) demonstrate likewise a broad distribution over the electric quantum number λ that finally results in an increase of the dielectronic recombination rate.

References

V.A. Astapenko, Impact ionization of atoms: calculation in the Born-Compton approximation. Laser Phys. **11**, 1336 (2001)

V.A. Astapenko, *Polarization Bremsstrahlung on Atoms, Plasmas, Nanostructures and Solids* (Springer, Berlin, 2013)

V.A. Astapenko, A. Eletskii, V. Kudrya, P. Ventzek, Calculation of the cross-sections for electron impact excitation of magnesium. Laser Phys. **10**, 1220 (2000)

V.A. Astapenko, V.S. Lisitsa, *Collisional Processes in low Temperature Plasma* (MIPT-Textbook, Moscow, 2007). (in Russian)

J.G. Baker, D.H. Menzel, Physical processes in gazeous nebulae. Astrophys. J. **88**, 52 (1938)

V.A. Bazylev, M.I. Chibisov, Excitation and ionization of multicharged ions by electron collisions. Phys.-Usp. **24**, 276 (1981)

I.L. Beigman, L.A. Vainshtein, B.N. Chichkov, Dielectron recombination. JETP **53**, 490 (1981)

I.L. Beigman, L.A. Vainshtein, M. Brix, A. Pospieszczyk, I. Bray, D.B. Fursa, YuV Ralchenko, Excitation and ionization cross-sections for HeI from normalized Born and K-matrix calculations: $\Delta S = 0$ transitions from $n = 2{,}3$ excited states. ADNDT **74**, 123 (2000)

V.B. Berestetskii, L.P. Pitaevskii, E.M. Lifshitz, *Quantum Electrodynamics* (Elsevier, Oxford, 1982)

H.A. Bethe, E.E. Salpeter, *Quantum Mechanics of One- and Two-Electron Atoms* (Plenum Publishing, New York, 1977)

D. Braunstein, S. Shuker, X-ray laser without inversion in a three-level ladder system. Phys. Rev. A **68**, 013812 (2003)

L.A. Bureyeva, T. Kato, V.S. Lisitsa, C. Namba, Quasiclassical representation of autoionization decay reates in parabolic coordinates. J. Phys. B: At. Mol. Opt. Phys. **34**, 3909 (2001)
L.A. Bureyeva, T. Kato, V.S. Lisitsa, C. Namba, Quasiclassical theory of dielecronic recombination in plasmas. Phys. Rev. A **65**, 032702 (2002)
A. Burgess, Dielectronic recombination and the temperature of the solar corona. Astrophys. J. **139**, 776 (1964)
H.K. Chung, C. Bowen, C.J. Fontes, S.B. Hansen, Yu. Ralchenko, Comparison and analysis of collisional-radiative models at the NLTE-7 workshop. HEDP **9**, 645 (2013)
J. Colgan, C.J. Fontes, H. Zhang, J. Abdallah Jr., Collisional-radiative modeling of Tungsten at temperatures of 1200–2400 eV. Atoms **3**, 76 (2015). https://doi.org/10.3390/atoms3020076
R.D. Cowan, Resonant-scattering (autoionisation) contributions to excitation rates in O^{4+} and similar ions. J. Phys. B: At. Mol. Opt. Phys. **13**, 1471 (1980)
R.D. Cowan, *The Theory of Atomic Structure and Spectra* (California University Press, 1981)
R.D. Cowan, Effects of autoionising levels in highly ionized atoms. Phys. Scr. **T3**, 200 (1983)
J. Davis, V.L. Jabobs, Effects of plasma microfields on radiative transitions from atomic levels above the ionization threshold. Phys. Rev. A **12**, 2017 (1975)
B. Deschaud, O. Peyrusse, F.B. Rosmej, Generalized atomic physics processes when intense femtosecond XUV- and X-ray radiation is interacting with solids. Europhys. Lett. **108**, 53001 (2014)
B. Deschaud, O. Peyrusse, F.B. Rosmej, Atomic kinetics for isochoric heating of solid aluminum under short intense XUV free electron laser irradiation. HEDP **15**, 22 (2015)
R.C. Elton, *X-Ray Lasers* (Academic Press, New York, 1990)
E. Fermi, Über die Theorie des Stoßes zwischen Atomen und elektrisch geladenen Teilchen. Zeitschrift für Physik **29**, 315 (1924)
V. Fischer, V. Bernshtam, H. Golten, Y. Maron, Electron-impact excitation cross-sections for allowed transitions in atoms. Phys. Rev. A **53**, 2425 (1996)
M.N. Gailitis, The use of the Bethe approximation in calculating the excitation cross-section of an ion by electrons (in Russian), in *Atomic Collisions* (Latvian State University, Riga, 1963), p. 93
J.N. Gau, R.J.W. Henry, Excitation of lithumlike ions by electron impact. Phys. Rev. A **16**, 986 (1977)
M. Gryzinski, Classical theory of electronic and ionic inelastic collisions. Phys. Rev. **115**, 374 (1959)
M. Gryzinski, Two particle collisions. I. General relations for collisions in the laboratory system. Phys. Rev. **138**, A305 (1965a)
M. Gryzinski, Two particle collisions. II. Coulomb collisions in the laboratory system of coordinates. Phys. Rev. **138**, A322 (1965b)
M. Gryzinski, Classical theory of atomic collisions. I. Theory of inelastic collisions. Phys. Rev. **138**, A336 (1965c)
Y. Hahn, Plasma density effects on the three-body recombination rate coefficients. Phys. Lett. A **231**, 82 (1997)
Y. Hahn, J. Li, Transient behavior of nonequilibrium plasma formed by merged beams. Zeitschrift für Physik **D36**, 85 (1996)
W. Heitler, *The Quantum Theory of Radiation* (Dover, New York, 1984)
J. Jackson, *Classical Electrodynamics*, 3rd ed. (Wiley, 1998), ISBN: 978-0-471-30932-1
V.J. Jacobs, J. Davis, Properties of Rydberg autoionizing states in electric field. Phys. Rev. A **19**, 776 (1979)
V.J. Jacobs, J. Davis, P.C. Kepple, Enhancement of dielectronic recombination by plasma electric microfields. Phys. Rev. Lett. **37**, 1390 (1976)
T. Kato, E. Asano, Comparison of recombination rate coefficients given by empirical formulas for ions from hydrogen through nickel. NIFS-DATA-54 (June 1999)
T. Kato, K. Masai, M. Arnaud, Comparison of ionization rate coefficients of ions from hydrogen through nickel. NIFS-DATA-14 (Sept. 1991)
D.P. Kilcrease, S. Brookes, Correction of near threshold behavior of electron collisional excitation cross-sections in the plane-wave Born approximation. HEDP **9**, 722 (2013)
Y.-K. Kim, M.E. Rudd, Binary-encounter-dipole model for electron-impact ionization. Phys. Rev. A **50**, 3954 (1994)

V.I. Kogan, A.B. Kukushkin, V.S. Lisitsa, Kramers electrodynamics and electron-atomic radiative-collisional processes. Phys. Rep. **213**, 1 (1992)

L.D. Landau, E.M. Lifschitz, *The Classical Theory of Fields* (Pergamon, Oxford, 2003)

L.D. Landau, E.M. Lifschitz, *Mechanics* (Elsevier, Oxford, 2005)

W. Lotz, Electron impact-ionization cross-sections for atoms up to $Z = 108$. Zeitschrift für Physik **232**, 101 (1970)

R. Loudon, *The Quantum Theory of Light*, 3rd ed. (Oxford Science Publications, 2000)

P. Mansbach, J. Keck, Monte Carlo Trajectory calculations of atomic excitation and ionization by thermal electrons. Phys. Rev. **181**, 275 (1969)

S.A. Mayorov, A.N. Tkachev, S.I. Yakovlenko, Metastable supercooled plasma. Physics Uspekhi **37**, 279 (1994)

NIST (2019). http://www.nist.gov

I.C. Percival, M.J. Seaton, The polarization of atomic line radiation excited by electron impact. Philos. Trans. Royal Soc. London Ser. A Math. Phys. Sci. **251**, 113 (1958)

F. Petitdemange, F.B. Rosmej, Dielectronic satellites and Auger electron heating: irradiation of solids by intense XUV-free electron laser radiation, in *New Trends in Atomic & Molecular Physics—Advanced Technological Applications*, vol. 76, ed. by Mohan (Springer, 2013), pp. 91–114, ISBN 978-3-642-38166-9

H. Van Regemorter, Rate of collisional excitation in stellar atmospheres. Astrophys. J. **136**, 906 (1962)

F. Robicheaux, M.S. Pindzola, Enhanced dielectronic recombination in crossed electric and magnetic fields. Phys. Rev. Lett. **79**, 2237 (1997)

F.B. Rosmej, Diagnostic properties of Be-like and Li-like satellites in dense transient plasmas under the action of highly energetic electrons. JQSRT **51**, 319 (1994)

F.B. Rosmej, The He_β-emission in dense non-Maxwellian plasmas. J. Phys. B Lett: At. Mol. Opt. Phys. **33**, L1 (2000)

F.B. Rosmej, X-ray emission spectroscopy and diagnostics of non-equilibrium fusion and laser produced plasmas, in *Highly Charged Ion Spectroscopic Research*, ed. by Y. Zou, R. Hutton (Taylor and Francis 2012), pp. 267–341, ISBN: 978-1-4200-7904-3

F.B. Rosmej, A.Ya. Faenov, T.A. Pikuz, F. Flora, P. Di Lazzaro, S. Bollanti, N. Lizi, T. Letardi, A. Reale, L. Palladino, O. Batani, S. Bossi, A. Bornardinello, A. Scafati, L. Reale, Line formation of high intensity He_β-Rydberg dielectronic satellites $1s3lnl'$ in laser produced plasmas. J. Phys. B Lett.: At. Mol. Opt. Phys. **31**, L921 (1998)

F.B. Rosmej, R. Stamm, V.S. Lisitsa, Convergent coupling of Helium to the H/D background in magnetically confined plasmas. Europhys. Lett. **73**, 342 (2006)

F.B. Rosmej, V.A. Astapenko, V.S. Lisitsa, L.A. Vainshtein, *Dielectronic recombination in non-LTE plasmas*, Matter and Radiation at Extremes (Review) **5**, 064601 (2020). https://doi.org/10.1063/5.0014158

J.G. Rubiano, R. Florido, C. Bowen, R.W. Lee, Yu. Ralchenko, Review of the 4th NLTE Code Comparison Workshop. HEDP **3**, 225 (2007)

V.P. Shevelko, L.A. Vainshtein, *Atomic Physics for Hot Plasmas* (IOP Publishing, Bristol, 1993)

I.I. Sobelman, A.V. Vinogradov, On the problem of extreme UV and X-ray lasers, in *Advances in Atomic and Molecular Physics*, vol. 20, 327, ed. by S.D. Bates, B. Bederson (Academic Press, New York, 1985)

I.I. Sobelman, L.A. Vainshtein, *Excitation of Atomic Spectra* (Alpha Science, 2006)

V.N. Thytovich, I.M. Oiringel, *Polarization Bremsstrahlung of Particles and Atoms* (Plenum, New York, 1991)

L.A. Vainshtein, V.P. Shevelko, *Program ATOM*, Preprint No. 43, (Lebedev Physical Institute, Moscow, 1996)

G.S. Voronov, A practical fit formula for ionization rate coefficients of atoms and ions by electron impact: $Z = 1$–28. ADNDT **65**, 1 (1997)

J.G. Wang, T. Kato, I. Murakami, in Dielectronic Recombination Rate Coefficients to Excited States of He from He^+. NIFS-DATA-53 (1999)

Chapter 6
Atomic Population Kinetics

Abstract This chapter introduces to the theory of atomic population kinetics and radiative properties of atomic and ionic bound–bound transitions. Particular attention is devoted to the general problems related to an extremely large number of kinetic equations describing populations of Rydberg and autoionization atomic states in plasmas. A new method of reduced kinetics for autoionizing states, the virtual contour shape kinetic theory (VCSKT), is described in details. The method is based on a probability method for LTE- and non-LTE-level populations that allows effective level reduction while preserving all detailed atomic transitions. The representation employs effective relaxation constants that have analytical solutions. The comparison with detailed level-by-level calculations demonstrates high accuracy and large efficiency of the VCSKT. In order to solve many states' kinetic problems for Rydberg atomic states, the quasi-classical representation of the system of kinetic equations is proposed. In particular, the two-dimensional radiative cascades between Rydberg atomic states are described by a purely classical motion of atomic electrons in a Coulomb field that lose energy and orbital momentum. The general collisional-radiative model for large principal quantum numbers is reduced to an effective diffusion in two-dimensional energy and orbital momentum space. The results of these new kinetic models are compared with standard collisional-radiative kinetics demonstrating an important reduction of computer times, the possibility to obtain scaling relations and to independently study the precision of complex quantum calculations for these many level kinetic problems.

6.1 Generalized Atomic Kinetics of Non-Equilibrium Plasmas Containing Ions of Various Charge States

6.1.1 Principles of Atomic Line Emission: The Two-Level Atom

Let us consider a two-level atom to understand the basic principles of atomic line emission. Figure 6.1 depicts the two-level atom and the related atomic physics

F. B. Rosmej et al., *Plasma Atomic Physics*, Springer Series on Atomic, Optical, and Plasma Physics 104, https://doi.org/10.1007/978-3-030-05968-2_6

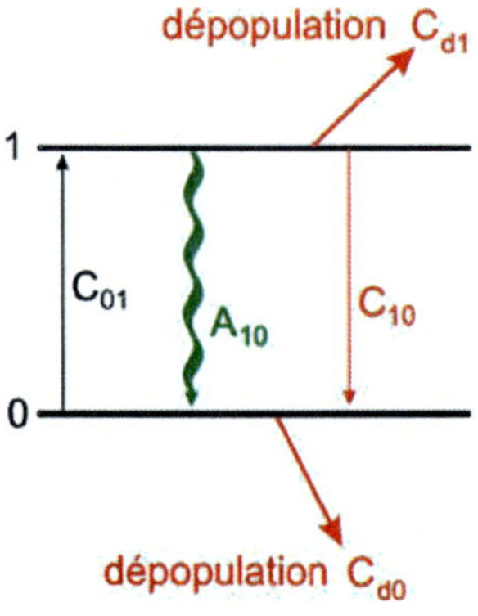

Fig. 6.1 Two-level atom of an open system

processes. For a two-level atom, the system of differential population equations takes the form:

$$\frac{dn_1}{dt} = n_0 n_e C_{01} - n_1 A_{10} - n_1 (n_e C_{10} + n_e C_{d1}). \tag{6.1}$$

n_1 is the upper level density, n_0 the lower level density, n_e is the electron density, the C's are the electron collisional rate coefficients and A is the spontaneous radiative decay rate. In stationary plasmas, $d/dt = 0$ and (6.1) can readily be solved for the upper-level density:

$$n_1 = \frac{n_0 n_e C_{01}}{A_{10} + n_e C_{10} + n_e C_{1d}}. \tag{6.2}$$

The intensity of the spectral line is then given by

$$I_{10} = \frac{\hbar\omega_{10}}{4\pi} n_0 n_e C_{01} \frac{A_{10}}{A_{10} + n_e C_{10} + n_e C_{1d}}. \tag{6.3}$$

In the high-density limit when $n_e C_{10} \gg A_{10}$ and $C_{d1} = C_{d0} = 0$ (due to the detailed balance of populating and depopulating collisions from and to levels not explicitly included in the two-level system), the intensity is proportional to the radiative decay rate:

$$I_{10} \propto A_{10}. \tag{6.4}$$

In the low-density limit, however, when $n_e C_{10} \ll A_{10}$ (Corona model), the intensity is given by

$$I_{10} = \frac{\hbar\omega_{10}}{4\pi} n_0 n_e C_{01}. \tag{6.5}$$

Equation (6.5) shows that the intensity is independent of the spontaneous radiative decay rate. How to understand this result? Let us imagine that we fill a

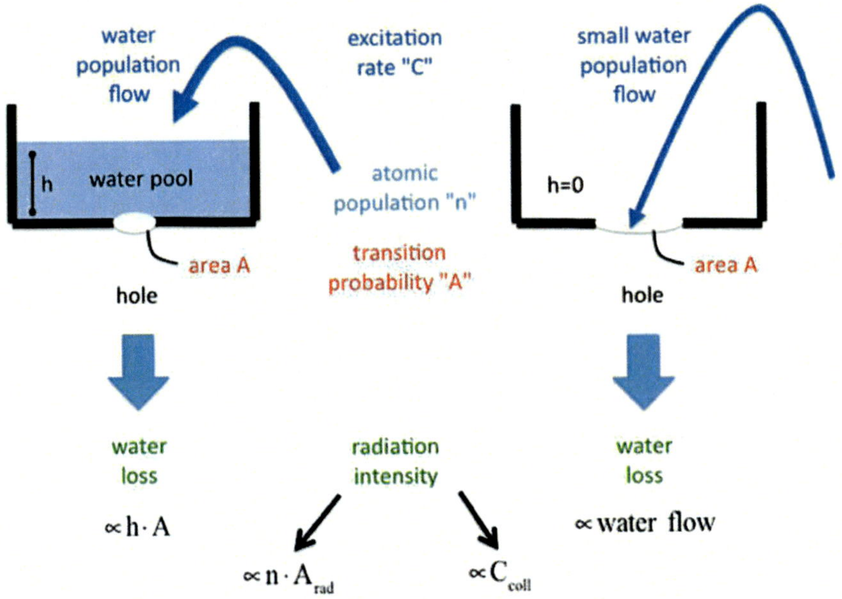

Fig. 6.2 Water pool model of atomic radiation emission. The collisional excitation rate C corresponds to the water population flow into the pool, the height h to the atomic population n, the size of the hole (area A) to the radiative decay A_{rad}, and the water flow out of the hole to the radiative emission

bottle with water that has a hole, see Fig. 6.2 left. The size of the hole (area A) corresponds to the radiative decay rate A_{rad}, the water that flows out of the hole corresponds to the line intensity, and the water flow into the bottle corresponds to the excitation rate C. If the hole is small, the water mounts in the bottle because it cannot escape quickly enough through the small hole.

Let us now imagine that we fill the bottle only with a tiny rate. In this case, the water escapes immediately through the whole and the water is not mounting in the bottle. Under these circumstances, we could increase the size of the hole without changing the amount of water that is escaping from the hole because for the small hole already all water escapes. This regime is equivalent to the case where the intensity does not depend on the radiative decay rate and corresponds to the Corona model. Equation (6.2) shows that in the limit of low densities the upper state population is given by

$$n_1 \approx \frac{n_0 n_e C_{01}}{A_{10}}. \tag{6.6}$$

If the radiative decay rate is small, the upper state population is large (so-called metastable level). This explains why we can observe in experimental spectra line emissions of forbidden transitions with intensities that are of the same order like those for resonance lines. Famous examples are the light emission from the Aurora

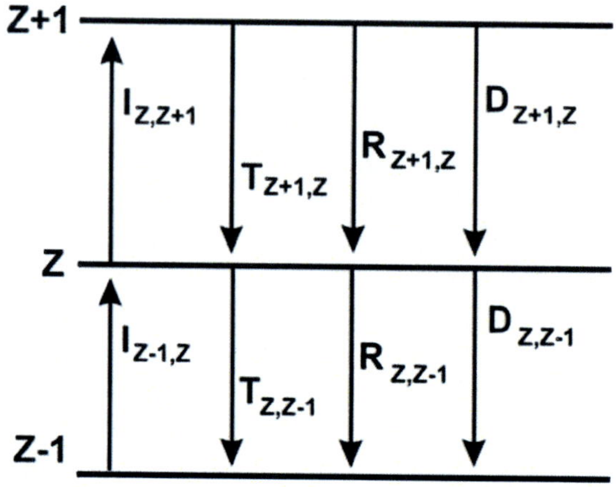

Fig. 6.3 Schematic ionic level system showing ionization (*I*) and recombination processes (*T*, *R*, *D*), *T* is the three-body recombination, *R* the radiative recombination, and *D* the dielectronic recombination

Borealis (green and red emission from atomic oxygen), the observation of the forbidden lines *X* and *Z* of He-like impurity ions in tokamaks (see Sect. 1.2.3) as well as the observation of the intercombination line *Y* of He-like ions in many dense laser-produced plasmas.

6.1.2 The Principles of Ionic Charge State Distributions in Plasmas

Atomic radiation in plasmas is rather complex as line emission from several ionization stages of the atom contribute at the same time. We therefore start our investigation with the so-called ionic charge state distribution in plasmas. In order to get some insight in the relevant physics, we consider an atomic level with population density n^Z and charge state "*Z*" that is linked to $(Z + 1)$ and $(Z - 1)$ via several elementary atomic processes, see Fig. 6.3: electron collisional ionization *I*, three-body recombination *T*, radiative recombination *R*, and dielectronic recombination *D*. These processes are defined as follows (see also Chap. 1 and Sect. 3.5):

Ionization: $X^Z + e \rightarrow X^{Z+1} + e + e$

Three-body recombination: $X^Z + e + e \rightarrow X^{Z-1} + e$

Radiative recombination: $X^Z + e \rightarrow X^{Z-1} + \hbar\omega_{\text{rad.recom.}}$

Dielectronic recombination: $X^Z + e \rightarrow X^{Z-1,**} \rightarrow X^{Z-1,*} + \hbar\omega_{\text{sat}} \rightarrow X^{Z-1} + \hbar\omega_{\text{l}}$

X^Z characterizes an atom "*X*" in charge state "*Z*", "*e*" is an electron in the continuum, $\hbar\omega_{\text{rad.recom.}}$ is the continuum radiation of the radiative recombination. $X^{Z,*}$ and $X^{Z,**}$ characterize single- and double-excited ions, $\hbar\omega_{\text{sat}}$ and $\hbar\omega_{\text{l}}$ indicate bound–bound radiation from atomic and ionic lines. The dielectronic recombination describes a multistep process: it starts from the so-called dielectronic capture of a continuum electron that forms first a double-excited atom $X^{Z-1,**}$ (means that the energy of the recombining originally free electron is used to excite another bound electron in the atom). De-excitation is via successive radiative decays between bound atomic levels, creating the photons $\hbar\omega_{sat}$ and $\hbar\omega_{l}$. The photon $\hbar\omega_{sat}$ originates from a double-excited state and is called "dielectronic satellite". As the atoms start from charge state "Z" and end up finally in charge state "Z − 1", effective recombination has occurred.

The evolution of the atomic populations can be described by the following system of differential rate equations:

$$\begin{aligned}\frac{\mathrm{d}n_Z}{\mathrm{d}t} &= -n_Z\left(n_e^2 T_{Z,Z-1} + n_e D_{Z,Z-1} + n_e R_{Z,Z-1} + n_e I_{Z,Z+1}\right)\\ &\quad + n_{Z+1}\left(n_e^2 T_{Z+1,Z} + n_e D_{Z+1,Z} + n_e R_{Z+1,Z}\right)\\ &\quad + n_{Z-1}\left(n_e I_{Z-1,Z}\right).\end{aligned} \tag{6.7}$$

Let us now consider explicit solutions of the set of (6.7). The stationary solution is given by

$$\frac{n_{Z+1}}{n_Z} = \frac{n_e I_{Z,Z+1}}{n_e R_{Z+1,Z} + n_e D_{Z+1,Z} + n_e^2 T_{Z+1,Z}}. \tag{6.8}$$

Due to the n_e^2-dependence of the three-body recombination, radiative recombination and dielectronic recombination are negligible compared to three-body recombination at high densities:

$$\lim_{n_e \to \infty}\left(\frac{n_{Z+1}}{n_Z}\right) = \frac{I_{Z,Z+1}}{n_e T_{Z+1,Z}}. \tag{6.9}$$

The ionization rate coefficient $I_{Z,Z+1}$ is related to the three-body recombination rate coefficient $T_{Z+1,Z}$ via the principle of microreversibility (see also Sects.7.7.2 and 10.6.5.4) that for a system containing Maxwellian electrons at temperature T_e takes the form ($E_{Z,Z+1}$ is the ionization energy from the charge state "Z" to charge state "Z + 1"):

$$T_{Z+1,Z} = I_{Z,Z+1}\frac{g_Z}{2g_{Z+1}}\left(\frac{2\pi\hbar^2}{m_e k T_e}\right)^{3/2} e^{+E_{Z,Z+1}/kT_e}. \tag{6.10}$$

With the help of (6.10), (6.9) can be rewritten as:

$$\lim_{n_e \to \infty}\left(\frac{n_{Z+1}}{n_Z}\right) = 2\left(\frac{m_e k T_e}{2\pi\hbar^2}\right)^{3/2}\frac{g_{Z+1}}{g_Z}\frac{e^{-E_{Z,Z+1}/kT_e}}{n_e}. \tag{6.11}$$

Equation (6.11) is equivalent to the famous Saha–Boltzmann equation. Note that (6.11) connects only two levels in charge states "Z" and "Z + 1", whereas the so-called Saha-equation connects all levels from charge state "Z" to all levels of charge state "Z + 1" with the help of their respective partition functions.

At low densities, three-body recombination is small compared to radiative and dielectronic recombinations:

$$\lim_{n_e \to 0} \left(\frac{n_{Z+1}}{n_Z} \right) = \frac{n_e I_{Z,Z+1}}{n_e R_{Z+1,Z} + n_e D_{Z+1,Z}}. \quad (6.12)$$

As ionization rate coefficients, radiative recombination rate coefficients, and dielectronic recombination rate coefficients depend on the electron temperature, (6.12) does not depend on density and is a function of electron temperature only:

$$\lim_{n_e \to 0} \left(\frac{n_{Z+1}}{n_Z} \right) = F_{Z,Z+1}(kT_e) = \frac{I_{Z,Z+1}}{R_{Z+1,Z} + D_{Z+1,Z}}. \quad (6.13)$$

The low-density limit according to (6.13) is called "Corona distribution". In the Corona limit, $F_{Z,Z+1}(kT_e)$ is a universal function of the electron temperature. As for every charge state a universal function can be obtained, the Corona limit describes a universal charge state distribution of all ions in a plasma. Even if the density changes by orders of magnitude, the charge state distribution does not change as long as for every charge state three-body recombination is negligible compared to the sum of radiative and dielectronic recombination.

Equation (6.8) demonstrates that the distribution of the ionic charge state populations is strongly dependent on elementary atomic processes. We therefore discuss in the following radiative recombination, dielectronic recombination, ionization, and three-body recombination in the context of their application for the calculation of the ionic charge state distribution.

In order to make practical use of the general solution of the charge state distribution according to (6.8), we need explicit expressions for the radiative recombination rate coefficient $R_{Z,Z+1}$, the dielectronic recombination rate coefficient $D_{Z,Z+1}$, the three-body recombination rate coefficient $T_{Z,Z+1}$, and the ionization rate coefficient $I_{Z,Z+1}$.

Let us begin with the Corona limit (6.13) and consider the schematic atomic level system depicted in Fig. 6.4. Radiative recombination takes place into the ground and all excited states:

$$X^{Z+1}(1) + e \to X^Z(n) + \hbar\omega_{\text{rad.recom.}}$$

After radiative recombination, the excited states $X^Z(n)$ can decay via spontaneous radiative emission as indicated by the red flashes in Fig. 6.4. In the Corona limit, radiative decay $A_{nn'}$ is much more important than collisional transfer processes between excited states $C_{nn'}$ because the electron density is low: $A_{nn'} \gg n_e C_{nn}$ '. This implies that excited state population is low compared to the ground state and effective ionization from excited states is very small compared to the ionization from the ground states (an exception might be metastable levels: radiative decay is

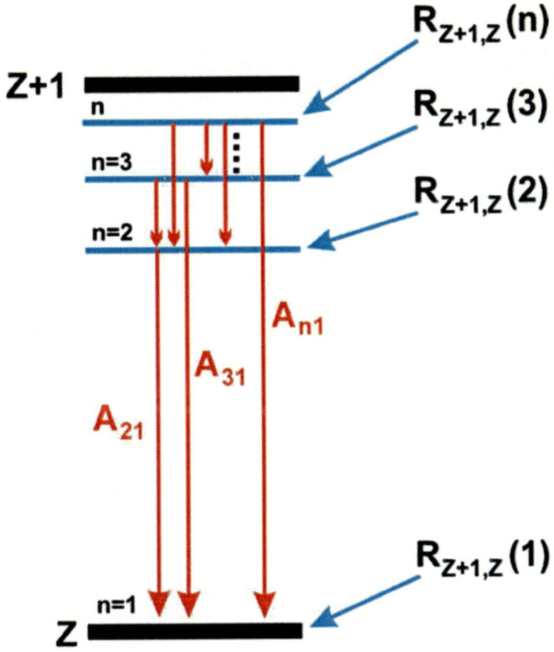

Fig. 6.4 Schematic ionic level system showing radiative recombination to ground (n = 1) and excited sates (n) followed by radiative cascades

low and population might be very high). Therefore, all radiative recombination ends up finally into the ground state. The total radiative recombination is therefore the sum of all radiative recombination into the ground and excited states (N_{max} is the largest principal quantum number to be taken into account):

$$R_{Z+1,Z} = \sum_{n=1}^{N_{max}} \sum_{l=0}^{n-1} R_{Z+1,Z}(nl). \tag{6.14}$$

In the optical electron model (hydrogenic approximation), the radiative recombination can be directly represented by a sum over the orbital l-quantum numbers

$$R_{Z+1,Z}(n) = \sum_{l=0}^{n-1} R_{Z+1,Z}(nl). \tag{6.15}$$

The rate coefficient $R(n)$ can be estimated with the formulas from (5.61) while the sum $R^{tot}(n \geq n_1)$ over the n-quantum numbers (n_1 is the principal quantum number from which the sum is taken, i.e., overall higher-lying excited states with $n > n_1$) can be directly approximated with (5.62).

In a similar manner, dielectronic recombination $D_{Z+1,Z}(\alpha_0 \to \alpha, nl)$ has to be summed over the excited state contribution to account for the total recombination due to cascading from excited levels:

$$D_{Z+1,Z} = \sum_{n}^{N_{\max}} \sum_{l=0}^{n-1} \sum_{\alpha_0} \sum_{\alpha} D_{Z+1,Z}(\alpha_0 \rightarrow \alpha, nl). \tag{6.16}$$

The sums in (6.16) can considerably be simplified with the help of the Burgess formula (see also Sect. 5.6.2) where the sums over the quantum numbers "*nl*" of the spectator electrons are explicitly taken into account:

$$D_{Z+1,Z}(\alpha_0 \rightarrow \alpha) = \sum_{n} \sum_{l=0}^{n-1} D_{Z+1,Z}(\alpha_0 \rightarrow \alpha, nl) \tag{6.17}$$

assuming that dielectronic recombination into the ground state is usually the most important one. In this case, the state α_0 coincides with the atomic ground state and the sum over α_0 can be suppressed (see also Sect. 5.6):

$$D_{Z+1,Z} \approx \sum_{\alpha} D_{Z+1,Z}(\alpha_0 \rightarrow \alpha), \tag{6.18}$$

$$D_{Z+1,Z}(\alpha_0 \rightarrow \alpha) = 4.8 \times 10^{-11} f_{\alpha_0\alpha} B_{\mathrm{d}} \beta^{3/2} e^{-\beta\chi_{\mathrm{d}}} \left[\mathrm{cm}^3\,\mathrm{s}^{-1}\right]. \tag{6.19}$$

The factor B_{d} is a so-called branching factor: after dielectronic capture, a double-excited state is formed that can decay via autoionization or radiative decay. For the dielectronic recombination, only the radiative decays contribute finally to recombination as autoionization only returns the original state. In the one-channel approximation, (6.19) can be estimated with the help of the Burgess and Cowan formulas from (5.138–5.143).

Due to multichannel autoionization and radiative decay and the complex configurations involved numerical calculations of the dielectronic recombination turn out to be very complex and the precision of the Burgess formula is difficult to estimate. This is one of the major reasons that up to present-day different atomic population models to calculate the ionization charge state distribution differ largely from each other, in particular for high-*Z* elements (Rubiano et al. 2007; Chung et al. 2013; Colgan et al. 2015).

The ionization rates involved in (6.7) are the ionizations from the ground state that can be directly estimated from the formulas (5.49) while radiative recombination and dielectronic recombination rates are given by (6.14), (6.16) and its approximations discussed in this chapter and in the Annex A.1.

As it has been discussed above for the radiative recombination and dielectronic recombination processes, also the three-body recombination rate into excited states followed by radiative cascades has to be taken into account:

$$T_{Z+1,Z} = \sum_{n=1}^{N_{\max}} \sum_{l=0}^{n-1} T_{Z+1,Z}(nl). \tag{6.20}$$

Summation over the orbital *l*-quantum numbers "*l*" provides:

$$T_{Z+1,Z}(n) = \sum_{l=0}^{n-1} T_{Z+1,Z}(nl). \qquad (6.21)$$

The three-body recombination rate $T_{Z+1,Z}(n)$ can be estimated from (5.50). The summations over principal quantum number "*n*" until N_{max} in (6.21) have to be taken out with care and follow the methods described in Sect. 5.3.2 and corresponding approximation formulas from (5.51–5.58).

6.1.3 Characteristics of the Ionic Charge State Distribution

Figure 6.5 shows the charge state distribution of Argon obtained from the collisional–radiative code MARIA (Rosmej 1997; 2001, 2006, 2012). The dominance of the shell structure in the distribution of different charge states is clearly visible: a rather wide existence over temperature of the Ne-like and He-like ions densities.

The dominance of closed shell configurations is a general feature and almost independent of the atom and the electron density. The large "high-temperature wings" of the Na-like and Li-like charge states are related to the dielectronic recombination that proceeds from the closed shell configurations $1s^2 2s^2 2p^6$ and $1s^2$. As one can see from Fig. 6.5, in general, only about 3–6 charge states are highly populated for a given temperature. This is a typical feature of plasmas with Maxwellian electron energy distributions. We note that in non-Maxwellian plasma,

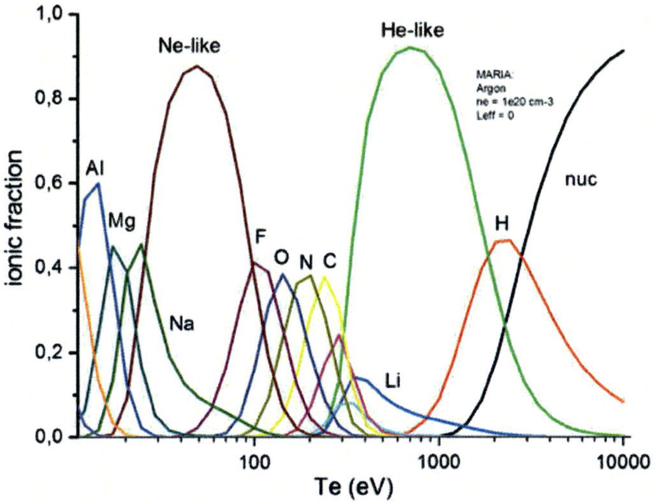

Fig. 6.5 Charge state distribution of Argon in dependence of the electron temperature calculated with the MARIA code, $n_e = 10^{20}$ cm^{-3}

however, qualitative changes appear (Rosmej 1997) that in turn witness the presence of suprathermal electrons.

6.1.4 Generalized Atomic Population Kinetics

In dense plasmas, collisional excitation results into an important population of excited states from which ionization processes may then proceed more efficiently. A particular important case is the ionization from a metastable state because radiative decay is low and population correspondingly high. At the same time, collisional and radiative processes are equally important. It is therefore desirable to consider ionic population and excited states on the same footing rather than calculating the ionic charge state distribution from a set of (6.7) and, separately from these, the corresponding excited states. A widely applied and very successful model (albeit only rates are considered) is the so-called collisional–radiative model (CRM) where all ionization states, ground states, and excited states are connected via elementary collisional radiative processes. The population equations are based on the rate equation principle (see Fig. 6.2) while the elementary processes are calculated with quantum mechanical, quasi-classical or classical methods. The CRM is also called the standard atomic kinetics. The time-dependent evolution of the atomic populations is given by a set of differential rate equations:

$$\frac{dn_{j_Z}}{dt} = -n_{j_Z} \sum_{Z'=0}^{Z_n} \sum_{i_{Z'}=1}^{N_{Z'}} W_{j_Z i_{Z'}} + \sum_{Z'=0}^{Z_n} \sum_{k_{Z'}=1}^{N_{Z'}} n_{k_{Z'}} W_{k_{Z'} j_Z}. \tag{6.22}$$

n_{j_Z} is the atomic population of level j in charge state Z, Z_n is the nuclear charge, $N_{Z'}$ is the maximum number of atomic levels in charge state Z, and $W_{j_Z i_{Z'}}$ is the population matrix which contains the rates of all elementary processes from level j of charge state Z to level i of charge state Z'.

In general, (6.22) is a system of nonlinear differential equations because the population matrix might contain the populations by itself (e.g., when radiation transport is included). Only for special cases, the population matrix W does not depend on the atomic populations and the set of equations becomes linear. Equations (6.22) provide N differential equations where the number of levels N is given by:

$$N = \sum_{Z=0}^{Z_n} N_Z. \tag{6.23}$$

Looking more carefully to the symmetry relations of (6.22), one finds that the system of equations contains only $(N - 1)$ independent equations for the N atomic populations. We are therefore seeking for a supplementary equation. Let us consider atomic populations in terms of a probability (like in quantum mechanics). In this case, the probability to find the atom in any state is equal to 1:

$$\sum_{Z=0}^{Z_n}\sum_{j_Z=1}^{N_Z} n_{j_Z} = 1\,. \tag{6.24}$$

Equation (6.24) is the desired Nth equation and is called the "boundary condition". The population matrix is given by:

$$W_{ij} = W_{ij}^{rad} + W_{ij}^{col}\,. \tag{6.25}$$

The matrix describing the radiative and autoionizing processes is given by

$$W_{ij}^{rad} = A_{ij} + \Gamma_{ij} + P_{ij}^{abs} + P_{ij}^{em} + P_{ij}^{rr} + P_{ij}^{iz}\,. \tag{6.26}$$

The collisional processes are described by

$$W_{ij}^{col} = n_e C_{ij} + n_e I_{ij} + n_e^2 T_{ij} + n_e R_{ij} + n_e D_{ij} + W_{ij}^{col-heavy}\,, \tag{6.27}$$

$$W_{ij}^{col-heavy} = Cx_{ij} + n_{HP} C_{ij}^{HP} + n_{HP} I_{ij}^{HP} \ldots\,, \tag{6.28}$$

where $W_{ij}^{col-heavy}$ describes the heavy-particle collisions, A_{ij} is the spontaneous radiative decay rate, Γ_{ij} the autoionization rate, P_{ij}^{abs} the stimulated photoabsorption, P_{ij}^{em} the stimulated photoemission, P_{ij}^{rr} the stimulated radiative emission, P_{ij}^{iz} the photoionization, C_{ij} the electron collisional excitation/de-excitation, I_{ij} the electron collisional ionization, T_{ij} the three-body recombination, R_{ij} the radiative recombination, D_{ij} the dielectronic capture, Cx_{ij} the charge exchange (see also Annex 1), C_{ij}^{HP} the excitation/de-excitation by heavy-particle collisions, and I_{ij}^{HP} the ionization by heavy-particle collisions.

In the framework of the general set of (6.22), the distribution of atomic populations over the various charge states is readily obtained from its detailed solution:

$$n_Z = \sum_{j_Z=1}^{N_Z} n_{j_Z}\,. \tag{6.29}$$

n_Z is the population for the charge state Z. Heavy-particle collisions are usually not very important in dense hot plasmas. However, there are a few important exceptions, e.g., the coupling of the H-like levels $2s_{1/2}$ and $2p_{1/2}$ via heavy-particle collisions that might change the line ratio of the Lyman-alpha doublet (Boiko et al. 1985) because the energy difference between the levels $2s_{1/2}$ and $2p_{1/2}$ is very small compared to the difference between the levels $2s_{1/2}$ and $2p_{3/2}$. Therefore, the coupling to the level $2p_{3/2}$ is inefficient. Another example is the proton collisional induced ionization of Rydberg levels in magnetic fusion plasmas (Rosmej and Lisitsa 1998).

6.1.5 Statistical Charge State Distribution Based on Average Occupation Numbers

It is evident from (6.22)–(6.29) that the calculation of the charge state distribution can be very complex, in particular, for mid-Z or more heavy elements. It is therefore of interest to develop simplified methods to estimate the charge state distribution over the various shells (in particular for the more complex shells L, M, N, O, P). For these purposes, a statistical model has been developed (Rosmej et al. 2002a) to calculate the probability of the charge state distribution based on an average occupation number:

$$f(k_n) = \left(\frac{P_n}{2n^2}\right)^{k_n} \cdot \left(1 - \frac{P_n}{2n^2}\right)^{2n^2-k_n} \cdot \frac{(2n^2)!}{(2n^2 - k_n)!k_n!}. \tag{6.30}$$

$f(k_n)$ is the probability to find k_n-electrons $(0 \leq k_n \leq 2n^2)$ in quantum shell n (K-shell: $n = 1$, L-shell: $n = 2$, M-shell: $n = 3$ etc.) if the average non-integer population is P_n. Figure 6.6 shows the charge state distribution for L-, M- and N-shell for various different averaged populations P_n.

It can be seen from Fig. 6.6 that if the average occupation number is $P_n = n^2$ the probabilities are centered around the maximum probability at $k_n = P_n$ and that the maxima are far from 1, e.g., for the L-shell, we find a maximum at 0.273, M-shell at 0.185, and N-shell at 0.141. At the same time, the charge state distribution becomes more wider from L-shell to M-shell to N-shell. The calculations for $P_n = 2n^2 - 1$ show that even at such high-shell occupation, the maximum fraction for the corresponding charge state is much below 1, only for the case of almost complete shell occupation, fractions near 1 are encountered (see calculations for $P_n = 2n^2 - 0.5$).

The charge state distribution can be visualized with the spectral distribution. This is demonstrated in Fig. 6.7 via the inner-shell X-ray transitions of type $1s^1 2s^n 2p^m \rightarrow 1s^2 2s^n 2p^{m-1} + \hbar\omega$ for copper. The spectral distribution $I(\omega)$ has been calculated from

$$I(\omega) = \sum_{k_n=0}^{2n^2} f(k_n) \cdot \sum_{i,j} \hbar\omega_{ji}^{(k_n)} \cdot g_j^{(k_n)} \cdot A_{ji}^{(k_n)} \cdot \varphi\left(\omega, \omega_{ji}^{(k_n)}\right), \tag{6.31}$$

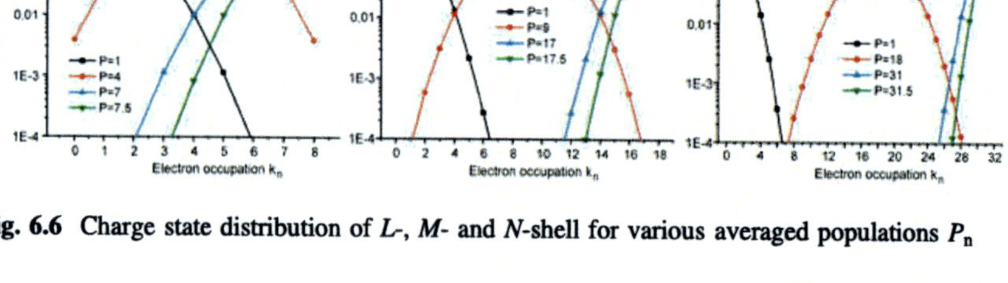

Fig. 6.6 Charge state distribution of L-, M- and N-shell for various averaged populations P_n

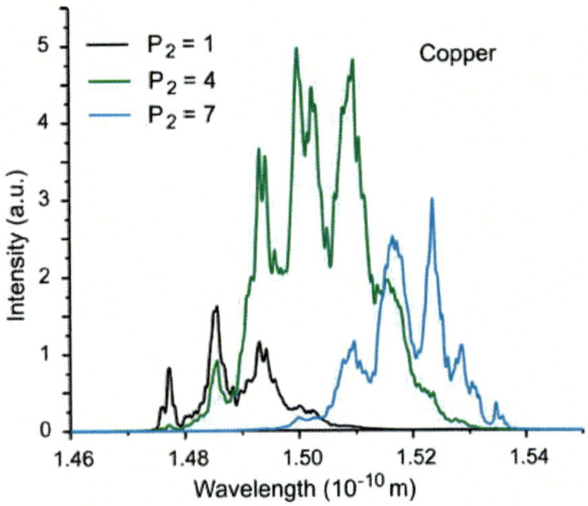

Fig. 6.7 Spectral distribution of inner-shell X-ray transitions $1s^1 2s^n 2p^m \to 1s^2 2s^n 2p^{m-1} + \hbar\omega$ for various averaged *L*-shell populations $P_2 = 1, 4, 7$

where $g_j^{(k_n)}$ is the statistical weight of level *j* in charge state k_n, $A_{ji}^{(k_n)}$ is the transition probability from level *j* to level *i* in charge state k_n and $\hbar\omega_{ji}^{(k_n)}$ is the corresponding transition energy.

6.2 Characteristic Time Scales of Atomic and Ionic Systems

The development of short-pulse lasers (optical and free electron lasers) allows to study systems that are highly out of equilibrium and it is therefore of great interest to study the general properties of the radiating atoms and ions for time-dependent perturbations. It turns out that two principle time scales can be identified: the characteristic time scale to establish an ionization balance and the characteristic time scale of photon emission.

6.2.1 Characteristic Times to Establish Ionization Balance

The time-dependent response properties can be studied in the framework of a two-level atom considering the level "Z + 1" and "Z" of Fig. 6.3. Equation (6.7) then takes the form:

$$\frac{\partial n_Z}{\partial t} = -n_Z n_e I_{Z,Z+1} + n_{Z+1}\left(n_e^2 T_{Z+1,Z} + n_e R_{Z+1,Z} + n_e D_{Z+1,Z}\right). \qquad (6.32)$$

For the two-level atom, the normalization condition (closure relation) reads

$$n_Z + n_{Z+1} = 1, \tag{6.33}$$

which means that the probability to find the atom either in state "Z" or in state "Z + 1" is equal to one. Inserting (6.33) into (6.32), we obtain:

$$\frac{\partial n_Z}{\partial t} = -n_Z a + b, \tag{6.34}$$

where

$$a = n_e I_{Z,Z+1} + n_e^2 T_{Z+1,Z} + n_e R_{Z+1,Z} + n_e D_{Z+1,Z}, \tag{6.35}$$

$$b = n_e^2 T_{Z+1,Z} + n_e R_{Z+1,Z} + n_e D_{Z+1,Z}. \tag{6.36}$$

If the rate coefficients and the electron density do not depend explicitly on time, the differential equation (6.34) has an analytical solution:

$$n_Z(t) = \alpha e^{\beta t} + \gamma. \tag{6.37}$$

Let us consider a rapid cooling process (e.g., a recombining plasma when the laser interaction is switched off) where all initial populations are in the state n_{Z+1}:

$$n_Z(t = 0) = 0, \tag{6.38}$$

$$n_{Z+1}(t = 0) = 1. \tag{6.39}$$

Inserting (6.37) into (6.34), we obtain for $t = 0$:

$$\alpha\beta = -a\alpha - a\gamma + b. \tag{6.40}$$

Inserting (6.38) into (6.37), it follows

$$\alpha + \gamma = 0. \tag{6.41}$$

An additional equation can be obtained remembering that at $t \to \infty$ a physical solution must be finite. Inserting (6.37) into (6.34), we obtain:

$$\alpha\beta e^{\beta t} = -a\left(\alpha e^{\beta t} + \gamma\right) + b. \tag{6.42}$$

In order to select finite solutions for $t \to \infty$, we must request $\beta < 0$:

$$\gamma = \frac{b}{a}. \tag{6.43}$$

From (6.40), (6.41), (6.43), we obtain all further integration constants:

$$\alpha = -\frac{b}{a}, \tag{6.44}$$

$$\beta = -a. \tag{6.45}$$

The final solution (6.37) is therefore:

$$n_Z(t) = \frac{b}{a}\left(1 - e^{-at}\right). \tag{6.46}$$

Equation (6.46) shows that the cooling process which populates the level n_Z has a characteristic time scale:

$$\frac{1}{a} = \tau_{Z,Z+1} = \frac{1}{n_e I_{Z,Z+1} + n_e^2 T_{Z+1,Z} + n_e R_{Z+1,Z} + n_e D_{Z+1,Z}}. \tag{6.47}$$

A similar result can be obtained for rapid heating. Therefore, even sudden cooling/heating processes do not lead to a sudden response of the atomic level populations. It is important to note that the time scale for the ionization process (Z) → (Z+1) is not given by the inverse rate of ionization itself but rather by the inverse of the sum of the ionization and all recombination process. This has important numerical consequences for the time-dependent charge state evolution. The physical reason is that equilibrium requests not only the equilibrium of the atomic state that is ionized but also the equilibrium of those levels that are populated by ionization. From these levels, however, recombination processes originate which request to be in equilibrium with these processes too.

In order to obtain more insight in the meaning of (6.47), let us rewrite the equation in the following form:

$$\tau_{Z,Z+1} = \frac{1}{n_e} \cdot \frac{1}{I_{Z,Z+1} + R_{Z+1,Z} + D_{Z+1,Z} + n_e T_{Z+1,Z}}. \tag{6.48}$$

If three-body recombination is negligible (Corona model), the characteristic time scale (6.48) is inversely proportional to the electron density:

$$\lim_{n_e \to 0} \tau_{Z,Z+1} = \frac{1}{n_e} \cdot \frac{1}{I_{Z,Z+1} + R_{Z+1,Z} + D_{Z+1,Z}} \propto \frac{1}{n_e}. \tag{6.49}$$

Therefore, the characteristic time scale for low-density plasmas can be very long. Although each ionization stage and each element has, in principle, its own characteristic time scale according to (6.48), numerical calculations demonstrate, however, that rather general time constants can be identified (Rosmej 1997; 2001; 2006). For example, the characteristic time constant of the *K*-shell of highly charged ions is given by

$$\tau_{Z,Z+1}(K - shell) \approx \frac{10^{12}\,\mathrm{cm}^{-3}\mathrm{s}}{n_e(\mathrm{cm}^{-3})} \tag{6.50}$$

that is rather insensitive of the temperature and the atomic element. This can be directly understood from (6.48) that contains the sum of the recombination and ionization processes: at high temperature ionization is dominating, whereas at low temperature recombination processes dominate so that the sum of all these processes is finally not strongly dependent on temperature.

6.2.2 Characteristic Times of Photon Emission

We consider now the transient evolution of photon emission according to Fig. 6.8. The relevant set of differential equations is given by

$$\frac{\partial n_j}{\partial t} = -n_j\left(A_{ji} + n_e C_{ji}\right) + n_i n_e C_{ij}, \tag{6.51}$$

$$n_i + n_j = 1, \tag{6.52}$$

which means that the probability to find the atom either in state "*i*" or in state "*j*" is equal to one. Inserting (6.52) in (6.51), we obtain:

$$\frac{\partial n_j}{\partial t} = -n_j a + b, \tag{6.53}$$

where

$$a = n_e C_{ij} + A_{ji} + n_e C_{ji}, \tag{6.54}$$

$$b = n_e C_{ij}. \tag{6.55}$$

The analytical solution of the differential equation (6.53)–(6.54) is given by

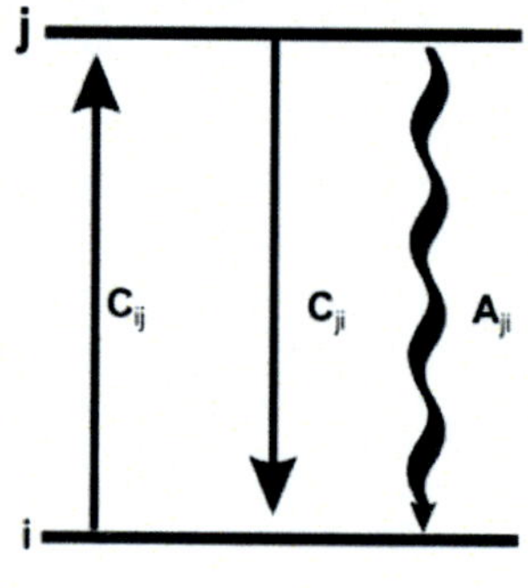

Fig. 6.8 Schematic of a two-level atom

$$n_j(t) = \alpha e^{\beta t} + \gamma. \tag{6.56}$$

Let us consider a rapid cooling process and an initial condition

$$n_j(t = 0) = 1. \tag{6.57}$$

The analytical solution of the differential equation (6.53)–(6.57) is then given by:

$$\gamma = \frac{b}{a}, \tag{6.58}$$

$$\alpha = 1 - \frac{b}{a}, \tag{6.59}$$

$$\beta = -a \tag{6.60}$$

and the time-dependent upper-level density is given by

$$n_j(t) = \left(1 - \frac{b}{a}\right) e^{-at} + \frac{b}{a}. \tag{6.61}$$

As can be seen from (6.61) the cooling process that populates the level n_j has a characteristic time scale:

$$\frac{1}{a} = \tau_j = \frac{1}{A_{ji} + n_e C_{ji} + n_e C_{ij}}. \tag{6.62}$$

Therefore, a sudden cooling does not lead to a sudden response of the atomic level populations and the radiative decay. At very low densities, the relaxation constant is given by

$$\tau_j \approx 1/A_{ji}. \tag{6.63}$$

Equation (6.63) is principally different from (6.48). Even for low densities, the relaxation time can be very small due to the radiative decay rate. The relaxation constant of allowed transitions between principal quantum numbers can be estimated from the following expression (n, m are principal quantum numbers, $m > n$) (Cowan 1981):

$$A_{mn} = A(m \to n) \approx \frac{1.57 \times 10^{10} Z_{eff}^4}{nm^3(m^2 - n^2)} \, [s^{-1}]. \tag{6.64}$$

Note that (6.64) is valid only for allowed dipole transitions without any change in spin quantum number. In a multilevel system, all radiative decay rates to the lower levels have to be considered for the relaxation constant (Sobelman and Vainshtein 2006):

$$A_{\mathrm{m}} = \sum_{\mathrm{n}} A(m \to n) \approx \frac{7.79 \times 10^9 Z_{\mathrm{eff}}^4}{m^5} \ln\left\{\frac{m^3 - m}{2}\right\} [\mathrm{s}^{-1}]. \tag{6.65}$$

6.2.3 Collisional Mixing of Relaxation Time Scales

Equations (6.61), (6.62) show that the population of levels which decay radiatively is strongly density-dependent if the rates of collisional processes are of the order of the radiative decay rate. In this case, the characteristic time scales for photon emission are strongly density-dependent. Moreover, in a multilevel system, collisions might transfer population from levels with different relaxation constants, the so-called Mixing of Relaxation Times (Rosmej and Rosmej 1996; Rosmej 2012). This can have very important impact on the time-dependent radiative properties. For example, in a multilevel system, a metastable level can "feed" a resonance emission for a long time via collisions. This phenomenon is demonstrated in Fig. 6.9 for a rapidly cooled argon plasma. The multilevel collisional radiative simulations are carried out with the MARIA code (Rosmej 1997, 2001, 2006, 2012) for $Z_n = 18$ at $n_e = 10^{21}$ cm^{-3} and rapid cooling from $kT_e = 2000$ eV to $kT_e = 500$ eV.

The shortest relaxation time is those of the He-like resonance line $\tau(W) \approx 9 \times 10^{-15}$ s (indicated by the arrow at the first step in Fig. 6.9). The next step is due to a collisional coupling between the levels $1s2p\ ^1P_1$ and $1s2s\ ^1S_0$.

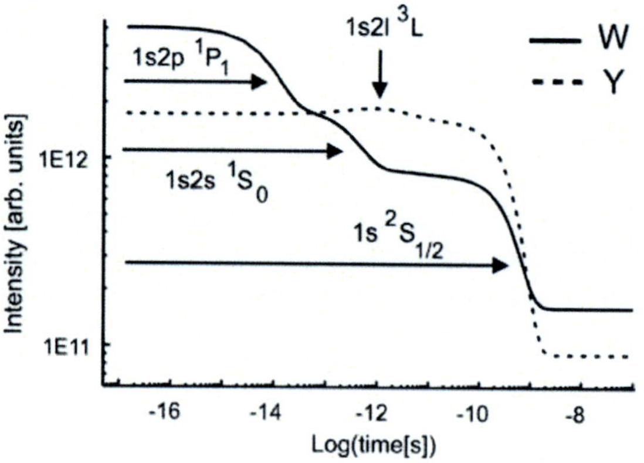

Fig. 6.9 Collisional mixing of relaxation times of the He-like levels $1s2s\ ^3S_1$, $1s2s\ ^1S_1$, $1s2p\ ^3P_2$, $1s2p\ ^3P_1$, $1s2p\ ^3P_0$, $1s2p\ ^1P_1$. The simulations show the collisional mixing of the relaxation times for the He-like resonance line $W = 1s^2\ ^1S_0\text{–}1s2p\ ^1P_1$ and the He-like intercombination line $Y = 1s^2\ ^1S_0\text{–}1s2p\ ^3P_1$. Simulations are carried out with the MARIA code (Rosmej 1997, 2001, 2006, 2012) for argon, $Z_n = 18$ at $n_e = 10^{21}$ cm^{-3} and rapid cooling from $kT_e = 2000$ eV to $kT_e = 500$ eV

The relaxation time of the $1s2s\ ^1S_0$-level is determined by the two-photon decay $\tau(2E1) \approx 3 \times 10^{-9}$ s as well as by collisions. At an electron density of $n_e = 10^{21}$ cm^{-3}, the relaxation time of the $1s2s\ ^1S_0$-level is determined by collisions (rate coefficient $C(1s2s\ ^1S_0\text{–}1s2p\ ^1P_1) \approx 2 \times 10^{-9}$ cm^3 s^{-1}). The effective relaxation time is therefore about $\tau(1s2s\ ^1S_0) \approx 4 \times 10^{-13}$ s as indicated by the arrow "$1s2s\ ^1S_0$" (giving rise to the second step at about $t = 10^{-13}\text{–}10^{-12}$ s). The last step is due to the establishment of ionization equilibrium: the recombination rate from the H-like to He-like ions at $kT_e = 500$ eV is about $R \approx 4 \times 10^{-12}$ cm^3 s^{-1}, giving a relaxation time of about $\tau(1s\ ^2S_{1/2}) \approx 3 \times 10^{-10}$ s. This is indicated by the arrow "$1s\ ^2S_{1/2}$". Almost stationary conditions are achieved at times larger than 1 ns, providing $\tau(1s\ ^2S_{1/2})\ n_e \approx 1 \times 10^{12}$ cm^{-3} s. These numerical results are in good agreement with (6.50).

Due to the strong Z-scaling of intercombination and forbidden transitions ($Z_{\text{eff}}^8 \ldots Z_{\text{eff}}^{10}$, contrary to the Z-scaling of allowed dipole transitions with Z_{eff}^4), the relaxation steps depicted in Fig. 6.9 may change by many orders of magnitude for different elements. Therefore, in transient dense plasmas, collisional processes do not lead only to a transfer of population but also to a mixing of relaxation times. This can result in a considerable prolongation of the radiation emission. Let us, for example, consider the intercombination line of He-like argon ions as an example: the radiative relaxation time is $\tau(Y = 1s^2\text{–}1s2p\ ^3P_1) \approx 6 \times 10^{-13}$ s, however, the fine structure $1s2l\ ^3L$ is metastable and decays by magnetic multipole transitions with very long relaxation times: $\tau(Z = 1s^2\text{–}1s2s\ ^3S_1) \approx 2 \times 10^{-7}$ s and $\tau(X = 1s^2\text{–}1s2p\ ^3P_2) \approx 3 \times 10^{-9}$ s. It is therefore possible that the intercombination line emission has a collisionally enhanced relaxation time by about five orders of magnitude compared to the radiative relaxation time of the Y-line itself (indicated by the vertical arrow "$1s2l\ ^3L$" in Fig. 6.9). This can lead to very long-lasting intercombination line emission in cooling plasmas like, e.g., in laser-produced plasmas and Z-pinch plasmas. This effect has been observed with X-ray streak camera measurements in a dense plasma focus experiment (Lebert et al. 1995): intercombination line emission over time scales of the order of some 0.1 ns are observed.

Figure 6.9 demonstrates, likewise, that the intercombination line intensity in the time interval of about $10^{-13} - 10^{-9}$ s is much stronger than those of the resonance line. This effect has likewise been observed in experiments (Lebert et al. 1995). It is important to note that inner-shell ionization ($1s^22l + e \rightarrow 1s2l\ ^{1,3}L + 2e$) may explain at maximum three times larger intensities of the intercombination line compared to the resonance line (due to the ratio of the statistical weights of the singlet and triplet systems) but is practically limited to about a factor of 2 due to charge state distribution effects.

Therefore, collisional mixing of relaxation times explains simultaneously up to order of magnitude different intensities in certain time intervals and very long-lasting emission. Both effects have been simultaneously observed in experiments of a dense argon pinch during its transition from the column to the micropinch mode (Lebert et al. 1995). The time-dependent measurement has been performed with the help of an X-ray streak camera coupled to a X-ray Bragg crystal, and time-dependent observation of the spectral distribution containing the X-ray intercombination and resonance line emissions $Y = 1s^2{-}1s2p\ ^3P_1$ and $W = 1s^2{-}1s2p\ ^1P_1$.

6.3 Reduced Atomic Kinetics

6.3.1 Ground States, Single-Excited and Autoionizing Levels: General Considerations

The atomic structure of multielectron atoms is rather complex, and the large number of levels is often prohibitive for numerical solution of the population kinetic equations. This is essentially due to the large number of autoionizing states that have to be explicitly involved in dense plasmas in order to reasonably approximate the dielectronic recombination to get right the ionic charge state distribution. It is important to underline that the dielectronic recombination rates calculated by, e.g., the Burgess formula and other similar approaches (see Chap. 5) are strictly only applicable in Corona plasmas, where density effects do not play an important role. There are principally two different density effects:

I. Due to collisional excitation also single-excited states are subjected to dielectronic capture (see Sect. 5.6.2.3 "Excited states driven dielectronic recombination", comparison of Tables 5.5 and 5.6).
II. In dense plasmas, electron collisions may redistribute population between the autoionizing levels, thereby changing the dielectronic recombination after dielectronic capture. This invalidates in general the assumption made for using branching factors [see (5.130)] that do not depend on density and therefore invalidates the use of the simple dielectronic recombination formulas. In order to take into account the density effects among the autoionizing states, all autoionizing levels have to be included explicitly in the population kinetics.

As the number of autoionizing levels is excessively larger than the number of ground and single-excited states, the numerical load to solve the population kinetic equations in dense hot plasmas is finally dominated by the number of the autoionizing states. Currently, there are essentially three different methods in use to handle a large number of levels (thousands up to millions of levels) in population kinetics:

(1) Averaged models of the Fermi type and its various modifications. These models, however, are not very useful for high-resolution spectroscopy and related plasma diagnostics. They are usually employed for equation of state and opacity simulations (Lieb and Simon 1977; Piron and Blenski 2011). So-called plasma atomic models (Demura et al. 2013) have recently been proposed to extend statistical models to plasma diagnostic precision (to be discussed in Chap. 9).

(2) The super-configuration methods where numerous levels are lumped together via to a certain coupling scheme (Bar-Shalom et al. 1995; Bauche et al. 2006; Abdallah and Sherrill 2008; Hansen et al. 2011). As details of the level structure are suppressed, high-resolution spectroscopic applications are very challenging (a discussion with respect to dielectronic satellite transitions can be found in (Petitdemange and Rosmej 2013).
(3) The virtual contour shape kinetic theory (VCSKT) that is based on a probability formalism to account for collisional–radiative effects in complex autoionizing configurations (Rosmej 2006). VCSKT allows for a maximum reduction of autoionzing levels in population kinetics (in the limit to one autoionizing level for a certain configuration instead of all detailed autoionizing levels, e.g., the 274 LSJ-split autoionizing levels $1s3l5l'$ are replaced by just one level) while maintaining the details of all transitions (e.g., means all detailed transitions originating from the 274 levels of the $1s3l5l'$-configuration) with respect to their existence and to their distribution of oscillator strengths over frequency. This allows maximum simplification in the population kinetics while maintaining maximum information in the spectral distribution (e.g., necessary for diagnostic applications).

6.3.2 The Virtual Contour Shape Kinetic Theory (VCSKT)

6.3.2.1 Exact and Reduced Kinetics

Due to the important practical difference between autoionizing states and single-excited states, it is convenient to first reformulate the population kinetics and corresponding spectral distribution explicitly with respect to autoionizing states. Let us start with the general expression for the spectral distribution:

$$I(\omega) = \sum_{i=1}^{N}\sum_{j=1}^{N} \hbar\omega_{\mathrm{ji}} \cdot n_{\mathrm{j}} \cdot A_{\mathrm{ji}} \cdot \varphi_{\mathrm{ji}}\left(\omega, \omega_{\mathrm{ji}}, \theta\right), \tag{6.66}$$

where the indexes i,j run over all ground, single, and autoionzing states from all charge states. N is the number of levels included in the model, n_{j} is the population of level j, ω_{ji} is the frequency of the transition $j \to i$, A_{ji} is the corresponding spontaneous transition probability (of any multipole order for electric and magnetic transitions), φ_{ji} is the line profile, and θ specifies the ensemble of parameters for the line profile calculation (e.g., the ionic temperature, electron density, ion density, etc.). The population n_{j} of level j is determined from

$$\frac{dn_j}{dt} = -n_j \cdot \sum_{i=1}^{N} W_{ji} + \sum_{k=1}^{N} n_k \cdot W_{kj} \tag{6.67}$$

with

$$N = \sum_{\alpha=1}^{N_a} N_{a,\alpha} + \sum_{\beta=1}^{N_b} N_{b,\beta}. \tag{6.68}$$

W_{ij} are the transition matrix elements [see also (6.25)–(6.28)] connecting the discrete levels [i, j, k in (6.67)] in all ionization stages. If a particular transition $j \rightarrow i$ cannot occur because of energy or symmetry considerations, $W_{ij} = 0$. N_a and N_b are the total numbers of autoionizing-state and bound-state manifolds, respectively, and $N_{a,\alpha}$ and $N_{b,\beta}$ are the numbers of levels in the individual autoionizing-state and bound-state manifolds $\{\alpha\}$ and $\{\beta\}$, respectively. These manifolds may be defined to include states with the same principal quantum numbers but different angular-momentum combinations, e.g., $\{\alpha\} = \{1s3l5l'\}$, $N_{a,\alpha} = 274$, $\{\beta\} = \{1s3l\}$, $N_{b,\beta} = 10$. The number of possible angular-momentum combinations $N_{a,\alpha}$ can be enormous. Consequently, it is necessary to consider many thousands, possibly millions of levels, even for combinations of only a few nl-configurations in the evaluation of the radiative emission $I_\alpha(\omega)$ from the manifold $\{\alpha\}$ of the autoionizing states:

$$I_\alpha(\omega) = \sum_{i=1}^{N} \sum_{j \in \{\alpha\}} \hbar\omega_{ji} \cdot n_j \cdot A_{ji} \cdot \varphi_{ji}(\omega, \omega_{ji}, \theta). \tag{6.69}$$

The difficulty associated with (6.66), (6.67) is that the retention of a reduced number of autoionizing levels $N_{a,\alpha} \rightarrow N^r_{a,\alpha}$ in order that

$$N^r = \sum_{\alpha=1}^{N_a} N^r_{a,\alpha} + \sum_{\beta=1}^{N_b} N^r_{b,\beta} \tag{6.70}$$

in the atomic-state kinetics leads to the omission of many emission lines in the evaluation of (6.69). The multitude of original, detailed emission lines is thereby replaced by a reduced set of artificial lines (with averaged intensities, line center positions, and broadening parameters). Consequently, important spectral features and plasma-parameter sensitivities can be lost as a result of this reduction procedure.

We are therefore led to inquire, if (6.69) is the only possible form for the determination of the spectral distribution. This is not only a fundamental question but also one of great practical importance: the exact treatments of (6.67), (6.69) present a severe challenge for practical integrated calculations, which are needed to provide spectroscopic/diagnostic accuracy for the radiation field generated by inertial fusion and other dense plasmas. It is therefore highly desirable to develop alternative methods.

6.3.2.2 The Probability Method for Boltzmann-like Populations

For the purpose of a more transparent presentation of the principal ideas, let us consider the case

$$N_{\mathrm{a},\alpha} \rightarrow N^{1}_{\mathrm{a},\alpha} = 1 \tag{6.71}$$

for which all autoionizing levels α are represented by only a single level in the population kinetics described by (6.67), with a density $n_{\alpha\Re}$ and a statistical weight $g_{\alpha\Re}$ $(i' \in N)$:

$$I_{\alpha\Re}(\omega) = \frac{n_{\alpha\Re}}{g_{\alpha\Re}} \sum_{i'=1}^{N} \sum_{j\in\{\alpha\}} \hbar\omega_{\mathrm{j}i'} \cdot \Re_{\mathrm{j}} \cdot A_{\mathrm{j}i'} \cdot \varphi_{\mathrm{j}i'}\left(\omega, \omega_{\mathrm{j}i'}, \theta\right). \tag{6.72}$$

Note that in (6.72) we have used the index $(i' \in N)$ instead of $(i \in N)$ because after level reduction the overall level identification changes. A generalization of (6.71) to several levels for the manifold $\{\alpha\}$ is straightforward. As one can see from the comparison of (6.69), (6.72), the dimensionless vector $\Re_{\mathrm{j}}$ transforms the averaged level $n_{\alpha\Re}$ to non-statistical individual populations n_{j}. Practically, we seek for a solution for $\Re_{\mathrm{j}}$ that continuously transforms the individual level populations from the Corona model to the Boltzmann case with increasing densities. For clarity of the physical meaning of $\Re_{\mathrm{j}}$, let us first consider the trivial solution of (6.72):

$$\Re_{\mathrm{j}}^{(T)} = \frac{g_{\alpha\Re}}{n_{\alpha\Re}} \cdot n_{\mathrm{j}}, \tag{6.73}$$

i.e., Equation (6.73) makes (6.72) equal to (6.69): in other words, $\Re_{\mathrm{j}}^{(T)}$ depends on the exact individual population vector n_{j}. A non-trivial solution for $\Re_{\mathrm{j}}$ does not invoke the exact solution for all n_{j} (6.67), (6.68) but employs only the reduced kinetics according to (6.67), (6.70), i.e.,

$$\Re_{\mathrm{j}} \simeq \Re_{\mathrm{j}}(n_{\alpha\Re}) \tag{6.74}$$

from which the approximate individual populations are obtained according to

$$n_{\mathrm{j}}^{(n)} \simeq \frac{n_{\alpha\Re}}{g_{\alpha\Re}} \cdot \Re_{\mathrm{j}}^{(n)}. \tag{6.75}$$

The upper index (n) for $n_j^{(n)}$ and $\Re_j^{(n)}$ indicates the approximate solutions of order "n". Equation (6.75) has a clear physical meaning. The first term in (6.75) for $n_j^{(n)}$ is an individual level population according to the statistical assumption (the total population is just divided by the total statistical weight), i.e., the population per statistical weight. The second term, namely $\Re_j^{(n)}$ provides a correction to the statistical population.

A solution of (6.75) can be obtained recalling that the radiation emission from autoionizing states is primarily produced by four individual atomic mechanisms: dielectronic recombination (D), inner-shell collisional excitation (C), collisional coupling among the autoionizing levels $\{\alpha\}$ (B), and couplings to all other levels retained in the atomic kinetics (A). We therefore split $\Re_j$ into the respective contributions ($j \in \alpha$, $k, l, q, s, t, u \in$ reduced set of bound levels retained in the population kinetics pertaining to the various excitation channels):

$$\Re_j = \sum_k \Re_{k,j}^{(D)} + \sum_l \Re_{l,j}^{(C)} + \Re_j^{(B)} + \sum_q \Re_{q,j}^{(A)}. \tag{6.76}$$

Within the manifold $\{\alpha\}$ $\Re_j^{(B)}$ describes collisions corresponding to no change in the principal quantum number n, whereas $\Re_{q,j}^{(A)}$ pertains to transitions with changes in n:

$$\Re_{q,j}^{(A,B)} \sim \int_{\Delta E}^{\infty} \sigma_{q,j}^{(A,B)}(E) \cdot F(E) \cdot \mathrm{d}E. \tag{6.77}$$

ΔE is the energy threshold, $F(E)$ is the electron energy distribution function and σ the cross section. For the majority of relevant transitions

$$\sigma_{\Delta n=0} \gg \sigma_{\Delta n > 0}. \tag{6.78}$$

We therefore neglect detailed collisional processes between different n-quantum numbers from and to the manifold $\{\alpha\}$ and approximate $\Re_{q,j}^{(A)}$ by

$$\Re_{q,j}^{(A)} \approx \sum_X \Re_{q,j}^{(X)}. \tag{6.79}$$

The symbol X denotes additional (to D and C type) processes, e.g., direct radiative recombination, three-body recombination, charge transfer, and ionization. Accordingly, (6.74) generalizes the standard processes of dielectronic recombination and inner-shell excitation (Gabriel 1972; Jacobs and Blaha 1980) to further excitation channels (X). Equations (6.74), (6.75) reduce the complex redistribution effects to collisional processes with the manifold $\{\alpha\}$ only. Processes (D), (C), and (X) are therefore decoupled from the detailed level populations according to (6.67), (6.70). This permits the derivation of an analytical solution for the elements $\Re_{\mathrm{j}}$ $(\gamma = D, C, X)$:

$$\Re_{\mathrm{j}}^{(B)} = g_{\mathrm{j}} \rho_{\mathrm{j}} b_{\mathrm{j}}, \tag{6.80}$$

$$\Re_{\mathrm{k}',\mathrm{j}}^{(\gamma)} = \left(1 - \rho_{\mathrm{j}}\right) g_{\mathrm{j}} R_{\mathrm{k}',\mathrm{j}}^{(\gamma)}. \tag{6.81}$$

In Maxwellian plasmas, b_{j} is the Boltzmann factor. ρ_{j} describes the degree of collisionality over radiative and autoionization processes and is given by

$$\rho_{\mathrm{j}} \simeq 1 - \frac{\sum_{i''=1}^{N} A_{\mathrm{j}i''} + \sum_{k} \Gamma_{\mathrm{k,j}}}{\nu_{\mathrm{j}}^{(\mathrm{redis})} + \sum_{i''=1}^{N} A_{\mathrm{j}i''} + \sum_{k} \Gamma_{\mathrm{k,j}}}, \tag{6.82}$$

where $\nu_{\mathrm{j}}^{(\mathrm{redis})}$ is a characteristic collision frequency for level j and $\Gamma_{\mathrm{k,j}}$ is the autoionization rate of level j via channel k. Taking into account all details of the atomic data via the index j in (6.72), (6.82) even metastable level features are recovered. The strengths $R_{\mathrm{k}',\mathrm{j}}^{(\gamma)}$ from (6.81) can be derived by considering the low-density limit. In this limit, the spectral distribution, which is defined by (6.69), can be exactly evaluated as the sum of the contributions from all individual excitation channels as follows:

$$I_{\alpha}(\omega) = \sum_{i'=1}^{N} \sum_{j \in \{\alpha\}} \sum_{\gamma, k'} \hbar \omega_{\mathrm{j}i'} n_{\mathrm{e}} n_{\mathrm{k}'}^{(\gamma)} \langle \gamma \rangle_{\mathrm{k}'j} \varphi_{\mathrm{j}i'}\left(\omega, \omega_{\mathrm{j}i'}, \theta\right) \cdot \frac{A_{\mathrm{j}i'}}{\sum_{i''=1}^{N} A_{\mathrm{j}i''} + \sum_{k} \Gamma_{\mathrm{k,j}}}, \tag{6.83}$$

where $n_{\mathrm{k}'}^{(\gamma)}$ are the population densities of the initial states in various excitation channels (γ) and $\langle \gamma \rangle_{\mathrm{k}'\mathrm{j}}$ are the corresponding individual collisional excitation rate coefficients, $k' \in k, l, q$, i.e., k' is an index in the reduced set of bound levels [see (6.76)]. The link of (6.83) to (6.72) can be accomplished via (6.76, 6.79) approximating $n_{\alpha,\Re}$ from (6.67), (6.70), (6.72) by

$$n_{\alpha\Re} \cdot \left\{ \sum_{s} \bar{A}_{\alpha\Re,s} + \sum_{t} \bar{\Gamma}_{\alpha\Re,t} + \sum_{X,u} \overline{\langle X \rangle}_{\alpha\Re,u} \right\} \simeq \sum_{k''} \sum_{\gamma'} n_{\mathrm{k}''}^{(\gamma')} \cdot \overline{\langle \gamma' \rangle}_{\mathrm{k}'',\alpha\Re}. \tag{6.84}$$

$\bar{A}_{\alpha\Re,s}$, $\bar{\Gamma}_{\alpha\Re,t}$ and $\overline{\langle X \rangle}_{\alpha\Re,u}$ are effective depopulation rates that decrease the level density $n_{\alpha\Re}$ due to radiative decay, autoionization (decay of level "$\alpha\Re$" to level "t"), and processes (*X*), while $\overline{\langle\gamma\rangle}_{\mathrm{k},\alpha\Re}$ are effective population rates that increase the level density $n_{\alpha\Re}$ due to the processes (γ):

$$\overline{\langle\gamma\rangle}_{\mathrm{k},\alpha\Re} = \sum_{j\in\{\alpha\}} \langle\gamma\rangle_{\mathrm{k,j}}, \tag{6.85}$$

$$\bar{A}_{\alpha\Re,s} = \frac{1}{g_{\alpha\Re}} \cdot \sum_{j\in\{\alpha\}} n_{\mathrm{j}} \cdot A_{\mathrm{j}s}, \tag{6.86}$$

$$\bar{\Gamma}_{\alpha\Re,t} = \frac{1}{g_{\alpha\Re}} \cdot \sum_{j\in\{\alpha\}} n_{\mathrm{j}} \cdot \Gamma_{\mathrm{t,j}}, \tag{6.87}$$

$$\overline{\langle X\rangle}_{\alpha\Re,u} = \frac{1}{g_{\alpha\Re}} \cdot \sum_{j\in\{\alpha\}} n_{\mathrm{j}} \cdot X_{\mathrm{j}u}. \tag{6.88}$$

As can be seen from (6.85), the effective population rate is given just by the sum of all detailed population rates. The depopulation rates are more complex as they request the knowledge of the individual populations that are expressed in terms of the vector $\Re_{\mathrm{j}}$ from (6.75). With the help of (6.86)–(6.88), we can now determine the strengths $R^{(\gamma)}_{\mathrm{k'},\mathrm{j}}$ from (6.81). Inserting (6.81) and (6.84) into (6.72) and equating the result with (6.83) we obtain

$$R^{(\gamma)}_{\mathrm{k'},\mathrm{j}} = \frac{n^{(\gamma)}_{\mathrm{k'}} \langle\gamma\rangle_{\mathrm{k'},j}}{\sum_{k''}\sum_{\gamma'} n^{(\gamma')}_{\mathrm{k''}} \overline{\langle\gamma'\rangle}_{\mathrm{k''},\alpha\Re}} \cdot \frac{g_{\alpha\Re}}{g_{\mathrm{j}}} \cdot \frac{\sum_s \bar{A}_{\alpha\Re,s} + \sum_t \bar{\Gamma}_{\alpha\Re,t} + \sum_{X,u} \overline{\langle X\rangle}_{\alpha\Re,u}}{\sum_{i''}^{N} A_{\mathrm{j}i''} + \sum_k \Gamma_{\mathrm{k,j}}}. \tag{6.89}$$

The strength parameter $R^{(\gamma)}_{\mathrm{k'},\mathrm{j}}$ has a clear physical meaning: it determines the strength to populate level *j* from level *k'* via the elementary process (γ) in the Corona limit while the strength parameter $\Re^{(\gamma)}_{\mathrm{k'},\mathrm{j}} = (1-\rho_{\mathrm{j}}) g_{\mathrm{j}} R^{(\gamma)}_{\mathrm{k'},\mathrm{j}}$ from (6.81) determines this strength for arbitrary density.

According to (6.80)–(6.82), the intermediate densities and corresponding redistribution among the individual levels are determined via a probability method: ρ_{j} is the probability for level *j* to be "Boltzmann-like" (see (6.80), while $1-\rho_{\mathrm{j}}$ is the probability for level *j* to be "non-Boltzmann-like" [see (6.81)]. Therefore, the redistribution among the levels from the manifold $\{\alpha\}$ that is a complex interplay between collisional–radiative and autoionization processes is replaced by the individual probabilities from (6.82). The system of equations is closed, if the individual densities n_{j} from (6.86)–(6.88) are replaced by the approximate individual populations from (6.75).

6.3.2.3 Maximum Recovery Properties and Convergence Properties

In order to solve the system of (6.67), (6.70), the effective depopulation and population rates from (6.85)–(6.88) have to be specified. The system (6.67), (6.70) can be initially set up assuming $\rho_j^{(0)} = 1$ (the upper index specifies the iteration number). According to (6.76), (6.79), (6.80), (6.81), this corresponds to:

$$\Re_j^{(B),(0)} = g_j \rho_j^{(0)} b_j = g_j b_j, \tag{6.90}$$

$$\Re_{k',j}^{(\gamma),(0)} = \left(1 - \rho_j^{(0)}\right) g_j R_{k',j}^{(\gamma)} = 0 \tag{6.91}$$

from which it follows [see (6.76)]

$$\Re_j^{(0)} = \Re_j^{(B),(0)} = g_j b_j. \tag{6.92}$$

According to (6.75), this corresponds to an individual level population of

$$n_j^{(0)} \simeq \frac{n_{\alpha\Re}}{g_{\alpha\Re}} \cdot \Re_j^{(0)} = n_{\alpha\Re} \cdot \frac{g_j}{g_{\alpha\Re}} \cdot b_j = n_{\alpha\Re} \cdot \frac{g_j}{g_{\alpha\Re}} \cdot \exp\left(-\Delta E_{j,\alpha\Re}/kT_e\right), \tag{6.93}$$

i.e., the Boltzmann population. The system of (6.67), (6.70) is therefore initially set up with statistical/Boltzmann averaged rate coefficients. The population densities $n_k^{(\gamma),(0)}$ are then used in (6.89) to calculate non-statistical vectors $R_j^{(1)}$ from (6.81), (6.82) and non-statistical depopulations rates from (6.86)–(6.88) and so on. The numerical calculations show extremely rapid convergence, in fact, already the 0-iteration (mean no iteration in the set of (6.67), (6.70) providing the first non-statistical approximation $R_j^{(1)}$) provides a very good approximation to the spectral distribution.

In order to demonstrate the maximum efficiency of the virtual contours shape kinetic theory (VCSKT), let us consider the extreme case

$$N_{a,\alpha} \rightarrow N_{a,\alpha}^{r} = 1 \tag{6.94}$$

for which all autoionizing levels $\{\alpha\}$ are represented by only a single level in the population kinetics described by (6.67), with a density $n_{\alpha\Re}$ and a statistical weight $g_{\alpha\Re}$ $(i' \in N)$. We consider also examples where (*D*) and (*C*) driven dielectronic satellite transitions are well separated: model inaccuracies are not masked by line overlapping and a stringent test for the accuracy of VCSKT is provided. We likewise chose parameter intervals so large that all experimental situations of interest are covered.

Figure 6.10 displays the spectral range of the He-like resonance line $W = He_\alpha = 1s2p\,^1P_1 \rightarrow 1s^2\,^1S_0$, intercombination line, $Y = 1s2p\,^3P_1 \rightarrow 1s^2\,^1S_0$ and Li-like dielectronic satellites $1s2l2l'\,LSJ \rightarrow 1s^2 2l\,L'S'J'$. This spectral range is of particular interest for spectroscopic diagnostics (Gabriel 1972; Boiko et al. 1985; Rosmej 1997; Rosmej 2012; Glenzer et al. 1998). The agreement is found to be very good: first, although large temperature variations are considered, the relative intensity between the He-like resonance line W and the satellite transitions is very well described. This demonstrates the correct connection between the reduced atomic kinetics description of (6.67), (6.70), (6.85)–(6.88) and the recovered spectral distribution of (6.72). Second, in Fig. 6.10a, the plasma density is too low for titanium to couple the autoionizing levels via $\Re_j^{(B)}$. Therefore, correct intensities driven by dielectronic recombination and inner-shell excitation show the correct distribution over excitation channels (6.76), (6.79), (6.81). Third, the intensity redistribution among the transitions due to collisions (Jacobs and Blaha 1980; Petitdemange and Rosmej 2013) is correct over many orders of magnitude. Therefore, the probability method for Boltzmann-like populations (6.80)–(6.82) provides a very satisfactory approximation.

Let us study the probability method for Boltzmann-like populations with another important example, namely the $1s2l3l'$-satellites near $He_\beta = W3 = 1s3p\,^1P_1 \rightarrow 1s^2\,^1S_0$ that have been employed in gas-bag experiments to control the uniformity of the compression toward near-solid density (Woolsey et al. 1997) and in dense laser-produced plasmas to characterize non-Maxwellian effects (Rosmej et al. 2001). Figure 6.11a shows a near-solid density case, Fig. 6.11b an intermediate density case, and Fig. 6.11c shows a low-density (corona) case. The agreements

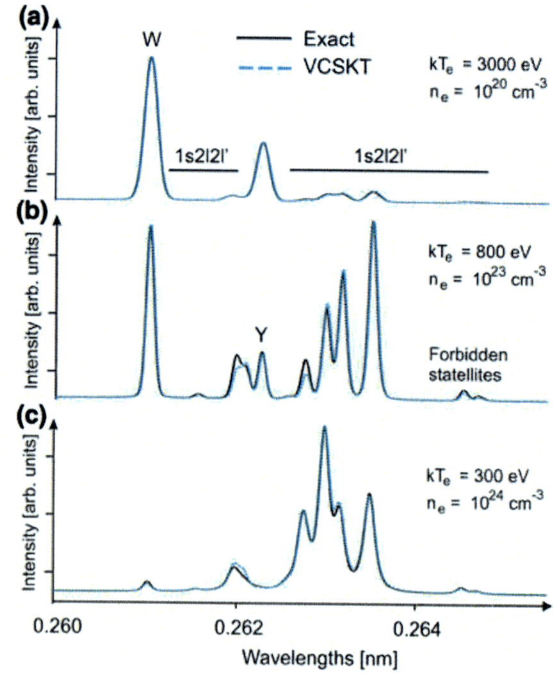

Fig. 6.10 Spectral distributions of the He-like resonance line W, intercombination line Y and Li-like $1s2l2l'$-satellites of titanium for various temperatures and densities. The simulations with the VCSKT with maximum reduction, i.e., $N^r_{a,\alpha} = 1$, show overall very good agreement with the exact solutions

between the results of the exact simulations and the predictions of VCSKT over many orders of magnitude in density are remarkable and demonstrate the efficiency of the probability method for Boltzmann-like populations.

6.3.2.4 Broadening Properties of Complex Emission Groups

Let us consider the broadening properties of the emission from the manifold $\{\alpha\}$. Unlike the broadening of a single line, the broadening of the total contour is determined by the broadening of a single transition from $\{\alpha\}$ and also by the number of transitions with their respective line center positions. In VCSKT, the last effect is treated exactly, because all transitions with their exact line center positions are retained in the summation, based on (6.83). In (Rosmej and Abdallah 1998), a Voigt profile representation was proposed for φ_{ij} with a Lorentz width given by

$$\Delta\omega_{ji}^{(L)} = \sum_{k}\left(A_{jk} + \Gamma_{jk} + C_{jk}\right) + \sum_{l}\left(A_{il} + \Gamma_{il} + C_{il}\right). \tag{6.95}$$

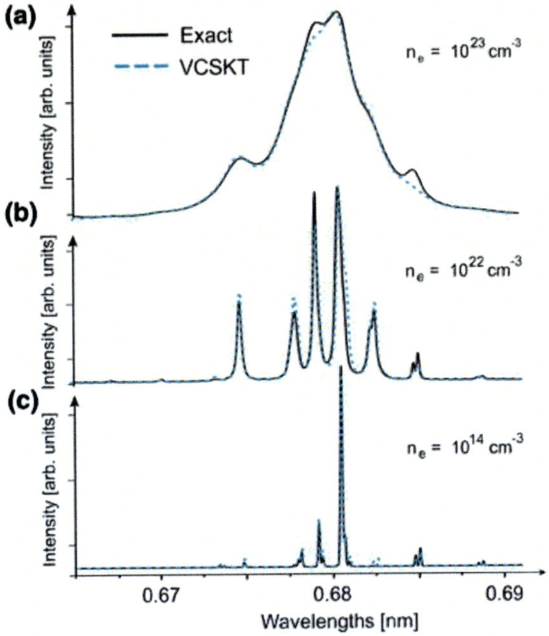

Fig. 6.11 Spectral distributions of the Li-like $1s2l3l'$-satellites (near He$_\beta$) of aluminum for various densities. The simulations with the VCSKT with maximum reduction, i.e., $N_{a,\alpha}^{r} = 1$, show overall very good agreement with the exact solutions

The inelastic collision rates C_{jk} can be approximated by a unique frequency $\nu_{\mathrm{eff}}^{(\alpha)}$, i.e.,

$$\Delta\omega_{\mathrm{ji}}^{(L)} \simeq \sum_{k}\left(A_{\mathrm{jk}}+\Gamma_{\mathrm{jk}}\right)+\sum_{l}\left(A_{\mathrm{il}}+\Gamma_{\mathrm{il}}\right)+\nu_{\mathrm{eff}}^{(\alpha)}. \tag{6.96}$$

Figure 6.11 shows the results of simulations using width expression according to (6.95) (solid curves) and (6.96) (dashed curves): $\nu_{\mathrm{eff}}^{(\alpha)}$ is found to provide a good agreement for the broadening of the total satellite contour. We note that the expression for the width according to (6.96) readily permits further sophistications via the introduction of additional effective width expression $\nu_{\mathrm{eff}}^{(\alpha)} \to \nu_{\mathrm{eff}}^{(\alpha)} + \nu_{\mathrm{eff},1}^{(\alpha)} + \ldots$. We note that also Stark broadening effects could be incorporated in this approach (Rosmej et al. 2002b).

6.3.2.5 Response Properties of VCSKT to Hot Electrons

We consider now the response properties of the VCSKT with respect to hot electrons that have important impact on the radiative properties of matter, in particular, in inertial confinement fusion ICF and high-intensity laser-produced plasmas. The hot electron fraction is defined as follows (Rosmej 1997):

$$f_{\mathrm{hot}} = \frac{n_{\mathrm{e,hot}}}{n_{\mathrm{e,hot}}+n_{\mathrm{e,bulk}}}, \tag{6.97}$$

where T_{hot} and T_{bulk} are the "bulk" and "hot" electron temperature, respectively.

Figure 6.12 shows the Lyman-alpha satellite emission $2l2l' \to 1s2l + \hbar\omega_{\mathrm{satellite}}$ of non-Maxwellian and optically thick argon plasmas. A group of transitions is appreciably populated by hot electrons via the inner-shell excitation process

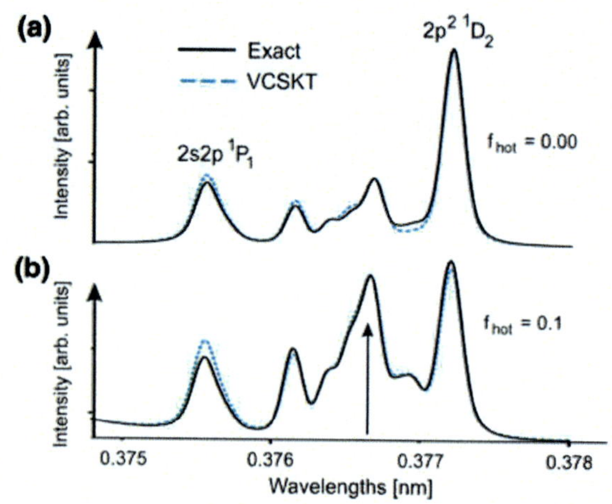

Fig. 6.12 Spectral distributions of the He-like 2l2l′-satellites (near Ly_α) of argon for dense plasmas containing hot electrons for $kT_{\mathrm{bulk}} = 500$ eV, $kT_{\mathrm{hot}} = 20$ keV, $n_{\mathrm{e,tot}} = n_{\mathrm{e,bulk}} + n_{\mathrm{e,hot}} = 10^{23}\ \mathrm{cm}^{-3}$, effective plasma size $L_{\mathrm{eff}} = 10\ \mu\mathrm{m}$. The simulations with the VCSKT with maximum reduction, i.e., $N_{\mathrm{a},\alpha}^{r} = 1$, show overall very good agreement with the exact solutions

$1s2l + e \to 2l2l' + e$ (indicated by the arrow in Fig. 6.12). The results of exact and analytical non-Maxwellian VCSKT simulations are found to be in very good agreement. This indicates that preferential population via single channels (γ) (e.g., the inner-shell excitation driven by hot electrons) is very well described by (6.81).

Equations (6.80), (6.81) can be regarded as providing a virtual contour shape $I_{\alpha\Re}$ and population kinetics description (VCSK): the strengths of channels (γ) are redistributed by the action of $\Re_{\mathrm{j}}^{(B)}$ and $\Re_{\mathrm{k'j}}^{(\gamma)}$ from (6.80, 6.81). The levels $\{\alpha\}$ are thereby decoupled from the atomic kinetics while retaining the details of all individual transitions according to (6.83).

An important property of the VCSKT is that (6.79)–(6.89) are exact in the high-density limit, as well as in the low-density limit. Consequently, VCSKT is applicable for all kinds of plasma conditions. Equation (6.83) together with (6.79)–(6.89) differs from the spectral distribution obtained from common reduction schemes, e.g., (Bar-Shalom et al. 1995; Bauche et al. 2006; Abdallah and Sherrill 2008; Hansen et al. 2011). In these schemes, the reduction of the atomic kinetics is also applied to the evaluation of (6.66), and therefore the reduced number of levels $N_{\mathrm{a},\alpha}^{(r)} = 1$ (e.g., the maximum reduction possible and applied for all examples of Figs. 6.10, 6.11, 6.12) would then result in the retention of only a single-line transition for each lower state i'. Practically, all information from the detailed spectral distribution would be lost. However, (6.83) together with (6.76)–(6.89) recovers all spectral details via the summation over the full manifold $\{\alpha\}$ from the reduced population $n_{\alpha\Re}$ via $\Re_{\mathrm{j}}(n_{\alpha\Re})$. VCSKT generates therefore a detailed, unreduced spectral distribution from a reduced description of atomic level population kinetics. This is of fundamental interest for the atomic radiative properties and also of great practical importance because VCSKT reduces the computational effort by orders of magnitude. VCSKT could therefore be especially promising for applications: fully integrated simulations with diagnostic accuracy for the most complex configurations (e.g., hollow atoms/ions) become feasible.

6.4 Two-Dimensional Radiative Cascades Between Rydberg Atomic States

Many physical applications require calculations of radiative cascade between highly excited atomic states. Examples include calculations of the level populations and line intensities of hydrogen and ionized He(II) in interstellar gas plasmas (nebulas) (Seaton 1959; Pengelly 1964; Summers 1977; Grin and Hirata 2010), spectral line calculations for highly stripped ions in hot rarefied plasmas whose levels are populated by the processes of charge transfer (Abramov et al. 1987), or dielectronic recombination (Sobelman and Vainshtein 2006) as well as natural lasing (Strelnitski et al. 1996; Messenger and Strelnitski 2010). Several analytical and numerical techniques for calculating the parameters of radiative cascades were developed and discussed (Seaton 1959; Pengelly 1964; Summers 1977; Biberman et al. 1982;

Kukushkin and Lisitsa 1985; Flannery and Vrinceanu 2003; Sobelman and Vainshtein 2006; Grin and Hirata 2010).

Many works deal with one-dimensional radiative cascades, in which the populations f_{nl} of atomic states with different orbital quantum numbers l are assumed to be determined by their statistical weights: $f_{nl} = f_n \cdot (2l+1)/n^2$, where the function f_n depends only on the principal quantum number n and corresponds to the total (with respect to l) level population. The radiative transitions in such a consideration thus occur between levels with a definite n, and the corresponding probabilities $W(n \rightarrow n')$ are obtained by averaging the probabilities $W(nl \rightarrow n'l')$ over l and l' (this is called the n-method). Pengelly (Pengelly 1964) and Summers (1977) have carried out numerical calculations for two-dimensional cascades, i.e., dealing with the populations of the individual nl-levels (this is called the nl-method). In the work of (Summers 1977) also collisional transitions are considered making it difficult to trace the role of radiative cascades using his data.

The amount of data and the complexity of the numerical calculations in the nl-method clearly increase with the number of levels considered. Moreover, even in numerical calculations, one ought to treat levels with large principal quantum number up to about $n \sim 10^2$ (cf., e.g., [Sobelman and Vainshtein 2006)]. For large principal and orbital momenta, scaling relations need to be invoked to calculate the cascade matrix and the error increases with the increase of n and l (Grin and Hirata 2010). It has been demonstrated (Pengelly 1964) that already for $n = 5$ considerable deviations are encountered. On the other hand, just for $n \gg 1$ and $l \gg 1$, the radiative transition probabilities could be accurately described by quasi-classical methods, and in particular by the Kramers Electrodynamics. This is realized due to the good agreement between quasi-classical results and quantum numerical calculations. We will show below that the description of radiative cascade based on the quasi-classical approach leads to manageable analytic solutions which are in good agreement with quantum numerical calculations. These solutions also allow identification of the parameters in terms of which the numerical data can be interpreted in a consistent, unified way without recourse to laborious numerical methods.

Apart from its practical significance, the study of radiative cascades between Rydberg states is of general physical interest: it can shed light on the relative importance of direct and cascade populations of atomic levels and on the interrelation between quantum mechanical and classical descriptions of electron motion along the atomic levels. Indeed, the problem can be solved in two extreme cases:

(1) The nl-state may be assumed to be populated directly by a source q_{nl}, after which it decays with a probability A_{nl} into all of the lower-lying states; the population will then be equal to q_{nl}/A_{nl} (this is the direct population model).
(2) One may assume that the electron can reach a certain nl-level only by downward cascading through all of the upper-lying states (the cascade population model).

The latter approach is closely related to the classical concept of motion in nl-space, in which the electron motion is associated with a gradual loss of energy $(E = -Ry/n^2)$ and angular momentum $[M = \hbar(l+1/2)]$ at a rate which is determined by corresponding classical quantities (Landau and Lifschitz 2000). This classical description has been employed (Belyaev and Budker 1958) for the treatment of radiative cascades; this method is equivalent to using the equation of continuity in phase space for the population $f(E, M)$. On the other hand, it was shown (Beigman and Gaisinsky 1982; Beigman 2001) that the classical "flow" description with respect to the energy variable E is invalid: the electron always moves in quantum-mechanical jumps. It is therefore of interest to examine the regions of nl-space within which the electron can be considered to move classically or by quantum jumps.

Of particular interest is the cascade population in the case of a photorecombination source of external population when the free electrons with an equilibrium (Maxwellian) energy distribution populate the bound atomic states, and the radiative transitions determine both the population source and the subsequent radiative cascade. It is noteworthy that the distribution of the atomic electrons with respect to the orbital quantum number l is by no means always proportional to statistical weights, even if the source of electrons populating the levels is in equilibrium (Pengelly 1964).

6.4.1 *Classical Kinetic Equation*

Following (Belayev and Budker 1958), we will use canonically conjugate action-angle variables to analyze the classical kinetic equation for the electron distribution function (DF) in an atom or ion. These variables are most convenient because the characteristic time of action variables variation for a radiating electron is appreciably larger than the period of electron motion (the latter is the characteristic time of the variation of the angles variables). That is why the DF may be regarded as independent of the angle variables. We shall take the initial kinetic equation to be the continuity equation in six-dimensional phase space. After averaging over the angle variables, this equation takes the form

$$\partial f/\partial t + \partial(\dot{I}_k f)/\partial I_k = q, \tag{6.98}$$

where I_k are the action variables,

$$I_1 = m\alpha^2/\sqrt{2E}, \quad I_2 = M, \quad I_3 = M_z, \quad \alpha \equiv Ze^2 \tag{6.99}$$

and the $\dot{I}_k$ are the corresponding generalized momenta (averaged over the angle variables):

$$\dot{I}_1 = |\partial I_1/\partial E|\dot{E}, \quad \dot{I}_1 = (1 - M^2/3I_1^2)me^{10}Z^4/c^3M^5, \tag{6.100}$$

$$\dot{I}_2 \equiv M = -2me^{10}Z^4/c^3M^2I_1^3, \quad \dot{I}_3 \equiv \dot{M}_z = M_z\dot{M}/M. \tag{6.101}$$

Here and below, $E > 0$ is the modulus of the total energy of the bound electron. Equations (6.100), (6.101) give the rate at which a classically radiating electron losses energy (I_1), angular momentum (I_2), and its z-component (I_3) (Landau and Lifschitz 2000).

We shall consider only the stationary case in what follows. The spherical symmetry of the Coulomb field implies that the DF f must be independent of M_z (we also assume that the source q is independent of M_z). Equation (6.98) thus simplifies to

$$\dot{E}(\partial f^{(3)}/\partial E) + \dot{M}(\partial f^{(3)}/\partial M) = q^{(3)}. \tag{6.102}$$

Here the superscript indicates the dimensionality of the space in which f is defined. We note that the variables E, M, and M_z satisfy the classical kinematic constraints

$$M \leq M_{\max}(E) \equiv (m\alpha^2/2E)^{1/2}, \quad |M_z| \leq M.$$

In deriving (6.102), we have used the important property

$$div^{(3)}\dot{I} = 0 \tag{6.103}$$

of the generalized momentum, which implies that the electron flux in the space E, M, M_z may be uniform ($f^{(3)} = const.$ satisfies (6.98) if $q = 0$). Solving (6.102) by the method of characteristics, we find

$$f^{(3)}(E,M) = \phi[M(\tau,E_0)] + \int_{E_0}^{E} dE' q^{(3)}[E', M(\tau,E')], \tag{6.104}$$

where $\varphi(M)$ is the boundary condition for (6.102) (we take the boundary to be the line $E = E_0$; the generalization to the case of an arbitrary boundary is evident),

$$\tau \equiv \tau(E,M) = M^{-3}(1 - 2EM^2/m\alpha^2) \equiv M^{-3}\varepsilon^2. \tag{6.105}$$

ε is the eccentricity of the electron orbit, and the dependence $M(\tau, E)$ in (6.104) is determined by (6.105). Using (6.104), we can rewrite the Green function for (6.102) in the form

$$\begin{aligned} G(E'M' \to EM) &= \frac{\eta(E - E')}{\dot{E}(E', M')}\delta[M' - M(\tau, E')] \\ &\equiv \frac{\eta(M' - M)}{|\dot{M}(E', M')|}\delta[E' - E(\tau, M')], \end{aligned} \tag{6.106}$$

where $\eta = 0$ for $x<0$ and $\eta = 1$ for $x > 0$. The δ-function in (6.106) corresponds to the classical motion of the radiating electron in the two-dimensional $\{E, M\}$-space; the trajectories coincide with the characteristic curves of (6.102) defined by the relation $\tau(E, M) = const$. Since the energy loss rate exceeds the angular momentum loss, ε decreases during the radiation emission process so that the orbits eventually become "rounder".

6.4.2 Quantum Kinetic Equation in the Quasi-classical Approximation

We will consider the quantum mechanical kinetic equation for the distribution function $f^{(2)}$ in the two-dimensional space $\{I_1, I_2\}$ and use the formulae

$$I_1 = \hbar n, \quad I_2 = \hbar(l + 1/2), \quad I_3 = \hbar m_z, \tag{6.107}$$

which relate the action variables to the quantum numbers n, l and m_z. Because $f^{(3)}$ is independent of M_z, $f^{(2)}$ and $f^{(3)}$ obey the simple relation

$$f^{(2)}(I_1, I_2) = 2Mf^{(3)}(I_1, I_2) \equiv (2l + 1)f^{(3)}(I_1, I_2). \tag{6.108}$$

The kinetic equation has the standard form ($\Gamma = \{nl\}$)

$$\sum_{n'=n+1}^{\infty} \sum_{l'=l\pm 1} f^{(2)}(\Gamma')W(\Gamma' \to \Gamma) + q(\Gamma) = A(\Gamma)f^{(2)}(\Gamma), \tag{6.109}$$

where we have allowed for cascades from all higher-lying states; W is the probability per unit time for a radiative transition $\Gamma' \to \Gamma$, q is the external population source, and A is the total rate of radiative decay from the Γ level:

$$A(\Gamma) = \sum_{n'=l+1}^{n-1} \sum_{l'=l\pm 1} W(\Gamma' \to \Gamma). \tag{6.110}$$

For $n \gg 1$, we can replace the sum in (6.109) by an integral, and for $l \gg 1$, $f(\Gamma')$ can be expanded in l near the state Γ. This leads to an integro-differential equation (we will henceforth write f in place of $f^{(2)}$ where no confusion may arise)

$$A(\Gamma)f(\Gamma) = q + \int_{n+1}^{\infty} \left[f(n', l)W(n' \to nl) + \frac{\partial f(n', l)}{\partial l} \sum_{\Delta l = \pm 1} (l' - l)W(\Gamma' \to \Gamma) \right] dn', \tag{6.111}$$

where

$$W(n' \to nl) = \sum_{\Delta l = \pm 1} W(\Gamma' \to \Gamma). \tag{6.112}$$

The quasi-classical kinetic (6.111) reduces to a simpler one-dimensional integral or two-dimensional differential equation, depending on the specific region in nl-space, and the solutions can be joint uniquely because the corresponding regions overlap.

Indeed, consider (6.111) for the region $l \ll n$, for which the Kramers approximation is valid for the radiative transition probabilities W. The radiative angular momentum loss ($\Delta l = \pm 1$) for $l \ll n$ is slower than the energy loss, because transitions with $\Delta n \gg 1$ (including those with $\Delta n \simeq 1$) are more likely to occur. If the DF is smooth enough we can therefore discard the differential term in (6.111), so that l appears to be merely a parameter of the resulting integral equation ($E = 1/2n^2$, $M = \hbar(l + 1/2)$):

$$\int_0^{x_m} G_0(x) f\left[E\left(1 - \frac{x}{x_m}\right), M\right] dx - f(E, M) \int_0^{\infty} G_0(x) dx = Q \equiv \frac{\pi}{\sqrt{3}} \cdot \frac{q(\Gamma)}{A(\Gamma)}, \tag{6.113}$$

where $x_m \equiv \dfrac{(l + 1/2)^3}{6n^2}$, $E = 1/2n^2$ (in atomic units), $M = \hbar(l + 1/2)$, and, as before f is normalized in Γ space. The function G_0 is related to the leading term in the expansion of the transition probability $W(n' \to nl)$ (6.112) with respect to $\hbar$ for $l \ll n$

$$G_0(x) = x \cdot \left[K_{1/3}^2(x) + K_{2/3}^2(x)\right]. \tag{6.114}$$

The function $A(\Gamma)$ is the total radiative decay rate for the level $\Gamma = \{nl\}$

$$A(\Gamma) = 4\left[\sqrt{3}\pi c^3 n^3 (l + 1/2)^2\right]^{-1}. \tag{6.115}$$

The first (cascade) integral in (6.113) is negligible for small x_m, so that the population of level Γ is determined by the external source q,

$$f(\Gamma) = q(\Gamma)/A(\Gamma). \tag{6.116}$$

The cascade term becomes important as x_m increases.

Since the Kramers' probability W depends only on the difference between the energies of the initial and final states, the integral (6.113) can be solved by taking Laplace transforms. The latter satisfy the equation

$$\bar{f}(s) = \bar{Q}(s)/s\bar{G}_2(s), \tag{6.117}$$

where s is the Laplace variable conjugate to x_{m},

$$G_2(x) = \int_x^{\infty} G_0(x')\mathrm{d}x' = xK_{1/3}(x)K_{2/3}(x), \tag{6.118}$$

$$s\bar{G}_2(s) = \bar{G}_0(0) - \bar{G}_0(s). \tag{6.119}$$

We can approximate G_2 to within 10% by the expression

$$G_2 \simeq \alpha \exp(-2x), \tag{6.120}$$

$$\bar{G}_2(s) = \alpha(s)/(s+2). \tag{6.121}$$

where the function $\alpha(s)$ is slowly varying, $\alpha(s=0) = \pi^2/6 = 1.64; \alpha(s=\infty) = \pi\sqrt{3} = 1.81$. If we set $\alpha = 1.7$, ensuring at most a 10% error in (6.120), (6.121), we obtain the approximate analytic expression

$$f(\Gamma) = q(\Gamma)/A(\Gamma) + \int_{n+1}^{\infty} \mathrm{d}n' q(n', l)/|\dot{n}(n', l)| \tag{6.122}$$

for an arbitrary source q; here, the quantity $\hbar\dot{n} \equiv \dot{I}_1$ is the rate of energy loss [see E in (6.100)] in Kramers' domain $l \ll n$.

To illuminate the essence of the approximation (6.120), (6.121) it should be pointed out that the exact relation between G_2 and G_0 takes the form (with account of (6.118), (6.119))

$$G_0(x) - 2G_2(x) = x[K_{1/3}(x) - K_{2/3}(x)]^2 \equiv D(x). \tag{6.123}$$

The correction $D(x)$ which is the "Bethe Rule Defect" is proportional to the Kramers' transition probability for a transition with $\Delta l = -sgn(\Delta n)$. Such transitions are suppressed (relative to the transitions with $\Delta l = sgn(\Delta n)$) the stronger the larger Δn. In the Kramers' domain, this leads to an approximate coincidence of the averaged Δl transition probability with the one corresponding to $\Delta l = sgn(\Delta n)$ transitions only. The transition to the limit of a classical trajectory (in Γ space) corresponds to the motion with averaged (over Δl) probabilities. That is why the transitions with $\Delta l = -sgn(\Delta n)$, in spite of their existence as an elementary, one-step transition, can, within the framework of the KrED, be neglected in multistep transitions.

The DF (6.122) satisfies (6.111) for $x_m \le 1$ (including $x_m \ll 1$), where both integrals in (6.113) are of the same order of magnitude. The integrals cancel each other for $x_m \gg 1$ that corresponds to the classical limit in (6.122). We can follow this limit by expanding $f(\Gamma')$ in the integrand with respect to both n and l (not only with respect to l as in the derivation of (6.111)). This expansion, which is valid for $x_m \gg 1$, leads to the two-dimensional differential equation

$$\dot{E}\partial f^{(2)}/\partial E + \dot{M}\partial f^{(2)}/\partial M - \dot{M}f^{(2)}/M = q^{(2)} \tag{6.124}$$

for $f^{(2)}(\Gamma)$. Recalling (6.108), we see that (6.124) is equivalent to (6.102).

We note that since the classical limit is consistent with the inequality $l \ll n$, it can be described in terms of Kramers transition probabilities. The contribution of the leading term in the $\hbar$-expansion for the transition probability, which is proportional to $\hbar^{-1}$, vanishes due to the aforementioned cancellation between the contribution of cascades from all upper levels to the nl-level under consideration and the contribution of cascades from the nl-level to all lower levels. This cancellation takes place (in the two-dimensional consideration) only for the leading terms of the $\hbar$-expansion for the contributions mentioned. The calculation of these contributions, with account of the quantum corrections to the leading term of the $\hbar$-expansion for W, gives the third term on the left-hand of (6.124). As $\hbar \to 0$, a continuous classical flow of electrons described by (6.124) thus replaces the discrete quantum mechanical "jumps" specified by the non-local coupling in the integral (6.113).

We will now consider how the quasi-classical and classical distributions (6.104) and (6.122) are to be matched. Comparison in the Kramers' domain $l \ll n$ shows that the first term in (6.104) (the contribution from the boundary condition for a classical differential equation) must be replaced by the contribution from the direct population. The resulting distribution function is valid for the entire quasi-classical domain of n and l, including the non-Kramers region $n \sim l$:

$$f(\Gamma) = q(\Gamma)/A(\Gamma) + M\int\limits_{n+1}^{\infty} \frac{q(n', l(\tau, n'))\mathrm{d}n'}{|\dot{n}(n', l(\tau, n'))|M(\tau, n')} = q/A + \hat{C}[q], \tag{6.125}$$

where $l(\tau, n)$ is given by (6.105) (note, that $M = \hbar(l+1/2)$). Indeed, the boundary condition contributes to the classical distribution function (6.104) mostly for large n and, respectively, small x_m, for which the purely classical description breaks down. We will carry out calculations for a specific (photorecombination) source and explicitly piece the solutions together. The results will prove the correctness of the quasi-classical expression (6.125).

6.4.3 Relationship of the Quasi-classical Solution to the Quantum Cascade Matrix. The Solution in the General Quantum Case

We will interpret the above result (6.125) by using the quantum cascade matrix formalism, in which the cascade matrix $C(\Gamma' \to \Gamma)$ plays the role of the Green function for the quantum mechanical equation (6.109). The DF obeying (6.109) can be expressed in the form (Sobelman and Vainshtein 2006; Seaton 1959; Pengelly 1964):

$$\begin{aligned} f(\Gamma) &= A^{-1}(\Gamma)\sum_{n'=n}^{\infty}\sum_{l'=0}^{n-1} C(\Gamma' \to \Gamma)q(\Gamma') \equiv \frac{q(\Gamma)}{A(\Gamma)} \\ &+ A^{-1}(\Gamma)\sum_{n'=n+1}^{\infty}\sum_{l'=0}^{n-1} C(\Gamma' \to \Gamma)q(\Gamma'). \end{aligned} \tag{6.126}$$

The matrix C can be regarded as the probability of a $\Gamma' \to \Gamma$ transition via all possible cascades $(C(\Gamma \to \Gamma) = 1)$ and obeys the two equivalent recursion formulae:

$$\begin{aligned} C(\Gamma' \to \Gamma) &= \sum_{n''=n}^{n'-1}\sum_{l''=l'\pm 1} \frac{W(\Gamma' \to \Gamma'')}{A(\Gamma')} C(\Gamma'' \to \Gamma) \\ &\equiv \sum_{n''=n+1}^{n'}\sum_{l''=l\pm 1} C(\Gamma' \to \Gamma'')\frac{W(\Gamma'' \to \Gamma)}{A(\Gamma'')}. \end{aligned} \tag{6.127}$$

Comparison of (6.126) with the quasi-classical function, (6.125) shows that the cascade population will be purely classical if f is smooth enough (so that $f(\Gamma')$ can be expanded in (6.109) as a Taylor series near the point $\Gamma' = \Gamma$). In the classical limit, the matrix C takes the form

$$C(\Gamma' \to \Gamma'') \propto MA(\Gamma)\delta(\tau - \tau'), \tag{6.128}$$

where the δ-function of the argument τ [cf. (6.105)] describes the classical trajectory. A similar expression for C also follows directly from (6.127) in the classical limit. If we let $\hbar \to 0$ as in the derivation of (6.124), we find that

$$C(\Gamma' \to \Gamma'') \propto MA(\Gamma)F(\tau - \tau'), \tag{6.129}$$

where the function F is arbitrary.We will now estimate the error in the classical description of cascades for an arbitrary source q (including a selective population source $q \propto \delta(\Gamma - \Gamma_0)$) by substituting the approximate solution (6.122) for the

Kramers' domain $l \ll n$ into the corresponding (6.113). The remaining term can be transformed into

$$\int_{o}^{x_m} \frac{q(E',M)}{A(E',M)}\left[G_0(x) - \frac{4\sqrt{3}}{\pi}G_2(x)\right]dx, \quad x = \left[(E' - E)M^3/3\right]a.u. \qquad (6.130)$$

The expression in square brackets in the integrand coincides with the above-defined "Bethe rule defect" to within 10%. Equation (6.130) implies that the terms in square brackets cancels only for those x for which the Bethe rule defect can be neglected. The distribution function f given by (6.125) cannot be used for sources q whose main contribution to the integral in (6.127) comes from small x, for which the terms in square brackets do not cancel.

Let us analyze the case of a δ-function source. Equation (6.125) is clearly not applicable if direct transitions from the level Γ' populated by the source to the level Γ are important (this corresponds to the leading term in the Bethe rule defect as $x \to 0$). In any case, such direct transitions will be important for levels Γ close to Γ', as well as for more remote levels that are populated solely by Bethe rule defect transitions, i.e., by electrons lying far from the classical trajectory. Classical cascade may occur between the levels which lie close to the classical trajectory provided they are sufficiently far from the levels Γ' populated directly by the source ($\Delta x_m \geq 1$).

The situation depicted (i.e., transition from the quantum direct population, in the domain close to an externally populated level, to the classical cascade population) can be described in terms of a modified classical cascade. For example, in Kramers' domain this gives [here x is the same as in (6.130)]:

$$f(\Gamma) = q(\Gamma)/A(\Gamma) + \int_{n+1}^{\infty} \frac{q(n',l)G_0(x)}{|\dot{n}(n',l)|2G_2(x)}dn'. \qquad (6.131)$$

However, there is an alternative, more systematic method for treating "the quantum mechanical properties" of the external source of population. This method exploits the fact that the form of the quantum mechanical kinetic equation remains unchanged if we subtract an arbitrary number of the leading terms in the expansion of the distribution function in powers of the number of the transitions in a cascade from the externally populated level Γ' to the investigated level Γ. Indeed, (6.109) continues to hold for the function $(f - q/A)$ if we replace q by

$$\langle q \rangle \equiv \sum_{n'=n+1}^{\infty} \sum_{\Delta l=\pm 1} q(\Gamma')\frac{W(\Gamma' \to \Gamma)}{A(\Gamma')}. \qquad (6.132)$$

We thus arrive at the distribution function [compare with (6.125)]

$$f = \langle f \rangle \equiv q/A + \langle q \rangle/A + \hat{C}[\langle q \rangle]. \tag{6.133}$$

The generalization of the result (6.125) in the case of an arbitrary number of averaging procedures for the source q gives the result

$$f = \langle f \rangle_{\mathrm{N}} \equiv q/A + A^{-1} \sum_{i=1}^{N} \langle q \rangle_{\mathrm{i}}/A + \hat{C}[\langle q \rangle_{\mathrm{N}}], \tag{6.134}$$

where the effective source $\langle q \rangle_{\mathrm{N}}$ describes the population of the level Γ by all possible N-step (i.e., N-photon) cascade transitions from all points of the source,

$$\begin{aligned} \langle q \rangle_{\mathrm{N}} = &\sum_{n_1=n+N}^{\infty} \sum_{n_2=n+N-1}^{\infty} \cdots \sum_{n_{\mathrm{N}}=n+1}^{\infty} \sum_{l_1,\ldots,l_{\mathrm{N}}=0}^{n_{\mathrm{i}}-1} q(\Gamma_1) \\ &\times \frac{W(\Gamma_1 \to \Gamma_2)}{A(\Gamma_1)} \cdots \frac{W(\Gamma_{\mathrm{N}} \to \Gamma_{\mathrm{N}+1})}{A(\Gamma_{\mathrm{N}})} \end{aligned} \tag{6.135}$$

and the appropriate selection rules for the radiative transition probabilities must be used in calculating (6.135). Each additional summation in (6.135) further smoothens the effective source and thus decreases the error caused by summing the remainder terms in the series (6.126) "classically" to $\simeq 10\%$. The error in the final result depends both on the specific form of the source q and on the values of quantum numbers n and l. The error will be small if the relative change of $f(\Gamma)$ is small due to subtracting one more term (corresponding to $(N+1)$-step transitions) out of the classical cascade.

The above algorithm can be used to calculate the distribution function f for radiative electron cascades between Rydberg atomic or ionic states for arbitrary sources and quantum numbers (in particular, n and l may be of the order of unity).

Note that the extent to which the population source q is of essentially "quantum-mechanical" character depends partly on the sharpness of its distribution in Γ-space (6.127) and partly on the range of values of n, l within which the source is concentrated. For example, the distribution function (DF) is of essentially quantum character if a smoothly distributed (i.e., "classical") source is concentrated in the "quantum" region $x_{\mathrm{m}} \ll 1$. On the other hand, the cascade population can be described quasi-classically even for a selective source if the latter is concentrated in the "classical" region $l \sim n$. Thus, if the levels with $l = n - 1$ are selectively populated by the external source, the population of the lower levels by cascades can be described purely classically and the result agrees with the exact quantum calculation. Specifically, if we use the quantum cascade matrix and recall the relations for this case ($l = n - 1$),

$$W(n, l \to n-1, l-1) = A(n,l) = |\dot{n}| = |\dot{l}| = \frac{2}{3}n^{-5} \quad (6.136)$$

we find from (6.126) that

$$f(n,l) = q(n_0, l_0)A^{-1}(n,l)\delta(n, n_0 - k)\delta(l, l_0 - k), \quad (6.137)$$

where δ is the Kronecker symbol and $k \geq 0$. The calculation using (6.125) leads to the same result.

6.4.4 Atomic Level Populations for a Photorecombination Source. Quasi-classical Scaling Laws

The general results of Sects. 6.4.2, 6.4.3 are now applied to the calculation of level population for a Rydberg atom externally populated by a photorecombination source which is of great interest for astrophysical applications. Since the source involves the same radiative transitions as the cascade between the atomic levels, the above approximations for the cascade can be also applied to the recombination source [see approximation (6.120), (6.121) for the error in the quasi-classical DF (6.125)] in a specific case.

The calculation of the DF (6.122) for the photorecombination source q provides ($x_m = E_n M^3/3 = (l{+}1/2)^3/6n^2$, $x_T = 3/TM^3$ in a.u.)

$$q(\Gamma) = \frac{2}{2+x_T}G_2(x_m) + \frac{x_T}{2+x_T}\psi \exp(E/T), \quad (6.138)$$

$$\psi = \int_{x_m}^{\infty} D(y)\exp(-yx_T)dy \quad (6.139)$$

and corresponds to including the Bethe rule defect contribution to the source q but neglecting it in the Green function for (6.113). It is worthwhile to express the result for the DF in terms of the equilibrium DF (as well as the corresponding ratio $b(\Gamma)$)

$$f(\Gamma) \equiv 2M\bar{A}\exp(E/T)b(\Gamma), \quad \bar{A} = (2\pi mT)^{-3/2}, \quad (6.140)$$

$$b(\Gamma) \equiv b_{nl} = \frac{2}{2+x_T}\exp(-E/T) + \frac{1}{\alpha}\psi(x_m, x_T), \quad (6.141)$$

where ψ and α are given by (6.139) and (6.120), (6.121), respectively (remind, that we use $E > 0$).

For $n, l \gg 1$, the second term in b_{nl} is of importance only for $x_m \ll 1$, $x_T \gg 1$; the DF is therefore independent on the energy E at the edge ($l \sim n$, or $x_m \geq 1$) of the Kramers' domain. This implies that the solution outside the Kramers' domain could be found from the first term in (6.141), regarded simply as a classical boundary condition. Because this term is independent of E, the resulting DF will be the same regardless of which line in nl-space is chosen as the boundary. If we then use (6.104) and (6.105) to continue the DF (6.140, 6.141) along the characteristic curves, we obtain the final result

$$b_{nl} = \frac{2}{2 + x_T \varepsilon^2} \exp(-E/T) + \frac{1}{\alpha} \psi(x_m, x_T), \tag{6.142}$$

which is valid for all quasi-classical values of n and l. It is legitimate to continue the solution in this way because the source (6.138) is concentrated in the Kramers' domain, so that there is no need to evaluate (6.125) directly (recall that A and dn/dt in (6.125) are the transition probabilities for an arbitrary ratio l/n). Indeed, a calculation using (6.125) for $(l/n - 1) \ll 1$ reveals that these states are populated solely by classical cascades; moreover, most of the contribution comes from the transitions whose initial state is far from the curve $l \sim n$. The latter result corresponds precisely to the classical behavior, in which the states near the boundary $M = M_{max}(E)$ can be populated by a source concentrated within a region with an eccentricity $\varepsilon \to 1$. For a recombination source, the Kramers' domain shrinks along the n axis as l increases (as the edge of the continuum is approached) and thus is effectively transformed into a boundary condition.

We will now show that the use of the algorithm discussed in Sect. 6.4.3 permits to incorporate some additional Bethe rule defect contributions to the DF. For a singly averaged source ($N = 1$), (6.132) gives

$$\langle Q \rangle \equiv \langle q \rangle / A = \frac{3}{\pi^2} \int_0^{x_m} G_0(x) Q(x_m - x) \mathrm{d}x \,. \tag{6.143}$$

Within 10% error (approximation of the coefficient α in (6.120, 6.121)), this gives

$$\langle f \rangle = f + \frac{3}{\pi^2} \int_0^{x_m} D(x) Q(x_m - x) \mathrm{d}x \tag{6.144}$$

for $\langle f \rangle$ in (6.133); where f is defined by (6.125). The calculation of $\langle f \rangle$ for the photorecombination source reveals that the corrections due to the Bethe rule defect contribution are smaller than the 10% error arising from the approximation (6.120), (6.121). Thus, if we include the linear correction to (6.141), in accordance with (6.144), we find that

$$\langle b\rangle = \frac{2}{2+x_{\mathrm{T}}}\left[1+\exp(-2x_{\mathrm{m}}) - \frac{\sqrt{3}}{\pi}G_2(x_{\mathrm{m}})\right]\exp(-E/T) + \frac{\psi}{\alpha}\left[1+\frac{x_{\mathrm{T}}}{(2+x_{\mathrm{T}})\alpha}\int_0^{x_{\mathrm{m}}} D(x)\mathrm{d}x\right]. \quad (6.145)$$

Since the factor multiplying ψ in (6.145) is significant only for $x_{\mathrm{m}} \ll 1$, $x_{\mathrm{T}} \gg 1$, we conclude, as in Sect. 6.4.3, that the accuracy of the DF (6.141) is the same as for (6.120), (6.121).

The quasi-classical DF (6.140, 6.142) derived above reveals approximate scaling laws for the exact quantum-level populations, while these laws are unknown from the results of complex quantum numerical calculations. These laws are a consequence of the fact that the quasi-classical DF (6.140, 6.142) depends on a lower number of variables than in the case for the quantum DF. Indeed, f_{nl} depends only on x_{m} and x_{T} for $x_{\mathrm{m}} \ll 1$, $x_{\mathrm{T}} \gg 1$. If one of the parameters x_{m} or $1/x_{\mathrm{T}}$ becomes ~ 1, the second term in (6.142) becomes much less than the first term and f depends only on x_{T}. Elsewhere in nl-space, f depends on the parameter $x_{\mathrm{T}}\varepsilon^2$. We thus have a smooth transition between three scaling laws for $n, l \gg 1$. Comparison of the quasi-classical DF (6.140, 6.142) with the results of numerical quantum calculations (Pengelly 1964) reveals that quasi-classical DF (6.142) can also be used for relatively small values of n and l. The validity of the scaling laws derived by the quasi-classical method can be verified by means of a corresponding transformation of quantum numerical data.

It is worthwhile to illustrate the relative importance of direct and cascade population in the Kramers' domain $l \ll n$ for a photorecombination source. We find from (6.122)

$$\begin{aligned} f_{\mathrm{C}} &\propto \frac{2}{2+x_{\mathrm{T}}}(1-\exp(-2x_{\mathrm{m}})) + \frac{2}{2+x_{\mathrm{T}}}\frac{\psi}{\alpha}\exp(E/T), \\ f_{\mathrm{D}} &\propto \frac{2}{2+x_{\mathrm{T}}}\exp(-2x_{\mathrm{m}}) + \frac{x_{\mathrm{T}}}{2+x_{\mathrm{T}}}\frac{\psi}{\alpha}\exp(E/T) \end{aligned} \quad (6.146)$$

for the cascade (f_{C}) and direct (f_{D}) populations. The contribution from f_{D} clearly decreases as x_{m} increases while the sum $(f_{\mathrm{C}}+f_{\mathrm{D}})$ coincides with (6.141). The numerical values of the ratio $f_{\mathrm{D}}/(f_{\mathrm{D}}+f_{\mathrm{C}})$ agree reasonably well with the data in (Pengelly 1964), e.g., for $n=6$ and $T=10^4$ K, (6.146) implies that this ratio is equal to 96% for $l=1$ and 87% for $l=2$; the corresponding values from (Pengelly 1964) are 81% and 73%, respectively.

It is important to note that the dependence of the integral over l population differs from the one found by the n-method. The major contribution to $f(n)$ comes from the first term in (6.142), which gives the following dependence on the parameter Tn^3:

$$f(n)n^{-2} \propto \int_0^1 x\left[1+\frac{3(1-x^2)}{2Tn^3x^3}\right]^{-1} \mathrm{d}x \approx \begin{cases} \dfrac{1}{2} & Tn^3 \gg 1 \\ \dfrac{Tn^3}{3}\ln\left(\dfrac{3}{Tn^3}\right), & Tn^3 \ll 1. \end{cases} \quad (6.147)$$

We note, that in the n-method the universal parameter is Tn^2.

The quasi-classical method for an analytic description of radiative cascades developed in this section gives the possibility of an approximate (to within 10%) calculation of the contribution of multistep cascade transitions to the atomic level populations. This calculation is known to be the most difficult part of the corresponding numerical calculations. Indeed, the δ-function properties of the cascade matrix that correspond to radiating electrons moving (in the nl-space) along the characteristics in the classical domain are difficult to reveal from the results of quantum numerical calculations. For example, the calculations in (Pengelly 1964) detected only the boundary characteristics corresponding to $l = l_{\max} = n-1$.

The algorithm in Sect. 6.4.3 for calculating populations in the general quantum case and for arbitrary sources can thus be used to correctly treat cascades through an arbitrary large number of Rydberg states. The number of quantum mechanical cascade transitions which cannot be described classically may be quite small in practice, particularly for the case of distributed sources. For example, the cascade population is purely classical (to within 10%) for a photorecombination source.

6.5 Two-Dimensional Collisional–Radiative Model of Highly Excited Atomic States

As outlined in the foregoing Sect. 6.4, the population of Rydberg atomic states determined by radiative–collisional cascades is the subject of many years' investigations (Sobelman and Vainshtein 2006; Beigman 2001; Griem 2005; Strelnitski et al. 1996; Grin and Hirata 2010). As a rule, the one-dimensional (1D) kinetics is used for the modeling. When one makes a transition to 2D (in principle n and orbital momentum l quantum numbers) the number of kinetic equations grows sharply. Really if, for example, $n = 100$, one needs to take into account $100 * 100 = 10^4$ equations for radiative–collisional transitions between two highly excited states multiplied by 100 transitions from the continuum to a specific atomic energy state which results in more than 10^6 kinetic equations with further account of cascade transitions to other Rydberg atomic states. Direct solution of such a large number of kinetic equations is a problem even for modern computers, but the most essential point is the estimation of the precision of such calculations.

The possible solution of the problem is the application of quasi-classical or pure classical model for description of highly excited atomic states. This was demonstrated above for radiative (*R*) cascade transitions. We generalize the model for general case of both radiative and collisional (*RC*) processes. For a specific application, we will use below experimental data for recombination radio lines of highly excited hydrogen atoms ($n > 100$) observed in astrophysical plasma with low electron density ($n_e \propto 10^3 - 10^4\,\mathrm{cm}^{-3}$) and moderate temperatures ($T_e \propto 1\,\mathrm{eV}$). One can apply the results for large densities in the case of highly charged ions with ion charge $Z \gg 1$ using the scaling between radiative and collisional processes, proportional to Z^7 [see also (5.53)].

6.5.1 Kinetic Model of Radiative–Collisional Cascades

Let us consider the transition from the quantum kinetic equation to the classical one. The structure of a quantum kinetic equation for a radiative–collisional (RC) cascade takes the form (Kadomtsev et al. 2007, 2008):

$$\left[\hat{L}_c + \hat{L}_r^q\right] f(nl) + q(nl) = 0, \tag{6.148}$$

where $\hat{L}_c$ is the collision transition operator, $\hat{L}_r^q$ is radiation transition operator, $f(nl)$ is population distribution function in two-dimensional space of principal n and orbital momentum l atomic quantum numbers, and $q(nl)$ is the population source of atomic energy states. The action of these operators on population distribution function takes the form:

$$\hat{L}_c f(n,l) = \Sigma[W(n',l';nl) f(n,l) - W(n,l;n',l') f(n',l')], \tag{6.149}$$

$$\hat{L}_r f(n,l) = A(n,l) f(n,l) - \Sigma A(n,l;n',l') f(n',l'). \tag{6.150}$$

Here, $W(n',l';nl)$ and $A(n,l;n',l')$ are the rates of collisional and radiative transitions between atomic states with different quantum numbers, and $A(n,l) = \Sigma_{n',l'} A(n,l;n',l')$ is the total radiation decay rate to all lower atomic states. The sums in (6.149), (6.150) go over all values of quantum numbers with account for corresponding selection rules.

For the case of highly excited (Rydberg) atomic states, it is possible to simplify the collision operator by its transformation to a diffusion operator in the space of principle and orbital momentum quantum numbers. The situation is complex due to the non-local connection between atomic states via radiative transitions. In fact, the radiation operator (6.150) is in reality an integral operator describing the possibilities of large changes (jumps) of principle quantum numbers in radiative transitions. It is possible, however, to show that in the domain $n > l \gg n^{2/3}$ the action of the radiation operator (6.150) can be reduced to a differential one describing the continuous motion (flux) of the electron in 2D space of quantum numbers with rates determined by the classical conservation laws with respect to energy and momentum:

$$\hat{L}_{\mathrm{r}}f = \dot{n}\partial f/\partial n + \dot{l}\partial f/\partial l \equiv \hat{L}_{\mathrm{r}}^{\mathrm{cl}}f. \tag{6.151}$$

Therefore, the quantum integral radiation operator $\hat{L}_{\mathrm{r}}$ reduces to the classical differential operator $\hat{L}_{\mathrm{r}}^{cl}$. However, this approach is only valid for population sources which are broad enough (for example, three-body recombination) so that the statistical weight of atomic states with small values of orbital momentum would be also small: $l_{\mathrm{eff}}^2 \propto n^{4/3} \ll n^2$. However, this is not the case for many others recombination sources (for example, radiative or dielectronic ones).

We will use the iteration approach to account for the non-local connection of 2D distributions used above for a radiative cascade where the distribution function is presented by

$$f = q/A + \langle q \rangle/A + \hat{C}[\langle q \rangle]. \tag{6.152}$$

The same iteration method can be developed accounting for the collision operator. The general scheme looks as follows. In the first step of the iteration procedure, we find f as the sum $f = f_0 + f_1$ where f_0 satisfies the zero-order approximation determined by populations from external sources, collisional diffusion and radiation decay to lower atomic states:

$$\hat{L}_{\mathrm{c}}f_0(n,l) - A(n,l) \cdot f_0(n,l) + q(n,l) = 0. \tag{6.153}$$

After the substitution $f = f_0 + \delta f$ into (6.148), we obtain the equation for δf:

$$\begin{aligned} 0 = {} & \hat{L}_{\mathrm{c}}\delta f(n,l) - A(n,l) \cdot \delta f(n,l) \\ & + \sum_{n'=n+1} \sum_{l'=l\pm 1} f_0(n',l') \cdot A(n',l' \to n,l) \\ & + \sum_{n'=n+1} \sum_{l'=l\pm 1} \delta f(n',l') A(n',l' \to n,l) \end{aligned} \tag{6.154}$$

or, in the operator form:

$$\left[\hat{L}_{\mathrm{c}} + \hat{L}_{\mathrm{r}}^{\mathrm{cl}}\right]\delta f(n,l) + \sum_{n'=n+1} \sum_{l'=l\pm 1} f_0(n',l') A(n',l' \to n,l) = 0. \tag{6.155}$$

It can be seen that the function δf is determined from an equation that is of similar type as for f (6.148), where the modified source $q_1(n, l)$ enters instead of $q(n, l)$:

$$q_1(n,l) = \sum_{n'=n+1} \sum_{l'=l\pm 1} f_0(n',l') A(n',l' \to n,l). \tag{6.156}$$

It corresponds to the population after the first quantum emission. Application of the same procedure like in (6.152) to the function δf we separate out the direct population $f_1(n,l)$ by the source q_1 (determined from (6.156)) and from the previous iteration for δf with f_0. Applying the iteration procedure step-by-step to the initial (6.148), we find the solution of f in the form of a series $f = f_0 + f_1 + f_2 + \cdots + f_k + \cdots$ that corresponds to a step-by-step quanta emission where every term f_k is determined from the equation

$$\hat{L}_c f_k(n,l) - A(n,l) f_k(n,l) + q_k = 0,$$
$$q_k \equiv \sum_{n'=n+1} \sum_{l'=l\pm1} f_{k-1}(n',l') A(n',l' \to n,l) \tag{6.157}$$

using the function f_{k-1} calculated at the previous step of (6.157), while f_0 can be found from (6.153) with initial source. From the physical point of view, the contributions f_k are determined by a collision transition kinetics between the steps k and $k+1$ with correspondingly emitted quanta. The convergence of the series is, in general, rather slow (logarithmic), so that higher-order contributions are of importance. From a mathematical point of view, the iterative solution corresponds to a von Neumann series for integral equations.

According to (6.157), the source q_k in a specific state nl is determined by different transitions between the atomic states $(n' > n)$ after the emission of a k-th-quantum. At every further step, the quantum becomes more "continuous" and from a specific iteration number on it is possible to make a transition from the quantum operator $\hat{L}_r$ to the classical differential operator $\hat{L}_r^{cl}$ [see (6.151)]. This is realized by changing in the further iteration series terms the quantum for the classical distribution function f_n^{cl}:

$$\left[\hat{L}_c + \hat{L}_r^{cl}\right] f_k^{cl}(n,l) + A f_{k-1}(n,l) = 0. \tag{6.158}$$

As a result, the total distribution function is determined by the sum

$$f(n,l) \approx \sum_{i=0}^{k-1} f_i(n,l) + f_k^{cl}(n,l). \tag{6.159}$$

The quasi-classical radiation operator has been discussed in the foregoing Sections and will be used in specific calculations below.

6.5.2 The Classical Collision Operator

The collision operator determining the diffusion in the space of principal and orbital momentum quantum numbers will be determined in a pure classical representation. For this purpose, the general kinetic equation was solved in the framework of a quasi-classical representation for radiation transitions and a pure classical collision integral. The specific form of the collision integral can be obtained by calculations of energy and orbital momentum transfer in classical Coulomb collision between an atomic electron on a Kepler orbit with a plasma particle. The diffusion coefficients in energy and orbital momentum space are then determined from the well-known averaged quadrates of energy and orbital momentum transfer. The general form of the kinetic equation is of Fokker–Planck-type (Belyaev and Budker 1958, Ecker 1972, Kadomtsev et al. 2008):

$$\partial f/\partial t = \frac{1}{2}\frac{\partial}{\partial I_\mathrm{k}}\langle \Delta I_\mathrm{k}\Delta I_\mathrm{j}\rangle \frac{\partial f}{\partial I_\mathrm{j}}. \tag{6.160}$$

Here $I_{\mathrm{k,j}}$ are in general arbitrary motion integrals. For a Coulomb field, a natural choice is to choose the motion integrals as the energy and orbital momentum integrals. The average values in (6.160) represent an integration over the parameters of the plasma particles parameters and the motion phases of the Kepler atomic electron.

Let us consider the cases of fast (inelastic) and slow (elastic) collisions for plasma electrons and ions correspondently. In the case of fast collisions, the motion of a plasma particle can be taken as rectangular with an impact parameter ρ and velocity υ_e:

$$|\Delta\vec{\upsilon}_\mathrm{e}| = \left|\int\limits_{-\infty}^{\infty} F(t)\mathrm{d}t\right| = \int\limits_{-\infty}^{\infty} \frac{\rho\ \mathrm{d}t}{\left(\rho^2+\upsilon_\mathrm{e}^2 t^2\right)^{3/2}} = \frac{2}{\rho\upsilon_\mathrm{e}}. \tag{6.161}$$

This results in the squared energy change:

$$\left\langle (\Delta E)^2\right\rangle = \left\langle (\Delta\vec{\upsilon}_\mathrm{e}\vec{\upsilon}_\mathrm{a})^2\right\rangle = \frac{1}{3}\left(\Delta\vec{\upsilon}_\mathrm{e}^2\right)\left\langle\vec{\upsilon}_\mathrm{a}^2\right\rangle, \tag{6.162}$$

where $\vec{\upsilon}_\mathrm{a}$ is the electron velocity on the Bohr orbit. After averaging over impact parameters, Maxwellian velocity distribution and multiplication by the perturbing particles' density n_e, one obtains the collision frequency ω (or the time between collisions):

$$\left\langle \Delta E^2\right\rangle \propto E_\mathrm{n}\omega\tau, \tag{6.163}$$

$$\omega = \frac{4\sqrt{2\pi}}{3} n_\mathrm{e} L \sqrt{\frac{1}{T}}, \tag{6.164}$$

where T is the temperature of the perturber particles, L is the Coulomb logarithm. In order to obtain independent diffusion operators in 2D scheme, it is useful to introduce dimensionless variables:

$$\xi = (-E/E_0)^{1/2} = n_0/n, \tag{6.165}$$

$$\eta = \left(\frac{-2EM^2}{mz_0^2e^4}\right)^{1/2} = l/n, \tag{6.166}$$

$$\zeta = \frac{M_z}{M}. \tag{6.167}$$

Equation (6.165) corresponds to the electron energy, (6.166) to the ratio l/n, and (6.167) is the orbital momentum projection (that is of no importance for spherically plasma electron distributions). The domains for changing of variables are: $0<\xi<\infty$, $0<\eta<1$, $-1<\zeta<1$. The energy parameter E_0 for the dimensionless energy variable in (6.167) is taken to be equal to one which corresponds to equal rates of classical collision and radiative decay rates:

$$\begin{aligned} E_0 &= [\sqrt{2\pi}Le^6z_0^2m^{3/2}c^3n_e/12\sqrt{T}]^{1/4}, \\ E_0 &= \frac{1}{2n_0^2}. \end{aligned} \tag{6.168}$$

It is obvious that such a choice corresponds to the transition of a collision cascade to a radiation one, if $\xi \approx 1$. The average squared of the new variable is equal to

$$\left\langle (\Delta\eta)^2 \right\rangle = \left\langle \left(\frac{\Delta l}{n} - \frac{l\Delta n}{n^2}\right)^2 \right\rangle = \frac{1}{3}\left(\Delta\vec{v}_e^2\right)\frac{5}{2}n^2(1-\eta^2). \tag{6.169}$$

It is convenient also to introduce a dimensionless distribution function according

$$\psi = f/n_e(2\pi mT)^{-3/2}. \tag{6.170}$$

The collision operator expressed in terms of the new variables takes the form:

$$\hat{L}_c\psi = \frac{\xi^4}{\eta}\cdot\frac{\partial}{\partial\xi}\left[\frac{\eta}{\xi^4}\cdot\frac{\partial\psi}{\partial\xi}\right] + \frac{\xi^4}{\eta}\cdot\frac{\partial}{\partial\eta}\left[\frac{5\eta(1-\eta^2)}{2\xi^6}\cdot\frac{\partial\psi}{\partial\eta}\right]. \tag{6.171}$$

Such representation of the collision integral was obtained in (Belyaev and Budker 1958) by direct application of the Landau collision integral to the collisions between plasma and atomic electrons. The elastic collisions correspond to the diffusion in the orbital momentum quantum number space. Most contributions originate from ion-atom collisions due to relatively small ion's velocity compared to the electron ones.

6.5.3 Numerical Solution for Delta-Function Source

The solution of the kinetic equation solution for a local (delta-function) source is of interest in order to demonstrate the effects of collisions on the radiative cascades. From the point of view of application, it is of interest for a selective (laser) population source. The calculation procedure includes the introduction of a point source and an iterative calculation procedure as described above.

The plasma parameters have been selected close to astrophysical experimental conditions (temperature of $kT_e = 1\,\text{eV}$, density $n_e = 2.5 \times 10^3\,\text{cm}^{-3}$), and two pairs of quantum numbers n, l: (100,55) and (50,20). The first one corresponds to the values presented in (Strelnitski et al. 1996) in order to check the results with quantum calculations. The second pair corresponds to the decrease of quantum numbers approximately two times in order to demonstrate the sharp change between radiation and collision processes when quantum numbers are changed. Note that the collision operator scaling is close to $\hat{L}_c \sim n_e \upsilon_e n^4$, whereas the radiation operator scaling is proportional to $\hat{L}_r \sim n^{-3} l^{-2}$. This results in different distributions in 2D space of quantum numbers.

Figure 6.13 shows the 2D population distribution function for the pure radiation cascade for a delta-function source for a populated atomic state $n, l = (50, 20)$. The radiation cascade follows the classical trajectory according to (6.151).

Figure 6.14 shows the 2D population distribution function for the collision–radiative cascade. Collisions result into an important distribution over principle and orbital momentum quantum numbers (compare with Fig. 6.13).

6.5.4 Radiation Recombination Population Source

Numerical calculations have been performed employing the general iteration scheme outlined above. It is of interest to compare 1D and 2D models. In order to

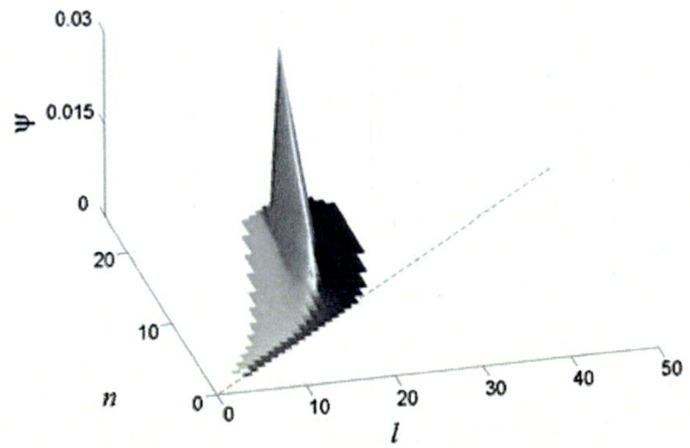

Fig. 6.13 2D population distribution function for the radiation cascade from a delta-function source supporting the population $\psi_0 = 1$ for $n, l = (50, 20)$ in the case when collisions are absent (note that the sharp maximum $\psi_0 = 1$ at n, $l = (50, 20)$ is not shown for better demonstration of the overall features of the distribution function)

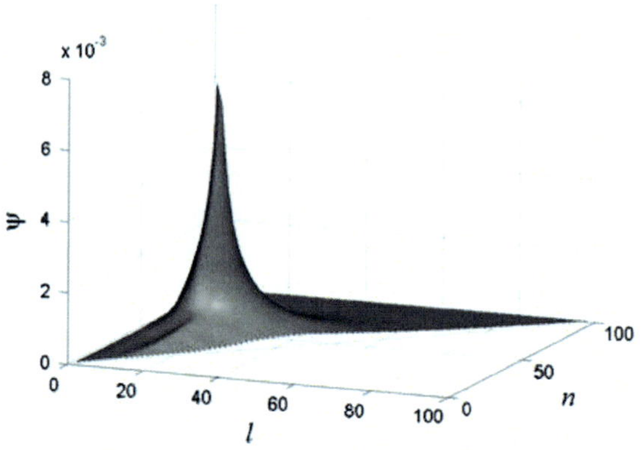

Fig. 6.14 2D population distribution function for the radiation–collision cascade for $n_e = 2500$ cm^{-3}, $kT_e = 1$ eV

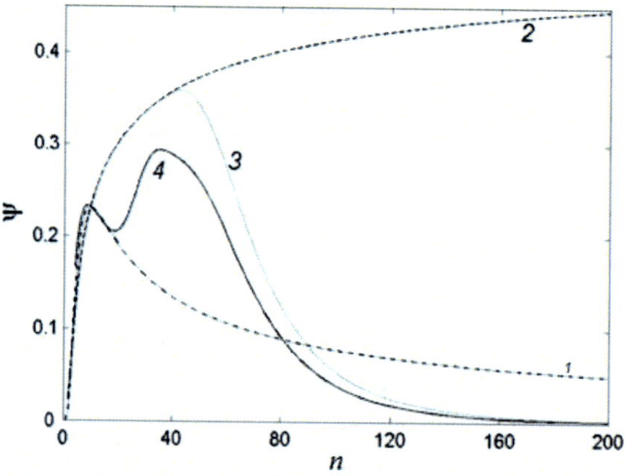

Fig. 6.15 Populations of atomic levels in direct population by a photorecombination source: 1—averaged two-dimensional calculations without collisions, 2—one-dimensional calculations without collisions, 3—one-dimensional calculations with allowance for collisions, 4—averaged two-dimensional calculations with allowance for collisions. The plasma parameters correspond to astrophysical conditions: $n_e = 2500$ cm^{-3}, $kT_e = 1$ eV

do so, the total population function was averaged over orbital momentum quantum numbers l, i.e.,

$$\overline{\psi} \equiv \langle \psi \rangle_l = 2 \int \psi l \, \mathrm{d}l / n^2. \tag{6.172}$$

Figure 6.15 presents a comparison between 1D and averaged 2D population distribution functions. One can see that in the absence of collisions, the two models differ strongly. It is due to the simple circumstance: the 1D model deals with the ratio of averaged rates, whereas the 2D model deals with the averaged ratio of the rates; such difference in the averaging procedure results in different dependencies of populations on principle quantum numbers. When collision frequencies increase, the difference between 1D and 2D populations decreases. It is due to the strong mixing over orbital momentum states driving their populations to statistical equilibrium which is just the initial starting point for the 1D kinetics.

6.5.5 Intensities of Rydberg Spectral Lines

The intensities of Rydberg spectral lines are calculated from the equation:

$$I_{nn'} = \hbar\omega_{nn'} \sum_{l,l'} N_{nl} A(nl; n'l'), \qquad (6.173)$$

where $\hbar\omega_{nn'}$ are the radiation transition energies, N_{nl} are populations of upper energy atomic levels calculated according to the scheme described above, $A(nl; n'l')$ are radiation transition probabilities. In experiments on radio-recombination lines (Biberman et al. 1982) the dependence of radiation transition intensities on the principle quantum number of the upper level is of interest for a fixed value of transition frequency ω. Figure 6.16 demonstrates the line intensities for the transitions from the energy levels n = 50–100 at the observed frequency near $\omega = 8 \cdot 10^{-6}$ a.u. The comparison between statistical equilibrium (large n) and non-equilibrium (low n) upper-level populations demonstrates the essential difference between the two curves in Fig. 6.16 (dashed curve show calculations for a statistical equilibrium, while the solid curve presents non-statistical equilibrium calculations). This is important for Rydberg spectra interpretation. The results allow also to judge the degree of non-equilibrium of the populations.

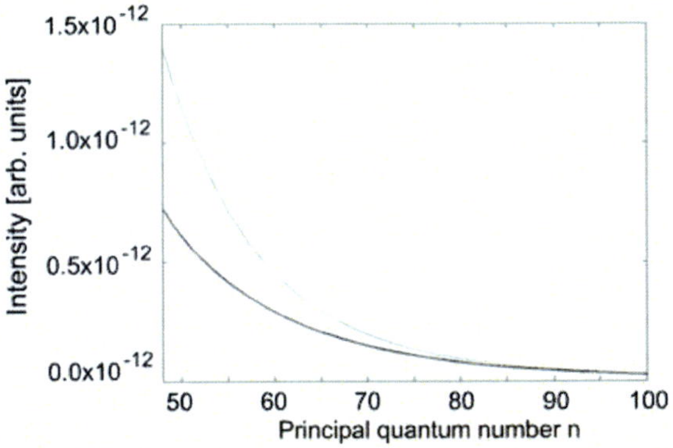

Fig. 6.16 Intensities of Rydberg spectral lines for the transition $n = 50 - 100$ at the observed frequency $\omega = 8 \times 10^{-6}$ a.u., corresponding to the transition with $\Delta n = 1$ between energy levels 50 and 49 for an electron density n_e = 2500 cm^{-3} and temperature kT_e = 1 eV

References

J. Abdallah Jr., M.E. Sherrill, The reduced detailed configuration accounting RDCA model for NLTE plasma calculations. HEDP **4**, 124 (2008)

V.A. Abramov, F.F. Baryshnikov, A.I. Kazanskii, I.V. Komarov, V.S. Lisitsa, M.I. Chibisov, in *Charge Exchange of Atoms on Multiply Charged Ions*, Reviews of Plasma Physics, vol. 12, ed. by M.A. Leantovich, B.B. Kadomtsev (Consultants Bureau, New York, London, 1987)

A. Bar-Shalom, J. Oreg, W.H. Goldstein, D. Shvarts, A. Zigler, Super-transition-arrays: a model for the spectral analysis of hot, dense plasma. Phys. Rev. A **40**, 6414 (1995)

J. Bauche, C. Bauche-Arnoult, O. Peyrusse, Effective temperatures in hot dense plasmas. JQSRT **99**, 55 (2006)

I.L. Beigman, Analytical methods for highly excited level populations in hot plasma, in *Astrophysics and Space Science Reviews*, OPA (2001). ISBN-9789058232410

I.L. Beigman, I.M. Gaisinsky, An analytical description of populations of highly excited levels. JQSRT **28**, 441 (1982)

S.T. Belyaev, G.I. Budker, Multiquantum recombination in ionized gases, in *Plasma Physics and the Problem of Controlled Thermonuclear Reactions*, ed. by M.A. Leontovich, vol. 3 (Pergamon Press, Oxford, 1958)

L.M. Biberman, V.S. Vorobiev, I.T. Yakubov, *Kinetics of a non-equilibrium low-temperature plasma* (Nauka, Moscow, 1982)

V.A. Boiko, A.V. Vinogradov, S.A. Pikuz, I.Y. Skobelev, A.Y. Faenov, X-ray spectroscopy of laser produced plasmas. J. Sov. Laser Res. **6**, 82 (1985)

H.K. Chung, C. Bowen, C.J. Fontes, S.B. Hansen, Yu. Ralchenko, Comparison and analysis of collisional-radiative models at the NLTE-7 workshop. HEDP **9**, 645 (2013)

J. Colgan, C.J. Fontes, H. Zhang, J. Abdallah Jr., Collisional-radiative modeling of tungsten at temperatures of 1200–2400 eV. Atoms **3**, 76 (2015). https://doi.org/10.3390/atoms3020076

R.D. Cowan, *The Theory of Atomic Structure and Spectra* (University of California Press, Berkely, 1981)

A.V. Demura, M.B. Kadomtsev, V.S. Lisitsa, V.A. Shurygin, Statistical model of radiation losses for heavy ions in plasmas. JETP Lett. **98**, 786 (2013)

G. Ecker, *Theory of Fully Ionized Plasmas* (Academic Press, New York 1972)

M.R. Flannery, D. Vrinceanu, Quantal and classical radiative cascade in Rydberg plasmas. Phys. Rev. A **68**, 030502(R) (2003)

A.H. Gabriel, Dielectronic satellite spectra for highly-charged he-like ion lines. Mon. Not. R. Astro. Soc. **160**, 99 (1972)

S.H. Glenzer, F.B. Rosmej, R.W. Lee, C.A. Back, K.G. Estabrook, B.J. MacCowan, T.D. Shepard, R.E. Turner, Measurements of suprathermal electrons in hohlraum plasmas with X-ray spectroscopy. Phys. Rev. Lett. **81**, 365 (1998)

H.R. Griem, On the narrowing of radio recombination lines at high principal quantum numbers. Astrophys. J. **620**, L133 (2005)

D. Grin, C.M. Hirata, Cosmological hydrogen recombination: the effect of extremely high-n states. Phys. Rev. D **81**, 083005 (2010)

S.B. Hansen, J. Bauche, C. Bauche-Arnoult, Superconfiguration widths and their effects on atomic models. HEDP **7**, 27 (2011)

V.L. Jacobs, M. Blaha, Effects of angular-momentum–changing collisions on dielectronic satellite spectra. Phys. Rev. A **21**, 525 (1980)

M.B. Kadomtsev, M.G. Levashova, V.S. Lisitsa, Universal two-dimensional kinetics of the populations of Rydberg atoms in plasmas. JETP Lett. **85**, 493 (2007)

M.B. Kadomtsev, M.G. Levashova, V.S. Lisitsa, Semiclassical theory of the radiative-collisional cascade in a Rydberg atom. JETP **106**, 635 (2008)

A.B. Kukushkin, V.S. Lisitsa, Radiative cascades between Rydberg atomic states. Sov. Phys. JETP **61**, 937 (1985)

L.D. Landau, E.M. Lifschitz, *The Classical Theory of Fields*, 4th edn. (Butterworth-Heineman, Oxford, 2000)

R. Lebert, A. Engel, W. Neff, Investigation on the transition between column and micropinch mode of plasma focus operation. J. Appl. Phys. **78**, 6414 (1995)

E.H. Lieb, B. Simon, The Thomas-Fermi theory of atoms, molecules and solids. Adv. Math. **23**, 22 (1977)

S.J. Messenger, V. Strelnitski, On the 1.7 mm FeII and other natural lasers. Mon. Not. R. Astron. Soc **404**, 1545 (2010)

F. Petitdemange, F.B. Rosmej, Dielectronic satellites and Auger electron heating: irradiation of solids by intense XUV-free electron laser radiation, in *New Trends in Atomic & Molecular Physics—Advanced Technological Applications*, ed. by M. Mohan, vol. 76 (Springer, Heidelberg, 2013), pp. 91–114. ISBN 978-3-642-38166-9

R.M. Pengelly, Recombination spectra—I. calculations for hydrogenic ions in the limit of low densities. Mon. Not. R. Astron. Soc **127**, 145 (1964)

R. Piron, T. Blenski, Variational-average-atom-in-quantum-plasmas VAAQP code and virial theorem: equation-of-state and shock-Hugoniot calculations for warm dense Al. Fe, Cu, and Pb, Phys. Rev. E **83**, 026403 (2011)

F.B. Rosmej, Hot electron X-ray diagnostics, J. Phys. B. Lett. At. Mol. Opt. Phys. **30**, L819 (1997)

F.B. Rosmej, A new type of analytical model for complex radiation emission of hollow ions in fusion and laser produced plasmas. Europhys. Lett. **55**, 472 (2001)

F.B. Rosmej, An alternative method to determine atomic radiation. Europhys. Lett. **76**, 1081 (2006)

F.B. Rosmej, X-ray emission spectroscopy and diagnostics of non-equilibrium fusion and laser produced plasmas, in *Highly Charged Ion Spectroscopic Research*, ed. by Y. Zou, R. Hutton, (Taylor and Francis, Abingdon, UK 2012), pp. 267–341. ISBN: 9781420079043. http://www.crcnetbase.com/isbn/9781420079050

F.B. Rosmej, O.N. Rosmej, Transient formation of forbidden lines, J. Phys. B Lett. At. Mol. Opt. Phys. **29**, L359 (1996)

F.B. Rosmej, J. Abdallah Jr., Blue satellite structure near He_α and He_β and redistribution of level populations. Phys. Lett. A **245**, 548 (1998)

F.B. Rosmej, V.S. Lisitsa, A self-consistent method for the determination of neutral density from X-ray impurity spectra. Phys. Lett. A **244**, 401 (1998)

F.B. Rosmej, A. Calisti, B. Talin, R. Stamm, D.H.H. Hoffmann, W. Süß, M. Geißel, A.Ya. Faenov, T.A. Pikuz, Observation of two-electron transitions in dense non-Maxwellian laser produced plasmas and their use as diagnostic reference lines. JQSRT **71**, 639 (2001)

F.B. Rosmej, R. More, O.N. Rosmej, J. Wieser, N. Borisenko, V.P. Shevelko, M. Geißel, A. Blazevic, J. Jacoby, E. Dewald, M. Roth, E. Brambring, K. Weyrich, D.H.H. Hoffmann, A.A. Golubev, V. Turtikov, A. Fertman, BYu. Sharkov, A.Ya. Faenov, T.A. Pikuz, A.I. Magunov, IYu. Skobelev, Methods of charge state analysis of fast ions inside matter based on their X-ray spectral distribution. Laser Part. Beams **20**, 479 (2002a)

F.B. Rosmej, H.R. Griem, R.C. Elton, V.L. Jacobs, J.A. Cobble, A.Ya. Faenov, T.A. Pikuz, M. Geißel, D.H.H. Hoffmann, W. Süß, D.B. Uskov, V.P. Shevelko, R.C. Mancini, Investigation of charge exchange induced formation of two electron satellite transitions in dense laser produced plasmas. Phys. Rev. E **66**, 056402 (2002b)

J.G. Rubiano, R. Florido, C. Bowen, R.W. Lee, Y. Ralchenko, Review of the 4th NLTE code comparison workshop. HEDP **3**, 225 (2007)

M.J. Seaton, The solution of capture-cascade equations for hydrogen. Mon. Not. R. Astron. Soc. **119**, 90 (1959)

I.I. Sobelman, L.A. Vainshtein, *Excitation of Atomic Spectra* (Alpha Science International Ltd., Oxford, U.K., 2006)

V.S. Strelnitski, V.O. Ponomarev, H.A. Smith, Hydrogen masers. I. Theory and prospects. Astrophys. J. **470**, 1118 (1996)

H.P. Summers, The recombination and level populations of ions—II; resolution of angular momentum states. Mon. Not. R. Astron. Soc. **178**, 101 (1977)

N.C. Woolsey, B.A. Hammel, C.J. Keane, A. Asfaw, C.A. Back, J.C. Moreno, J.K. Nash, A. Calisti, C. Mossé, R. Stamm, B. Talin, L. Klein, R.W. Lee, Evolution of electron temperature and electron density in indirectly driven spherical implosions. Phys. Rev. E **56**, 2314 (1997)

Chapter 7
Quantum Atomic Population Kinetics in Dense Plasmas

Abstract The atomic populations are the fundamental quantities for various different disciplines in science and applications. Although the rate equation model has widely been employed, it has principal deficiencies as it considers only populations related to the squared of the wave functions, while all mixed populations (coherences) are missing. In plasmas, however, large effects of strong plasma electric microfields on atomic (usually highly charged ions) energy levels are encountered that require more advanced descriptions, e.g., energy level evolution under the action of a quasi-static electric ion field including coherence effects. For real systems, these phenomena cannot be described consistently in the framework of the Schrödinger picture but request the density matrix description in contrast to the standard population balance rate equations. The general system of density matrix equations, however, is extremely challenging. It is demonstrated that the system can be transformed to a standard form of population kinetic rate equations with additional terms describing the effect of the ion electric field, the so-called quantum F-matrix theory (QFMT) allowing to construct real (large) closed model systems. The density matrix approach results in a new understanding of the fundamental physical problem connected with the transition of atomic populations to the Boltzmann equilibrium in dense plasmas. The standard transition is provided by electron collisions only satisfying the principle of detailed balance related to a statistical population of energy levels over magnetic quantum numbers. However, the effect of static electric fields on the atomic population is selective in magnetic quantum numbers resulting in a destruction of statistical Boltzmann populations. The interplay between the statistical effect of electrons and non-statistical effect of quasi-static ions is illustrated in detail for autoionizing atomic states of highly charged ions in dense laser-produced plasmas.

7.1 Rate Equation and Quantum Populations

The atomic populations are the fundamental quantity for various different disciplines in science and applications, like the equation of state in thermodynamics, absorption, emission and scattering properties of matter, lasing, radiation transport,

F. B. Rosmej et al., *Plasma Atomic Physics*, Springer Series on Atomic, Optical, and Plasma Physics 104, https://doi.org/10.1007/978-3-030-05968-2_7

radiative cooling and energy loss, diagnostic and spectroscopy employing the radiative properties of matter, astrophysics, planetary science, radiation sources, and fusion science (Loudon 2000; McWhirter 1965; Griem 1964, 1974, 1997; Sobelman 2006; Salzman 1988; Boiko et al. 1985; Michelis and Mattioli 1981; Lisitsa 1994; Bureyeva and Lisitsa 2000; Unsöld 1955; Mihalas 1978; Mihalas and Weibel-Mihalas 1999; Bekefi 1966; Drake 2018; Zel'dovich and Raizer 2002). The most frequently employed approach is the rate equation theory, where the atomic populations are considered as classical quantities while the various elementary population and depopulation processes are described by cross sections calculated with quantum mechanical and/or classical methods. In the framework of the rate equation theory, the atomic populations of the levels are given by a set of differential equations (see also Chap. 6):

$$\frac{\mathrm{d}n_\mathrm{j}}{\mathrm{d}t} = -n_\mathrm{j}\sum_{k=1}^{N} W_\mathrm{jk} + \sum_{l=1}^{N} n_\mathrm{l} W_\mathrm{lj}, \tag{7.1}$$

where n_j is the population density (cm^{-3}) of level j and N is the number of levels included in the model (e.g., ground and excited states from various charge states). The matrix W_ij describes the rates (s^{-1}) for the various elementary processes from level i to level j (e.g., radiative decay, collisional excitation/de-excitation, ionization, three-body recombination, photoionization, radiative recombination, autoionization, dielectronic capture, stimulated emission and absorption, and stimulated radiative recombination). If a certain transition between levels does not exist (e.g., due to selection rules), the matrix element is zero.

Although the rate model has widely been employed, it has principal deficiencies. It can consider only populations related to the squared of wave functions, i.e., $n_\mathrm{j} \propto |\psi_\mathrm{i}|^2$, while all mixed populations (coherences) related to the terms $\psi_\mathrm{i}\psi_\mathrm{j}^*$ (i / $= j$) are entirely absent in the theory.

The absence of the mixed populations drives two important drawbacks compared to the fully quantum approach: First, coherences cannot be described by the model system (7.1), and second, the influence of the non-diagonal populations on the diagonal ones is likewise entirely absent. In other words, even if coherences and/or polarizations are not of interest, the absence of the coupling remains an uncertainty in the theory for the determination of populations. Although the principal mechanisms of negligible non-diagonal populations on the diagonal ones are well identified (e.g., broadband illumination of the electromagnetic perturbation potential, elastic collision rates much larger than radiative decay rates) (Loudon 2000), their quantitative impact cannot be studied within the framework of the system (7.1) and remains an open question up to present days (in particular, in applications to dense correlated systems, e.g., dense plasmas, high-energy-density physics, astrophysics, etc.). Moreover, the Schrödinger picture does not allow to consistently describe relaxation constants in the amplitude equations and is therefore extremely challenging to study a quantum atom immersed into a dense collisional environment.

It is the purpose of the present quantum kinetic approach, to close this gap by developing a framework of a quantum kinetics theory for atoms in dense plasmas that permits to study the impact of the non-diagonal populations on the diagonal ones for realistic (large) model systems. This is of great practical importance, because at present only the rate equation approach permits model systems that are large enough to be realistically considered as closed systems. This is in strong contrast to the fully quantum approach: the complexity due to the coherences is so large that only rather restricted model systems have been considered, e.g., for applications in nonlinear quantum optics.

7.2 Schrödinger Picture

The Schrödinger picture of quantum dynamics describes the probability distributions for atomic states via the wave function amplitudes

$$|\Psi\rangle = \sum_{\mathrm{n}} a_{\mathrm{n}} |\psi_{\mathrm{n}}\rangle, \tag{7.2}$$

where $|\Psi\rangle$ is an arbitrary state vector, $|\psi_{\mathrm{n}}\rangle$ are the basis vectors, and a_{n} are the corresponding amplitudes. The probability amplitudes

$$|a_{\mathrm{n}}|^2 = |\langle\psi_{\mathrm{n}} \mid \Psi\rangle|^2 \tag{7.3}$$

provide the probability that a system known to be in a state described by $|\Psi\rangle$ has attributes associated with $|\psi_{\mathrm{n}}\rangle$. It is convenient to define the basis vectors $|\psi_{\mathrm{n}}\rangle$ as the solution of the Hamiltonian $\hat{H}_0$ in the absence of any externally applied forces, i.e.,

$$\hat{H}_0 \psi_{\mathrm{n}} = E_{\mathrm{n}}^{(0)} \psi_{\mathrm{n}}, \tag{7.4}$$

where the energies $E_{\mathrm{n}}^{(0)}$ are the unperturbed energies of the eigenstates $|\psi_{\mathrm{n}}\rangle$. An atom that is known to be in an eigenstate of the Hamiltonian $\hat{H}_0$ will remain infinitely in that states until external forces destroy this repose and cause transitions between states. If we disturb the isolated atom by a time-dependent interaction energy represented by the operator $\hat{V}(t)$, the full Hamiltonian becomes

$$\hat{H} = \hat{H}_0 + \hat{V}(t). \tag{7.5}$$

As $\hat{V}(t)$ acts on the basis vectors, the interaction produces new linear combinations of basis vectors, i.e.,

$$\hat{V}(t)\psi_{\mathrm{m}} = \sum_{\mathrm{n}} \psi_{\mathrm{n}} V_{\mathrm{nm}} \tag{7.6}$$

with

$$V_{\mathrm{nm}} = \langle n|\hat{V}|m\rangle \tag{7.7}$$

while the time evolution is given by the time-dependent Schrödinger equation:

$$i\hbar\frac{\partial\Psi(t)}{\partial t} = \hat{H}(t)\Psi(t). \tag{7.8}$$

It is evident that the Schrödinger picture of (7.8) is extremely challenging for correlated systems, e.g., dense hot plasmas where the fluctuating electric micro-fields constitute an important part of the interaction $\hat{V}(t)$, while numerous time-dependent collisional excitation and de-excitation processes drive a complex population kinetics of each amplitude related to the basis vectors $|\psi_{\mathrm{n}}\rangle$. Due to this large complexity, model systems are rather restricted if the Schrödinger description of (7.8) is chosen for numerical solutions. Consequently, a variety of processes may eventually place the atom into a quantum state that is not included in the restricted model system. Such events must diminish the probability of finding the atom within the limited model system. Obviously, any detailed description of such processes must introduce an enlarged model system to account for the additional quantum transitions. In order to take into account effects from the enlarged system but retain the limited model system itself, we can suppress the requirement that the probability of the limited model must be conserved. The decline from the probability conservation represents therefore the probability lost as the atom undergoes transitions into quantum states that are not included in the quantum vector basis of the limited model. As the probability loss is irreversible, it cannot be consistently incorporated into the quantum mechanical amplitude equations (Rautian and Shalagin 1991). However, these processes can consistently be described via incoherent relaxation processes (random phase approximation) into the atomic density matrix equations.

7.3 Atomic Density Matrix: Open and Closed Systems

The density matrix equation of quantum populations in a dense collisional plasma consisting of multicharged ions and electrons is given by Anufrienko et al. (1990), Rautian and Shalagin (1991) and Lisitsa (1994):

$$\begin{aligned}\frac{\mathrm{d}\tilde{\rho}_{\alpha\beta}}{\mathrm{d}t} &= -i\omega_{\alpha\beta}\tilde{\rho}_{\alpha\beta} - \frac{i}{\hbar}\sum_{\gamma}\left(V_{\alpha\gamma}\tilde{\rho}_{\gamma\beta} - \tilde{\rho}_{\alpha\gamma}V_{\gamma\beta}\right) + S_{\alpha\beta} \\ &\quad - \left(\frac{\gamma_\alpha}{2} + \frac{\gamma_\beta}{2}\right)\tilde{\rho}_{\alpha\beta} + R_{\alpha\beta} + A_{\alpha\beta} + Q_{\alpha\beta}.\end{aligned} \tag{7.9}$$

$\rho_{\alpha\beta}$ is the density matrix element of the states α and β, i.e., $\rho_{\alpha\beta} = a_\alpha a_\beta^*$ that is related to the reduced representation according to

$$\tilde{\rho}_{\alpha\beta} = \rho_{\alpha\beta} \cdot \exp\left[-\frac{i}{\hbar}\left(E_\alpha - E_\beta\right)t\right]. \tag{7.10}$$

Note that for the diagonal elements $\tilde{\rho}_{\alpha\beta} = \rho_{\alpha\beta}$. The first term on the right-hand side of (7.9) is related to the oscillatory behavior of the non-diagonal matrix elements with frequency $\omega_{\alpha\beta} = \frac{1}{\hbar}\left(E_\alpha - E_\beta\right)$, and the second term accounts for the mixing of populations $\rho_{\alpha\gamma}$ due to the potential $V_{\gamma\beta}$. $S_{\alpha\beta}$ is the change of the density matrix in time due to collisions, and $\Phi_{\alpha\beta}^{\alpha'\beta'}$ are the matrix elements of the collision operator (Griem 1964, 1974, 1997; Sholin et al. 1973; Lisitsa 1977, 1994; Rautian and Shalagin 1991; Anufrienko et al. 1993; Sobelman and Vainshtein 2006):

$$S_{\alpha\beta} = \sum_{\alpha'\beta'} \Phi_{\alpha\beta}^{\alpha'\beta'} \cdot \tilde{\rho}_{\alpha'\beta'}. \tag{7.11}$$

The collision matrix elements $\Phi_{\alpha\beta}^{\alpha'\beta'}$ describe the influence of the density matrix element $\rho_{\alpha'\beta'}$ on the matrix element $\rho_{\alpha\beta}$ due to collisions.

The fourth term accounts for the loss of populations due to transitions into quantum states that are not explicitly included in the set of the density matrix equations. In order to distinguish the levels from the system of the density matrix equations from quantum states not explicitly included, we introduce the following conventions: $\alpha, \beta, \gamma \in \{\text{density matrix}\}$, $n, m \notin \{\text{density matrix}\}$. With this designation, the loss rates γ_α and γ_β are given by

$$\gamma_\alpha = \sum_n \gamma_{\alpha n}, \tag{7.12}$$

$$\gamma_\beta = \sum_m \gamma_{\beta m}. \tag{7.13}$$

The coefficients $\gamma_{\alpha n}$ and $\gamma_{\beta m}$ describe the transition rates (number of transitions per unit time) from the quantum state α to the quantum state n. These depopulating transitions to quantum levels $n, m \notin \{\text{density matrix}\}$ might be due to various elementary processes, e.g., radiative decay, autoionization, collisional excitation, ionization, dielectronic capture, radiative recombination, three-body recombination, etc. Note that the collisional depopulation from the levels $\alpha, \beta \in \{\text{density matrix}\}$ to the levels $\alpha', \beta' \in \{\text{density matrix}\}$ is described by the collisional operator of (7.11).

The fifth term describes the inverse process of the fourth term, namely the population of the quantum states $\alpha, \beta, \gamma \in \{\text{density matrix}\}$ from the quantum states $n, m \notin \{\text{density matrix}\}$:

$$R_{\alpha\beta} = \delta_{\alpha\beta} \sum_{\mathrm{m}} \gamma_{\mathrm{m}\alpha} \rho_{\mathrm{mm}}. \tag{7.14}$$

In open model systems, the fifth term (7.14) is absent and the model has therefore serious drawbacks as it does not maintain, e.g., the principle of detailed balance.

The sixth term $A_{\alpha\beta}$ describes radiative transitions between the quantum levels included in the density matrix model (Rautian and Shalagin 1991)

$$A_{\alpha\beta} = -\left(\frac{\gamma_{\alpha}^{\mathrm{rad}}}{2} + \frac{\gamma_{\beta}^{\mathrm{rad}}}{2}\right)\tilde{\rho}_{\alpha\beta} + \delta_{\alpha\beta} \sum_{\alpha'} \gamma_{\alpha'\alpha}^{\mathrm{rad}} \tilde{\rho}_{\alpha'\alpha'} \tag{7.15}$$

with

$$\gamma_{\alpha}^{\mathrm{rad}} = \sum_{\alpha'} \gamma_{\alpha\alpha'}^{\mathrm{rad}}, \tag{7.16}$$

$$\gamma_{\beta}^{\mathrm{rad}} = \sum_{\beta'} \gamma_{\beta\beta'}^{\mathrm{rad}}. \tag{7.17}$$

It is convenient to introduce also explicitly the pumping rates $Q_{\alpha\beta}$, i.e.,

$$Q_{\alpha\beta} = \delta_{\alpha\beta} Q_{\alpha}. \tag{7.18}$$

We note that the pumping rates could formally also be described via the matrix elements $R_{\alpha\beta}$. This, however, is rather challenging in open model systems due to the absence of such terms.

The summation over the quantum states $n, m \notin \{\text{density matrix}\}$ in (7.12), (7.13) can be consistently performed as they concern only summations over the cross sections of respective elementary processes. This is quite different from (7.14) where all elementary processes request the level populations ρ_{mm} that are not known within the limited density matrix system $\alpha, \beta, \gamma \in \{\text{density matrix}\}$. In dense plasmas, however, a considerable part of the population flow originates from the quantum levels $n, m \notin \{\text{density matrix}\}$; examples are the radiative recombination cascading as well as the spontaneous radiative cascading in triplet systems.

At present, there are no general and consistent solutions available: either the population flows from the unknown populations densities ρ_{mm} are neglected or the density matrix approach is at all suppressed and replaced by the set of (7.1) with all their aforementioned drawbacks. Therefore, the introduction of relaxation terms (rates) in the atomic density matrix enables to keep track of some effects originating from a variety of processes that place the atom into a quantum state that is not explicitly included in the restricted model system. But in either case this cannot consistently account for the backflow (7.14) of the populations that is very important in dense plasmas to be consistent with the principle of detailed balance and the possibility that the model system approaches correctly the thermodynamic limit.

It is therefore of general interest (fundamental and application) to develop a new theory of atomic population kinetic equations that permit model systems to be large enough to be considered to be closed, i.e.,

$$\sum_{\alpha=1}^{N} \rho_{\alpha\alpha} = 1, \tag{7.19}$$

while maintaining at the same time the coupling of the non-diagonal populations $\rho_{\alpha\beta}$ on the diagonal ones $\rho_{\alpha\alpha}$, i.e., to incorporate quantum kinetic effects.

7.4 The Electron Collisional Operator $\hat{\Phi}$

Apart its traditional applications in quantum optics and nonlinear laser physics (Rautian and Shalagin 1991), the density matrix approach is likewise of fundamental interest in atomic plasma physics (Lisitsa 1994). Apart usual collisional radiative elementary processes that are described by the *W*-matrix of (7.1), we need to specify the matrix elements $V_{\alpha\beta}$ of the field interaction as well as those of the collisional operator $\Phi_{\alpha\beta}^{\alpha'\beta'}$. The explicit determination of the field operators $\hat{V}$ and $\hat{\Phi}$ is a very complex undertaking because these operators describe a quantum mechanical many-body correlated charged particle system.

In dense hot plasmas, electrons are usually of much greater mobility than ions (due to the large electron ion mass ratio). We therefore can assume ions to be quasi-static ions and responsible for the field operator $\hat{V}$ (that results into a mixing of levels due to a quasi-static electric ion field), while the electron collisions are described by the operator $\hat{\Phi}$ that acts on a "quasi-static atomic Stark split structure." This approximation is the so-called impact approximation and has originally been developed in line shape theory (Baranger 1958; Griem and Shen 1959; Griem 1964, 1968, 1974; Griem et al. 1979; Smith and Hooper 1967; Cooper and Oertel 1969; Sholin et al. 1973; Lisitsa 1977, 1994). In order to obtain explicit expressions for $\hat{V}$ and $\hat{\Phi}$ for quantum kinetics applications in the framework of the impact approximation, we apply first- and second-order quantum mechanical perturbation theory.

7.4.1 Scattering Matrix Representation

In order to obtain explicit expressions for the evolution of the system, let us restrict our interest to times Δt that are large compared to the ionic collision time ρ/υ_i (υ_i is the ion velocity and ρ the mean ion distance). At the same time, Δt must be small compared to γ^{-1} where γ is the width of the atomic level, so the change of the system evolution operator in the time interval Δt is still small. These conditions correspond approximately to the assumption that collisions are binary. Let us therefore assume that the evolution of the density matrix can be cast into the following form:

$$\frac{\mathrm{d}\hat{\rho}}{\mathrm{d}t} = -\frac{i}{\hbar}\left[\hat{H}_0\hat{\rho} - \hat{\rho}\hat{H}_0\right] - \frac{i}{\hbar}\left[\hat{V}\hat{\rho} - \hat{\rho}\hat{V}\right] + \left(\frac{\mathrm{d}\hat{\rho}}{\mathrm{d}t}\right)_{\text{coll}}, \tag{7.20}$$

where the term $\left(\frac{\mathrm{d}\hat{\rho}}{\mathrm{d}t}\right)_{\text{coll}}$ describes the evolution of the operator $\hat{\rho}$ due to collisions. If the wave function before collision is designated as Ψ and after collision as Ψ', they are related to each other via the collision matrix $\hat{S}$:

$$\Psi' = \hat{S}\Psi. \tag{7.21}$$

The corresponding change of the density matrix is then given by

$$\Delta\hat{\rho} = \hat{S}^{+}\hat{\rho}\hat{S} - \hat{\rho}. \tag{7.22}$$

Therefore

$$\left(\frac{\mathrm{d}\hat{\rho}}{\mathrm{d}t}\right)_{\text{coll}} = N\cdot\int_0^\infty\int_0^\infty \upsilon\cdot F(\upsilon)\,\mathrm{d}\upsilon\cdot 2\pi\rho\mathrm{d}\rho\left[\hat{S}^{+}\hat{\rho}\hat{S} - \hat{\rho}\right]. \tag{7.23}$$

N is the density of the perturbing particles, $F(\upsilon)$ the particle velocity distribution function, and ρ the impact parameter. The matrix elements of the collisional part of the density matrix are designated as in (7.9), i.e.,

$$S_{\alpha\beta} = \langle\alpha|\left(\frac{\mathrm{d}\hat{\rho}}{\mathrm{d}t}\right)_{\text{coll}}|\beta\rangle \tag{7.24}$$

from which it follows (using $S^{+}_{\alpha\beta} = S^{*}_{\beta\alpha}$)

$$S_{\alpha\beta} = \sum_{\alpha',\beta'}\rho_{\alpha'\beta'}\cdot N\cdot\int_0^\infty\int_0^\infty \upsilon\cdot F(\upsilon)\,\mathrm{d}\upsilon\cdot 2\pi\rho\mathrm{d}\rho\left[S^{*}_{\alpha'\alpha}\cdot S_{\beta'\beta} - \delta_{\alpha\alpha'}\delta_{\beta'\beta}\right]. \tag{7.25}$$

According to (7.11), the matrix elements of the collisional operator $\hat{\Phi}$ are defined as

$$\Phi^{\alpha'\beta'}_{\alpha\beta} = N\cdot\int_0^\infty\int_0^\infty \upsilon\cdot F(\upsilon)\,\mathrm{d}\upsilon\cdot 2\pi\rho\mathrm{d}\rho\left[S^{*}_{\alpha'\alpha}\cdot S_{\beta'\beta} - \delta_{\alpha\alpha'}\delta_{\beta'\beta}\right]. \tag{7.26}$$

In dense hot plasmas, electron–ion collisions are usually the most important ones for the excitation of levels and ionization of atoms and ions and we shall therefore assume in the following that the collisional operator $\hat{\Phi}$ describes electron collisions, i.e.,

$$\Phi_{\alpha\beta}^{\alpha'\beta'} = n_e \cdot \int_0^{\infty}\int_0^{\infty} \upsilon \cdot F(\upsilon)\, d\upsilon \cdot 2\pi\rho d\rho \left[S_{\alpha'\alpha}^{*} \cdot S_{\beta'\beta} - \delta_{\alpha\alpha'}\delta_{\beta\beta'}\right], \tag{7.27}$$

where n_e is the electron density. Due to the long-range character of the Coulomb field, the scattering matrix $\hat{S}$ can be expanded in a perturbation-theory series:

$$\hat{S}(t) = 1 + \left(-\frac{i}{\hbar}\right)\int_{t_0}^{t} \hat{V}^{(I)}(t')dt' + \left(-\frac{i}{\hbar}\right)^2 \int_{t_0}^{t} \hat{V}^{(I)}(t')dt' \int_{t_0}^{t'} \hat{V}^{(I)}(t'')dt'' + \cdots, \tag{7.28}$$

where the operator $\hat{V}^{(I)}$ is given by

$$\hat{V}^{(I)} = \exp\left[\frac{i}{\hbar}\hat{H}_0 t\right] \cdot \hat{V} \cdot \exp\left[-\frac{i}{\hbar}\hat{H}_0 t\right]. \tag{7.29}$$

$\hat{V}$ represents the field interaction operator.

7.4.2 Electron Collisional Operator in Second-Order Perturbation Theory

In the classical nonrelativistic approximation, the interaction of perturbers with charge Z_p with a charge distribution $\rho(\vec{r})$ is given by

$$\hat{H}_{\text{int}}(t) = -\sum_{p} Z_p \cdot e^2 \int_{\text{vol}} \frac{\rho(\vec{r})d\vec{r}}{|\vec{r}_p - \vec{r}|}. \tag{7.30}$$

The charge state distribution $\rho(\vec{r})$ describes the radiating ions including bound electrons, and the perturbers are assumed to act as point charges. The radius vectors of the perturbers $\vec{r}_p$ and the vector $\vec{r}$ of the "interaction charge $\rho(\vec{r})$" are measured from the nucleus of the radiator. The term $1/|\vec{r}_p - \vec{r}|$ can be developed in a Taylor series:

$$\frac{1}{|\vec{r}_{\mathrm{p}}-\vec{r}|}=\sum_{n=0}^{\infty}\frac{1}{n!}\cdot\left(-\vec{r}\cdot\nabla_{\mathrm{p}}\right)^{n}\frac{1}{r_{\mathrm{p}}}=\frac{1}{r_{\mathrm{p}}}-\vec{r}\cdot\nabla_{\mathrm{p}}\frac{1}{r_{\mathrm{p}}}+\frac{1}{2}\vec{r}\cdot\vec{r}\cdot\nabla_{\mathrm{p}}\nabla_{\mathrm{p}}\frac{1}{r_{\mathrm{p}}}+\cdots$$
$$=\frac{1}{|\vec{r}_{\mathrm{p}}|}+\frac{\vec{r}_{\mathrm{p}}\cdot\vec{r}}{|\vec{r}_{\mathrm{p}}|^{3}}+\frac{1}{2}\cdot\frac{\vec{r}_{\mathrm{p}}\cdot\vec{r}_{\mathrm{p}}}{|\vec{r}_{\mathrm{p}}|^{5}}\cdot\left(3\cdot\vec{r}\cdot\vec{r}-|\vec{r}|^{2}\cdot\hat{1}\right)+\cdots, \tag{7.31}$$

where $\hat{1}$ is the unit matrix. Assuming that all perturbers have the same charge Z_{p}, the interaction Hamiltonian can be represented as a multipole expansion:

$$\begin{aligned}\hat{H}_{\mathrm{int}}=-Z_{\mathrm{p}}\cdot e^{2}\cdot\sum_{p}\Bigg\{&\frac{1}{|\vec{r}_{\mathrm{p}}|}\cdot\int_{vol}\rho(\vec{r})\mathrm{d}\vec{r}\\&+\frac{\vec{r}_{\mathrm{p}}}{|\vec{r}_{\mathrm{p}}|^{3}}\cdot\int_{vol}\vec{r}\cdot\rho(\vec{r})\mathrm{d}\vec{r}\\&+\frac{\vec{r}_{\mathrm{p}}\cdot\vec{r}_{\mathrm{p}}}{|\vec{r}_{\mathrm{p}}|^{5}}\cdot\int_{vol}\left[3\vec{r}\cdot\vec{r}-|\vec{r}|^{2}\cdot\hat{1}\right]\rho(\vec{r})\mathrm{d}\vec{r}+\cdots\Bigg\}.\end{aligned} \tag{7.32}$$

The first term in (7.31) corresponds to the monopole, the second one is the dipole, and the third one is the quadrupole. The corresponding moments are

$$Z:=\int_{vol}\rho(\vec{r})\mathrm{d}\vec{r}, \tag{7.33}$$

$$\vec{d}=\int_{vol}\vec{r}\cdot\rho(\vec{r})\mathrm{d}\vec{r}, \tag{7.34}$$

$$\hat{Q}=\int_{vol}\left[3\vec{r}\cdot\vec{r}-|\vec{r}|^{2}\cdot\hat{1}\right]\rho(\vec{r})\mathrm{d}\vec{r}. \tag{7.35}$$

Note that the dipole moment Z is a scalar, the dipole moment $\vec{d}$ is a vector, and the quadrupole moment $\hat{Q}$ is a matrix.

In atomic physics, the overwhelming part of radiative and collisional processes is very well described by the first three terms in the series development of (7.32), while the dipole term is dominating in most cases. Let us therefore specify the interaction Hamiltonian in the dipole approximation for the present case of electron collisions; i.e., an electron collides with the electrons of the radiator and induces elastic and inelastic transitions (note that due to the orthogonality of the wave functions, the monopole term vanishes for transitions unless initial and final quantum numbers are not identical). In collision theory, the radius vector $\vec{r}_{\mathrm{p}}$ of the perturber particles of (7.32) is usually called the impact parameter. Let $\vec{\rho}$ and $\vec{\upsilon}$ the impact parameter and velocity vectors of the colliding electron (perturber) of closest approach, the classical interaction potential is given by

$$V(t) = e \cdot \frac{\vec{d} \cdot (\vec{\rho} + \vec{\upsilon} \cdot t)}{(\rho^2 + \upsilon^2 \cdot t^2)^{3/2}}, \tag{7.36}$$

where ρ is the impact parameter of the colliding electron, $R = \sqrt{\rho^2 + \upsilon^2 \cdot t^2}$ is the distance between the perturbing particle and the atom, and υ the relative velocity of the perturbing particle and the atom. In the dipole approximation

$$\hat{H}_{\mathrm{int}} \approx \hat{H}_{\mathrm{int}}^{(\mathrm{dipole})} = \hat{V} = e \cdot \frac{\vec{d} \cdot (\vec{\rho} + \vec{\upsilon} \cdot t)}{(\rho^2 + \upsilon^2 \cdot t^2)^{3/2}}, \tag{7.37}$$

where the electron charge state distribution of a certain configuration is determined from the squared of the wave functions of a specific quantum mechanical state, i.e.,

$$\vec{d} = e \cdot \int_{vol} \psi_\alpha^*(\vec{r}) \cdot \vec{r} \cdot \psi_\beta(\vec{r}) d\vec{r} = \vec{d}_{\alpha\beta}. \tag{7.38}$$

If we develop the scattering matrix up to second order and integrate from $t = -\infty$ to $t = \infty$ (i.e., over the whole duration of the collisional process), we obtain from (7.28)

$$\hat{S}(t) \approx 1 + \left(-\frac{i}{\hbar}\right) \int_{-\infty}^{\infty} \hat{V}^{(I)}(t) \mathrm{d}t + \left(-\frac{i}{\hbar}\right)^2 \int_{-\infty}^{\infty} \hat{V}^{(I)}(t) \mathrm{d}t \int_{-\infty}^{t} \hat{V}^{(I)}(t') \mathrm{d}t'. \tag{7.39}$$

Physically, a development of the collisional operator up to second order is equivalent to the fact that the main contributions to collisions originate from distant collisions where the impact parameters are essentially larger than the Weisskopf radius (i.e., weak collisions).

Assuming that a transition occurs between a level a and a level b consisting of the states $(\alpha, \alpha', \ldots)$ and $(\beta, \beta', \ldots)$, respectively, the scattering matrix contains then both the second-order terms $\vec{r}_{\mathrm{a}} \cdot \vec{r}_{\mathrm{a}}$ and $\vec{r}_{\mathrm{b}} \cdot \vec{r}_{\mathrm{b}}$ and cross-terms originating from the product of the first-order terms $\vec{r}_{\mathrm{a}}$ and $\vec{r}_{\mathrm{b}}$ where $\vec{r}_{\mathrm{a}}$ and $\vec{r}_{\mathrm{b}}$ are the radius vectors of the atomic electrons in states "a" and "b", respectively. With the help of (7.31) we obtain the requested matrix elements for the collisional operator of (7.27):

$$
\begin{aligned}
S^*_{\alpha'\alpha} &= \langle\alpha' \mid \alpha\rangle^* + \langle\alpha'|\left(-\frac{i}{\hbar}\right) \int\limits_{-\infty}^{\infty} \hat{V}_{\rm a}^{(I)}(t)\mathrm{d}t|\alpha\rangle^* \\
&+ \langle\alpha'|\left(-\frac{i}{\hbar}\right)^2 \int\limits_{-\infty}^{\infty} \hat{V}_{\rm a}^{(I)}(t)\mathrm{d}t \int\limits_{-\infty}^{t} \hat{V}_{\rm a}^{(I)}(t')\mathrm{d}t'|\alpha\rangle^*
\end{aligned}
\tag{7.40}
$$

and

$$
\begin{aligned}
S_{\beta'\beta} &= \langle\beta' \mid \beta\rangle + \langle\beta'|\left(-\frac{i}{\hbar}\right) \int\limits_{-\infty}^{\infty} \hat{V}_{\rm b}^{(I)}(t)\mathrm{d}t|\beta\rangle \\
&+ \langle\beta'|\left(-\frac{i}{\hbar}\right)^2 \int\limits_{-\infty}^{\infty} \hat{V}_{\rm b}^{(I)}(t)\mathrm{d}t \int\limits_{-\infty}^{t} \hat{V}_{\rm b}^{(I)}(t')\mathrm{d}t'|\beta\rangle.
\end{aligned}
\tag{7.41}
$$

According to (7.27), the collisional operator contains the product of the matrix elements; therefore, the product of the second-order approximations (7.40), (7.41) contain third- and fourth-order terms that need to be dropped, i.e.,

$$
\begin{aligned}
S^*_{\alpha'\alpha}S_{\beta'\beta} &\approx \langle\alpha' \mid \alpha\rangle^*\langle\beta' \mid \beta\rangle + \langle\alpha' \mid \alpha\rangle^*\langle\beta'|\left(-\frac{i}{\hbar}\right) \int\limits_{-\infty}^{\infty} \hat{V}_{\rm b}^{(I)}(t)\mathrm{d}t|\beta\rangle \\
&+ \langle\alpha' \mid \alpha\rangle^*\langle\beta'|\left(-\frac{i}{\hbar}\right)^2 \int\limits_{-\infty}^{\infty} \hat{V}_{\rm b}^{(I)}(t)\mathrm{d}t \int\limits_{-\infty}^{t} \hat{V}_{\rm b}^{(I)}(t')\mathrm{d}t'|\beta\rangle \\
&+ \langle\beta' \mid \beta\rangle\langle\alpha'|\left(-\frac{i}{\hbar}\right) \int\limits_{-\infty}^{\infty} \hat{V}_{\rm a}^{(I)}(t)\mathrm{d}t|\alpha\rangle^* \\
&+ \langle\beta' \mid \beta\rangle\langle\alpha'|\left(-\frac{i}{\hbar}\right)^2 \int\limits_{-\infty}^{\infty} \hat{V}_{\rm a}^{(I)}(t)\mathrm{d}t \int\limits_{-\infty}^{t} \hat{V}_{\rm a}^{(I)}(t')\mathrm{d}t'|\alpha\rangle^* \\
&+ \langle\alpha'|\left(-\frac{i}{\hbar}\right) \int\limits_{-\infty}^{\infty} \hat{V}_{\rm a}^{(I)}(t)\mathrm{d}t|\alpha\rangle^*\langle\beta'|\left(-\frac{i}{\hbar}\right) \int\limits_{-\infty}^{\infty} \hat{V}_{\rm b}^{(I)}(t)\mathrm{d}t|\beta\rangle.
\end{aligned}
\tag{7.42}
$$

In order to apply the collisional operator to dense hot plasmas, we need to average (7.27) over all directions of the impact vector $\vec{\rho}$ and the velocity vector $\vec{\upsilon}$, i.e.,

$$
\Phi_{\alpha\beta}^{\alpha'\beta'} = \left\{ n_{\rm e} \cdot \int\limits_0^{\infty}\int\limits_0^{\infty} \upsilon \cdot F(\upsilon)\,\mathrm{d}\upsilon \cdot 2\pi\rho\mathrm{d}\rho\left[S^*_{\alpha'\alpha} \cdot S_{\beta'\beta} - \delta_{\alpha\alpha'}\delta_{\beta\beta'}\right] \right\}_{\rm av}. \tag{7.43}
$$

As we are interested only in times Δt that are large compared to the mean ionic collision time $\rho/\upsilon_{\mathrm{i}}$, the evolution terms $S^*_{\alpha'\alpha} \cdot S_{\beta'\beta}$ are independent from $\delta_{\alpha\alpha'}\delta_{\beta\beta'}$ and the averaging procedure can be carried out for these terms independently. If we assume a random distribution of directions, the second and fourth terms in (7.42) vanish due to the specific form of the interaction potential (7.36), i.e.,

$$\begin{aligned}\left\{\left[S^*_{\alpha'\alpha}S_{\beta'\beta} - \delta_{\alpha'\alpha}\delta_{\beta'\beta}\right]\right\}_{\mathrm{av}} \approx\ & \delta_{\alpha'\alpha}\left\{\langle\beta'|\left(-\frac{i}{\hbar}\right)^2 \int_{-\infty}^{\infty} \hat{V}_{\mathrm{b}}^{(I)}(t)\mathrm{d}t \int_{-\infty}^{t} \hat{V}_{\mathrm{b}}^{(I)}(t')\mathrm{d}t'|\beta\rangle\right\}_{\mathrm{av}} \\ & + \delta_{\beta'\beta}\left\{\langle\alpha'|\left(-\frac{i}{\hbar}\right)^2 \int_{-\infty}^{\infty} \hat{V}_{\mathrm{a}}^{(I)}(t)\mathrm{d}t \int_{-\infty}^{t} \hat{V}_{\mathrm{a}}^{(I)}(t')\mathrm{d}t'|\alpha\rangle^*\right\}_{\mathrm{av}} \\ & + \left\{\langle\alpha'|\left(-\frac{i}{\hbar}\right) \int_{-\infty}^{\infty} \hat{V}_{\mathrm{a}}^{(I)}(t)\mathrm{d}t|\alpha\rangle^* \langle\beta'|\left(-\frac{i}{\hbar}\right) \int_{-\infty}^{\infty} \hat{V}_{\mathrm{b}}^{(I)}(t)\mathrm{d}t|\beta\rangle\right\}_{\mathrm{av}}.\end{aligned} \tag{7.44}$$

The second-order expansion indicates that the electronic collisions of the upper and lower levels are not independent from each other; i.e., interference effects may take place. We are now left to calculate the average of products of the interaction potential $\left\{\hat{V}^{(I)}(t)\hat{V}^{(I)}(t')\right\}_{\mathrm{av}}$. With the help of (7.37), we obtain:

$$\left\{\hat{V}(t)\hat{V}(t')\right\}_{\mathrm{av}} = e^4 \cdot \frac{\vec{r}\cdot\vec{r}}{3} \cdot \frac{\rho^2 + \upsilon^2 tt'}{(\rho^2 + \upsilon^2 \cdot t^2)^{3/2} \cdot (\rho^2 + \upsilon^2 \cdot t'^2)^{3/2}}. \tag{7.45}$$

The first term in (7.44) provides

$$\begin{aligned}&\left\{\langle\beta'|\left(-\frac{i}{\hbar}\right)^2 \int_{-\infty}^{\infty} \hat{V}_{\mathrm{b}}^{(I)}(t)\mathrm{d}t \int_{-\infty}^{t} \hat{V}_{\mathrm{b}}^{(I)}(t')\mathrm{d}t'|\beta\rangle\right\}_{\mathrm{av}} \\ &= -\frac{e^4}{3\hbar^2} \cdot \sum_{\beta''} \langle\beta'|\vec{r}_{\mathrm{b}}|\beta''\rangle\langle\beta''|\vec{r}_{\mathrm{b}}|\beta\rangle \\ &\times \int_{-\infty}^{\infty} \mathrm{d}t \int_{-\infty}^{t} \mathrm{d}t' \frac{(\rho^2 + \upsilon^2 tt') \cdot \exp(i\omega_{\beta'\beta''}t) \cdot \exp(i\omega_{\beta''\beta}t')}{(\rho^2 + \upsilon^2 \cdot t^2)^{3/2} \cdot (\rho^2 + \upsilon^2 \cdot t'^2)^{3/2}}\end{aligned} \tag{7.46}$$

and correspondingly for the second term in (7.44), we have

$$\left\{\langle\alpha'|\left(-\frac{i}{\hbar}\right)^2\int\limits_{-\infty}^{\infty}\hat{V}_{\mathrm{a}}^{(I)}(t)\mathrm{d}t\int\limits_{-\infty}^{t}\hat{V}_{\mathrm{a}}^{(I)}(t')\mathrm{d}t'|\alpha\rangle^*\right\}_{\mathrm{av}}$$
$$= -\frac{e^4}{3\hbar^2}\cdot\sum_{\alpha''}\langle\alpha'|\vec{r}_{\mathrm{a}}|\alpha''\rangle^*\langle\alpha''|\vec{r}_{\mathrm{a}}|\alpha\rangle^* \tag{7.47}$$
$$\times\int\limits_{-\infty}^{\infty}\mathrm{d}t\int\limits_{-\infty}^{t}\mathrm{d}t'\frac{(\rho^2+\upsilon^2tt')\cdot\exp(-i\omega_{\alpha'\alpha''}t)\cdot\exp(-i\omega_{\alpha''\alpha}t')}{(\rho^2+\upsilon^2\cdot t^2)^{3/2}\cdot(\rho^2+\upsilon^2\cdot t'^2)^{3/2}}.$$

The third term in (7.44) is a "cross-term" and given by

$$\left\{\langle\alpha'|\left(-\frac{i}{\hbar}\right)\int\limits_{-\infty}^{\infty}\hat{V}_{\mathrm{a}}^{(I)}(t)\mathrm{d}t|\alpha\rangle^*\langle\beta'|\left(-\frac{i}{\hbar}\right)\int\limits_{-\infty}^{\infty}\hat{V}_{\mathrm{b}}^{(I)}(t)\mathrm{d}t|\beta\rangle\right\}_{\mathrm{av}}$$
$$= \frac{e^4}{3\hbar^2}\cdot\langle\alpha'|\vec{r}_{\mathrm{a}}|\alpha\rangle^*\langle\beta'|\vec{r}_{\mathrm{b}}|\beta\rangle \tag{7.48}$$
$$\times\int\limits_{-\infty}^{\infty}\mathrm{d}t\int\limits_{-\infty}^{\infty}\mathrm{d}t'\frac{(\rho^2+\upsilon^2tt')\cdot\exp(-i\omega_{\alpha'\alpha}t)\cdot\exp\left(i\omega_{\beta'\beta}t'\right)}{(\rho^2+\upsilon^2\cdot t^2)^{3/2}\cdot(\rho^2+\upsilon^2\cdot t'^2)^{3/2}}.$$

Let us introduce the dimensionless parameters (note that $\omega_{\alpha\beta} = -\omega_{\beta\alpha}$)

$$z_1 = \frac{\rho}{\upsilon}\cdot\omega_{\beta'\beta''}, \tag{7.49}$$

$$z_2 = \frac{\rho}{\upsilon}\cdot\omega_{\beta\beta''}, \tag{7.50}$$

$$z_1' = \frac{\rho}{\upsilon}\cdot\omega_{\alpha'\alpha''}, \tag{7.51}$$

$$z_2' = \frac{\rho}{\upsilon}\cdot\omega_{\alpha\alpha''}, \tag{7.52}$$

$$z_3 = \frac{\rho}{\upsilon}\cdot\omega_{\beta'\beta}, \tag{7.53}$$

$$z_4 = \frac{\rho}{\upsilon}\cdot\omega_{\alpha'\alpha}, \tag{7.54}$$

$$x_1 = \frac{\upsilon}{\rho}, \tag{7.55}$$

$$x_2 = \frac{\upsilon'}{\rho}. \tag{7.56}$$

With the help of the dimensionless parameters (7.49)–(7.56) and (7.46)–(7.48), (7.44) can be represented in the form

$$\begin{aligned}
&\left\{\left[S^{*}_{\alpha'\alpha}S_{\beta'\beta}-\delta_{\alpha'\alpha}\delta_{\beta'\beta}\right]\right\}_{\mathrm{av}}\\
&\quad=-\frac{2e^4}{3\hbar^2}\cdot\frac{1}{\rho^2 v^2}\times\left(\delta_{\alpha'\alpha}\cdot\sum_{\beta''}\langle\beta'|\vec{r}_{\mathrm{b}}|\beta''\rangle\langle\beta''|\vec{r}_{\mathrm{b}}|\beta\rangle\cdot\frac{1}{2}\int\limits_{-\infty}^{\infty}\mathrm{d}x_1\right.\\
&\quad\times\int\limits_{-\infty}^{x_1}\mathrm{d}x_2\frac{(1+x_1x_2)\cdot\exp[i(z_1x_1-z_2x_2)]}{\left(1+x_1^2\right)^{3/2}\cdot\left(1+x_2^2\right)^{3/2}}\\
&\quad+\delta_{\beta'\beta}\cdot\sum_{\alpha''}\langle\alpha'|\vec{r}_{\mathrm{a}}|\alpha''\rangle^{*}\langle\alpha''|\vec{r}_{\mathrm{a}}|\alpha\rangle^{*}\cdot\frac{1}{2}\int\limits_{-\infty}^{\infty}\mathrm{d}x_1\int\limits_{-\infty}^{x_1}\mathrm{d}x_2\frac{(1+x_1x_2)\cdot\exp\left[-i\left(z_1'x_1-z_2'x_2\right)\right]}{\left(1+x_1^2\right)^{3/2}\cdot\left(1+x_2^2\right)^{3/2}}\\
&\quad\left.+\langle\alpha'|\vec{r}_{\mathrm{a}}|\alpha\rangle^{*}\langle\beta'|\vec{r}_{\mathrm{b}}|\beta\rangle\cdot\frac{1}{2}\int\limits_{-\infty}^{\infty}\mathrm{d}x_1\int\limits_{-\infty}^{\infty}\mathrm{d}x_2\frac{(1+x_1x_2)\cdot\exp[-i(z_3x_1+z_4x_2)]}{\left(1+x_1^2\right)^{3/2}\cdot\left(1+x_2^2\right)^{3/2}}\right).
\end{aligned}\tag{7.57}$$

The integrals can be expressed in terms of Bessel functions of the first kind J_{n}, modified Bessel functions of the second kind K_{n} and hypergeometric functions ${}_{\mathrm{n}}F_m$ (Deutsch et al. 1969a, b). The integral of the third term in (7.57) is given by

$$\begin{aligned}
&\frac{1}{2}\cdot\int\limits_{-\infty}^{\infty}\mathrm{d}x_1\int\limits_{-\infty}^{\infty}\mathrm{d}x_2\frac{(1+x_1x_2)\cdot\exp[-i(z_3x_1+z_4x_2)]}{\left(1+x_1^2\right)^{3/2}\cdot\left(1+x_2^2\right)^{3/2}}\\
&\quad:=2\cdot[|z_3|\cdot|z_4|\cdot K_1(|z_3|)\cdot K_1(|z_4|)+z_3\cdot z_4\cdot K_0(|z_3|)\cdot K_0(|z_4|)].
\end{aligned}\tag{7.58}$$

As the second integral in (7.57) reduces to the first one when $z_1'=-z_1$ and $z_2'=-z_2$, we obtain

$$\frac{1}{2}\cdot\int\limits_{-\infty}^{\infty}\mathrm{d}x_1\int\limits_{-\infty}^{x_1}\mathrm{d}x_2\frac{(1+x_1x_2)\cdot\exp[i(z_1x_1-z_2x_2)]}{\left(1+x_1^2\right)^{3/2}\cdot\left(1+x_2^2\right)^{3/2}}:=A(z_1,z_2)+iB(z_1,z_2)\tag{7.59}$$

and

$$\frac{1}{2}\cdot\int\limits_{-\infty}^{\infty}\mathrm{d}x_1\int\limits_{-\infty}^{x_1}\mathrm{d}x_2\frac{(1+x_1x_2)\cdot\exp\left[-i\left(z_1'x_1-z_2'x_2\right)\right]}{\left(1+x_1^2\right)^{3/2}\cdot\left(1+x_2^2\right)^{3/2}}=A\left(-z_1',-z_2'\right)+iB\left(-z_1',z_2'\right)\tag{7.60}$$

with

$$A(z_1, z_2) = |z_1| \cdot |z_2| \cdot K_1(|z_1|) \cdot K_1(|z_2|) + z_1 \cdot z_2 \cdot K_0(|z_1|) \cdot K_0(|z_2|) \tag{7.61}$$

and

$$\begin{aligned} B(z_1, z_2) = {} & z_1 \left[{}_1F_2\left(1; \frac{1}{2}, \frac{3}{2}; \frac{1}{4} z_1^2\right) - \frac{1}{4}\pi |z_1| {}_1F_2\left(\frac{3}{2}; 2, \frac{3}{2}; \frac{1}{4} z_1^2\right)\right] |z_2| K_1(|z_2|) \\ & - \left[{}_1F_2\left(1; \frac{1}{2}, \frac{1}{2}; \frac{1}{4} z_1^2\right) - \frac{1}{2}\pi |z_1| {}_1F_2\left(\frac{3}{2}; 1, \frac{3}{2}; \frac{1}{4} z_1^2\right)\right] z_2 K_1(|z_2|) \\ & + z_2 \int_0^\infty \mathrm{d}t \int_0^\infty \mathrm{d}t' \frac{t \cdot J_0(t)}{t^2 + z_2^2} \cdot \frac{t'(t+t') \cdot J_0(t')}{(t+t')^2 + (z_1 - z_2)^2} \\ & + (z_2 - z_1) \int_0^\infty \mathrm{d}t \int_0^\infty \mathrm{d}t' \frac{t^2 \cdot J_0(t)}{t^2 + z_2^2} \cdot \frac{t' \cdot J_0(t')}{(t+t')^2 + (z_1 - z_2)^2} \\ & + z_2 \int_0^\infty \mathrm{d}t \int_0^\infty \mathrm{d}t' \frac{t \cdot J_1(t)}{t^2 + z_2^2} \cdot \frac{t'(t+t') \cdot J_1(t')}{(t+t')^2 + (z_1 - z_2)^2} \\ & + (z_2 - z_1) \int_0^\infty \mathrm{d}t \int_0^\infty \mathrm{d}t' \frac{t^2 \cdot J_1(t)}{t^2 + z_2^2} \cdot \frac{t' \cdot J_1(t')}{(t+t')^2 + (z_1 - z_2)^2}. \end{aligned} \tag{7.62}$$

We note the symmetry relations (Deutsch et al. 1969a, b): $A(z_1, z_2) = A(-z_1, -z_2)$ and $B(z_1, z_2) = -B(-z_1, -z_2)$. The B-function is rather complex, but can be evaluated numerically without great difficulties. In the straight path approximation, the B-function reduces to Deutsch (1969a, b), Griem (1974), Sobelman and Vainshtein (2006) and Sahal-Bréchot et al. (2015)

$$B^{(\text{straight})}(z_1, z_2) = \pi \cdot z_1 \cdot z_2 \cdot [K_0(z_1) I_0(z_2) - K_1(z_1) I_1(z_2)]. \tag{7.63}$$

In the Born limit, when the dimensionless parameters z of (7.49)–(7.56) are very small, i.e., $z_1 \ll 1$ and $z_2 \ll 1$, the A-function takes a very simple form:

$$A^{(\text{Born})}(z_1, z_2) \to 1 + z_1 \cdot z_2. \tag{7.64}$$

In the straight path approximation, also the B-function takes a simple form:

$$B^{(\text{Born})}_{\text{straight}}(z_1, z_2) \to \pi \cdot z_1 \cdot z_2. \tag{7.65}$$

The approximations (7.64), (7.65) are of great practical interest because for $z < 0.1$ the A-value deviates less than 10% from its asymptotic value and the B-value is smaller than about 0.1. Asymptotic expansions for large values of z-parameters, i.e., $z_1 \gg 1$ and $z_2 \gg 1$, have been considered by (Deutsch et al. 1969b).

With the help of (7.57)–(7.62), the matrix elements of the collisional operator (7.43) can therefore be cast into the following form

$$\Phi_{\alpha\beta}^{\alpha'\beta'} = -n_e \cdot \frac{4\pi e^4}{3\hbar^2} \cdot \int_0^\infty \frac{f(\upsilon)}{\upsilon} \mathrm{d}\upsilon \int_{\rho_{min}}^{\rho_{max}} \frac{\mathrm{d}\rho}{\rho} \cdot \left\{ \sum_{\beta''} \phi_{(1),\alpha\beta}^{(M),\alpha'\beta'} \cdot [A(z_1,z_2)+iB(z_1,z_2)] \right.$$
$$\left. + \sum_{\alpha''} \phi_{(2),\alpha\beta}^{(M),\alpha'\beta'} \cdot [A(-z_{1'},-z_{2'})+iB(-z_{1'},-z_{2'})] - \phi_{(3),\alpha\beta}^{(M),\alpha'\beta'} \cdot [A(z_3,z_4)] \right\}. \tag{7.66}$$

with (note, that the third term does not contain a sum over the quantum states and that the upper index "(M)" remembers that the collisional operator matrix elements are resolved in magnetic quantum number)

$$\phi_{(1),\alpha\beta}^{(M),\alpha'\beta'} = \delta_{\alpha'\alpha} \cdot \langle\beta'|\vec{r}_b|\beta''\rangle\langle\beta''|\vec{r}_b|\beta\rangle, \tag{7.67}$$

$$\phi_{(2),\alpha\beta}^{(M),\alpha'\beta'} = \delta_{\beta'\beta} \cdot \langle\alpha''|\vec{r}_a|\alpha''\rangle^*\langle\alpha''|\vec{r}_a|\alpha\rangle^*, \tag{7.68}$$

$$\phi_{(3),\alpha\beta}^{(M),\alpha'\beta'} = 2 \cdot \langle\alpha'|\vec{r}_a|\alpha\rangle^*\langle\beta'|\vec{r}_b|\beta\rangle, \tag{7.69}$$

$f(\upsilon)$ is the electron velocity distribution function. Assuming a Maxwellian electron energy distribution function, $f(\upsilon)$ is given by

$$f(\upsilon) = \sqrt{\frac{2}{\pi}} \cdot \left(\frac{m_e}{kT_e}\right)^{3/2} \cdot \upsilon^2 \cdot \exp\left[-\frac{m_e\upsilon^2}{2kT_e}\right]. \tag{7.70}$$

υ is the electron velocity, m_e the electron mass, k the Boltzmann constant, and T_e the electron temperature. The integrals [A- and B-functions, (7.61), (7.62)] describe the non-degenerate case and non-isolated line ($z_1 = z_2 = z$ for isolated lines) case because of the possibility $z_1 \neq z_2$. This general case, accompanied with a rigorous numerical calculation of the integrals of (7.58)–(7.60), is well adopted for applications to quantum atomic population kinetics of arbitrary level structure.

If the quasi-static Stark splitting between the levels is neglected, i.e., $z_i = 0$ and correspondingly $A(0,0) = 1, B(0,0) = 0$ and $\hat{V}^{(I)} = \hat{V}$ [see (7.29)], we encounter the following relations

$$\int_{-\infty}^{\infty} \mathrm{d}t \int_{-\infty}^{t} \mathrm{d}t' \hat{V}_a(t)\hat{V}_a(t') = \frac{1}{2}\left[\int_{-\infty}^{\infty} \mathrm{d}t \cdot \hat{V}_a(t)\right]^2 \tag{7.71}$$

and

$$\hat{\Phi}_{\alpha\beta}^{\alpha'\beta'} = -n_{\mathrm{e}} \cdot \frac{4\pi e^4}{3\hbar^2} \cdot \int_0^{\infty} \frac{f(\upsilon)}{\upsilon} \mathrm{d}\upsilon \int_{\rho_{\min}}^{\rho_{\max}} \frac{\mathrm{d}\rho}{\rho} \cdot \left\{ \sum_{\beta''} \phi_{(1),\alpha\beta}^{(M),\alpha'\beta'} + \sum_{\alpha''} \phi_{(2),\alpha\beta}^{(M),\alpha'\beta'} - \phi_{(3),\alpha\beta}^{(M),\alpha'\beta'} \right\} \tag{7.72}$$

Note that (7.71), (7.72) are a particular case of the general (7.66)–(7.69).

In dense plasmas, the integral over the impact parameters deserves particular attention (Smith and Hooper 1967; Sholin et al. 1973; Griem 1974, 1997; Griem et al. 1979; Sobelman and Vainshtein 2006; Anufrienko et al. 1993). Other ions and electrons surround the ion that is considered to be the center of electron scattering. This results in a screening of the ionic field and is of the order of the Debye sphere at large distances:

$$R_{\mathrm{D}} = \sqrt{\frac{kT}{4\pi e^2 n_{\mathrm{i}}(1+Z^2)}}, \tag{7.73}$$

where n_{i} is the ion density, Z is the ionic charge, and T is the plasma temperature (assuming that electrons and ions do have the same temperature, i.e., $T = T_{\mathrm{e}} = T_{\mathrm{i}}$). Therefore, the upper limit of integration for the impact integral (7.66), (7.72) is given by $\rho_{\max}$ rather than by an infinite impact parameter. For low values of impact parameters, (7.36), (7.37) become invalid; however, the region of small impact parameters with $\rho < \rho_{\mathrm{W}}$ where ρ_{W} is the Weisskopf radius (Weisskopf 1933; Kogan et al. 1973; Lisitsa 1977; Sobelman and Vainshtein 2006) gives a comparatively small contribution. It is therefore possible to approximate the integral over the impact parameters in (7.26) according to

$$\int_0^{\infty} 2\pi\rho \mathrm{d}\rho \left[S_{\alpha'\alpha}^* \cdot S_{\beta'\beta} - \delta_{\alpha\alpha'}\delta_{\beta'\beta}\right] \simeq \pi\rho_{\mathrm{W}}^2 + \int_{\rho_{\mathrm{W}}}^{\infty} 2\pi\rho \mathrm{d}\rho \left[S_{\alpha'\alpha}^* \cdot S_{\beta'\beta} - \delta_{\alpha\alpha'}\delta_{\beta'\beta}\right] \tag{7.74}$$

Finally, it should be noted that the impact approximation can only be used if the duration of the collision fulfills the condition $\rho/\upsilon_i > \Delta\omega^{-1}$ where $\Delta\omega$ is the width of the level. Therefore, the upper limit should be chosen as the minimum value of the two values R_{D} and $\upsilon_i/\Delta\omega$. The integrals in (7.57) can be approximated by the following G-functions (Griem and Shen 1959; Griem et al. 1979; Smith and Hooper 1967):

$$\begin{aligned} \hat{\Phi}_{\alpha\beta}^{\alpha'\beta'} \approx\ & \delta_{\alpha'\alpha} \cdot \sum_{\beta''} \langle\beta'|\vec{r}_{\mathrm{b}}|\beta''\rangle\langle\beta''|\vec{r}_{\mathrm{b}}|\beta\rangle \cdot G(-\Delta\omega_{\alpha\beta''}) \\ & + \delta_{\beta'\beta} \cdot \sum_{\alpha''} \langle\alpha'|\vec{r}_{\mathrm{a}}|\alpha''\rangle^* \langle\alpha''|\vec{r}_{\mathrm{a}}|\alpha\rangle^* \cdot G(\Delta\omega_{\alpha''\beta}) \\ & - \langle\alpha'|\vec{r}_{\mathrm{a}}|\alpha\rangle^* \langle\beta'|\vec{r}_{\mathrm{b}}|\beta\rangle \cdot \left[G(\Delta\omega_{\alpha\beta'}) + G(-\Delta\omega_{\alpha'\beta})\right] \end{aligned} \tag{7.75}$$

with ($H_0^{(2)}$ is the Hankel function of second kind) (Smith and Hooper 1967)

$$G(\Delta\omega_{\alpha\beta}) = -n_e \cdot \frac{\pi e^4 n}{3\hbar^2} \cdot \sqrt{\frac{8\pi m_e}{kT_e}} \cdot \exp\left(\frac{\hbar\Delta\omega_{\alpha\beta}}{2kT_e}\right) \cdot H_0^{(2)}\left(-i\frac{\hbar\Delta\omega_{\alpha\beta}}{2kT_e}\right). \quad (7.76)$$

The G-function can be approximated by Griem et al. (1979)

$$G(\Delta\omega_{\alpha\beta}) = -n_e \cdot \frac{4\pi}{3} \cdot \sqrt{\frac{2m_e}{\pi kT_e}} \cdot \left(\frac{\hbar}{m_e}\right)^2 \cdot \left(C_n + \frac{1}{2} \cdot \int\limits_y^\infty \frac{e^{-x}}{x} dx\right), \quad (7.77)$$

$$y = \left(\frac{n^2}{2 \cdot Z}\right)^2 \cdot \frac{\hbar^2\Delta\omega_S^2 + \hbar^2\omega_p^2 + \hbar^2\Delta\omega^2}{Ry \cdot kT_e}, \quad (7.78)$$

$$\Delta\omega = \omega_{\alpha\beta} - \omega, \quad (7.79)$$

$$\omega_p = \sqrt{\frac{4\pi n_e^2}{m_e}}. \quad (7.80)$$

$\Delta\omega$ is the frequency separation from the unperturbed line, ω_p is the electron plasma frequency, $\Delta\omega_S$ is the Stark shift of the upper levels (i.e., $\Delta\omega_S = \omega_{\alpha\alpha'}(F) - \omega_{\alpha\alpha'}(F = 0)$), n is the principal quantum number of the upper level, $Ry = 13.6057$ eV, and Z is the charge of the radiating ion. C_n is a constant that accounts for the so-called strong collisions at small impact parameters (where the perturbation theory fails). This constant can roughly be approximated by

$$C_n \approx \frac{5}{2n} + 0.2. \quad (7.81)$$

The exponential integral of (7.77) can conveniently be approximated by

$$\int\limits_y^\infty \frac{e^{-x}}{x} dx \approx e^{-y} \cdot \ln\left[1 + \frac{e^{-\gamma} + 1.4y}{y \cdot (1 + 1.4y)}\right], \quad (7.82)$$

where $\gamma = 0.577216$ is the Euler constant.

For simple estimates [that include the approximations of (7.63)–(7.65)], the following approximations for the integrals in (7.72) can be employed (Sholin et al. 1973; Lisitsa 1977; Akhmedov et al. 1985; Anufrienko et al. 1993):

$$\hat{\Phi}_{\alpha\beta}^{\alpha'\beta'} \approx \tilde{G}(T_e, \rho_{\min}, \rho_{\max}) \cdot \left[\delta_{\alpha'\alpha} \cdot \sum_{\beta''} \langle \beta' | \vec{r}_b | \beta'' \rangle \langle \beta'' | \vec{r}_b | \beta \rangle \right. \\ \left. + \delta_{\beta'\beta} \cdot \sum_{\alpha''} \langle \alpha' | \vec{r}_a | \alpha'' \rangle^* \langle \alpha'' | \vec{r}_a | \alpha \rangle^* - 2 \cdot \langle \alpha' | \vec{r}_a | \alpha \rangle^* \langle \beta' | \vec{r}_b | \beta \rangle \right], \tag{7.83}$$

$$\tilde{G}(T_e, \rho_{\min}, \rho_{\max}) = -n_e \cdot \frac{4\pi}{3} \cdot \left(\frac{e^2}{\hbar} \right)^2 \cdot \left\langle \frac{1}{\upsilon} \right\rangle \cdot \Lambda, \tag{7.84}$$

$$\left\langle \frac{1}{\upsilon} \right\rangle = \frac{4}{\pi \langle \upsilon \rangle} = \frac{4}{\pi} \cdot \sqrt{\frac{\pi m_e}{8kT_e}}, \tag{7.85}$$

$$\Lambda \approx \ln\left(\frac{\rho_{\max}}{\rho_{\min}} \right) + 0.215 \tag{7.86}$$

with $\rho_{\max} = R_D$, $\rho_{\min} = \max\{\rho_W, h/m\upsilon, Ze^2/m\upsilon^2, a_n\}$ where $\lambda_D = h/m_e\upsilon$ is the electron De Broglie wavelength, h is the Planck constant, $a_n = n^2 a_0/Z$ is the generalized Bohr radius, and

$$\rho_W = (\alpha_x C_x/\upsilon)^{1/(x-1)} \tag{7.87}$$

is the Weisskopf radius. C_x is the Stark constant: For $x = 2$, we encounter the linear Stark effect while for the quadratic Stark effect $x = 4$. The Stark constant can in some cases be approximated by the reduced matrix elements $\langle \alpha \| r^\kappa \| \beta \rangle$ that can easily be obtained from atomic structure calculations. For the quadratic Stark effect, we have

$$C_4 \approx \frac{1}{2J_\alpha + 1} \cdot \frac{\left| \langle \gamma_\alpha J_\alpha \| r \| \gamma_\beta J_\beta \rangle \right|^2}{\omega_{\alpha\beta}} \tag{7.88}$$

(note that in this expression the reduced dipole matrix element and the transition frequency $\omega_{\alpha\beta}$ are to be employed in atomic units). The constants α_x can be expressed in terms of the Γ-function:

$$\alpha_x = \sqrt{\pi} \cdot \Gamma[(x-1)/2]/\Gamma(x/2) \tag{7.89}$$

(note that $\alpha_{x=2} = \pi$ and $\alpha_{x=4} = \pi/2$). The velocity can be approximated by the mean velocity of the Maxwellian average, i.e.,

$$\upsilon \approx \langle \upsilon \rangle = \sqrt{8kT_e/\pi m_e}. \tag{7.90}$$

We note that improvements of the collisional operator concern not only use of the exact integrals for the A- and B-functions of (7.59)–(7.62) and higher-order terms in the series development of the scattering matrix (7.28) but in particular also the inclusion of the monopole and quadrupole terms in the interaction potential (7.30) itself (Deutsch et al. 1969a, b; Griem 1974).

7.4.3 Reduced Matrix Element Representation

The matrix elements of the radius vectors of (7.67)–(7.69) that enter into the expression for the collisional operator $\hat{\Phi}$ are very inconvenient for numerical evaluation of the matrix elements $\Phi_{\alpha\beta}^{\alpha'\beta'}$. They can, however, conveniently be transformed with the help of irreducible tensor operators and the Wigner-Eckhart theorem (Cowan 1981). The radius vector matrix elements

$$r_{\alpha\beta} = a_0 \langle \alpha | \vec{r} | \beta \rangle \tag{7.91}$$

can be considered as a tensor operator $\hat{T}_{\mathrm{q}}^{(k)}$ of rank one, i.e.,

$$r_{\alpha\beta} = a_0 \cdot \sum_q \langle \alpha | r_{\mathrm{q}}^{(1)} | \beta \rangle := \frac{d_{\alpha\beta}}{e}, \tag{7.92}$$

where $d_{\alpha\beta}$ is the dipole matrix element. The radius vector matrix element can be then transformed with the help of the Wigner–Eckart theorem according to (r is in atomic units, i.e., in units of the Bohr radius a_0)

$$\begin{aligned} \langle \alpha | r_{\mathrm{q}}^{(1)} | \beta \rangle &= \langle \gamma_\alpha J_\alpha M_\alpha | r_{\mathrm{q}}^{(1)} | \gamma_\beta J_\beta M_\beta \rangle \\ &= (-1)^{J_\alpha - M_\alpha} \begin{pmatrix} J_\alpha & 1 & J_\beta \\ -M_\alpha & q & M_\beta \end{pmatrix} \langle \gamma_\alpha J_\alpha \| r \| \gamma_\beta J_\beta \rangle . \end{aligned} \tag{7.93}$$

The state $\langle \alpha |$ is described by a set of quantum numbers $\gamma_\alpha J_\alpha M_\alpha$ where γ_α is a short notation of all quantum numbers except J_α and except M_α. $\begin{pmatrix} J_\alpha & 1 & J_\beta \\ -M_\alpha & q & M_\beta \end{pmatrix}$ is the 3j symbol that vanishes unless the relations $\Delta J = J_\alpha - J_\beta = 0, \pm 1$ and $-M_\alpha + q + M_\beta = 0$ are not fulfilled (with the restriction that $J_\alpha = J_\beta = 0$ is not allowed). The q-parameter specifies linear ($q = 0$) and circular ($q = \pm 1$) polarization. $\langle \gamma_\alpha J_\alpha \| r \| \gamma_\beta J_\beta \rangle$ is the reduced matrix element that does not depend anymore on the magnetic quantum numbers. It is very convenient to use reduced matrix elements in atomic units because atomic structure codes almost exclusively provide these matrix elements in atomic units.

Let us begin with the transformation of the matrix element $\phi^{(M),\alpha'\beta'}_{(1),'\alpha\beta} = \delta_{\alpha'\alpha} \cdot \langle\beta'|\vec{r}_\mathrm{b}|\beta''\rangle\langle\beta''|\vec{r}_\mathrm{b}|\beta\rangle$ from (7.67). Direct application of the Wigner–Eckart theorem according to (7.93) would be very inconvenient, as it would result in a product of a sum over different q-terms. However, the matrix elements of $(\vec{r}\cdot\vec{r})$ can be conveniently evaluated considering the product of the radius vectors as a product of tensor operators of rank one that transforms according to

$$\langle\alpha|\vec{r}\cdot\vec{r}|\beta\rangle = \sum_q (-1)^q \cdot \langle\alpha|r^{(1)}_\mathrm{q} \cdot r^{(1)}_{-\mathrm{q}}|\beta\rangle. \tag{7.94}$$

Therefore (see also (7.67))

$$\begin{aligned}
\phi_1 &= \delta_{\alpha'\alpha} \cdot \langle\beta'|\vec{r}_\mathrm{b}\cdot\vec{r}_\mathrm{b}|\beta\rangle \\
&= \delta_{\alpha'\alpha} \cdot \sum_q (-1)^q \cdot \left\langle \beta' \left\| r^{(1)}_{\mathrm{b,q}} \cdot r^{(1)}_{\mathrm{b,-q}} \right\| \beta \right\rangle \\
&= \delta_{\alpha'\alpha} \cdot \sum_{q,\beta''} (-1)^{q+J_{\beta'}-M_{\beta'}+J_{\beta''}-M_{\beta''}} \cdot \langle\gamma_{\beta'}J_{\beta'}\|r_\mathrm{b}\|\gamma_{\beta''}J_{\beta''}\rangle \cdot \langle\gamma_{\beta''}J_{\beta''}\|r_\mathrm{b}\|\gamma_\beta J_\beta\rangle \\
&\quad \times \begin{pmatrix} J_{\beta'} & 1 & J_{\beta''} \\ -M_{\beta'} & q & M_{\beta''} \end{pmatrix} \cdot \begin{pmatrix} J_{\beta''} & 1 & J_\beta \\ -M_{\beta''} & -q & M_\beta \end{pmatrix}; \\
\phi_2 &= \delta_{\beta'\beta} \cdot \langle\alpha'|\vec{r}_\mathrm{a}\cdot\vec{r}_\mathrm{a}|\alpha\rangle \\
&= \delta_{\alpha'\alpha} \cdot \sum_q (-1)^q \cdot \left\langle \alpha' \left\| r^{(1)}_{\mathrm{a,q}} \cdot r^{(1)}_{\mathrm{a,-q}} \right\| \alpha \right\rangle \\
&= \delta_{\alpha'\alpha} \cdot \sum_{q,\alpha''} (-1)^{q+J_{\alpha'}-M_{\alpha'}+J_{\alpha''}-M_{\alpha''}} \cdot \langle\gamma_{\alpha'}J_{\alpha'}\|r_\mathrm{a}\|\gamma_{\alpha''}J_{\alpha''}\rangle \cdot \langle\gamma_{\alpha''}J_{\alpha''}\|r_\mathrm{a}\|\gamma_\alpha J_\alpha\rangle \\
&\quad \times \begin{pmatrix} J_{\alpha'} & 1 & J_{\alpha''} \\ -M_{\alpha'} & q & M_{\alpha''} \end{pmatrix} \cdot \begin{pmatrix} J_{\alpha''} & 1 & J_\alpha \\ -M_{\alpha''} & -q & M_\alpha \end{pmatrix}
\end{aligned} \tag{7.95}$$

and with the help of the symmetry properties of the 3j symbols

$$\begin{pmatrix} J_1 & J_2 & J \\ M_1 & M_2 & M \end{pmatrix} = \begin{pmatrix} J & J_1 & J_2 \\ M & M_1 & M_2 \end{pmatrix} = (-1)^{J+J_1+J_2} \begin{pmatrix} J & J_2 & J_1 \\ M & M_2 & M_1 \end{pmatrix} \tag{7.96}$$

and

$$\begin{pmatrix} J_1 & J_2 & J \\ M_1 & M_2 & M \end{pmatrix} = (-1)^{J+J_1+J_2} \begin{pmatrix} J_1 & J_2 & J \\ -M_1 & -M_2 & -M \end{pmatrix} \tag{7.97}$$

it follows

$$\begin{aligned}&\begin{pmatrix} J_{\beta'} & 1 & J_{\beta''} \\ -M_{\beta'} & q & M_{\beta''} \end{pmatrix} \cdot \begin{pmatrix} J_{\beta''} & 1 & J_{\beta} \\ -M_{\beta''} & -q & M_{\beta} \end{pmatrix} \\ &= (-1)^{2J_{\beta'}+2J_{\beta''}+2} \begin{pmatrix} J_{\beta''} & 1 & J_{\beta'} \\ -M_{\beta''} & -q & M_{\beta'} \end{pmatrix} \cdot \begin{pmatrix} J_{\beta''} & 1 & J_{\beta} \\ -M_{\beta''} & -q & M_{\beta} \end{pmatrix} \\ &= \begin{pmatrix} J_{\beta''} & 1 & J_{\beta'} \\ -M_{\beta''} & -q & M_{\beta'} \end{pmatrix} \cdot \begin{pmatrix} J_{\beta''} & 1 & J_{\beta} \\ -M_{\beta''} & -q & M_{\beta} \end{pmatrix}.\end{aligned} \tag{7.98}$$

The last relation in (7.98) follows from the fact that $J_{\beta'} + 1 + J_{\beta''}$ is integral. Because

$$\langle \gamma_{\beta''} J_{\beta''} \| r_b \| \gamma_\beta J_\beta \rangle = (-1)^{J_{\beta''}-J_\beta} . \langle \gamma_\beta J_\beta \| r_b \| \gamma_{\beta''} J_{\beta''} \rangle^* \tag{7.99}$$

we obtain with (7.96), (7.97), (7.98) from (7.95)

$$\begin{aligned}\phi_1 = \delta_{\alpha'\alpha} \cdot &\sum_{\gamma_{\beta''}, J_{\beta''}, q, M_{\beta''}} (-1)^{q+J_{\beta'}-M_{\beta'}+J_{\beta''}-M_{\beta''}+J_{\beta''}-J_\beta} \\ &\times \langle \gamma_\beta J_\beta \| r_b \| \gamma_{\beta''} J_{\beta''} \rangle^* \cdot \langle \gamma_{\beta'} J_{\beta'} \| r_b \| \gamma_{\beta''} J_{\beta''} \rangle \\ &\times \begin{pmatrix} J_{\beta''} & 1 & J_{\beta'} \\ -M_{\beta''} & -q & M_{\beta'} \end{pmatrix} \cdot \begin{pmatrix} J_{\beta''} & 1 & J_{\beta} \\ -M_{\beta''} & -q & M_{\beta} \end{pmatrix}.\end{aligned} \tag{7.100}$$

Because $-M_{\beta''} - q + M_{\beta'} = 0$ and $2J_{\beta''} - 2M_{\beta''}$ is even the alternating sign of (7.100) transforms into

$$(-1)^{-M_{\beta''}+J_{\beta'}+J_{\beta''}-M_{\beta''}+J_{\beta''}-J_\beta} = (-1)^{2J_{\beta''}-2M_{\beta''}+J_{\beta'}-J_\beta} = (-1)^{J_{\beta'}-J_\beta}. \tag{7.101}$$

Equation (7.100) therefore transforms into

$$\begin{aligned}\phi_1 = \delta_{\alpha'\alpha} \cdot &\sum_{\gamma_{\beta''}, J_{\beta''}} (-1)^{J_{\beta'}-J_\beta} . \langle \gamma_\beta J_\beta \| r_b \| \gamma_{\beta''} J_{\beta''} \rangle^* \cdot \langle \gamma_{\beta'} J_{\beta'} \| r_b \| \gamma_{\beta''} J_{\beta''} \rangle \\ &\times \sum_{q, M_{\beta''}} \begin{pmatrix} J_{\beta''} & 1 & J_{\beta'} \\ -M_{\beta''} & -q & M_{\beta'} \end{pmatrix} \cdot \begin{pmatrix} J_{\beta''} & 1 & J_{\beta} \\ -M_{\beta''} & -q & M_{\beta} \end{pmatrix}.\end{aligned} \tag{7.102}$$

The last sum in (7.102) can be evaluated with the 3j symbol sum rule because the alternating sign is independent of the summation indexes:

$$\sum_{M_1, M_2} \begin{pmatrix} J_1 & J_2 & J \\ M_1 & M_2 & M \end{pmatrix} \begin{pmatrix} J_1 & J_2 & J' \\ M_1 & M_2 & M' \end{pmatrix} = \frac{\delta_{JJ'} \cdot \delta_{MM'} \cdot \delta(J_1, J_2, J)}{2J+1}, \tag{7.103}$$

where $\vec{J} = \vec{J}_1 + \vec{J}_2$ (i.e., $J = J_1 + J_2, J_1 + J_2 - 1, \ldots, |J_1 - J_2|$). The delta function $\delta(J_1, J_2, J)$ indicates the fulfillment of the triangular relation, i.e., $\delta(J_1, J_2, J) = 1$ if the triangular relation is satisfied and $\delta(J_1 J_2 J) = 0$ otherwise. Therefore

$$\sum_{q,M_{\beta''}} \begin{pmatrix} J_{\beta''} & 1 & J_{\beta'} \\ -M_{\beta''} & -q & M_{\beta'} \end{pmatrix} \cdot \begin{pmatrix} J_{\beta''} & 1 & J_{\beta} \\ -M_{\beta''} & -q & M_{\beta} \end{pmatrix} = \frac{\delta_{J_{\beta'}J_\beta} \cdot \delta_{M_{\beta'}M_\beta} \cdot \delta\left(J_{\beta''}, 1, J_{\beta'}\right)}{2J_\beta + 1}. \tag{7.104}$$

With (7.102), (7.104) the first term of the collisional operator from (7.66) is then given by

$$\begin{aligned} \sum_{\beta''} \phi^{(M),\alpha'\beta'}_{(1),\alpha\beta} \cdot [A(z_1, z_2) + iB(z_1, z_2)] = {} & \frac{\delta_{\alpha'\alpha} \cdot \delta_{J_{\beta'}J_\beta} \cdot \delta_{M_{\beta'}M_\beta} \cdot \delta\left(J_{\beta''}, 1, J_\beta\right)}{2J_\beta + 1} \\ & \times \sum_{\gamma_{\beta''}, J_{\beta''}} \langle \gamma_\beta J_\beta \| r_\mathrm{b} \| \gamma_{\beta''} J_{\beta''} \rangle^* \cdot \langle \gamma_{\beta'} J_{\beta'} \| r_\mathrm{b} \| \gamma_{\beta''} J_{\beta''} \rangle \\ & \cdot [A(z_1, z_2) + iB(z_1, z_2)]. \end{aligned} \tag{7.105}$$

A similar transformation with the help of the Wigner–Eckart theorem and tensor operators for ϕ_2 from (7.95) provides

$$\begin{aligned} \sum_{\alpha''} \phi^{(M),\alpha'\beta'}_{(2),\alpha\beta} \cdot [A(-z_1', -z_2') + iB(-z_1', -z_2')] = {} & \frac{\delta_{\beta'\beta} \cdot \delta_{J_{\alpha'}J_\alpha} \cdot \delta_{M_{\alpha'}M_\alpha} \cdot \delta(J_{\alpha''}, 1, J_\alpha)}{2J_\alpha + 1} \\ & \times \sum_{\gamma_{\alpha''}, J_{\alpha''}} \langle \gamma_\alpha J_\alpha \| r_\mathrm{a} \| \gamma_{\alpha''} J_{\alpha''} \rangle \cdot \langle \gamma_{\alpha'} J_{\alpha'} \| r_\mathrm{a} \| \gamma_{\alpha''} J_{\alpha''} \rangle^* \\ & \cdot \left[A(-z_1', -z_2') + iB(-z_1', -z_2')\right]. \end{aligned} \tag{7.106}$$

The cross-term $\phi^{(M),\alpha'\beta'}_{(3),\alpha\beta}$ from (7.69) may be transformed with the help of the tensor product transformation (Sobelman 2006)

$$T^{(k)} \cdot U^{(k)} = \sum_q (-1)^q \cdot T^{(k)}_\mathrm{q} \cdot U^{(k)}_{-\mathrm{q}}. \tag{7.107}$$

We can transform the term $\langle \alpha' | \vec{r}_\mathrm{a} | \alpha \rangle^* \langle \beta' | \vec{r}_\mathrm{b} | \beta \rangle$ with the help of (7.107) if the tensors $T^{(k)}$ and $U^{(k)}$ operate in the same space. According to (7.23)–(7.25), the vectors $|\alpha'\rangle$ and $|\beta'\rangle$ are introduced as a complete orthonormal functional set, i.e., $\hat{1} = \sum |\alpha'\rangle\langle\alpha'|$ and $\hat{1} = \sum |\beta'\rangle\langle\beta'|$, while the same holds true for the vectors $|\alpha\rangle$ and $|\beta\rangle$ itself. Therefore (note that $\langle \alpha' | \vec{r}_\mathrm{a} | \alpha \rangle^* = \langle \alpha | \vec{r}_\mathrm{a} | \alpha' \rangle$)

$$
\begin{aligned}
\langle\alpha'|\vec{r}_{\mathrm{a}}|\alpha\rangle^{*}\langle\beta'|\vec{r}_{\mathrm{b}}|\beta\rangle &= \langle\alpha\beta'|\vec{r}_{\mathrm{a}}\cdot\vec{r}_{\mathrm{b}}|\alpha'\beta\rangle \\
&= \sum_{q}(-1)^{q}\cdot\langle\alpha\beta'|r^{(1)}_{\mathrm{a,q}}\cdot r^{(1)}_{\mathrm{b,-q}}|\alpha'\beta\rangle \\
&= \sum_{\alpha'',\beta''}\sum_{q}(-1)^{q}\cdot\langle\alpha\beta'|r^{(1)}_{\mathrm{a,q}}|\alpha''\beta''\rangle\cdot\langle\alpha''\beta''|r^{(1)}_{\mathrm{b,-q}}|\alpha'\beta\rangle.
\end{aligned}
\tag{7.108}
$$

We can carry out the integration of the first matrix element in (7.108) with respect to the vectors $|\alpha'\rangle$ and $|\alpha''\rangle$. After integration, the radial dependence vanishes and provides finally the term $\langle\alpha\beta'|r^{(1)}_{\mathrm{a,q}}|\alpha''\beta''\rangle = \langle\alpha|r^{(1)}_{\mathrm{a,q}}|\alpha''\rangle\langle\beta' \mid \beta''\rangle$. Likewise we can carry out the first integration for the second matrix element with respect to the vectors $|\beta''\rangle$ and $|\beta\rangle$ providing $\langle\alpha''\beta''|r^{(1)}_{\mathrm{b,q}}|\alpha'\beta\rangle = \langle\beta''|r^{(1)}_{\mathrm{b,q}}|\beta\rangle\langle\alpha'' \mid \alpha'\rangle$. Equation (7.108) therefore takes the form

$$
\langle\alpha|\vec{r}_{\mathrm{a}}|\alpha'\rangle\langle\beta'|\vec{r}_{\mathrm{b}}|\beta\rangle = \delta_{\beta'\beta''}\cdot\delta_{\alpha''\alpha'}\sum_{\alpha'',\beta''}\sum_{q}(-1)^{q}\cdot\langle\alpha|r^{(1)}_{\mathrm{a,q}}|\alpha''\rangle\cdot\langle\beta''|r^{(1)}_{\mathrm{b,q}}|\beta\rangle. \tag{7.109}
$$

We apply now the Wigner–Eckart theorem for each matrix element:

$$
\begin{aligned}
\langle\alpha|\vec{r}_{\mathrm{a}}|\alpha'\rangle\langle\beta'|\vec{r}_{\mathrm{b}}|\beta\rangle &= \delta_{\gamma_{\beta'},\gamma_{\beta''}}\cdot\delta_{J_{\beta'},J_{\beta''}}\cdot\delta_{M_{\beta'},M_{\beta''}}\cdot\delta_{\gamma_{\alpha''},\gamma_{\alpha'}}\cdot\delta_{J_{\alpha''},J_{\alpha'}}\cdot\delta_{M_{\alpha''},M_{\alpha'}} \\
&\times\sum_{\gamma_{\alpha''},J_{\alpha''},M_{\alpha''}}\;\sum_{\gamma_{\beta''},J_{\beta''},M_{\beta''}}\;\sum_{q}(-1)^{q}\cdot(-1)^{J_\alpha-M_\alpha+J_{\beta'}-M_{\beta'}} \\
&\times\langle\gamma_\alpha J_\alpha\|r_{\mathrm{a}}\|\gamma_{\alpha''}J_{\alpha''}\rangle\cdot\begin{pmatrix} J_\alpha & 1 & J_{\alpha''} \\ -M_\alpha & q & M_{\alpha''}\end{pmatrix} \\
&\times\langle\gamma_{\beta''}J_{\beta''}\|r_{\mathrm{b}}\|\gamma_\beta J_\beta\rangle\cdot\begin{pmatrix} J_{\beta''} & 1 & J_\beta \\ -M_{\beta''} & -q & M_\beta\end{pmatrix}.
\end{aligned}
\tag{7.110}
$$

Due to the δ-functions and the fact that $2\cdot\left(J_{\beta''}+1+J_{\beta'}\right)$ is even, (7.110) can be simplified:

$$
\begin{aligned}
\langle\alpha|\vec{r}_{\mathrm{a}}|\alpha'\rangle\langle\beta'|\vec{r}_{\mathrm{b}}|\beta\rangle &= \sum_{q}(-1)^{q}\cdot(-1)^{J_\alpha-M_\alpha+J_{\beta'}-M_{\beta'}+J_\alpha-J_{\alpha'}} \\
&\times\langle\gamma_{\alpha'}J_{\alpha'}\|r_{\mathrm{a}}\|\gamma_\alpha J_\alpha\rangle^{*}\cdot\begin{pmatrix} J_\alpha & 1 & J_{\alpha'} \\ -M_\alpha & q & M_{\alpha'}\end{pmatrix} \\
&\times\langle\gamma_{\beta'}J_{\beta'}\|r_{\mathrm{b}}\|\gamma_\beta J_\beta\rangle\cdot\begin{pmatrix} J_\beta & 1 & J_{\beta'} \\ -M_\beta & q & M_{\beta'}\end{pmatrix}.
\end{aligned}
\tag{7.111}
$$

Therefore, we obtain finally for the cross-term $\phi^{(M),\alpha'\beta'}_{(3),\alpha\beta}$ (with -M$_\alpha$ + q + M$_{\alpha'}$ = 0):

$$\phi_3 = \phi^{(M),\alpha'\beta'}_{(3),\alpha\beta} = 2 \cdot \sum_q (-1)^{2J_\alpha - M_{\alpha'} - M_{\beta'} + J_{\beta'} - J_{\alpha'}} \cdot \langle \gamma_{\alpha'} J_{\alpha'} \| r_{\mathrm{a}} \| \gamma_\alpha J_\alpha \rangle^* \cdot \langle \gamma_{\beta'} J_{\beta'} \| r_{\mathrm{b}} \| \gamma_\beta J_\beta \rangle$$
$$\times \begin{pmatrix} J_\alpha & 1 & J_{\alpha'} \\ -M_\alpha & q & M_{\alpha'} \end{pmatrix} \cdot \begin{pmatrix} J_\beta & 1 & J_{\beta'} \\ -M_\beta & q & M_{\beta'} \end{pmatrix}. \tag{7.112}$$

7.4.4 *Symmetry Properties*

The interest in symmetry properties is related to the principle of microreversibility and the principle of detailed balance which are fundamental conditions that systems can reach the thermodynamic limit. We are therefore in particular interested in the symmetry properties of the electron collisional operator because this operator is the "driving force" in the system of density matrix equations to reach the thermodynamic limit. Although general symmetry properties exist for quantum mechanical operators, it is of practical importance to verify the final expressions explicitly in order to clarify if certain important symmetry properties are maintained in the course of the various approximations made (second-order perturbation theory, dipole approximation, etc.).

There are no general symmetry properties related to the collisional operator matrix elements $S_{\alpha\beta}$ from (7.11) because it explicitly contains all the populations $\tilde{\rho}_{\alpha'\beta'}$ and energy differences. We are therefore interested in the symmetry properties of some particular matrix elements $\phi_1; \phi_2; \phi_3; \Phi^{\alpha'\beta'}_{\alpha\beta}$ that follow directly from the symmetry properties of the functions $\phi^{(M),\alpha'\beta'}_{(1),\alpha\beta}$, $\phi^{(M),\alpha'\beta'}_{(2),\alpha\beta}$ and $\phi^{(M),\alpha'\beta'}_{(3),\alpha\beta}$ from (7.67)–(7.69), (7.112). For $\alpha' = \beta' = \alpha$ we have $\phi^{(M),\alpha\alpha}_{(3),\alpha\beta} = 0$ while $\phi^{(M),\alpha\alpha}_{(1),\alpha\beta}$ and $\phi^{(M),\alpha\alpha}_{(2),\alpha\beta}$ are symmetric to each other as can be directly seen from (7.95), i.e.,

$$\phi_1 = \delta_{\alpha'\alpha} \cdot \langle \beta' | \vec{r}_{\mathrm{b}} \cdot \vec{r}_{\mathrm{b}} | \beta \rangle \tag{7.113}$$

and correspondingly

$$\phi_2 = \delta_{\beta'\beta} \cdot \langle \alpha' | \vec{r}_{\mathrm{a}} \cdot \vec{r}_{\mathrm{a}} | \alpha \rangle \tag{7.114}$$

Therefore,

$$\sum_{\beta''} \phi^{(M),\alpha\alpha}_{(1),\alpha\beta} + \sum_{\alpha''} \phi^{(M),\alpha\alpha}_{(2),\alpha\beta} - \phi^{(M),\alpha\alpha}_{(3),\alpha\beta} = \sum_{\beta''} \phi^{(M),\alpha\alpha}_{(1),\beta\alpha} + \sum_{\alpha''} \phi^{(M),\alpha\alpha}_{(2),\beta\alpha} - \phi^{(M),\alpha\alpha}_{(3),\beta\alpha} \tag{7.115}$$

and likewise

$$\sum_{\beta''}\phi^{(M),\alpha\beta}_{(1),\alpha\alpha}+\sum_{\alpha''}\phi^{(M),\alpha\beta}_{(2),\alpha\alpha}-\phi^{(M),\alpha\beta}_{(3),\alpha\alpha}=\sum_{\beta''}\phi^{(M),\beta\alpha}_{(1),\alpha\alpha}+\sum_{\alpha''}\phi^{(M),\beta\alpha}_{(2),\alpha\alpha}-\phi^{(M),\beta\alpha}_{(3),\alpha\alpha}. \tag{7.116}$$

For $\alpha'=\alpha$ and $\beta'=\beta$ we have $\phi^{(M),\alpha\beta}_{(3),\alpha\beta}=0$ while $\phi^{(M),\alpha\beta}_{(1),\alpha\beta}$ and $\phi^{(M),\alpha\beta}_{(2),\alpha\beta}$ are also symmetric to each other, i.e.,

$$\sum_{\beta''}\phi^{(M),\alpha\beta}_{(1),\alpha\beta}+\sum_{\alpha''}\phi^{(M),\alpha\beta}_{(2),\alpha\beta}-\phi^{(M),\alpha\beta}_{(3),\alpha\beta}=\sum_{\beta''}\phi^{(M),\beta\alpha}_{(1),\beta\alpha}+\sum_{\alpha''}\phi^{(M),\beta\alpha}_{(2),\beta\alpha}-\phi^{(M),\beta\alpha}_{(3),\beta\alpha}. \tag{7.117}$$

For $\alpha'=\beta$ and $\beta'=\alpha$ we have $\phi^{(M),\beta\alpha}_{(1),\alpha\beta}=0$ and $\phi^{(M),\beta\alpha}_{(2),\alpha\beta}=0$. If we apply (7.99) two times, we obtain

$$\langle\gamma_\beta J_\beta\|r\|\gamma_\alpha J_\alpha\rangle^*\cdot\langle\gamma_\alpha J_\alpha\|r\|\gamma_\beta J_\beta\rangle=\langle\gamma_\alpha J_\alpha\|r\|\gamma_\beta J_\beta\rangle\cdot\langle\gamma_\beta J_\beta\|r\|\gamma_\alpha J_\alpha\rangle^* \tag{7.118}$$

from which it follows

$$\sum_{\beta''}\phi^{(M),\beta\alpha}_{(1),\alpha\beta}+\sum_{\alpha''}\phi^{(M),\beta\alpha}_{(2),\alpha\beta}-\phi^{(M),\beta\alpha}_{(3),\alpha\beta}=\sum_{\beta''}\phi^{(M),\alpha\beta}_{(1),\beta\alpha}+\sum_{\alpha''}\phi^{(M),\alpha\beta}_{(2),\beta\alpha}-\phi^{(M),\alpha\beta^*}_{(3),\beta\alpha}. \tag{7.119}$$

For $\beta=\alpha$ and $\alpha'=\beta'=\beta$ we have $\phi^{(M),\beta\beta}_{(1),\alpha\alpha}=0$ and $\phi^{(M),\beta\beta}_{(2),\alpha\alpha}=0$. Because of the selection rule $-2M_\alpha=-2M_\beta+2q$ and the fact that $2J_\alpha-2M_\alpha$ and $2q$ are even, we find

$$\sum_{\beta''}\phi^{(M),\beta\beta}_{(1),\alpha\alpha}+\sum_{\alpha''}\phi^{(M),\beta\beta}_{(2),\alpha\alpha}-\phi^{(M),\beta\beta}_{(3),\alpha\alpha}=\sum_{\beta''}\phi^{(M),\alpha\alpha}_{(1),\beta\beta}+\sum_{\alpha''}\phi^{(M),\alpha\alpha}_{(2),\beta\beta}-\phi^{(M),\alpha\alpha}_{(3),\beta\beta}. \tag{7.120}$$

The symmetry relations for $\left\{\sum_{\beta''}\phi^{(M),\alpha'\beta'}_{(1),\alpha\beta}+\sum_{\alpha''}\phi^{(M),\alpha'\beta'}_{(2),\alpha\beta}-\phi^{(M),\alpha'\beta'}_{(3),\alpha\beta}\right\}$ do in general not coincide with the symmetry relations for $\Phi^{\alpha'\beta'}_{\alpha\beta}$ because the matrix elements $\Phi^{\alpha'\beta'}_{\alpha\beta}$ contain the integral over the electron distribution function and different energies. Therefore, e.g., different lower threshold energies have to be taken into account when exchanging the indexes. This circumstance corresponds to the physical phenomena that in thermodynamic equilibrium the direct and inverse rates are related to each other via the Boltzmann factor and not only via the statistical weights (note that in the high temperature limit—or Born limit—the Boltzmann factor vanishes and the differences in energies, i.e z-terms in (7.49)–(7.54) vanish too).

We note that for the case $\alpha' = \alpha$ and $\beta' = \beta$, and $A = 1, B = 0$ we encounter also symmetric integrals because $\phi^{(M),\alpha\beta}_{(3),\alpha\beta} = 0$ while $\phi^{(M),\alpha\beta}_{(1),\alpha\beta}$ and $\phi^{(M),\alpha\beta}_{(2),\alpha\beta}$ enter only as a sum in $\Phi^{\alpha\beta}_{\alpha\beta}$. Therefore (for $A = 1$ and $B = 0$)

$$\Phi^{\alpha\beta}_{\alpha\beta} = \Phi^{\beta\alpha}_{\beta\alpha}. \tag{7.121}$$

A similar circumstance is encountered for $\alpha' = \beta' = \alpha$. Therefore (for $A = 1$ and $B = 0$)

$$\Phi^{\alpha\alpha}_{\alpha\beta} = \Phi^{\alpha\alpha}_{\beta\alpha}. \tag{7.122}$$

7.5 Matrix Elements and Atomic Physics Processes

7.5.1 Line Strengths and Oscillator Strengths

With the help of the Wigner–Eckart theorem, the M-quantum average over the relaxation constants can be easily obtained from the reduced matrix elements of atomic structure calculations. They can be expressed in terms of the dipole strength, the squared dipole matrix element, the spontaneous transition probability, emission and absorption oscillator strengths or directly in terms of cross sections and rate coefficients. The dipole line strength (in atomic units, i.e., in units of $e^2a_0^2$) is defined by

$$S_{J_\alpha J_\beta} = \left|\langle \gamma_\alpha J_\alpha \| r \| \gamma_\beta J_\beta \rangle\right|^2, \tag{7.123}$$

i.e., the line strength is symmetric

$$S_{J_\alpha J_\beta} = S_{J_\beta J_\alpha}. \tag{7.124}$$

Reduced matrix elements and line strengths are given in atomic units because numerical atomic structure calculations provide these quantities almost exclusively in atomic units. Note that $a_0 = \hbar^2/m_e e^2, Ry = m_e e^4/2\hbar^2$ and with $\hbar = 1.054572 \times 10^{-27}$ erg s, $e = 4.803204 \times 10^{-10}$ esu, $m_e = 9.109383 \times 10^{-28}\,g$ we obtain $a_0 = 5.291772 \times 10^{-9}$ cm, $Ry = 2.179870 \times 10^{-11}$ erg, $1\,\text{eV} = 1.602176 \times 10^{-12}$ erg in cgs-units while $Ry = 13.605693$ eV. The spontaneous transition probability from $|\gamma_\alpha J_\alpha M_\alpha\rangle \to |\gamma_\beta J_\beta M_\beta\rangle$ is given by (see also (7.92))

$$A_{\alpha\beta} = \frac{4e^2 a_0^2 \omega_{\alpha\beta}^3}{3c^3\hbar} \sum_q \left| \langle \gamma_\alpha J_\alpha M_\alpha | r_q^{(1)} | \gamma_\beta J_\beta M_\beta \rangle \right|^2 = \frac{4\omega_{\alpha\beta}^3}{3c^3\hbar} \left| d_{\alpha\beta} \right|^2. \tag{7.125}$$

With the help of the 3*j* symbol sum rule (assuming that the triangular relation is fulfilled, see also (7.103))

$$(2J_\beta + 1) \sum_{M_\alpha, q} \begin{pmatrix} J_\alpha & 1 & J_\beta \\ -M_\alpha & q & M_\beta \end{pmatrix}^2 = 1 \tag{7.126}$$

we obtain for the squared of the dipole matrix element summed over all final states M_β (see also (7.93) and (7.96))

$$\begin{aligned} \left| d_{\alpha J_\beta} \right|^2 &= \sum_{M_\beta} \left| d_{\alpha\beta} \right|^2 = e^2 a_0^2 \left| \langle \gamma_\alpha J_\alpha \| r \| \gamma_\beta J_\beta \rangle \right|^2 \sum_{M_\beta, q} \begin{pmatrix} J_\alpha & 1 & J_\beta \\ -M_\alpha & q & M_\beta \end{pmatrix}^2 \\ &= e^2 a_0^2 \cdot \frac{\left| \langle \gamma_\alpha J_\alpha \| r \| \gamma_\beta J_\beta \rangle \right|^2}{2J_\alpha + 1}. \end{aligned} \tag{7.127}$$

The line strength $S_{J_\alpha J_\beta}$ and absorption oscillator strength $f_{\alpha J_\beta}$ are related via

$$f_{\alpha J_\beta} = \frac{2m_e}{3\hbar} \cdot \omega_{\alpha\beta} \cdot \frac{a_0^2 \cdot \left| \langle \gamma_\alpha J_\alpha \| r \| \gamma_\beta J_\beta \rangle \right|^2}{(2J_\alpha + 1)} = \frac{2m_e a_0^2 \omega_{\alpha\beta}}{3\hbar} \cdot \frac{S_{J_\alpha J_\beta}(e^2 a_0^2)}{(2J_\alpha + 1)}, \tag{7.128}$$

where the argument $(e^2 a_0^2)$ indicates that the line strength is in atomic units (see also (7.123)). The oscillator strength $f_{\alpha J_\beta}$ is an absorption oscillator strength from a specific lower state $|\alpha\rangle = |\gamma_\alpha J_\alpha M_\alpha\rangle$ resolved in M_α to all upper states M_β, i.e., summed over all M_β. The corresponding expression for the spontaneous transition probability is likewise obtained from the sum over all final states M_β (note, that here the index α is the upper state and the index β is the lower state):

$$\begin{aligned} A_{\alpha J_\beta} &= \sum_{M_\beta} A_{\alpha\beta} = \frac{4\omega_{\alpha\beta}^3}{3c^3\hbar} \sum_{M_\beta} \left| d_{\alpha\beta} \right|^2 = \frac{4e^2 a_0^2 \omega_{\alpha\beta}^3}{3c^3\hbar} \cdot \frac{\left| \langle \gamma_\alpha J_\alpha \| r \| \gamma_\beta J_\beta \rangle \right|^2}{2J_\alpha + 1} \\ &= \frac{4e^2 a_0^2 \omega_{\alpha\beta}^3 \cdot S_{J_\alpha J_\beta}}{3c^3\hbar \cdot (2J_\alpha + 1)} \\ &= 1.063076 \times 10^6 \cdot \Delta E_{\alpha\beta}^3(\mathrm{eV}) \cdot \frac{S_{J_\alpha J_\beta}(e^2 a_0^2)}{(2J_\alpha + 1)} \left[\mathrm{s}^{-1} \right]. \end{aligned} \tag{7.129}$$

In the last expression: (easily verified in atomic units using the unit of time $2.418884 \cdot 10^{-17}$ sec and the number 137.036 for the speed of light) the constant is given for the transition energy in (eV) (positive: note that binding energies are negative; therefore, e.g., the transition energy of Ly_α is given by the indexes

$\alpha = 2p, \beta = 1s$ and $\Delta E_{\alpha\beta} = (-0.25 \cdot Ry) - (-1 \cdot Ry) = +0.75 \cdot Ry$ with $Ry = 13.605693\,\mathrm{eV}$), the line strengths in atomic units, and the transition probability in (s^{-1}). From (7.128), (7.129) it follows

$$f_{\beta J_\alpha} = 1.49922 \times 10^4 \cdot \lambda_{\beta\alpha}^2(m) \cdot A_{\beta J_\alpha}. \tag{7.130}$$

The constant in (7.130) is given for the emission oscillator strength $f_{\beta J_\alpha}$, transition wavelength $\lambda_{\beta\alpha}$ in (m), and the spontaneous transition probability $A_{\beta J_\alpha}$ in (s^{-1}).

In order to relate emission and absorption oscillator strength to each other as well as with corresponding spontaneous transition probabilities, an average over initial and final M-quantum numbers has to be performed:

$$\left|d_{J_\alpha J_\beta}\right|^2 = \frac{1}{2J_\alpha + 1} \cdot \sum_{M_\alpha, M_\beta} \left|d_{\alpha\beta}\right|^2 = e^2 a_0^2 \cdot \frac{\left|\langle \gamma_\alpha J_\alpha \| r \| \gamma_\beta J_\beta \rangle\right|^2}{2J_\alpha + 1}. \tag{7.131}$$

The last expression follows from the fact that according to (7.127) the dipole matrix element summed over final M-quantum numbers M_β does not depend any more on M_α: Therefore, the subsequent sum over the initial M-quantum numbers M_α provides $2J_\alpha + 1$ identical terms and one factor $1/(2J_\alpha + 1)$ cancels. For the transition probability, we obtain

$$A_{J_\alpha J_\beta} = \frac{1}{2J_\alpha + 1} \cdot \sum_{M_\alpha, M_\beta} A_{\alpha\beta} = \frac{4e^2 a_0^2}{3c^3 \hbar} \cdot \omega_{\alpha\beta}^3 \cdot \frac{S_{J_\alpha J_\beta}(e^2 a_0^2)}{2J_\alpha + 1}. \tag{7.132}$$

Equation (7.132) is therefore identical to (7.129). Likewise, we have

$$\begin{aligned} f_{J_\alpha J_\beta} &= \frac{1}{2J_\alpha + 1} \cdot \sum_{M_\alpha, M_\beta} f_{\alpha\beta} = \frac{2m_e a_0^2 \omega_{\alpha\beta}}{3\hbar} \cdot \frac{S_{J_\alpha J_\beta}(e^2 a_0^2)}{(2J_\alpha + 1)} \\ &= \frac{2m_e a_0^2 \omega_{\alpha\beta}}{3\hbar} \cdot \frac{\left|\langle \gamma_\alpha J_\alpha \| r \| \gamma_\beta J_\beta \rangle\right|^2}{(2J_\alpha + 1)}. \end{aligned} \tag{7.133}$$

Because the line strength is symmetric, the absorption oscillator strengths follow directly from (7.133) changing the initial and final states:

$$\begin{aligned} f_{J_\beta J_\alpha} &= \frac{1}{2J_\beta + 1} \cdot \sum_{M_\beta, M_\alpha} f_{\beta\alpha} = \frac{2m_e a_0^2 \omega_{\beta\alpha}}{3\hbar} \cdot \frac{S_{J_\beta J_\alpha}(e^2 a_0^2)}{(2J_\beta + 1)} \\ &= -\frac{2m_e a_0^2 \omega_{\alpha\beta}}{3\hbar} \cdot \frac{S_{J_\alpha J_\beta}(e^2 a_0^2)}{(2J_\beta + 1)} \\ &= -\frac{2m_e a_0^2 \omega_{\alpha\beta}}{3\hbar} \cdot \frac{\left|\langle \gamma_\alpha J_\alpha \| r \| \gamma_\beta J_\beta \rangle\right|^2}{(2J_\beta + 1)}. \end{aligned} \tag{7.134}$$

Emission and absorption oscillator strengths are therefore related via the expression

$$(2J_\alpha + 1) \cdot f_{J_\alpha J_\beta} = -(2J_\beta + 1) \cdot f_{J_\beta J_\alpha}. \tag{7.135}$$

Comparing (7.132) (exchanging the indexes α and β to account for the fact that α is the index of the upper state for the transition probability in (7.132) while α corresponds to the lower state for the absorption oscillator strengths in 7.133)) and (7.133), we find, with (7.135) and $\omega_{\alpha\beta} = -\omega_{\beta\alpha}$

$$-f_{J_\alpha J_\beta} = \frac{c^3 \cdot m_e}{2e^2 \cdot \omega_{\alpha\beta}^2} \cdot \frac{2J_\beta + 1}{2J_\alpha + 1} \cdot A_{J_\beta J_\alpha} = \frac{c \cdot m_e}{8\pi^2 e^2} \cdot \lambda_{\alpha\beta}^2 \cdot \frac{2J_\beta + 1}{2J_\alpha + 1} \cdot A_{J_\beta J_\alpha} \tag{7.136}$$

and

$$f_{J_\beta J_\alpha} = \frac{c^3 \cdot m_e}{2e^2 \cdot \omega_{\beta\alpha}^2} \cdot A_{J_\beta J_\alpha} = \frac{c \cdot m_e}{8\pi^2 e^2} \cdot \lambda_{\beta\alpha}^2 \cdot A_{J_\beta J_\alpha}. \tag{7.137}$$

In convenient units, (7.137) reads (compare with (7.130))

$$f_{J_\beta J_\alpha} = 1.49922 \times 10^4 \cdot \lambda_{\beta\alpha}^2(m) \cdot A_{J_\beta J_\alpha}. \tag{7.138}$$

With the above equations, all M-quantum averages for radius vector dipole matrix elements are specified.

The matrix elements are not only useful for the calculation of radiative transitions, but also for the calculation of collisional processes and the study of field effects; detailed studies of different methods of calculations have been performed (Rosmej et al. 2013). Note, that the detailed calculation of energies, wavelengths, oscillator strengths and reduced matrix elements including sign for HI and HeII are presented in Annexes A.2. and A.3.

7.5.2 Reduced Matrix Elements and Cross Sections

Let us consider the dipole allowed electron excitation cross section $\sigma(J \rightarrow J')$ in the Born limit. Because the Bessel function in the scattering matrix element can be represented via a dipole matrix element for high energies, the dipole allowed cross section can be expressed via reduced matrix elements or oscillator strengths $(f(J_\alpha \rightarrow J_\beta) = f_{J_\alpha J_\beta})$ too (Cowan 1981; Sobelman and Vainshtein 2006):

$$\begin{aligned}
\sigma(J_\alpha \rightarrow J_\beta) &= \frac{4e^4}{3\hbar^2} \cdot \int\limits_{\rho_{min}}^{\rho_{max}} 2\pi\rho d\rho \cdot \left[\frac{1}{\rho^2 \upsilon^2} \cdot \frac{\left|\langle \gamma_\alpha J_\alpha \| r \| \gamma_\beta J_\beta \rangle\right|^2}{2J_\alpha + 1}\right] \\
&= \frac{8\pi e^4}{3\hbar^2} \cdot \frac{1}{\upsilon^2} \cdot \left\{\frac{\left|\langle \gamma_\alpha J_\alpha \| r \| \gamma_\beta J_\beta \rangle\right|^2}{2J_\alpha + 1}\right\} \cdot g \\
&= 8\pi a_0^2 \cdot \frac{Ry}{E} \cdot \frac{Ry}{\Delta E_{\alpha\beta}} \cdot f(J_\alpha \rightarrow J_\beta) \cdot g,
\end{aligned} \tag{7.139}$$

where [see also (7.86)]

$$g = \ln\left(\frac{\rho_{\max}}{\rho_{\min}}\right). \tag{7.140}$$

The corresponding transition rate coefficient [in units of ($\mathrm{cm}^3\,\mathrm{s}^{-1}$)] averaged over a Maxwellian distribution function follows directly from (7.139), (7.140), i.e.,

$$\begin{aligned}\langle \upsilon\sigma(J_\alpha \to J_\beta)\rangle &= \int \upsilon\sigma(J_\alpha \to J_\beta) f(\upsilon)\mathrm{d}\upsilon \\ &= 16\sqrt{\pi}a_0^2\upsilon_0 \cdot f(J_\alpha \to J_\beta)\cdot\left(\frac{Ry}{\Delta E_{\alpha\beta}}\right)^{3/2}\cdot\sqrt{\beta}\cdot\exp\left(-\frac{\Delta E_{\alpha\beta}}{kT_e}\right)\cdot\bar{g},\end{aligned} \tag{7.141}$$

where

$$\bar{g} = \overline{\ln\left(\frac{\rho_{\max}}{\rho_{\min}}\right)}. \tag{7.142}$$

a_0 and υ_0 are the Bohr radius and velocity and the bar over the Coulomb logarithm indicates that maximum and minimum impact parameters are to be taken for the thermal velocity [for a more detailed discussion of the Coulomb logarithm the interesting reader is referred to Kogan et al. (1973)] with electron temperature kT_e in (eV), $Ry = 13.6057\,\mathrm{eV}$, $\Delta E_{\alpha\beta}$ is the transition energy (positive) in (eV) and

$$\beta = \frac{\Delta E_{\alpha\beta}}{kT_e}. \tag{7.143}$$

For closely spaced levels, the exponential factor in (7.141) is close to 1.

Numerous expressions for (7.142) have been proposed in the literature (Gaunt-factor discussion) that all practically differ not too much from each other (Sobelman and Vainshtein 2006). The following closed expression can be employed for somewhat more general estimates of dipole allowed transitions in ions:

$$\langle \upsilon\sigma(J_\alpha \to J_\beta)\rangle = 3.15\times 10^{-7}\cdot f(J_\alpha \to J_\beta)\left(\frac{Ry}{\Delta E_{\alpha\beta}}\right)^{3/2}\frac{e^{-\beta}}{\sqrt{\beta}}\,\bar{g}(\beta)\,\left[\mathrm{cm}^3\,\mathrm{s}^{-1}\right]. \tag{7.144}$$

With ($\gamma = 0.577216$ is the Euler constant)

$$\bar{g}(\beta) \approx 0.2757\cdot e^{-1.3\beta}\left(\beta - \frac{\beta^2}{4} - \ln(\beta) - \gamma\right) + 0.2\cdot\left(1 - e^{-4.5\beta}\right). \tag{7.145}$$

We note that the Gaunt-factor approximation of (7.145) is valid for the whole range of β-parameters, i.e., exceeds the application range of the pure Born limit.

7.6 Magnetic Quantum Number Averages

7.6.1 The Rate Equation Case

Although in the overwhelming number of applications M-quantum number resolved structure is usually employed, the set of equations (7.1) also allows an M-resolved description of the populations. In practice, the M-resolved descriptions are essentially limited to applications in polarization spectroscopy (Fujimoto and Imawae 2008; Degl'Ionnocenti and Landolfi 2004). As the M-quantum number averages in density matrix approach are rather complex due to the double index for quantum populations, let us first consider the general principles of M-quantum number averages in the rate equation approach, i.e., in the framework of the populations (7.1).

As before, we designate M-quantum-number-resolved levels with the greek indices $\alpha, \beta, \gamma, \ldots$ while M-quantum-number-averaged levels are designated with latin indices $a, b, c, \ldots$. The M-quantum-number-resolved set of (7.1) takes therefore the form

$$\frac{\mathrm{d}n_\alpha}{\mathrm{d}t} = -n_\alpha \sum_\beta W_{\alpha\beta} + \sum_\gamma n_\gamma W_{\gamma\alpha}, \tag{7.146}$$

while the corresponding M-quantum-number-averaged equation is given by

$$\frac{\mathrm{d}n_\mathrm{a}}{\mathrm{d}t} = -n_\mathrm{a} \sum_b W_\mathrm{ab} + \sum_g n_\mathrm{g} W_\mathrm{ga}. \tag{7.147}$$

The summation of (7.146) over M-quantum numbers provides

$$\sum_{M_\alpha} \left(\frac{\mathrm{d}n_\alpha}{\mathrm{d}t} \right) = -\sum_{M_\alpha} \left(n_\alpha \sum_\beta W_{\alpha\beta} \right) + \sum_{M_\alpha} \left(\sum_\gamma n_\gamma W_{\gamma\alpha} \right). \tag{7.148}$$

The left-hand side defines the M-quantum-number-summed populations, i.e.

$$\sum_{M_\alpha} \left(\frac{\mathrm{d}n_\alpha}{\mathrm{d}t} \right) = \frac{\mathrm{d}n_\mathrm{a}}{\mathrm{d}t}. \tag{7.149}$$

We note that at this point, (7.148) and (7.149) are still exact. However, in order that (7.147) are useful, no direct reference to the M-quantum-number-resolved set of populations (7.146) should be made; otherwise, there would be no useful simplification if the averaged populations are just calculated from (7.149).

In order to find a non-trivial solution to the set of population (7.147), additional approximations need to be involved. The most common one is the hypothesis of a statistical population among the M-quantum-number-resolved levels γJM (J is the total angular momentum and γ specifies all quantum numbers of a level except J and M), i.e.

$$n_{\alpha}^{(\mathrm{approx})} = \frac{1}{2J_{\alpha}+1} n_{\mathrm{a}} \tag{7.150}$$

(Note that usually no Boltzmann exponential factor is applied to (7.150) because the M-quantum levels are degenerated). Injecting (7.150) into (7.148) we obtain

$$\begin{aligned}
\sum_{M_\alpha}\left(\frac{\mathrm{d}n_\alpha}{\mathrm{d}t}\right) &= -\sum_{M_\alpha}\left(\frac{1}{2J_\alpha+1}n_{\mathrm{a}}\sum_{\beta}W_{\alpha\beta}\right)+\sum_{M_\alpha}\left(\sum_{\gamma}\frac{1}{2J_\gamma+1}n_{\mathrm{g}}W_{\gamma\alpha}\right)\\
&= -\sum_{M_\alpha}\left(\frac{1}{2J_\alpha+1}n_{\mathrm{a}}\sum_{J_\beta}\sum_{M_\beta}W_{\alpha\beta}\right)+\sum_{M_\alpha}\left(\sum_{J_\gamma}\sum_{M_\gamma}\frac{1}{2J_\gamma+1}n_{\mathrm{g}}W_{\gamma\alpha}\right)\\
&= -n_{\mathrm{a}}\sum_{J_\beta}\left\{\frac{1}{2J_\alpha+1}\sum_{M_\alpha}\sum_{M_\beta}W_{\alpha\beta}\right\}+\sum_{J_\gamma}n_{\mathrm{g}}\left\{\frac{1}{2J_\gamma+1}\sum_{M_\gamma}\sum_{M_\alpha}W_{\gamma\alpha}\right\},
\end{aligned} \tag{7.151}$$

where the summation over J_β and J_γ in the second and third line indicates the summation over all levels that are not resolved in M-quantum number. The comparison of (7.147) and (7.148) with (7.151) indicates that the M-quantum-number-averaged rates of the elementary atomic physics processes should be

$$W_{\mathrm{ab}} = \left\{\frac{1}{2J_\alpha+1}\sum_{M_\alpha}\sum_{M_\beta}W_{\alpha\beta}\right\}. \tag{7.152}$$

With the help of (7.152), (7.147) represents a non-trivial approximate solution of the M-quantum-number-averaged set of population equations, i.e.

$$\frac{\mathrm{d}n_{\mathrm{a}}^{(\mathrm{approx})}}{\mathrm{d}t} = -n_{\mathrm{a}}^{(\mathrm{approx})}\cdot\sum_{b}W_{\mathrm{ab}}+\sum_{g}n_{\mathrm{g}}^{(\mathrm{approx})}\cdot W_{\mathrm{ga}}. \tag{7.153}$$

7.6.2 Formal Solution for the Density Matrix Equations

Let us now investigate the M-quantum average of the general set of density matrix (7.9):

$$\frac{\mathrm{d}\tilde{\rho}_{\alpha\beta}}{\mathrm{d}t} = -i\omega_{\alpha\beta}\tilde{\rho}_{\alpha\beta} - \frac{i}{\hbar}\sum_{\gamma}\left(V_{\alpha\gamma}\tilde{\rho}_{\gamma\beta} - \tilde{\rho}_{\alpha\gamma}V_{\gamma\beta}\right) + \sum_{\alpha'\beta'}\Phi_{\alpha\beta}^{\alpha'\beta'}\cdot\tilde{\rho}_{\alpha'\beta'}$$
$$- \left(\frac{\gamma_\alpha}{2} + \frac{\gamma_\beta}{2}\right)\tilde{\rho}_{\alpha\beta} + R_{\alpha\beta} + A_{\alpha\beta} + Q_{\alpha\beta}. \tag{7.154}$$

Because $\sum_{M_\alpha}\sum_{M_\beta}\psi_{\gamma_\alpha J_\alpha M_\alpha}\cdot\psi^*_{\gamma_\beta J_\beta M_\beta} = \left(\sum_{M_\alpha}\psi_{\gamma_\alpha J_\alpha M_\alpha}\right)\cdot\left(\sum_{M_\beta}\psi^*_{\gamma_\beta J_\beta M_\beta}\right)$ the M-quantum-number-summed density matrix elements are given by

$$\tilde{\rho}_{\mathrm{ab}} := \sum_{M_\alpha,M_\beta}\tilde{\rho}_{\alpha\beta}. \tag{7.155}$$

The M-quantum-number-summed density matrix elements of (7.155) request the solution of the full M-quantum-number-resolved (7.154). The practical interest in the M-quantum-number-averaged equations is therefore related to a presentation that employs from the very beginning only averaged quantities; otherwise, the average would not provide any simplifications. We are therefore looking for a set of density matrix equations that can be written in the form

$$\frac{\mathrm{d}\tilde{\rho}_{\mathrm{ab}}}{\mathrm{d}t} = -i\omega_{\mathrm{ab}}\tilde{\rho}_{\mathrm{ab}} - \frac{i}{\hbar}\sum_{g}\left(V_{\mathrm{ag}}\tilde{\rho}_{\mathrm{gb}} - \tilde{\rho}_{\mathrm{ag}}V_{\mathrm{gb}}\right) + \sum_{a'b'}\Phi_{\mathrm{ab}}^{\mathrm{a'b'}}\cdot\tilde{\rho}_{a'b'}$$
$$- \left(\frac{\gamma_{\mathrm{a}}}{2} + \frac{\gamma_{\mathrm{b}}}{2}\right)\tilde{\rho}_{\mathrm{ab}} + R_{\mathrm{ab}} + A_{\mathrm{ab}} + Q_{\mathrm{ab}}, \tag{7.156}$$

where the indexes a, b instead of α, β are used to explicitly indicate M-quantum number averages. Equation (7.156) represents therefore a density matrix equation that is resolved in γJ-quantum numbers only (for designation of γJ see Sect. 7.5).

In order that (7.156) is useful, the averaged relaxation constants $R_{\mathrm{ab}}, A_{\mathrm{ab}}, Q_{\mathrm{ab}}$, $\gamma_{\mathrm{a}}, \gamma_{\mathrm{b}}$, the collisional operator $\Phi_{\mathrm{ab}}^{\mathrm{a'b'}}$, the ionic field interaction V_{ag}, and the transition frequencies ω_{ab} need to be specified without making use of the solution of the M-quantum-number-resolved populations $\tilde{\rho}_{\alpha\beta}$.

Because energies and relaxation constants do not contain interference terms, their average is readily performed like in the rate equation case outlined above. Because $\omega_{\alpha\alpha} = 0$, we can define an average frequency via

$$\hbar\omega_{\mathrm{ab}} = \left(\frac{1}{2J_\alpha + 1}\cdot\sum_{M_\alpha}E_\alpha\right) - \left(\frac{1}{2J_\beta + 1}\cdot\sum_{M_\beta}E_\beta\right), \tag{7.157}$$

while for relaxation constants (that do not contain interferences) a similar procedure like for (7.152) can be adopted:

$$\gamma_{\mathrm{a}} = \frac{1}{2J_\alpha + 1} \cdot \sum_{M_\alpha} \gamma_\alpha, \tag{7.158}$$

$$\gamma_{\mathrm{b}} = \frac{1}{2J_\beta + 1} \cdot \sum_{M_\beta} \gamma_\beta, \tag{7.159}$$

$$Q_{\mathrm{ab}} = \frac{\delta_{\mathrm{ab}}}{2J_\alpha + 1} \sum_{M_\alpha} Q_\alpha, \tag{7.160}$$

$$\gamma_{\mathrm{a}}^{\mathrm{rad}} = \frac{1}{2J_\alpha + 1} \cdot \sum_{M_\alpha} \sum_{\alpha'} \gamma_{\alpha\alpha'}, \tag{7.161}$$

$$\gamma_{\mathrm{b}}^{\mathrm{rad}} = \frac{1}{2J_\beta + 1} \cdot \sum_{M_\beta} \sum_{\beta'} \gamma_{\beta\beta'}, \tag{7.162}$$

$$A_{\mathrm{ab}} = -\left(\frac{\gamma_{\mathrm{a}}^{\mathrm{rad}}}{2} + \frac{\gamma_{\mathrm{b}}^{\mathrm{rad}}}{2} \right) \tilde{\rho}_{\mathrm{ab}} + \delta_{\mathrm{ab}} \sum_{a'} \gamma_{a'a}^{\mathrm{rad}} \tilde{\rho}_{a'a'}, \tag{7.163}$$

$$R_{\mathrm{ab}} = \delta_{\mathrm{ab}} \cdot \sum_{M_\alpha} \sum_{m} \gamma_{\mathrm{m}\alpha} \rho_{\mathrm{mm}}. \tag{7.164}$$

For the collisional operator $\Phi_{\mathrm{ab}}^{\mathrm{a'b'}}$ and the ionic field interaction V_{ag}, the situation is different because here, the two different indexes of the density matrix are relevant (interference terms). Summation of (7.154) over M_α and M_β and comparison with (7.156) provides formally

$$\sum_{M_\alpha} \sum_{M_\beta} \left\{ -\frac{i}{\hbar} \sum_{\gamma} \left(V_{\alpha\gamma} \tilde{\rho}_{\gamma\beta} - \tilde{\rho}_{\alpha\gamma} V_{\gamma\beta} \right) \right\} = -\frac{i}{\hbar} \sum_{g} \left(V_{\mathrm{ag}} \tilde{\rho}_{\mathrm{gb}} - \tilde{\rho}_{\mathrm{ag}} V_{\mathrm{gb}} \right) \tag{7.165}$$

and

$$\sum_{M_\alpha} \sum_{M_\beta} \left\{ \sum_{\alpha'\beta'} \Phi_{\alpha\beta}^{\alpha'\beta'} \cdot \tilde{\rho}_{\alpha'\beta'} \right\} = \sum_{a'b'} \Phi_{\mathrm{ab}}^{\mathrm{a'b'}} \cdot \tilde{\rho}_{\mathrm{a'b'}}. \tag{7.166}$$

From (7.165), we obtain

$$\sum_{M_\alpha}\sum_{M_\beta}\left\{-\frac{i}{\hbar}\sum_{\gamma_\gamma J_\gamma M_\gamma}\left(V_{\alpha\gamma}\tilde{\rho}_{\gamma\beta}-\tilde{\rho}_{\alpha\gamma}V_{\gamma\beta}\right)\right\}=-\frac{i}{\hbar}\sum_{\gamma_\gamma J_\gamma}\left(V_{\rm ag}\tilde{\rho}_{\rm gb}-\tilde{\rho}_{\rm ag}V_{\rm gb}\right) \quad (7.167)$$

from which it follows

$$\sum_{M_\alpha}\sum_{M_\beta}\sum_{M_\gamma}\left(V_{\alpha\gamma}\cdot\tilde{\rho}_{\gamma\beta}\right)=V_{\rm ag}\tilde{\rho}_{\rm gb}. \quad (7.168)$$

We therefore formally write

$$V_{\rm ag}=F_{\rm V}^{\alpha\beta\gamma}\cdot\sum_{M_\alpha}\sum_{M_\beta}\sum_{M_\gamma}V_{\alpha\gamma}, \quad (7.169)$$

where $F_{\rm V}^{\alpha\beta\gamma}$ is a function that fulfills (7.168), i.e.

$$\sum_{M_\alpha}\sum_{M_\beta}\sum_{M_\gamma}\left(V_{\alpha\gamma}\cdot\tilde{\rho}_{\gamma\beta}\right)=\left(F_{\rm V}^{\alpha\beta\gamma}\cdot\sum_{M_\alpha}\sum_{M_\beta}\sum_{M_\gamma}V_{\alpha\gamma}\right)\cdot\tilde{\rho}_{\rm gb}. \quad (7.170)$$

Similar, from (7.166) it follows

$$\sum_{M_\alpha}\sum_{M_\beta}\left\{\sum_{\gamma_{\alpha'}J_{\alpha'}M_{\alpha'},\gamma_{\beta'}J_{\beta'}M_{\beta'}}\Phi_{\alpha\beta}^{\alpha'\beta'}\cdot\tilde{\rho}_{\alpha'\beta'}\right\}=\sum_{\gamma_{\alpha'}J_{\alpha'},\gamma_{\beta'}J_{\beta'}}\Phi_{\rm ab}^{\rm a'b'}\cdot\tilde{\rho}_{a'b'} \quad (7.171)$$

from which we deduce

$$\sum_{M_\alpha}\sum_{M_\beta}\sum_{M_{\alpha'}}\sum_{M_{\beta'}}\left(\Phi_{\alpha\beta}^{\alpha'\beta'}\cdot\tilde{\rho}_{\alpha'\beta'}\right)=\Phi_{\rm ab}^{\rm a'b'}\cdot\tilde{\rho}_{a'b'}. \quad (7.172)$$

We therefore formally write

$$\Phi_{\rm ab}^{\rm a'b'}=F_{\Phi}^{\alpha\beta\alpha'\beta'}\cdot\sum_{M_\alpha}\sum_{M_\beta}\sum_{M_{\alpha'}}\sum_{M_{\beta'}}\Phi_{\alpha\beta}^{\alpha'\beta'}, \quad (7.173)$$

where $F_{\Phi}^{\alpha\beta\alpha'\beta'}$ is a function that fulfills (7.172), i.e.

$$\sum_{M_\alpha}\sum_{M_\beta}\sum_{M_{\alpha'}}\sum_{M_{\beta'}}\left(\Phi_{\alpha\beta}^{\alpha'\beta'}\cdot\tilde{\rho}_{\alpha'\beta'}\right)=\left(F_{\Phi}^{\alpha\beta\alpha'\beta'}\cdot\sum_{M_\alpha}\sum_{M_\beta}\sum_{M_{\alpha'}}\sum_{M_{\beta'}}\Phi_{\alpha\beta}^{\alpha'\beta'}\right)\cdot\tilde{\rho}_{a'b'}. \quad (7.174)$$

As already outlined for the rate equation case (7.151), the averaged density matrix (7.156) is only useful, if reasonable approximations can be involved for the matrix elements (7.166), (7.168) so that these equations do not depend any more on the populations (neither on the M-resolved ones $\tilde{\rho}_{\alpha\beta}$ nor on the M-averaged ones $\tilde{\rho}_{\mathrm{ab}}$). Therefore, at this point (7.167)–(7.170) and (7.171)–(7.174) are only formal solutions that request to invoke additional relations and/or further approximations to be practically useful.

7.6.3 *The Failure of the Rate Equation Approach for Quantum Averages*

It is instructive to investigate first what happens if we employ an averaging procedure for the density matrix elements that corresponds to (7.150), (7.152), i.e., a simple statistical hypothesis. For the diagonal elements of the density matrix, the averaging procedure to be found should then exactly provide the same result as (7.150). Due to the symmetry relation of the density matrix elements, i.e., $\tilde{\rho}_{\beta\alpha} = \tilde{\rho}^{*}_{\beta\alpha}$ we would likewise request this symmetry relation for the approximate solution. This means $\tilde{\rho}^{(\mathrm{approx})}_{\alpha\beta} = \tilde{\rho}^{(\mathrm{approx})*}_{\beta\alpha}$ should be likewise preserved for all matrix elements. These two conditions (limit for diagonal elements, symmetry relation for all density matrix elements) could be fulfilled, e.g., by

$$\tilde{\rho}^{(\mathrm{approx})}_{\alpha\beta} \stackrel{?}{=} \frac{1}{\sqrt{2J_{\alpha}+1}} \cdot \frac{1}{\sqrt{2J_{\beta}+1}} \cdot \tilde{\rho}_{\mathrm{ab}}. \tag{7.175}$$

We note, that (7.175) is not necessarily correct, but it fullfills the above mentioned two conditions and will be used below to demonstrate its failure. For $\alpha = \beta$, we obtain from (7.175)

$$\tilde{\rho}^{(\mathrm{approx})}_{\alpha\alpha} = \frac{1}{2J_{\alpha}+1} \cdot \tilde{\rho}_{\mathrm{aa}}. \tag{7.176}$$

Equation (7.176) is identical to (7.150) because $n^{(\mathrm{approx})}_{\alpha} = \tilde{\rho}^{(\mathrm{approx})}_{\alpha\alpha}$ and $n_{\mathrm{a}} = \tilde{\rho}_{\mathrm{aa}}$. The symmetry relation can be verified as follows. From (7.155) it follows

$$\tilde{\rho}_{\mathrm{ab}} = \tilde{\rho}^{*}_{\mathrm{ba}} \tag{7.177}$$

because $\tilde{\rho}_{\alpha\beta} = \tilde{\rho}^{*}_{\beta\alpha}$. Inserting (7.176) into (7.177), we obtain

$$\sqrt{2J_{\alpha}+1} \cdot \sqrt{2J_{\beta}+1} \cdot \tilde{\rho}^{(\mathrm{approx})}_{\alpha\beta} = \sqrt{2J_{\beta}+1} \cdot \sqrt{2J_{\alpha}+1} \cdot \tilde{\rho}^{(\mathrm{approx})*}_{\beta\alpha}. \tag{7.178}$$

Therefore

$$\tilde{\rho}_{\alpha\beta}^{(\text{approx})} = \tilde{\rho}_{\beta\alpha}^{(\text{approx})*} \tag{7.179}$$

as it should be.

We now study the averaging produce for the electron collisional operator. If we insert (7.175) in (7.172), we obtain

$$\frac{1}{\sqrt{2J_{\alpha'}+1}} \cdot \frac{1}{\sqrt{2J_{\beta'}+1}} \cdot \tilde{\rho}_{a'b'} \cdot \sum_{M_\alpha} \sum_{M_\beta} \sum_{M_{\alpha'}} \sum_{M_{\beta'}} \left(\Phi_{\alpha\beta}^{\alpha'\beta'} \right) = \Phi_{\text{ab}}^{\text{a'b'}} \cdot \tilde{\rho}_{a'b'} \tag{7.180}$$

from which it follows

$$\Phi_{\text{ab}}^{\text{a'b'}} = \frac{1}{\sqrt{2J_{\alpha'}+1}} \cdot \frac{1}{\sqrt{2J_{\beta'}+1}} \cdot \sum_{M_\alpha} \sum_{M_\beta} \sum_{M_{\alpha'}} \sum_{M_{\beta'}} \left(\Phi_{\alpha\beta}^{\alpha'\beta'} \right). \tag{7.181}$$

With the help of (7.173), (7.174) we obtain

$$F_{\Phi}^{\alpha\beta\alpha'\beta'} = \frac{1}{\sqrt{2J_{\alpha'}+1}} \cdot \frac{1}{\sqrt{2J_{\beta'}+1}}. \tag{7.182}$$

Let us apply (7.181), (7.182) to the particular term $\Phi_{\text{ab}}^{\text{a'b'}} \cdot \tilde{\rho}_{a'b'} = \Phi_{\text{aa}}^{\text{aa}} \cdot \tilde{\rho}_{\text{aa}}$ in (7.156). In the population rate equation picture and also in the density matrix approach, this term in (7.156) corresponds to the electron collisional loss rate from level *a*. First we note that $\phi_{(3),\alpha\alpha}^{(M),\alpha\alpha} = 0$ because the reduced dipole matrix elements are zero for identical indexes [see (7.112)]. Due to the product of the δ-functions $\delta_{\alpha'\alpha} \cdot \delta_{J_{\beta'}J_\beta} \cdot \delta_{M_{\beta'}M_\beta} \cdot \delta(J_{\beta'}, 1, J_\beta)$ and $\delta_{\beta'\beta} \cdot \delta_{J_{\alpha'}J_\alpha} \cdot \delta_{M_{\alpha'}M_\alpha} \cdot \delta(J_{\alpha'}, 1, J_\alpha)$ in the expressions for $\phi_{(1),\alpha\beta}^{(M),\alpha'\beta'}$ and $\phi_{(2),\alpha\beta}^{(M),\alpha'\beta'}$ of (7.67), (7.68), (7.105), (7.107) the summations over $M_{\alpha'}$ and $M_{\beta'}$ provide only a single term, so that only summations over M_α and M_β effectively remain in (7.181). Assuming $A = 1$ and $B = 0$, we obtain

$$\begin{aligned} \phi_{(1),\text{ab}}^{(av),\text{a'b'}} &= \frac{1}{\sqrt{2J_{\alpha'}+1}} \cdot \frac{1}{\sqrt{2J_{\beta'}+1}} \cdot \sum_{M_\alpha} \sum_{M_\beta} \sum_{M_{\alpha'}} \sum_{M_{\beta'}} \left(\sum_{\gamma'', J_{\beta''}} \phi_{(1),\alpha\beta}^{(M),\alpha'\beta'} \right) \\ &= \frac{1}{\sqrt{2J_\alpha+1}} \cdot \frac{1}{\sqrt{2J_\beta+1}} \cdot \sum_{M_\alpha} \sum_{M_\beta} \left(\sum_{\gamma'', \beta''} \phi_{(1),\alpha\beta}^{(M),\alpha\beta'} \right). \end{aligned} \tag{7.183}$$

For $\phi_{(1),\mathrm{aa}}^{(av),\mathrm{aa}}$, we obtain formally from (7.183) and (7.105):

$$\phi_{(1),\mathrm{aa}}^{(av),\mathrm{aa}} == \frac{1}{2J_\alpha+1}\cdot\sum_{M_\alpha}\sum_{M_{\beta=\alpha}}\left\{\frac{1}{2J_{\beta=\alpha}+1}\cdot\sum_{\gamma_{\beta''},J_{\beta''}}\left|\left\langle\gamma_{\beta=\alpha}J_{\beta=\alpha}\|r\|\gamma_{\beta''}J_{\beta''}\right\rangle\right|^2\right\}. \tag{7.184}$$

As the term in parenthesis of (7.184) does not any more depend on M_α nor on $M_{\beta=\alpha}$, the summation over M_α provides therefore $2J_\alpha+1$ and the summation over $M_{\beta=\alpha}$ $2J_\beta+1=2J_\alpha+1$. Therefore, (7.184) results finally into

$$\phi_{(1),\mathrm{aa}}^{(av),\mathrm{aa}} = (2J_\alpha+1)\cdot\frac{1}{2}\cdot\sum_{\gamma_{\beta''},J_{\beta''}}\left\{\frac{2\cdot\left|\left\langle\gamma_\alpha J_\alpha\|r\|\gamma_{\beta''}J_{\beta''}\right\rangle\right|^2}{2J_\alpha+1}\right\} \tag{7.185}$$

and correspondingly

$$\phi_{(2),\mathrm{aa}}^{(av),\mathrm{aa}} = (2J_\alpha+1)\cdot\frac{1}{2}\cdot\sum_{\gamma_{\beta''},J_{\beta''}}\left\{\frac{2\cdot\left|\left\langle\gamma_\alpha J_\alpha\|r\|\gamma_{\beta''}J_{\beta''}\right\rangle\right|^2}{2J_\alpha+1}\right\}. \tag{7.186}$$

Therefore (note, that α'' and β'' run over the same functional space)

$$\phi_{(1),\mathrm{aa}}^{(av),\mathrm{aa}}+\phi_{(3),\mathrm{aa}}^{(av),\mathrm{aa}}+\phi_{(3),\mathrm{aa}}^{(av),\mathrm{aa}} = (2J_\alpha+1)\cdot\sum_{\gamma_{\beta''},J_{\beta''}}\left\{\frac{2\cdot\left|\left\langle\gamma_\alpha J_\alpha\|r\|\gamma_{\beta''}J_{\beta''}\right\rangle\right|^2}{2J_\alpha+1}\right\}. \tag{7.187}$$

According to (7.66)–(7.69), (7.139), the term in parenthesis of (7.187) is proportional to cross sections, i.e.

$$\sigma\left(J_\alpha\to J_\beta\right)\propto\frac{2\cdot\left|\left\langle\gamma_\alpha J_\alpha\|r\|\gamma_\beta J_\beta\right\rangle\right|^2}{2J_\alpha+1}. \tag{7.188}$$

Therefore, the sum in (7.187) represents the total loss cross section multiplied with the statistical weight, i.e.

$$\Phi_{\mathrm{aa}}^{\mathrm{aa}} = [(2J_\alpha+1)]\cdot\left[-n_\mathrm{e}\cdot\left\{\sum_{J_{\alpha''}}\langle\upsilon\sigma(J_\alpha\to J_{\alpha''})\rangle\right\}\right]. \tag{7.189}$$

Due to the prefactor $(2J_\alpha+1)$ in (7.189), the electron collisional operator matrix element $\Phi_{\mathrm{aa}}^{\mathrm{aa}}$ is *not* proportional to a loss rate as it should be!

A similar problematic is faced, e.g., for the term $\Phi_{\mathrm{ab}}^{\mathrm{a'b'}}\cdot\tilde{\rho}_{\mathrm{a'b'}}=\Phi_{\mathrm{bb}}^{\mathrm{aa}}\cdot\tilde{\rho}_{\mathrm{aa}}$ in (7.156). In the population rate equation picture and also in the density matrix approach, this term in (7.156) corresponds to the electron collisional transfer rate from level $a \to b$. First we note that $\phi^{(M),\alpha\alpha}_{(1),\beta\beta} = 0$ and $\phi^{(M),\alpha\alpha}_{(2),\beta\beta} = 0$ due to the δ-functions in the expressions for $\phi^{(M),\alpha'\beta'}_{(1),\alpha\beta}$ and $\phi^{(M),\alpha'\beta'}_{(2),\alpha\beta}$, see (7.67), (7.68), (7.105), (7.106). For $\phi^{(M),\alpha'\beta'}_{(3),\alpha\beta}$ we obtain from (7.181)

$$\phi^{(av),a'b'}_{(3),ab} = \frac{1}{\sqrt{2J_{\alpha'}+1}} \cdot \frac{1}{\sqrt{2J_{\beta'}+1}} \cdot \sum_{M_\alpha} \sum_{M_\beta} \sum_{M_{\alpha'}} \sum_{M_{\beta'}} \left(\phi^{(M),\alpha'\beta'}_{(3),\alpha\beta} \right). \tag{7.190}$$

For $\phi^{(av),aa}_{(3),bb}$, we obtain formally from (7.190) and (7.112):

$$\phi^{(av),aa}_{(3),bb} = \frac{1}{\sqrt{2J_{\alpha'=\alpha}+1}} \cdot \frac{1}{\sqrt{2J_{\beta'=\alpha}+1}} \times \sum_{M_{\alpha=\beta}} \sum_{M_\beta} \sum_{M_{\alpha'=\alpha}} \sum_{M_{\beta'=\alpha}} \left\{ \sum_q 2 \cdot \left| \langle \gamma_\alpha J_\alpha \| r \| \gamma_\beta J_\beta \rangle \right|^2 \cdot \begin{pmatrix} J_\beta & 1 & J_\alpha \\ -M_\beta & q & M_\alpha \end{pmatrix}^2 \right\}. \tag{7.191}$$

Due to the sum rule of the $3j$ symbol, the sum over M_α and q provides finally

$$\phi^{(av),aa}_{(3),bb} = \frac{1}{2J_\alpha+1} \cdot \sum_{M_{\alpha=\beta}} \sum_{M_{\beta'=\alpha}} \sum_{M_{\alpha'=\alpha}} \left\{ \frac{2 \cdot \left| \langle \gamma_\alpha J_\alpha \| r \| \gamma_\beta J_\beta \rangle \right|^2}{2J_\alpha+1} \right\}. \tag{7.192}$$

Because the term in parenthesis of (7.192) does not any more depend on $M_{\alpha=\beta}, M_{\beta'=\alpha}$ and $M_{\alpha'=\alpha}$, the summation over the 3 identical sums in (7.192) would formally provide $(2J_\alpha+1)^3$. As the term in parenthesis of (7.192) is proportional to a cross section [see also relation (7.188)], (7.192) would *not* be proportional to a cross section because

$$\phi^{(av),aa}_{(3),bb} = (2J_\alpha+1)^2 \cdot \frac{2 \cdot \left| \langle \gamma_\alpha J_\alpha \| \| \gamma_\beta J_\beta \rangle \right|^2}{2J_\alpha+1}. \tag{7.193}$$

Therefore (note, that α'' and β'' run over the same functional space)

$$\phi^{(av),aa}_{(1),bb} + \phi^{(av),aa}_{(3),bb} + \phi^{(av),aa}_{(3),bb} = (2J_\alpha+1)^2 \cdot \frac{2 \cdot \left| \langle \gamma_\alpha J_\alpha \| r \| \gamma_\beta J_\beta \rangle \right|^2}{2J_\alpha+1}. \tag{7.194}$$

Correspondingly, the collisional operator matrix element

$$\Phi^{aa}_{bb} = [(2J_\alpha+1)]^2 \cdot [n_e \cdot \langle V\sigma(J_\alpha \to J_\beta) \rangle] \tag{7.195}$$

would *not* be proportional to a collisional rate from level "*a*" to level "*b*" as it should be.

For completeness, we shortly consider the case for the field matrix elements $V_{\alpha\beta}$. Injecting (7.175) into (7.168), we obtain

$$\frac{1}{\sqrt{2J_\gamma+1}}\cdot\frac{1}{\sqrt{2J_\beta+1}}\cdot\tilde{\rho}_{\mathrm{gb}}\cdot\sum_{M_\alpha}\sum_{M_\beta}\sum_{M_\gamma}\left(V_{\alpha\gamma}\right)=V_{\mathrm{ag}}\cdot\tilde{\rho}_{\mathrm{gb}} \tag{7.196}$$

and, with the help of (7.169)

$$F_V^{\alpha\beta\gamma}=\frac{1}{\sqrt{2J_\gamma+1}}\cdot\frac{1}{\sqrt{2J_\beta+1}}. \tag{7.197}$$

As there is no "simple cross section" analog for the field matrix element as for the cases discussed above in relation with the collisional operator elements, it is therefore difficult to discuss at this point the failure or correctness of (7.196), (7.197) and need instead discuss the averaged field matrix elements in relation with the ion quasi-static field.

7.6.4 The Cross Section Method for Quantum Averages

As can be seen from (7.184), (7.187), (7.189) and (7.192), (7.194), (7.195), the failure of (7.189) and (7.195) to obtain the physical rates is related to summations over M-quantum numbers with identical indexes. In other words: in (7.184) one needs to sum two times over M_α providing the wrong additional factor $(2J_\alpha+1)$ in (7.189), and, in (7.192) one needs to sum three times over M_α providing the wrong additional factor $(2J_\alpha+1)^2$ in (7.195). This suggests, to consider a solution to the problem the other way around and ask, what prefactors $F_\Phi^{\alpha\beta\alpha'\beta'}$ have to be applied to obtain the physical cross sections and rates.

The failure is connected with a wrong combination of statistical weights in a particular way due to multiple equivalent summations; namely, concerning the elements $\phi_{(1)}^{(\mathrm{av})}$ and $\phi_{(2)}^{(\mathrm{av})}$ an additional wrong factor $(2J_\alpha+1)$ is encountered, while processes concerning the element $\phi_{(3)}^{(\mathrm{av})}$ an additional wrong factor $(2J_\alpha+1)^2$ is encountered. This suggests also not to look for a solution of a single prefactor $F_\Phi^{\alpha\beta\alpha'\beta'}$ (as discussed before) but for a prefactor for each $\phi_{(1)}^{(\mathrm{av})}$, $\phi_{(2)}^{(\mathrm{av})}$, and $\phi_{(3)}^{(\mathrm{av})}$, namely $F_{\phi_1}^{\alpha\beta\alpha'\beta'}$, $F_{\phi_2}^{\alpha\beta\alpha'\beta'}$, and $F_{\phi_3}^{\alpha\beta\alpha'\beta'}$, respectively:

$$\phi_{(1),\mathrm{ab}}^{(av),\mathrm{a}'\mathrm{b}'}=F_{\phi_1}^{\alpha\beta\alpha'\beta'}\cdot\sum_{M_\alpha}\sum_{M_\beta}\sum_{M_{\alpha'}}\sum_{M_{\beta'}}\left(\sum_{\gamma_{\beta''},J_{\beta''}}\phi_{(1),\alpha\beta}^{(M),\alpha'\beta'}\right), \tag{7.198}$$

$$\phi_{(2),\mathrm{ab}}^{(av),\mathrm{a'b'}} = F_{\phi_2}^{\alpha\beta\alpha'\beta'} \cdot \sum_{M_\alpha}\sum_{M_\beta}\sum_{M_{\alpha'}}\sum_{M_{\beta'}}\left(\sum_{\gamma_{\alpha''},J_{\alpha''}} \phi_{(2),\alpha\beta}^{(M),\alpha'\beta'}\right), \tag{7.199}$$

$$\phi_{(3),\mathrm{ab}}^{(av),\mathrm{a'b'}} = F_{\phi_3}^{\alpha\beta\alpha'\beta'} \cdot \sum_{M_\alpha}\sum_{M_\beta}\sum_{M_{\alpha'}}\sum_{M_{\beta'}}\left(\phi_{(3),\alpha\beta}^{(M),\alpha'\beta'}\right), \tag{7.200}$$

with

$$F_{\phi_1}^{\alpha\beta\alpha'\beta'} = \frac{1}{\sqrt{2J_\alpha+1}}\cdot\frac{1}{\sqrt{2J_{\alpha'}+1}}\cdot\frac{1}{\sqrt{2J_\beta+1}}\cdot\frac{1}{\sqrt{2J_{\beta'}+1}}, \tag{7.201}$$

$$F_{\phi_2}^{\alpha\beta\alpha'\beta'} = \frac{1}{\sqrt{2J_\alpha+1}}\cdot\frac{1}{\sqrt{2J_{\alpha'}+1}}\cdot\frac{1}{\sqrt{2J_\beta+1}}\cdot\frac{1}{\sqrt{2J_{\beta'}+1}}, \tag{7.202}$$

$$F_{\phi_3}^{\alpha\beta\alpha'\beta'} = \frac{1}{\sqrt{2J_\alpha+1}}\cdot\frac{1}{\sqrt{2J_{\alpha'}+1}}\cdot\frac{1}{\sqrt{2J_\beta+1}}\cdot\frac{1}{\sqrt{2J_{\beta'}+1}}\cdot\frac{1}{\sqrt{2J_{\alpha'}+1}}\cdot\frac{1}{\sqrt{2J_{\beta'}+1}}. \tag{7.203}$$

Let us now consider the combinations of indexes that are physically related to cross sections or rates.

7.6.4.1 The Total Loss Rate from a Quantum Level: $\Phi_{\mathrm{aa}}^{\mathrm{aa}}$

For the case $b = a, a' = a$ and $b' = a$, it follows from (7.112) that $\phi_{(3),\alpha\beta}^{(M),\alpha'\beta'} = \phi_{(3),\alpha\alpha}^{(M),\alpha\alpha} = 0$ and therefore

$$\phi_{(3),\mathrm{aa}}^{(av),\mathrm{aa}} = 0. \tag{7.204}$$

For $\phi_{(1),\alpha\alpha}^{(M),\alpha\alpha}$ and $\phi_{(2),\alpha\alpha}^{(M),\alpha\alpha}$, it follows from (7.67), (7.68), (7.105), (7.106), (7.198), (7.199), (7.201), (7.202):

$$\begin{aligned}\phi_{(1),\mathrm{aa}}^{(av),\mathrm{aa}} &= F_{\phi_1}^{\alpha\alpha\alpha\alpha} \cdot \sum_{M_\alpha}\sum_{M_{\beta=\alpha}}\sum_{M_{\alpha'=\alpha}}\sum_{M_{\beta'=\alpha}}\left(\sum_{\gamma_{\beta''},J_{\beta''}} \phi_{(1),\alpha\alpha}^{(M),\alpha\alpha}\right)\\ &= \left(\frac{1}{\sqrt{2J_\alpha+1}}\right)^4 \cdot \sum_{M_\alpha}\sum_{M_{\beta=\alpha}}\sum_{M_{\alpha'=\alpha}}\sum_{M_{\beta'=\alpha}}\left(\sum_{\gamma_{\beta''},J_{\beta''}} \frac{\left|\langle\gamma_\alpha J_\alpha\|r_b\|\gamma_{\beta''}J_{\beta''}\rangle\right|^2}{2J_\alpha+1}\right)\\ &= \sum_{\gamma_{\beta''},J_{\beta''}} \frac{\left|\langle\gamma_\alpha J_\alpha\|r_b\|\gamma_{\beta''}J_{\beta''}\rangle\right|^2}{2J_\alpha+1}\end{aligned} \tag{7.205}$$

and likewise

$$\begin{aligned}\phi_{(2),\mathrm{aa}}^{(av),\mathrm{aa}} &= F_{\phi_2}^{\alpha\alpha\alpha\alpha}\cdot\sum_{M_\alpha}\sum_{M_{\beta=\alpha}}\sum_{M_{\alpha'=\alpha}}\sum_{M_{\beta'=\alpha}}\left(\sum_{\gamma_{\alpha''},J_{\alpha''}}\phi_{(2),\alpha\alpha}^{(M),\alpha\alpha}\right)\\ &= \left(\frac{1}{\sqrt{2J_\alpha+1}}\right)^4\cdot\sum_{M_\alpha}\sum_{M_{\beta=\alpha}}\sum_{M_{\alpha'=\alpha}}\sum_{M_{\beta'=\alpha}}\left(\sum_{\gamma_{\alpha''},J_{\alpha''}}\frac{|\langle\gamma_\alpha J_\alpha\|r_\mathrm{b}\|\gamma_{\alpha''}J_{\alpha''}\rangle|^2}{2J_\alpha+1}\right)\\ &= \sum_{\gamma_{\alpha''},J_{\alpha''}}\frac{|\langle\gamma_\alpha J_\alpha\|r_\mathrm{b}\|\gamma_{\alpha''}J_{\alpha''}\rangle|^2}{2J_\alpha+1}.\end{aligned} \tag{7.206}$$

Because the final sums in (7.205), (7.206) run over the same functional space, (7.205) and (7.206) are identical and we have:

$$\phi_{(1),\mathrm{aa}}^{(av),\mathrm{aa}}+\phi_{(2),\mathrm{aa}}^{(av),\mathrm{aa}}+\phi_{(3),\mathrm{aa}}^{(av),\mathrm{aa}} = \sum_{\gamma_{\beta''},J_{\beta''}}\frac{2\cdot|\langle\gamma_\alpha J_\alpha\|r_\mathrm{b}\|\gamma_{\beta''}J_{\beta''}\rangle|^2}{2J_\alpha+1}. \tag{7.207}$$

Therefore, we obtain with the help of (7.72), (7.139)

$$\Phi_{\mathrm{aa}}^{\mathrm{aa}} = -n_\mathrm{e}\cdot\left\{\sum_{J_{\alpha''}}\langle\upsilon\sigma(J_\alpha\to J_{\alpha''})\rangle\right\}. \tag{7.208}$$

The matrix element $\Phi_{\mathrm{aa}}^{\mathrm{aa}}$ corresponds therefore to the decay of the diagonal matrix element $\tilde{\rho}_{\mathrm{aa}}$, i.e., to the total loss rate of the atomic population of level "a" due to electron–ion collisions.

7.6.4.2 The Loss Rate of Coherences: $\Phi_{\mathrm{ab}}^{\mathrm{ab}}$

For the case $a' = a$ and $b' = b$, it follows from (7.112) that $\phi_{(3),\alpha\beta}^{(M),\alpha'\beta'} = \phi_{(3),\alpha\beta}^{(M),\alpha\beta} = 0$ and therefore

$$\phi_{(3),\mathrm{ab}}^{(av),\mathrm{ab}} = 0. \tag{7.209}$$

For $\phi_{(1),\alpha\beta}^{(M),\alpha\beta}$ and $\phi_{(2),\alpha\beta}^{(M),\alpha\beta}$, it follows from (7.67), (7.68), (7.105), (7.106), (7.198), (7.199), (7.201), (7.202), (7.206):

$$
\begin{aligned}
\phi_{(1),\mathrm{ab}}^{(av),\mathrm{ab}} &= F_{\phi_1}^{\alpha\beta\alpha\beta}\cdot\sum_{M_\alpha}\sum_{M_\beta}\sum_{M_{\alpha'}=\alpha}\sum_{M_{\beta'}=\beta}\left(\sum_{\gamma_{\beta''},J_{\beta''}}\phi_{(1),\alpha\beta}^{(M),\alpha\beta}\right)\\
&= \left(\frac{1}{\sqrt{2J_\alpha+1}}\right)^2\cdot\left(\frac{1}{\sqrt{2J_\beta+1}}\right)^2\cdot\sum_{M_\alpha}\sum_{M_\beta}\sum_{M_{\alpha'}=\alpha}\sum_{M_{\beta'}=\beta}\left(\sum_{\gamma_{\beta''},J_{\beta''}}\frac{\left|\langle\gamma_\beta J_\beta\|r_\mathrm{b}\|\gamma_{\beta''}J_{\beta''}\rangle\right|^2}{2J_\beta+1}\right)\\
&= \sum_{\gamma_{\beta''},J_{\beta''}}\frac{\left|\langle\gamma_\beta J_\beta\|r_\mathrm{b}\|\gamma_{\beta''}J_{\beta''}\rangle\right|^2}{2J_\beta+1}
\end{aligned}
\tag{7.210}
$$

and likewise

$$
\begin{aligned}
\phi_{(2),\mathrm{ab}}^{(av),\mathrm{ab}} &= F_{\phi_2}^{\alpha\beta\alpha\beta}\cdot\sum_{M_\alpha}\sum_{M_\beta}\sum_{M_{\alpha'}=\alpha}\sum_{M_{\beta'}=\beta}\left(\sum_{\gamma_{\alpha''},J_{\alpha''}}\phi_{(2),\alpha\beta}^{(M),\alpha\beta}\right)\\
&= \left(\frac{1}{\sqrt{2J_\alpha+1}}\right)^2\cdot\left(\frac{1}{\sqrt{2J_\beta+1}}\right)^2\cdot\sum_{M_\alpha}\sum_{M_\beta}\sum_{M_{\alpha'}=\alpha}\sum_{M_{\beta'}=\beta}\left(\sum_{\gamma_{\beta''},J_{\beta''}}\frac{\left|\langle\gamma_\alpha J_\alpha\|r_\mathrm{b}\|\gamma_{\beta''}J_{\beta''}\rangle\right|^2}{2J_\alpha+1}\right)\\
&= \sum_{\gamma_{\alpha''},J_{\alpha''}}\frac{\left|\langle\gamma_\alpha J_\alpha\|r_\mathrm{b}\|\gamma_{\beta''}J_{\beta''}\rangle\right|^2}{2J_\alpha+1}.
\end{aligned}
\tag{7.211}
$$

Because the final sums in (7.210), (7.211) run over the same functional space, (7.210) and (7.211) are similar and we have:

$$
\begin{aligned}
\phi_{(1),\mathrm{aa}}^{(av),\mathrm{aa}}+\phi_{(2),\mathrm{aa}}^{(av),\mathrm{aa}}+\phi_{(3),\mathrm{aa}}^{(av),\mathrm{aa}} &= \sum_{\gamma_{\beta''},J_{\beta''}}\frac{\left|\langle\gamma_\beta J_\beta\|r_\mathrm{b}\|\gamma_{\beta''}J_{\beta''}\rangle\right|^2}{2J_\beta+1}\\
&\quad+\sum_{\gamma_{\alpha''},J_{\alpha''}}\frac{\left|\langle\gamma_\alpha J_\alpha\|r_\mathrm{b}\|\gamma_{\beta''}J_{\beta''}\rangle\right|^2}{2J_\alpha+1}.
\end{aligned}
\tag{7.212}
$$

Therefore, we obtain with the help of (7.72), (7.139)

$$
\Phi_{\mathrm{ab}}^{\mathrm{ab}} = -n_\mathrm{e}\cdot\left\{\frac{1}{2}\sum_{J_{\beta''}}\langle\upsilon\sigma(J_\beta\to J_{\beta''})\rangle+\frac{1}{2}\sum_{J_{\alpha''}}\langle\upsilon\sigma(J_\alpha\to J_{\alpha''})\rangle\right\}. \tag{7.213}
$$

Equation (7.213) represents therefore the loss rate of coherences of the upper and lower levels as it should be [see also (7.156)].

7.6.4.3 The Transfer Rate Between Two Levels: $\Phi_{\rm aa}^{\rm bb}$

As can be seen from (7.67), (7.68), (7.105), (7.106), for the case $\beta = \alpha, \alpha' = \beta$ and $\beta' = \beta$, $\phi_{(1),\alpha\beta}^{(M),\alpha'\beta'} = \phi_{(1),\alpha\alpha}^{(M),\beta\beta} = 0$ and $\phi_{(2),\alpha\beta}^{(M),\alpha'\beta'} = \phi_{(2),\alpha\alpha}^{(M),\beta\beta} = 0$, therefore

$$\phi_{(1),\rm aa}^{(av),\rm bb} = 0, \tag{7.214}$$

$$\phi_{(2),\rm aa}^{(av),\rm bb} = 0. \tag{7.215}$$

As concerns $\phi_{(3),\rm aa}^{(av),\rm bb}$ for the selection of these indexes, the alternating sign in (7.112) transforms to $(-1)^{2J_\alpha - M_{\alpha'} - M_{\beta'} + J_{\beta'} - J_{\alpha'}} = (-1)^{2J_\beta - 2M_\alpha + J_\alpha - J_\alpha} = (-1)^{2J_\beta - 2M_\alpha}$. From the selection rule $-M_\beta + q + M_\alpha = 0$, it follows that $-2M_\beta + 2q = -2M_\alpha$; therefore, $(-1)^{2J_\beta - 2M_\beta + 2q} = +1$ because $2J_\alpha - 2M_\alpha$ and $2q$ are even. The alternating sign therefore vanishes, and we are left with

$$\phi_{(3),\rm aa}^{(M),\rm bb} = \sum_q 2 \cdot |\langle \gamma_\alpha J_\alpha \| r \| \gamma_\beta J_\beta \rangle|^2 \cdot \begin{pmatrix} J_\beta & 1 & J_\alpha \\ -M_\beta & q & M_\alpha \end{pmatrix}^2. \tag{7.216}$$

For $\phi_{(3),\alpha\alpha}^{(M),\beta\beta}$, it follows from (7.112), (7.200), (7.203):

$$\begin{aligned}
\phi_{(3),\rm aa}^{(av),\rm bb} &= F_{\phi_3}^{\alpha\alpha\beta\beta} \cdot \sum_{M_\alpha} \sum_{M_{\beta=\alpha}} \sum_{M_{\alpha'=\beta}} \sum_{M_{\beta'=\beta}} \left(\phi_{(3),\alpha\alpha}^{(M),\beta\beta} \right) \\
&= \left(\frac{1}{\sqrt{2J_\alpha + 1}} \right)^2 \cdot \left(\frac{1}{\sqrt{2J_\beta + 1}} \right)^4 \\
&\times \sum_{M_\alpha} \sum_{M_{\beta=\alpha}} \sum_{M_{\alpha'=\beta}} \sum_{M_{\beta'=\beta}} \left(\sum_q 2 \cdot |\langle \gamma_\alpha J_\alpha \| r \| \gamma_\beta J_\beta \rangle|^2 \cdot \begin{pmatrix} J_\beta & 1 & J_\alpha \\ -M_\beta & q & M_\alpha \end{pmatrix}^2 \right) \\
&= \left(\frac{1}{\sqrt{2J_\alpha + 1}} \right)^2 \cdot \left(\frac{1}{\sqrt{2J_\beta + 1}} \right)^4 \cdot \sum_{M_\alpha} \sum_{M_{\alpha'=\beta}} \sum_{M_{\beta'=\beta}} \left(\frac{2 \cdot |\langle \gamma_\alpha J_\alpha \| r \| \gamma_\beta J_\beta \rangle|^2}{2J_\alpha + 1} \right) \\
&= \frac{2 \cdot |\langle \gamma_\alpha J_\alpha \| r \| \gamma_\beta J_\beta \rangle|^2}{2J_\alpha + 1}
\end{aligned} \tag{7.217}$$

and we have:

$$\phi_{(1),\rm aa}^{(av),\rm bb} + \phi_{(2),\rm aa}^{(av),\rm bb} + \phi_{(3),\rm aa}^{(av),\rm bb} = \frac{2 \cdot |\langle \gamma_\alpha J_\alpha \| r \| \gamma_\beta J_\beta \rangle|^2}{2J_\alpha + 1}. \tag{7.218}$$

Therefore, we obtain with the help of (7.72), (7.139)

$$\Phi_{\mathrm{aa}}^{\mathrm{bb}} = n_{\mathrm{e}} \cdot \langle \upsilon\sigma(J_\beta \to J_\alpha) \rangle \tag{7.219}$$

corresponding to the electron collisional transfer from level "*b*" to level "*a*" and determines therefore the influence of the density matrix element $\tilde{\rho}_{\mathrm{bb}}$ on the density matrix element $\tilde{\rho}_{\mathrm{aa}}$ due to collisions. $\Phi_{\mathrm{bb}}^{\mathrm{aa}}$ and $\Phi_{\mathrm{aa}}^{\mathrm{bb}}$ correspond therefore to direct and inverse transitions due to electron collisions. If the electron collisions are assumed to be random (i.e., a Maxwellian electron energy distribution function), these direct and inverse matrix elements provide the possibility to obtain the thermodynamic limit for the atomic populations, i.e., a Boltzmann population in quantum kinetics.

Therefore, expressions (7.198)–(7.203) provide the physically requested cross sections and rates for the electron collisional operator elements $\Phi_{\mathrm{aa}}^{\mathrm{aa}}$ from (7.208), $\Phi_{\mathrm{ab}}^{\mathrm{ab}}$ from (7.213) and $\Phi_{\mathrm{aa}}^{\mathrm{bb}}$ from (7.219). The particular interesting point in this kind of cross section normalization is that it concerns all terms $\phi_{(1),\mathrm{ab}}^{(av),\mathrm{a'b'}}$, $\phi_{(1),\mathrm{ab}}^{(av),\mathrm{a'b'}}$, and $\phi_{(1),\mathrm{ab}}^{(av),\mathrm{a'b'}}$ of the electron collisional operator. We therefore assume (7.198)–(7.203) as the *M*-quantum number averages for all combinations of indexes.

7.7 About the Boltzmann Limit in Quantum Kinetics

7.7.1 *The Two-Level* M-*Quantum-Number-Averaged Level System*

Let us consider a closed two-level system and specify explicitly diagonal and non-diagonal density matrix elements. From (7.156), we obtain with $a = 1, 2$ and $b = 1, 2$:

$$\frac{\mathrm{d}\tilde{\rho}_{22}}{\mathrm{d}t} = -\tilde{\rho}_{22}\left(A_{21} - \Phi_{22}^{22}\right) + \Phi_{22}^{11}\tilde{\rho}_{11} - \frac{i}{\hbar}\left(V_{21}\tilde{\rho}_{12} - \tilde{\rho}_{21}V_{12}\right) + \Phi_{22}^{12}\tilde{\rho}_{12} + \Phi_{22}^{21}\tilde{\rho}_{21}, \tag{7.220}$$

$$\frac{\mathrm{d}\tilde{\rho}_{11}}{\mathrm{d}t} = -\tilde{\rho}_{11}\left(-\Phi_{11}^{11}\right) + \left(\Phi_{11}^{22} + A_{21}\right)\tilde{\rho}_{22} - \frac{i}{\hbar}\left(V_{12}\tilde{\rho}_{21} - \tilde{\rho}_{12}V_{21}\right) + \Phi_{11}^{12}\tilde{\rho}_{12} + \Phi_{11}^{21}\tilde{\rho}_{21}, \tag{7.221}$$

$$\begin{aligned}\frac{\mathrm{d}\tilde{\rho}_{12}}{\mathrm{d}t} = &-\tilde{\rho}_{12}\left(\frac{A_{21}}{2} - \Phi_{12}^{12}\right) - i\omega_{12}\tilde{\rho}_{12} + \Phi_{12}^{21}\tilde{\rho}_{21} \\ &- \frac{i}{\hbar}\left(V_{12}\tilde{\rho}_{22} - \tilde{\rho}_{11}V_{12}\right) + \Phi_{12}^{11}\tilde{\rho}_{11} + \Phi_{12}^{22}\tilde{\rho}_{22},\end{aligned} \tag{7.222}$$

$$\frac{d\tilde{\rho}_{21}}{dt} = -\tilde{\rho}_{21}\left(\frac{A_{21}}{2} - \Phi_{21}^{21}\right) - i\omega_{21}\tilde{\rho}_{21} + \Phi_{21}^{12}\tilde{\rho}_{12} - \frac{i}{\hbar}(V_{21}\tilde{\rho}_{11} - \tilde{\rho}_{22}V_{21}) + \Phi_{21}^{11}\tilde{\rho}_{11} + \Phi_{21}^{22}\tilde{\rho}_{22}. \quad (7.223)$$

From (7.105), (7.106), (7.112), (7.198)–(7.203), it can be seen that for a two-level system

$$\Phi_{22}^{12} = 0, \quad (7.224)$$

$$\Phi_{11}^{12} = 0, \quad (7.225)$$

$$\Phi_{22}^{21} = 0, \quad (7.226)$$

$$\Phi_{11}^{21} = 0, \quad (7.227)$$

$$\Phi_{12}^{11} = 0, \quad (7.228)$$

$$\Phi_{12}^{22} = 0, \quad (7.229)$$

$$\Phi_{21}^{11} = 0, \quad (7.230)$$

$$\Phi_{21}^{22} = 0. \quad (7.231)$$

The elements Φ_{12}^{21} and Φ_{21}^{12} are mixed terms as described by (7.112). With the help of (7.224)–(7.231), (7.220)–(7.223) reduce to:

$$\frac{d\tilde{\rho}_{22}}{dt} = -\tilde{\rho}_{22}\left(A_{21} - \Phi_{22}^{22}\right) + \Phi_{22}^{11}\tilde{\rho}_{11} - \frac{i}{\hbar}(V_{21}\tilde{\rho}_{12} - \tilde{\rho}_{21}V_{12}), \quad (7.232)$$

$$\frac{d\tilde{\rho}_{11}}{dt} = -\tilde{\rho}_{11}\left(-\Phi_{11}^{11}\right) + \left(\Phi_{11}^{22} + A_{21}\right)\tilde{\rho}_{22} - \frac{i}{\hbar}(V_{12}\tilde{\rho}_{21} - \tilde{\rho}_{12}V_{21}), \quad (7.233)$$

$$\frac{d\tilde{\rho}_{12}}{dt} = -\tilde{\rho}_{12}\left(\frac{A_{21}}{2} - \Phi_{12}^{12}\right) - i\omega_{12}\tilde{\rho}_{12} + \Phi_{12}^{21}\tilde{\rho}_{21} - \frac{i}{\hbar}(V_{12}\tilde{\rho}_{22} - \tilde{\rho}_{11}V_{12}), \quad (7.234)$$

$$\frac{d\tilde{\rho}_{21}}{dt} = -\tilde{\rho}_{21}\left(\frac{A_{21}}{2} - \Phi_{21}^{21}\right) - i\omega_{21}\tilde{\rho}_{21} + \Phi_{21}^{12}\tilde{\rho}_{12} - \frac{i}{\hbar}(V_{21}\tilde{\rho}_{11} - \tilde{\rho}_{22}V_{21}). \quad (7.235)$$

The collisional operator elements $\Phi_{ab}^{a'b'}$ are proportional to the electron density (in the binary approximation), i.e., $\Phi_{ab}^{a'b'} \sim n_e$, while the field matrix elements V_{ab} are either essentially constant/oscillating external applied fields. For the case of quasi-static ion fields in plasmas V_{ab} scales approximately like the classical Holtsmark field, i.e., $V_{ab} \sim n_i^{2/3}$. From this it follows that at high densities, the electron collisional operator elements dominate all other terms (in particular the radiative decay A_{ba} and the imaginary terms $i\omega_{ab}$) in (7.222), (7.223). Therefore, in the stationary case at very high densities (7.234), (7.235) reduce to

$$\tilde{\rho}_{12} \cdot \left(\Phi_{12}^{12} \cdot \Phi_{21}^{21} - \Phi_{12}^{21} \cdot \Phi_{21}^{12}\right) = 0 \tag{7.236}$$

and

$$\tilde{\rho}_{21} \cdot \left(\Phi_{21}^{21} \cdot \Phi_{12}^{12} - \Phi_{21}^{12} \cdot \Phi_{12}^{21}\right) = 0. \tag{7.237}$$

Relations (7.236), (7.237) can only be valid in general if

$$\tilde{\rho}_{21} = 0 \tag{7.238}$$

and

$$\tilde{\rho}_{12} = 0. \tag{7.239}$$

Due to (7.238), (7.239), the field terms in (7.232), (7.233) vanish and we are left with

$$0 = \tilde{\rho}_{22}\Phi_{22}^{22} + \Phi_{22}^{11}\tilde{\rho}_{11}, \tag{7.240}$$

$$0 = \tilde{\rho}_{11}\Phi_{11}^{11} + \Phi_{11}^{22}\tilde{\rho}_{22} \tag{7.241}$$

from which we obtain

$$\frac{\tilde{\rho}_{22}}{\tilde{\rho}_{11}} = -\frac{\Phi_{22}^{11}}{\Phi_{22}^{22}}, \tag{7.242}$$

$$\frac{\tilde{\rho}_{22}}{\tilde{\rho}_{11}} = -\frac{\Phi_{11}^{11}}{\Phi_{11}^{22}}. \tag{7.243}$$

From (7.219), it follows from (7.242), (7.243):

$$\frac{\tilde{\rho}_{22}}{\tilde{\rho}_{11}} = -\frac{n_e\langle v\sigma(J_1 \to J_2)\rangle}{(-n_e\langle v\sigma(J_2 \to J_1)\rangle)} = \frac{\langle v\sigma(J_1 \to J_2)\rangle}{\langle v\sigma(J_2 \to J_1)\rangle}, \tag{7.244}$$

$$\frac{\tilde{\rho}_{22}}{\tilde{\rho}_{11}} = -\frac{(-n_e\langle v\sigma(J_1 \to J_2)\rangle)}{n_e\langle v\sigma(J_2 \to J_1)\rangle} = \frac{\langle v\sigma(J_1 \to J_2)\rangle}{\langle v\sigma(J_2 \to J_1)\rangle}. \tag{7.245}$$

Equations (7.244) and (7.245) are therefore identical as it should be. Physically they correspond to the principal of detailed balance where the flow of each elementary process is exactly balanced by its inverse one, i.e.

$$n_e \cdot \langle \upsilon\sigma(J_1 \to J_2) \rangle \cdot \tilde{\rho}_{11} = n_e \cdot \langle \upsilon\sigma(J_2 \to J_1) \rangle \cdot \tilde{\rho}_{22}. \tag{7.246}$$

If the relations between the cross sections or rate coefficients for the direct and inverse process are known, (7.246) provides readily the population ratio $\tilde{\rho}_{22}/\tilde{\rho}_{11}$.

7.7.2 The Principle of Detailed Balance and Microreversibility

Let us consider a system in thermodynamic equilibrium where the principle of detailed balance holds. The principle of detailed balance holds true not only for a total elementary process but also for each energy interval dE. For the electron collisional excitation, we can formally write

$$X_i + e(E) \to X_j + e(E'), \tag{7.247}$$

where X_i and X_j designate the atom in the quantum state "*i*" and "*j*", respectively, and the electron energy E and E' are the energies before and after scattering. These energies are related to each other via the excitation energy ΔE_{ij}:

$$E = E' + \Delta E_{ij}. \tag{7.248}$$

The detailed balance for the energy interval dE then reads:

$$n_e n_i \sigma_{ij}(E)\upsilon(E)F(E)\mathrm{d}E = n_e n_j \sigma_{ji}(E')\upsilon(E')F(E')\mathrm{d}E', \tag{7.249}$$

where σ_{ij} is the excitation cross section, σ_{ji} the de-excitation cross section, $\upsilon(E)$ the electron velocity at electron impact energy E, and $F(E)$ the electron energy distribution function. From (7.248), it follows

$$\mathrm{d}E = \mathrm{d}E'. \tag{7.250}$$

In thermodynamic equilibrium, the electron energy distribution function is the Maxwell distribution

$$F(E) = \frac{2}{\sqrt{\pi}} \cdot \sqrt{E} \cdot \frac{\exp(-E/kT_e)}{(kT_e)^{3/2}} \tag{7.251}$$

and the atomic populations are the Boltzmann populations

$$\frac{n_j}{n_i} = \frac{g_j}{g_i} \cdot \exp(-\Delta E_{ij}/kT_e). \tag{7.252}$$

Assuming a nonrelativistic relation between energy and velocity, i.e.

$$\upsilon = \sqrt{\frac{2E}{m_e}} \tag{7.253}$$

we obtain from (7.249) with the help of (7.250)–(7.253)

$$\begin{aligned} & n_e \sigma_{ij}(E) \sqrt{\frac{2E}{m_e}} \frac{2}{\sqrt{\pi}} \sqrt{E} \cdot \frac{\exp(-E/kT_e)}{(kT_e)^{3/2}} \cdot \mathrm{d}E' \\ & = n_e \frac{g_j}{g_i} \cdot \exp\left(-\Delta E_{ij}/kT_e\right) \sigma_{ji}(E') \sqrt{\frac{2E'}{m_e}} \frac{2}{\sqrt{\pi}} \sqrt{E'} \frac{\exp(-E'/kT_e)}{(kT_e)^{3/2}} \mathrm{d}E'. \end{aligned} \tag{7.254}$$

From (7.254), it follows with (7.248)

$$\begin{aligned} \sigma_{ji}(E') &= \frac{g_i}{g_j} \sigma_{ij}(E) \frac{E}{E'} \\ &= \frac{g_i}{g_j} \sigma_{ij}\left(E' + \Delta E_{ij}\right) \frac{E' + \Delta E_{ij}}{E'}. \end{aligned} \tag{7.255}$$

Although thermodynamic equilibrium, Maxwell energy distribution, and Boltzmann relation have been used for the derivation of (7.255), the final expression does not contain anymore the temperature; i.e., the relation (7.255) is independent from the energy distribution function.

From cross sections, we can readily derive the rate coefficients:

$$\langle \upsilon \sigma_{ij} \rangle = \int_{\Delta E_{ij}}^{\infty} \upsilon(E) \cdot F(E) \cdot \sigma_{ij}(E) \cdot \mathrm{d}E. \tag{7.256}$$

Equation (7.256) can be transformed with the help of (7.248), (7.250) into

$$\langle \upsilon \sigma_{ij} \rangle = \int_0^{\infty} \upsilon\left(E' + \Delta E_{ij}\right) \cdot F\left(E' + \Delta E_{ij}\right) \cdot \sigma_{ij}\left(E' + \Delta E_{ij}\right) \cdot \mathrm{d}E'. \tag{7.257}$$

Injecting (7.255) into (7.257), we obtain

$$\langle \upsilon \sigma_{ij} \rangle = \frac{g_j}{g_i} \int_0^{\infty} \upsilon\left(E' + \Delta E_{ij}\right) \cdot F\left(E' + \Delta E_{ij}\right) \cdot \frac{E'}{E' + \Delta E_{ij}} \cdot \sigma_{ji}(E') \cdot \mathrm{d}E'. \tag{7.258}$$

Assuming a Maxwellian energy distribution function (7.251), it follows from (7.258):

$$\begin{aligned}\langle \upsilon\sigma_{ij}\rangle &= \frac{g_j}{g_i}\cdot\int_0^\infty \sqrt{\frac{2}{m_e}}\cdot\frac{2}{\sqrt{\pi}}\cdot E'\cdot\frac{\exp\left(-(E'+\Delta E_{ij})/kT_e\right)}{(kT_e)^{3/2}}\cdot\sigma_{ji}(E')\cdot dE'\\ &= \frac{g_j}{g_i}\cdot\exp\left(-\Delta E_{ij}/kT_e\right)\cdot\int_0^\infty\sqrt{\frac{2E'}{m_e}}\cdot\frac{2}{\sqrt{\pi}}\cdot\sqrt{E'}\cdot\frac{\exp(-E'/kT_e)}{(kT_e)^{3/2}}\cdot\sigma_{ji}(E')\cdot dE'\\ &= \frac{g_j}{g_i}\cdot\exp\left(-\Delta E_{ij}/kT_e\right)\cdot\langle\upsilon\sigma_{ji}\rangle,\end{aligned} \tag{7.259}$$

i.e.,

$$\langle\upsilon\sigma_{ij}\rangle = \frac{g_j}{g_i}\cdot\exp\left(-\Delta E_{ij}/kT_e\right)\cdot\langle\upsilon\sigma_{ji}\rangle. \tag{7.260}$$

Would we inject (7.260) into (7.245) (note that $i=1, j=2$, $\sigma_{12}=\sigma_{12}(J_1\to J_2)$ and $\sigma_{21}=\sigma_{21}(J_2\to J_1)$ because the levels are M-quantum number averaged) we would recover the Boltzmann relation for the level populations of the density equations. This conclusion, however, is ill-defined because it is based on the relation (7.255) that has been derived under the assumption of thermodynamic equilibrium (i.e., detailed balance, Maxwellian energy distribution, Boltzmann relation). In other words, the conclusion that (7.245) is equivalent to the Boltzmann relation for populations had already been used during the course of derivation [to obtain (7.255)]. Equation (7.260) should therefore *not* be considered as a result but rather as a simple recovery of a relation being already used before.

A correct line of reasoning has to start from the CPT-invariance properties of the Hamiltonian that describes the scattering process, i.e., from the principle of microreversibility. It can be shown (Muirhead 1965; Dawydow 1981) that the direct and inverse cross sections for the process (7.247) are related via

$$\sigma_{ji}(E')\big|_{CPT} = \frac{g_i}{g_j}\sigma_{ij}\left(E'+\Delta E_{ij}\right)\frac{E'+\Delta E_{ij}}{E'}. \tag{7.261}$$

Equation (7.261) is the famous Klein–Rosseland formula. This formula is formally identical to (7.255) but is derived without any assumptions of thermodynamic equilibrium, Maxwellian energy distribution nor Boltzmann population. It is therefore conceptionally different from (7.255) and for this reason, we have inserted the index "CPT" in (7.261). We now can calculate directly the rate coefficients of the direct and inverse processes from the integrals of (7.258) without using the assumptions of detailed balance and Boltzmann distribution. Inserting (7.261) into (7.258), we obtain:

$$\langle \upsilon \sigma_{ij} \rangle = \frac{g_j}{g_i} \int_0^{\infty} \upsilon\left(E' + \Delta E_{ij}\right) \cdot F\left(E' + \Delta E_{ij}\right) \cdot \frac{E'}{E' + \Delta E_{ij}} \cdot \sigma_{ji}\left(E'\right)\Big|_{CPT} \cdot dE' \quad (7.262)$$

from which it follows with the help of the Maxwellian distribution function

$$\begin{aligned} \langle \upsilon \sigma_{ij} \rangle &= \frac{g_j}{g_i} \cdot \int_0^{\infty} \sqrt{\frac{2}{m_e}} \cdot \frac{2}{\sqrt{\pi}} \cdot E' \cdot \frac{\exp\left(-\left(E' + \Delta E_{ij}\right)/kT_e\right)}{\left(kT_e\right)^{3/2}} \cdot \sigma_{ji}\left(E'\right)\Big|_{CPT} \cdot dE' \\ &= \frac{g_j}{g_i} \cdot \exp\left(-\Delta E_{ij}/kT_e\right) \cdot \int_0^{\infty} \sqrt{\frac{2E'}{m_e}} \cdot \frac{2}{\sqrt{\pi}} \cdot \sqrt{E'} \cdot \frac{\exp\left(-E'kT_e\right)}{\left(kT_e\right)^{3/2}} \cdot \sigma_{ji}\left(E'\right)\Big|_{CPT} \cdot dE'. \end{aligned} \quad (7.263)$$

The integral in (7.263) corresponds to the rate coefficient for the inverse process, i.e.

$$\int_0^{\infty} \sqrt{\frac{2E'}{m_e}} \cdot \frac{2}{\sqrt{\pi}} \cdot \sqrt{E'} \cdot \frac{\exp\left(-E'/kT_e\right)}{\left(kT_e\right)^{3/2}} \cdot \sigma_{ji}\left(E'\right)\Big|_{CPT} \cdot dE' = \langle \upsilon \sigma_{ji} \rangle. \quad (7.264)$$

Because

$$\langle \upsilon \sigma_{ij} \rangle = \langle \upsilon \sigma_{12} \rangle = \langle \upsilon \sigma(J_1 \to J_2) \rangle \quad (7.265)$$

and

$$\langle \upsilon \sigma_{ji} \rangle = \langle \upsilon \sigma_{21} \rangle = \langle \upsilon \sigma(J_2 \to J_1) \rangle \quad (7.266)$$

(7.263) takes the form

$$\langle \upsilon \sigma(J_1 \to J_2) \rangle = \frac{g_j}{g_i} \cdot \exp\left(-\Delta E_{ij}/kT_e\right) \cdot \langle \upsilon \sigma(J_2 \to J_1) \rangle. \quad (7.267)$$

Inserting (7.267) into (7.245), we finally obtain

$$\frac{\tilde{\rho}_{22}}{\tilde{\rho}_{11}} = \frac{\langle \upsilon \sigma(J_1 \to J_2) \rangle}{\langle \upsilon \sigma(J_2 \to J_1) \rangle} = \frac{g_j}{g_i} \cdot \exp\left(-\Delta E_{ij}/kT_e\right). \quad (7.268)$$

Therefore, (7.242)–(7.245) and the corresponding electron collisional operator do show that the diagonal elements of the density matrix equations approach the Boltzmann ratio at high density.

It should be emphasized that the correct thermodynamic limit can only be obtained when the interference term of the electron collisional operator from (7.66), (7.69), (7.112) is included in its general form. Therefore, a quantum kinetic approach that is consistent with the thermodynamic limit requests a rigorous and consistent application of the second-order perturbation theory for the scattering matrix and the system evolution operator [see also (7.27), (7.28), (7.36)] that contain all possible matrix elements (i.e., all matrix elements that can be formed with a given set of wave function vectors that is not restricted to either upper or lower levels). This is different from usual broadening considerations. Here, the interference term constitutes a correction to the dominating first term for the upper level.

It should be noted that the second-order perturbation theory of the scattering matrix has not to be confused with the multipole expansion of the interaction potential (see 7.31)–(7.35) in terms of monopole, dipole, quadrupole, etc. These multipoles arise from the Taylor expansion of the interaction potential. We note, that (7.246) corresponds to the general result of the first-order Born approximation in quantum mechanical perturbation theory where detailed balance holds true for all systems (in systems that are spin dependent we have to average over the spin projections of the initial and final states, respectively, sometimes to be referred to as semi-detailed balance).

7.7.3 Comments to a Two-Level Quantum Kinetics Resolved in M-Quantum Number

If the two-level system is resolved in M-quantum number [here the indexes refer to $\alpha = 1, 2$ and $\beta = 1, 2$ while in the case of M-quantum averages $a = 1, 2$ and $b = 1, 2$, [see (7.220)], we have to start our investigations from (7.9). In order to facilitate the discussion, let us assume $A = 1$ and $B = 0$ in (7.66) for the electron collisional operators Φ_{11}^{11} and Φ_{11}^{22}, i.e.

$$\hat{\Phi}_{11}^{11} = -n_e \cdot \frac{4\pi e^4}{3\hbar^2} \cdot \int_0^{\infty} \frac{f(\upsilon)}{\upsilon} d\upsilon \int_{\rho_{\min}}^{\rho_{\max}} \frac{d\rho}{\rho} \cdot \left\{ \phi_{(1),11}^{(M),11} + \phi_{(2),11}^{(M),11} \right\}, \tag{7.269}$$

$$\hat{\Phi}_{11}^{22} = -n_e \cdot \frac{4\pi e^4}{3\hbar^2} \cdot \int_0^{\infty} \frac{f(\upsilon)}{\upsilon} d\upsilon \int_{\rho_{\min}}^{\rho_{\max}} \frac{d\rho}{\rho} \cdot \left\{ \phi_{(3),11}^{(M),22} \right\} \tag{7.270}$$

with [see also (7.105), (7.106), (7.112)]

$$\phi_{(1),11}^{(M),11} = \frac{\delta_{11} \cdot \delta_{J_1 J_1} \cdot \delta_{M_1 M_1} \cdot \delta(J_1, 1, J_2)}{2J_1 + 1} \cdot |\langle \gamma_1 J_1 \| r \| \gamma_2 J_2 \rangle|^2, \tag{7.271}$$

$$\phi_{(2),11}^{(M),11} = \frac{\delta_{11} \cdot \delta_{J_1 J_1} \cdot \delta_{M_1 M_1} \cdot \delta(J_1, 1, J_2)}{2J_1 + 1} \cdot |\langle \gamma_1 J_1 \| r \| \gamma_2 J_2 \rangle|^2, \tag{7.272}$$

$$\phi_{(3),11}^{(M),22} = 2 \cdot \frac{|\langle \gamma_2 J_2 \| r \| \gamma_1 J_1 \rangle|^2}{2J_2+1} \cdot \sum_q (2J_2+1) \cdot \begin{pmatrix} J_1 & 1 & J_2 \\ -M_1 & q & M_2 \end{pmatrix}^2. \tag{7.273}$$

Equation (7.273) shows that the assumption of a framework of a two-level system that is resolved in M-quantum number is inconsistent. This can be seen directly from the $3j$ symbol of (7.273). In the two-level case, there is only one upper level resolved in M-quantum number and one lower level also resolved in M-quantum number. Therefore, $J_1 = 0$ in order to have only one level with $M_1 = 0$ and likewise $J_2 = 0$ with $M_2 = 0$. The $3j$ symbol, however, indicates that the transition $\Delta J = 0$ is forbidden for the case $J_1 = 0$ and $J_2 = 0$; i.e., the $3j$ symbol is zero. The same holds true for the other matrix elements from (7.271), (7.272): the triangular relation is not fulfilled, i.e., $\delta(J_1, 1, J_2) = 0$ for the case $\Delta J = 0$ with $J_1 = 0$ and $J_2 = 0$. Therefore, no collisional transition exists between these two levels. From this, it follows that one cannot study consistently the Boltzmann limit induced by random electron collisions at high densities in the framework of a two-level system that is resolved in M-quantum number.

7.8 The Field Perturbation Operator $\hat{V}$

7.8.1 The Quasi-classical Electric Field Perturbation

The perturbation of the Hamiltonian due to an electric field can be described within the well-established quasi-classical approach, i.e.

$$\hat{V} = -\vec{d} \cdot \vec{F}, \tag{7.274}$$

where $\vec{d}$ is the electric dipole moment and $\vec{F}$ is the electric field. The corresponding matrix element of the field perturbation can be expressed via the reduced dipole matrix element (see (7.93)):

$$\begin{aligned} V_{\alpha\beta} &= \langle \gamma_\alpha J_\alpha M_\alpha | \hat{V} | \gamma_\beta J_\beta M_\beta \rangle \\ &= -\sum_q F \cdot (-1)^{q+J_\alpha - M_\alpha} \begin{pmatrix} J_\alpha & 1 & J_\beta \\ -M_\alpha & q & M_\beta \end{pmatrix} \langle \gamma_\alpha J_\alpha \| d \| \gamma_\beta J_\beta \rangle . \end{aligned} \tag{7.275}$$

If we align the z-axis along with the electric field vector $\vec{F}$, (7.275) simplifies to

$$\begin{aligned} V_{\alpha\beta} &= -F \cdot (-1)^{J_\alpha - M_\alpha} \begin{pmatrix} J_\alpha & 1 & J_\beta \\ -M_\alpha & 0 & M_\beta \end{pmatrix} \langle \gamma_\alpha J_\alpha \| d \| \gamma_\beta J_\beta \rangle \\ &= -e a_0 F (-1)^{J_\alpha - M_\alpha} \begin{pmatrix} J_\alpha & 1 & J_\beta \\ -M_\alpha & 0 & M_\beta \end{pmatrix} \langle \gamma_\alpha J_\alpha \| r \| \gamma_\beta J_\beta \rangle . \end{aligned} \tag{7.276}$$

In order to replace the reduced dipole matrix element with the expressions (7.133) for the oscillator strength, it is necessary to consider first the conjugate complex of (7.276) rather than inserting the square root of the oscillator strength (because the latter procedure destroys the correct quantum mechanical phase). From (7.276), we find

$$\begin{aligned} V_{\alpha\beta}^{*} &= -F \cdot (-1)^{J_\alpha - M_\alpha} \begin{pmatrix} J_\alpha & 1 & J_\beta \\ -M_\alpha & 0 & M_\beta \end{pmatrix} \langle \gamma_\alpha J_\alpha \| d \| \gamma_\beta J_\beta \rangle^{*} \\ &= -ea_0 F \cdot (-1)^{J_\alpha - M_\alpha} \begin{pmatrix} J_\alpha & 1 & J_\beta \\ -M_\alpha & 0 & M_\beta \end{pmatrix} \langle \gamma_\alpha J_\alpha \| r \| \gamma_\beta J_\beta \rangle^{*} \end{aligned} \tag{7.277}$$

from which it follows with (7.133) and the fact that $2J_\alpha - 2M_\alpha$ is even

$$\begin{aligned} V_{\alpha\beta} \cdot V_{\alpha\beta}^{*} &= F^2 \cdot (-1)^{2J_\alpha - 2M_\alpha} \begin{pmatrix} J_\alpha & 1 & J_\beta \\ -M_\alpha & 0 & M_\beta \end{pmatrix}^2 \left| \langle \gamma_\alpha J_\alpha \| d \| \gamma_\beta J_\beta \rangle \right|^2 \\ &= \frac{3e^2\hbar^2}{2m_e} \cdot F^2 \cdot \begin{pmatrix} J_\alpha & 1 & J_\beta \\ -M_\alpha & 0 & M_\beta \end{pmatrix}^2 \cdot \frac{(2J_\alpha + 1) f_{J_\alpha J_\beta}}{\hbar\omega_{\alpha\beta}} \\ &= 3e^2 a_0^2 Ry \cdot F^2 \cdot \begin{pmatrix} J_\alpha & 1 & J_\beta \\ -M_\alpha & 0 & M_\beta \end{pmatrix}^2 \cdot \frac{(2J_\alpha + 1) f_{J_\alpha J_\beta}}{\hbar\omega_{\alpha\beta}}. \end{aligned} \tag{7.278}$$

7.8.2 The Ionic Field Mixing

If particles are moving sufficiently slowly so that the frequencies characterizing the actual time dependence of the perturbing electric field produced in the vicinity of the radiator ("jumping frequency") are much smaller than the resulting Stark shifts $\Delta\omega_S(F) = \Delta\omega_{\alpha\beta}(F)$ (where $\Delta\omega_{\alpha\beta}(F) = \omega_{\alpha\beta}(F) - \omega_{\alpha\beta}(F = 0)$) the electric field is quasi-static, i.e.

$$\left| \frac{\mathrm{d}F(t)/\mathrm{d}t}{F(t)} \right| \ll |\Delta\omega_S(F)|. \tag{7.279}$$

The Stark shifts can be estimated from perturbation theory:

$$\Delta\omega_{\alpha\beta}(F) = C_{\alpha\beta}^{(n)} \cdot F^n. \tag{7.280}$$

For the linear Stark effect, (H-like ions) we have $n = 1$ while for the quadratic Stark effect (non-hydrogen-like ions) $n = 2$, $C_{\alpha\beta}$ is the Stark constant.

The essential features of the quasi-static approximation can be estimated with the electric field produced by the nearest perturber. This can be seen as follows. The characteristic ionic field in a plasma is given by the Holtsmark characteristic field (Unsöld 1955):

$$F_0 = 2\pi \cdot \left(\frac{4}{15}\right)^{2/3} \cdot Z_p e \cdot n_i^{2/3} = 2.6031 \cdot Z_p e \cdot n_i^{2/3}. \tag{7.281}$$

Taking the ion sphere radius R_i according to

$$n_i = \frac{1}{\frac{4}{3}\pi R_i^3} \tag{7.282}$$

as the mean interaction radius for the distance between perturber and radiator, i.e.

$$r \approx R_i = \left(\frac{3}{4\pi}\right)^{1/3} \cdot n_i^{-1/3} = 0.62035 \cdot n_i^{-1/3} \tag{7.283}$$

we find

$$F \approx \frac{Z_p e}{r^2} \approx \left(\frac{4\pi}{3}\right)^{2/3} \cdot Z_p e \cdot n_i^{2/3} = 2.5985 \cdot Z_p e \cdot n_i^{2/3}. \tag{7.284}$$

Comparing (7.281), (7.284), we find that the binary approximation is very close to the statistical mean Holtsmark field. Estimating now the relative field perturbation from (7.279) via

$$\left|\frac{\mathrm{d}F(t)/\mathrm{d}t}{F(t)}\right| \approx \frac{\upsilon_p}{r} \tag{7.285}$$

(υ_p is the velocity of the ionic perturbers), we find within the binary approximation (7.284)

$$\Delta\omega_{\alpha\beta}(F) \gg \left(\frac{1}{C_{\alpha\beta}^{(n)}}\right)^{2n-1} \cdot \left(\frac{\upsilon_p^2}{Z_p e}\right)^{\frac{n}{2n-1}}. \tag{7.286}$$

The linear Stark effect can be estimated from the first-order perturbation theory:

$$\hbar C^{(1)} \approx \frac{3}{2} \cdot \frac{\hbar^2}{Z \cdot e \cdot m_e} \cdot n(n_1 - n_2), \tag{7.287}$$

where n is the principal quantum number of the atomic level and n_1 and n_2 are the parabolic quantum numbers and Z is the ion charge (e.g., $Z = 1$ for hydrogen, $Z = 2$ for ionized helium). The outermost levels correspond to $n_1 = n - 1, n_2 = 0$ and

$n_1 = 0, n_2 = n - 1$. Therefore, the energy difference between these outermost levels is

$$\Delta E = \frac{3ea_0}{Z} \cdot n(n-1) \cdot F. \tag{7.288}$$

According to (7.287), for large quantum numbers, the linear Stark constant can be estimated from

$$\hbar C^{(1)}_{\alpha\beta} \approx \frac{3\hbar^2}{2 \cdot Z \cdot e \cdot m_e} \cdot \left(n_\alpha^2 - n_\beta^2\right) \approx \hbar C^{(1)}_{ab}, \tag{7.289}$$

where n_α and n_β are the principal quantum numbers of the upper and lower levels, respectively. The last relation in (7.289) indicates the relation to an averaged Stark constant (averaged over magnetic quantum numbers). In atomic units, (7.289) takes the form (note that in this case the field strengths F have to be measured in atomic units too, i.e., in units of e/a_0^2):

$$\left[\hbar C^{(1)}_{ab}\right]_{\text{a.u.}} \approx \frac{3}{2 \cdot Z} \cdot \left(n_\alpha^2 - n_\beta^2\right). \tag{7.290}$$

For the quadratic Stark effect, the Stark constant can be estimated from the second-order perturbation theory:

$$\hbar C^{(2)} \approx \sum_{\gamma' J'} \frac{\left|\langle \gamma JM|\vec{d}|\gamma' J' M'\rangle\right|^2}{\hbar\omega_{\gamma J,\gamma' J'}}. \tag{7.291}$$

With the help of the Wigner–Eckart theorem, the matrix element in (7.291) can be expressed in terms of the reduced matrix element. As before, the z-coordinate is along the electric field vector [see also (7.276)] and we obtain:

$$\begin{aligned} \hbar C^{(2)} &\approx \sum_{\gamma' J' M'} \begin{pmatrix} J & 1 & J' \\ -M & 0 & M' \end{pmatrix}^2 \cdot \frac{|\langle \gamma J\|d\|\gamma' J'\rangle|^2}{\hbar\omega_{\gamma J,\gamma' J'}} \\ &= \sum_{\gamma' J'} \frac{1}{2J+1} \cdot \frac{|\langle \gamma J\|d\|\gamma' J'\rangle|^2}{\hbar\omega_{\gamma J,\gamma' J'}} = e^2 a_0^2 \cdot \sum_{\gamma' J'} \frac{1}{2J+1} \cdot \frac{|\langle \gamma J\|r\|\gamma' J'\rangle|^2}{\hbar\omega_{\gamma J,\gamma' J'}}. \end{aligned} \tag{7.292}$$

With the help of (7.134), the quadratic Stark constant can be expressed in terms of oscillator strengths:

$$\hbar C^{(2)} \approx \frac{3}{2} \cdot \frac{\hbar^2 e^2}{m_e} \sum_{\gamma' J'} \frac{f_{\gamma J,\gamma' J'}}{\left(\hbar\omega_{\gamma J,\gamma' J'}\right)^2}. \tag{7.293}$$

In atomic units, the quadratic Stark constant takes likewise a very simple form:

$$\hbar C^{(2)} \approx \frac{3}{2} \cdot \sum_{\gamma' J'} \frac{f_{\gamma J,\gamma' J'}}{\Delta E^2_{\gamma J,\gamma' J'}(a.u.)}. \tag{7.294}$$

Taking into account only the most closely lying level in the summation of (7.294), we obtain (in atomic units)

$$\left[\hbar C_{\mathrm{ab}}^{(2)}\right]_{\mathrm{a.u.}} \approx \frac{3}{2} \cdot \frac{f_{\mathrm{ab}}}{\Delta E^2_{\mathrm{ab}}(a.u.)}. \tag{7.295}$$

With the help of (7.279), (7.283), (7.285), (7.286), the condition for the quasi-static regime can finally be formulated as follows:

$$n_{\mathrm{i}} \gg \left[\frac{\upsilon_{\mathrm{p}}}{\left(\frac{3}{4\pi}\right)^{1/3} \cdot (2\pi)^n \cdot \left(\frac{4}{15}\right)^{2n/3} \cdot (Z_{\mathrm{p}} e)^n \cdot C_{\mathrm{ab}}^{(n)}}\right]^{\frac{3}{2n-1}}, \tag{7.296}$$

where the perturber particle velocity is given by

$$\upsilon_{\mathrm{p}} = \sqrt{\frac{3kT_{\mathrm{p}}}{\mu}}. \tag{7.297}$$

μ is the reduced mass of the perturber and radiator and T_{p} is the temperature of the perturbers. For the linear Stark effect, we obtain

$$\begin{aligned} n_{\mathrm{i}}^{(1)} &\gg \frac{\upsilon_{\mathrm{p}}^3}{6\pi^2 \cdot \left(\frac{4}{15}\right)^2 \cdot Z_{\mathrm{p}}^3 e^3 \cdot \left[C_{\mathrm{ab}}^{(1)}\right]^3} \\ &\approx 7.5 \times 10^{17} \cdot \left(\frac{kT_{\mathrm{p}}(\mathrm{eV})}{\mu(amu)}\right)^{3/2} \cdot \frac{1}{Z_{\mathrm{p}}^3 \cdot \left[\hbar C_{\mathrm{ab}}^{(1)}\right]^3_{\mathrm{a.u.}}} \left[\mathrm{cm}^{-3}\right], \end{aligned} \tag{7.298}$$

while for the quadratic Stark effect

$$\begin{aligned} n_{\mathrm{i}}^{(2)} &\gg \frac{\upsilon_{\mathrm{p}}}{4\pi^2 \cdot \left(\frac{3}{4\pi}\right)^{1/3} \cdot \left(\frac{4}{15}\right)^{4/3} \cdot Z_{\mathrm{p}}^2 e^2 \cdot C_{\mathrm{ab}}^{(2)}} \\ &\approx 1.2 \times 10^{22} \cdot \sqrt{\frac{kT_{\mathrm{p}}(\mathrm{eV})}{\mu(amu)}} \cdot \frac{1}{Z_{\mathrm{p}}^2 \cdot \left[\hbar C_{\mathrm{ab}}^{(2)}\right]_{\mathrm{a.u.}}} \left[\mathrm{cm}^{-3}\right]. \end{aligned} \tag{7.299}$$

The conditions (7.298), (7.299) set a lower density limit for the quasi-static approximation. It should however be noted that the notion of the quasi-static approximation according to (7.279) is related to the framework of the line broadening theory and not to the level mixing phenomena that are described by (7.9), (7.274), and it is therefore necessary to distinguish between broadening and energy level mixing.

In line broadening the condition of static perturbation is a simple phase modulation of the atomic energy levels due to phase shift fluctuations induced by the quadratic Stark effect. If these fluctuations are slow (static), the effect of line broadening is described by the static approximation. In the opposite case, when the fluctuations are fast, the broadening effect is described by the adiabatic impact approximation where only elastic collisions are taken into account and level mixing is negligible. In both cases, the energy level mixing effect is the same and determined by statistic electric fields that depend parametrically on time: the field direction can be considered as constant and aligned along z-direction according to (7.276). In other words: The failure of the quasi-static condition (7.279) for the quadratic Stark effect does not mean a randomization of the ionic field direction.

A static condition for the level populations means that the field fluctuation rate is small compared to the sum of the decay rate τ_α^{-1} that can be estimated (lower limit) with the radiative and Auger decay (see also Chap. 6), i.e.

$$\left|\frac{\mathrm{d}F(t)/\mathrm{d}t}{F(t)}\right| \ll \tau_\alpha^{-1} \approx \sum_\beta A_{\alpha\beta} + \sum_k \Gamma_{\alpha k}. \tag{7.300}$$

Employing the approximation (7.285) as before, we obtain from (7.300) with (7.283), (7.285), (7.297)

$$n_{\mathrm{i}}^{(k)} \ll \frac{1}{\tau_\alpha^3}\cdot\frac{3}{4\pi}\cdot\left(\frac{\mu}{3kT_{\mathrm{p}}}\right)^{3/2}. \tag{7.301}$$

We note that in a more generalized density matrix theory (Rautian and Shalagin 1991; Anufrienko et al. 1993) the density matrix elements appear as functions of the frequency, i.e., $\rho_{\alpha\beta} = \rho_{\alpha\beta}(\omega)$: Line broadening and atomic populations are therefore coupled and generalized in a unified framework. This presentation, however, requests the introduction of the quantized electromagnetic field in the density matrix representation that is outside the scope of the present introductory chapter.

Let us now consider the condition for the application of the perturbation theory, i.e., the condition

$$\left|V_{\alpha\beta}\right| \ll \left|\hbar\omega_{\alpha\beta}\right|. \tag{7.302}$$

Substituting the mean Holtsmark field F_0 from (7.281) into (7.276), (7.277), we obtain

$$\begin{aligned}|V_{\alpha\beta}| &= e^2 a_0\sqrt{3}\cdot 2\pi\cdot\left(\frac{4}{15}\right)^{2/3}\cdot Z_\mathrm{p} n_\mathrm{i}^{2/3} \\ &\times\left|(-1)^{J_\alpha - M_\alpha + 1}\begin{pmatrix} J_\alpha & 1 & J_\beta \\ -M_\alpha & 0 & M_\beta\end{pmatrix}\right|\cdot\sqrt{\frac{(2J_\alpha+1)f_{J_\alpha J_\beta}}{\hbar\omega_{\alpha\beta}/Ry}}\end{aligned} \tag{7.303}$$

or, in convenient units

$$\frac{|V_{\alpha\beta}|}{\hbar} = 19.253\cdot Z_\mathrm{p}\cdot n_\mathrm{i}^{2/3}\cdot|Y|\cdot\sqrt{\frac{(2J_\alpha+1)f_{J_\alpha J_\beta}}{\Delta E_{\alpha\beta}\,(\mathrm{eV})}}\left[\mathrm{s}^{-1}\right] \tag{7.304}$$

with

$$Y = (-1)^{J_\alpha - M_\alpha + 1}\begin{pmatrix} J_\alpha & 1 & J_\beta \\ -M_\alpha & 0 & M_\beta\end{pmatrix}. \tag{7.305}$$

For estimations of the order of magnitude of the interaction energy, we may approximate $|Y|\approx 1$ and obtain directly from (7.277), (7.281)

$$\begin{aligned}|V_{\alpha\beta}| &\approx 2\pi\cdot\left(\frac{4}{15}\right)^{2/3}\cdot e^2 a_0\cdot Z_\mathrm{p}\cdot n_\mathrm{i}^{2/3}\cdot\left|\langle\gamma_\alpha J_\alpha\|r\|\gamma_\beta J_\beta\rangle\right| \\ &\approx 2.0\times 10^{-15}\cdot Z_\mathrm{p}\cdot n_\mathrm{i}^{2/3}\cdot\left|\langle\gamma_\alpha J_\alpha\|r\|\gamma_\beta J_\beta\rangle\right|\,[\mathrm{eV}].\end{aligned} \tag{7.306}$$

Combining (7.302), (7.306), we find

$$n_\mathrm{i} \ll 1.1\times 10^{22}\cdot\frac{\Delta E_{\alpha\beta}^{3/2}(\mathrm{eV})}{Z_\mathrm{p}^{3/2}\cdot\left|\langle\gamma_\alpha J_\alpha\|r\|\gamma_\beta J_\beta\rangle\right|^{3/2}}\left[\mathrm{cm}^{-3}\right]. \tag{7.307}$$

The application of the perturbation theory sets therefore an upper limit for the density that can be quite severe according to (7.307). This limitation is incompatible with atomic kinetics because the correct asymptotic Boltzmann limit is only achieved for high densities. Moreover, for closely lying levels the condition (7.302) for the application of perturbation theory makes no sense. In fact, in the framework of a more exact theory, the diagonalization of the perturbed Hamiltonian shows a minimum level separation that is given by the field splitting itself. Therefore, the level crossing is avoided due to the interaction itself (the avoided crossing is related to the Landau–Zener effect where the transition occurs essentially in the neighborhood of the minimum distance between the levels).

It is therefore necessary to develop a density matrix approach including level mixing that circumvents the limitations of perturbation theory while maintaining a moderate complexity in order to be useful for practical applications where large open systems have to be handled. This theory will be discussed below (Sect. 7.9) within the framework of the quantum mechanical F-matrix theory QFMT.

7.8.3 Magnetic Quantum Number Averages and Symmetry Properties

In order to proceed to the consideration of large open systems, let us finish before the discussion of the M-quantum number averages for the ionic field perturbation too. The M-quantum number averages (see Sect. 7.6) concern elementary processes that are represented by the product of reduced matrix elements. As the field operator $\hat{V}$ from (7.274), (7.275) is only proportional to the matrix element itself, we define (with respect to Sects. 7.6.3 and 7.6.4) an M-quantum average according to

$$V_{\text{ab}} = \frac{1}{\sqrt{2J_\alpha + 1}} \cdot \sum_{M_\alpha, M_\beta} V_{\alpha\beta}. \tag{7.308}$$

The summation in (7.308) can be carried out with the help of (7.275), and we obtain the general expression (means for all possible directions of q):

$$V_{\text{ab}} = \frac{1}{\sqrt{2J_\alpha + 1}} \cdot \sum_{M_\alpha, M_\beta} V_{\alpha\beta} = ea_0 F \frac{\langle \gamma_\alpha J_\alpha \| r \| \gamma_\beta J_\beta \rangle}{\sqrt{2J_\alpha + 1}} \cdot X(\tilde{q}) \tag{7.309}$$

with

$$X(\tilde{q}) = \sum_{\substack{M_\alpha, M_\beta, q \\ |q| \le \tilde{q}}} (-1)^{J_\alpha - M_\alpha + 1} \begin{pmatrix} J_\alpha & 1 & J_\beta \\ -M_\alpha & q & M_\beta \end{pmatrix}. \tag{7.310}$$

The parameter $\tilde{q}$ specifies the direction, i.e., $\tilde{q} = 0$ for $q = 0$ and $\tilde{q} = 1$ for $q = 0, \pm 1$. Physically, the factor X determines the quantum number properties of the electric field with respect to the population kinetics. For the random case $q = 0, \pm 1$ (like considered above for the electron collisional operator) while for the quasi-static case $q = 0$ (if the ions are quasi-static, the Z-axis can be aligned along the direction of the electric field $\vec{F}$).

As (7.310) cannot be readily simplified (e.g., with the help of $3j$-symmetry properties or sum rules), let us calculate for a few relevant cases the numerical factor X; see Table 7.1. Concerning a quasi-static ion field as discussed above, the summation over the magnetic quantum numbers in (7.308) requests particular attention in view of the density matrix (7.9) that contains simultaneously the field matrix elements $V_{\alpha\beta}$ and the collisional operator matrix elements $\Phi_{\alpha\beta}^{\alpha'\beta'}$. The electron

Table 7.1 Numerical values of the X-function from (7.310) for a few relevant cases of initial (J_α) and final (J_β) J-quantum numbers, J_α and J_β, respectively

J_α	J_β	$X(q = 0)$	$X(q = 0, \pm 1)$
0	0	0.00000000	0.00000000
	1	0.57735027	1.73205081
1	0	−0.57735027	0.57735027
	1	0.00000000	0.00000000
	2	0.99760390	2.88963500
2	1	-0.99760390	0.89442719
	2	0.00000000	0.00000000
	3	1.28125825	3.66573376
3	2	−1.28125825	1.10321726
	3	0.00000000	0.00000000
	4	1.50969547	4.29288472
4	3	−1.50969547	1.27349379
	4	0.00000000	0.00000000
	5	1.70631492	4.83451992
5	4	−1.70631492	1.42189008
	5	0.00000000	0.00000000
	6	1.88161878	5.31873093

Note that due to the selection rule $\Delta J = J_\alpha - J_\beta = 0, \pm 1$ only the corresponding values of the X-function are nonzero

motion is assumed to be random, and we have to sum over all q-components in the $3j$ symbol. During the characteristic time of electron interaction, the ions are quasi-static, providing the possibility to align the ion field in Z-direction (which is equivalent to a $q = 0$ component only). The summation in (7.308) contains therefore only the summation over initial and final magnetic quantum numbers M_α and M_β, respectively, but does not contain anymore a summation over the q-directions.

It can be seen from Table 7.1, that for J-quantum numbers below 6 (covering the most relevant cases for typical applications in plasma atomic physics) the absolute values of the factors X for $q = 0$ are of the order of unity albeit different from one.

Let us now determine the symmetry properties of the V-operator. We have to start from the symmetry relation of $V_{\alpha\beta}$ from (7.276). With the help of the symmetry properties of the $3j$ symbol and the tensor matrix elements (7.107), we obtain:

$$\begin{aligned} V_{\beta\alpha} &= -ea_0 F \cdot (-1)^{J_\beta - M_\beta} \begin{pmatrix} J_\beta & 1 & J_\alpha \\ -M_\beta & 0 & M_\alpha \end{pmatrix} \langle \gamma_\beta J_\beta \| r \| \gamma_\alpha J_\alpha \rangle \\ &= -ea_0 F (-1)^{J_\beta - M_\beta + J_\beta - J_\alpha} \begin{pmatrix} J_\alpha & 1 & J_\beta \\ -M_\alpha & 0 & M_\beta \end{pmatrix} \langle \gamma_\alpha J_\alpha \| r \| \gamma_\beta J_\beta \rangle^* . \end{aligned} \tag{7.311}$$

Because $2J_\alpha + 2 + 2J_\beta$ and $4J_\beta$ are even, and $M_\alpha = M_\beta$ (due to the selection rules), we have $(-1)^{J_\beta - M_\beta + J_\beta - J_\alpha} = (-1)^{4J_\beta + J_\alpha - M_\beta} = (-1)^{J_\alpha - M_\alpha}$ from which it follows

$$V_{\beta\alpha} = V_{\alpha\beta}^{*}. \tag{7.312}$$

Therefore, the approximation of the quasi-static ion field conserves well the usual relation of matrix elements and its conjugate complex. It should be noted that not only the symmetry property of (7.312) is of interest, but also the symmetry of the product $V_{\alpha\beta} \cdot V_{\beta\alpha}$. In fact, this product arises in the solution of the density matrix (7.9) for the populations $\tilde{\rho}_{\alpha\alpha}$. This will be considered below in a dedicated paragraph in the framework of the construction of large closed model systems.

7.9 The Quantum Mechanical *F*-Matrix Theory QFMT

7.9.1 Rate Equation and Quantum Populations

Let us consider the equivalent of (7.9) in the framework of the rate equation theory of (7.1), i.e.

$$\begin{aligned} \frac{dn_\alpha}{dt} = &-n_\alpha \sum_{\beta'} \left(\gamma_{\alpha\beta'} + \gamma_{\alpha\beta'}^{\mathrm{rad}} + \Phi_{\alpha\beta'}\right) \\ &+ \sum_{\alpha'} n_{\alpha'} \left(\gamma_{\alpha'\alpha} + \gamma_{\alpha'\alpha}^{\mathrm{rad}} + \Phi_{\alpha'\alpha}\right) + Q_\alpha. \end{aligned} \tag{7.313}$$

We are now seeking for a solution for an *F-matrix* model system

$$\begin{aligned} \frac{dn_\alpha}{dt} = &-n_\alpha \sum_{\beta'} \left(\gamma_{\alpha\beta'} + \gamma_{\alpha\beta'}^{\mathrm{rad}} + \Phi_{\alpha\beta'} + F_{\alpha\beta'}\right) \\ &+ \sum_{\alpha'} n_{\alpha'} \left(\gamma_{\alpha'\alpha} + \gamma_{\alpha'\alpha}^{\mathrm{rad}} + \Phi_{\alpha'\alpha} + F_{\alpha'\alpha}\right) + Q_\alpha \end{aligned} \tag{7.314}$$

such that

$$n_\alpha = \rho_{\alpha\alpha}, \tag{7.315}$$

where the $\rho_{\alpha\alpha}$ are the populations from the system (7.10) of the coupled diagonal and non-diagonal density matrix elements.

A solution of (7.315) would certainly provide an expression (albeit very complex) for the *F*-matrix in (7.314); however, this would be equivalent to directly solve the density matrix (7.9). In addition, this type of solution (7.315) would not met the physical motivation of (7.314), to employ a rate equation framework of atomic levels (7.1) in order that:

(a) the system can be realized large enough to be considered as a closed system and,

(b) seek for an approximate solution for the F-matrix in (7.314) in terms of a function of the matrix elements $W_{\alpha\beta}$ from (7.1) only, i.e.

$$F_{\alpha\beta} = F\left(W_{\alpha\beta}\right), \tag{7.316}$$

while (7.315) is still well approximated, i.e.

$$n_\alpha \approx \rho_{\alpha\alpha}. \tag{7.317}$$

(c) go beyond the perturbation theory for the field interaction in order that the system (7.314) approaches the correct Boltzmann limit at high densities.

It is the purpose of the Quantum-F-Matrix theory QFMT to solve this general problem related to (7.313)–(7.317). Note, that the Greek indexes α and β in (7.316), (7.317) indicate that both equations are resolved in M-quantum number. Note that a classical system of population equations like those of (7.1) could either be formulated M-quantum-number-resolved or M-quantum-number-averaged.

7.9.2 The Open Two-Level System

In order to proceed toward a solution for the F-matrix according to (7.316), (7.317), let us consider an open two-level system that is schematically depicted in Fig. 7.1. For the system in Fig. 7.1a, the rate equation (7.313) takes the form

$$\frac{dn_2}{dt} = -n_2(\gamma_2 + A_{21} + \Phi_{21}) + n_1(\Phi_{12}) + Q_2, \tag{7.318}$$

$$\frac{dn_1}{dt} = -n_1(\gamma_1 + \Phi_{12}) + n_2(A_{21} + \Phi_{21}) + Q_1. \tag{7.319}$$

From (7.318), (7.319), we obtain in the stationary case (i.e., $d/dt = 0$):

$$Q_1 + Q_2 = n_1\gamma_1 + n_2\gamma_2. \tag{7.320}$$

Physically (7.320) means that all pumping rates are balanced by all loss rates irrespective of the collisional transfer between the levels "1" and "2".

For the density matrix system of Fig. 7.1b (7.9)–(7.18), we obtain in the dipole approximation for field operator $\hat{V} = -\vec{d} \cdot \vec{F}$ (i.e., $V_{\alpha\beta} \neq 0$ if $\alpha \neq \beta$ and $V_{\alpha\beta} = 0$ if $\alpha = \beta$ where $\vec{d}$ is the dipole moment and $\vec{F}$ the electric field strength; see (7.220)–(7.223)) the following set of equations:

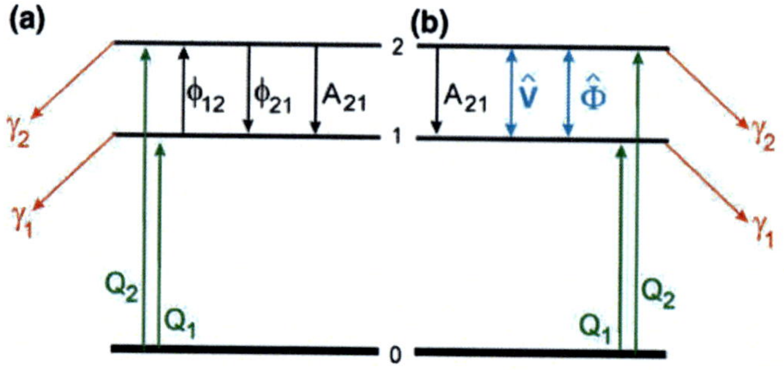

Fig. 7.1 Open two-level system for the rate equation approach (**a**) and in the framework of the density matrix approach (**b**). The field operator $\hat{V}$ describes the electric-field-induced level mixing due to the quasi-static ionic field while the collisional operator $\hat{\Phi}$ describes the electronic impact on the level populations. The pumping rates Q, the loss rates γ and the radiative decay rate A are relaxation constants, see text

$$\begin{aligned}\frac{\mathrm{d}\tilde{\rho}_{22}}{\mathrm{d}t} &= -\frac{i}{\hbar}\left(V_{21}\tilde{\rho}_{12} - \tilde{\rho}_{21}V_{12}\right) + \Phi_{22}^{11}\tilde{\rho}_{11} + \Phi_{22}^{12}\tilde{\rho}_{12} + \Phi_{22}^{21}\tilde{\rho}_{21}\\ &\quad - \tilde{\rho}_{22}\left(\gamma_2 + A_{21} - \Phi_{22}^{22}\right) + Q_2,\end{aligned} \tag{7.321}$$

$$\begin{aligned}\frac{\mathrm{d}\tilde{\rho}_{11}}{\mathrm{d}t} &= -\frac{i}{\hbar}\left(V_{12}\tilde{\rho}_{21} - \tilde{\rho}_{12}V_{21}\right) + \left(\Phi_{11}^{22} + A_{21}\right)\tilde{\rho}_{22} + \Phi_{11}^{12}\tilde{\rho}_{12} + \Phi_{11}^{21}\tilde{\rho}_{21}\\ &\quad - \tilde{\rho}_{11}\left(\gamma_1 - \Phi_{11}^{11}\right) + Q_1,\end{aligned} \tag{7.322}$$

$$\begin{aligned}\frac{\mathrm{d}\tilde{\rho}_{12}}{\mathrm{d}t} &= -i\omega_{12}\tilde{\rho}_{12} - \frac{i}{\hbar}\left(V_{12}\tilde{\rho}_{22} - \tilde{\rho}_{11}V_{12}\right) + \Phi_{12}^{11}\tilde{\rho}_{11} + \Phi_{12}^{22}\tilde{\rho}_{22} + \Phi_{12}^{21}\tilde{\rho}_{21}\\ &\quad - \tilde{\rho}_{12}\left(\frac{\gamma_1}{2} + \frac{\gamma_2}{2} + \frac{A_{21}}{2} - \Phi_{12}^{12}\right),\end{aligned} \tag{7.323}$$

$$\begin{aligned}\frac{\mathrm{d}\tilde{\rho}_{21}}{\mathrm{d}t} &= -i\omega_{21}\tilde{\rho}_{21} - \frac{i}{\hbar}\left(V_{21}\tilde{\rho}_{11} - \tilde{\rho}_{22}V_{21}\right) + \Phi_{21}^{11}\tilde{\rho}_{11} + \Phi_{21}^{22}\tilde{\rho}_{22} + \Phi_{21}^{12}\tilde{\rho}_{12}\\ &\quad - \tilde{\rho}_{21}\left(\frac{\gamma_1}{2} + \frac{\gamma_2}{2} + \frac{A_{21}}{2} - \Phi_{21}^{21}\right).\end{aligned} \tag{7.324}$$

From (7.321), (7.322), we obtain an equivalent expression to (7.320) in the framework of the density matrix approach:

$$\begin{aligned}Q_1 + Q_2 &= \tilde{\rho}_{11}\gamma_1 + \tilde{\rho}_{22}\gamma_2 + \left\{\tilde{\rho}_{11}\left(-\Phi_{11}^{11} - \Phi_{22}^{11}\right) + \tilde{\rho}_{22}\left(-\Phi_{22}^{22} - \Phi_{11}^{22}\right)\right\}\\ &\quad + \left\{\tilde{\rho}_{12}\left(-\Phi_{22}^{12} - \Phi_{11}^{12}\right) + \tilde{\rho}_{21}\left(-\Phi_{22}^{21} - \Phi_{11}^{21}\right)\right\}.\end{aligned} \tag{7.325}$$

With the help of (7.224)–(7.227), we can proceed toward an explicit identification of the collision operator elements occurring in (7.325):

$$\left\{\tilde{\rho}_{12}\left(-\Phi_{22}^{12}-\Phi_{11}^{12}\right)+\tilde{\rho}_{21}\left(-\Phi_{22}^{21}-\Phi_{11}^{21}\right)\right\}=0. \tag{7.326}$$

Therefore, in a two-level system, the non-diagonal elements do not influence on the loss rates of the open system that is balanced by the pumping rates. Let us now consider the contributions to the loss rates via the elements of the collisional operator that are proportional to the diagonal elements of the density matrix [see (7.325)]:

$$\Phi_{11}^{11}=-\Phi_{12}, \tag{7.327}$$

$$\Phi_{11}^{22}=\Phi_{21}. \tag{7.328}$$

Physically, the elements $-\Phi_{11}^{11}$ and $-\Phi_{22}^{22}$ of the collisional operator represent the total electron-induced collisional depopulation rates from the levels "1" and "2", respectively. In a two-level system, the total rates are identical to the transition rates between the levels. According to Sect. 7.6.4.3, the elements [see also (7.219)]

$$\Phi_{22}^{11}=\Phi_{12}, \tag{7.329}$$

$$\Phi_{22}^{22}=-\Phi_{21} \tag{7.330}$$

correspond to the transfer rates from one level to another. Injecting (7.327)–(7.330) into the term with the first parenthesis of (7.325), we find

$$\left\{\tilde{\rho}_{11}\left(-\Phi_{11}^{11}-\Phi_{22}^{11}\right)+\tilde{\rho}_{22}\left(-\Phi_{22}^{22}-\Phi_{11}^{22}\right)\right\}=0 \tag{7.331}$$

and with the help of (7.131), (7.326) finally

$$Q_1+Q_2=\tilde{\rho}_{11}\gamma_1+\tilde{\rho}_{22}\gamma_2. \tag{7.332}$$

Therefore, in a two-level system, the pumping is directly balanced by the loss rates as it is the case in the rate equation approach (see (7.320)).

The identification of the electron collisional operator elements (neglecting the imaginary part in (7.66), (7.105), (7.106) with cross sections (or rate coefficients)) in (7.327)–(7.330) can be completed with (7.213) (see also Sect. 7.6.4.2), i.e.

$$\Phi_{12}^{12}=-\frac{\Phi_{12}}{2}-\frac{\Phi_{21}}{2}, \tag{7.333}$$

$$\Phi_{21}^{21}=-\frac{\Phi_{21}}{2}-\frac{\Phi_{12}}{2}. \tag{7.334}$$

7.9.3 The Exact QFMT Solution for a Two-Level System

Let us now consider the solution for the system containing the matrix elements $F_{\alpha\beta}$ from (7.314). Instead of (7.318), (7.319), we encounter the following system of equations:

$$\frac{dn_2}{dt} = -n_2(\gamma_2 + A_{21} + \Phi_{21} + F_{21}) + n_1(\Phi_{12} + F_{12}) + Q_2, \tag{7.335}$$

$$\frac{dn_1}{dt} = -n_1(\gamma_1 + \Phi_{12} + F_{12}) + n_2(A_{21} + \Phi_{21} + F_{21}) + Q_1. \tag{7.336}$$

We are now seeking for a solution of the density matrix (7.321), (7.322) and are looking for an explicit expression for $\tilde{\rho}_{22}$ that can be compared with the rate equation solution of n_2 from the system of (7.335), (7.336). In order to be general for later discussions, let us not use the relations (7.224)–(7.231) that are specific for a two-level system. For this purpose, we consider the quasi-stationary case (i.e., $d/dt = 0$) and solve (7.323), (7.324) for $\tilde{\rho}_{12}$ and $\tilde{\rho}_{21}$. For this purpose, we rewrite (7.321) in a more convenient form:

$$\frac{d\tilde{\rho}_{22}}{dt} = \tilde{\rho}_{11}\Phi_{22}^{11} - \tilde{\rho}_{22}\left(\gamma_2 + A_{21} - \Phi_{22}^{22}\right) + \tilde{\rho}_{12}a_1 + \tilde{\rho}_{21}a_2 + Q_2, \tag{7.337}$$

$$\frac{d\tilde{\rho}_{11}}{dt} = -\tilde{\rho}_{11}\left(\gamma_1 - \Phi_{11}^{11}\right) + \tilde{\rho}_{22}\left(\Phi_{11}^{22} + A_{21}\right) + \tilde{\rho}_{12}a_3 + \tilde{\rho}_{21}a_4 + Q_1, \tag{7.338}$$

$$\frac{d\tilde{\rho}_{12}}{dt} = \tilde{\rho}_{11}a_5 + \tilde{\rho}_{22}a_6 - \tilde{\rho}_{12}(a + i\omega_{12}) + \tilde{\rho}_{21}\Phi_{12}^{21}, \tag{7.339}$$

$$\frac{d\tilde{\rho}_{21}}{dt} = \tilde{\rho}_{11}a_7 + \tilde{\rho}_{22}a_8 + \tilde{\rho}_{12}\Phi_{21}^{12} - \tilde{\rho}_{21}(b + i\omega_{21}) \tag{7.340}$$

with

$$a_1 = -\frac{i}{\hbar}V_{21} + \Phi_{22}^{12}, \tag{7.341}$$

$$a_2 = \frac{i}{\hbar}V_{12} + \Phi_{22}^{21}, \tag{7.342}$$

$$a_3 = \frac{i}{\hbar}V_{21} + \Phi_{11}^{12}, \tag{7.343}$$

$$a_4 = -\frac{i}{\hbar}V_{12} + \Phi_{11}^{21}, \tag{7.344}$$

$$a_5 = \frac{i}{\hbar} V_{12} + \Phi^{11}_{12}, \tag{7.345}$$

$$a_6 = -\frac{i}{\hbar} V_{12} + \Phi^{22}_{12}, \tag{7.346}$$

$$a_7 = -\frac{i}{\hbar} V_{21} + \Phi^{11}_{21}, \tag{7.347}$$

$$a_8 = \frac{i}{\hbar} V_{21} + \Phi^{22}_{21}, \tag{7.348}$$

$$a = \frac{\gamma_1}{2} + \frac{\gamma_2}{2} + \frac{A_{21}}{2} - \Phi^{12}_{12}, \tag{7.349}$$

$$b = \frac{\gamma_1}{2} + \frac{\gamma_2}{2} + \frac{A_{21}}{2} - \Phi^{21}_{21}, \tag{7.350}$$

$$\omega_{12} = -\omega_{21}. \tag{7.351}$$

From (7.338), (7.339) it follows

$$\tilde{\rho}_{21} = -\tilde{\rho}_{11} \cdot \frac{A}{C} - -\tilde{\rho}_{22} \cdot \frac{B}{C}, \tag{7.352}$$

$$\tilde{\rho}_{12} = \frac{1}{a + i\omega_{12}} \cdot \left\{ \tilde{\rho}_{11} \left[a_5 - \Phi^{21}_{12} \cdot \frac{A}{C} \right] + \tilde{\rho}_{22} \left[a_6 - \Phi^{21}_{12} \cdot \frac{B}{C} \right] \right\} \tag{7.353}$$

with

$$A = a_7 + \frac{a_5 \Phi^{12}_{21}}{a + i\omega_{12}}, \tag{7.354}$$

$$B = a_8 + \frac{a_6 \Phi^{12}_{21}}{a + i\omega_{12}}, \tag{7.355}$$

$$C = \frac{\Phi^{21}_{12} \cdot \Phi^{12}_{21}}{a + i\omega_{12}} - (b - i\omega_{12}). \tag{7.356}$$

Assuming $A = 1$ and $B = 0$, we encounter a the symmetry relation (7.121) and we have from (7.349), (7.350)

$$a = b. \tag{7.357}$$

The solutions (7.352), (7.353) are then injected into (7.337) with the help of (7.341)–(7.351), (7.357). The resulting equation is then compared with (7.335) to obtain the *F*-matrix elements. Let us split the matrix elements F_{21} in two parts according $F_{21} = X_{21} + iY_{21}$ and $F_{12} = X_{12} + iY_{12}$ and we obtain for the *X*-part of F_{21}

$$
\begin{aligned}
X_{21} = {} & +2\frac{V_{21}\cdot V_{12}}{\hbar^2}\cdot G_1\cdot G_2-\Phi_{21}-\Phi_{22}^{22}\\
& -G_1\cdot G_2\left\{\frac{V_{12}\cdot V_{12}}{\hbar^2}\cdot\frac{\Phi_{21}^{12}}{a}+\frac{V_{21}\cdot V_{21}}{\hbar^2}\cdot\frac{\Phi_{12}^{21}}{a}\right\}\\
& -G_1\cdot G_2\cdot\left\{\Phi_{22}^{12}\cdot\Phi_{12}^{22}+\Phi_{22}^{21}\cdot\Phi_{21}^{22}+\frac{\Phi_{21}^{12}}{a}\cdot\Phi_{22}^{21}\cdot\Phi_{12}^{22}+\frac{\Phi_{12}^{21}}{a}\cdot\Phi_{22}^{12}\cdot\Phi_{21}^{22}\right\}\\
& +G_1\cdot G_2\cdot\frac{\omega}{a}\cdot\left\{\frac{V_{12}}{\hbar}\left(\Phi_{21}^{22}+\Phi_{22}^{12}\right)+\frac{V_{21}}{\hbar}\left(\Phi_{12}^{22}+\Phi_{22}^{21}\right)\right\}
\end{aligned}
\tag{7.358}
$$

and for the Y-part of F_{21}:

$$
\begin{aligned}
Y_{21} = {} & -G_1\cdot G_2\cdot\frac{\omega_{12}}{a}\cdot\left\{\Phi_{22}^{21}\cdot\Phi_{21}^{22}-\Phi_{22}^{12}\cdot\Phi_{12}^{22}\right\}\\
& -G_1\cdot G_2\cdot\frac{V_{12}}{\hbar}\cdot\left\{\Phi_{21}^{22}-\Phi_{22}^{12}+\frac{\Phi_{21}^{12}}{a}\cdot\Phi_{12}^{21}-\frac{\Phi_{21}^{12}}{a}\cdot\Phi_{22}^{21}\right\}\\
& -G_1\cdot G_2\cdot\frac{V_{21}}{\hbar}\cdot\left\{\Phi_{22}^{21}-\Phi_{12}^{22}+\frac{\Phi_{12}^{21}}{a}\cdot\Phi_{22}^{12}-\frac{\Phi_{12}^{21}}{a}\cdot\Phi_{21}^{22}\right\}.
\end{aligned}
\tag{7.359}
$$

Likewise we obtain for the X-part of F_{12}

$$
\begin{aligned}
X_{12} = {} & +2\frac{V_{21}\cdot V_{12}}{\hbar^2}\cdot G_1\cdot G_2-\Phi_{12}-\Phi_{22}^{11}\\
& -G_1\cdot G_2\left\{\frac{V_{12}\cdot V_{12}}{\hbar^2}\cdot\frac{\Phi_{21}^{12}}{a}+\frac{V_{21}\cdot V_{21}}{\hbar^2}\cdot\frac{\Phi_{12}^{21}}{a}\right\}\\
& +G_1\cdot G_2\cdot\left\{\Phi_{22}^{12}\cdot\Phi_{12}^{11}+\Phi_{22}^{21}\cdot\Phi_{21}^{11}+\frac{\Phi_{12}^{21}}{a}\cdot\Phi_{22}^{12}\cdot\Phi_{21}^{11}+\frac{\Phi_{21}^{12}}{a}\cdot\Phi_{22}^{21}\cdot\Phi_{12}^{11}\right\}\\
& +G_1\cdot G_2\cdot\frac{\omega}{a}\cdot\left\{\frac{V_{12}}{\hbar}\left(\Phi_{22}^{11}-\Phi_{21}^{11}\right)+\frac{V_{21}}{\hbar}\left(\Phi_{22}^{21}-\Phi_{12}^{11}\right)\right\}
\end{aligned}
\tag{7.360}
$$

and for the Y-part

$$
\begin{aligned}
Y_{12} = {} & -G_1\cdot G_2\cdot\frac{\omega_{12}}{a}\cdot\left\{\Phi_{22}^{12}\cdot\Phi_{12}^{11}-\Phi_{22}^{21}\cdot\Phi_{21}^{11}\right\}\\
& -G_1\cdot G_2\cdot\frac{V_{12}}{\hbar}\cdot\left\{\Phi_{22}^{11}+\Phi_{21}^{11}-\frac{\Phi_{21}^{12}}{a}\Phi_{12}^{11}-\frac{\Phi_{21}^{12}}{a}\Phi_{22}^{21}\right\}\\
& -G_1\cdot G_2\cdot\frac{V_{21}}{\hbar}\cdot\left\{-\Phi_{12}^{11}-\Phi_{22}^{21}+\frac{\Phi_{12}^{21}}{a}\Phi_{21}^{11}+\frac{\Phi_{12}^{21}}{a}\Phi_{22}^{12}\right\}
\end{aligned}
\tag{7.361}
$$

with

$$G_1 = \frac{1}{\frac{\omega_{12}^2}{a} + a}, \tag{7.362}$$

$$G_2 = \frac{a^2 + \omega_{12}^2}{a^2 + \omega_{12}^2 - \Phi_{12}^{21} \cdot \Phi_{21}^{12}}, \tag{7.363}$$

$$a = \frac{\gamma_1}{2} + \frac{\gamma_2}{2} + \frac{A_{21}}{2} - \Phi_{12}^{12}. \tag{7.364}$$

Note that in a two-level system the operator Φ_{12}^{12} can be approximated in terms of cross sections (see (7.333), (7.334)) under assumptions leading to (7.357). With the help of relations (7.224)–(7.231) and (7.327)–(7.330), we obtain

$$\begin{aligned} X_{21} = &+2\frac{V_{21} \cdot V_{12}}{\hbar^2} \cdot G_1 \cdot G_2 \\ &- G_1 \cdot G_2 \left\{ \frac{V_{12} \cdot V_{12}}{\hbar^2} \cdot \frac{\Phi_{21}^{12}}{a} + \frac{V_{21} \cdot V_{21}}{\hbar^2} \cdot \frac{\Phi_{12}^{21}}{a} \right\}, \end{aligned} \tag{7.365}$$

$$Y_{21} = 0, \tag{7.366}$$

$$\begin{aligned} X_{12} = &+2\frac{V_{21} \cdot V_{12}}{\hbar^2} \cdot G_1 \cdot G_2 \\ &- G_1 \cdot G_2 \left\{ \frac{V_{12} \cdot V_{12}}{\hbar^2} \cdot \frac{\Phi_{21}^{12}}{a} + \frac{V_{21} \cdot V_{21}}{\hbar^2} \cdot \frac{\Phi_{12}^{21}}{a} \right\}, \end{aligned} \tag{7.367}$$

$$Y_{12} = 0. \tag{7.368}$$

7.9.4 Successive Pair Coupling of Quantum Effects

One can see from (7.365)–(7.368) that the F-matrix elements for F_{12} and F_{21} are symmetric and contain only the X-functions (see (7.366), (7.368)). We therefore might approximate the F-matrix solution of a multilevel system by the solution from the two-level system with general indexes, i.e.

$$\begin{aligned} F_{\alpha\beta} \approx &+2\frac{V_{\alpha\beta} \cdot V_{\beta\alpha}}{\hbar^2} \cdot G_{1,\alpha\beta} \cdot G_{2,\alpha\beta} - G_{1,\alpha\beta} \\ &\cdot G_{2,\alpha\beta} \left\{ \frac{V_{\alpha\beta} \cdot V_{\alpha\beta}}{\hbar^2} \cdot \frac{\Phi_{\beta\alpha}^{\alpha\beta}}{a_{\alpha\beta}} + \frac{V_{\beta\alpha} \cdot V_{\beta\alpha}}{\hbar^2} \cdot \frac{\Phi_{\alpha\beta}^{\beta\alpha}}{a_{\alpha\beta}} \right\} \end{aligned} \tag{7.369}$$

with

$$G_{1,\alpha\beta} = \frac{1}{\dfrac{\omega_{\alpha\beta}^2}{a_{\alpha\beta}} + a_{\alpha\beta}}, \tag{7.370}$$

$$G_{2,\alpha\beta} = \frac{a_{\alpha\beta}^2 + \omega_{\alpha\beta}^2}{a_{\alpha\beta}^2 + \omega_{\alpha\beta}^2 - \Phi_{\alpha\beta}^{\beta\alpha} \cdot \Phi_{\beta\alpha}^{\alpha\beta}}, \tag{7.371}$$

$$a_{\alpha\beta} = \frac{\gamma_\alpha}{2} + \frac{\gamma_\beta}{2} + \frac{A_{\alpha\beta} + A_{\beta\alpha}}{2} - \Phi_{\alpha\beta}^{\alpha\beta}. \tag{7.372}$$

Note that in (7.372) we have replaced the term $\frac{A_{21}}{2}$ in (7.364) by $\frac{A_{\alpha\beta}+A_{\beta\alpha}}{2}$ in order to avoid that the contribution of the radiative decay to the loss of the non-diagonal matrix element vanishes if we exchange the index as in a two-level system there exists only one nonvanishing radiative decay from the upper to the lower level. Therefore, if α designates the upper level and β the lower one, $A_{\alpha\beta} \geq 0$ while $A_{\beta\alpha} = 0$. On the other hand, if β designates the upper level and α the lower one, then $A_{\beta\alpha} \geq 0$ while $A_{\alpha\beta} = 0$. The radiative term $\frac{A_{\alpha\beta}+A_{\beta\alpha}}{2}$ in (7.372) therefore generalizes the radiative contribution for arbitrary level combinations in a multilevel system. Physically, the $F_{\alpha\beta}$-matrix elements from (7.369) connect a particular selected level α with all other levels β of the multilevel system. This is schematically depicted in Fig. 7.2.

Figure 7.2 illustrates that the levels α and β are not only coupled to each other via $F_{\alpha\beta}$ (see green flashes) but also to all other levels γ in the system via $F_{\alpha\gamma}$ and $F_{\beta\gamma}$ (blue flashes). Moreover, in the proposed multilevel scheme (7.369)–(7.372) there is even recovered some influence of the coupling between the levels γ (red flashes) and the coupling of the levels γ to the levels α and β (see red dashed flashes).

There are therefore important advantages of the application of (7.369)–(7.372) in a multiple-pair approach to a multilevel system:

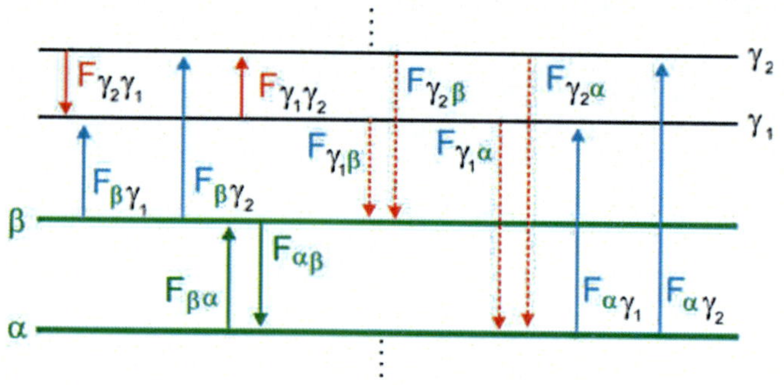

Fig. 7.2 F-matrix coupling in a multilevel system. Apart its direct coupling between the levels α and β, it couples likewise all other system levels γ to α and to β

(a) Due to the solution from the two-level system, no approximation with respect to the magnitude of the V-parameter has been made (in other words, (7.369)–(7.372) are valid even for large values of V-matrix elements). In other words, no perturbation theory with limiting conditions like (7.302) is employed. This is important when considering high densities and the thermodynamic limit: in plasmas, the thermodynamic limit is achieved for high densities, i.e., for large values of V-matrix elements (because these matrix elements increase with the density [see (7.276), (7.281)]. Therefore, (7.369)–(7.372) constitute a well-adopted compromise: It includes the interferences in a coupled two-level approach (see discussion concerning Fig. 7.2), while it is appropriate for large values of V-matrix elements.

(b) The two-level approximation according to (7.369)–(7.372) conserves also the symmetry properties with respect to a multilevel system. This can be seen as follows. According to (7.121), the operator elements $\Phi_{\alpha\beta}^{\alpha\beta}$ are symmetric with respect to an exchange in indexes α and β. Therefore, we encounter a symmetry (see also (7.357) in the parameter $a_{\alpha\beta}$ from (7.372):

$$a_{\alpha\beta} = a_{\beta\alpha}. \tag{7.373}$$

Because $\Phi_{\alpha\beta}^{\beta\alpha} \cdot \Phi_{\beta\alpha}^{\alpha\beta}$, and $\omega_{\alpha\beta}^2$ are symmetric with respect to an exchange of indexes, we encounter that $G_{1,\alpha\beta}$ and $G_{2,\alpha\beta}$ are likewise symmetric:

$$G_{1,\alpha\beta} = G_{1,\beta\alpha}, \tag{7.374}$$

$$G_{2,\alpha\beta} = G_{2,\beta\alpha}. \tag{7.375}$$

As $V_{\alpha\beta} \cdot V_{\beta\alpha}$ is trivially symmetric with respect to its indexes, we obtain a symmetry for the F-matrix from (7.369):

$$F_{\alpha\beta} = F_{\beta\alpha}. \tag{7.376}$$

7.9.5 QFMT and Statistical Boltzmann Populations

An M-quantum-number-averaged model system (7.1) is well adopted to realize large model systems that include not only a large number of excited levels, but consider at the same time the ionic distribution. The system is therefore resolved in nLSJ-quantum numbers for each possible ionization stage, hereafter referred as Z-nLSJ-resolved population kinetics. Such a model system may contain hundreds up to millions of levels that realize a closed model system. It is therefore of great interest, to study the F-matrix approach averaged over M-quantum numbers. This question is strongly related to the fundamental question, whether the F-matrix supports Boltzmann relations or not. We are therefore interested to seek for a solution for the F-matrix for the following set of equations [instead of (7.314), (7.315)]

$$\begin{aligned}\frac{dn_a}{dt} &= -n_a \sum_{b'} \left(\gamma_{ab'} + \gamma_{ab'}^{rad} + \Phi_{ab'} + F_{ab'}\right) \\ &\quad + \sum_{a'} n_{a'} \left(\gamma_{a'a} + \gamma_{a'a}^{rad} + \Phi_{a'a} + F_{a'a}\right) + Q_a\end{aligned} \tag{7.377}$$

and (see also (7.137))

$$n_a = \rho_{aa}, \tag{7.378}$$

where the ρ_{aa} are the M-quantum-number-averaged populations from the system of density matrix (7.156). In analogy to (7.369)–(7.372), we obtain

$$\begin{aligned}F_{ab} &\approx +2\frac{V_{ab}\cdot V_{ba}}{\hbar^2}\cdot G_{1,ab}\cdot G_{2,ab} - G_{1,ab} \\ &\quad \cdot G_{2,ab}\left\{\frac{V_{ab}\cdot V_{ab}}{\hbar^2}\cdot\frac{\Phi_{ba}^{ab}}{a_{ab}} + \frac{V_{ba}\cdot V_{ba}}{\hbar^2}\cdot\frac{\Phi_{ab}^{ba}}{a_{ab}}\right\}\end{aligned} \tag{7.379}$$

with

$$G_{1,ab} = \frac{1}{\dfrac{\omega_{ab}^2}{a_{ab}} + a_{ab}}, \tag{7.380}$$

$$G_{2,ab} = \frac{a_{ab}^2 + \omega_{ab}^2}{a_{ab}^2 + \omega_{ab}^2 - \Phi_{ab}^{ba}\cdot\Phi_{ba}^{ab}}, \tag{7.381}$$

$$a_{ab} = \frac{\gamma_a}{2} + \frac{\gamma_b}{2} + \frac{A_{ab} + A_{ba}}{2} - \Phi_{ab}^{ab}. \tag{7.382}$$

According to (7.213), the collisional operator matrix element Φ_{ab}^{ab} contains two symmetric sums over all collisional rate coefficients. Therefore Φ_{ab}^{ab} is symmetric, i.e.

$$\Phi_{ab}^{ab} = \Phi_{ba}^{ba}. \tag{7.383}$$

From (7.362), it follows immediately that

$$a_{ab} = a_{ba}, \tag{7.384}$$

$$G_{1,ab} = G_{1,ba}, \tag{7.385}$$

$$G_{2,\mathrm{ab}} = G_{2,\mathrm{ba}}. \tag{7.386}$$

As $V_{\mathrm{ab}} \cdot V_{\mathrm{ba}}$ is evidently symmetric with respect to an exchange of indexes, we encounter finally a symmetric M-quantum-number-averaged F-matrix too, i.e.

$$F_{\mathrm{ab}} = F_{\mathrm{ba}}. \tag{7.387}$$

According to (7.377) the matrix elements F_{ab} and F_{ba} act like effective processes that induce transitions from state a to state b and vice versa, e.g., like the electron collisional matrix elements Φ_{ab} and Φ_{ba} do. Physically, it can be interpreted that interferences are interrupted by relaxation which can result into an effective loss rate. Let us consider the detailed balance for these two processes. For the electron collisions, the "collisional flow" from level n_{a} is given by $n_{\mathrm{a}}\Phi_{\mathrm{ab}}$ and the detailed balance reads

$$n_{\mathrm{a}}\Phi_{\mathrm{ab}} = n_{\mathrm{b}}\Phi_{\mathrm{ba}}. \tag{7.388}$$

Because [see also (7.139)–(7.142)]

$$\Phi_{\mathrm{ab}} \propto \frac{\left|\langle \gamma_\alpha J_\alpha \| r \| \gamma_\beta J_\beta \rangle\right|^2}{2J_\alpha + 1} \cdot \exp\left(-\frac{\Delta E_{\mathrm{ab}}}{kT_{\mathrm{e}}}\right) \tag{7.389}$$

and

$$\Phi_{\mathrm{ba}} \propto \frac{\left|\langle \gamma_\beta J_\beta \| r \| \gamma_\alpha J_\alpha \rangle\right|^2}{2J_\beta + 1} \tag{7.390}$$

we encounter with the help of the symmetry relation of the line strengths [see also (7.124)], i.e.

$$\left|\langle \gamma_\alpha J_\alpha \| r \| \gamma_\beta J_\beta \rangle\right|^2 = \left|\langle \gamma_\beta J_\beta \| r \| \gamma_\alpha J_\alpha \rangle\right|^2 \tag{7.391}$$

the following relation between the direct and inverse electron collisional processes:

$$\Phi_{\mathrm{ba}} = \Phi_{\mathrm{ab}} \cdot \frac{2J_\alpha + 1}{2J_\beta + 1} \cdot \exp\left(\frac{\Delta E_{\mathrm{ab}}}{kT_{\mathrm{e}}}\right). \tag{7.392}$$

Injecting (7.392) into (7.388), we find

$$\frac{n_{\mathrm{b}}}{n_{\mathrm{a}}} = \frac{2J_\beta + 1}{2J_\alpha + 1} \cdot \exp\left(-\frac{\Delta E_{\mathrm{ab}}}{kT_{\mathrm{e}}}\right). \tag{7.393}$$

According to (7.393), the electron collisions support Boltzmann populations. In fact, it is due to the dominating role in the population kinetics of electron collisions (that are proportional to the electron density) that populations approach for high density the statistical Boltzmann limit.

A corresponding detailed balance equation for the field operator $\hat{V}$ is difficult to identify directly from the density matrix (7.154), (7.156), because the operator $\hat{V}$ acts via an imaginary term $\frac{i}{\hbar}\hat{V}_{ag}\tilde{\rho}_{gb}$ in the equations. It is therefore necessary to consider the F-matrix solution of (7.379) to obtain its real contribution to the populations (i.e., its influence on the populations of the diagonal elements). A corresponding detailed balance for the $\hat{V}$-operator should be defined as

$$n_a F_{ab} = n_b F_{ba}. \tag{7.394}$$

According to the symmetry relation of (7.387), we obtain from (7.394)

$$\frac{n_b}{n_a} = 1. \tag{7.395}$$

Therefore, the $\hat{V}$-field-induced transition rates do *not* support Boltzmann populations. This is a very important result for high-density plasmas, because unlike spontaneous radiative rates, the $\hat{V}$-field-induced transition rates increase with density and therefore reduce the efficiency of electron collisions with respect to their capacity to establish Boltzmann populations (and thereby changing the so-called critical density, i.e., the minimum density for achieving Boltzmann populations).

Equation (7.395) indicates that ionic field rates have the tendency to drive the system to equal populations, while electron collisions have the tendency to establish a statistical population [except the Boltzmann factor, see (7.389)]. This is related to the fact that for the electron collisional operator, a random distribution of impact vectors and velocity vectors has been assumed that has suppressed terms in the electron collisional operator that are only proportional to one reduced matrix element [see (7.42)–(7.44)]. Therefore, from the very beginning, the average of the scattering matrix (7.43) has suppressed "non-statistical terms." For the ion field, this is different. Ions have been assumed to be quasi-static and no random average over the field contributions in the kinetic equations [see also (7.9), (7.20)] has been made. We note that the relation (7.395) is independent from the assumption of the q-components; different assumptions for the q-values result in different numerical factors (see, e.g., Table 7.1) but do not alter the principle circumstance that ionic fields do not support the Boltzmann populations.

We note that the density dependence of the influence of the ionic field effect is not trivial, because the ionic field increases also with density (as do the random electron collisions). For large densities, we encounter the following scaling relations related to (7.379)–(7.382):

$$a_{ab} \propto n_e, \tag{7.396}$$

$$G_{1,ab} \propto n_e^{-1}, \tag{7.397}$$

$$G_{2,ab} \propto n_e^0, \tag{7.398}$$

$$V_{ab} V_{ba} \propto n_e^{4/3}, \tag{7.399}$$

$$V_{ab} V_{ab} \propto n_e^{4/3}, \tag{7.400}$$

$$V_{ba} V_{ba} \propto n_e^{4/3}, \tag{7.401}$$

$$\frac{\Phi_{ab}^{ba}}{a_{ab}} \propto n_e^0, \tag{7.402}$$

$$\frac{\Phi_{ba}^{ab}}{a_{ab}} \propto n_e^0. \tag{7.403}$$

From the scaling relations (7.396)–(7.403), it follows the scaling for the F-matrix:

$$F_{ab} \propto n_e^{1/3}. \tag{7.404}$$

The F-matrix scaling is less strong than the scaling of the electron collisional processes, i.e.

$$\Phi_{ab} \propto n_e, \tag{7.405}$$

therefore, the relative importance of field effects vanishes for large densities according to the scaling

$$\frac{F_{ab}}{\Phi_{ab}} \propto n_e^{-2/3}. \tag{7.406}$$

The vanishing ratio $\frac{F_{ab}}{\Phi_{ab}}$ for large densities allows electron collisions to be dominant for very high densities and to establish Boltzmann populations. As the influence of the F-matrix vanishes for very large densities, a possible failure of perturbation theory will have little impact on the final result. It is therefore of great interest to apply the density matrix equations as well as the F-matrix approach to a level system of practical interest (e.g., in plasma spectroscopy) to get more practical insight into the changes induced by the ionic fields and their dependence with density.

7.10 Application to Autoionizing Levels of Highly Charged Ions

7.10.1 Dielectronic Satellites Near H-like Lyman-Alpha

Let us apply the density matrix approach to the autoionizing levels $2p^2\,{}^1D_2$, $2p^2\,{}^1S_0$ and $2s2p\,{}^1P_1$ that are of great interest for dense plasma diagnostic to measure density and temperature (Boiko et al. 1985; Renner et al. 2009; Renner and Rosmej 2019; Rosmej and Lee 2007; Rosmej et al. 2015; Michelis and Mattioli 1981; Rosmej 2012a, b). This diagnostic is typically realized with a high-resolution X-ray Bragg crystal to obtain a highly resolved *X*-spectral distribution of the transitions $2l2l' \rightarrow 1s2l$ (the so-called dielectronic satellites of He-like Lyman-alpha) whose qualitative changes (or line ratios) are then indicative of changes in temperature and density.

Tables 7.2 and 7.3 present the relevant atomic properties of the levels and the related transitions obtained from the Flexible Atomic Code FAC (Gu 2008).

The density matrix equations have been solved for the autoionizing levels $2p^2\,{}^1D_2$, $2p^2\,{}^1S_0$, and $2s2p\,{}^1P_1$ that are resolved in LSJ-quantum numbers but not in *M*-quantum number. For these purposes, (7.156) and (7.377), (7.378) have been numerically solved for $q = 0$. As dielectronic capture is the dominating population channel for the $2l2l$ levels for highly charged ions, dielectronic capture has been assumed for the pumping rates $Q_{\alpha\beta}$.

Likewise, the loss rates of the autoionizing levels are dominated by the radiative losses γ_A from the levels $\alpha, \beta, \gamma \in \{\text{density matrix}\}$ to the levels $n, m \notin \{\text{density matrix}\}$ and the autoionizing rates Γ. Collisional rates from levels $\alpha, \beta, \gamma \in \{\text{density matrix}\}$ to the levels $n, m \notin \{\text{density matrix}\}$ have therefore to be compared with the sum of the loss rates $\gamma = \gamma_A + \Gamma$ given in Table 7.2 and with the rates already included in the density matrix system. Calculations for typical

Table 7.2 Statistical weights g, autoionizing rates Γ, dielectronic capture energies E_s, radiative loss rates γ_A, and total loss rates γ for aluminum

Level	g	Γ (s^{-1})	γ_A (s^{-1})	E_s (eV)	γ (s^{-1})
$2p^2\,{}^1S_0$	1	1.26×10^{13}	2.73×10^{13}	1237.683	3.99×10^{13}
$2s2p\,{}^1P_1$	3	1.70×10^{14}	1.63×10^{13}	1219.174	1.86×10^{14}
$2p^2\,{}^1D_2$	5	3.13×10^{14}	3.18×10^{13}	1217.566	3.45×10^{14}

Table 7.3 Transition energies ΔE, reduced dipole matrix elements $\langle \gamma_\alpha J_\alpha \| r \| \gamma_\beta J_\beta \rangle$, absorption oscillator strengths *f*, and transition probabilities *A* for aluminum

Transition	ΔE (eV)	$\langle \gamma_\alpha J_\alpha \| r \| \gamma_\beta J_\beta \rangle$	$f_{J_\alpha J_\beta}$	$A_{J_\beta J_\alpha}$ (s^{-1})
$a = 2p^2\,{}^1D_2 \rightarrow b = 2s2p\,{}^1P_1$	1.6078	−0.7615	4.57×10^{-3}	8.54×10^5
$a = 2s2p\,{}^1P_1 \rightarrow b = 2p^2\,{}^1S_0$	18.506	−0.5782	5.05×10^{-2}	2.55×10^9

temperatures show ($kT_e = 500\,\mathrm{eV}$) that ionization rate coefficients $2l2l' + e \to 2l + e + e$ are of the order of $I(2l2l' \to 2l) \approx 10^{-10}\,\mathrm{cm^3\,s^{-1}}$. Therefore, even for the highest densities considered ionization loss rates turn out not to be very important.

Concerning electron excitation rates, the most dominant ones are those within the same group, i.e., collisions from levels $\alpha = 2l2l'$ $\alpha \in \{\text{density matrix}\}$ to the levels $n = 2l2l'$ while $n \notin \{\text{density matrix}\}$. The calculations show that the rate $C(2p^2\,{}^1D_2{-}2s2p\,{}^1P_1) \approx 6 \times 10^{-9}\,\mathrm{cm^3\,s^{-1}}$ is the dominating one for the level $2p^2\,{}^1D_2$. This rate, however, is included in the calculations as both levels belong to the system $\alpha, \beta \in \{\text{density matrix}\}$. A similar situation holds true for the $2p^2\,{}^1S_0$ level: $C(2p^2\,{}^1S_0{-}2s2p\,{}^1P_1) \approx 9 \times 10^{-9}\,\mathrm{cm^3\,s^{-1}}$. The situation is only marginally different for the $2s2p\,{}^1P_1$-level. The strongest rate coefficient is $C(2s2p\,{}^1P_1{-}2p^2\,{}^1S_0) \approx 1 \times 10^{-8}\,\mathrm{cm^3\,s^{-1}}$ and concerns the levels included in the density matrix system, while the transition $C(2s2p\,{}^1P_1 - 2s^2\,{}^1S_0) \approx 1 \times 10^{-9}\,\mathrm{cm^3\,s^{-1}}$ concerns a loss rate to a level $n, m \notin \{\text{density matrix}\}$.

For further physical understanding, let us consider a few critical densities for typical conditions of dense hot aluminum plasma: ion temperature of $kT = 100\,\mathrm{eV}$, ion charge $Z = 12$, a mean perturber charge of $Z_p = 11$, and reduced mass of $\mu = 27/2$. Because the $2l2l'$-levels are essentially subject to the quadratic Stark effect, we have to use (7.295) in order to estimate the quasi-static conditions for the quadratic Stark effect. From the data presented in Table 7.1, we obtain:

$$\left[\hbar C^{(2)}_{{}^1D_2-{}^1P_1}\right]_{\mathrm{a.u.}} \approx 1.96 \times 10^0, \tag{7.407}$$

$$\left[\hbar C^{(2)}_{{}^1P_1-{}^1S_0}\right]_{\mathrm{a.u.}} \approx 1.64 \times 10^{-1}. \tag{7.408}$$

With the help of (7.407), (7.408), the lower density limit for the quasi-static broadening condition (7.299) can be estimated:

$$n_i^{(2)}\left({}^1D_2 \to {}^1P_1\right) \gg 1.4 \times 10^{20}\,\mathrm{cm^{-3}}, \tag{7.409}$$

$$n_i^{(2)}\left({}^1P_1 \to {}^1S_0\right) \gg 1.7 \times 10^{21}\,\mathrm{cm^{-3}}. \tag{7.410}$$

The comparison of estimates (7.409) and (7.410) shows that due to the much smaller level separation of the *D–P* transition (see Table 7.3), the quasi-static regime starts at about one order of magnitude lower density compared to the *P–S* transition. In this regime, we encounter a strong level mixing where the transition from spherical quantization to parabolic one has occurred. At densities lower than those given by (7.409), (7.410), the field mixing is weak. In both cases, the energy level mixing effect is the same and determined by statistic electric fields that will be considered in Sects. 7.10.2 and 7.10.3 in the low-frequency approximation of the plasma microfield.

Table 7.4 Dielectronic satellite X-ray transitions and corresponding atomic data obtained from the MZ code (Vainshtein and Shevelko 1986)

Transition	λ (Å)	A (s^{-1})	Γ (s^{-1})	K	Q (s^{-1})
$2p^2\ ^1D_2 \rightarrow 1s2p\ ^1P_1$	7.2759	3.38×10^{13}	3.12×10^{14}	9.77×10^{-2}	1.52×10^{14}
$2s2p\ ^1P_1 \rightarrow 1s2s\ ^1S_0$	7.2316	1.74×10^{13}	1.74×10^{14}	9.09×10^{-2}	4.75×10^{13}
$2p^2\ ^1S_0 \rightarrow 1s2p\ ^1P_1$	7.1929	2.65×10^{13}	1.27×10^{13}	6.75×10^{-1}	8.60×10^{12}

λ is the wavelengths of the X-ray transition, Γ the autoionizing rate, K the branching factor between radiative decay rate and total radiative + autoionizing decay rates, Q is the dielectronic satellite intensity factor

We note that the two different transitions from the strongly mixed levels $2p^2\ ^1D_2$ and $2s2p\ ^1P_1$ can be well observed because the lower levels are quite different ($1s2p\ ^1P_1$ and $1s2s\ ^1S_0$) resulting is a large separation in the spectral distribution, i.e., quite different wavelengths (see Table 7.4).

Let us now consider the density limit for the application of perturbation theory as provided by (7.307):

$$n_i\left(^1D_2 \rightarrow {}^1P_1\right) \ll 9.3 \times 10^{20}\,\text{cm}^{-3}, \tag{7.411}$$

$$n_i\left(^1P_1 \rightarrow {}^1S_0\right) \ll 5.5 \times 10^{22}\,\text{cm}^{-3}. \tag{7.412}$$

In the framework of the present QFMT, these density estimates are not quite relevant because the QFMT solution does not rely on the assumption of small field perturbation.

We estimate now the critical density for the quasi-static condition in atomic kinetics from (7.301). The critical density (employing the sum of the radiative decay and autoionizing rates from the *P–S* and *D–P* transition, Table 7.4) is given by

$$n_i^{(k)} \ll 2 \times 10^{22}\,\text{cm}^{-3}. \tag{7.413}$$

With the help of the relation $n_e \approx Z_p \cdot n_i$ the estimations (7.409)–(7.413) can be transformed to electron densities if perturber ions and radiating ions are identical and originate from the same one component ionized atoms.

We note that the density estimate according to (7.413) is near-solid density. If the fluctuation rates are faster than the characteristic atomic times scales (i.e., if the density estimate (7.413) is seriously invalidated), the time-dependent density matrix equations for the fluctuating microfield have to be solved. Alternatively, one can apply the frequency fluctuation method or the model microfield method that enables to keep the static solutions of the density matrix while introducing a jumping frequency (Brissaud and Fritsch 1971; Kosarev and Lisitsa 1996). This approach maintains the quasi-static consideration presented below even in the case of dynamic microfields.

7.10.2 The Low-Frequency Plasma Microfield

The ion microfield distribution has been taken into account in the following way. The density matrix equations have been solved for each microfield value F. The resulting populations $\tilde{\rho}_{ab}(F)$ have then been summed up attributing a probability to each population that corresponds to the probability $W(F)$ of the microfield distribution, i.e.

$$\tilde{\rho}_{ab} = \int_0^\infty W(F) \cdot \tilde{\rho}_{ab}(F) \cdot dF. \tag{7.414}$$

The simplest estimate of the microfield distribution function can be performed assuming that the ions are statistically independent (i.e., the ideal gas approximation). In this approximation, the correlation of the positions of the ions is neglected; i.e., each ion is found at any point in the plasma independently of how all the other ions are located. This situation is usually characterized by a small Coulomb ion–ion coupling parameter (dimensionless) that is defined as the ratio of the Coulomb energy and the mean distance of the ions:

$$\Gamma_{ii} = \frac{Z_p^2 \cdot e^2}{R_i \cdot kT_i}. \tag{7.415}$$

Substituting the mean ion sphere radius (7.283) into (7.415), we obtain

$$\begin{aligned}\Gamma_{ii} &= \left(\frac{4\pi}{3}\right)^{1/3} \cdot \frac{Z_p^2 \cdot e^2 \cdot n_i^{1/3}}{kT_i} \\ &= 2.321 \times 10^{-7} \cdot \frac{Z_p^2 \cdot n_i^{1/3}\,(\mathrm{cm}^{-3})}{kT_i\,(\mathrm{eV})}.\end{aligned} \tag{7.416}$$

A microfield distribution valid for $\Gamma_{ii} \ll 0.1$ has first been derived by Holtsmark (Unsöld 1955; Hooper 1966, 1968; Griem 1974; Sobelman 2006) (see also Sect. 3.2):

$$W_H = \frac{2}{\sqrt{\pi}} \cdot \frac{1}{\beta} \cdot \int_0^\infty x \cdot \sin(x) \cdot \exp\left[-\left(\frac{x}{\beta}\right)^{3/2}\right] dx, \tag{7.417}$$

where

$$\beta = \frac{F}{F_0} \tag{7.418}$$

with

$$F_0 = 2\pi \cdot \left(\frac{4}{15}\right)^{2/3} \cdot Z_p e \cdot n_i^{2/3}. \tag{7.419}$$

Z_p is the ionic perturber charge, and n_i the perturber density. Numerical calculations show that the maximum of the Holtsmark distribution function is located at

$$F_{max} = 1.607 \cdot F_0. \tag{7.420}$$

The Holtsmark distribution has a well-defined asymptotic behavior for small and large electric field values:

$$W_H \approx \begin{cases} 1.496 \cdot \beta^{-5/2} \cdot \left(1 + 5.107 \cdot \beta^{-3/2} + 14.93 \cdot \beta^{-3} + \cdots\right) & \text{if } \beta \gg 1 \\ 0.4244 \cdot \beta^2 \cdot \left(1 - 0.463 \cdot \beta^2 + 0.1227 \cdot \beta^4 + \cdots\right) & \text{if } \beta \ll 1. \end{cases} \tag{7.421}$$

7.10.3 The Screened Effective Pair Potential Method in Strongly Coupled Plasmas

Important deviations from the Holtsmark distribution are encountered, if the Coulomb ion–ion coupling parameter (7.415), (7.416) is $\Gamma > 0.1$. In fact, in dense plasmas, the electric microfield is related to the probability to find an electric field that is equal to the total electric field composed from the sum of the electron and screened single ion contribution.

Numerous modifications of the Holtsmark distribution to account for ion correlations have been proposed in the framework of the Debye–Hückel theory (Griem 1974, Demura 2010). The simulations are very complex (Demura 2010), and for the sake of convenience, the so-called "Adjustable Parameter Exponential Expansion—APEX" has been developed to take into account ion correlation in dense strongly coupled plasmas.

For the present work, we use the APEX method as proposed by Potekhin et al. (2002) that is based on a screened potential of Yukawa form:

$$V(r) = \frac{Z_p^2 \cdot e^2}{r} \cdot \exp\left(-s \cdot \frac{r}{R_i}\right) = \frac{Z_p^2 \cdot e^2}{r} \cdot \exp(-k_s \cdot r). \tag{7.422}$$

s is a screening parameter, R_i the ion sphere radius (7.283), and k_s is an effective screening wave vector. The low-frequency ion microfield is approximated by

$$W_{APEX} \approx \frac{\beta^2}{S_N} \cdot \left[A \cdot \exp(-a \cdot \beta^{\alpha}) + B \cdot \exp(-b\beta^{\gamma}) + \frac{\exp(-\Gamma_{ii} \cdot \sqrt{\beta})}{1 + c \cdot \beta^{9/2}} \right], \quad (7.423)$$

where the parameter β is a dimensionless field strengths defined by

$$\beta = \frac{F}{\tilde{F}_0} \quad (7.424)$$

with

$$\tilde{F}_0 = \frac{Z_p \cdot e}{R_i^2} = \left(\frac{4\pi}{3}\right)^{2/3} \cdot Z_p \cdot e \cdot n_i^{2/3} = 2.598518 \cdot Z_p \cdot e \cdot n_i^{2/3}. \quad (7.425)$$

Note that $\tilde{F}_0$ is slightly different from the characteristic Holtsmark field strength F_0 (see (7.281) and discussions related to (7.281)–(7.284)). The parameter S_N is a normalization constant:

$$S_N = A \cdot \frac{\Gamma(3/\alpha)}{\alpha \cdot a^{3/\alpha}} + B \cdot \frac{\Gamma(3/\gamma)}{\alpha \cdot b^{3/\gamma}} + \frac{1}{\Gamma_{ii}^6} \cdot F\left(\frac{c}{\Gamma_{ii}^9}\right), \quad (7.426)$$

where $\Gamma(x)$ is the Gamma function and

$$F(y) = \int_0^{\infty} \frac{x^2 \cdot \exp(-\sqrt{x})}{1 + y \cdot x^{9/2}} \cdot \mathrm{d}x. \quad (7.427)$$

The integral (7.427) may be approximated within a precision of a few percent by

$$F(y) \approx \frac{1 + \frac{4\pi}{9 \cdot \sqrt{3}} \cdot y^{1/9}}{\frac{1}{240} + 0.849 \cdot y^{1/3} + 3.2 \cdot y^{5/9} + 2.43 \cdot y^{2/3} + y^{7/9}}. \quad (7.428)$$

The parameters A, B, a, b, c, α and γ are expressed in terms of the ion–ion coupling parameter Γ_{ii} and the potential screening parameter s:

$$A = A_1 \cdot \frac{1 + A_4 \cdot \Gamma_{ii}^{1/2}}{1 + A_2 \cdot \Gamma_{ii}^2 + A_3 \cdot \Gamma_{ii}^4}, \quad (7.429)$$

$$A_1 = 0.59 + 2540 \cdot s^4 + 3 \cdot s^{14}, \quad (7.430)$$

$$A_2 = \frac{0.55 + 10 \cdot s^{0.5} + 2 \cdot s^{4.5}}{1 + 20 \cdot s^{0.5}}, \tag{7.431}$$

$$A_3 = 2.17 \times 10^{-3} \cdot s^5, \tag{7.432}$$

$$A_4 = \frac{14.8}{1 + 117 \cdot s^{3.5}}, \tag{7.433}$$

$$B = \frac{B_1}{1 + B_2 \cdot \Gamma_{ii}^2 + B_3 \cdot \Gamma_{ii}^4}, \tag{7.434}$$

$$B_1 = 0.386 + 300 \cdot s^2 + 1.1 \cdot s^{9.5}, \tag{7.435}$$

$$B_2 = 0.038 + 0.79 \cdot s^{0.75}, \tag{7.436}$$

$$B_3 = \frac{3.7 \times 10^{-3} \cdot s^{5.5}}{1 + 4 \times 10^{-3} \cdot s^9}, \tag{7.437}$$

$$a = 0.5 \cdot \Gamma_{ii} + 1.15 + 2 \cdot s^{1.8}, \tag{7.438}$$

$$b = 0.25 \cdot \Gamma_{ii} + \frac{1 + 0.54 \cdot s^{2.5}}{1 + 0.07 \cdot s}, \tag{7.439}$$

$$c = \frac{0.097}{1 + 210 \cdot s^{2.5} \cdot \exp(-1.3 \cdot s^{1.5})}, \tag{7.440}$$

$$\alpha = \frac{\alpha_1 + 2 \cdot \alpha_2 \cdot \Gamma_{ii}^{1/2}}{1 + \alpha_2 \cdot \Gamma_{ii}^{1/2}}, \tag{7.441}$$

$$\alpha_1 = 0.1 + \frac{1.1}{1 + 0.145 \cdot s^3}, \tag{7.442}$$

$$\alpha_2 = \frac{5.4}{1 + 20 \cdot s^2} + \frac{1.1}{1 + 14 \cdot s^{0.35}}, \tag{7.443}$$

$$\gamma = \frac{\gamma_1 + 1.5 \cdot \gamma_2 \cdot \Gamma_{ii}^{1/2}}{1 + \gamma_2 \cdot \Gamma_{ii}^{1/2}}, \tag{7.444}$$

$$\gamma_1 = 0.1 + \frac{1.1}{1 + 0.174 \cdot s^{2.5}}, \tag{7.445}$$

$$\gamma_2 = \frac{5.4}{1 + 21 \cdot s^{1.5}} + \frac{1.1}{1 + 19 \cdot s^{0.16}}. \tag{7.446}$$

The range of validity of the fit functions (7.429)–(7.446) concerning the ion–ion coupling parameters is $\Gamma_{ii} \leq 100$ and the screening parameter $s \leq 2$. These ranges cover most cases of interest in dense plasma atomic physics.

Let us now determine the effective screening lengths of the potential (7.422) that is an effective pair potential of Debye–Hückel form. A test charge Z_p at position $\vec{r}_p$ creates a perturbation of the electron density. In thermodynamic equilibrium, the perturbed electron density is given by

$$n_e(r) = n_e^{(0)} \cdot \exp\left(-\frac{e\phi}{kT_e}\right), \tag{7.447}$$

where $n_e^{(0)}$ is the unperturbed electron density and ϕ is the potential induced by the test charge. This potential induces changes to the chemical potential of the electrons; i.e., the chemical potential becomes a function of the potential ϕ. Within the framework of the grand canonical ensemble theory, the induced charge due to the potential can therefore be written

$$\rho(\vec{r}) = -e\left[n_e\left(\mu^{(0)} + e\phi\right) - n_e\left(\mu^{(0)}\right)\right], \tag{7.448}$$

where $\mu^{(0)}$ is the chemical potential of the unperturbed electrons. The potential ϕ is defined in such a manner that $\phi(\vec{r}) = 0$ at the points $\vec{r}$, where the plasma is charge neutral and correspondingly $\mu^{(0)}$ is defined as the electron chemical potential at points $\vec{r}$ where the plasma is charge neutral. If we linearize (7.448) (i.e., the chemical potential does not vary very much in space), we obtain for the induced charge

$$\rho(\vec{r}) \approx -e^2\phi(\vec{r}) \cdot \left.\frac{\partial n_e}{\partial \mu}\right|_{\mu=\mu^{(0)}}. \tag{7.449}$$

The potential can then be determined from the Poisson equation:

$$\nabla^2\phi = -4\pi Z_p \delta\left(\vec{r}_p\right) - 4\pi e^2 \phi(\vec{r}) \cdot \left.\frac{\partial n_e}{\partial \mu}\right|_{\mu=\mu^{(0)}}. \tag{7.450}$$

If we rewrite (7.450) like

$$\left[\nabla^2 + 4\pi e^2 \left.\frac{\partial n_e}{\partial \mu}\right|_{\mu=\mu^{(0)}}\right]\phi(\vec{r}) = -4\pi Z_p \delta\left(\vec{r}_p\right) \tag{7.451}$$

the second term on the left-hand side can be interpreted as an effective screening of the pure ionic Coulomb potential. Considering the Fourier transform of (7.451), an effective electron screening wavenumber k_s can be defined according to

$$k_s^2 = 4\pi e^2 \frac{\partial n_e}{\partial \mu}\bigg|_{\mu=\mu^{(0)}}. \tag{7.452}$$

The chemical potential μ of the electron Fermi–Dirac distribution

$$F_{\mathrm{FD}} := \frac{\sqrt{2}\cdot m_e^{3/2}}{\hbar^3 \pi^2}\cdot\frac{1}{n_e}\cdot\frac{\sqrt{\varepsilon}}{e^{(\varepsilon-\mu)/kT_e}+1} \tag{7.453}$$

has to be obtained numerically from the normalization condition

$$1 = \int_0^\infty F_{\mathrm{FD}}(E)\cdot \mathrm{d}E. \tag{7.454}$$

The integral (7.454) can be conveniently transformed to

$$1 = \frac{\sqrt{2}\cdot (m_e kT)^{3/2}}{\hbar^3 \pi^2}\cdot\frac{1}{n_e}\cdot\frac{\sqrt{\pi}}{2}\cdot f_{3/2}(z) \tag{7.455}$$

via the functions $f_x(z)$

$$\begin{aligned} f_x(z) &:= \frac{1}{\Gamma(x)}\int_0^\infty \frac{t^{x-1}\cdot \mathrm{d}t}{z^{-1}e^t+1} \quad \text{for } 0\le z\le 1,\ x\in\Re \\ &= \sum_{k=1}^{\infty}(-1)^{k-1}\cdot\frac{z^k}{k^x} \end{aligned} \tag{7.456}$$

that are well known in statistical physics. We note that also powerful closed analytical approximations have been developed (Antia 1993). The functions $f_x(z)$ obey a simple recurrence relation

$$\frac{\partial}{\partial z} f_x(z) = \frac{1}{z}\cdot f_{x-1}(z). \tag{7.457}$$

Due to the exponential dependence of the fugacity z from the chemical potential, it is in some cases more convenient to develop the functions $f_n(z)$ in terms of a logarithmic dependence from the fugacity (Sommerfeld 1971):

$$\begin{aligned} f_x(z) = \frac{(\ln z)^x}{\Gamma(x+1)}\cdot\Bigg\{1 + 2x\cdot\sum_{j=1}^{\infty}\binom{n-1}{2j-1} \\ \cdot(\ln z)^{-2j}\cdot\Gamma(2j)\cdot\zeta(2j)\cdot\left(1-\frac{1}{2^{2j-1}}\right)\Bigg\}, \end{aligned} \tag{7.458}$$

where ζ is the Rieman zeta function. With the help of (7.456), the *f*-function in (7.455) takes the form

$$f_{3/2}(z) = \frac{2}{\sqrt{\pi}} \cdot \int_0^{\infty} \frac{t^{1/2} \cdot dt}{z^{-1} \cdot e^t + 1}, \tag{7.459}$$

$$z = \exp(y), \tag{7.460}$$

$$y = \frac{\mu}{kT_e}. \tag{7.461}$$

In certain ranges of temperature and density, the parameter y characterizes the degeneracy of the electron gas, i.e., y is large, if the density is high and the temperature low. From (7.455)–(7.457), (7.459)–(7.461), we can determine the derivative $\partial n_e / \partial \mu$ that appears in (7.452):

$$\begin{aligned}\frac{\partial n_e}{\partial \mu} &= \sqrt{\frac{\pi}{2}} \cdot \frac{(m_e kT_e)^{3/2}}{\pi^2 \hbar^3} \cdot \frac{\partial f_{3/2}(z)}{\partial z} \cdot \frac{\partial z}{\partial y} \frac{\partial y}{\partial \mu} \\ &= \sqrt{\frac{\pi}{2}} \cdot \frac{(m_e kT_e)^{3/2}}{\pi^2 \hbar^3} \cdot \frac{1}{z} f_{1/2}(z) \cdot \frac{z}{kT_e} \\ &= \sqrt{\frac{\pi}{2}} \cdot \frac{m_e^{3/2} (kT_e)^{1/2}}{\pi^2 \hbar^3} \cdot f_{1/2}(z),\end{aligned} \tag{7.462}$$

where

$$f_{1/2}(z) = \frac{1}{\sqrt{\pi}} \cdot \int_0^{\infty} \frac{t^{-1/2} \cdot dt}{z^{-1} \cdot e^t + 1}. \tag{7.463}$$

From (7.452), (7.462), (7.463), the effective screening wave vector is given by

$$k_s^2 = (2m_e)^{3/2} \cdot \frac{e^2}{\pi \hbar^3} \cdot (kT)^{1/2} \cdot \sqrt{\pi} \cdot f_{1/2}(z). \tag{7.464}$$

For an ideal electron gas, the chemical potential can be obtained from (7.455) taking into account only the first expansion term in (7.456):

$$\mu^{(\text{ideal})} = -kT_e \cdot \ln\left\{ \frac{2}{n_e} \cdot \left(\frac{mkT_e}{2\pi\hbar^2} \right)^{3/2} \right\}. \tag{7.465}$$

In the limit of weak coupling, we can likewise replace $f_{1/2}(z)$ by the first expansion term in (7.456), i.e.,

$$f_{1/2}(z) \to z = \frac{n_e}{2} \cdot \left(\frac{2\pi\hbar^2}{m_e k T_e} \right)^{3/2} . \tag{7.466}$$

Inserting (7.466) in (7.462) we obtain

$$k_s^2 \Big|_{\text{weak coupling}} \to \frac{4\pi e^2 n_e}{kT_e} . \tag{7.467}$$

The effective screening wave vector for small coupling parameters is therefore the Debye length, i.e.,

$$k_s \Big|_{\text{weak coupling}} \to \frac{1}{\lambda_D} \tag{7.468}$$

with

$$\lambda_D = \sqrt{\frac{kT_e}{4\pi e^2 n_e}} = 7.4339 \times 10^2 \cdot \sqrt{\frac{n_e\,(\text{cm}^{-3})}{kT_e\,(\text{eV})}}\,[\text{cm}]. \tag{7.469}$$

Equation (7.455) shows that at high densities and low temperatures, i.e., for large coupling parameters the function $f_{3/2}(z)$ should become very large and can therefore be approximated by the first expansion term in (7.458):

$$1 \approx \frac{\sqrt{2} \cdot (m_e k T_e)^{3/2}}{\hbar^3 \pi^2} \cdot \frac{1}{n_e} \cdot \frac{\sqrt{\pi}}{2} \cdot \frac{4}{3\sqrt{\pi}} \cdot (\ln z)^{3/2} . \tag{7.470}$$

From (7.460), (7.461), we obtain an analytical expression for the chemical potential

$$1 \approx \frac{(2 m_e k T_e)^{3/2}}{3\hbar^3 \pi^2} \cdot \frac{1}{n_e} \cdot \left(\frac{\mu}{kT_e} \right)^{3/2} . \tag{7.471}$$

Therefore, the chemical potential is independent from the temperature and given by

$$\mu \approx \frac{3^{2/3} \pi^{4/3} \hbar^2}{2 m_e} n_e^{2/3} = \varepsilon_F , \tag{7.472}$$

where ε_F is the Fermi energy

$$\varepsilon_F = \frac{3^{2/3} \pi^{4/3} \hbar^2}{2 m_e} \cdot n_e^{2/3} = 3.6464 \times 10^{-15} \cdot n_e^{2/3}\left(\text{cm}^{-3}\right) [\text{eV}], \tag{7.473}$$

which is identical to the chemical potential at zero temperature. From (7.464), we obtain with the help of (7.472) the expression

$$k_s^2 = \frac{4\pi e^2 n_e}{kT_e} \cdot \left(\frac{kT_e}{\varepsilon_F}\right)^{3/2} \cdot \frac{3}{4} \cdot \sqrt{\pi} \cdot f_{1/2}(z). \tag{7.474}$$

In the limit of strong coupling, we replace for the screening wave vector likewise $f_{1/2}(z)$ by the first expansion term in (7.458):

$$k_s^2 \approx \frac{4\pi e^2 n_e}{(2/3) \cdot \varepsilon_F} \tag{7.475}$$

and

$$k_s^2\Big|_{\text{strong coupling}} \to \frac{4\pi e^2 n_e}{(2/3) \cdot \varepsilon_F}. \tag{7.476}$$

Comparing the relations (7.467) and (7.475) we see that in the limit of strong coupling the electron temperature kT_e is replaced by $(2/3) \cdot \varepsilon_F$. The physical reason is that at very low temperature, the pressure of a Fermi gas is due to the Pauli principle while thermal pressure is negligible. The asymptotic relations (7.467) and (7.476) suggest therefore a simple approximation via a match of the asymptotes:

$$k_s^2 \approx \frac{4\pi e^2 n_e}{\left[(kT_e)^\alpha + (2 \cdot \varepsilon_F/3)^\alpha\right]^{1/\alpha}}. \tag{7.477}$$

As the asymptotes are matched by (7.477) irrespective of the value of the fitting parameter α, we may determine this parameter for an intermediate case where the electron temperature is of the order of the Fermi energy. If we take

$$kT_e = \frac{2}{3} \cdot \varepsilon_F \tag{7.478}$$

we obtain from (7.477)

$$k_s^2\left(kT_e = \frac{2}{3}\varepsilon_F\right) = \frac{4\pi e^2 n_e}{2^{1/\alpha} kT_e}. \tag{7.479}$$

Combining (7.464), (7.472) with (7.479) provides

$$\begin{aligned} \frac{1}{2^{1/\alpha}} &= \frac{3}{4}\sqrt{\pi} \cdot \left(\frac{kT}{\varepsilon_F}\right)^{3/2} \cdot f_{1/2}(z) \\ &= \frac{3}{4}\sqrt{\pi} \cdot \left(\frac{2}{3}\right)^{3/2} \cdot f_{1/2}(z) = \sqrt{\frac{\pi}{6}} \cdot f_{1/2}(z) \end{aligned} \tag{7.480}$$

from which it follows

$$\alpha = \frac{-\ln 2}{\ln\left(\sqrt{\frac{\pi}{6}} \cdot f_{1/2}(z)\right)}. \tag{7.481}$$

Numerical calculations carried out for the chemical potential show that for $kT_e = \frac{2}{3} \cdot \varepsilon_F$ we have $\mu/kT_e \approx 0.80284$ and for $\sqrt{\pi} \cdot f_{1/2}(z) = F_{-1/2}(\mu/kT_e) \approx 1.6653$ (Cloutman 1989). According to (7.481), this results in $\alpha \approx 1.7963$. With the help of (7.481), (7.478) and $k_s = s/R_i$ (7.422), we can determine the screening constant s:

$$s \approx \left(\frac{3}{4\pi}\right)^{1/3} \cdot n_i^{-1/3} \cdot \sqrt{\frac{4\pi e^2 n_e}{\left[(kT_e)^\alpha + (2 \cdot \varepsilon_F/3)^\alpha\right]^{1/\alpha}}}. \tag{7.482}$$

Setting $n_e = Z_p \cdot n_i$, we obtain from (7.482) the screening parameter in convenient units:

$$s \approx 8.3448 \times 10^{-4} \cdot \frac{n_e^{1/6}\,(\mathrm{cm}^{-3})}{Z_p^{1/3}} \cdot \frac{1}{\left[(kT_e\,(\mathrm{eV}))^\alpha + (2 \cdot \varepsilon_F\,(\mathrm{eV})/3)^\alpha\right]^{1/2\alpha}}. \tag{7.483}$$

For example, with $n_e = 10^{21}\,\mathrm{cm}^{-3}$, $kT_e = 100\,\mathrm{eV}$, and $Z_p = 11$, we obtain for the ion–ion coupling parameter $\Gamma_{ii} = 1.3$, $\varepsilon_F = 0.3646\,\mathrm{eV}$, and $s = 0.119$ while, e.g., for $n_e = 10^{24}\,\mathrm{cm}^{-3}$, $kT_e = 2\,\mathrm{eV}$ and $Z_p = 3$ we obtain for the ion–ion coupling parameter $\Gamma_{ii} = 72$, $\varepsilon_F = 36.46\,\mathrm{eV}$ and $s = 1.17$. Therefore, the interval $s = [0, 2]$ covers almost all cases of interest.

The comparison of the Holtsmark distribution (7.417) with the APEX (7.423) shows that for high ion–ion coupling parameters, the microfield distribution function becomes narrower and the maximum shifts to lower field values. The maximum of the field distribution can be determined from the following fitting formula (Potekhin et al. 2002):

$$F_{\max}^{(\mathrm{APEX})} = \frac{\tilde{F}_0}{0.622 + 0.25 \cdot s \cdot e^s} \cdot \left[1 + \frac{\Gamma_{ii}^{1/4} + \Gamma_{ii}}{0.774 + 0.54 \cdot s \cdot e^s}\right]^{-1/2}. \tag{7.484}$$

The maximum of (7.484) has to be compared with the maximum of the Holtsmark distribution:

$$r = \frac{F_{\max}^{(\mathrm{APEX})}}{F_{\max}^{(\mathrm{Holtsmark})}} \approx \frac{0.622}{0.622 + 0.25 \cdot s \cdot e^s} \cdot \left[1 + \frac{\Gamma_{ii}^{1/4} + \Gamma_{ii}}{0.774 + 0.54 \cdot s \cdot e^s}\right]^{-1/2}. \tag{7.485}$$

For the examples given above, we obtain $r(\Gamma_{ii} = 1.3) \approx 0.49$ and $r(\Gamma_{ii} = 72) \approx 0.076$. Therefore, even for moderately coupled plasmas, the deviations from the Holtsmark distribution are considerable and the microfield distribution for high coupling parameters is strongly shifted to smaller field values.

7.10.4 The Relaxation Rate Approximation of QFMT: QFMT-W

The F-matrix from (7.369)–(7.372) has been further simplified in order to meet also the requirements of (7.316), (7.317) rather than only those of (7.315):

$$F_{\mathrm{ab}} \approx 2\frac{V_{\mathrm{ab}} \cdot V_{\mathrm{ba}}}{\hbar^2} \cdot G \tag{7.486}$$

with

$$G = \frac{a_{\mathrm{ab}}}{\omega_{\mathrm{ab}}^2 + a_{\mathrm{ab}}^2}, \tag{7.487}$$

$$a_{\mathrm{ab}} = \frac{\gamma_{\mathrm{a}}}{2} + \frac{\gamma_{\mathrm{b}}}{2} + \frac{A_{\mathrm{ab}} + A_{\mathrm{ba}}}{2} - \Phi_{\mathrm{ab}}^{\mathrm{ab}}, \tag{7.488}$$

where the operator $\Phi_{\mathrm{ab}}^{\mathrm{ab}}$ is represented in terms of collisional rates, see (7.213), i.e. $\Phi_{\mathrm{ab}}^{\mathrm{ab}} = \Phi_{\mathrm{ab}}/2 + \Phi_{\mathrm{ba}}/2$. The loss rates γ_{a} and γ_{b} (see Table 7.2) are approximated by the autoionizing rates and the radiative decay rates $2l2l' \rightarrow 1s2l$. This approximation neglects collisional transfer and ionization loss rates to other levels which is well justified as discussed above.

Figures 7.3, 7.4 and 7.5 show population ratios and effective field rates (Rosmej et al. 2019). Figure 7.3 shows the population ratio of the states $2p^2\ {}^1D_2$ and

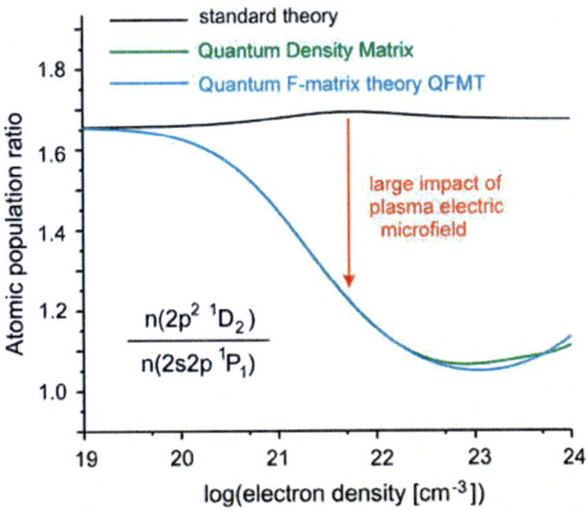

Fig. 7.3 Population ratios of autoionizing levels $n(2p^2\ {}^1D_2)/n(2s2p\ {}^1P_1)$ in dependence of the electron density. The microfield calculations are averaged over the APEX ion distribution function. The Quantum F-matrix Theory employed in the quasi-classical approach provides very good agreement with the quantum mechanical atomic density matrix theory

$2s2p\ ^1P_1$. The black curve presents the calculation of the standard theory (7.1), the green curve is the solution of the density matrix (7.156) and the blue curve is the Quantum F-Matrix Theory QFMT from (7.377). It is observed that for the standard theory, the population ratio does not depend much on density because of the large autoionization rates of the $2p^2\ ^1D_2$ and $2s2p\ ^1P_1$ states that strongly populates these levels by dielectronic capture (see Table 7.2). Therefore, collisions do not change much the populations even for high density. The solution of the density matrix results, however, in a strong decrease of the ratio starting from densities of about $10^{20}\ \text{cm}^{-3}$. Although the QFMT approach of (7.486)–(7.488) is a rather simplified one, it provides a very good approximation to the density matrix solution.

Figures 7.3 demonstrates also that due to the plasma electric microfield the population ratio approaches 1 at about $5 \times 10^{22}\ \text{cm}^{-3}$. This strong deviation from the standard ratio (of about 1.7) is due to the non-random character of the quasi-static ion field (see also (7.394), (7.395) and related discussion) that does not support the Boltzmann limit but has the tendency to equilibrate the atomic states to equal populations (7.395). The regime from low density to densities where the level populations approach approximately 1 is characterized by a transition from spherical quantization to parabolic one. We note, that the quantum density matrix calculations of Fig. 7.3 are not resolved in magnetic quantum number: whether this allows analyzing previously unexplained data (e.g. Renner et al. 2009) remains an open question.

Figure 7.4 shows the population ratio of the states $2p^2\ ^1D_2$ and $2p^2\ ^1S_0$. The black curve presents the calculation of the standard theory (7.1), the green curve is the solution of the density matrix (7.156) and the blue curve is the quantum F-matrix theory from (7.377), (7.486)–(7.488). It is observed that for the standard theory, the population ratio depends on density because the autoionization rate of the $2p^2\ ^1S_0$ state is rather small (see Table 7.2). Therefore, electron collision between the autoionizing levels transfers population from the levels $2p^2\ ^1D_2$ and $2s2p\ ^1P_1$ to the $2p^2\ ^1S_0$ level leading to a decrease of the ratio with increasing densities. At electron

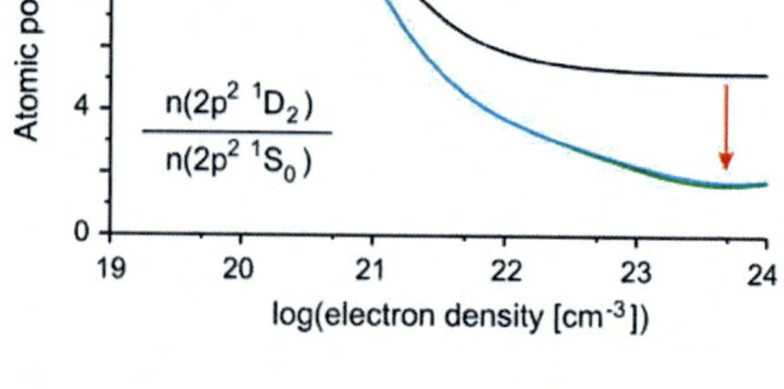

Fig. 7.4 Same like Fig. 7.4, however showing the population ratios of the autoionizing levels $n(2p^2\ ^1D_2)/n(2p^2\ ^1S_0)$ in dependence of the electron density

densities of about 10^{22} cm^{-3}, the ratio stabilizes because the Boltzmann limit is reached. The exact solution of the density matrix results, however, in a strong decrease of the ratio starting from densities of about 10^{21} cm^{-3}. Although the QFMT approach of (7.486)–(7.488) is a rather simplified one, it provides also for this ratio a very good approximation to the density matrix solution.

As the approximate QFMT solution according to (7.486)–(7.488) have provided very good agreement with the quantum mechanical density matrix solution, this QFMT represents an efficient approximation as it can be also included in the general rate equation atomic kinetics. We therefore name hereafter this approximation QFMT-W. The particular efficiency of QFMT-W arises from the fact that it can be represented only with rate coefficients and energies, that are already included in the *W*-matrix elements of the usual rate equation approach.

Figure 7.5 explores the impact of the field perturbation in atomic population kinetics comparing the field-induced rate F_{ab} with the usual electron collisional rate C_{ab}. It can be seen that for higher plasma densities, the field effects may have even larger values than the standard electron collisional rates. This is the origin of the strong deviation from the standard rate equation kinetics shown in Figs. 7.3 and 7.4. At very high densities, the electron collisions exceed the field effects because their scaling with electron density is $\propto n_e$ (see also (7.405)) while the field effects scale with $\propto n_e^{1/3}$ [see also (7.404)]. The field effects therefore vanish for very high densities (see also (7.406) and discussion of scaling relations). However, as the calculations of Figs. 7.3 and 7.4 demonstrate, this limit is difficult to achieve. In fact, the field effect stay important even for near-solid densities, thereby changing the usual Boltzmann limit of the standard rate equation approach.

Figure 7.5 demonstrates likewise that the *F*-matrix approach takes into account the different action of the electric field mixing due to different energy separation. Due to the much smaller level separation between the levels $2p^2\,{}^1D_2$ and $2s2p\,{}^1P_1$ (1.6 eV, see Table 7.3) than between the levels $2s2p\,{}^1P_1$ and $2p^2\,{}^1S_0$ (18.5 eV, see

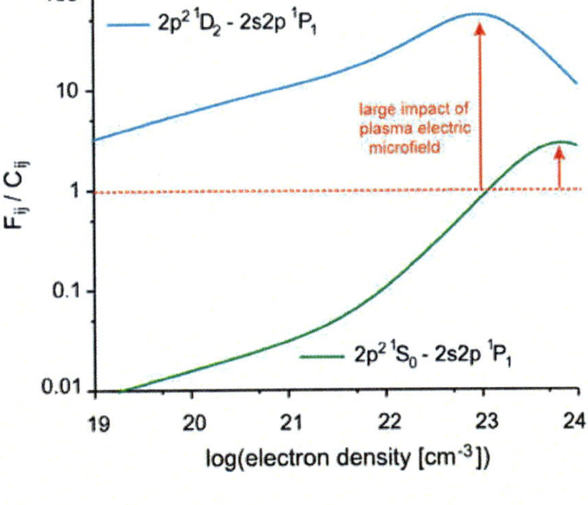

Fig. 7.5 Comparison of the effective electric field rate F_{ab} employed in the atomic rate population kinetics with the corresponding electron collisional rate C_{ab}. The impact of the plasma electric microfield results in strong perturbations with effective rates that can be much larger than the standard electron collisional rates (see red flashes)

Table 7.3), the field interaction is much more important. This is clearly visible in Fig. 7.5 where the ratio of the effective fields rates to the standard collisional rates is much larger for the transition $2p^2\,{}^1D_2$–$2s2p\,{}^1P_1$ than for the transition $2p^2\,{}^1S_0$–$2s2p\,{}^1P_1$. This behavior is quite different from usual collisional processes employed in standard kinetics: For closely spaced levels, the collisional transfer rates are almost all in the Born limit and no important different effects are emerging from different levels separations of the $2l2l'$-configurations (an exception might be the consideration of proton collisions that are very sensitive to the level spacing due to typically low velocity in plasmas: Well-known examples are the proton collisions in the fine structure of Lyman-alpha).

In conclusion, the new quantum F-matrix theory QFMT opens up the possibility to consider large closed population kinetic dense collisional model systems that could not be treated consistently in the Schrödinger picture nor efficiently in the framework of the quantum mechanical atomic density matrix theory (due to its prohibitive complexity). The QFMT is based on a successive pair coupling of quantum effects, while all pair combinations are considered at the same time when solving the large kinetic model system. QFMT allows likewise including quantum mechanical interference effects in standard rate equations. It introduces field- and level-separation-dependent effective rates in standard collisional radiative modeling and therefore incorporates the main quantum effects in population kinetics. QFMT-W represents an efficient approximation to QFMT as it employs only energies and rate coefficients, that are usually implemented in the transition matrix W of the standard rate equations. QFMT-W allows therefore developing consistently quantum kinetics approximations for large closed model systems of dense plasmas that are also numerically manageable.

References

A.N. Antia, Rational function approximations for Fermi-Dirac integrals. Astrophys. J. Suppl. Ser. **84**, 101 (1993)

A.N. Anufrienko, A.L. Godunov, A.V. Demura, Y.K. Zemtsov, V.S. Lisitsa, A.N. Starostin, M.D. Taran, V.A. Shchipakov, Nonlinear interference effects in Stark broadening of ion lines in a dense plasma. Sov. Phys. JETP **71**, 728 (1990)

A.N. Anufrienko, A.E. Bulyshev, A.L. Godunov, A.V. Demura, Y.K. Zemtsov, V.S. Lisitsa, A.N. Starostin, Nonlinear interference effects and ion dynamics in the kinetic theory of Stark broadening of the spectral lines of multicharged ions in a dense plasma. Sov. Phys. JETP **76**, 219 (1993)

E.K. Akhmedov, A.L. Godunov, Y.K. Zemtsov, V.A. Makhrov, A.N. Starostin, M.D. Taran, Profile of the Ly_α line of hydrogen-like ions in a dense plasma with allowance for the fine structure and the Lamb shift. Sov. Phys JETP **62**, 266 (1985)

M. Baranger, Problem of overlapping lines in the theory of pressure broadening. Phys. Rev. **111**, 494 (1958)

G. Bekefi, *Radiation Processes in Plasmas* (Wiley, New York, 1966)

V.A. Boiko, A.V. Vinogradov, S.A. Pikuz, I.Y. Skobelev, A.Ya. Faenov, X-ray spectroscopy of laser produced plasmas. J. Sov. Laser Research **6**, 82 (1985)

A. Brissaud, U. Fritsch, Theory of Stark Broadening II. - Exact Line Profile with Model Microfield. JQSRT **11**, 1767 (1971)

L.A. Bureyeva, V.S. Lisitsa, A Perturbed Atom. In: *Astrophysics and Space Science Reviews* **11** (2000)

L.D. Cloutman, Numerical evaluation of the Fermi-Dirac integrals. Astrophys. J. Suppl. Ser. **71**, 677 (1989)

J. Cooper, G.K. Oertel, Electron impact broadening of isolated lines of neutral atoms in a plasma. Phys. Rev. **180**, 286 (1969)

R.D. Cowan, *The Theory of Atomic Structure and Spectra* (University of California Press, Berkely, 1981)

A.S. Dawydow, *Quantum Mechanics* (1981)

C. de Michelis, M. Mattioli, Soft-X-ray spectroscopy of laboratory plasmas. Nucl. Fusion **21**, 677 (1981)

E.L. Degl'Ionnocenti, M. Landolfi, *Polarization in Spectral Lines* (Kluwer Academic Publishers, Dordrecht, 2004)

A.V. Demura, Physical models of plasma microfield. Int. J. Spectrosc. Rev. **42** (2010). https://doi.org/10.1155/2010/671073 (Article ID 671073)

C. Deutsch, L. Herman, H.-W. Drawin, Electron-impact broadening of overlapping HeI lines in plasmas. Phys. Rev. **178**, 261 (1969a)

C. Deutsch, L. Herman, H.-W. Drawin, Asymptotic behavior of generalized width and shift functions in the electron-impact broadening theory of neutral spectral lines in plasmas. Phys. Rev. **186**, 204 (1969b)

R.P. Drake, *High-Energy-Density Physics* (Springer, 2018)

T. Fujimoto, A. Iwamae, *Plasma Polarization Spectroscopy* (Springer, Berlin, 2008)

H.R. Griem, *Plasma Spectroscopy* (McGraw-Hill Book Company, New York, 1964)

H.R. Griem, Semiempirical formulas for the electron-impact widths and shifts of isolated ion lines in plasmas. Phys. Rev. **165**, 258 (1968)

H.R. Griem, *Spectral Line Broadening by Plasmas* (Academic Press, New York, London, 1974)

H.R. Griem, *Principles of Plasma Spectroscopy* (Cambridge University Press, New York, 1997)

H.R. Griem, M. Blaha, P.C. Kepple, Stark-profile calculations for Lyman-series lines of one-electron ions in dense plasmas. PRA **19**, 2421 (1979)

H.R. Griem, K.Y. Shen, Stark broadening of hydrogen lines in a plasma. Phys. Rev. **116**, 4 (1959)

M.F. Gu, The flexible atomic code. Can. J. Phys. **86**, 675 (2008)

C.F. Hooper, Electric microfield distributions in plasmas, J. Phys. Rev. **149**, 77 (1966)

C.F. Hooper Jr., Low-frequency component electric microfield distributions in plasmas. Phys. Rev. **165**, 215 (1968)

I.N. Kosarev, V.S. Lisitsa, Model mircofield method for atomic state kinetics in dense plasmas: analytical solutions. J. Phys. B: At. Mol. Opt. Phys. **29**, 1183 (1996)

V.I. Kogan, V.S. Lisitsa, A.D. Selidovkin, Two-level system with damping in a plasma. Sov. Phys. JETP **38**, 76 (1973)

V.S. Lisitsa, Stark broadening of hydrogen lines in plasmas. Sov. Phys. Usp. **20**, 603 (1977)

V.S. Lisitsa, *Atoms in Plasmas* (Springer, New York, 1994)

R. Loudon, *The Quantum Theory of Light*, 3rd edn. (Oxford Science Publications, 2000)

R.W.P. McWhirter, Spectral intensities, in *Plasma Diagnostic Techniques*, ed. by R.H. Huddelstone, S.L. Leonard (Academic Press, New York, 1965)

D. Mihalas, *Stellar Atmospheres*, 2nd edn. (W.H. Freeman, San Francisco, 1978)

D. Mihalas, B. Weibel-Mihalas, *Foundations of Radiation Hydrodynamics* (Dover Publications Inc., New York, 1999)

H. Muirhead, *The Principles of Elementary Particles* (1965)

A.Y. Potekhin, G. Chabrier, D. Gilles, Electric microfield distributions in electron-ion plasmas. Phys. Rev. E **65**, 036412 (2002)

S.G. Rautian, A.M. Shalagin, *Kinetic Problems of Non-linear Spectroscopy* (North-Holland, 1991)

O. Renner, R. Liska, F.B. Rosmej, Laser-produced plasma-wall interaction. Laser Part. Beams **27**, 725 (2009)

O. Renner, F.B. Rosmej, Challenges of X-ray spectroscopy in investigations of matter under extreme conditions. Matter Radiat. Extrem. Rev. **4**, 024201 (2019)

F.B. Rosmej, Exotic states of high density matter driven by intense XUV/X-ray free electron lasers, in *Free Electron Laser, InTech 2012*, ed. by S. Varró (2012a), pp. 187–212. ISBN 978-953-51-0279-3

F.B. Rosmej, X-ray emission spectroscopy and diagnostics of non-equilibrium fusion and laser produced plasmas, in *Highly Charged Ion Spectroscopic Research*, ed. by Y. Zou, R. Hutton (Taylor and Francis, 2012b), pp. 267–341. ISBN: 978-1-4200-7904-3

F.B. Rosmej, R.W. Lee, D.H.G. Schneider, Fast X-ray emission switches driven by intense X-ray free electron laser radiation. High Energy Density Phys. **3**, 218 (2007)

F.B. Rosmej, B. Deschaud, K. Bennadji, P. Indelicato, J.P. Marquès, Study of electric dipole matrix elements of He-like ions for X-ray line shape calculations. Phys. Rev. A **87**, 022515 (2013)

F.B. Rosmej, R. Dachicourt, B. Deschaud, D. Khaghani, M. Dozières, M. Smîd, O. Renner, Exotic X-ray emission from dense plasmas. J. Phys. B: Rev. Spec. Top. **48**, 224005 (2015)

F.B. Rosmej, R.W. Lee, Hollow ion emission driven by pulsed X-ray radiation fields. Europhys. Lett. **77**, 24001 (2007)

F.B. Rosmej, Y. Aouad, S. Ongala-Edoumou, A. Demura, V.S. Lisitsa, *Private communications* 2008–2019

D. Salzmann, *Atomic Physics in Hot Plasmas* (Oxford University Press, New York, 1988)

S. Sahal-Bréchot, M.S. Dimitrijevic, N.B. Nessib, Width and shifts of isolated lines of neutral and ionized atoms perturbed by collisions with electrons and ions: an outline of the semiclassical perturbation (SCP) method and of the approximations used for the calculations. Atoms **2**, 225 (2015)

G.V. Sholin, A.V. Demura, V.S. Lisitsa, Theory of stark broadening of hydrogen lines in plasma. Sov. Phys. JETP **37**, 1057 (1973)

E.W. Smith, C.F. Hooper, Relaxation theory of spectral line broadening in plasmas. Phys. Rev. **157**, 126 (1967)

I.I. Sobelman, *Theory of Atomic Spectra* (Alpha Science International Ltd., Oxford, UK, 2006)

I.I. Sobelman, L.A. Vainshtein, *Excitation of Atomic Spectra* (Alpha Science International Ltd., Oxford, UK, 2006)

A. Sommerfeld, *Thermodynamics and Statistical Mechanics*. Lectures on Theoretical Physics V (Academic Press, New York, 1971)

A. Unsöld, *Physik der Sternatmosphären* (Springer, Berlin, 1955)

L.A. Vainshtein, V.P. Shevelko, The structure and characteristics of ions in hot plasmas. Phys. Techn. Spektrosc. (1986) (in Russian)

V. Weisskopf, Die Streuung des Lichts an ageregten Atomen. Z. Phys. **85**, 451 (1933)

Y.B. Zel'dovich, Y.P. Raizer, *Physics of Shock Waves and High-Temperature Hydrodynamic Phenomena* (Dover Publications Inc., 2002)

Chapter 8
Ionization Potential Depression

Abstract In a low-density environment, where atoms and ions are essentially free, atomic population kinetics of gases and plasmas has been very successful in many different scientific and technical disciplines. As density increases, the free atom model breaks down resulting in a perturbation of the atomic energy levels and a corresponding ionization potential depression IPD. The IPD is of great fundamental interest, for thermodynamic applications and also for the understanding of the various radiative properties (emission, absorption, scattering). Different IPD models are discussed including the finite temperature ion sphere FTIS model and the Atomic-Solid-Plasma ASP model. The ASP model accounts for the difference between real atomic ionization potentials in solids and free atoms taking into account the structure of the valence band and the Fermi energy. The FTIS model accounts for self-consistent screening effects of both bound and free electrons inside the ion sphere with effective radius depending on plasma density. Different regimes of ionization potential depression are considered as well as plasma polarization shifts of X-ray spectral lines. Finally, Fermi surface rising in much above solid density compressed matter is discussed leading to increased K-edge energies rather than decreased ones.

8.1 The Atomic-Solid-Plasma ASP Model

As was pointed out in Chap. 7, a correct description of the population of a quantum mechanical state is of fundamental importance across a wide range of disciplines. In a low-density environment, where atoms and ions are essentially free, atomic population kinetics of gases and plasmas (see Chap. 6) has been very successful in many different scientific and technical disciplines. For example, it resulted in a unique characterization of matter via dense atomic plasma physics methods and in the development of plasma spectroscopy (Rautian et al. 1991; Griem 1997; Kunze 2009; Fujimoto 2004; Sobelman et al. 1995; Lisitsa 1994; Salzman 1998; Unsöld 1955; Mihalas 1978; Mihalas and Weibel-Mihalas 1999; Zeldovich and Raizer 2002; Drake 2006). As density increases, the free atom model breaks down and numerous corrections to the free atom picture have been developed among which

F. B. Rosmej et al., *Plasma Atomic Physics*, Springer Series on Atomic, Optical, and Plasma Physics 104, https://doi.org/10.1007/978-3-030-05968-2_8

the famous ionization potential depression IPD has attracted a broad interest since many decades (Inglis and Teller 1939; Ecker and Weizel 1956; Ecker and Kröll 1963; Kohn and Majumdar 1965; Zimmermann and Moore 1980; More 1981; Nguyen et al. 1986; Massacrier and Dubeau 1990; Hummer and Mihalas 1988; Shimamura and Fujimoto 1990; Blenski and Ishikawa 1995; Stewart and Pyatt 1996; Rosmej et al. 2011; Li and Rosmej 2012, 2020; Chung et al. 2013; Li et al. 2019). We also note that the IPD is strongly related to the long-standing discussion of the divergence of the canonical partition function in thermodynamics (Blinder 1995).

The continuity of bound and unbound states is a fundamental question and has stimulated sophisticated high-resolution X-ray line shift measurements in very dense plasmas (Griem 1997; Renner et al. 1997b, 1998; Saemann et al. 1999; Rosmej et al. 2000; Dervieux et al. 2015; Hansen et al. 2017; Beiersdorfer et al. 2019). Among a variety of studies, those performed with the vertical Johann X-ray spectrometer scheme (Renner et al. 1997a) for the H-like Lyman series of aluminum (Renner et al. 1997b, 1998) attracted particular interest (Renner and Rosmej 2019).

With the emergence of the X-ray Free Electron Lasers, also measurements of the ionization potential depression have been attempted and the data seem to question our present understanding of IDP (Ciricosta et al. 2012). It was claimed that the performed measurements support the early model of Ecker and Kröll (Ecker and Kröll 1963), whereas the well-accepted model of Stewart and Pyatt (More 1981; Stewart and Pyatt 1996) is in disagreement with the data. On the other hand, subsequently performed ionization potential depression measurements at laser-driven compression experiments (Hoarty et al. 2013; Fletcher et al. 2014) confirmed the validity of the Stewart and Pyatt model and demonstrated a worse agreement with the Ecker and Kröll model. Density functional theory (DFT) calculations (Vinko et al. 2014) renewed the claims made in (Ciricosta et al. 2012) while the analysis of the XFEL data presented in Ciricosta et al. (2012) has recently again been questioned (Iglesias and Sterne 2013; Iglesias 2014; Son et al. 2014). Shortly later similar IPD data from XFEL had been re-published (Ciricosta et al. 2016). Also the pseudo-potential assumptions in the DFT calculations of (Vinko et al. 2010, 2014) had been subjects to critics (Iglesias 2011, Karasiev and Hu 2021). In addition, very recent line shift measurements of the He-like resonance line of aluminum in a dense laser-produced plasma experiment (Stillman et al. 2017) turned out to agree well with the analytic finite temperature ion sphere model proposed in (Li et al. 2006; Li and Rosmej 2012, 2020). It was found empirically (Stillman et al. 2017) that better agreement is obtained scaling down the ion sphere radius of the ab initio analytic 2nd-order model (Li and Rosmej 2012) by about 10% while numerically, scaling properties have been investigated in the framework of the FTIS by (Rosmej et al. 2011; Rosmej 2018; Li et al. 2019; Li and Rosmej 2020). Moreover, recent measurements in strongly compressed iron plasmas indicate that ionization energies did not decrease but increase (Hansen et al. 2017).

In view of the continued and controversy discussion of the ionization potential depression, the atomic-solid-plasma (ASP) model (Rosmej 2018) has been developed for high-density matter where crystal-like structure exists. ASP was shown to provide good agreement to a large variety of data and explains observed scaling relations that remained unexplored. Moreover, ASP also predicts increased ionization energies for much above solid density compressed matter.

Figure 8.1 visualizes via magnesium schematically the principles of the solid and atomic-solid-plasma model. The cold solid (or "usual" solid) is characterized by the core states $1s^2$, $2s^2$, $2p^6$, and the valence band $(\mathrm{VB})^2$ contains 2 electrons and the corresponding Fermi energy is $\varepsilon_{\mathrm{F}}^{(2)}$. As the electrons in the VB are essentially free, the ASP model considers the states $1s^2$, $2s^2$, $2p^6$ as bound states and the states above as continuum states (indicated as ASP continuum in Fig. 8.1). Within this framework, the *K*-edge of the solid (indicated with energy $\Delta E_{\mathrm{K}}^{(\mathrm{solid})}$ in Fig. 8.1), the related transition in the free atom (indicated with energy $\Delta E_{\mathrm{K}}^{(\mathrm{free})}$ in Fig. 8.1), and the Fermi energy are connected via the relation

$$\Delta E_{\mathrm{K}}^{(\mathrm{free})} - \Delta E_{\mathrm{K}} = \Delta E_{\mathrm{K}}^{(\mathrm{solid})} - \varepsilon_{\mathrm{F}}^{(2)}, \tag{8.1}$$

where the upper index for the function ε_{F} indicates the number of electrons in the VB. At zero temperature, the *K*-edge of the cold solid is the energy to transfer an electron from the *K*-shell to the upper bound of the Fermi energy. A link from the cold solid to the atomic-solid-plasma model is accomplished if the energy ΔE_{K} is considered to be the ionization potential depression for the isolated atom or ion, i.e.,

$$\Delta E_{\mathrm{K}} \approx \Delta E_{\mathrm{IPD}}. \tag{8.2}$$

We therefore obtain for the ionization potential depression

$$\Delta E_{\mathrm{IDP}} = \Delta E_{\mathrm{K}}^{(\mathrm{free})}(2) - \Delta E_{\mathrm{K}}^{(\mathrm{solid})}(2) + \varepsilon_{\mathrm{F}}^{(2)}. \tag{8.3}$$

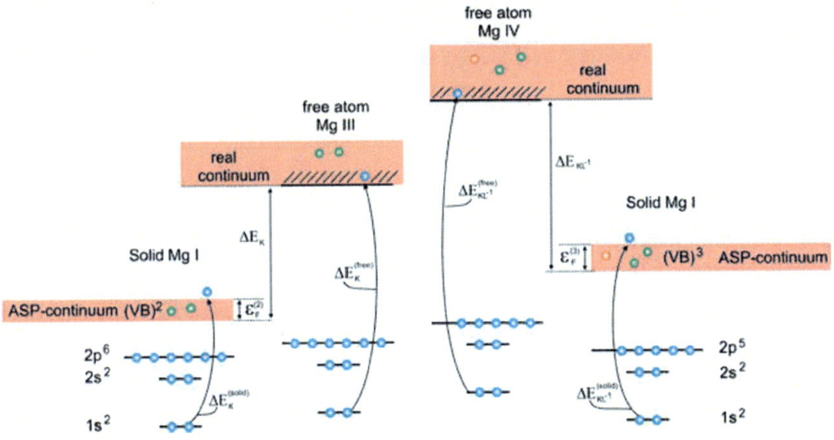

Fig. 8.1 Schematic of the atomic-solid-plasma (ASP) model visualized with magnesium and different ionization stages of the free atom. The electrons (green) in the VB form an effective continuum (ASP continuum), while the ionized free atoms form a usual continuum

Figure 8.1 (right part) presents the corresponding example for a single electron vacancy in the $2p$-shell.

Equation (8.3) can be rewritten for arbitrary core hole vacancies:

$$\Delta E_{\mathrm{IDP}}\left(2, C_{\mathrm{N}}^{(i)}\right) = \Delta E_{\mathrm{K}}^{(\mathrm{free})}\left(2, C_{\mathrm{N}}^{(i)}\right) - \Delta E_{\mathrm{K}}^{(\mathrm{solid})}\left(2, C_{\mathrm{N}}^{(i)}\right) + \varepsilon_{\mathrm{F}}^{(2+N)}, \quad (8.4)$$

where N is the number of core hole vacancies and $C_{\mathrm{N}}^{(i)}$ is the particular core hole vacancy configuration (i) for N-vacancies. In the ASP model, the Fermi energy is determined by the number of electrons of the valence band plus the number of vacancies (i.e., $2+N$ for the case of Fig. 8.1). We can generalize (8.4) to the case of a heated solid, remembering that for finite temperature, quantum states are available even below the Fermi edge ε_{F}:

$$\Delta E_{\mathrm{IDP}}(2, C_{\mathrm{N}}^{(i)}) = \Delta E_{\mathrm{K}}^{(\mathrm{free})}(2, C_{\mathrm{N}}^{(i)}) - \Delta E_{\mathrm{K}}^{(\mathrm{solid})}(2, C_{\mathrm{N}}^{(i)}) + p(T)\varepsilon_{\mathrm{F}}^{(2+N)}. \quad (8.5)$$

$p(T)$ is a function that accounts for finite temperatures in the VB. At zero temperature, no places are free below the Fermi energy due to the Pauli principle; therefore, the chemical potential $\mu(T) = \varepsilon_{\mathrm{F}}$ and $p(T) = 1$. For high temperatures (i.e., $k_{\mathrm{B}}T_{\mathrm{e}} \gg \varepsilon_{\mathrm{F}}$), an important number of places are offered just at the "bottom" of the VB and $p(T) = 0$. In this case, the chemical potential is close to $\mu(T) \approx -1.5 \cdot kT_{\mathrm{e}} \cdot \ln[(m_{\mathrm{e}}kT_{\mathrm{e}})/(2\pi\hbar^2 n_{\mathrm{e}}^{2/3})]$. These two limits correspond to the "cold solid" ASP-CS and "hot solid" ASP-HS, respectively. $p(T)$ accounts in an averaged manner for the probability of free places below the Fermi edge ε_{F} (Rosmej 2018; Gournay and Rosmej 2022):

$$p(T) \approx \frac{1}{\varepsilon_{\mathrm{F}}} \cdot \int_0^{\varepsilon_{\mathrm{F}}} \frac{d\varepsilon}{\exp[\beta \cdot (\varepsilon - \mu(T))] + 1} \approx 1 - \frac{1}{\left(\frac{100}{(kT_{\mathrm{e}}/\varepsilon_{\mathrm{F}})}\right)^{0.79} + 1}. \quad (8.6)$$

The last expression in (8.6) recovers the correct asymptotes for low and high temperatures and is accurate within 10% for intermediate values (Gournay and Rosmej 2022). The Fermi energies for any core hole configuration $C_{\mathrm{N}}^{(i)}$ depend only on the number of core hole vacancies but not on the details of the core hole configurations itself:

$$\varepsilon_{\mathrm{F}}^{(N)} = \frac{3^{2/3}\pi^{4/3}\hbar^2}{2m_{\mathrm{e}}} \cdot n_{\mathrm{e}}^{2/3}(N) = 3.6464 \times 10^{-15} \cdot n_{\mathrm{e}}^{2/3}(N)\left(\mathrm{cm}^{-3}\right)[\mathrm{eV}]. \quad (8.7)$$

The Fermi energies at zero temperature are well known from solid-state physics (Ashcroft and Mermin 1976; Hyperphysics 2021).

8.2 Approximate Solid-State Core Hole Configuration Energies

The calculation of the K-shell ionization energies $\Delta E_{\mathrm{K}}^{(\mathrm{solid})}$ of solid matter for different core hole vacancies is rather challenging: complex, very time-consuming, not very well established, and available experimental data are very rare. In order to obtain with reasonable effort the full set of requested data for the application of the ASP model [in particular, expressions similar to those of (8.5)], an empirical method is proposed that fits naturally to the model depicted in Fig. 8.1. It is based on the single atom multiconfiguration Hartree–Fock approach, where successive core hole vacancies are transformed into valence band bound electrons to the next higher allowed nl-quantum states. If the shell occupation is designated according to

$$(n_1 l_1)^{w_1}(n_2 l_2)^{w_2}\ldots(n_\mathrm{q} l_\mathrm{q})^{w_\mathrm{q}}, \tag{8.8a}$$

where w_i is the number of equivalent electrons in subshell i, the sum over all subshells provides

$$\sum_{i=1}^{q} w_\mathrm{i} = Z_\mathrm{n}. \tag{8.8b}$$

The hollow ion configurations that include a number of electrons equal to the nuclear charge (8.8a) are then numerically solved including the exact exchange terms $A_\mathrm{i}(r)$ and $B_\mathrm{ij}(r)$:

$$\left[-\frac{\mathrm{d}^2}{\mathrm{d}r^2} + \frac{l_\mathrm{i}(l_\mathrm{i}+1)}{r^2} - \frac{2Z_\mathrm{n}}{r} + \sum_{j=1}^{q}\left(w_\mathrm{j} - \delta_\mathrm{ij}\right)\cdot\int_0^\infty \frac{2}{r_>}P_\mathrm{j}^2(r')\mathrm{d}r' - (w_\mathrm{i}-1)A_\mathrm{i}(r)\right]P_\mathrm{i}(r)$$
$$= \varepsilon_\mathrm{ii}P_\mathrm{i}(r) + \sum_{j=1\, j\neq i}^{q} w_\mathrm{j}\left[\delta_{l_\mathrm{i}l_\mathrm{j}}\varepsilon_\mathrm{ij} + B_\mathrm{ij}(r)\right]P_\mathrm{j}(r), \tag{8.9}$$

$$A_\mathrm{i}(r) = \frac{2l_\mathrm{i}+1}{4l_\mathrm{i}+1}\sum_{k>0}\begin{pmatrix} l_\mathrm{i} & k & l_\mathrm{j} \\ 0 & 0 & 0 \end{pmatrix}^2 \int_0^\infty \frac{2r_<^k}{r_>^k}P_\mathrm{i}^2(r')\mathrm{d}r', \tag{8.10}$$

$$B_\mathrm{ij}(r) = \frac{1}{2}\sum_{k}\begin{pmatrix} l_\mathrm{i} & k & l_\mathrm{j} \\ 0 & 0 & 0 \end{pmatrix}^2 \int_0^\infty \frac{2r_<^k}{r_>^k}P_\mathrm{j}(r')P_\mathrm{i}(r')\mathrm{d}r'. \tag{8.11}$$

$r_<$ and $r_>$ are the lesser and greater of r and r', respectively, $P_\mathrm{i}(r)$ are the radial functions for each subshell and ε_ii and ε_ij are the energies of equivalent and

Table 8.1 *K*-edge energies [in eV] as calculated with the AS-HF and AS-HF-AC method and comparison with reference data from (Desclattes et al. 2003; Thomson et al. 2009; NIST 2021)

Element	AS-HF	AS-HF-AC	Reference
Ne	865		867
Na	1073		1072
Mg	1306		1303
Al	1562		1560
Si	1841		1839
P	2144		2146
S	2470		2471
Cl	2819		2820
Ar	3208		3206
Ti	4968	4965	4965
Fe	7129	7109	7111

non-equivalent electrons. In addition, relativistic corrections to the radial wave functions are included in the Pauli approximation.

The *K*-edge is then naturally defined as the difference of the multielectron averaged energies between the atom in the ground state and the hollow atom configurations (this respects also the Pauli principle). Let us consider a few relevant examples to estimate the precision of this empirical "Atomic-Solid Hartree–Fock" method (shortly designated as AS-HF) with the help of reference data (Desclattes et al. 2003; Thomson et al. 2009; NIST 2021), Table 8.1. One observes a quite good match with precisions of the order of some eV.

The precision might be improved employing the Hartree–Fock wave functions from the numerical solution of the system of integro-differential equations (8.9)–(8.11) and performing a subsequent detailed angular coupling calculation, shortly designated as AS-HF-AC (details are described elsewhere (Cowan 1981). Table 8.1 shows examples of Ti and Fe in a simple 4-configuration approach (e.g., for Ti, $x = 2$ and Fe, $x = 6$ taking into account the configurations $1s^2 2s^2 2p^6 3s^2 3p^6 3d^x 4s^2$, $1s^2 2s^2 2p^6 3s^2 3p^6 3d^x 4s^1 4p^1$, $1s^1 2s^2 2p^6 3s^2 3p^6 3d^{x+1} 4s^2$, $1s^1 2s^2 2p^6 3s^2 3p^6 3d^x 4s^2 4p^1$). Although the more complex AS-HF-AC calculations provide better agreement with the reference data (and could be further improved involving more sophisticated configuration interactions and intermediate couplings), it should be remembered the present context: It concerns ionization potential depression energies of up to several 100 eV and experimental error bars of about ±10 eV. Moreover, different ionization potential depression theories differ much more than the experimental error bars and show parameter dependences that are even not qualitatively reproduced. Therefore, the precision of the AS-HF method seems to be quite appropriate for the present purposes.

Table 8.2 presents the *K*-edge energies of aluminum for different *L*- and *K*-shell vacancies, assuming $p(T) = 0$ (note that under typical experimental situation of XFEL irradiation of solids, the sample is heated up to temperatures of several 100 eV). As can be seen from the comparison with currently available data

Table 8.2 *K*-shell ionization energies in solid aluminum in dependence of different core hole vacancies

Z_n–N_{K+L}	Core configuration	Experiment Ciricosta et al. (2016)	Present AS-HF No shift	Vinko et al. (2014), Ciricosta et al. (2016) Calibration shifts +35…40 eV	Son et al. (2014) Calibration shift +21,5 eV	Desclattes et al. (2003), Thomson et al. (2009), NIST (2021)
3	$1s^2\ 2s^2\ 2p^6$	1556 ± 10	1562	1556	1552	1559,53
4	$1s^2\ 2s^2\ 2p^5$	1582 ± 10	1588	1583	1578	–
5	$1s^2\ 2s^2\ 2p^4$	1607 ± 10	1616	1612	1612	–
6	$1s^2\ 2s^2\ 2p^3$	1644 ± 10	1644	1642	1652	–
7	$1s^2\ 2s^2\ 2p^2$	1673 ± 10	1675	1673	1684	–
8	$1s^2\ 2s^2\ 2p^1$	1711 ± 10	1719	1706	–	–
9	$1s^2\ 2s^2$	1750 ± 10	1755	1738*	–	–
10	$1s^2\ 2s^1$	–	1781	–	–	
11	$1s^2$	–	1808	–	–	
12	$1s^1$	–	1980	–	–	

The present AS-HF model includes the exact exchange term for all electrons in the Hamiltonian and does not include any free parameters neither a calibration shift. On the other hand, the methods employed in (Vinko et al. 2014; Son et al. 2014; Ciricosta et al. 2016) employ potential approximations and requested a calibration shift of 35–40 eV (Vinko et al. 2014; Ciricosta et al. 2016) and 21.5 eV (Son et al. 2014) to match the measured standard *K*-edge in cold solid aluminum. Note that, e.g., for aluminum, the core configuration of a cold solid is $1s^2 2s^2 2p^6$ and the corresponding number Z_n–N_{K+L} = 13–(2 + 8) = 3

(Ciricosta et al. 2016), excellent agreement over the whole range of multiple core hole vacancies is obtained. Concerning the calculations in the framework of the density functional theory (DFT) as presented in (Vinko et al. 2014), it should be mentioned that the use of pseudo-potentials and approximations for the exchange potential of the few-electron system with core hole states are challenging (Engel and Vosko 1993; Iglesias 2011; Karasiev and Hu 2021). In addition, the final results (Vinko et al. 2014; Ciricosta et al. 2016) have artificially been shifted by $+35-40\,\text{eV}$ (Vinko et al. 2014; Ciricosta et al. 2016) in order to match the experimental data for IPD. For the $1s^2 2s^2$-configuration (indicated with * in Table 8.2), the much larger discrepancies despite of calibration shifts might indicate difficulties in the DFT method (Ciricosta et al. 2012, 2016; Vinko et al. 2014) to correctly describe the subshell structure where the Pauli principle is important.

Therefore, the present ASP-HF method provides results that for the given context are not better or worse than any other calculations currently available. Moreover, the present AS-HF itself does not contain any pseudo- and exchange potential approximations: The AS-HF includes the exact exchange terms, and all wave functions are subject to a self-consistent field iteration procedure that includes configuration interaction.

8.3 Ionization Potential Depression Formulas

The K-shell ionization energies of the various free ion cores $\Delta E_{\mathrm{K}}^{(\mathrm{free})}(2, C_{\mathrm{N}}^{(i)})$ are calculated as described above; however, core hole vacancies are not transferred to the VB band but correspond to an ionized atom (see Fig. 8.1 for qualitative illustration).

Figure 8.2 shows the ionization potential depression energies for Mg in dependence of various core hole vacancies according to (8.5) for $p(T) = 1$ (corresponding to a cold solid, indicated with blue symbols, designated as "ASP-CS") and for $p(T) = 0$ (corresponding to a hot solid, indicated with red symbols, designated as "ASP-HS"). The ASP-HS calculations are in very good agreement with the data (black squares in Fig. 8.2) of (Ciricosta et al. 2016). We note that the ASP model provides likewise good agreement with the data for double core hole states (e.g., configurations of the type $K^0L^XM^Y$, i.e., configurations of hollow crystals) (Gournay and Rosmej 2022).

Also presented in Fig. 8.2 are the widely applied model from Stewart and Pyatt (More 1981; Stewart and Pyatt 1996) (green symbols designated with "SP") and the Ecker–Kröll model (Ecker and Kröll 1963) (orange symbols designated with "EK") for an electron temperature of $kT_e = 100$ eV and free electron density of $n_e(N = 3) = 8.61 \times 10^{22}$ cm^{-3} (electron density of the Fermi energy for the "usual" solid). For the Stewart–Pyatt model, assuming $n_e = n_i \cdot \langle Z \rangle$ (n_i is the ion density and $\langle Z \rangle$ is the average ion charge), the ionization potential depression can be estimated from:

$$\Delta E_{\mathrm{IPD}}^{(\mathrm{SP})} \approx 2.16 \times 10^{-7} \cdot \frac{Z}{R_i} \cdot \left[\left(1 + \frac{\lambda_D^3}{R_i^3}\right)^{2/3} - \left(\frac{\lambda_D}{R_i}\right)^2 \right] \quad [\mathrm{eV}], \tag{8.12}$$

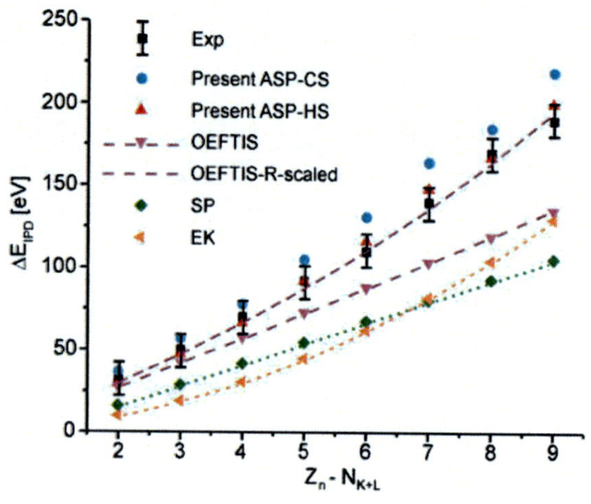

Fig. 8.2 Comparison of different models of ionization potential depression with the present ASP and OEFTIS models. Experimental data are for Mg in dependence of successive K- and L-shell vacancies in near solid density matter. Z_n is the nuclear charge, and N_{K+L} is the number of K- and L-shell electron vacancies (core hole vacancies)

$$\lambda_{\mathrm{D}} \approx 7.43 \times 10^{2} \cdot \sqrt{\frac{kT_{\mathrm{e}}(\mathrm{eV})}{n_{\mathrm{e}}(\mathrm{cm}^{-3}) \cdot (1 + \langle Z \rangle)}}\,[\mathrm{cm}], \tag{8.13}$$

$$R_{\mathrm{i}} \approx 0.620 \cdot \left(\frac{\langle Z \rangle}{n_{\mathrm{e}}(\mathrm{cm}^{-3})} \right)^{1/3} [\mathrm{cm}]. \tag{8.14}$$

For the Ecker–Kröll model, the ionization potential depression formulas can be roughly summarized as follows:

$$\Delta E_{\mathrm{IPD}}^{(\mathrm{EK})} \approx 1.44 \times 10^{-7}\,\mathrm{eV} \cdot (1 + \langle Z \rangle) \cdot \begin{cases} \dfrac{1}{\tilde{\lambda}_{\mathrm{D}}} & \text{if } n_{\mathrm{i}} \cdot (1 + \langle Z \rangle) \leq n_{\mathrm{C}} \\ \dfrac{C}{\tilde{R}_{\mathrm{i}}} & \text{if } n_{\mathrm{i}} \cdot (1 + \langle Z \rangle) \geq n_{\mathrm{C}} \end{cases}, \tag{8.15}$$

$$\tilde{\lambda}_{\mathrm{D}} \approx 7.43 \times 10^{2} \cdot \sqrt{\frac{kT_{\mathrm{e}}(\mathrm{eV})}{\langle Z \rangle \cdot n_{\mathrm{i}}(\mathrm{cm}^{-3}) \cdot (1 + \langle Z \rangle)}}\,[\mathrm{cm}], \tag{8.16}$$

$$\tilde{R}_{\mathrm{i}} \approx 0.620 \cdot \left(\frac{1}{\langle Z \rangle \cdot n_{\mathrm{i}}\,(\mathrm{cm}^{-3})} \right)^{1/3} [\mathrm{cm}], \tag{8.17}$$

$$C = \frac{\left[\tilde{R}_{\mathrm{i}}\right]_{n_{\mathrm{C}} = n_{\mathrm{i}} \cdot (1 + \langle Z \rangle)}}{\left[\tilde{\lambda}_{\mathrm{D}}\right]_{n_{\mathrm{C}} = n_{\mathrm{i}} \cdot (1 + \langle Z \rangle)}}, \tag{8.18}$$

$$n_{\mathrm{C}} = 8.00 \times 10^{19} \cdot \sqrt{\frac{kT_{\mathrm{e}}}{Z_{\max}^{2}}}\,[\mathrm{cm}^{-3}]. \tag{8.19}$$

For example, for aluminum at $kT_{\mathrm{e}} = 58$ eV, with $Z = 6$, $\langle Z \rangle = 6$, $Z_{\max} = 13$, $n_{\mathrm{e}} = n_{\mathrm{i}} \cdot \langle Z \rangle = 3.6 \times 10^{23}\,\mathrm{cm}^{-3}$, we obtain $n_{\mathrm{C}} = 4.69 \times 10^{19}\mathrm{cm}^{-3} \leq n_{\mathrm{i}} \cdot (1 + \langle Z \rangle)$, $\tilde{R}_{\mathrm{i}}\big|_{n_{\mathrm{i}} = n_{\mathrm{C}}} = 1.72 \times 10^{-7}\,\mathrm{cm}$, $\tilde{\lambda}_{\mathrm{D}}\big|_{n_{\mathrm{e}} = \langle Z \rangle \cdot n_{\mathrm{C}}} = 3.37 \times 10^{-7}\,\mathrm{cm}$, $C = 0.51$ and finally $\Delta E_{\mathrm{IPD}}^{(\mathrm{EK})} = 62$ eV and $\Delta E_{\mathrm{IPD}}^{(\mathrm{SP})} = 78$ eV.

It can clearly be seen from Fig. 8.2 that the SP and EK models not only fail to describe the absolute ionization depression energies but fail also to describe the Z-dependence of the data (black symbols designated as "Exp" in Fig. 8.2).

It should be noted that although Fig. 8.2 employs only particular data for Mg, it demonstrates the important failure in Z-dependence that requests particular attention. Also shown in Fig. 8.2 is the optical electron finite temperature ion sphere (OEFTIS) model that will be described in more detail below.

8.4 Optical Electron Finite Temperature Ion Sphere Model OEFTIS

8.4.1 Plasma Polarization Shift and Level Disappearance in Dense Hot Plasmas

The optical electron finite temperature ion sphere (OEFTIS) model (Rosmej 2018) is based on an analytical potential description of the arbitrary perturbation potential method APPM (Rosmej et al. 2011). In OEFTIS, an effective potential is calculated from an effective charge Z_{eff} seen by the optical electron. The ionization potential depression can then be cast into an entirely analytical description that only needs energies and quantum numbers of the free atom/ion configuration. In the second order approximation in terms of matrix $\langle i|r^{\beta}|j\rangle$ elements of integer power β (that have analytical solutions in terms of principal and orbital quantum number) we have:

$$\Delta E_{\mathrm{IPD}} = 2Ry \cdot \frac{Z}{R_0} \cdot \frac{F(\langle r\rangle, \langle r^2\rangle)}{F\left(R_0, R_0^2\right)}, \tag{8.20}$$

$$F(x,y) \approx \frac{R_0^2}{2} - \frac{y}{6} + \frac{4}{3\sqrt{\pi}} \cdot R_0^{3/2} \cdot \sqrt{\frac{2Ry \cdot Z}{k_B T_e}} - \frac{8}{15\sqrt{\pi}} \cdot \left(\frac{4}{\sqrt{\pi}} \cdot x + \frac{1}{10} \cdot y\right) \cdot \sqrt{\frac{2Ry \cdot Z}{k_B T_e}}, \tag{8.21}$$

$$\langle r\rangle = \frac{1}{2Z_{\mathrm{eff}}}\left[3n^2 - l(l+1)\right], \tag{8.22}$$

$$\langle r^2\rangle = \frac{n^2}{2Z_{\mathrm{eff}}^2}\left[5n^2 + 1 - 3l(l+1)\right], \tag{8.23}$$

$$Z_{\mathrm{eff}} = n\sqrt{|E_0|/Ry}, \tag{8.24}$$

$$R_0 \approx 1.1723 \times 10^8 \cdot (Z/n_i\langle Z\rangle)^{1/3}. \tag{8.25}$$

$Ry = 13.606$ eV, ΔE_{IPD} is the ionization potential depression energy in [eV], E_0 is the ionization potential ([eV]) of the optical electron with principal/orbital quantum numbers nl that move in an effective Coulomb potential with charge Z_{eff} in the free atom/ion picture, z is the spectroscopic symbol i.e., $z = Z + 1$, where Z is the ion charge, R_0 is the ion sphere radius in atomic units (vanishing wave function at $r = R_0$ has been assumed) and $\langle Z\rangle$ is the average charge of the plasma, $k_B T_e$ is the electron temperature [eV]. It should be noted, that the 2nd-order potential function $F(x,y)$ of (8.21) should only be considered as a convenient fit function for particular applications (estimates for ionization potential depression and line shifts in intermediate density and temperature range, analysis of scaling relations) rather than an overall consistent approximation of the finite temperature ion sphere theory.

In this context, the critics of (Iglesias 2019) concerning the limited precision of the second order expression (e.g. the neutrality condition) represent a rather artificial discussion in particular as the necessary matrix elements $\langle i|r^{\beta}|j\rangle$ of non-integer power β of the APPM (Rosmej et al. 2011) (that matches well the neutrality condition and high precision line shift data) can be represented by the well-known Γ-function, see Annex A.4. Moreover, improved precision of OEFTIS has been demonstrated with a 4th-order approximation (of integer power β) (Li and Rosmej 2020) that matches very well high precision X-ray spectroscopic line shift data and the exact Γ-function representation and has in addition the advantage of an explicit analytical description in terms of principal and orbital quantum numbers (see Annex A.4 for further details).

In general, it is somewhat difficult to determine the range of validity for the various models of the rather complex IPD phenomenon but some insight can roughly be estimated from the validity of the first order perturbation theory: $k_{\mathrm{B}}T_{\mathrm{e}} > 3 \cdot \varepsilon_{\mathrm{F}}, n_{\mathrm{e}} \ll 2 \times 10^{24} \cdot Z^3 \cdot \langle Z \rangle / n^6\,[\mathrm{cm}^{-3}]$.

As the plasma polarization shift (Griem 1997; Lisitsa 1994; Nguyen et al. 1986; Massacrier and Dubeau 1990) of each state is nl-quantum number-dependent, (8.20)–(8.25) can also readily be employed to calculate line shifts. This provides critical tests (Griem 1997; Salzman et al. 1998; Nguyen et al. 1986; Massacrier and Dubeau 1990; Renner et al. 1997b, 1998; Saeman et al. 1999; Rosmej et al. 2000; Dervieux et al. 2015; Hansen et al. 2017). Table 8.3 compares measured line shifts of H-like and He-like aluminum lines with the present 2nd-order OEFTIS [columns (a)–(d)]. As can be seen, quite good agreement is obtained over different ranges of quantum numbers, temperatures, and densities. For the He-α transition, the agreement is not very good. However, the extraction of the He-α shift is rather complex because contributions of high-order satellites need to be extracted with a complex procedure (Renner et al. 2001). The remaining discrepancies might be ascribed to the radiation transport effects in differentially moving plasmas (see Chap. 1) that lead to significant He-α line asymmetries (and therefore to the line center shifts). Finally we note, that the 2nd-order approximation leads in general to too large values while the 4th-order provides almost exact results (for details see Annex A.4).

As concerns the measurements of the line disappearance due to ionization potential depression (Hoarty et al. 2013), the OEFTIS provides equally good agreement with the data. For case (e) and (f), the ionization potential depressions for He-α and Ly-α are much lower than the ionization potential. Both lines should therefore be clearly observed in good agreement with the data. For He-β and Ly-β and case (e), the ionization potential depressions are larger than the ionization potential: Lines should therefore not be observable. This is in good agreement with the experimental data. Case (f) is a very challenging one, because He-β is just marginally observed. This is likewise in good agreement with OEFTIS: The IPD values are close to the ionization potential for the $1s3l$-levels.

Table 8.3 Line shift measurements (first values, in eV) and OEFTIS calculations (second values, in eV) of H-like and He-like aluminum lines in dense laser-produced plasmas, Ly series (a) of Renner et al. at for $n_e = 1.5 \times 10^{22}$ cm^{-3} and $k_BT_e = 1400$ eV (Renner et al. 1997b, 1998), Ly-α (b) of Saemann et al. at $n_e = 5 \times 10^{23}$ cm^{-3} and $k_BT_e = 280$ eV (Saeman et al. 1999), He-β (b) of Saemann et al. at $n_e = 8 \times 10^{23}$ cm^{-3} and $k_B\mathrm{T}_e = 220$ eV (Saeman et al. 1999), He-β (c) of Dervieux et al. at $n_e = 5 \times 10^{23}$ cm^{-3} and $k_B\mathrm{T}_e = 560$ eV (Dervieux et al. 2015), He-α (d) of Stillman et al. at $n_e = 2.2 \times 10^{23}$ cm^{-3} and $k_B\mathrm{T}_e = 330$ eV (Stillman et al. 2017)

	(a) Renner et al. (1997b, 1998)	(b) Saeman et al. (1999)	(c) Dervieux et al. (2015)	(d) Stillman et al. (2017)	(e) Hoarty et al. (2013)	(f) Hoarty et al. (2013)
He-α	2.2 ± 0.6/ 1		–	2.4 ± 0.3/ 1.9 ± 0.7[a]	487.7/ 283–309[a] observed	487.7/ 205–226[a] observed
He-β	–	20 ± 8/ 25	6 ± 2/ 6.8	–	217.3/ 243–259[a] not observed	217.3/ 196–204[a] marginally observed
Ly-α	–	3.7 ± 0.7/ 4.1	–	–	575.5/ 310–339[a] observed	574.5/ 223–247[a] observed
Ly-β	0.2 ± 0.1/ 0.30	–	–	–	255.8/ 271–290[a] not observed	255.8/ 207–225[a] observed
Ly-γ	0.5 ± 0.2/ 0.76	–	–	–		
Ly-δ	1.1 ± 0.2/ 1.1	–	–	–		
Ly-ε	2.0 ± 0.5/ 2.4	–	–	–		

Also given the data of Ly-α, Ly-β, He-α, and He-β disappearance/existence of Hoarty et al. (Hoarty et al. 2013), (e) $n_e = 2.0$–2.5×10^{24} cm^{-3}, $k_BT_e = 700$ eV, (f) $n_e = 8.6$–11×10^{23} cm^{-3}, $k_BT_e = 700$ eV: First values are ionization potentials [eV], second ones ionization potential depression from OEFTIS [eV], and third entry indicates whether the line is observed or not
[a]The error bars for the OEFTIS calculations correspond to the density uncertainties given in (Hoarty et al. 2013, Stillman et al. 2017)

The study of X-ray line disappearance for dense plasma physics investigations is most conveniently demonstrated via a comparison of spectra from laser-produced plasmas originating from high contrast and from prepulse. Figure 8.3 demonstrates the potential of high-contrast high-intensity lasers to contribute to the challenging field of dense plasma atomic physics. A Nd-Glas laser with pulse duration of 400 fs, 1 J energy, wavelength 0.53 nm, focal spot diameter 10 μm, and intensity of 5×10^{18} W/cm^3 was brought to interaction with a solid magnesium target at normal incidence for different irradiation conditions: (1) high contrast 1:10^{10} (solid blue curve), (2) a prepulse with energy of 0.03 J separated from the main pulse by 150 ps (solid black curve) (Rosmej et al. 2000). Figure 8.3 shows the spectral window containing the He-like resonance lines $\mathrm{He}_\beta = 1s3p\ ^1P_1 – 1s^2\ ^1S_0$,

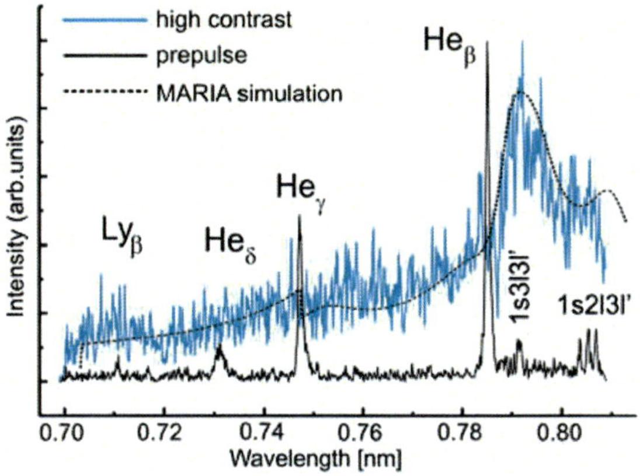

Fig. 8.3 *K*-shell X-ray emission of He$_\beta$ until Ly$_\beta$ of magnesium for different irradiation conditions: extremely high contrast 10^{10}–10^{11} and prepulse. The MARIA simulations indicate near solid density $n_e = 3 \times 10^{23}$ cm^{-3}, $kT_e = 0.2$ keV and expansion velocities of the order of $V = 3 \times 10^7$ cm/sec. For high contrast, the plasma density is near solid density resulting in a disappearance of the He$_\gamma$ and He$_\delta$ lines whereas these lines are clearly visible in the low-density spectra obtained with prepulse

He$_\gamma$ = $1s4p\ ^1P_1$–$1s^2\ ^1S_0$, He$_\delta$ = $1s5p\ ^1P_1$–$1s^2\ ^1S_0$, and the H-like Ly$_\beta$ = $3p\ ^2P_{1/2,\,3/2}$–$1s\ ^2S_{1/2}$ lines, and the Li-like dielectronic satellites $1s3l3l'$ and $1s2l3l'$. For the case of prepulse, resonance line emission is dominating, the He-like Rydberg series is well developed, and the dielectronic satellite intensities are relatively weak. With high contrast, however, the situation is dramatically different. A new type of spectra develops: Resonance lines seem to disappear, and the dielectronic satellites become the most pronounced emission features in the spectrum. This phenomenon is known as dielectronic satellite accumulation DSA and has been described in Sect. 1.4.3.

Also shown in Fig. 8.3 the simulations carried out with the MARIA code (Rosmej 1997, 2001, 2006) indicating near solid density plasmas: electron density $n_e \approx 3 \times 10^{23}$cm^{-3}, electron temperature $kT_e \approx 0.2$ keV, effective plasma source size of $L_{eff} = 3$ μm and expansion velocities of $V \approx 3 \times 10^7$ cm/s (see also discussion in Chap. 1 concerning Fig. 1.5). The high-density results in a high ion–ion coupling parameter ($\Gamma \approx 3$) and manifests itself in a strongly developed DSA (clearly seen for the He$_\beta$-emission structure), line broadening, line shift, and line disappearance (as can be clearly seen for the He$_\gamma$ and He$_\delta$ lines). Also seen in the simulations is the continuum edge shift of He-like ions near 0.75 nm along with the line disappearance of He$_\delta$ and even He$_\gamma$. These kinds of measurements provide critical data for theoretical modeling of the ionization potential depression and line shifts, but high precision experiments are very challenging (Renner and Rosmej 2019).

8.4.2 Scaled Ion Sphere Radii and Lattice Structure

Despite the quite good agreement of OEFTIS for line disappearances and even for line shift analysis (Table 8.3), the OEFTIS model is not really in good agreement with the data presented in Fig. 8.2 (purple triangles, $Z_{core} = Z_n - N_{K+L} = 2\text{–}9$, $\langle Z \rangle = Z_{core}$, $kT_e = 100$ eV, $n_i = 4.31 \times 10^{22}$ cm^{-3}). But OEFTIS provides a significant better agreement with the data compared to the SP and EK models. Empirically, it is observed that a scaled ion sphere radius (Z_n is the nuclear charge and N_{VB} is the number of electrons in the VB, e.g., for Mg $Z_n = 12$, $N_{VB} = 2$ and $Z_{core} = 2\ldots10$)

$$R_{scaled} = R_0 \cdot f(Z_{core}, Z_n, N_{VB}), \tag{8.26}$$

where the function $f(Z_{core}, Z_n, N_{VB})$ is approximated with a simple linear dependence in Z_{core} according to

$$f(Z_{core}, Z_n, N_{VB}) \approx 0.95 \cdot \{1 - 0.35 \cdot (Z_{core} - N_{VB})/Z_n\} \tag{8.27}$$

provides a good agreement even with these data (purple dashed curve indicated as OEFTIS-R-scaled in Fig. 8.2). Likewise in a recent different laser compression experiment, it was empirically found that a scaled down ion sphere radius describes better the observed line shift of Helium-α (Stillman et al. 2017). However, in dense laser-produced plasmas, higher-order dielectronic satellites of the type $1s2lnl'\text{–}1s^2nl'$ are overlapping with the He-α resonance line that makes analysis challenging. Moreover, as has been discovered recently (Rosmej 2012), these satellite transitions are subject to important quantum interference effects that lead to a narrowing of the satellite group emission (see also Sect. 1.5.3). Densities might therefore be underestimated (Rosmej 2012).

Finally, we mention recent scaling investigations of the plasma screening potential in the framework of the self-consistent ion sphere model. Based on fully quantum mechanical self-consistent finite temperature ion sphere calculations, an analytical potential approximation has been developed (Li et al. 2019):

$$\phi_p(r) = \frac{Z_n \cdot e}{r} - \frac{N_f \cdot e}{R} \cdot \left\{1 + \frac{1}{x-1} - \frac{1}{x-1} \cdot \left(\frac{r}{R}\right)^{x-1}\right\}, \quad r < R, \tag{8.28}$$

$$\phi_p(r) = \frac{(Z_n - N_f) \cdot e}{r}, \quad r \geq R, \tag{8.29}$$

$$x = 3 - b \cdot \frac{e}{\pi} \cdot \sqrt{\frac{N_f}{R \cdot kT_e}}, \tag{8.30}$$

$$N_f = Z_n - N_b, \tag{8.31}$$

$$\frac{4}{3}\pi R^3 \cdot n_e = N_f. \tag{8.32}$$

N_f is the free electron number inside the ion sphere with radius R, n_e is the electron density, and Z_n is the nuclear charge of the ion (for detailed application of (8.28–8.32) in convenient units see Annex A.4). Note that the second term in (8.28) represents the potential of the free electrons inside the ion sphere and that (8.28), (8.29) have a continuous derivative. Comparison with numerical simulations has shown (Li et al. 2019) that excellent agreement with the self-consistent ion sphere calculations is achieved setting $b = 2$. The great advantage of the b-potential is that it substitutes the complex self-consistent ion sphere calculations via a perturbation potential in the Hamiltonian while maintaining spectroscopic precision (e.g. to describe high-precision X-ray line shift measurements) (Li and Rosmej 2020). Finite temperature atomic structure calculations are therefore greatly simplified and the Li-Rosmej b-potential has therefore attracted growing interest in dense plasma atomic physics (Chen et al. 2021; Karasiev and Hu 2021; Ma et al. 2020; Shing et al. 2020).

For infinite temperature, $x = 3$ and the analytical potential turns into the well-known uniform electron gas model (UEGM) as it should be:

$$\phi_p^{(UEGM)}(r) = \frac{Z_n \cdot e}{r} - \frac{(Z_n - N_b) \cdot e}{2 \cdot R} \cdot \left\{3 - \left(\frac{r}{R}\right)^2\right\}, \quad r < R. \tag{8.33}$$

The analytical potential can easily be applied to estimate the shift ΔE_K of the K-edge via the $1s$-electron energy shift ΔE_{1s} setting $r = 0$:

$$\Delta E_K \approx \Delta E_{1s} \approx \frac{N_f \cdot e^2}{R} \cdot \left\{1 + \frac{1}{x-1}\right\}. \tag{8.34}$$

It has been empirically found (Li et al. 2019) that calculations using the analytical b-potential setting approximately $b \approx 4$ are in better agreement with some recent data obtained from experiments where XFEL radiation interacts with solids (Ciricosta et al. 2012). In addition, it has been demonstrated (Li and Rosmej 2020) that the unexplained X-ray line shift data of He_α (Stillman et al. 2017) are in excellent agreement employing the analytical b-potential with $b = 4$. These findings suggest that the b4-potential has the capacity to imitate lattice effects in ionization potential depression and line shifts in near solid density heated matter (note, that a higher b-value makes the free electron potential more smooth inside the ion sphere). We note, that also effective analytical formulas can be developed for arbitrary b-parameters, see Annex A.4. To what extend the b-potential method could imitate lattice effects for $b > 2$ at temperatures much larger than the Fermi temperature remains an open question and is subject to active research. The same holds true for scaled ion sphere radii (correspondingly, (8.26–8.34) are only given to guide an eventual future discussion). In this context we note, that for near solid density plasmas with temperatures near the Fermi temperature, the Atomic-Solid-Plasma Model ASP (as discussed above) has provided excellent agreement with the data.

8.5 Strongly Compressed Matter and Fermi Surface-Rising

Recent quantum molecular dynamics (QMD) simulations (Hu 2017) of much above solid density compressed carbon predicted unusually increased *K*-edge energies, while existing models failed to describe the QMD predictions. Let us consider this situation in the framework of the present ASP model that is readily extended to compressed solids, i.e., instead of (8.3), we have

$$\Delta E_{\mathrm{K}}^{(\mathrm{compressed})} = \Delta E_{\mathrm{K}}^{(\mathrm{solid})} + \varepsilon_{\mathrm{F}}^{(\mathrm{compressed})} - \Delta E_{\mathrm{IDP}}. \tag{8.35}$$

The ionization potential depression is now an input to (8.35) rather than a result like in (8.4), (8.5). The "low-density" limit of (8.35) corresponds to the "usual" solid, i.e., $\Delta E_{\mathrm{K}}^{(\mathrm{compressed})} = \Delta E_{\mathrm{K}}^{(\mathrm{solid})}$. In this case, the corresponding ionization potential depression should correspond to the Fermi energy of the "usual" solid, i.e., $\varepsilon_{\mathrm{F}}^{(\mathrm{solid})} = \Delta E_{\mathrm{IDP}}$. Within the framework of ASP, we approximate the ionization potential depression values for carbon from ΔE_{K} (calculated with the ASP-HF method as described above) assuming that the $2p^2$-electrons form a VB with a corresponding Fermi energy of $\varepsilon_{\mathrm{F}}^{(\mathrm{solid})} = 13.56\,\mathrm{eV}$ at $\rho^{(\mathrm{solid})} = 2.26\mathrm{g/cm}^3$. The ΔE_{K}-values are extended to the low-density case (solid) with a simple linear expression

$$\Delta E_{\mathrm{IDP}} \approx \varepsilon_{\mathrm{F}}^{(\mathrm{solid})} \cdot \frac{\varepsilon_{\mathrm{F}}^{(\mathrm{solid})}}{\varepsilon_{\mathrm{F}}^{(\mathrm{compressed})}} + \Delta E_{\mathrm{K}} \cdot \left(1 - \frac{\varepsilon_{\mathrm{F}}^{(\mathrm{solid})}}{\varepsilon_{\mathrm{F}}^{(\mathrm{compressed})}}\right). \tag{8.36}$$

Equation (8.36) employs the solid density state energies ΔE_{K}, solid density Fermi energies $\varepsilon_{\mathrm{F}}^{(\mathrm{solid})}$, and Fermi energies $\varepsilon_{\mathrm{F}}^{(\mathrm{compressed})}$ of compressed matter. According to its construction with Fermi energies, (8.35), (8.36) are applicable only for temperatures not much larger than the Fermi energy itself.

If we estimate, e.g., the ionization potential depression for the critical He-β line of the Hoarty experiment (Hoarty et al. 2013) (see Table 8.3 case (f) and corresponding discussion) with the help of (8.5) for $p(T) = 0$, we find $\Delta E_{\mathrm{IPD}} \approx 204$ eV. As Table 8.3 shows, this value is not out of order. More detailed discussions concerning the disappearance/appearance of lines for different densities as observed in Hoarty et al. (2013) would be, however, inappropriate as the ASP model is not valid in the temperature range of this experiment (Hoarty et al. 2013) because $kT_{\mathrm{e}} \gg \varepsilon_{\mathrm{F}}$ (rather than $kT_{\mathrm{e}} < \varepsilon_{\mathrm{F}}$ for the ASP model).

Figure 8.4 compares the K-edge energies from the ASP model of (8.35), (8.36) with the recent QMD simulations of (Hu 2017) for a temperature of 1.35 eV (i.e., $kT_e \ll \varepsilon_F$). It is observed that the ASP model correctly predicts decreasing values of K-edge energies up to about $\rho \approx 10$ g/cm^3, while for even higher densities much increased K-edge energies are obtained. According to (8.35), (8.36), the physical reason for the much increased K-edge energies is the energy term $\varepsilon_F^{(\text{compressed})}$ that effectively rises the Fermi surface (like depicted in Fig. 8.1) for the core hole ionization. The decrease of the K-edge for intermediate values is due to the potential depression of the optical electron that is still more effective than the rise of the Fermi surface energy. This is in excellent agreement with the findings from the recent QMD simulations (Hu 2017), while currently employed models fail to describe these features. Also shown in Fig. 8.4 the SP model in its original version (blue curve) and a modified one that includes the Fermi energy (green curve SP-F).

For very high densities, the ASP model predicts somewhat higher values compared to the QMD simulations but is still in agreement with QMD within 20% up to 100 times compressed solid. This somewhat larger deviation for the highest compression values is not very surprising: In Fig. 8.4, we use for all densities the energies ΔE_K from AS-HF that are free atom/ion energy values. To obtain more advanced values would probably need modifications in (8.9) (e.g., a free electron term in the Wigner–Seitz cell) which is not an evident task and outside the frame of the present studies. We note that the QMD simulations have artificially been shifted

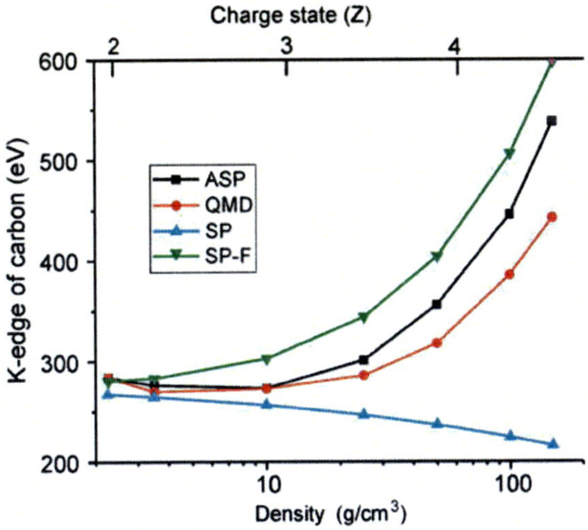

Fig. 8.4 ASP-predicted K-edges of carbon plasmas (black curve) as a function of mass density (or ion charge state Z), in comparison with quantum molecular dynamics (QMD) modeling (red curve) (Hu 2017). The blue curve indicates the continuum lowing model of Stewart and Pyatt (SP) (More 1981; Stewart and Pyatt 1996) while the green curve (SP-F) adds the Fermi energy to the SP model. The ASP model is in good agreement with complex QMD simulations and predicts well decreased K-edge energies for intermediate densities and increased ones for much above solid density compression

by about 12 eV in order to match experimental data for $\rho \approx 3.5$ g/cm^3 and this constant shift has then been applied for all densities of the QMD simulations [for more details, see (Hu 2017)]. Correspondingly, an error bar in the QMD simulations of near 12 eV should be attributed for the K-edge energies. Therefore, within the QMD error bars, the present ASP model matches very well with the QMD up to about ten times compressed solid carbon and is in agreement within 20% up to 100 times compressed solid carbon.

Finally we note that very recently it was claimed (Hansen et al. 2017) that the above-predicted increased K-edge energies have been observed in a dense Z-pinch plasma (Hansen et al. 2017): increased K-edge energy for iron of $\Delta E_{\mathrm{K}}^{(\mathrm{compressed})} - \Delta E_{\mathrm{K}}^{(\mathrm{solid})} \approx +5$ eV for $kT_{\mathrm{e}} = 10$ eV and $n_{\mathrm{e}} = 2.3 \times 10^{24}$ cm^{-3}. Quantitative analysis of these data, however, is challenging as the experiment in (Hansen et al. 2017) employed a gas mixture (Fe-impurity in Be). In addition, different methods of calculation delivered different average charges Z^* for the iron impurity: $Z^* = 3.5$ (SCRAM model) and $Z^* = 5.5$ (DFT model). Applying under these circumstances the present ASP model to the Be–Fe mixture Z-pinch experiment (Hansen et al. 2017), we need to align ionization of the Fe-impurity to the outer shell electrons (this is drastically different from the above-discussed XFEL experiment where different ionization is correlated with L-shell core hole ionization, Tables 8.1 and 8.2). In addition, as the iron impurity is rather small (180 ppm) the Fermi energy of (8.7) should be those of Be, i.e., 14.3 eV. Using (8.35), (8.36) with corresponding energies of Fe obtained from the AS-HF method, we obtain for $Z^* = 4$ a shifted K-edge of $\left[\Delta E_{\mathrm{K}}^{(\mathrm{compressed})} - \Delta E_{\mathrm{K}}^{(\mathrm{solid})}\right] \approx +17\,\mathrm{eV}$, while for $Z^* = 5$ and $Z^* = 6$ the ASP model gives +3.1 and −21.7 eV, respectively. The ASP model predicts therefore increased Fe K-edge energies in the relevant parameter range. More detailed conclusions are difficult to derive: The parameter uncertainties are too large and the observed shifts are too small due to the mid-Z-element employed.

Finally, we note that the experiment is subject to important transient evolution that drives many overlapping transitions form numerous different charge states. The K_β-analysis is therefore particularly challenging: Detailed atomic structure calculations of open $3d$-shell electrons predict negative screening (Condamine et al. 2019; Smid et al. 2019) that strongly obscures line and group shifts' measurements because dense plasma effects are equivalent to positive screening. It should further be noted that calibration shifts of up to 104 eV have been applied to the density functional theory (DFT) simulations (Hansen et al. 2017) to extract various data.

8.6 Discussion of Different Regimes of Ionization Potential Depression

"Why do well-established models like SP, EK, and the finite temperature ion sphere models that have proved to provide good agreement with numerous data do not match the XFEL data of K-edge energies in dependence of the number of L-shell core hole vacancies (Fig. 8.2)?" Figure 8.5 explores possible pathways. Let us separate the parameter space into five qualitative different regions with four boundaries: (1) the cold solid, (2) the hot solid, (3) the cold plasma, and (4) the hot plasma. The yellow-colored region between the cold and the hot solid corresponds to a state of matter, where the lattice structure of the solid density matter is still present and (8.5) applies. According to Fig. 8.5, the blue dashed line is therefore the "natural" ionization potential depression of a solid in the ASP picture, where the Fermi energy $\varepsilon_F^{(\mathrm{solid})}$ fully contributes to the K-shell ionization energy. If the solid crystal structure still prevails while electrons in the conduction band are heated, the Fermi energy contributes less and less to the ionization energy until the limit of the hot solid [$p(T) = 0$ according to (8.5)] is obtained (red dashed line in Fig. 8.5).

In the ASP model, the nonlinear dependence with respect to core hole vacancies is essentially a result of the nonlinear dependence of the energies predicted by the AS-HF method described above (in particular, when changing the subshell from one to another orbital quantum number; see, e.g., the change in Fig. 8.5 between $Z_n - N_{K+L} = 7$ and $Z_n - N_{K+L} = 8$). Within the framework of AS-HF, we find that the IPD energies $\Delta E_K = \Delta E_K^{(\mathrm{free})} - \Delta E_K^{(\mathrm{solid})}$ from (8.2), (8.3) scale for a variety of elements roughly like

$$\Delta E_K \propto (Z+1)^{4/3}. \tag{8.37}$$

The excellent agreement of the ASP model with the data (see Fig. 8.2) therefore suggests that the IPD data correspond to a heated solid where crystal structure essentially remained while VB electrons are strongly heated.

Debye models, ions sphere models, or interpolations between them [e.g., the widely applied SP model (More 1981; Stewart and Pyatt 1996)] are constructed

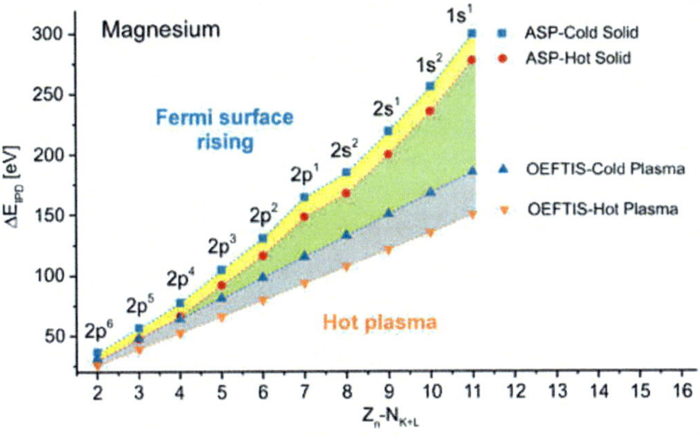

Fig. 8.5 Visualization of the different physical regimes of ionization potential depression (for Mg ions) in the framework of the present ASP and OEFTIS models. The abscissa indicates the dependence on different L-shell vacancies

with respect to an atom plasma point of view. The corresponding boundaries of interest in Fig. 8.5 are indicated with "cold plasma" and "hot plasma" (approximate boundaries have been visualized with the OEFTIS model at solid density for 10 and 1000 eV). The intermediate region is indicated in gray color. Figure 8.2 demonstrates that Debye and ion sphere models provide essentially a linear dependence on charge (see curves for OEFTIS and SP in Fig. 8.2), while their absolute values appear to be too low to match the data. This twofold failure might therefore be attributed to the absence of lattice structure, Fermi, and solid-state energies.

On the other hand, SP (More 1981; Stewart and Pyatt 1996), finite temperature ion sphere model (Li and Rosmej 2012, 2020), and the OEFTIS discussed above provide good agreement with previous data (see, e.g., Table 8.3). In the framework of Fig. 8.5, this is not a contradiction. Let us consider the green area in Fig. 8.5 that is located between the hot solid and the cold plasma. This region corresponds to matter, where crystal structure disappears when going from the upper part of the green area to the lower one. In fact, in typical optical laser-driven experiments [e.g., Dervieux et al. 2015; Renner et al. 1997b; Fletcher et al. 2014)], the pulse duration is much longer than in XFEL-driven plasmas (Ciricosta et al. 2016) (about a factor 10–100) and disintegration of cold/hot solid crystal structure is important. Correspondingly, the data (Hoarty et al. 2013) might be more close to the gray region than the extremely short pulse data from XFEL (Ciricosta et al. 2016). This point of view might be supported by the observation [(8.26), (8.27) and (Stillman et al. 2017)] that better agreement with the data is obtained, when scaling down the ion sphere radius. This moves the ionization potential depression to the region of the ASP model, i.e., to the green and yellow regions of Fig. 8.5 where some lattice structure is still present. Whether scaled down ion sphere radii and b4-potentials can imitate lattice structure effects remain currently an open question (see also discussions in the Annex A.4 and in [Li et al. 2019; Li and Rosmej 2020]).

In conclusion, Fig. 8.5 permits also a rough recommendation for the application of different theories and models on the basis of a physical landscape rather than limiting the discussion to coupling and degeneracy parameters only (see further discussions below): 1) the yellow region (cold or heated solid) is well described by the ASP-model (that can be generalized to any chemical element and ionization stage) while the region above (Fermi surface rising) might be approximated by "ASP-compressed" (according (8.36)); 2) the green region, where lattice structure starts to disappear (from top to down) the b-potential and its analytical representation in terms of Γ-functions (with b-parameters larger than 2, for details see Annex A.4) provides a reasonable description; 3) the grey region of a dense structure less plasma is very well described by OEFTIS in 4th-order analytical approximation or by the b-potential method with b = 2 (for details see Annex A4).

Let us now discuss timescales that are relevant for the application of the various theories. Models that are based on the minimization of the free energy [e.g., the Ecker–Kröll model (Ecker and Kröll 1963)] request thermodynamic equilibrium and the obtained potential reduction is a reduction that corresponds to the Saha equation. In fs-laser-produced plasmas (e.g., XFEL or UHI produced plasmas), these preconditions are questionable. In addition, the resulting potential reduction in the free energy minimization procedure specifies à priori the energy reduction that is consistent with a particular ionization and/or conductivity but is not necessarily identical with the data from spectroscopic measurements (line shifts, line disappearance in the continuum, etc.). This last point has historically already been stressed by Ecker and Weizel themselves more than 60 years ago (Ecker and Weizel 1956).

The discussions of the Ecker–Kröll model in the context of fs-XFEL data (Ciricosta et al. 2012) are therefore likewise questionable, and the replacement of the lattice and polarization functions in the Ecker–Kröll model (Ecker and Weizel 1956; Ecker and Kröll 1963) by a simple constant (Ciricosta et al. 2012, 2016; Vinko et al. 2014) is more a particular fit to specific data rather than a modified model. Moreover, as discussed above the AS-HF ionization potential depression energies $\Delta E_{\mathrm{K}} = \Delta E_{\mathrm{K}}^{(\mathrm{free})} - \Delta E_{\mathrm{K}}^{(\mathrm{solid})}$ scale for a variety of elements roughly like $\Delta E_{\mathrm{K}} \propto (Z+1)^{4/3}$. This scaling is an accidental coincidence with the EK model that shows a similar scaling for the ionization potential depression in certain parameter regimes. A simple fitting constant could therefore accidentally match some *Z*-scaled data.

Finally we note that the existence of a lattice structure and corresponding application of models is not only a question of temperature but also of the timescale: If the heating is almost instantaneous, electrons could be heated to high temperatures (i.e., $kT_{\mathrm{e}} \gg \varepsilon_{\mathrm{F}}$) while the ions are still immobile. Therefore, the ASP model can be even applied for strongly heated electrons in the XFEL experiments: AS-HF energies and Fermi energies are well described. In this respect, Fig. 8.5 classifies the overall experimental and physical landscape that is otherwise difficult to be cast into coupling and degeneracy parameters only. We note that the ASP model provides likewise excellent agreement for ionization potential depression in hollow crystals, i.e., for double core hole configurations (Gournay and Rosmej 2022).

The present ASP model has also favorable properties via timescale effects. ASP does not contain the total plasma particle thermodynamic equilibrium assumption nor an equilibrium assumption between the ionized ion and the polarized plasma (like in the Debye theory): It is essentially based on the solid and ionic state energies (Fig. 8.1). At maximum, an equilibrium assumption for the conduction band electrons is requested (for $0 < p(T) < 1$ calculations). This is, however, a much less restrictive condition as electrons at near solid density equilibrate on fs-time scales. Similar arguments hold true for the finite temperature ion sphere model OEFTIS. We note, that recent developments of non-equilibrium ion-sphere models have been made to account for non-thermalized ionized electrons (Li et al. 2022).

References

N.W. Ashcrof, N.D. Mermin, *Solid State Physics* (Saunders, 1976)

P. Beiersdorfer, G. Brown, A. McKelvey, R. Shepherd, D. Hoarty, C. Brown, M. Hill, L. Hobbs, S. James, J. Morton, L. WIlson, *High resolution measurements of Cl^{15+} line shifts in hot, solid-density plasmas*, Phys. Rev. A **100**, 012511 (2019)

T. Blenski, K. Ishikawa, Pressure ionization in the spherical ion-cell model of dense plasma and a pressure formula in the relativistic Pauli approximation. Phys. Rev. E **51**, 4869 (1995)

S.M. Blinder, Canonical partition function for the hydrogen atom via the Coulomb propagator, Journal of Mathematical Physics **36**, 1208 (1995)

Z.B. Chen, Y.S. Tian, Y.M. Yin, Y.Y. Qi, G.P. Zhao, X.Z. Shen, K. Wang, *Theoretical determination of level delocalizations, plasma shifts and radiative properties of fusion relevant Ni XXII in finite temperature dense plasmas using a generalized analytical b-potential*, JQSRT **266** 107570 (2021)

O. Ciricosta, S.M. Vinko, H.-K. Chung, B.I. Cho, C.R.D. Brown, T. Burian, J. Chalupsky, K. Engelhorn, R.W. Falcone, C. Graves, V. Hajkova, A. Higginbotham, L. Juha, J. Krzywinski, H.J. Lee, M. Messerschmid, C.D. Murphy, Y. Ping, D.S. Rackstraw, A. Scherz, W. Schlotter, S. Toleikis, J.J. Turner, L. Vysin, T. Wang, B. Wu, U. Zastrau, D. Shu, R.W. Lee, P. Heimann, B. Nagler, J.S. Wark, Direct measurements of the ionization potential depression in a dense plasma. Phys. Rev. Lett. **109**, 065002 (2012)

O. Ciricosta, S.M. Vinko, B. Barbrel, D.S. Rackstraw, T.S. Preston, T. Burian, J. Chalupsky, B.I. Cho, H.-K. Chung, G.L. Dakovski, K. Engelhorn, V. Hajkova, P. Heiman, M. Holmes, L. Juha, J. Krzywinski, R.W. Lee, S. Toleikis, J.J. Turner, U. Zastrau, J.S. Wark, Measurements of continuum lowering in solid-density plasmas created from elements and compounds. Nat. Comm. **7**, 11713 (2016)

F.P. Condamine, E. Filippov, P. Angelo, S.A. Pikuz, O. Renner, F.B. Rosmej, *High-spatial and spectral resolution spectroscopic study of Copper K-alpha open M-shell complex driven by hot electrons in dense laser produced plasmas*, HEDP **32**, 89 (2019)

R.D. Cowan, *The Theory of Atomic Structure and Spectra* (University of California Press, 1981)

V. Dervieux, B. Loupias, S. Baton, L. Lecherbourg, K. Glize, C. Rousseaux, C. Reverdin, L. Gremillet, C. Blancard, V. Silvert, J.-C. Pain, C.R.D. Brown, P. Allan, M.P. Hill, D.J. Hoarty, P. Renaudin, Characterization of near-LTE, high-temperature and high-density aluminum plasmas produced by ultra-high intensity lasers. HEDP **16**, 12 (2015)

R.D. Desclattes, E.Jr. Kessler, P. Indelicato, L. de Billy, E. Lindroth, J. Anton, X-ray transition energies: new approach to a comprehensive evaluation. Rev. Mod. Phys. **75**, 35 (2003)

R.P. Drake, *High-Energy-Density Physics* (Springer, 2006)

G. Ecker, W. Kröll, Lowering of the ionization energy for a plasma in thermodynamic equilibrium. Physics Fluids **6**, 62 (1963)

G. Ecker, W. Weizel, Zustandssumme und effective Ionisierungsspannung eines Atoms im Inneren des Plasmas. Ann. Phys. **6**, 126 (1956)

E. Engel, S.H. Vosko, Exact exchange-only potentials and the virial relation as microscopic criteria for generalized gradient approximations, Phys. Rev. B. **47**, 13164 (1993)

L.B. Fletcher, A.L. Kritcher, A. Pak, T. Ma, T. Döppner, C. Fortmann, L. Divol, O.S. Jones, O.L. Landen, H.A. Scott, J. Vorberger, D.A. Chapman, D.O. Gericke, B.A. Mattern, G.T. Seidler, G. Gregori, R.W. Falcone, S.H. Glenzer, Observations of continuum depression in warm dense matter with X-ray thomson scattering. Phys. Rev. Lett. **112**, 145004 (2014)

T. Fujimoto, *Plasma Spectroscopy* (Oxford Science Publications, Oxford, 2004)

L. Gournay, F.B. Rosmej, *Continuum Lowering of Hollow Crystals within the Atomic-Solid-Plasma Picture* (in preparation, 2022)

H.R. Griem, *Principles of Plasma Spectroscopy* (Cambridge University Press, 1997)

R.D. Inglis, E. Teller, Ionic depression of series limits in one electron spectra. Astr. Phys. J. **90**, 439 (1939)

V.V. Karasiev, S.X. Hu, *Unraveling the intrinsic atomic physics behind x-ray absorption line shifts in warm dense silicon plasmas*, Phys. Rev. E **103**, 033202 (2021)

W. Kohn, C. Majumdar, Continuity between bound and unbound states in a fermi gas. Phys. Rev. **138**, A1617 (1965)

H.-J. Kunze, *Introduction to Plasma Spectroscopy* (Springer, Berlin, 2009)

S.B. Hansen, E.C. Harding, P.F. Knapp, M.R. Gomez, T. Nagayama, J.E. Bailey, Changes in the electronic structure of highly compressed iron revealed by X-ray fluorescence lines and absorption edges. HEDP **24**, 39 (2017)

D.J. Hoarty, P. Allan, S.F. James, C.R.D. Brown, L.M.R. Hobbs, M.P. Hill, J.W.O. Harris, J. Morton, M.G. Brookes, R. Shepherd, J. Dunn, H. Chen, E. von Marley, P. Beiersdorfer, H.-K. Chung, R.W. Lee, G. Brown, J. Emig, Observation of the effect of ionization-potential depression in hot dense plasma. Phys. Rev. Lett. **110**, 265003 (2013)

S.X. Hu, Continuum lowering and fermi-surface rising in strongly coupled and degenerate plasmas. Phys. Rev. Lett. **119**, 065001 (2017)

D.G. Hummer, D. Mihalas, The equation of state for stellar envelopes: an occupation probability formalism for the truncation of internal partition functions. Astr. J. **331**, 794 (1988)

Hyperphysics: http://hyperphysics.phy-astr.gsu.edu/hbase/Tables/fermi.html (2021)

C.A. Iglesias, Comment on "Free-free opacity in warm aluminum". HEDP **7**, 38 (2011)

C.A. Iglesias, A plea for a reexamination of ionization potential depression measurements. HEDP **12**, 5 (2014)

C.A. Iglesias, *On spectral line shifts from analytic fits to the ion-sphere model potential*, HEDP **30**, 41 (2019)

C.A. Iglesias, P.A. Sterne, Fluctuations and the ionization potential in dense plasmas. HEDP **9**, 103 (2013)

X. Li, F.B. Rosmej, Quantum number dependent energy level shifts of ions in dense plasmas: a generalized analytical approach. Europhys. Lett. **99**, 33001 (2012)

X. Li, F. B. Rosmej, *Analytical approach to level delocalization and line shifts in finite temperature dense plasmas*, Phys. Lett A **384**, 126478 (2020)

X. Li, F.B. Rosmej, V.A. Astapenko, V.S. Lisitsa, An analytical plasma screening potential based on the self-consistent-field ion-sphere model. Phys. Plasmas **26**, 03301 (2019)

X. Li, F.B. Rosmej, V.A. Astapenko, V.S. Lisitsa, *Generalized free-bound approach for ions embedded in dense plasmas* (in preparation, 2019)

X. Li, Z. Xu, F.B. Rosmej, Exchange energy shifts in dense plasma conditions. J. Phys. B: At. Mol. Opt. Phys. **39**, 3373 (2006)

V.S. Lisitsa, *Atoms in Plasmas* (Springer, 1994)

K. Ma, Y. Chu, Z.B. Chen, *Analytical Calculation of* Cl^{15+} *Ion Immersed in Dense Plasmas*, Few-Body Systems **61**, 42 (2020)

G. Massacrier, J. Dubeau, A theoretical approach to n-electron ionic structure under dense plasma conditions: I. blue and red shift. J. Phys. B: At. Mol. Opt. Phys. **23**, 24595 (1990)

D. Mihalas, *Stellar Atmospheres* (W. H. Freeman and Company, San Francisco, 1978)

D. Mihalas, B. Weibel-Mihalas, *Foundations of Radiation Hydrodynamic* (Dover Publications, 1999)

R.M. More, *Atomic Physics in Inertial Confinement Fusion, LLNL technical report*, no. ucrl-84991 (unpublished, 1981)

H. Nguyen, M. Koenig, D. Benredjem, M. Caby, G. Coulaud, Atomic structure and polarization line shift in dense and hot plasmas. Phys. Rev. A **33**, 1279 (1986)

NIST-data base: https://physics.nist.gov/PhysRefData/XrayTrans/Html/search.html (2021)

S.G. Rautian, A.M. Shalagin, *Kinetic Problems of Non-Linear Spectroscopy* (North-Holland, Amsterdam, 1991)

O. Renner, F.B. Rosmej, Challenges of X-ray spectroscopy in investigations of matter under extreme conditions. Matter Radiat. Extrem. Rev. **4**, 024201 (2019)

O. Renner, T. Missalla, P. Sondhauss, E. Krousky, E. Förster, C. Chenais-Popovics, O. Rancu, High-luminosity, high-resolution, X-ray spectroscopy of laser-produced plasma by vertical-geometry Johann spectrometer. Rev. Sci. Instrum. **68**, 2393 (1997a)

O. Renner, P. Sondhauss, D. Salzmann, A. Djaoui, M. Koenig, E. Förster, Measurement of the polarization shifts in hot and dense aluminum plasma. JQSRT **58**, 851 (1997b)

O. Renner, D. Salzmann, P. Sondauss, A. Djaoui, E. Krousky, E. Förster, Experimental evidence for plasma shifts in Lyman series of aluminum. J. Phys. B At. Mol. Opt. Phys. **31**, 1379 (1998)

O. Renner, F.B. Rosmej, E. Krouský, P. Sondhauss, M.P. Kalachnikov, P.V. Nickles, I. Uschmann, E. Förster, Overcritical density plasma diagnosis inside the laser-produced craters. Appl. Phys. Lett. **79**, 177 (2001)

F.B. Rosmej, Hot electron X-ray diagnostics. J. Phys. B. Lett.: At. Mol. Opt. Phys. **30**, L819 (1997)

F.B. Rosmej, A new type of analytical model for complex radiation emission of hollow ions in fusion and laser produced plasmas. Europhys. Lett. **55**, 472 (2001)

F.B. Rosmej, An alternative method to determine atomic radiation. Europhys. Lett. **76**, 1081 (2006)

F.B. Rosmej, Exotic states of high density matter driven by intense XUV/X-ray Free Electron Lasers, in *Free Electron Laser, InTech 2012*, ed. by S. Varró, ISBN 978-953-51-0279-3 (2012), p. 187–212. Free download: http://www.intechopen.com/books/free-electron-lasers/exotic-states-of-high-density-matter-driven-by-intense-xuv-x-ray-free-electron-lasers

F.B. Rosmej, Ionization potential depression in an atomic-solid-plasma picture. Letter J. Phys. B. **51**(9), LT01 (2018)

F.B. Rosmej, K. Bennadji, V.S. Lisitsa, Dense plasmas effects on exchange energy shifts in highly charged ions: an alternative approach for arbitrary perturbation potentials. Phys. Rev. A **84**, 032512 (2011)

F.B. Rosmej, U.N. Funk, M. Geißel, D.H.H. Hoffmann, A. Tauschwitz, A.Ya. Faenov, T.A. Pikuz, I.Yu. Skobelev, F. Flora, S. Bollanti, P.Di. Lazzaro, T. Letardi, A. Grilli, L. Palladino, A. Reale, A. Scafati, L. Reale, T. Auguste, P. D'Oliveira, S. Hulin, P. Monot, A. Maksimchuk, S.A. Pikuz, D. Umstadter, M. Nantel, R. Bock, M. Dornik, M. Stetter, S. Stöwe, V. Yakushev, M. Kulisch, N. Shilkin, X-ray radiation from ions with K-shell vacancies. JQSRT **65**, 477 (2000)

A. Saemann, K. Eidmann, I.E. Golovkin, R.C. Mancini, E. Andersson, E. Förster, K. Witte, Isochoric heating of solid aluminum by ultrashort laser pulses focused on a tampered target. Phys. Rev. Lett. **82**, 4843 (1999)

D. Salzman, *Atomic Physics in Hot Plasmas* (Oxford University Press, 1998)

I. Shimamura, T. Fujimoto, State densities and ionization equilibrium of atoms in dense plasmas. Phys. Rev. A **42**, 2346 (1990)

A.K. Shing, D. Dawra, M. Dimri, A.K.S. Jha, R.K. Pandey, M. Mohan, *Plasma screening effects on the atomic structure of He-like ions embedded in strongly coupled plasma*, Phys. Lett. A **384**, 126369 (2020)

M. Smid, O. Renner, A. Colaitis, V.T. Tikhonchuk, T. Schlegel, F.B. Rosmej, *Characterization of suprathermal electrons inside laser accelerated solid density matter via axially-resolved K_α-emission,* Nature Comm. **10**, 4212 (2019)

I.I. Sobelman, L.A. Vainshtein, E.A. Yukov, *Excitation of Atoms and Broadening of Spectral Lines* (Springer, 1995)

S.-K. Son, R. Thiele, Z. Jurek, B. Ziaja, R. Santra, Quantum-mechanical calculation of ionization-potential lowering in dense plasmas. Phys. Rev. X **4**, 031004 (2014)

J. Stewart, K. Pyatt, Lowering of ionization potentials in plasmas. Astrophys. J. **144**, 1203 (1996)

C.R. Stillman, P.M. Nilson, S.T. Ivancic, I.E. Golovkin, C. Mileham, I.A. Begishev, D.H. Froula, Picosecond time-resolved measurements of dense plasma line shifts. PRE **95**, 063204 (2017)

A. Thompson, I. Lindau, D. Attwood, Y. Liu, E. Gulliksn, P. Pianetta, M. Howells, A. Robinson, K.-J. Kim, J. Scofield, J. Kirz, J. Underwood, J. Kortright, G. Williams, H. Winick, *X-ray Data Booklet*, Center for X-ray Optics and Advanced Light Source, LBNL/PUB-490 Rev. 3 (2009)

A. Unsöld, *Physik der Sternatmosphären* (Springer, 1955)

S.M. Vinko, O. Ciricosta, J.S. Wark, Density functional theory calculations of continuum lowering in strongly coupled plasmas. Nature Comm. **5**, 3533 (2014)

S.M. Vinko, G. Gregori, B. Nagler, T.J. Whitcher, J.S. Wark, U. Zastrau and E. Förster, S. Mazevet, J. Andreasson, S. Bajt, R.R. Fäustlin, S. Toleikis, and T. Tschentscher, J. Chalupsky, J. Cihelka, V. Hajkova, L. Juha, H. Chapman, T. Dzelzainis, D.Riley, E. Galtier, F.B. Rosmej, P. A. Heimann, M. Jurek, J. Krzywinski, R.W. Lee, A.J. Nelson, R. Sobierajski, Electronic structure of an XUV photo-generated solid-density aluminum plasma. Physical Review Letters **104**, 225001 (2010)

Ya.B. Zeldovich, Yu.P. Raizer, *Physics of Shock Waves and High-Temperature Hydrodynmaic Phenomena* (Dover Publications, 2002)

G.B. Zimmermann, R.M. Moore, Pressure ionization in laser fusion target simulation. JQSRT **23**, 417 (1980)

Chapter 9
The Plasma Atom

Abstract In this chapter the statistical description of atomic radiative-collisional processes for complex heavy ions in electron atomic collisions is developed. The local plasma frequency model and the Thomas-Fermi electron density distribution are applied for the description of collisional processes making it possible to express the atomic characteristics (energy structure and oscillator strengths) as well as transition probabilities in terms of a functional of the electron density distribution inside atoms and ions. The Fermi method of equivalent photons allows to express the collisional rates in terms of photo-excitation or ionization cross sections. The statistical description is applied for efficient calculations of ionization cross sections and rates for different highly charged ions demonstrating a very good correspondence with detailed quantum mechanical calculations. Likewise, the dielectronic recombination rates obtained from statistical models are compared with quantum results for different ionization states of many chemical elements. The statistical method is in very good agreement with sophisticated detailed level-by-level quantum calculations and is of much higher precision than the usually applied Burgess formula. Finally, the statistical approach is applied for calculations of radiative energy losses of tungsten ions in hot thermonuclear plasmas. The results for the low-density case (coronal condition) of magnetically confined plasmas demonstrate a rather good correspondence with more detailed numerical calculations and measurements. In addition, the transition from the low-density corona condition to the high-density Boltzmann limit can be described via a simple application of detailed balance in the two-state approximation. In general, quite reasonable precision of the statistical model for different kinds of radiative-collisional processes is demonstrated. Moreover, general formulae and scaling relations can be obtained from the statistical approach that would otherwise difficult to obtain.

9.1 The Thomas–Fermi Statistical Approach

From the earliest days of quantum mechanics, it has been clear that one could not hope to solve exactly most of the physically interesting systems, especially those with three or more particles. This has stimulated the development of a large variety

F. B. Rosmej et al., *Plasma Atomic Physics*, Springer Series on Atomic, Optical, and Plasma Physics 104, https://doi.org/10.1007/978-3-030-05968-2_9

of approximate methods such as the time-dependent and time-independent perturbation theory, variational methods, and the Hartree method. The improvements of the Hartree method, namely the Hartree–Fock–Slater and the Dirac–Hartree–Fock–Slater (including the multiconfiguration Hartree–Fock (MCHF) method that incorporates configuration interaction and intermediate coupling, or its relativistic version MCDF—multiconfiguration Dirac–Fock method), are currently the most general and widely used methods to study the atomic structure of atoms and ions (Cowan 1981; Grant et al. 1980).

The theoretical description of multielectron systems and in particular the structure of heavy atoms and ions is still challenging, and one of the traditional approaches to this problem is the use of the Thomas–Fermi statistical theory (Fermi 1928; Gombas 1943, 1949, 1963; Lieb and Simon 1977; Kemister and Nordholm 1982). The Thomas–Fermi model represents the simplest way to take into account not only the Pauli principle, but also the mutual electrostatic repulsion of the electrons, at least in a general way, in a many-electron system. The starting point is the only approximately correct idea that there is a fixed potential well and that it is the same for all electrons. The model therefore gives a similar electron density for all atoms. Although the model does not permit to provide very detailed information about the atomic structure, it provides general insight into the properties of heavy atoms, e.g., the electron density distribution, ionization energies, size of the atom/ion, polarizability. The Thomas–Fermi model is also used to describe the equation of state of highly compressed and ionized matter, has stimulated the development of the density functional theory, and provides often a good starting point for more complex self-consistent field calculations.

Apart the atomic structure itself, the study of the interaction of multielectron atoms and ions with an electromagnetic field is of great practical interest due to applications in material science, atomic physics, plasma physics, radiative properties of matter, spectroscopy. The statistical model provides the possibility of a universal description of elementary processes (Astapenko et al. 2002, 2003) and radiative properties (Demura et al. 2013, 2015a, b) and to extract general scaling laws for all nuclear charges using the Thomas–Fermi density distribution $n(r)$. This is of particular interest for the fusion science: In magnetic fusion, high-Z divertor material (tungsten) is employed while in inertial fusion, high-Z materials (gold) are used as a hohlraum material. The determination of the detailed atomic structure (MCHF or MCDF) and corresponding radiative properties of these high-Z elements is, however, very challenging. Therefore, approximate and/or general methods are of great interest to derive the variety of requested properties. Moreover, more general methods and scaling relations are particularly useful for implementation of heavy element atomic physics in integrated simulations.

In the framework of the Thomas–Fermi model, the electron density distribution of a particular element and charge state is given by

$$n(x, q, Z_n) = \frac{32}{9\pi^3} \cdot Z_n^2 \cdot \left[\frac{\varphi(x, q)}{x}\right]^{3/2}, \tag{9.1}$$

where

$$x = \frac{r}{r_{TF}}, \tag{9.2}$$

$$r_{TF} = \left(\frac{9\pi^3}{128}\right)^{1/3} \cdot \frac{1}{Z_n^{1/3}} = 0.8853 \cdot Z_n^{-1/3}, \tag{9.3}$$

$$q = \frac{Z}{Z_n}. \tag{9.4}$$

Z_n is the nuclear charge, Z the ion charge, q characterizes the ionization degree and r_{TF} is the Thomas–Fermi radius. The Thomas–Fermi function $\varphi(x, q)$ can be approximated by the Sommerfeld method (Sommerfeld 1932; Gombas 1949) which is an exact particular solution of the Thomas–Fermi differential equation:

$$\varphi(x, q) = \varphi(x) \cdot \left[1 - \left(\frac{1 + z(x)}{1 + z_0(x)}\right)^{\lambda_1/\lambda_2}\right], \tag{9.5}$$

$$z(x) = \left(\frac{x}{144^{3/2}}\right)^{\lambda_2}, \tag{9.6}$$

$$z_0(x) = \left(\frac{x_0(q)}{144^{3/2}}\right)^{\lambda_2}, \tag{9.7}$$

$$\varphi_0(x) = \frac{1}{(1 + z(x))^{\lambda_1/2}}, \tag{9.8}$$

$$\lambda_1 = 0.5 \cdot \left(7 + \sqrt{73}\right) = 7.77200, \tag{9.9}$$

$$\lambda_2 = 0.5 \cdot \left(-7 + \sqrt{73}\right) = 0.77200. \tag{9.10}$$

The reduced radius $x_0(q)$ is determined from the boundary condition

$$x_0 \frac{d\varphi(x_0)}{dx} = -q. \tag{9.11}$$

In high-temperature plasma, i.e., when the ionization degree $q = Z/Z_n$ is not too low, the reduced radius can be approximated by

$$x_0(q) = \begin{cases} 2.96 \cdot \left(\dfrac{1-q}{q}\right)^{2/3} & if \quad 0.2 < q \leq 1 \\ 6.84 \cdot \dfrac{1}{q^3} & if \quad q < 0.05 \end{cases}. \tag{9.12}$$

The ionization energy of an atom or ion is then given by

$$I_Z = Z_n^2 Ry \cdot \left\{ \left(\frac{128}{9\pi^2}\right)^{1/3} \cdot \frac{2 \cdot Z}{Z_n^{5/3} \cdot x_0(q, Z_n)} \right\}. \tag{9.13}$$

As can be seen from (9.13), the hydrogenic approximation $Z_n^2 \cdot Ry$ of the ionization potential of an ion with charge Z_n is corrected via the Thomas–Fermi electron density distribution that depends on nuclear charge and ionic charge [factor in parenthesis in (9.13)]. The comparison of the ionization energies obtained from (9.13) with detailed Hartree–Fock calculations shows a reasonable agreement for heavy elements over a wide range of ionization degrees (Demura 2015a, b). We note that certainly more accurate descriptions of the ionization potentials can be obtained from a direct fit to the vast amount of ionization potentials in dependence of Z and Z_n (Kirillow et al. 1975):

$$I_Z \approx 0.221 \cdot Ry \cdot \frac{(1+Z)^{4/3}}{1 - 0.96 \cdot \left(\dfrac{1+Z}{Z_n}\right)^{0.257}}. \tag{9.14}$$

Many modifications of the Thomas–Fermi model have been proposed in order to include shell structure, improve ionization energies, and in particular to approach the Hartree–Fock results for the electron density distribution (Dmitrieva and Plindov 1984; Fromy et al. 1996; Dyachkov et al. 2016). Also, modifications to derive the average degree of ionization in a dense plasma have been proposed (Ying and Kalman 1989). In addition, in order to improve the studies of the interaction of multielectron atoms with an electromagnetic field the classical kinetic Vlasov equations with self-consistent field has been proposed (Vinogradov and Tolstikhin 1989). It leads to improvement for the calculation of elementary processes, e.g., photoionization cross sections, and permits the calculation of the real part of the polarizability.

In the further developments to improve the statistical approach, one must not lose sight of the requirement that the fundamental equations of the statistical model of atoms, including the various corrections terms, should not be too complicated and, in any case, not more complicated than the basic equations of the quantum mechanical many-body approximation, e.g., the multiconfiguration Hartree–Fock methods. One must always bear in mind that the statistical theory of atoms is only a rough approximation of the quantum atom, and its advantage is its extreme simplicity both in structure and application to determine the electron and potential distributions of atoms, derive elementary processes in collisional–radiative regimes, to shed light into the detailed atomic structure calculations (in particular for heavy atoms) and to derive general scaling laws that could be hardly obtained otherwise. It is this practical philosophy that we bear in mind when we consider (9.1)–(9.13) for the statistical framework of the atom/ion and the framework of the local plasma frequency (to be discussed further below) in order to enlarge the standard statistical Thomas–Fermi model also to elementary collisional–radiative processes and to the radiative properties of heavy atoms and ions in plasmas.

9.2 The Local Plasma Frequency Approximation

9.2.1 Oscillator Strengths Distribution and Photoabsorption

The response of an atom to an external field of given frequency ω can be conveniently discussed in terms of the properties of its differential oscillator strength distribution $f(\omega)$ that is directly related to the photoabsorption cross section of the atom:

$$\sigma(\omega) = \frac{2\pi^2 e^2}{m_e c} \cdot f(\omega). \tag{9.15}$$

The function $f(\omega)$ may be considered to comprise all the fundamental information on the quantum dynamics of atoms, but its quantum mechanical calculation is rather challenging and laborious. The distribution of the local atomic density determines a variety of elementary excitations with the classical plasma frequency. Concerning the Z_n- and frequency-dependence, we can identify three regions of interest. In the low-frequency range, where

$$0 \le \frac{\hbar\omega}{Ry} \le 1, \tag{9.16}$$

the function $f(\omega)$ essentially consists of the sharp lines familiar from optical spectroscopy separated by frequency ranges of low absorption while it changes irregularly with Z_n and reflects in its details the atomic binding. In the high-frequency range, where

$$\hbar\omega \ge Z_n^2 Ry, \tag{9.17}$$

the function $f(\omega)$ exhibits characteristic X-ray absorption edges. In the intermediate frequency range, where

$$1 \le \frac{\hbar\omega}{Ry} \le Z_n^2, \tag{9.18}$$

the contribution from intermediate shells of the atom is expected to overlap strongly and $f(\omega)$ becomes a smooth function on frequency. In this regime, excitations from intermediate atomic shells can be coupled rather strongly: new collective resonances of the atom as a whole become possible and a statistical approximation should apply best to the dynamics of the atom.

In the framework of the statistical approximation, the spectral distribution function $f(\omega)$ is derived from the general dynamic equations describing the density fluctuations induced in the atom by an external field. Using a local form of this framework, it is found that coherences between the motion in different parts of the atom causes modifications in $f(\omega)$ that can be formulated in terms of a dispersion denominator that identifies enhanced absorption as collective resonances of the atom as a whole (Brandt and Lundqvist 1965). The oscillation frequency is determined by the well-known formula for the electron plasma frequency

$$\omega(r) = \omega_{\mathrm{p}}(r) = \sqrt{\frac{4\pi e^2 n_{\mathrm{e}}(r)}{m_{\mathrm{e}}}}, \tag{9.19}$$

where $n_{\mathrm{e}}(r)$ is the local atomic electron density. $\omega_{\mathrm{p}}(r)$ is called the local plasma frequency LPF (see also Sect. 2.6). Comparison with the more general Vlasov approach shows that the LPF model does not take into account the polarization field induced by the external perturbation of the atomic electron density distribution (Vinogradov and Tolstikhin 1989). However, it turns out that the discrepancy between the LPF-statistical model for the photoionization cross sections is within the accuracy of calculations of radiative and collisional processes for multielectron ions by standard quantum mechanical codes and population kinetics (to be discussed below).

9.2.2 *Fermi Equivalent Photon Method and Local Plasma Oscillator Strength*

The interaction of the plasma electrons with a heavy atom can be considered within the framework of the Fermi approximation of equivalent photons (see Sect. 5.1). The electric field of the equivalent photon flux is determined by the Fourier expansion of the electric field of an electron, moving along with the classical trajectory in the field of the ion being excited. In this formulation, for example, the excitation of bound electrons in a multielectron ion is expressed in terms of the photoabsorption cross section of (9.15).

In quantum mechanics of atoms, almost all plasma physics relevant atomic characteristics can be approximated or expressed via the dipole oscillator strengths (see also Chaps. 2 and 7). It is therefore of particular interest for the generalization of the applications of the plasma atom to derive an "effective plasma oscillator strengths." The determination of the classical oscillator strength follows from the correspondence principle (see also Chap. 2). In fact, if we consider the atom as a classical oscillator with an eigenfrequency equal to the local plasma frequency, we can readily derive its response on periodic (harmonic) perturbations. This determines the dynamic response of the classical oscillator and via correspondence with the quantum radiation emission we can identify a "plasma oscillator strength."

The simplest relation between the induced dipole moment and the local electrical field is given by

$$\vec{p}_{\text{induced}} = \alpha \cdot \vec{E}_{\text{local}} \tag{9.20}$$

(Note that the polarizability according to (9.20) is related to the quadratic Stark constant, see Sects. 7.4.2, 7.8.2). In the classical description, the frequency dependence of the oscillating atom under the action of a local electric field is given by

$$m_{\text{e}} \cdot \ddot{\vec{x}} + m_{\text{e}} \cdot \gamma \cdot \dot{\vec{x}} + m_{\text{e}} \cdot \omega_0^2 \cdot \vec{x} = q \cdot \vec{E}_{\text{local}} \cdot \exp(-i\omega t), \tag{9.21}$$

where $\vec{x}$ is the amplitude, m the mass, γ the damping constant, ω_0 the eigenfrequency of the oscillator, i.e., transition frequency, q the electric charge, and ω is the frequency of the local oscillating electric field. The stationary solution of (9.21) is given by

$$\vec{x}(t) = \frac{1}{\omega_0^2 - \omega^2 - i\gamma\omega} \cdot \frac{q}{m_{\text{e}}} \cdot \vec{E}_{\text{local}} \cdot \exp(-i\omega t). \tag{9.22}$$

Because the induced dipole moment is given by

$$\vec{p}_{\text{induced}} = q \cdot \vec{x}(t), \tag{9.23}$$

we obtain with the help of (9.20), (9.22), (9.23) for the dynamic polarizability

$$\alpha(\omega) = \frac{q^2}{m_{\text{e}}} \cdot \frac{1}{\omega_0^2 - \omega^2 - i\gamma\omega}. \tag{9.24}$$

Generalizing (9.24) to several oscillation frequencies, we obtain

$$\alpha(\omega) = \frac{q^2}{m_{\text{e}}} \cdot \sum_n \frac{1}{\omega_{0n}^2 - \omega^2 - i\gamma_{0n}\omega}. \tag{9.25}$$

If we separate the dynamic polarizability (9.24) in the real and imaginary part, i.e.,

$$\alpha(\omega) = \Re e(\alpha) + i\Im m(\alpha), \tag{9.26}$$

we obtain

$$\Re e(\alpha) = \frac{q^2}{m_e} \cdot \sum_n \frac{\omega_{0n}^2 - \omega^2}{\left(\omega_{0n}^2 - \omega^2\right)^2 + \gamma_{0n}^2 \omega^2}, \tag{9.27}$$

$$\Im m(\alpha) = \frac{q^2}{m_e} \cdot \sum_n \frac{\gamma \cdot \omega}{\left(\omega_{0n}^2 - \omega^2\right)^2 + \gamma_{0n}^2 \omega^2}. \tag{9.28}$$

For $\omega = 0$, the real part of the polarizability (9.25) corresponds to the static polarizability because the imaginary part (9.28) vanishes. In the local plasma frequency model, (9.25) has to be transformed to the continuous case (see also Sects. 2.3–2.6). In the spherical shell of thickness dr, we encounter a confined charge of quantity d$q = e \cdot 4\pi n_e(r)\, r^2dr$ and we obtain instead of (9.25):

$$\alpha(\omega) = \frac{4\pi e^2}{m_e} \cdot \int_0^{R_{\text{atom}}} \frac{n_e(r) r^2 \mathrm{d}r}{\omega_{0n}^2 - \omega^2 - i\gamma_{0n}\omega}, \tag{9.29}$$

where R_{atom} is the size of the atom. From (9.29), it follows with (9.19)

$$\alpha(\omega) = \int_0^{R_{\text{atom}}} \frac{\omega_p^2 \cdot r^2 \cdot \mathrm{d}r}{\omega_{0n}^2 - \omega^2 - i\gamma_{0n}\omega}. \tag{9.30}$$

Equation (9.29) can be readily compared with the quantum mechanical result for the polarizability, i.e.,

$$\alpha(\omega) = \frac{e^2}{m_e} \cdot \sum_n \frac{f_{0n}}{\omega_{0n}^2 - \omega^2 - i\gamma_{0n}\omega} \tag{9.31}$$

from which it follows that the term $4\pi n_e(r) r^2\, \mathrm{d}r$ can be interpreted as a local strengths in the LPF model, i.e., the local plasma oscillator strength

$$f_p(r) = 4\pi n_e(r) r^2\, \mathrm{d}r = f_{ij}. \tag{9.32}$$

The plasma oscillator strengths $f_p(r)$ are a central plasma atomic property that allows to deal with collisional–radiative elementary processes that are usually expressed in terms of the oscillator strengths f_{ij} for the transition $i \to j$. The Regemorter formula of electron collisional excitation according to (5.90)–(5.94) is a typical example. In the LPF model, the indexes i, j are represented by the radial coordinate r and the radial interval dr [see right part of (9.32)]. From (9.32), it follows immediately that the plasma oscillator strengths fulfills the sum rule, i.e.,

$$\int \mathrm{d}f_{\mathrm{ij}} = Z_{\mathrm{n}} - Z = N_{\mathrm{e}}^{(\mathrm{bound})}. \tag{9.33}$$

9.3 Radiative Losses

9.3.1 General Relations

At low-density conditions, collisional excitation from the ground state to excited states decays readily by spontaneous radiative transitions (Corona model, see also Chap. 6) and the radiation losses are therefore determined by the collisional excitation rates itself. In the framework of the Fermi equivalent photon method, the radiation loss can therefore be expressed in terms of the photoexcitation rates in the field of equivalent photons (Demura et al. 2013, 2015a):

$$\begin{aligned} Q &= n_{\mathrm{e}}^{(\mathrm{free})} \int_0^{I/2Z_{\mathrm{n}}Ry} \mathrm{d}sRy \cdot \sigma_{\mathrm{ph}}(s) \cdot \left\langle \frac{\mathrm{d}I^{(\mathrm{Coulomb})}(s)}{\mathrm{d}s} \right\rangle_{\mathrm{E}} \\ &= n_{\mathrm{e}}^{(\mathrm{free})} \frac{4a_0 c\, Ry^2}{\sqrt{3}\pi \cdot e^2} \cdot \sqrt{\frac{Z_{\mathrm{n}}^2 Ry}{kT_{\mathrm{e}}}} \cdot \int_0^{I/2Z_{\mathrm{n}}Ry} \mathrm{d}s \cdot \sigma_{\mathrm{ph}}(s) \cdot \int_{\frac{2RyZ_{\mathrm{n}}}{kT_{\mathrm{e}}} \cdot s}^{\infty} \mathrm{d}u \cdot e^{-u} \cdot g(s,u) \end{aligned} \tag{9.34}$$

with

$$s = \frac{\hbar\omega}{2Z_{\mathrm{n}} \cdot Ry}, \tag{9.35}$$

$$u = \frac{E}{kT_{\mathrm{e}}}, \tag{9.36}$$

where a_0 is the Bohr radius, c the speed of light, e the electron charge, Ry the Rydberg energy, kT_{e} the thermal electron energy, $n_{\mathrm{e}}^{(\mathrm{free})}$ is the free electron density (note, that the atomic electron density is designated with $n_{\mathrm{e}}(r)$), I is the ionization potential of the ion with charge Z, and nuclear charge Z_{n}, σ_{ph} is the photoexcitation–photoionization cross section, $\left\langle \mathrm{d}I^{(\mathrm{Coulomb})}(s)/\mathrm{d}s \right\rangle_{\mathrm{E}}$ is the intensity of equivalent photon flux per unit of reduced frequency interval ds averaged over the energy E of the electron projectile scattered by the target and $g(s,u)$ is the Gaunt factor, describing the curvature of electron trajectory under its motion in the self-consistent field of the heavy ion or atom. In the Coulomb approximation, the Gaunt factor becomes a function of the unique variable (Kogan et al. 1992):

$$\nu = Z_{\text{eff}} \cdot Z_{\text{n}} \cdot \left(\frac{Ry}{kT_{\text{e}}}\right)^{3/2} \cdot \frac{s}{u^{3/2}}, \tag{9.37}$$

$$g(\nu) = \frac{\pi\sqrt{3}}{4} \cdot \left\{ i\nu H_{i\nu}^{(1)'}(i\nu) \cdot H_{i\nu}^{(1)}(i\nu) \right\} \approx \frac{\sqrt{6}}{\pi} \cdot \ln\left[\left(\frac{2}{\gamma\nu}\right)^{1/\sqrt{2}} + \exp\left(\pi/\sqrt{6}\right) \right], \tag{9.38}$$

where $H_{i\nu}^{(1)}(i\nu)$ and $H_{i\nu}^{(1)'}(i\nu)$ are Hankel functions and its first derivative with the argument $i\nu$, γ is the Euler constant ($\gamma = \exp(C) \approx 1.78$).

In the local plasma frequency model, the effective charge Z_{eff} from (9.37) is determined from the condition of equality of the Thomas–Fermi potential and the local Coulomb potential at the point $r_{\text{s}} = x_{\text{s}} \cdot r_{\text{TF}}$ (r_{TF} is the Thomas–Fermi radius from (9.3)) that corresponds to the resonance condition of (9.19) expressed in terms of the reduced frequency s of (9.35):

$$Z_{\text{eff}} = Z_{\text{n}} \cdot \left\{ \varphi(x_{\text{s}}, q) + \frac{q x_{\text{s}}}{x_0} \right\}, \tag{9.39}$$

where $\varphi(x_{\text{s}}, q)$, q and x are defined in (9.1)–(9.11). For low frequencies, the effective charge is equal to the ion charge Z, while in the high-frequency limit the effective charge is the nuclear charge Z_{n}. These limits are typically approximately approached for $s < 0.1$ (low-frequency limit) and $s > 30$ (high-frequency limit).

Taking into account only bound states, the first integration over frequencies in (9.34) extends up to the ionization threshold of the ion with charge Z while the second integration over energies of the incident electron corresponds to the excitation thresholds of atomic transitions in the statistical model. In the LPF approximation, the photoabsorption cross section is given by (see also (3.136) and (Rosmej et al. 2020a))

$$\sigma_{\text{ph}}(\omega) = \frac{2\pi^2 e^2}{m_{\text{e}} c} \cdot 4\pi r_\omega^2 \cdot \frac{n_{\text{e}}(r_\omega)}{\left|\frac{d\omega_{\text{p}}(r)}{dr}\right|} \tag{9.40}$$

or, expressed in terms of the Thomas–Fermi electron density:

$$\sigma_{\text{ph}}(\omega) = \pi a_0^2 \cdot \frac{3\pi^3 e^2}{16\hbar c} \cdot \frac{s \cdot x_{\text{s}}^2 \cdot \varphi(x_{\text{s}}, q)}{\left|\varphi'(x_{\text{s}}, q) - \varphi(x_{\text{s}}, q)/x_{\text{s}}\right|}. \tag{9.41}$$

Inserting (9.41) into (9.34), we obtain the final expression for the radiation loss in the Corona limit:

$$Q = n_{\mathrm{e}}^{(\mathrm{free})} \cdot \frac{\pi^3\sqrt{3\pi} \cdot a_0^3 Ry^2}{4} \cdot \sqrt{\frac{Z_{\mathrm{n}}^2 Ry}{kT_{\mathrm{e}}}} \cdot \int\limits_0^{1/2Z_{\mathrm{n}}Ry} \frac{s \cdot x_{\mathrm{s}}^2 \cdot \varphi(x_{\mathrm{s}}, q) \cdot \mathrm{d}s}{|\varphi'(x_{\mathrm{s}}, q) - \varphi(x_{\mathrm{s}}, q)/x_{\mathrm{s}}|} \cdot \int\limits_{\frac{2RyZ_{\mathrm{n}}}{kT_{\mathrm{e}}} \cdot s}^{\infty} \mathrm{d}u \cdot e^{-u} \cdot g(s, u). \tag{9.42}$$

The radiation losses determined by (9.42) have been compared with detailed collisional–radiative modeling of tungsten over a wide range of temperatures (Demura 2015a), and it is found that (9.42) reproduces the rise of the radiation loss at small temperatures as well as the minima and maxima in the emission (that corresponds to the shell structure) at intermediate temperatures and the decrease at very high temperature. As concerns the absolute values of radiation loss, it is found that reasonable agreement is obtained setting $g(s, u) \approx 2$. At low temperatures, however, the agreement is limited because the Corona model is not quite valid due to the proximity of levels in low charged ions. Finally, we note that in order to compare (9.42) with the detailed collisional–radiative modeling it is necessary, to sum up the emission from (9.42) for every charge state and multiply with the respective relative population (normalized to one) of this charge state.

9.3.2 Density Effects

The Corona approximation of the radiation loss (9.42) described above is valid in the low-density high-temperature limit. As density increases and temperature decreases, the Corona approximation becomes invalid and has to be replaced by the general collisional–radiative model described in Chap. 6. In order to establish more general expressions for the radiation loss in the framework of the local plasma frequency approximation, let us first consider the Boltzmann limit (i.e., high-density limit). In the Boltzmann limit, direct and inverse processes are related to the principle of detailed balance that contains in the local plasma frequency approximation an equivalent Boltzmann exponential factor with the plasma frequency, i.e., $\exp(\hbar\omega_{\mathrm{P}}/kT_{\mathrm{e}})$.

Let us begin with the radiative terms expressed via the Einstein coefficients and the Fermi equivalent photon method. The Einstein coefficients describing the emission probability in terms of the emission oscillator strengths f_{if} of particular transitions $i \to j$ are given by

$$A_{\mathrm{ij}} = -2\frac{e^2}{mc^3}\,\omega^2 f_{\mathrm{ij}}. \tag{9.43}$$

The oscillator strengths in emission $f_{\mathrm{if}} < 0$ and absorption $f_{\mathrm{ji}} > 0$ are connected by the well-known relation

$$-g_i f_{ij} = g_j f_{ji}. \tag{9.44}$$

We also need to evaluate the contribution of induced emission or absorption, using correspondingly the following relations for the Einstein coefficients of induced emission

$$B_{ij} = \frac{\pi^2 c^3}{\hbar\omega^3} A_{ij} \tag{9.45}$$

and induced absorption

$$B_{ji} = \frac{g_i}{g_j} \cdot \frac{\pi^2 c^3}{\hbar\omega^3} A_{ij}. \tag{9.46}$$

Let $N_{EQP}(\omega)$ the EQP number for a given frequency ω per unit frequency interval, $B_{ij} \cdot N_{EQP}(\omega)$ and $B_{ji} \cdot N_{EQP}(\omega)$ are the stimulated de-excitation and excitation rates corresponding to the EQP flux. Then from the equality for the direct and reverse processes, we obtain

$$B_{ij} N_{EQP}(\omega) N_i = B_{ji} N_{EQP}(\omega) N_j, \tag{9.47}$$

where N_i, N_j are the populations of levels i, j. For the Boltzmann distribution of level populations, i.e.,

$$N_i = N_j \exp[-\hbar\omega_{ij}/kT_e] \tag{9.48}$$

we obtain the following relation between direct and reverse processes

$$B_{ij} N_{EQP}(\omega) \exp[-\hbar\omega_{ij}/kT_e] = B_{ji} N_{EQP}(\omega). \tag{9.49}$$

On the other hand, the excitation rate under the influence of the EQP flux could be represented as the photoexcitation rate in terms of the photoabsorption cross section. The probability of induced radiation is determined by the product of the Einstein coefficient for induced radiation $B_{ij}(\omega)$ and the radiation energy density $U_\sigma(\omega)$ (erg/cm^3) with the polarization σ. It is connected with the integral over solid angles Ω of the radiation spectral intensity $I_\sigma(\omega, k)$ with polarization σ in the direction, determined by the wave vector $\vec{k}$ and divided by the speed of light c:

$$\begin{aligned} \int d\omega \cdot U_\sigma(\omega) \cdot B_{ij}(\omega) &= \int d\omega \cdot U_\sigma(\omega) \cdot \frac{\pi^2 c^3}{\hbar\omega^3} \cdot A_{ij}(\omega) \\ &= \int d\omega \left(\frac{1}{c}\int d\Omega\, I_\sigma(\omega, \vec{k})\right) \cdot \frac{\pi^2 c^3}{\hbar\omega^3} \cdot A_{ij}(\omega). \end{aligned} \tag{9.50}$$

Assuming that the electrons have an unpolarized isotropic Maxwellian velocity distribution, the intensity of the EQP flux in the frequency interval $ds = d\omega \cdot (\hbar/2Z_n \cdot Ry)$ (see (9.35)) produced in the elastic scattering of the electron flux $n_e^{(\text{free})} \cdot \upsilon_e$ (υ_e is the electron thermal velocity of the free electrons) by the Coulomb center and averaged over the electron energy distribution could be expressed as

$$\left[\int \mathrm{d}\Omega\, I_\sigma(\omega,\vec{k})\right] \cdot \mathrm{d}\omega = \left[\int \mathrm{d}\Omega\, I(\omega)\right] \cdot \mathrm{d}\omega$$
$$= \left[n_e^{(\text{free})} \cdot \frac{4a_0 c Ry^2}{\sqrt{3\pi} \cdot e^2} \left(\frac{Z_n^2 Ry}{kT_e}\right)^{1/2} \int\limits_{\frac{2RyZ_n}{kT_e} \cdot s}^{\infty} \mathrm{d}u\, \exp[-u]\, g(s,u) \right] \cdot \mathrm{d}s. \tag{9.51}$$

The photoabsorption cross section in the plasma model is given by (see also (3.135))

$$\sigma_{\text{abs}}(\omega) = \frac{2\pi^2 e^2}{m_e c} \int n_e(r) \cdot \delta(\omega - \omega_p(r)) \cdot \mathrm{d}^3 r. \tag{9.52}$$

The integral in (9.52) is the sum over the oscillator strengths of all transitions, while a separate transition could be represented through the differential $\mathrm{d}^3 r$ [see also comments related to (9.31)]:

$$\mathrm{d}\sigma_{\text{abs}}(\omega) = \frac{2\pi^2 e^2}{m_e c} n_e(r)\, \delta(\omega - \omega_p(r))\, \mathrm{d}^3 r. \tag{9.53}$$

Let us now derive the probability of EQP induced absorption within the LPF model with the help of (9.43)–(9.53). We first transform the Einstein coefficient for induced radiation to the LPF model with the help of (9.19), (9.33), (9.43)–(9.51):

$$B_{ij} = \frac{\pi^2 c^3}{\hbar\omega^3} A_{ij} = \frac{2\pi^2 e^2}{m_e \omega} n_e(r)\, \mathrm{d}^3 r \cdot \delta[\omega - \omega_p(r)] \cdot \mathrm{d}\omega. \tag{9.54}$$

From (9.54), we obtain the Einstein coefficient of spontaneous emission:

$$A_{ij}(\omega) = 2\, \frac{e^2}{m_e c^3}\, \omega^2\, n_e(r)\, \delta[\omega - \omega_p(r)]\, \mathrm{d}^3 r \cdot \mathrm{d}\omega. \tag{9.55}$$

Therewith all Einstein coefficients could be represented in the statistical model in terms of the integral operators like in (9.52)–(9.55).

Then, in the two-state approximation for each pair of levels, we derive a population density balance equation for arbitrary free electron density $n_e^{(\text{free})}$, equating the excitation rate via the EQP photoabsorption (from the lower state *j*) to the de-excitation rate via the spontaneous and EQP-induced radiative decay (from the upper state *i*):

$$N_j(r)\cdot\left[n_e^{(\mathrm{free})}\cdot\frac{2\pi^2a_0^3}{\sqrt{3\pi}}\left(\frac{Ry}{kT_e}\right)^{1/2}\frac{\omega_a}{\omega_p(r)}\int\limits_{u_{\min}(r)}^{\infty}e^{-u}\,g\left(s_p(r),u\right)\cdot du\right]$$

$$=N_i(r)\cdot\left[n_e^{(\mathrm{free})}\cdot\frac{2\pi^2a_0^3}{\sqrt{3\pi}}\left(\frac{Ry}{kT_e}\right)^{1/2}\frac{\omega_a}{\omega_p(r)}e^{u_{\min}(r)}\int\limits_{u_{\min}(r)}^{\infty}e^{-u}\,g\left(s_p(r),u\right)\cdot du\right.$$
$$\left.+\frac{2e^2}{m_ec^3}\cdot\frac{\omega_p^2(r)}{\omega_a}\right]$$

with

$$s_p(r)=\frac{\omega_p(r)}{Z_n\omega_a},\tag{9.57}$$

$$\omega_a=\frac{2Ry}{\hbar},\tag{9.58}$$

$$u_{\min}(r)=\frac{\hbar\omega_p(r)}{kT_e}.\tag{9.59}$$

Note that $n_e^{(\mathrm{free})}$ is the total electron density of the free electrons that are scattered by the atom while $n_e(r)$ is the bound atomic electron density. Then the excited state population N_i depends on density and is expressed via the lower state one N_j as follows

$$N_i(r)=N_j(r)\cdot\frac{\int_{u_{\min}(r)}^{\infty}e^{-u}g\left(s_p(r),u\right)du}{e^{u_{\min}(r)}\int_{u_{\min}(r)}^{\infty}e^{-u}g\left(s_p(r),u\right)du+\frac{\sqrt{3\pi kT_e}}{n_e^{(\mathrm{free})}\pi^2a_0^3\sqrt{Ry}}\cdot\frac{e^2}{m_ec^3}\cdot\frac{\omega_p^3(r)}{\omega_a^2}}.\tag{9.60}$$

From (9.60), we can see that it generalizes (9.48) for arbitrary densities. If the free electron density $n_e^{(\mathrm{free})}$ is very large, the second term in the denominator of (9.60) is very small compared with the first term and we obtain

$$N_i(r)=N_j(r)\cdot\exp(-u_{\min}(r))=N_j(r)\cdot\exp\left(-\frac{\hbar\omega_p(r)}{kT_e}\right).\tag{9.61}$$

According to (9.19), the plasma frequency depends on the atomic density; therefore, according to (9.61) the Boltzmann relation $N_i(r)/N_j(r)$ of levels depends on radius. The radiation losses due to the transitions $i\rightarrow j$ (corresponding to the local emission in the LPF model) can then be presented as an integral over frequencies using (9.55):

$$\begin{aligned}\int Q_{ij}(\omega, r)\cdot \mathrm{d}\omega &= \int \hbar\omega \cdot N_i(r)\cdot A_{ij}(\omega, r)\cdot \mathrm{d}\omega \\ &= \frac{2\hbar e^2}{m_e c^3}\cdot N_i(r)\cdot \omega_p^3(r)\cdot n_e(r)\cdot \mathrm{d}^3 r.\end{aligned} \tag{9.62}$$

In order to obtain the radiation loss for one ion with nuclear charge Z_n and charge state Z, we express the excited state density in terms of the ground state density to relate the radiative emission to a certain number of ions in the spherical shell at radius r with thickness $\mathrm{d}r$. Assuming that the excited state densities are negligible compared to the ground state densities, we can assume that $\sum_j N_j \approx N_Z$ where N_Z is the total density of ions with nuclear charge Z_n and charge state Z. Next, the total radiation loss $Q(Z_n, Z)$ is obtained from the sum of the contributions from all possible transitions. In the statistical model, the summation over contributions from the different transitions $i \rightarrow j$ consists in summing over the level populations N_j of different levels and integration over $\mathrm{d}^3 r$. Therefore, the total radiation losses $Q(Z_n, Z)$ in the framework of an effective two-state approximation plasma model and Coulomb center effective charge take the form:

$$Q(Z_n, Z) = \frac{2\hbar e^2}{m_e c^3}\cdot \sum_i \int_0^{R_{\mathrm{atom}}} N_i(r)\cdot \omega_p^3(r)\cdot n_e(r)\cdot \mathrm{d}^3 r. \tag{9.63}$$

Substituting (9.60) into (9.62) and switching to the Thomas–Fermi dimensionless reduced radius x and expressing $n_e(x)$ and $\omega_p(x)$ via the Thomas–Fermi function $\varphi(x, q)$ (e.g., (9.5)), we obtain a generalized analytical formula for the total radiation loss per ion for arbitrary free electron density $n_e^{(\mathrm{free})}$ and electron temperature T_e:

$$\frac{Q(Z_n, Z)}{N_Z \cdot n_e^{(\mathrm{free})}} = \frac{4RyZ^3\omega_a}{n_e^{(\mathrm{free})}}\left(\frac{e^2}{\hbar c}\right)^3\left(\frac{128}{9\pi^2}\right)^{3/2}\cdot \int_0^{x_0(q)} x^2 \cdot \left(\frac{\varphi(x, q)}{x}\right)^{15/4}\cdot G(x)\cdot R(x)\,\mathrm{d}x \tag{9.64}$$

with

$$G(x) = \int_{u_{\min}(x)}^{\infty} e^{-u}\cdot g(x, u)\cdot \mathrm{d}u, \tag{9.65}$$

$$R(x) = \frac{1}{\int_{u_{\min}(x)}^{\infty} e^{u_{\min}(x)-u} \cdot g(x,u) \cdot du + D\left(x, n_e^{(\text{free})}, T_e, Z_n, q\right)}, \tag{9.66}$$

$$D\left(x, n_e^{(\text{free})}, T_e, Z_n, q\right) = \frac{1}{n_e^{(\text{free})}} \cdot \left(\frac{128}{9\pi^2}\right)^{3/2} \cdot \left(\frac{e^2}{\hbar c}\right)^3 \cdot \frac{Z_n^3}{\pi^2 a_0^3} \cdot \sqrt{\frac{3\pi k T_e}{Ry}} \cdot \left(\frac{\varphi(x,q)}{x}\right)^{9/4}, \tag{9.67}$$

$$g(x,u) = g\left(\sqrt{\frac{Z_{\text{eff}}^2 Ry}{kT_e}} \cdot \frac{u_{\min}(x)}{2u^{3/2}}\right), \tag{9.68}$$

$$u_{\min}(x) = Z_n \cdot \left(\frac{128}{9\pi^2}\right)^{1/2} \cdot \left(\frac{\varphi(x,q)}{x}\right)^{3/4} \cdot \left(\frac{2Ry}{kT_e}\right). \tag{9.69}$$

Note that in (9.64)–(9.69), we have employed for consistency the Thomas–Fermi model for the ionization potential (9.13). In the limit of low free electron densities, (9.64)–(9.69) reproduce the result for the corona equilibrium; while in the opposite limit of high densities, the result corresponds to the Boltzmann distribution of atomic level populations.

The radiation losses of tungsten ions calculated from (9.64)–(9.69) are presented in Fig. 9.1 for different values of free electron density: 10^{14}, 10^{16}, 10^{18}, 10^{20} cm^{-3}. It is seen in Fig. 9.1 that these radiation losses are strongly suppressed with increase of plasma density in the region of low temperatures. This actually corresponds to the Boltzmann distribution of populations of excited states. When the second term

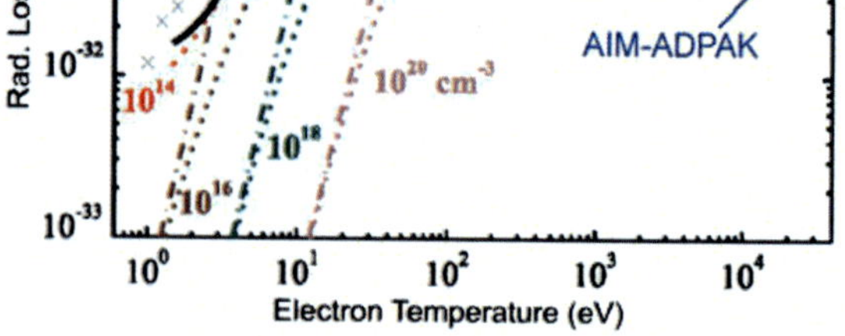

Fig. 9.1 Comparison of radiation losses from tungsten ions within the universal statistical approach with the numerical data from the AIM-ADPAK and ADPAK models

$D\left(x, n_e^{(\text{free})}, T_e, Z_n, q\right)$ (9.67) in $R(x)$ from (9.66) (that corresponds to spontaneous emission), is much larger than the first one (corresponding to collisional de-excitation), we return to the results of the corona limit, shown in Fig. 9.1 by the solid black curve. In the opposite limit, when the de-excitation rate is much larger than the rate of spontaneous emission, (9.62)–(9.67) take the form:

$$\frac{Q(Z_n, Z)}{N_Z \cdot n_e^{(\text{free})}} = \frac{4RyZ^3\omega_a}{n_e^{(\text{free})}} \left(\frac{e^2}{\hbar c}\right)^3 \left(\frac{128}{9\pi^2}\right)^{3/2} \cdot \int_0^{x_0(q)} x^2 \cdot \left(\frac{\varphi(x,q)}{x}\right)^{15/4} \cdot e^{-u_{\min}(x)}\, dx. \tag{9.70}$$

This limit in the statistical approach is represented in Fig. 9.1 by "dashed-dot" curves for various densities 10^{14}, 10^{16}, 10^{18}, 10^{20} cm^{-3}. Such dependence corresponds to a near exponential increase of Q in the Boltzmann limit versus increasing electron temperature T_e. Asymptotically the "dotted" curves approach the Boltzmann limit at lower temperatures and the coronal limit at higher temperatures. This behavior is physically transparent. At low temperatures, the collisional de-excitation rate coefficients are rather high. Then, with density increase the de-excitation rates become larger than the spontaneous radiative decay, establishing the Boltzmann type of equilibrium. On the other hand, for large temperatures, the de-excitation rate coefficients decrease and then, with sufficiently low densities, the de-excitation rates become smaller than the spontaneous radiative decay rate. In this way, the coronal distribution of atomic populations is restored.

Let us now compare the results for the radiation loss calculated with the statistical approach with the detailed collisional–radiative model (see Chap. 6). Figure 9.1 depicts the results from collisional–radiative model calculations for tungsten (Kogan et al. 1992; Summers 1994; Post et al. 1977, 1995), blue crosses and solid blue curve indicated as "AIM-ADPAK". As already discussed in relation with the corona approximation of (9.42), the comparison of the total radiation losses with different models requests to sum up the line emission for every charge state. This implies a need for the calculation of the ionic charge state distribution (see also Chap. 6) to properly weight the radiation emission for each charge stage. In order to compare the statistical model of (9.64)–(9.69) with detailed collisional–radiative models, we employ the charge state distribution proposed by Post (1977). It can be seen that the corona limit of the universal statistical approach (9.64)–(9.69) provides a reasonable approximation of the radiation losses within a factor of two in the temperature interval from about 50–30.000 eV. For lower temperatures, the emission peak near about 20 eV is not well reproduced. In this temperature range, the atomic configurations are very complex and important deviations from the corona model are encountered as discussed in relation with (9.70). It should be noted that the precise determination of the ionic populations and the total loss rates for heavy elements in dependence of temperature and density is a very complex problem that is controversy discussed up to present days in particular if M- and N-shells are involved (Beiersdorfer et al. 2012; Scott and Hansen 2010; Piron et al. 2017). For example, average charge states $\langle Z \rangle$ of gold at about 1 keV temperature at densities around 10^{21} cm^{-3} may differ as much as 10 for different calculations (Scott and Hansen 2010) while it should be remembered that a change in average charge state of only $\langle \Delta Z \rangle = 1$ results in an entirely different spectral distribution. Moreover, high precision experiments are likewise very difficult, as independent temperature, density, and charge state measurements are requested. For demonstration, Fig. 9.1 contains the experimental measurement of the radiation loss of tungsten (Pütterich et al. 2010). The experimental error bars are of the same order as the differences of different model calculations (Demura et al. 2015a; Pütterich et al. 2010).

Radiation loss related to radiative recombination and dielectronic recombination is much lower than the total radiation loss in the depicted temperature interval (Pütterich et al. 2010). Although the direct radiation loss related to dielectronic recombination is not very important for the total radiation loss, it is very important for the correct account of the ionization charge state distribution that in turn influences on the total radiation loss.

The statistical model for arbitrary density according to (9.64)–(9.69) represents therefore a useful approximation of the radiation loss for each charge state. Due to the Thomas–Fermi model approximation, (9.64)–(9.69) provide a generalized unified analytical description of the radiation loss for any heavy ion and allow to study in a transparent manner several physical properties of the line radiation loss. However, currently there exists no well-established theory to extend the statistical model to a generalized analytical description that includes also the ionic charge state distribution (resulting in a self-consistent total radiation loss calculation). Below, we discuss their essential ingredients, namely the statistical approach to ionization and dielectronic recombination.

Finally we note that for very high densities, the Thomas–Fermi atom size can become comparable of the ionic interparticle distance, such that the boundary condition at the periphery of the atomic electron density distribution could change. This would change the behavior of the distribution itself. This effect is strongly related to the ionization potential depression discussed in Chap. 8 and Annex A.4 and not specific to the statistical model but related to all types of collisional–radiative modeling.

9.4 Statistical Ionization Cross Sections and Rates

We now show that the statistical approach allows one to obtain the expressions for the total electron impact single ionization cross sections of multielectron ions and related ionization rates. Indeed, instead of the intensity of equivalent photons we can operate with the number of equivalent photons $\frac{\mathrm{d}N(\omega)}{\mathrm{d}(\omega/\omega_{\mathrm{a}})}$ at given frequency ω

in the unit frequency interval $d\omega$ in unit time interval at the fixed incident electron energy E, which in the dimensionless form could be determined by the expression (compare with (9.40), (9.41))

$$\frac{dN(\omega)}{d(\omega/\omega_a)} = \left(\frac{c\hbar}{e^2}\right) \cdot \frac{1}{2\sqrt{3}} \cdot \frac{\omega_a}{\omega} \cdot \left(\frac{Ry}{E}\right) \cdot g\left(Z, \left(\frac{Ry}{E}\right)^{3/2}, \frac{\omega}{\omega_a}, Z_n\right). \tag{9.71}$$

Multiplying the number of EQP (9.51) by the photoionization cross section and integrating over s from the reduced ionization potential $I_Z/(2RyZ_n)$ up to the reduced energy of electron projectile $E/(2RyZ_n)$, we arrive in the Coulomb approximation (9.37), (9.38) to the following expression for the electron ionization cross section

$$\sigma_i(E) = \frac{\pi^4 a_0^2 \sqrt{3}}{32} \cdot \frac{Ry}{E} \cdot \int\limits_{I_i/(2RyZ_n)}^{E/(2RyZ_n)} g\left[Z_{eff} Z_n \left(\frac{Ry}{E}\right)^{3/2} \cdot s\right] \cdot \frac{x_s^2 \varphi(x_s, q) \cdot ds}{|\varphi'(x_s, q) - \varphi(x_s, q)/x_s|}. \tag{9.72}$$

The corresponding statistical ionization rates are obtained by averaging the ionization cross section of (9.72) over the Maxwellian energy distribution. This can conveniently be performed by changing the sequence of integration and firstly integrate over energies of the incident plasma electrons. This average concerns in fact only the total flux of EQP number $n_e^{eff} \upsilon_e \frac{dN(\omega)}{d\omega}$ ($n_e^{(free)}$ is the plasma electron density, and υ_e is the corresponding thermal electron velocity). Then, the ionization rate could be expressed as

$$R_i(Z_n, q, T_e) = n_e^{(free)} \frac{a_0^3 \omega_a \pi^3 \sqrt{3\pi}}{16} \sqrt{\frac{Ry}{kT_e}} \cdot \left\{ \int\limits_{\frac{I_Z}{2Z_n Ry}}^{\infty} \frac{x_s^2 \varphi(x_s, q) \cdot ds}{|\varphi'(x_s, q) - \varphi(x_s, q)/x_s|} \right\} \cdot \left\{ \int\limits_{\frac{2Z_n Ry}{kT_e} \cdot s}^{\infty} e^{-u} \cdot g(v) \cdot du \right\}, \tag{9.73}$$

where v is defined by (9.37).

Figure 9.2 shows the ionization cross section for an open 4*f*-shell of a heavy element, the tungsten W^{9+} ion, calculated with the statistical model of (9.72). Also presented in Fig. 9.2 are the experimental data from (Stenke et al. 1995) and relativistic excitation–autoionization distorted wave calculations (Loch et al. 2005).

Figure 9.2 demonstrates that the statistical model describes the experimental data within a factor of two from threshold to high energies. The threshold for heavy elements requests particular attention because excitation–autoionization as well as excitation from metastable levels is of importance. It is very cumbersome to include all necessary excitation–autoionization channels in a fully quantum approach, as branching factors (for radiative and autoionization decay) have to be involved. In this respect, the statistical model is very convenient because the statistical model includes both direct and indirect contributions to ionization. Moreover, it is likewise difficult in experiment, to measure pure direct excitation cross sections because the electron beam results likewise in the excitation of metastable levels (highly populated) from which ionization can also proceed.

Let us now consider ionization rates from open 4*d*-shell and open 4*p*-shell configurations of Xe^{12+} and Xe^{20+}, respectively, and compare the statistical ionization rate coefficients of (9.73) with quantum mechanical calculations in the Coulomb–Born exchange approximation of the direct ionization rate using Vainshtein's ATOM code (Vainshtein and Shevelko 1986; Sobelman and Vainshtein 2006; Povyshev et al. 2001).

As can be seen from Fig. 9.3, also quite good agreement between the statistical model and the quantum calculation are obtained. For low electron temperatures, i.e., when the effective ionization cross sections are near threshold, the quantum cross sections provide systematically lower rates than the statistical model because the depicted quantum calculations do include only the direct cross section.

Numerous comparisons with experimental data and different methods of rather complex quantum mechanical ionization cross sections have been performed (Demura et al. 2015b) and it is found that the statistical approach to ionization cross sections constitutes an efficient and rather precise method for heavy elements and permits easily inclusion of direct and indirect ionization contributions.

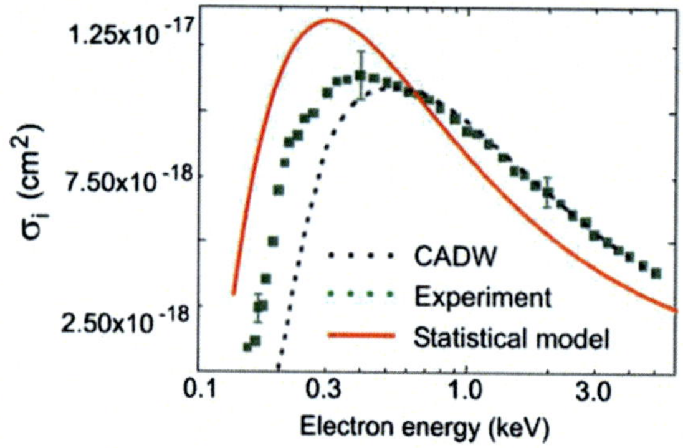

Fig. 9.2 Comparison of the ionization cross section of tungsten W^{9+} ions with experimental data (Stenke et al. 1995) and complex collisional excitation–autoionization distorted wave calculations (Loch et al. 2005)

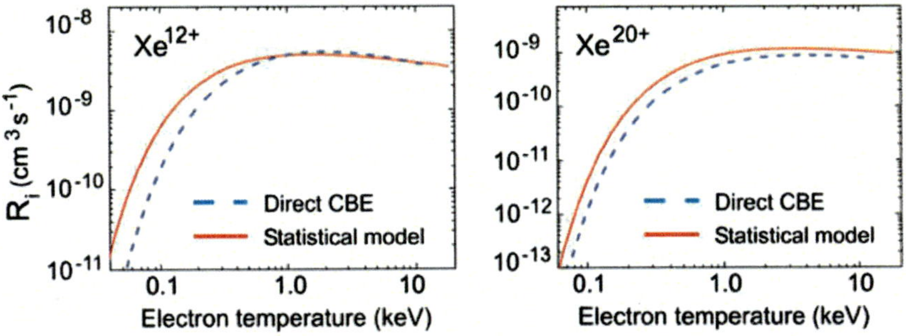

Fig. 9.3 Comparison of the ionization cross section of Xeon ions with the quantum mechanical calculations in Coulomb–Born exchange approximation for the direct ionization rate

9.5 Statistical Dielectronic Recombination Rates

Dielectronic recombination (see also Chap. 5 and review (Rosmej et al. 2020b)) is the most effective recombination channel in electron–heavy ion collisions. Due to the complex electronic structure of multielectron ions, the proper account of all necessary channels is a very difficult task, in particular for open shell configurations. In addition, in dense plasmas, dielectronic capture might effectively proceed from excited states (see also Chap. 5) that considerably increases the quantum channels for the dielectronic capture. Moreover, in heavy ions, numerous metastable states may play the role of excited states even in rather low-density plasmas, thereby increasing the numerical complexity of fully quantum calculations considerably. At present, the dielectronic recombination of heavy ions is still under controversial discussion and is one of the main sources of discrepancies between different methods of calculations for radiation loss and ionic charge state distributions. It is therefore of great interest to develop different methods for the calculation of the dielectronic recombination in heavy ions that permit more general studies including the analysis of scaling laws. As has been demonstrated in the foregoing paragraphs, the statistical approach provides reasonable approximations not only for the atomic structure, but for the calculations of elementary processes too. We are therefore seeking to extend the approach of the local plasma frequency approximation also to the dielectronic recombination.

9.5.1 General Formula

The general formula for the total dielectronic recombination rates DR can be written as Sobelman and Vainshtein (2006)

$$Q_{\mathrm{DR}}(T) = \left(\frac{4\pi Ry}{kT_{\mathrm{e}}}\right)^{3/2} a_0^3 \frac{g_{\mathrm{f}}}{g_{\mathrm{i}}} \cdot W_{\mathrm{R}} \cdot \sum_{n,l} \left\{ \frac{W_{\mathrm{A}}(n,l)}{W_{\mathrm{R}} + W_{\mathrm{A}}(n,l)} \cdot \exp\left(-\frac{\hbar\omega}{kT_{\mathrm{e}}} + \frac{Z_{\mathrm{i}}^2 Ry}{n^2 kT_{\mathrm{e}}} \right) \right\}, \tag{9.74}$$

where kT_{e} is the electronic temperature in [eV], $Ry = 13.61\,\mathrm{eV}$, g_{i} and g_{f} are the statistical weights of the initial and final states of the atomic core, respectively. W_{R} is the radiative transition probability inside the core, W_{A} is the autoionization decay rate of an excited atomic energy level, $\hbar\omega$ is the transition energy inside the core, Z_{i} is the ion charge, a_0 is the Bohr radius, and n, l are the principle and orbital quantum numbers of the captured electron, respectively. The radiative decay rate is expressed simply in terms of the oscillator strength f_{ij} for the transition inside the core (c is the speed of light):

$$W_{\mathrm{R}} = \frac{2\omega^2 f_{\mathrm{if}}}{c^3}. \tag{9.75}$$

In order to obtain the expression for the autoionization decay rate $W_{\mathrm{A}}(n,l)$, we use a relationship between the decay rate $W_{\mathrm{A}}(n,l)$ and the partial electron excitation cross section $\sigma_{\mathrm{ex}}(l)$ at threshold in the semiclassical representation. The quantities $W_{\mathrm{A}}(n,l)$ and $\sigma_{\mathrm{ex}}(l)$ describe the mutually inverse processes, so the relationship between them can be obtained from the detailed balance (see also Sect. 7.7.2) between ions $X^{Z_{\mathrm{i}}+1}$ and $X^{Z_{\mathrm{i}}}$. Thus, we obtain

$$(2l+1) g_{\mathrm{f}} W_{\mathrm{A}}(n,l) = \frac{Z_i^2}{n^3} \omega g_{\mathrm{i}} \frac{\sigma_{\mathrm{ex}}(l)}{\pi^2 a_0^2}. \tag{9.76}$$

The electron excitation cross section in the semiclassical approximation takes the form

$$\sigma_{\mathrm{exc}}(l) = \frac{8\pi}{3} \left(\frac{\hbar}{m_e v_{\mathrm{e}}}\right)^2 \frac{g_{\mathrm{f}}}{g_{\mathrm{i}}} f_{\mathrm{if}} Z_{\mathrm{i}}^{-2} (l+1/2)^2 G\left(\frac{\omega(l+1/2)^3}{3Z_{\mathrm{i}}^2}\right), \tag{9.77}$$

where the function $G(u)$ is given by ($K_{1/2}$ and $K_{3/2}$ are the Mcdonald functions)

$$G(u) = u \cdot \left(K_{1/3}^2(u) + K_{2/3}^2(u)\right). \tag{9.78}$$

Taking into account that the essential values of the argument of the $G(u)$ function are never close to zero, it is possible to replace $G(u)$ by its asymptotic expansion:

$$G(u) \approx 3.4 \cdot \exp(-2u). \tag{9.79}$$

Within these approximations, the autoionizing decay rate takes the form:

$$W_A(n,l) = 0.72 \cdot \frac{\omega(l+0.5)f_{ij}}{n^3} \cdot \exp\left(-\frac{2\omega(l+0.5)^3}{3Z_i^2}\right). \tag{9.80}$$

The sum of the oscillator strengths satisfies the Thomas–Reiche–Kuhn sum rule, i.e.,

$$N_e^{bound} = \sum_f f_{if}. \tag{9.81}$$

As discussed in the foregoing paragraphs, the oscillator strengths in the statistical model are expressed in terms of the atomic electron density $n_e(r,q,Z_n)$ (9.32) and the statistical sum rule is given by (N_e^{bound} is the total number of bound electrons)

$$N_e^{bound} = \int n_e(r,q,Z_n)dV. \tag{9.82}$$

The application of the semiclassical statistical model to the general formula (9.72) for the total DR is achieved by application of the relationships

$$\sum_f f_{if} \to \int_0^{r_0} dr \cdot 4\pi r^2 n_e(r,q,Z_n) \tag{9.83}$$

and

$$E_{if} \to \omega = \sqrt{4\pi n_e(r,q,Z_n)}. \tag{9.84}$$

After all substitutions, we obtain for the DR rates:

$$Q_{DR}(cm^3/s) = 0.61 \times 10^{-8} \cdot Q_{DR}(a.u), \tag{9.85}$$

$$Q_{DR}(a.u.) = \frac{54.5}{T_e^{3/2}}\left(\frac{Z_n}{Z_i}\right)^2 \cdot \int_0^{x_0} dx\, x^2 \left(\frac{\varphi(x)}{x}\right)^{9/4} \times \int_1^{\infty} dt\, \exp\left(-\frac{\omega(x)}{T_e}\left(1-\frac{1}{t^2}\right)\right) \int_0^{l_{max}=t \cdot n_1 - 1} dl \frac{(l+0.5)\exp\left[-2\omega(x)(l+0.5)^3/3Z_i^2\right]}{t^3 + A(x,l)}, \tag{9.86}$$

$$A(x,l) = 5.2 \times 10^6 (l+0.5) \frac{\exp\left[-2\omega(x)(l+0.5)^3/3Z_i^2\right]}{Z_i^3\sqrt{\omega(x)}}, \tag{9.87}$$

$$\omega(x) = 1.2 \cdot Z_n \cdot \left(\frac{\varphi(x)}{x}\right)^{3/4} \tag{9.88}$$

with $T_e(a.u.) = T_e(\mathrm{eV})/27.21$ and $t = n/n_1$, where n_1 is a minimal possible quantum number. n_1 is the lowest level, on which electron capture is possible and corresponds to an energy of an incident electron E_i

$$E_i = \omega - \frac{Z_i^2}{2n^2} \tag{9.89}$$

that is equal to zero, i.e.,

$$0 = \omega - \frac{Z_i^2}{2n_1^2}. \tag{9.90}$$

From (9.90), it follows

$$n_1 = \frac{Z_i}{\sqrt{2\omega}}. \tag{9.91}$$

9.5.2 Orbital Quantum Number Averaged Dielectronic Recombination Rates

In the simplest version of the statistical model, the atomic density, excitation energies, and oscillator strengths do not depend on the orbital momentum quantum number l. If we average the branching factor over orbital momentum we obtain for the total dielectronic recombination rate:

$$Q_{\mathrm{DR}}(a.u.) = \frac{39.2}{T_e^{3/2}}\left(\frac{Z_n}{Z_i}\right)^2 \frac{Z_n}{Z_i^2} \int_0^{x_0} \mathrm{d}x \cdot x^2 \left(\frac{\varphi(x)}{x}\right)^3 \int_1^{\infty} \frac{\mathrm{d}t}{t^2} \exp\left(-\frac{\omega(x)}{T}\left(1-\frac{1}{t^2}\right)\right)$$
$$\times \int_0^{l_{\max}=t \cdot n_1 - 1} \mathrm{d}l \frac{(l+0.5)^2 \exp\left[-2\omega(x)(l+0.5)^3/3Z_i^2\right]}{t^3 + A(x,l)}, \tag{9.92}$$

where the function $A(x, l)$ is given by (9.87), (9.88). Instead of averaging over the branching factor, we may investigate averaging the autoionization decay rate [3] from (9.80) over orbital quantum number, i.e.,

$$\langle W_{\mathrm{A}}(n,l)\rangle = 1.7\frac{f_{\mathrm{if}}Z_{\mathrm{i}}^{2}}{\pi n^{5}\omega}. \tag{9.93}$$

For the corresponding total dielectronic recombination rate, we then obtain

$$Q_{\mathrm{DR}}(a.u.) = \frac{0.86\times 10^{2}}{T_{\mathrm{e}}^{3/2}}\left(\frac{Z_{\mathrm{n}}}{Z_{\mathrm{i}}}\right)^{2}\int_{0}^{x_0}\mathrm{d}x\cdot x^{2}\left(\frac{\varphi(x,q)}{x}\right)^{9/4} \times \int_{1}^{\infty}\mathrm{d}t\frac{\exp\left[-\frac{1.2Z}{T_{\mathrm{e}}}\left(\frac{\varphi(x,q)}{x}\right)^{3/4}\left(1-\frac{1}{t^{2}}\right)\right]}{t^{5}+A(x)}, \tag{9.94}$$

$$A(x) = \frac{4.56\times 10^{6}}{Z_{\mathrm{i}}^{3}\sqrt{Z_{\mathrm{n}}}\left(\frac{\varphi(x,q)}{x}\right)^{3/8}}. \tag{9.95}$$

9.5.3 *Statistical Burgess Formula*

It is of great interest to apply the statistical approach to the Burgess-Mertz formula (Burgess 1964; Cowan 1981) for dielectronic recombination because this formula is widely employed and cast into an entirely analytical expression (z is the spectroscopic symbol $z = Z_n - N_e^{(\mathrm{bound})} + 1$, while $Z_i = Z_n - N_e^{(\mathrm{bound})}$ is the ion charge, see also eqs. (5.138)–(5.143)):

$$Q_{\mathrm{DR}}\ (\mathrm{cm}^{3}/\mathrm{s}) = 10^{-13}B_{\mathrm{d}}\cdot\beta^{3/2}\cdot e^{-\beta\cdot\chi_{\mathrm{d}}}, \tag{9.96}$$

$$B_{\mathrm{d}} = 480\,f_{\mathrm{if}}\left[\frac{z\chi}{z^{2}+13.4}\right]^{3/2}\left[1+0.105(z+1)\chi+0.015(z+1)^{2}\chi^{2}\right]^{-1}, \tag{9.97}$$

$$\beta = \frac{(z+1)^{2}Ry}{kT_{\mathrm{e}}}, \tag{9.98}$$

$$\chi_{\mathrm{d}} = \frac{\chi}{1+0.015\frac{z^{3}}{(z+1)^{2}}}, \tag{9.99}$$

$$\chi = \frac{E_{\mathrm{if}}}{(z+1)^{2}Ry}. \tag{9.100}$$

The statistical version of the Burgess-Mertz formula is obtained, going from the oscillator strength in (9.97) to the electron density and from the energy difference in (9.100) to the plasma frequency employing (9.83) and (9.84), respectively. After transformation to dimensionless variables, we obtain:

$$Q_{\mathrm{DR}}(\mathrm{cm}^3/\mathrm{s}) = 10^{-13}\beta^{3/2}\int_0^{x_0} \mathrm{d}x\cdot x^2 B_{\mathrm{d}}(x)e^{-\beta\chi_{\mathrm{d}}(x)}, \tag{9.101}$$

$$B_{\mathrm{d}}(x) = \frac{135\pi^3}{Z_n} n(x,q,Z_n)\left[\frac{z\chi(x)}{z^2+13.4}\right]^{1/2}\frac{1}{A(x)}, \tag{9.102}$$

$$A(x) = 1 + 0.105(z+1)\chi(x) + 0.015(z+1)^2\chi^2(x), \tag{9.103}$$

$$\chi(x) = \frac{2}{(z+1)^2}\sqrt{4\pi n_{\mathrm{e}}(x,q,Z)}, \tag{9.104}$$

$$\chi_{\mathrm{d}}(x) = \frac{\chi(x)}{1+0.015\dfrac{z^3}{(z+1)^2}}. \tag{9.105}$$

9.5.4 Statistical Vainshtein Formula

Several improvements to the Burgess formula have been proposed in the literature (see also Chap. 5). In this respect, it should be remembered that the Burgess formula is of interest due to its generality and great simplicity. The main drawback of the Burgess formula is the single channel approach that could result in considerable overestimation of the total dielectronic recombination rate (see Sect. 5.6.2.1). The multichannel approach requests usually complex atomic structure calculations that are very difficult to realize for heavy atoms. One of the most efficient general improvements of the Burgess formula including a simplified multichannel approach has been proposed by Vainshtein (Beigman et al. 1981; Sobelman and Vainshtein 2006) (see also Chap. 5). In the single channel approach, the Vainshtein formula can be summarized as follows:

$$Q_{\mathrm{DR}}\ (\mathrm{cm}^3/s) = 10^{-13}B_{\mathrm{d}}\beta^{3/2}e^{-\dfrac{\Delta E_{\mathrm{fi}}}{T}}, \tag{9.106}$$

$$B_{\mathrm{d}} = C'\frac{z}{n_1^4}f_{\mathrm{if}}\sum_{n>n_1} e^{\delta\beta}\sum_{l<n}\frac{2l+1}{1+\left(\dfrac{n}{n_{\mathrm{s}}}\right)^3}, \tag{9.107}$$

$$n_s = n_s(l) = 137\left[\frac{n_1^2\sigma_{\mathrm{if}}(l)}{\pi^2 a_0^2(2l+1)f_{\mathrm{if}}}\right]^{1/3}, \tag{9.108}$$

$$\delta\beta = \frac{z^2 Ry}{n^2 kT_e}, \tag{9.109}$$

$$C' = 10^{13}\frac{4\pi^{3/2}a_0\hbar}{(137)^3 m_e}\left(\frac{z}{z+1}\right)^3 = 0.53\left(\frac{z}{z+1}\right)^3, \tag{9.110}$$

where ΔE_{fi} is the energy difference between initial i and final f levels, and $\sigma_{\mathrm{if}}(l)$ is the partial excitation cross section at threshold. Implementation of the multichannel approach results into a modification of the B_{d} factor, i.e.,

$$B_{\mathrm{d}}^{(\mathrm{multi-channel})} = C'\frac{z}{n_1^4}f_{\mathrm{if}}\sum_{n>n_1} e^{\delta\beta}\sum_{l<n}\frac{2l+1}{B+\left(\dfrac{n}{n_s}\right)^3} \tag{9.111}$$

with

$$B = \sum_{f'}\frac{\Delta E_{\mathrm{ff'}}}{\Delta E_{\mathrm{fi}}}\cdot\frac{g_{\mathrm{f'}}}{g_{\mathrm{i}}}\cdot\frac{\sigma_{\mathrm{f'f}}(l)}{\sigma_{\mathrm{if}}(l)} \tag{9.112}$$

that results into an effective reduction of the branching factor (Beigman et al. 1981). Undertaking the substitutions of (9.83), (9.84) and replacing the sum over nl by integrations, we obtain from (9.106)–(9.110):

$$Q_{\mathrm{DR}}(a.u.) = \frac{13.6}{T_e^{3/2}}\left(\frac{Z_n}{Z_i}\right)^2\int_0^{x_0}\mathrm{d}x\cdot x^2\left(\frac{\varphi(x)}{x}\right)^{9/4}\int_1^{\infty}\mathrm{d}t\,\exp\left(-\frac{\omega(x)}{T_e}\left(1-\frac{1}{t^2}\right)\right)$$
$$\times\int_0^{l_{\max}=t\cdot n_1-1}\mathrm{d}l\frac{(l+0.5)^2\exp\left(-\dfrac{2\omega(x)(l+0.5)^3}{3Z_i^2}\right)}{t^3+A(x,l)}, \tag{9.113}$$

$$A(x,l) = 1.3\times 10^6(l+0.5)\frac{\exp\left[-2\omega(x)(l+0.5)^3/3Z_i^2\right]}{Z_i^3\sqrt{\omega(x)}}, \tag{9.114}$$

$$\omega(x) = 1.2Z_n\left(\frac{\varphi(x)}{x}\right)^{3/4}. \tag{9.115}$$

If we average (9.102)–(9.114) over the orbital angular electron momentum l and replace the sum over l by an integration, we obtain:

$$Q_{\mathrm{DR}}(a.u.) = \frac{98.1}{T_{\mathrm{e}}^{3/2}} \left(\frac{Z_{\mathrm{n}}}{Z_{\mathrm{i}}}\right)^2 \frac{Z_{\mathrm{n}}}{Z_{\mathrm{i}}^2} \int_0^{x_0} \mathrm{d}x \cdot x^2 \left(\frac{\varphi(x)}{x}\right)^3 \int_1^{\infty} \frac{\mathrm{d}t}{t^2} \exp\left(-\frac{\omega(x)}{T}\left(1-\frac{1}{t^2}\right)\right)$$
$$\times \int_0^{l_{\max}=t \cdot n_1 - 1} \mathrm{d}l \frac{(l+0.5)^3 \exp\left(-\frac{2\omega(x)(l+0.5)^3}{3Z_{\mathrm{i}}^2}\right)}{t^3 + A(x,l)}, \tag{9.116}$$

$$A(x,l) = 5.2 \times 10^6 (l+0.5) \frac{\exp\left[-2\omega(x)(l+0.5)^3/3Z_{\mathrm{i}}^2\right]}{Z_{\mathrm{i}}^3 \sqrt{\omega(x)}}, \tag{9.117}$$

$$\omega(x) = 1.2 Z_n \left(\frac{\varphi(x)}{x}\right)^{3/4}. \tag{9.118}$$

9.5.5 Numerical Comparison of Different Dielectronic Recombination Models

For heavy ions, the quantum mechanical level-by-level calculations are very complex and have so far mainly been carried out for closed shell configurations. Only recently, also open shell configurations have been considered (Balance et al. 2010; Wu et al. 2015). In open shell configuration (such as the open $4p$-, $4d$-, $4f$-shells or even higher ones like the $5p$-, $5d$-, $5f$-, $5g$-shells), excitation–autoionization channels are very complex and the overall completeness of quantum mechanical level-by-level calculations should still be considered with care. The analysis shows that order of magnitude disagreements can be excepted for low temperatures while for high temperatures, different level-by-level quantum mechanical models differ by about a factor of 2 while the Burgess-Mertz approach may deviate by many orders of magnitude providing also an entirely inadequate temperature dependence (Behar et al. 1996).

Let us compare first the various statistical approaches with detailed quantum level-by-level calculations. Figure 9.4 shows statistical dielectronic recombination rates nl-resolved according to (9.85)–(9.88), curve designated as "Statistical nl"; the statistical DR rate with orbital average of the branching factor according to (9.92), curve designated as "Statistical n"; the statistical approach with orbital average of the autoionization rate according to (9.94), (9.95), curve designated as "Statistical $W(n)$"; the Burgess DR rate according to (9.96)–(9.100), (9.101)–(9.105), curve

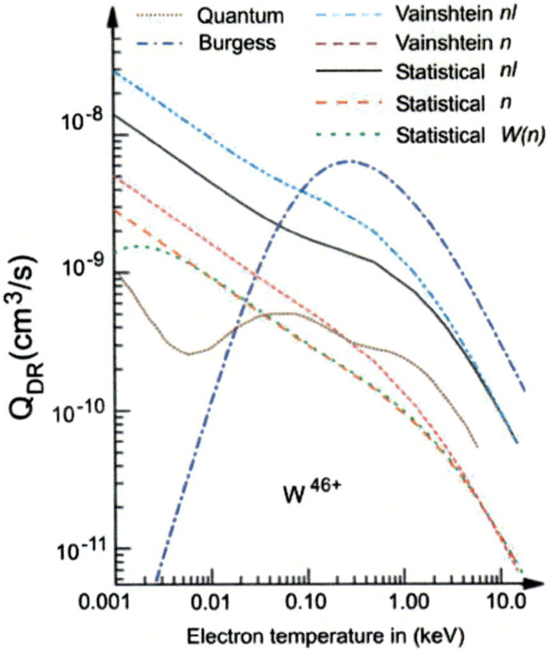

Fig. 9.4 Comparison of different statistical approaches with quantum level-by-level calculations for the Ni-like sequence $3s^23p^63d^{10}$ of tungsten W^{46+}

designated as "Burgess"; the Vainshtein approach according to (9.106)–(9.110), (9.113)–(9.115), curve designated as "Vainshtein *nl*"; the Vainshtein formula averaged over orbital quantum number according to (9.116)–(9.118), curve designated as "Vainshtein *n*"; detailed quantum level-by-level calculations (Behar et al. 1996), curve designated as "quantum." It can be seen that the statistical orbital averaged method compares quite well with the detailed quantum calculations, as does the *l*-averaged Vainshtein approach, in particular, these approaches describe very well the critical low-temperature dielectronic recombination while the Burgess-Mertz approach entirely fails in this region. Figure 9.4 demonstrates also that the difference between the *l*-averaged branching factor and autoionization rate in the statistical approach is rather small; only at very low temperatures, some difference becomes visible.

The great advantage of the statistical model is its generality that also provides an easy means to study scaling laws. Figure 9.5 shows the analysis of the scaling in nuclear charge number Z_n for fixed $kT_e = 100\,\text{eV}$ for the total dielectronic recombination rate obtained from the various approaches discussed in Fig. 9.4. It can be seen that the statistical *l*-averaged approaches as well as the *l*-averaged Vainshtein approach provide a reasonable agreement with the numerical data from complex quantum calculations.

As Figs. 9.4 and 9.5 demonstrate, the l-averaged approaches of the statistical and the Vainshtein approach seem to correspond better with the complex quantum mechanical level-by-level calculations, while the statistical and Vainshtein

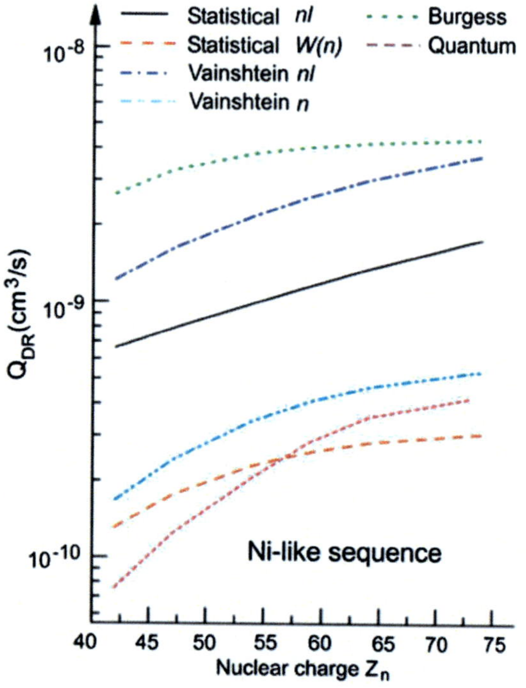

Fig. 9.5 Scaling properties in dependence of nuclear charge Z_n for $kT_e = 100$ eV for the various statistical approaches compared to quantum level-by-level calculations for the Ni-like sequence $3s^2 3p^6 3d^{10}$

nl-approach provides systematically larger total dielectronic recombination rates. The physical origin of this observation is difficult to explore and has not yet been discussed in the literature. In fact, in the current statistical approaches (Leontyev and Lisitsa 2016) multichannels as depicted by (9.111), (9.112) are not included. Multichannels, however, as have been demonstrated in detail in Sect. 5.6.2.2 may result in a decrease of total dielectronic recombination rate.

Below, we compare the different approaches with detailed quantum mechanical level-by-level calculations of the dielectronic recombination rates. Figure 9.6 shows the total dielectronic recombination rates of Xenon Xe^{26+} and Gold Au^{51+} (Ni-like $3s^2 3p^6 3d^{10}$-configuration into which dielectronic capture proceeds) calculated with the *l*-averaged statistical model from (9.85), (9.93), (9.94) that employs the Thomas–Fermi model of (9.1)–(9.13), the Burgess formula from (9.101)–(9.105) and the quantum level-by-level calculations of Behar et al. (1996).

The statistical model compares quite well (within a factor of two) over a very large temperature interval until very low temperatures while the Burgess approach entirely fails to describe the total dielectronic recombination rate of heavy ions. Similar observations are made for other isoelectronic sequences. Figure 9.7 shows a comparison of the Sr-like and Zn-like dielectronic recombination of tungsten and a comparison with detailed quantum mechanical level-by-level calculations (Wu et al. 2015).

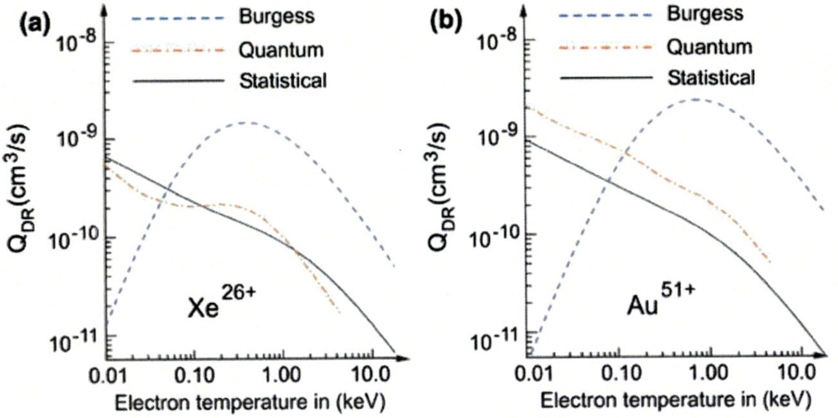

Fig. 9.6 Comparison of the l-averaged statistical approach with the Burgess and quantum level-by-level calculations for the Ni-like sequence $3s^2 3p^6 3d^{10}$ of Xenon Xe^{26+} and Gold Au^{51+}

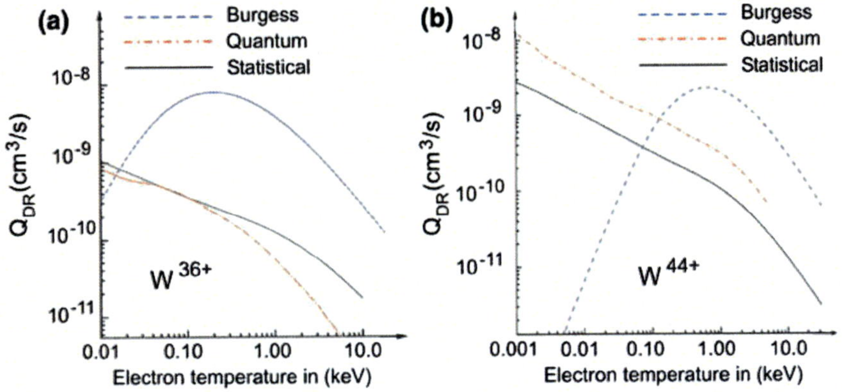

Fig. 9.7 Comparison of the l-averaged statistical approach with the Burgess and quantum level-by-level calculations for the Sr-like sequence $4s^2 4p^6 4d^2$ of Tungsten W^{36+} and the Zn-like sequence $4s^2$ of tungsten W^{44+}

It is particular impressive that the statistical model provides a rather good approximation of the total dielectronic recombination rate in the low-temperature region that is numerically exceedingly difficult to treat by fully quantum level-by-level calculations. Therefore, the statistical model in its simplest version of (9.1)–(9.13), (9.85)–(9.91), (9.92)–(9.95) seems to provide even the possibility to estimate the order of magnitude correctness of very complex quantum level-by-level calculations. Moreover, it should be remembered that currently even the most sophisticated quantum level-by-level calculations (Wu et al. 2015) have been obtained only in the low-density limit (corona approximation) where the branching factors are entirely determined by radiative and autoionization decay rates while dielectronic capture is proceeding from the respective ground states of the various charge states only. In high-density plasmas, however, collisional depopulation due to electron collisional ionization or collisional transfer to other levels (in particular of high nl-levels) strongly influences on the branching factors (see Sect. 5.6). In addition, excited states are highly populated from which very efficient channels of dielectronic recombination may proceed (Rosmej et al. 2020b). This may entirely change the properties of the total dielectronic recombination because dielectronic capture into excited states can be even more important than corresponding capture to the ground state (Sect. 5.6.3.2). This effect has explicitly been confirmed by high-resolution X-ray spectroscopy of dense laser-produced plasmas where it was shown that dielectronic recombination into excited states might exceed by many orders of magnitude the corresponding dielectronic recombination into ground states (Rosmej et al. 1998; Petitdemange and Rosmej 2013; Rosmej et al. 2020b). As for high-Z elements and open M-, N- and O-shells, excited states might be highly populated even at rather moderate electron densities. Therefore, all current detailed quantum level-by-level calculations to determine the dielectronic recombination have to be considered with care with respect to the particular application. In this respect, the properties and the innovation potential of the statistical model look very advantageous for the determination of the total dielectronic recombination rates for heavy elements.

Finally we note that the inclusion of more levels in the detailed quantum mechanical level-by-level calculations may not necessarily result only in an increase of the dielectronic recombination rate, but can also lead to a decrease as discussed in Sect. 5.6.2.1 and described by (9.111), (9.112). Therefore, at present, the simple above-presented statistical method compares quite well with other available much more complex methods of calculations and has the advantage of generality and ease of application. In addition, there is much room for improvements of the statistical model via improvements of the Thomas–Fermi model (ionization energies, l-quantum number dependence, Vlasov approach instead of the local plasma frequency, etc.).

References

V.A. Astapenko, L.A. Bureyeva, V.S. Lisitsa, Polarization mechanism for bremsstrahlung and radiative recombination in a plasma with heavy ions. Plasma Phys. Rep. **28**, 303 (2002)

V.A. Astapenko, L.A. Bureyeva, V.S. Lisitsa, Plasma models of atom and radiative-collisional processes, in *Reviews of Plasma Physics*, vol. 23, 1, ed. by V.D. Shafranov (Kluwer Academic/Plenum Publishers, New York, 2003); ISBN 0-306-11069-5

C.P. Balance, S.D. Loch, M.S. Pindzola, D.C. Griffin, Dielectronic recombination of W^{35+}. J. Phys. B: At. Mol. Opt. Phys. **43**, 205201 (2010)

P. Beiersdorfer, M.J. May, J.H. Scofield, S.B. Hansen, Atomic physics and ionization balance of high-Z ions: critical ingredients for characterizing and understanding high-temperature plasma. HEDP **8**, 271 (2012)

E. Behar, P. Mandelbaum, J.L. Schwob, A. Bar-Shalom, J. Oreg, W.H. Goldstein, Dielectronic recombination rate coefficients for highly-ionized Ni-like atoms. Phys. Rev. A **54**, 3070 (1996)

I.L. Beigman, L.A. Vainshtein, B.N. Chichkov, Dielectronic recombination. JETP **53**, 490 (1981)

W. Brandt, S. Lundqvist, Atomic oscillations in the statistical approximation. Phys. Rev. **139**, A612 (1965)

A. Burgess, Dielectronic Recombination and the Temperature of the Solar Corona. Astrophys. J. **139**, 776 (1964)

R.G. Cowan, *The Theory of Atomic Structure and Spectra* (California University Press, Berkeley, 1981)

A.V. Demura, M.B. Kadomtsev, V.S. Lisitsa, V.A. Shurygin, Statistical model of radiation losses for heavy ions in plasmas. JETP Lett. **98**, 786 (2013)

A.V. Demura, M.B. Kadomtsev, V.S. Lisitsa, V.A. Shurygin, Universal statistical approach to radiative and collisional processes with multielectron ions in plasmas. HEDP **15**, 49 (2015a)

A.V. Demura, M.B. Kadomtsev, V.S. Lisitsa, V.A. Shurygin, Statistical model of electron impact ionization of multielectron ions. J. Phys. B.: At. Molec. Opt. Phys. **48**, 055701 (2015b)

I.K. Dmitrieva, G.I. Plindov, The improved Thomas-Fermi model: chemical and ionization potentials in atoms. J. Physique **45**, 85 (1984)

S. Dyachkov, P. Leavshov, D. Minakov, Region of validity of the Thomas-Fermi model with corrections. Phys. Plasmas **23**, 112705 (2016)

E. Fermi, Eine statistische Methode zur Bestimmung einiger Eigenschaften des Atoms und ihre Anwendung auf die Theorie des periodischen Systems der Elemente. Zeitschrift für Physik **48**, 73 (1928)

P. Fromy, C. Deutsch, G. Maynard, Thomas-Fermi-like and average atom models for dense and hot matter. Phys. Plasmas **3**, 714 (1996)

P. Gombas, Erweiterung der Statistischen Theroy des Atoms. Zeitschrift für Physik **121**, 523 (1943)

P. Gombas, *Die statistische theorie des Atoms und ihre Anwendungen* (Springer, Wien, 1949)

P. Gombas, Present state of the statistical theory of atoms. Rev. Mod. Phys. **35**, 512 (1963)

I.P. Grant, B.J. McKenzie, P.H. Norrington, D.F. Mayers, N.C. Pyper, An atomic multiconfigurational Dirac-Fock package. Comput. Phys. Commun. **21**, 207 (1980)

G. Kemister, S. Nordholm, A radially restricted Thomas–Fermi theory for atoms. J. Chem. Phys. **76**, 5043 (1982)

V.D. Kirillow, B.A. Trubnikov, S.A. Trushin, Role of impurities in anomalous plasma resistance. Sov. J. Plasma Phys. **1**, 117 (1975)

V.I. Kogan, A.B. Kukushkin, V.S. Lisitsa, Kramers electrodynamics and electron-atomic radiative collisional processes. Phys. Rep. **213**, 1 (1992)

D.S. Leontyev, V.S. Lisitsa, Statistical model of dielectronic recombination of heavy ions in plasmas. Contrib. Plasma Phys. **56**, 846 (2016)

E.H. Lieb, B. Simon, The Thomas–Fermi theory of atoms, molecules and solids. Adv. Math. **23**, 22 (1977)

S.D. Loch, J.A. Ludlow, M.S. Pindzola, A.D. Whiteford, D.C. Griffin, Electron-impact ionization of atomic ions in the W isonuclear sequence. Phys. Rev. A **72**, 052716 (2005)

F. Petitdemange, F.B. Rosmej, Dielectronic satellites and Auger electron heating : irradiation of solids by intense XUV-Free Electron Laser radiation, in *New Trends in Atomic & Molecular Physics—Advanced Technological Applications*, vol. 76, ed. by M. Mohan (Springer 2013), pp. 91–114, ISBN 978-3-642-38166-9

R. Piron, F. Gilleron, Y. Aglitskiy, H.-K. Chung, C.J. Fontes, S.B. Hansen, O. Marchuk, H.A. Scott, E. Stambulchik, Yu. Ralchenko, Review of the 9th NLTE code comparison workshop. HEDP **23**, 38 (2017)

D.E. Post, R.V. Jensen, C.B. Tarter, W.H. Grasverger, W.A. Lokke, Steady-state raditiative coling rates for low-density high-temperature plasmas. ADNDT **20**, 397 (1977)

D. Post, J. Abdallah, R.E.H. Clark, N. Putvinskaya, Calculations of energy losses due to atomic processes in tokamaks with applications to the international thermonuclear experimental reactor divertor. Phys. Plasmas **2**, 2328 (1995)

V.M. Povyshev, A.A. Sadovoy, V.P. Shevelko, G.D. SHirkov, E.G. Vasina, V.V. Vatulin, *Electron-Impact Ionization Cross Sections of H, He, N, O, Ar, Xe, Au, Pb Atoms and their Ions in the Electron Energy Range from the Threshold up to 200 keV*, Preprint JINR E9-2001-148, Dubna 2001

T. Pütterich, R. Neu, R. Dux, A.D. Whiteford, M.G. O'Mullane, H.P. Summers, ASDEX upgrade team, calculation and experimental test of the cooling factor of tungsten. Nuc. Fusion **50**, 025012 (2010)

F.B. Rosmej, A.Ya. Faenov, T.A. Pikuz, F. Flora, P. Di Lazzaro, S. Bollanti, N. Lizi, T. Letardi, A. Reale, L. Palladino, O. Batani, S. Bossi, A. Bornardinello, A. Scafati, L. Reale, Line Formation of High Intensity He_β-Rydberg Dielectronic Satellites $1s3lnl'$ in Laser Produced Plasmas. J. Phys. B Lett.: At. Mol. Opt. Phys. **31,** L921 (1998)

F.B. Rosmej, L.A. Vainshtein, V.A. Astapenko, V.S. Lisitsa, Statistical and quantum photoionization cross sections in plasmas: analytical approaches for any configurations including inner shells, Matter and Radiation at Extremes (Review) **5**, 064202 (2020a) open access: https://aip.scitation.org/doi/10.1063/5.0022751

F.B. Rosmej, V.A. Astapenko, V.S. Lisitsa, L.A. Vainshtein, Dielectronic recombination in non-LTE plasmas, Matter and Radiation at Extremes (Review) **5**, 064201 (2020b) open access: https://doi.org/10.1063/5.0014158

H.A. Scott, S.B. Hansen, Advances in NLTE modeling for integrated simulations. HEDP **6**, 39 (2010)

I.I. Sobelman, L.A. Vainshtein, *Excitation of Atomic Spectra* (Alpha Science, 2006); ISBN 978-1842652336

A. Sommerfeld, Integrazione asintotica dell'equazione differentiale di Thomas–Fermi. Rend. R. Accademia dei Lincei **15**, 293 (1932)

M. Stenke, K. Aichele, D. Harthiramani, G. Hofmann, M. Steidl, R. Volpel, E. Salzborn, Electron-impact single-ionization of singly and multiply charged tungsten ions. J. Phys. B: At. Mol. Opt. Phys. **28**, 2711 (1995)

H.P. Summers, *ADAS Users Manual*, JET—IR 06 (Abingdon: JET Joint Undertaking, 1994)

L.A. Vainshtein, V.P. Shevelko, The structure and characteristics of ions in hot plasmas, in *Physika i Technika Spektroscopii* 1986 (in Russian)

A.V. Vinogradov, O.I. Tolstikhin, Plasma approach to the theory of photoabsorption and polarizability of complex atoms. Sov. Phys. JETP **69**, 683 (1989)

Z. Wu, Y. Fu, X. Ma, M. Li, L. Xie, J. Jiang, C. Dong, Electronic impact excitation and dielectronic recombination of highly charged Tungsten ions. ATOMS **3**, 474 (2015)

R. Ying, G. Kalman, Thomas–Fermi model for dense plasmas. Phys. Rev. A **40**, 3927 (1989)

Chapter 10
Applications to Plasma Spectroscopy

Abstract Applications to plasma spectroscopy are presented for different types of plasmas that are currently of great interest for science and applications: low-density tokamak plasmas, dense optical laser-produced plasmas, high-current Z-pinch plasmas, and X-ray Free Electron Lasers interacting with solids. The general principles of plasma electron temperature and density measurements as well as the characterization of suprathermal (hot electrons) and non-equilibrium phenomena are presented. Particular attention is based on the innovative concepts of dielectronic satellite and hollow ion X-ray emission. The effect of a neutral background that is coupled to plasma ions via charge exchange is considered in the framework of nonlinear atomic kinetics. Transient phenomena in the start-up phase, impurity diffusion, sawtooth oscillations, and superthermal electrons are discussed for magnetic fusion plasmas. For dense laser-produced plasmas and charge exchange coupling of colliding plasmas, the dynamics of fast ions in space and energy distribution functions are presented. The interaction of XFEL with solids is considered in the framework of a new kinetic plasma theory, where generalized atomic processes provide a link from the cold solid until the hot diluted plasma. Three-body recombination and Auger electrons constitute a generalized three-body recombination that is identified to play also a new role as a direct heating mechanism. General principles and new theories are illustrated along with detailed comparisons with experimental data.

10.1 The Emission of Light and Plasma Spectroscopy

The emission of light is one of the most fascinating phenomena in nature. Everybody feels the beauty while looking at the colors appearing at sunset, when a bolt of lightning illuminates the night, or when the emission of the aurora moves like magic in the dark heaven. And every day, we are looking at something in order to read information from a computer screen, to drive not into but around an obstacle, to look into the eyes of the child to understand that it tries to hide that it just burned off fathers' stamp collection in an unlucky physical experiment.

F. B. Rosmej et al., *Plasma Atomic Physics*, Springer Series on Atomic, Optical, and Plasma Physics 104, https://doi.org/10.1007/978-3-030-05968-2_10

In general terms, we all use light to obtain information, to diagnose something, to control or optimize a process, or to understand what is true and what is right. And since the discovery of the spectral analysis, no one doubted that the problems of describing atoms and matter would be solved once we had learned to understand the language of atomic spectra and the emission of light.

Light also transports energy, and it is the flow of energy from sunlight that has so much impacted on the evolution of our blue planet.

The "*Radiative properties of Matter*" is the related basic science, and their analysis for diagnostic purposes is called "Spectroscopy". Questions like "*Why the heaven is blue?*", "*What is the temperature in the flame of a candle?*", and "*What makes the sun burning so wonderful?*" have been the historical origin of scientific activity. We also might ask what light is by itself? This is a difficult question: Although almost everybody has some imagination what light is, it is difficult to say what it really is.

The light emission is accompanied by the transport of energy and the well-known formula

$$E_\omega = \hbar\omega \tag{10.1}$$

(where $\hbar$ is the Planck constant and ω the photon angular frequency) that manifests in a scientific manner the double role of light: the beauty of photons when looking at their various colors and the energy that is carried by them with their own velocity – the speed of light c.

That any radiating source loses consequently energy via its own radiation has a large impact on the evolution of the system itself. In general terms, the radiation losses influence on the energy balance of the system that is coupled to the motion of the particles from which the system is composed.

An impressive example of the importance of radiation losses is known from the early days of the fusion research based on the magnetic confinement invented in the 50s by the two Russians Igor Tamm and Andrei Sakharov: The term tokamak was likewise created by the Russians: "тороидальная камера с магнитными катушками" (***to**roïdalnaïa **ka**mera s **ma**gnitnymi **k**atushkami = toroidal chamber with magnetic coils*). The radiation emission from the impurities in the tokamak had been so large that it overcompensated the energy input leading to a plasma disruption. Only after generations of very intense and successful material research, the number of impurity atoms and ions could drastically be reduced to maintain tokamak discharges for more than minutes. A milestone in the tokamak research can be attributed to the year 1969: The Kurchatov Institute in Russia, Moscow, has announced to have obtained a tokamak discharge with an electron temperature of $kT_e > 1$ keV (means that the level of the right order of magnitude was reached where fusion reactions become probable). The scientific western community did not believe these results, and in 1969, an experimental group from England went to the Kurchatov Institute to measure independently the electron temperature with Thomson scattering. This group could only confirm the results announced by the Russians, and in the following years, many countries followed the Russian way of magnetic fusion research building many tokamaks all over the world.

Plasma radiation plays likewise an important role in the inertial confinement fusion (ICF). Here, fusion of the DT capsule is envisaged by means of powerful laser installations like NIF, MEGAJOULE, and GEKKO. In the indirect drive, the radiation field in the hohlraum (initiated by megajoule laser irradiation of the inner surface of the hohlraum composed of an heavy element like gold) is supposed to deliver energy to the capsule to evaporate in a homogenous manner (assuming that the hohlraum radiation is very close to the blackbody radiation) the ablator surface that in turn compresses the fusion material to ignition relevant parameters. In the direct drive approach, laser radiation is directly interacting with the capsule. Today, many other schemes of compression are investigated like fast ignition or shock ignition.

In astrophysics, radiation phenomena are strongly connected with the energy transport in stars from the fusion source in the inner part of the star to its surface. Therefore, so-called radiation transport plays an important role in star evolution.

Today, the study of the radiative properties of matter has proven to be one of the most powerful methods to understand various physical phenomena. Plasma spectroscopy (Griem 1964, 1974, 1997; McWhirter 1965; Lochte-Holtgreven 1968; Michelis and Mattioli 1981; Boiko et al. 1985; Lisitsa 1994; Fujimoto 2004; Sobelman and Vainshtein 2006; Kunze 2009, Rosmej 2012a) provides essential information about basic parameters, like temperature, density, chemical composition, velocities, and relevant physical processes (*"Why the aurora is green at low altitudes but red at larger ones?", "Why can we look with X-rays into the human body but not with visible light ?",...*).

The accessible parameter range of spectroscopy covers orders of magnitude in temperature and (especially) density, because practically all elements of particular, selected isoelectronic sequences can be used for diagnostic investigations. These elements can occur as intrinsic impurities or may be intentionally injected in small amounts (so-called tracer elements). This makes plasma spectroscopy also a very interdisciplinary science.

The rapid development of powerful laser installations including X-ray Free Electron Lasers, intense heavy ion beams, and the fusion research (magnetic and inertial fusion) enable the creation of matter under extreme conditions never achieved in laboratories so far. An important feature of these extreme conditions is the non-equilibrium nature of the matter, e.g., a solid is heated by a fs-laser pulse and undergoes a transformation from a cold solid to warm dense matter to strongly coupled plasma and then to a highly ionized gas while time is elapsing. We might think about using time-dependent detectors to temporally resolve the light emission in the hope to have then resolved the problem. However, this is not so simple: There are not only serious technical obstacles but also basic physical principals to respect. A simple technical reason is that for 10 fs-laser radiation interacting with matter, we do not have any X-ray streak camera available (the current technical limit is about 0.5 ps). A simple principal reason is that the atomic system from which light originates has a characteristic time constant (see also Sect. 6.2) that might be much longer than 10 fs. The atomic system is therefore "shocked," and any light emission is highly out of equilibrium even if the experimental observation is time resolved. Note, that only recently general studies of shocked atomic systems have begun (Deschaud et al. 2020).

In high-energy-density physics, laser-produced plasmas, and fusion research, the light emission in the X-ray spectral range is of particular interest: Only the X-ray emission is in general able to exit the volume without essential photoabsorption. Therefore, X-ray plasma spectroscopy is of particular interest to obtain information from objects that are not well understood or to control certain processes that are useful for applications (Renner and Rosmej 2019).

We are therefore looking to develop X-ray radiative properties and the related atomic and plasma physics to study non-equilibrium states of matter. This is a challenging field of activity for research and applications: Non-equilibrium atomic kinetics (see Chap. 6) involve fascinating topics in atomic physics of dense plasmas, like the study of exotic dense states of matter (like hollow ions and hollow crystals) with XFEL (Rosmej 2012a, b; Deschaud et al. 2020), the discovery by high-resolution spectroscopy of a new heating mechanism like Auger electron heating (Galtier et al. 2011; Rosmej 2012b; Petitdemange and Rosmej 2013), and three-body recombination assisted heating (Deschaud et al. 2014), see also Sect. 10.6.4.6. Radiative properties have likewise strong links and important impact to magnetic confinement fusion (MCF), inertial confinement fusion (ICF), and particular X-ray spectroscopy, and related atomic physics is a key element to study matter irradiated by X-ray Free Electron X-ray Lasers (XFEL).

It is often not quite clear what is meant with the word "Equilibrium." Does this mean that the plasma parameters do not change in time, which means for a parameter X:

$$\partial X/\partial t = 0 \tag{10.2}$$

or does the word "*Equilibrium*" mean that the particle statistics follows certain laws? Or something else?

In thermodynamics (Huang 1963; Reif 1965; Alonso and Finn 1968), the "*Equilibrium*" of an isolated system is characterized by the maximum entropy[1]:

$$\delta S = \delta\{k \ln W\} = 0. \tag{10.3}$$

k is the Boltzmann constant, and W is the microscopic probability (probability means "Wahrscheinlichkeit" in german) of a certain configuration which means the probability $P(N, N_1, N_2 \ldots N_M)$ to distribute N particles over M states with respective populations $N_1, N_2 \ldots N_M$.

It is important to realize that conditions (10.2) and (10.3) are entirely different: Thermodynamic equilibrium is not determined by $\partial/\partial t = 0$. In other words, the Saha–Boltzmann equation, the Planck radiation, the Maxwellian energy distribution

[1]Note that for a system with controlled constant temperature and volume, the Helmholtz free energy $F = U - TS$ is minimum at thermodynamic equilibrium and for a system with controlled constant temperature and pressure the Gibbs free energy $G = U - TS + pV$ is minimum. Note also the notions of thermal equilibrium, pressure equilibrium, and diffusive equilibrium (identity of chemical potentials) to characterize "equilibrium" between two systems.

function of free electrons,... are consequences of (10.3), and these laws cannot be obtained from (10.2). These laws are the solution of (10.3) for different particle statistics.

In most cases, laws obtained from (10.3) concern isolated systems, which means that external forces do not play a role for the particle statistics. What does it mean that external forces do not play an important role? In atomic physics, this means that external forces have to be compared with the "*Atomic Forces.*" Let us consider an example to illustrate this: the interaction of a high-intensity laser with a plasma. The external force is the laser electric field E_{laser}:

$$I_{\text{laser}} = c\varepsilon_0 \langle E_{\text{laser}}^2 \rangle. \tag{10.4}$$

I_{laser} is the laser intensity, c the speed of light, and ε_0 the electrical permittivity. Rewriting (10.4) in convenient units, we obtain:

$$E_{\text{laser}} \approx \sqrt{\langle E^2 \rangle} = \sqrt{\frac{I}{c\varepsilon_0}} = 19.4\sqrt{I\left(\frac{W}{cm^2}\right)}\left[\frac{\text{V}}{\text{cm}}\right]. \tag{10.5}$$

The relevant forces for an ion in a plasma are the atomic ones. They can be estimated with the simple Bohr atomic model:

$$E_{\text{Atom}} \approx \frac{1}{4\pi\varepsilon_0} \frac{Z \cdot e}{(a_0 n^2/Z)^2} = \frac{e}{4\pi\varepsilon_0 a_0^2} \frac{Z^3}{n^4} \tag{10.6}$$

or, expressed in convenient units

$$E_{\text{Atom}} \approx 5.14 \times 10^9 \frac{Z^3}{n^4} \left[\frac{\text{V}}{\text{cm}}\right]. \tag{10.7}$$

The standard atomic field, i.e., $E_{\text{a.u.}} = 5.14 \times 10^9$ V/cm (see also Annex A.5) is therefore obtained for a laser intensity of $I_{\text{a.u.}} = 7.0 \times 10^{16}$ W/cm^2. This means, e.g., that the 1s level of atomic hydrogen is weakly perturbed for laser intensities being many orders of magnitudes lower than $I_{\text{a.u.}}$.

Let us now consider the influence of the electrical laser field on the radiative properties of atoms and ions in a plasma. Due to the interaction of the laser electric field with the free electrons in the plasma, suprathermal electrons (or hot electrons) are generated. They seriously alter the light emission. First, there is an enhanced ionization due to hot electrons because ionization is more effective. The second change concerns the qualitative distortion of ionic charge stage distribution (Rosmej 1997). The visualization of the distortion of the charge stage distribution by high-resolution X-ray spectroscopy has been applied to inertial fusion hohlraums to determine the time-dependent hot electron fraction and the relevant mechanism of hot electron generation (Glenzer et al. 1998). Let us define the suprathermal electron fraction according to:

$$f_{\text{hot}} = \frac{n_e(hot)}{n_e(bulk) + n_e(hot)}. \tag{10.8}$$

$n_{e,hot}$ is the hot electron density, $n_{e,bulk}$ is the bulk electron density. $n_{e,hot} + n_{e,bulk}$ is therefore the total electron density, and (10.8) fulfills the normalization condition. It is important to realize that even rather small amounts of hot electrons (low as of the order of 10^{-5}) might have a considerable influence on the radiative properties, especially if the bulk electron temperature is low. This is due to the exponential temperature dependence of the excitation (C) and ionization (I) rate coefficients (see also Sects. 5.3.1 and 5.4.3), i.e.,

$$C \propto \frac{\exp(-\Delta E/kT_e)}{\sqrt{kT_e}}, \tag{10.9}$$

$$I \propto \frac{\exp(-E_i/kT_e)}{\sqrt{kT_e}}. \tag{10.10}$$

ΔE is the excitation energy of an atomic transition, E_i is the ionization energy of a particular atomic level, and T_e is the electron temperature.

It should be realized that small fractions of hot electrons can be produced already with laser intensities of 10^{15} W/cm^2 (Gitomer et al. 1986; Beg et al. 1997) and therefore the value of the laser electric field above which an important influence on the radiative properties is expected might be well below $I_{a.u.} = 7.0 \times 10^{16}$ W/cm^2. We note that suprathermal electrons generated in ICF hohlraums (e.g., generated by SBS = Stimulated Brillouin Scattering) may lead to a preheat of the DT target which in turn prevents efficient compression necessary to reach ignition. Suprathermal electrons (hot electrons) are therefore a very actual problem in the laser-driven inertial fusion ignition campaigns (Lindl 1995, Lindl et al. 2004, 2014; Atzeni 2009). We note that in the more recently discussed shock ignition scheme (Betti et al. 2007), suprathermal electrons impact on the fusion performance as an important fraction of laser energy is coupled to hot electrons.

In tokamaks, much lower electric fields lead to the generation of suprathermal electrons: Due to the low electron density, the collisional drag is small and even electric field values of the order of some V/cm (the so-called Dreicer field, Wesson 2004) lead to runaway electrons: The collisional drag is insufficient to compensate the electron acceleration due to the electric field and numerous circulations in the tokamak may then lead to electron energies up to MeV. These MeV electrons seriously influence on the fusion performance (e.g., electrons accelerated by lower hybrid waves, investigation of suitable current drives). These two foregoing examples show that the importance of an external force is not specified by an absolute value but rather by the comparison of the external force with the relevant "internal" one.

Let us now consider the principle idea for spectroscopic diagnostics that is based on line intensity ratios. Having once calculated the non-LTE level populations according to (6.22), all combinations of line intensity ratios can be obtained:

$$\frac{I_{ji}}{I_{j'i'}} = \frac{\omega_{ji}}{\omega_{j'i'}} \frac{A_{ji}}{A_{j'i'}} \frac{n_j}{n_{j'}}. \tag{10.11}$$

Of particular interest are those intensity ratios that depend only on one plasma parameter. The ideal case of a temperature diagnostic is therefore given by

$$\frac{I_{ji}}{I_{j'i'}} = G_{jij'i'}(T_e), \tag{10.12}$$

whereas the ideal case of a density diagnostic is represented by the relation

$$\frac{I_{kl}}{I_{k'l'}} = \gamma_{klk'l'}(n_e). \tag{10.13}$$

The functions G and γ are obtained from the solution of the system of rate equations (6.22). Having measured these intensity ratios with appropriate line emissions, the application of (10.12), (10.13) provides readily temperature and density. However, the solution of (6.22) shows that in general, the intensity ratio depends both on temperature and density:

$$\frac{I_{ji}}{I_{j'i'}} = \chi_{jij'i'}(T_e, n_e). \tag{10.14}$$

One aim of spectroscopic research is to find line ratios whose dependence is close to those of the ideal (10.12), (10.13). The difficulty in doing so lies in the fact that (10.14) has multiple solutions, which means that for different sets of density and temperature the same line intensity ratio is obtained (note that the two-parameter dependence is a simplified case and opacity, hot electrons and transient plasma evolution might considerably increase the complexity). It is therefore necessary to employ several line ratios at the same time to avoid misleading parameter information from single line ratios.

10.2 Dielectronic Satellite Emission

10.2.1 Electron Temperature

10.2.1.1 Satellite to Resonance Lines

Gabriel has introduced the dielectronic satellite transitions (see also Chap. 5) as a sensitive method to determine the electron temperature in hot plasmas (Gabriel 1972) that is based on dielectronic capture and dielectronic recombination (see also Sect. 5.6 and review [Rosmej et al. 2020a]). In low-density plasmas, this method approaches the ideal picture of a temperature diagnostic according to (10.12). Numerical calculations show (see also Sect. 6.3.2 and following Figures 10.2 and 10.3) that also in high-density plasmas, this method is still applicable and one of the most powerful methods for electron temperature determination of hot dense plasmas.

Let us therefore consider the basic principles via an example: The dielectronic satellites $2l2l'$ near the Lyman-alpha line of H-like ions (Fig. 5.1 show the relevant energy level diagram). As the He-like states $2l2l'$ are located above the ionization limit, a non-radiative decay to the H-like ground state (autoionization) is possible:

$$\text{autoionization}: 2l2l' \rightarrow 1s + e. \tag{10.15}$$

By first quantum mechanical principles, the reverse process, so-called dielectronic capture, must exist:

$$\text{dielectronic capture}: 1s + e \rightarrow 2l2l'. \tag{10.16}$$

The radiative decay reads

$$\text{radiative decay}: 2l2l' \rightarrow 1s2l + \hbar\omega_{\text{satellite}}. \tag{10.17}$$

The emitted photon is called a "satellite." The satellite transition is of similar nature like the resonance transition $Lyman_{\alpha} = 2p \rightarrow 1s + \hbar\omega_{Ly_{\alpha}}$ except the circumstance that an additional electron is present in the quantum shell $n = 2$, the so-called spectator electron. As the spectator electron screens the nuclear charge, the satellite transitions are essentially located on the long wavelength side of the corresponding resonance line. However, due to intermediate coupling effects and configuration interaction, also satellites on the short wavelengths side are emitted (see Fig. 10.1), so-called blue satellites (Rosmej and Abdallah 1998).

As the number of possible angular momentum couplings increases rapidly with the number of electrons, usually numerous satellite transitions are located near the resonance line (which often cannot be resolved spectrally even with high-resolution methods). Figure 10.1 shows an example of the $Lyman_{\alpha}$ satellite transitions obtained in a dense laser-produced magnesium plasma. The experiment shows also higher-order satellites where the spectator electrons are located in quantum shells $n > 2$ (configurations $2lnl'$).

Let us now proceed to the genius idea of Gabriel to obtain the electron temperature from satellite transitions. In a low-density plasma, the intensity of the resonance line is given by

$$I^{\text{res}}_{\text{k}',\text{j}'\text{i}'} = n_{\text{e}} n_{\text{k}'} \frac{A_{\text{j}'\text{i}'}}{\sum_{\text{l}'} A_{\text{j}'\text{l}'}} \langle C_{\text{k}'\text{j}'} \rangle, \tag{10.18}$$

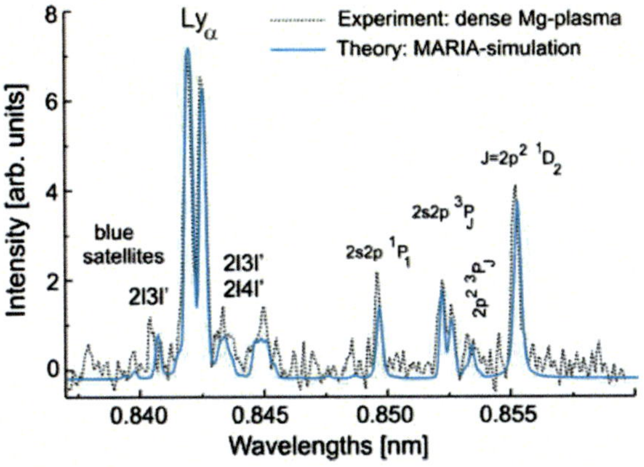

Fig. 10.1 Dielectronic satellite emission near Lyman-alpha of H-like Mg ions in a dense laser-produced plasma (50 J, 15 ns, 1.064 μm). Spectral simulation of optically thick plasma has been carried out with the MARIA code for an electron temperature of $kT_e = 210$ eV, electron density of $n_e = 3 \times 10^{20}$ cm^{-3}, effective photon path length $L_{eff} = 500$ μm, inhomogeneity parameter $s = 1.3$

where n_e is the free electron density, $n_{k'}$ the ground state density from which electron collisional excitation proceeds (k' is the 1s level in our example), $A_{j'i'}$ is the transition probability of the resonance transition $j' \rightarrow i'$ (the sum over A in the denominator accounts for possible branching ratio effects), and $\langle C_{k'j'} \rangle$ is the electron collisional excitation rate coefficient from level k' to level j'. The intensity of a satellite transition with a large autoionizing rate (and negligible collisional channel) is given by

$$I^{sat}_{k,ji} = n_e n_k \frac{A_{ji}}{\sum_l A_{jl} + \sum_m \Gamma_{jm}} \langle D_{kj} \rangle. \tag{10.19}$$

A_{ji} is the transition probability of the particular satellite transition, and $\langle D_{kj} \rangle$ is the dielectronic capture rate coefficient from level k to the level j. The sums over the radiative decay rates and autoionizing rates account for possible branching ratio effects (in our simple example, only $m = k$ exist, a particular upper-level $2l2l'$ may have more than one radiative decay possibilities $j \rightarrow l$). We note that already for the He_β satellites, numerous autoionizing channels exist which are very important in dense plasmas (Rosmej et al. 1998). As both intensities of (10.18), (10.19) are proportional to the electron density n_e and to the same ground state density ($k' = k$), the intensity ratio is a function of the electron temperature only, because the rate coefficients $\langle C \rangle$ and $\langle D \rangle$ depend only on the electron temperature but not on the density:

$$\frac{I_{\mathrm{k,ji}}^{\mathrm{sat}}}{I_{\mathrm{k'j'i'}}^{\mathrm{res}}} = G(T_{\mathrm{e}}). \tag{10.20}$$

The dielectronic capture rate (see also Chap. 5) is an analytical function and given by

$$\langle D_{\mathrm{kj}} \rangle = \alpha \Gamma_{\mathrm{jk}} \frac{g_{\mathrm{j}}}{g_{\mathrm{k}}} \frac{\exp(-E_{\mathrm{kj}}/kT_{\mathrm{e}})}{(kT_{\mathrm{e}})^{3/2}}. \tag{10.21}$$

$\alpha = 1.6564 \times 10^{-22}\ \mathrm{cm}^3\ \mathrm{s}^{-1}$, g_{j} and g_{k} are the statistical weights of the states j and k, Γ_{jk} is the autoionizing rate in $[\mathrm{s}^{-1}]$, E_{kj} is the dielectronic capture energy in [eV] (see also Fig. 5.1), and kT_{e} is the electron temperature [eV]. The intensity of a satellite transition can therefore be written as

$$I_{\mathrm{k,ji}}^{\mathrm{sat}} = \alpha n_{\mathrm{e}} n_{\mathrm{k}} \frac{Q_{\mathrm{k,ji}}}{g_{\mathrm{k}}} \frac{\exp(-E_{\mathrm{kj}}/kT_{\mathrm{e}})}{(kT_{\mathrm{e}})^{3/2}}. \tag{10.22}$$

$Q_{\mathrm{k,ji}}$ is the so-called dielectronic satellite intensity factor and given by

$$Q_{\mathrm{k,ji}} = \frac{g_{\mathrm{j}} A_{\mathrm{ji}} \Gamma_{\mathrm{jk}}}{\sum_{\mathrm{l}} A_{\mathrm{jl}} + \sum_{m} \Gamma_{\mathrm{jm}}}. \tag{10.23}$$

The calculation of the dielectronic satellite intensity factors $Q_{\mathrm{k,ji}}$ requests rather complicated multiconfiguration relativistic atomic structure calculations which have to include intermediate coupling effects as well as configuration interaction.

For the ease of applications, we provide an analytical set of all necessary formulas for the most important cases to apply the temperature diagnostic via dielectronic satellite transitions near Ly_{α} and He_{α} of highly charged ions. For the dielectronic satellite intensity factor, the following formula can be employed:

$$Q = 10^{10}\,\mathrm{s}^{-1} \frac{C_1 (Z_{\mathrm{n}} - C_2)^4}{C_3 Z_{\mathrm{n}}^{C_4} + 1}. \tag{10.24}$$

Table 10.1 provides the fitting parameters for the J-satellite near Ly_{α} as well as for the k-satellite and the j-satellite near He_{α} for all elements with nuclear charge $6 < Z_{\mathrm{n}} < 30$. We note that the k- and j-satellites are treated separately, as line

Table 10.1 Z_{n}-scaled fitting parameters of dielectronic satellite intensity factors Q according to (10.24), the range of validity is $6 < Z_{\mathrm{n}} < 30$

Satellite	C_1	C_2	C_3	C_4	Max. error (%)
$J = 2p^2\ {}^1D_2$–$1s2p\ {}^1P_1$	5.6696E−1	1.4374E−8	5.8934E0	2.2017E−2	1.5
$j = 1s^1 2p^2\ {}^2D_{5/2}$–$1s^2 2p\ {}^2P_{3/2}$	3.4708E−1	1.5569E−7	4.9939E0	8.6347E−1	1.5
$k = 1s^1 2p^2\ {}^2D_{3/2}$–$1s^2 2p\ {}^2P_{3/2}$	2.4072E−1	6.7212E−9	5.9468E0	1.1362E0	3

Table 10.2 Fitting parameters for Z- and β-scaled electron collisional excitation rates of H-like Ly_α and He-like He_α, $1/32 < \beta < 32$

Transition	A	χ	D	Max. error (%)
$Lyman_\alpha = 1s\ ^2S_{1/2}$–$2p\ ^2P_{1/2,\ 3/2}$	24.1	0.145	−0.120	4
$He_\alpha = 1s^2\ ^1S_0$–$1s2p\ ^1P_1$	24.3	0.198	1.06	6

overlapping may request their separate analysis. Note that $g_k = 2$ for the Ly_α-satellites and $g_k = 1$ for the He_α-satellites in (10.22). The dielectronic capture energies can be approximated by

$$E_{kj} \approx \delta(Z_n + \sigma)^2 Ry. \tag{10.25}$$

For the Ly_α-satellites $2l2l'$, $\delta = 0.5$, $\sigma \approx 0.5$, for the He_α-satellites $1s2l2l'$, $\delta = 0.5$, $\sigma \approx 0.1$, and $Ry = 13.6$ eV. The electron collisional excitation rate coefficients have been calculated with the Coulomb–Born exchange method including intermediate coupling effects and effective potentials (using Vainshtein's ATOM code (Vainshtein and Shevelko 1986; Sobelman and Vainshtein 2006)) and fitted into a simple Z- and β-scaled expression:

$$\langle C, T_e \rangle \approx \frac{10^{-8}\,\mathrm{cm^3\,s^{-1}}}{Z^3}\left(\frac{E_u}{E_l}\right)^{3/2}\sqrt{\beta}A\,\frac{\beta+1+D}{\beta+\chi}\exp\{-(E_l - E_u)/kT_e\} \tag{10.26}$$

with

$$\beta = \frac{Z^2 Ry}{kT_e}. \tag{10.27}$$

Z is the spectroscopic symbol ($Z = Z_n + 1 - N$ where Z_n is the nuclear charge and N the number of bound electrons), the fitting parameters A, χ, and D are given in Table 10.2.

E_l and E_u are the ionization energies of lower and upper states. If not particularly available, they can be approximated by the simple expression

$$E_j \approx \delta(Z_n - \sigma)^2 Ry. \tag{10.28}$$

For the $1s$-level, $\delta = 1$, $\sigma \approx -0.05$; for the $2p$ levels, $\delta = 0.25$, $\sigma \approx -0.05$; for the $1s^2$ level, $\delta = 1$, $\sigma \approx 0.6$, and for the $1s2p\ ^1P_1$ level, $\delta = 0.25$, $\sigma \approx 1$.

10.2.1.2 Rydberg Satellites

Higher-order satellites, namely $2lnl'$ and $1s2lnl'$, provide further possibilities for plasma diagnostics even if single transitions are not resolved. A rather tricky variant of electron temperature measurement which employs only satellite transitions has been proposed in (Renner et al. 2001):

$$\frac{I_{\mathrm{n}}^{\mathrm{sat}}}{I_2^{\mathrm{sat}}} \approx \frac{Q_{\mathrm{n}}}{Q_2}\exp\left\{-\frac{(Z_{\mathrm{n}}-0.6)^2 Ry}{4kT_{\mathrm{e}}}\left(1-\frac{4}{n^2}\right)\right\}. \tag{10.29}$$

Q_{n} and Q_2 are the total dielectronic satellite intensity factors for the $2lnl' \rightarrow 1snl'$ and $2l2l' \rightarrow 1s2l'$ transitions, respectively. The considerable advantage of this method is that it is even applicable, when the resonance line is absent due to high photoabsorption or due to very low electron temperatures—a typical situation in dense strongly coupled plasmas (Rosmej et al. 1997, 1998, 2000, 2003; Renner et al. 2001).

We note that another important excitation channel for satellite transitions is via electron collisional excitation from inner-shells. Concerning the above-discussed example of satellite transitions near Ly_{α}, this excitation channel reads

$$\text{inner-shell excitation} : 1s2l + e \rightarrow 2l2l' + e.$$

This excitation channel is important for satellite transitions with low autoionizing rates but high radiative decay rates. It drives satellite intensities, which allow an advanced characterization of the plasma (determination of charge exchange effects in tokamaks, characterization of suprathermal electrons, to be discussed below). For electron temperature measurements, the inner-shell excitation channel should be avoided.

Figure 10.2 shows the simulations of the spectral distribution near Ly_{α} carried out with the MARIA code (Rosmej 1997, 1998, 2001, 2006, 2012a). Dielectronic satellites $2l2l'$ as well as $2l3l'$-satellites are included in the simulations for a dense plasma: $n_{\mathrm{e}} = 10^{21}\ \mathrm{cm}^{-3}$. Several $2l3l'$-satellites are located at the blue wavelength side of Ly_{α}. For these particular transitions, LS-coupling effects are as important as the screening effect originating from the spectator electron. As can be seen, numerous satellites are located at the blue wavelengths wing of the resonance line, so-called blue satellites (Rosmej and Abdallah 1998).

Figure 10.3 shows the MARIA simulations of the spectral distribution near He_{α}, dielectronic satellites $1s2l2l'$, $1s2l3l'$, $1s2l4l'$, and $1s2l5l'$ which are included in the simulations. In all cases (Figs. 10.2 and 10.3), a strong sensitivity to electron temperature is seen from dominating until vanishing dielectronic satellite contribution. The blue curve in Fig. 10.3 shows the impact of the higher-order satellite emission ($n > 3$) on the intensity near the resonance line Helium-alpha. It can clearly be seen that higher-order satellites may still contribute considerably to the overall line emission.

Figure 10.1 shows also the fitting of the experimental spectrum obtained in a dense laser-produced plasma experiment taking into account opacity effects (important only for the Ly_{α}-line). A good match to the experimental data is obtained for $kT_{\mathrm{e}} = 210$ eV and $n_{\mathrm{e}} = 3 \times 10^{20}\ \mathrm{cm}^{-3}$. The effective photon path length was

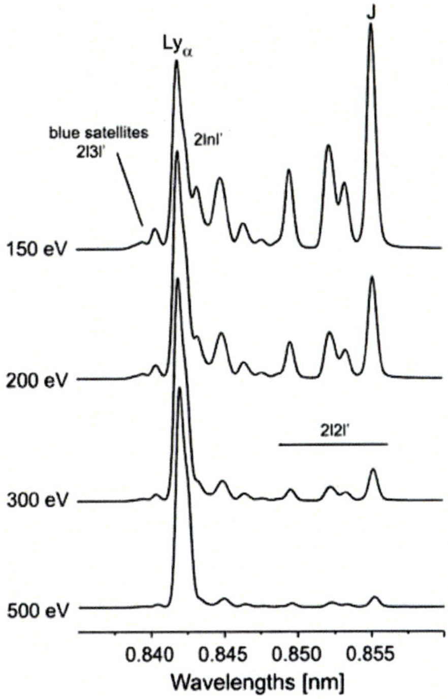

Fig. 10.2 MARIA simulations of the dielectronic satellite emission near Lyman-alpha of H-like Mg ions in dependence of electron temperature, $n_e = 10^{21}$ cm^{-3}

$L_{\text{eff}} = 500$ μm (determined from the width of the Lyman-alpha lines as well as the intensity ratio of the Lyman-alpha components), and the inhomogeneity parameter (see (1.42)) was $s = 1.3$ (determined from the dip between the Lyman-alpha components). An ion temperature of $kT_{\text{i}} = 100$ eV is assumed, and a convolution with an apparatus function $\lambda/\delta\lambda = 5000$ has been made. We note that the opacity broadening of Ly$_\alpha$ has been used to stabilize the fitting of the radiation transport. In this case, the line center optical thicknesses of Lyman-alpha lines are $\tau_0(\text{Ly}_{\alpha 1/2}) \approx 6$, $\tau_0(\text{Ly}_{\alpha 3/2}) \approx 12$; those of the satellites are of the order of $\tau_0(2l2l') \approx 2 \times 10^{-2}$. Figures 10.2 and 10.3 demonstrate that even in high-density plasmas, the temperature diagnostic via dielectronic satellite transitions works very well.

10.2.2 Ionization Temperature

Gabriel has also introduced the "ionization temperature T_Z" to plasma spectroscopy in order to characterize ionizing and recombining plasmas (Gabriel 1972). In general terms, the ionization temperature is the temperature used to solve (6.7), (6.22) for a certain density setting the left-hand side to zero (stationary and non-diffusive). This provides a certain set of ionic populations n_Z. If in an

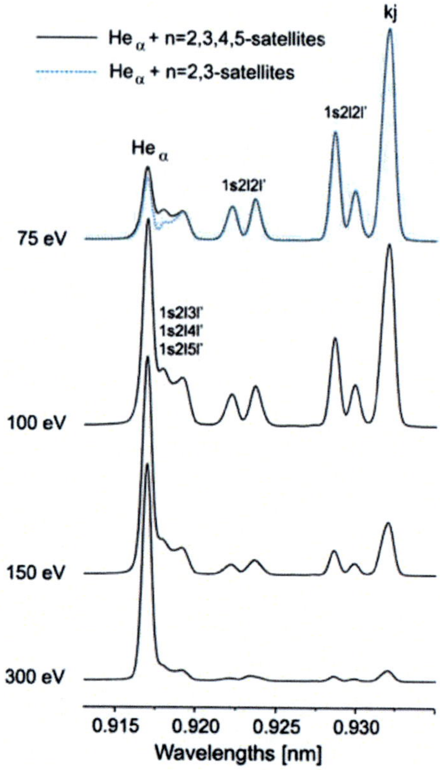

Fig. 10.3 MARIA simulations of the dielectronic satellite emission near Helium-alpha of He-like Mg ions in dependence of electron temperature, $n_e = 10^{21}$ cm^{-3}. The difference between the blue and black curve near Helium-alpha shows the impact of higher-order satellites $1s2lnl'$ with $n > 3$

experiment the electron temperature is known (e.g., by means of the dielectronic satellite method described above) and if, e.g., the ratio of the determined ionic populations n_{Z+1}/n_Z is smaller than it would correspond to the solution of (6.7), (6.22) (left-hand side is zero), the plasma is called ionizing. If n_{Z+1}/n_Z is larger, the plasma is called recombining. The physical picture behind this is as follows: Let us assume a rapid increase of the electron temperature that results in a subsequent plasma heating (e.g., a massive target is irradiated by a laser). Due to the slow relaxation time according to (6.48), the ionic populations need a considerable time to adopt their populations to the corresponding electron temperature. In the initial phase, the ionic populations are lagging behind the electron temperature and the plasma is called ionizing. Only after a rather long time (order of $\tau_{Z,Z+1}$), the ionic populations correspond to the electron temperature. The simulations of Fig. 6.9 provide detailed insight for this example. At an electron density of 10^{21} cm^{-3}, only after 1 ns the ionic populations have been stabilized. It is important to note that not the absolute time is important for the rapidity of the ionization but the inverse of the rates that are density dependent (see (6.48)). In more general terms, the ionic populations have stabilized after $t > 10^{12}$ cm^3 s/n_e for the K-shell of highly charged ions (see (6.50)).

Let us now assume that the electron temperature is rapidly switched off. Also in this case, the ionic populations need the time according (6.48) to decrease the plasma ionization. The plasma is therefore called recombining because higher charge states disappear successively until the ionic populations correspond to the decreased electron temperature.

In the original work of Gabriel, the radiation emission of the Li-like $1s2l2l'$-satellite transitions which had strong inner-shell excitation channels but low dielectronic capture (e.g., the *qr*-satellites) and strong dielectronic capture but low inner-shell excitation channel (e.g., the *jk*-satellites) have been employed to determine the ionic populations of the Li-like and He-like ions (note that the dielectronic capture channel for the Li-like $1s2l2l'$-satellites is connected to the He-like ground state $1s^2\ ^1S_0$, whereas the inner-shell excitation channel is connected to the Li-like states $1s^22l$). In the work (Yamamoto et al. 2005), satellite transitions near Ly_α have been employed to characterize the plasma regime. Also other emission lines can be used in order to characterize the ionizing/recombining nature of a plasma. The use of Rydberg line emission is another important example: In recombining plasmas, the Rydberg series emission is enhanced whereas in ionizing plasma, high *n*-members of the Rydberg series are barely visible.

The long time scale (6.48) to establish equilibrium in the ionic populations does not permit to employ standard temperature diagnostics which are based on the intensity ratio of resonance lines originating from different ionization stages, e.g., the line intensity ratio of the H-like Ly_α and the He-like He_α. For example, in ionizing plasmas, the intensity of the He-like He_α is enhanced due to ionization that is lagging behind the electron temperature, i.e., $T_Z < T_e$. Therefore, the electron temperature is underestimated if the transient evolution is not taken properly into account (if the time scale of characteristic changes of plasma parameters is much shorter than the characteristic time scale).

10.2.3 Relaxation Times

For the temperature diagnostic based on dielectronic satellite transitions (as discussed above), the obstacle of the long relaxation times according to (6.48) does practically *not* exist, because the employed line ratios concern only one ionization stage which then cancels in the line ratio method. Therefore, independent of any plasma regime (stationary, ionizing, recombining), the dielectronic satellite method allows to access the electron temperature and this is yet another reason why Gabriel's idea to employ satellite intensities for the temperature diagnostic is really a genius one.

Moreover, the response time of satellite transitions is much faster than for resonance lines according to (6.62). The reason is connected with the large autoionizing rate that has a characteristic time scale of the order of some 1..10 fs for *L*-shell electrons. For atomic transitions of multiple excited states, (6.62) has therefore to be modified according to (see also discussion of 1.105)

$$\tau_{\text{ji}} = \frac{1}{A_{\text{ji}} + C_{\text{ji}} + C_{\text{ji}} + \sum_k \Gamma_{\text{jk}}}. \tag{10.30}$$

This means that satellite transitions respond on a time scale of about some fs irrespective of any population mixing by collisional processes (Sect. 6.2.3). As the dielectronic capture population channel is proportional to an exponential temperature dependence (see (10.21), (10.22)), low electron temperatures are practically cut off because the dielectronic capture energy (e.g., (10.25)) is very large for highly charged ions:

$$I_{\text{ji}}^{\text{Sat}}\left(high\, \Gamma_{\text{kj}}\right) \propto \frac{\exp\left(-E_{\text{kj}}/kT_{\text{e}}\right)}{\left(kT_{\text{e}}\right)^{3/2}}. \tag{10.31}$$

In consequence, satellite transitions inherently cut off the low-density, low-temperature recombining regime. This is an extremely important and useful property in high-density plasma research as almost all high-density plasmas are very short living. This effect can clearly be seen from Fig. 1.11: The satellite transitions are confined near the target surface, whereas the He-like resonance and intercombination lines (*W* and *Y*, respectively) exist also far from the target surface.

10.2.4 Spatially Confined Emission

Inspection of the dielectronic capture channel and the correspondingly induced satellite line intensity (10.22) shows that the intensity is proportional to the square of the electron density (because the ground state n_{k} is proportional to the electron density):

$$I^{\text{sat}}(high\, \Gamma) \propto n_{\text{e}}^2. \tag{10.32}$$

Together with (10.31), the emission is therefore confined to high-density high temperature plasma areas. This effect is clearly seen on Figs. 1.11 and 1.12: Satellite transitions are visible just around the laser spot size. Line-of-sight integration effects are therefore minimized, as (10.31), (10.32) act like a "local emission source."

For He_β $1s3l3l'$-satellite transitions, an even stronger density dependence is expected. In high-density plasmas, their dominant excitation channel is dielectronic capture from the $1s2l$-states (Rosmej et al. 1998) and even density dependences up to $\propto n_{\text{e}}^3$ are possible. Figure 1.12 shows this effect on a space-resolved X-ray image of Si. In the spectral range around the He-like He_β-line, the $1s3l3l'$-satellites are much more confined to the target surface than the $1s2l3l'$-satellites (the *Z*-direction is the direction of the expanding plasma).

There is yet another wonderful property of satellite transitions which minimizes line-of-sight integration effects with respect to photon–plasma interaction: Their line center opacity (see Sect. 1.1.4) is small because the absorbing ground states for, e.g., the $2l2l'$ satellites transitions are the excited states $1s2l$ and not the atomic ground state $1s^2$ (like it is the case for the He-like resonance line). The population ratio $n(1s2l)/1s^2$ is rather small even in high-density plasmas and the maximum upper limit can be estimated from the Boltzmann relation. This results in a corresponding very low line center opacity of the satellite transitions.

We note that radiation transport effects in satellite transitions have been observed for Li-like $1s2l2l'$ transitions (Kienle et al. 1995; Elton et al. 2000; Rosmej et al. 2002a). This, however, is an exceptional case because their absorbing ground states coincide with the atomic ground and first excited states of the Li-like ions, namely the $1s^22l$ configuration. Also these obstacles can be avoided: employing higher-order satellite transitions from multiple excited states, other multiple excited configurations or even transitions from hollow ions (see also Sects. 10.6.4.2 and 10.6.4.3).

10.2.5 Electron Density

10.2.5.1 Collisional Redistribution

In dense plasmas, where electron collisions between the autoionizing levels become of increasing importance (compared to the radiative decay rates and autoionizing rates), population is effectively transferred between the autoionizing levels of a particular configuration (e.g., the $2l2l'$- and $1s2l2l'$-configuration). These angular momentum changing collisions (Vinogradov et al. 1977; Jacobs and Blaha 1980) result in characteristic changes of the satellite spectral distribution, i.e., their total contour (see also Sect. 5.6.3.3). In low-density plasmas, only those autoionizing levels are strongly populated which have a high autoionizing rate because in this case the dielectronic capture rate is large. This results in a high intensity of satellite transitions that do have high autoionizing rates and high radiative decay rates. Contrary, satellite transitions with high radiative decay rates but low autoionizing rates have small intensities (because the dielectronic capture is small). In high-density plasmas, population can be transferred via angular momentum changing collisions from highly populated levels to low populated ones, resulting in a density-dependent change of satellite line intensity. These characteristic changes of the spectral distribution can then be used for density diagnostics.

Figure 10.4 shows the effect of angular momentum changing collisions ("Density effect") on the satellite transitions near Ly_α of highly charged Mg ions. The simulations have been carried out with the MARIA code taking into account an extended level structure: LSJ-split levels of different ionization stages for ground, single, and multiple excited states have simultaneously been included. Strong density effects are indicated by red errors. Not only the $2l2l'$-satellites show strong

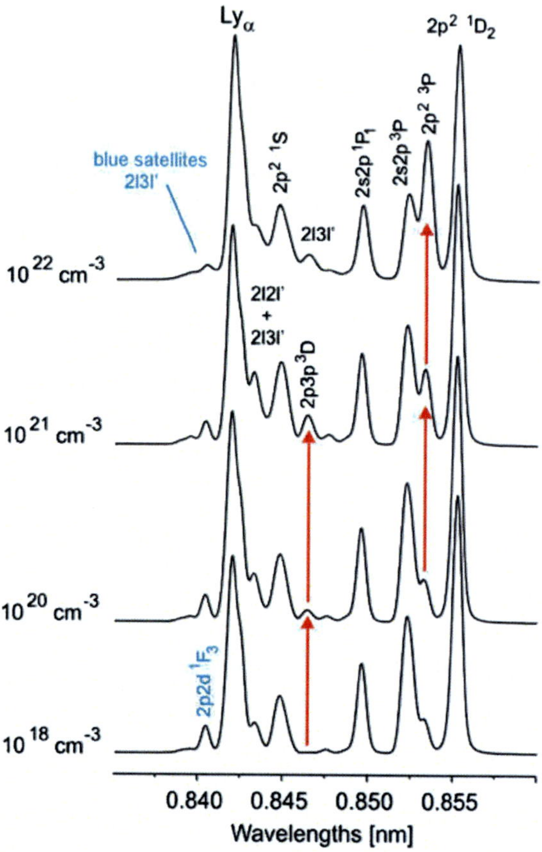

Fig. 10.4 MARIA simulations of the dielectronic satellite emission near Lyman-alpha of H-like Mg ions in dependence of electron density at $kT_e = 100$ eV. The red flashes indicate the intensity rise of particular satellite transitions with density. Blue satellites have effective negative screening due to strong angular coupling effects

density effects near $\lambda \approx 0.853$ nm, but also the $2l3l'$-satellites near $\lambda \approx 0.847$ nm. The density sensitivity of the $2l3l'$-satellites starts for lower densities, because the collisional rates between the $2l3l'$-configurations are in general larger than those for the $2l2l'$-configuration (collisional rates $C(2lnl'–2lnl'')$ increase with principal quantum number n), whereas corresponding radiative rates ($A \propto 1/n^3$) and non-radiative rates (autoionization rate $\Gamma \propto 1/n^3$) are smaller. Also indicated the so-called blue satellite emission located on the blue wing of the resonance line Lyman-alpha. These satellite transitions have negative screening (Rosmej and Abdallah 1998) that is due to the strong effect of angular momentum coupling (F-states). As can be seen from the Fig. 10.4, angular momentum changing collisions have little effect on blue satellites.

Figure 10.5 shows the MARIA simulations of the spectral distribution for the Li-like satellites near He-like Helium-alpha in dependence of the electron density. Strong density effects are visible near $\lambda \approx 0.930$ nm. Higher-order satellite transitions originating from the $1s2l3l'$-, $1s2l4l'$-, and $1s2l5l'$-configurations have been

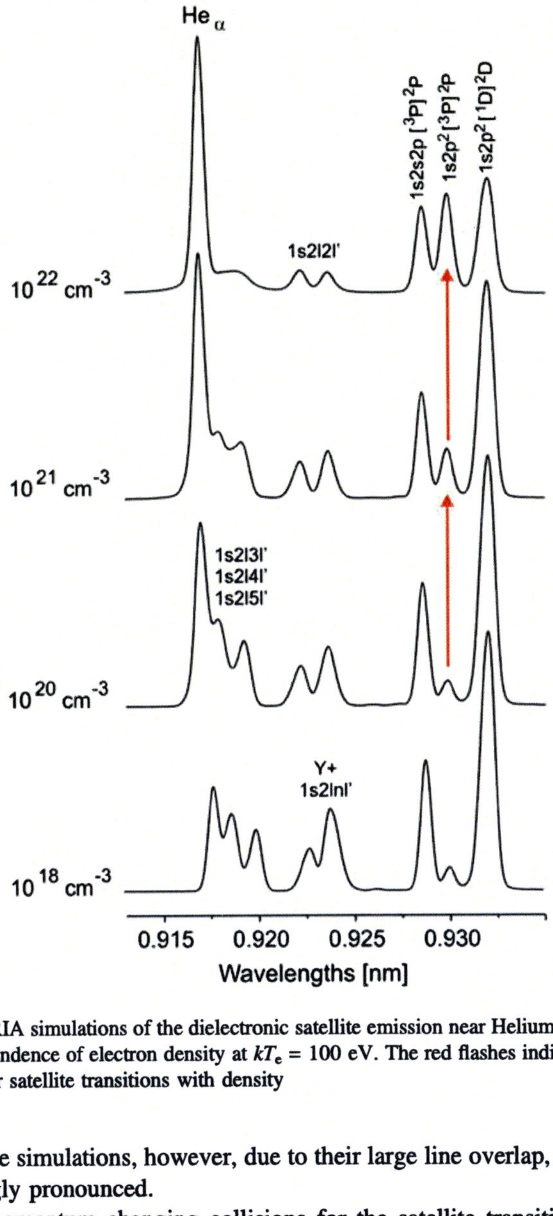

Fig. 10.5 MARIA simulations of the dielectronic satellite emission near Helium-alpha of He-like Mg ions in dependence of electron density at $kT_e = 100$ eV. The red flashes indicate the intensity rise of particular satellite transitions with density

included in the simulations, however, due to their large line overlap, density effects are not strongly pronounced.

Angular momentum changing collisions for the satellite transitions $1s2l3l' \rightarrow 1s^2 2l^1 + h\nu$ near the He_β-line (Rosmej and Abdallah 1998, Petitdemange and Rosmej 2013) are very useful: For aluminum, their density sensitivity is located in a very convenient interval of about 10^{19}–10^{22} cm^{-3} (corresponding to the critical

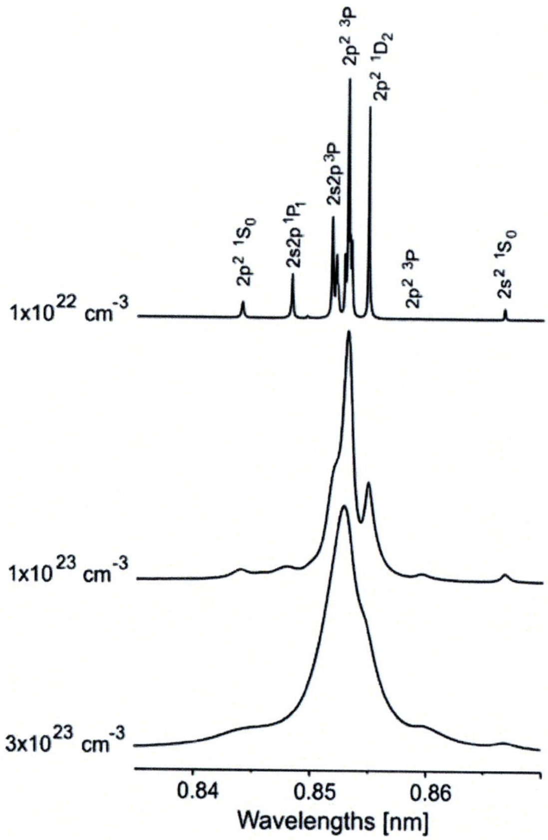

Fig. 10.6 Stark broadening simulations of the Lyman-alpha dielectronic satellite emission of He-like Mg ions in dependence of electron density at $kT_e = 100$ eV

density of almost all optical laser systems). Note, as radiative decay and autoionizing rates of $1s2l2l'$-satellites are higher while angular momentum changing collisions are smaller, their density sensitivity starts only at considerably higher densities.

Even lower densities can be accessed via Be-like satellites (Rosmej 1994, 1995a): Dielectronic capture is not only coupled to the Li-like ground state $1s^2 2s$ but likewise to the first excited state $1s^2 2p$ (see also Sect. 5.6.2.3). As the population of the $1s^2 2p$-states increases with density, the spectral distribution of the dielectronic capture reflects likewise this density dependence. The critical density for the $1s^2 2p$-states (i.e., when the radiative decay rate is equal to the collisional rate) can be estimated according to (Rosmej 1994) with the following simple analytical expression:

$$n_e^{(\mathrm{crit})} \approx 2 \times 10^{14} \cdot \frac{10^{0.05 \cdot Z_n} \cdot (Z_n - 2)^3 \cdot (\beta + 1.53)}{\beta^{0.5} \cdot (\beta + 4.3)} \ [\mathrm{cm}^{-3}], \tag{10.33}$$

$$\beta = \frac{(Z_n - 2)^2 \cdot Ry}{kT_e}. \tag{10.34}$$

Z_n is the nuclear charge, kT_e the electron temperature in [eV] and $Ry = 13.6\,\text{eV}$. For example, for aluminum at $kT_e = 100\,\text{eV}$, we obtain $n_e^{(\text{crit})} \approx 2.5 \times 10^{17}\,\text{cm}^3$.

10.2.5.2 Stark Broadening of Dielectronic Satellites

In very-high-density plasmas (near solid density), the Stark broadening analysis of satellites is very useful and has firstly been demonstrated for the $2l2l'$- and $1s2l2l'$-satellites (Woltz et al. 1991).

Figure 10.6 shows the Stark broadening simulations for the $2l2l'$-satellites of Mg carried out with the PPP code (Talin et al. 1995, 1997) assuming a statistical population between the autoionizing levels. It can clearly be seen that strong density sensitivities are obtained only for densities $n_e > 10^{22}\,\text{cm}^{-3}$.

In order to access lower electron densities via Stark broadening analysis, Rydberg-satellite transitions of the type $1s2lnl' \to 1s^2 2l^1 + h\nu$ have been studied in dense laser-produced plasma experiments with high spectral and spatial resolution (Rosmej et al. 2001a; Skobelev et al. 2002). This has stimulated Stark broadening calculations of Rydberg-satellite transitions (Rosmej et al. 2003) (see also discussion in Sect. 1.5.2).

10.2.5.3 Stark Broadening of Hollow Ions

As discussed in Sect. 1.5.4, a hollow ion (HI) is an ion, where one or more internal shells are entirely empty whereas higher shells are filled with 2 or more electrons. The hollow ion configurations are multiple excited configurations and are therefore also autoionizing configurations. Hollow ion transitions originating from the configurations K^0L^N of highly charged ions, i.e., $K^0L^N \to K^1L^{N-1} + h\nu_{\text{hollow}}$, are of particular interest for dense plasmas research: The hollow ion X-ray transitions $K^0L^N \to K^1L^{N-1} + h\nu_{\text{hollow}}$ can be easily identified as they are well separated from other transitions and, due to the large autoionizing rate, they do have very small opacity, very short emission time scale, and are sensitive to suprathermal electrons and radiation fields (Rosmej et al. 2015). It is therefore of interest to supplement the forgoing discussion (Sect. 1.5.4) with corresponding Stark broadening calculations.

Despite these outstanding properties for advanced diagnostics, hollow ion emission is rather complex: The large number of levels and transitions does not really permit ab initio simulations with a LSJ-split level structure to achieve spectroscopic precision. When employing usual reduction methods, e.g., the super-configuration method (Bar-Shalom et al. 1989) or a hydrogen-like approximation, the number of levels is reduced to a manageable number; however, the number of transitions is also strongly reduced. This reduction considerably modifies the total contour of the hollow ion transitions (e.g., important for Stark broadening analysis, see below) due to an average of transitions and other atomic data (transition probabilities, autoionizing rates, line center positions, etc.). It is therefore very difficult, to obtain a spectroscopic precision (high-resolution analysis of the spectral distribution) with the traditional super-configuration method. This reduction problem of the traditional super-configuration method has recently been solved by the "Virtual Contour Shape Kinetic Theory VCSKT" (Rosmej 2006) that has been discussed in detail in Sect. 6.3.

Figure 10.7 shows detailed Stark broadening calculations (carried out with the PPP code) for the hollow Mg ion X-ray transitions $K^0L^N \rightarrow K^1L^{N-1} + h\nu$, $kT_e = 100$ eV for $N = 1$–5. Line intensities within one configuration K^0L^N have been calculated assuming a statistical population for all LSJ-split levels in order not to mask the Stark broadening with population effects for different plasma densities. All hollow ion electric dipole transitions and all energy levels have been included in

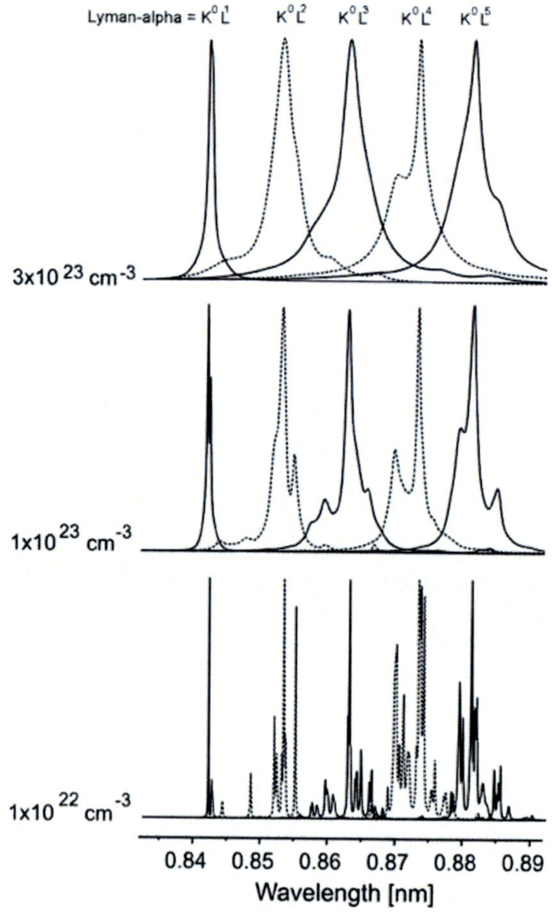

Fig. 10.7 Stark broadening simulations of the hollow ion X-ray transitions $K^0L^X \rightarrow K^1L^{X-1} + \hbar\omega_{HI}$ in magnesium (normalized to peak) in dependence of electron density at $kT_e = 100$ eV

the simulations (note that the minimum number of levels/transitions is 17/48 for the $N = 2$ configuration, 34/246 for the $N = 3$ configuration, 60/626 for the $N = 4$ configuration, and 65/827 for the $N = 5$ configuration; the number of Stark transitions is of the order of 10^6). Transitions from different charge states have been normalized to maximum peak intensity. It can be seen from Fig. 10.7 that the emission from different ionization stages is essentially separated and that strong changes of the total contours emerge for near solid density plasmas. For densities less than 10^{22} cm^{-3}, numerous single transitions are resolved (lower spectrum in Fig. 10.7). The low-density simulation indicates that the broadening of the total contour is not only determined by the Stark broadening of single transitions but also importantly by the oscillator strengths distribution over wavelengths. VCSKT provides also an appropriate answer here (see also Sect. 6.3.2.4): All line transitions are included in the simulations with their correct line center positions and oscillator strengths distribution over wavelengths (opposite to the traditional super-configuration method where new artificial line center positions are calculated from certain averages of LSJ-levels).

10.2.5.4 Interference Effects in Stark Broadening of Hollow Ions

Let us finish the Stark broadening analysis of HI with a discussion of interference effects (Griem 1964, 1974, 1997; Sobelman and Vainshtein 2006). As the lower states of the hollow ion configurations are autoionizing states by itself (states K^1L^{N-1}), the number of lower levels is also large and interference effects between upper and lower levels become important (see also Sect. 1.5.3).

Figure 10.8 compares Stark profile simulations for the hollow ion X-ray transitions $K^0L^3 \rightarrow K^1L^2 + h\nu$ with and without taking into account interference effects (intensities are normalized to peak). It can clearly be seen that interference effects lead to a considerable narrowing of the total contour as well as to a shift of the intensity peak of the total contour. Note that line narrowing effects due to

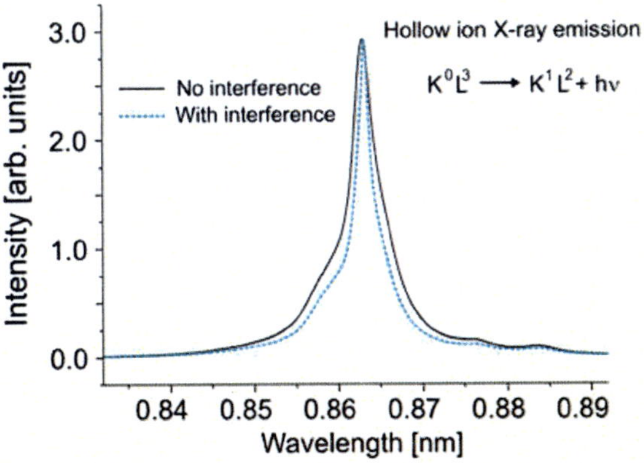

Fig. 10.8 Stark broadening simulations of the hollow ion X-ray transitions $K^0L^3 \rightarrow K^1L^2 + \hbar\omega_{HI}$ in magnesium (normalized to peak) showing the impact of the interference effects on the total contour, $n_e = 3 \times 10^{23}$ cm^{-3}, $kT_e = 100$ eV

interferences have originally been discussed for non-autoionizing levels (Aleseyev and Sobelman 1969).

10.2.5.5 Non-statistical Line Shapes

The traditional method of line shape calculations employs the so-called statistical lines shapes where the atomic level population of the corresponding configurations is assumed to be in statistical equilibrium (Griem 1974, 1997). In dense plasmas, however, the use of intercombination lines or other forbidden lines is of interest due to their advantageous properties with respect to opacity because despite of their low oscillator strengths, non-statistical effects in level populations (see Chap. 6) might drive intensities that are of the order of usual resonance lines. It is therefore of great interest to study non-statistical effects for the line shape calculations (so-called dynamical line shapes).

Figure 10.9 demonstrates the effect of so-called dynamical line shapes for the He-like resonance and intercombination lines of aluminum when the non-statistical populations of the $1s2l$-levels are taken into account. The line shape calculations have been performed with the PPP code; the dynamical properties of the level populations have been calculated with the MARIA code employing a relativistic atomic structure (LSJ-split), multipole transitions, cascading and ionization balance. Figure 10.9 demonstrates the case for He-like aluminum (spectral range of the

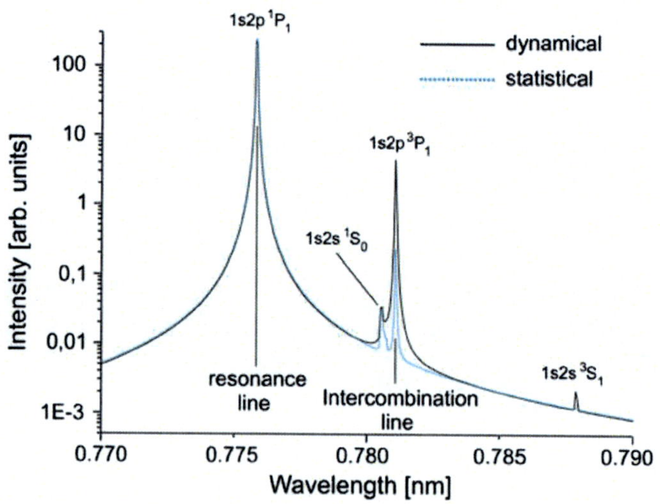

Fig. 10.9 Comparison of statistical and dynamical line shapes of the X-ray transitions $K^1L^1 \rightarrow K^2 + \hbar\omega$ in He-like aluminum for $n_e = 10^{21}$ cm^{-3}, $kT_e = 100$ eV. MARIA simulations of the dynamical level populations include LSJ-split level structure, electric and magnetic multipole transitions, and ionization balance calculation

He-like resonance line $W = 1s2p\ ^1P_1 – 1s^2\ ^1S_0$ and intercombination line $Y = 1s2p\ ^3P_1 – 1s^2\ ^1S_0$) for an electron density of $n_e = 10^{21}$ cm^{-3} and an electron temperature of $kT_e = 100$ eV. The simulations show that the intercombination line shape (Y) is essentially modified: Intensity and line wings are enhanced by about an order of magnitude providing a larger diagnostic potential as believed in the framework of the statistical line shape approach only. The two smaller peaks near 0.781 and 0.788 nm are due to Stark-induced transitions from the $1s2s\ ^1S_0$ and $1s2s\ ^3S_1$ levels, respectively. Note that the PPP code does not include multipole transitions and the intensity of the transition originating from the $1s2s\ ^3S_1$ level is therefore entirely due to the Stark mixing but not due to the magnetic quadrupole contribution (see discussion in Sect. 1.2.2).

10.3 Magnetic Fusion

10.3.1 Neutral Particle Background and Self-consistent Charge Exchange Coupling to Excited States

The confinement of the plasma is one of the most important issues in magnetic fusion research, and intensive efforts have therefore been devoted to the understanding of the particle transport. However, the physical processes that underlie plasma transport in toroidally confined plasmas are not so well understood. The plasma transport induced by Coulomb collisions (so-called classical or neo-classical transport) is often much less than what is actually observed (Engelhardt 1982; Hulse 1983; Pasini et al. 1990) and thus the transport is called anomalous.

Methods which determine the particle transport *independent* of theoretical plasma models are therefore of fundamental importance in the magnetic fusion research. Spectroscopic methods have turned out to be very effective, and one of the most powerful methods is based on the space- and time-resolved observation of the line emission from impurity ions (Engelhardt 1982; Hulse 1983; Pasini 1990). Emission spectroscopic methods (so-called passive methods) receive a renewed interest in view of the future installation ITER (International Thermonuclear Experimental Reactor, construction has begun in 2010 at Cadarache in France (ITER 2019)) because the strong radiation hazard during fusion operation combined with the large minor radius will not allow efficient use of many diagnostics (in particular active ones) that are currently in use at mid-sized tokamaks.

The radiation emission of the impurities (and also those from the neutral H/D/T) is simulated from an atomic physics model (see also Chap. 6):

$$\begin{aligned}\frac{\partial n_Z}{\partial t} + \nabla\left(\vec{\Gamma}_Z\right) &= -n_Z\left(I_{Z,Z+1} + T_{Z,Z-1} + R_{Z,Z-1} + D_{Z,Z-1}\right)\\ &\quad + n_{Z-1}\left(I_{Z-1,Z}\right)\\ &\quad + n_{Z+1}\left(T_{Z+1,Z} + R_{Z+1,Z} + D_{Z+1,Z}\right).\end{aligned} \tag{10.35}$$

$\vec{\Gamma}_Z$ is the particle flux (Z indicates the charge of the ion). With given temperature and density profiles, one tries to match the experimental observations by a best fit of $\vec{\Gamma}_Z$. For these purposes, it turned out to be convenient to split the flux into a diffusive and convective term according to $\vec{\Gamma}_Z = -D_Z \nabla n_Z + \vec{V}_Z n_Z$, D_Z is the diffusion coefficient (note that D_Z is the diffusion coefficient whereas $D_{Z,Z-1}$ is the dielectronic recombination rate coefficient connecting the charge states "Z" and "$Z - 1$") and $\vec{V}_Z$ is the convective velocity. These parameters are then varied in a numerical procedure in order to best fit the spectral emission data. The importance in this type of analysis lies in the fact that it provides a plasma simulation-independent information (independent from, e.g., turbulence models) for the diffusion coefficient and the convective velocity (Hulse 1983).

Under real experimental conditions of magnetically confined fusion plasmas, the impurity ions do interact with the plasma background H/D/T via charge exchange. This in turn leads to a change of the radial charge state distribution of the impurity ions, an effect which has a large impact for the analysis and the interpretation of possible particle transport: Diffusion in space (particle transport) and diffusion in charge states (charge exchange) are of similar nature in the framework of the traditional particle transport analysis (via diffusion coefficients D and convective velocities V (Rosmej and Lisitsa 1998; Rosmej et al. 1999a, Shurygin 2004)). This can easily be seen from the more generalized equation

$$\begin{aligned}\frac{\partial n_Z}{\partial t} + \nabla\left(\vec{\Gamma}_Z\right) = &-n_Z\left(I_{Z,Z+1} + T_{Z,Z-1} + R_{Z,Z-1} + D_{Z,Z-1} + \mathrm{Cx}_{Z,Z-1}\right) \\ &+ n_{Z-1}\left(I_{Z-1,Z} + \mathrm{Cx}_{Z-1,Z}\right) \\ &+ n_{Z+1}\left(T_{Z+1,Z} + R_{Z+1,Z} + D_{Z+1,Z} + \mathrm{Cx}_{Z+1,Z}\right).\end{aligned} \tag{10.36}$$

$\mathrm{Cx}_{Z,\,Z-1}$ etc., indicate possible charge exchange processes between the radiating test element (e.g., intrinsic impurities) and other species (namely, hydrogen, deuterium, tritium, and helium). Let us assume that the partial derivative is zero and integrate the set of (10.36) over space. The integration over space transforms the diffusion term into the so-called tau-approximation. Note that the tau-approximation is a rather powerful method of particle transport analysis which even permits to study details of the line emission not only of resonance lines but from forbidden lines too (Rosmej et al. 1999a; Rosmej and Lisitsa 1998). In the "tau-approximation" (10.36) takes the form

$$\begin{aligned} &n_Z\left(I_{Z,Z+1}+T_{Z,Z-1}+R_{Z,Z-1}+D_{Z,Z-1}\right)+n_Z\left(\mathrm{Cx}_{Z,Z-1}+\frac{1}{\tau_{Z,Z+1}}\right)\\ &\quad=n_{Z-1}\left(I_{Z-1,Z}\right)+n_{Z-1}\left(\mathrm{Cx}_{Z-1,Z}+\frac{1}{\tau_{Z-1,Z}}\right)\\ &\qquad+n_{Z+1}\left(T_{Z+1,Z}+R_{Z+1,Z}+D_{Z+1,Z}\right)+n_{Z+1}\left(\mathrm{Cx}_{Z+1,Z}+\frac{1}{\tau_{Z+1,Z}}\right). \end{aligned} \tag{10.37}$$

$\tau_{Z,\ Z+1}$ etc., are the respective diffusion times. It is clearly seen that diffusion/transport (represented by the tau-terms in (10.37)) are of the same origin as charge exchange processes (Cx-terms in (10.37)). It is therefore difficult to characterize the particle transport on the basis of (10.35): If the charge exchange is a free parameter as well as diffusion D_Z and convective velocity V_Z, their significance is not so evident as charge exchange (diffusion in charge states) and particle transport (diffusion in space) are overlapping effects.

In order to circumvent this difficulty, a self-consistent analysis has been proposed (Rosmej et al. 2006a, b) to eliminate the free parameters for the charge exchange: The coupling is a self-consisted excited states coupling of the tracer (impurity) kinetics to the plasma background (H,D,T) via atomic physics processes (charge exchange). The matrix coupling elements $M_{ji}(H,D,T,X)$ can schematically be written

$$M_{ji}(H,D,T,X)=n_j^{H,D,T}n_i^X\left\langle\sigma_{ji}^{Cx}V_{rel}\right\rangle. \tag{10.38}$$

H, D, T indicate the hydrogen, deuterium, tritium, and X is a spectroscopic tracer element (e.g., He, an intrinsic impurity or any other element intentionally introduced for diagnostic purposes), $n_j^{H,D,T}$ is the population density of the elements (H, D,T) in state "*j*", n_i^X is the population density of the tracer element in state "*i*", σ_{ji}^{Cx} is the charge exchange cross section from state "*j*" to state "*i*" between the elements (H, D, T) and X, V_{rel} is the relative particle velocity, and the brackets indicate an average over the particle energy distribution functions. As the coupling matrix elements according to (10.38) contain the product of different population densities, the system of equations (H, D, T) and (X) is nonlinear (even in the optically thin plasma approximation). The self-consistent numerical simulation of multi-ion multilevel (LSJ-split) non-LTE atomic kinetic systems coupled by charge exchange processes via the excited states coupling matrix (10.38) has been realized in the numerical code "SOPHIA" (Rosmej et al. 2006a; Rosmej 2012a).

The coupling matrix approach according to (10.38) lies in the fact that the selection rules for the charge exchange processes are respected: Charge transfer from excited states is *directly* coupled to excited states. Therefore, the population flow due to charge exchange is consistently treated without any free parameter along with the population flow of usual collisional–radiative processes. The excited states coupling also avoids critical divergences which arise from the strong scaling of the charge exchange cross sections with principal quantum number "*n*":

Table 10.3 Population density n_H multiplied by the 4th power of the principal quantum number

Principal quantum number	$n_H(n) \cdot n^4$	$n_H(n) \cdot n^4 / n_H(n = 1)$
$n = 1$	1.16D−03	1.00D+00
$n = 2$	1.24D−06	1.07D−03
$n = 3$	7.18D−07	6.21D−04
$n = 4$	3.28D−06	2.84D−03
$n = 5$	1.17D−05	1.01D−02
$n = 10$	6.66D−04	5.77D−01
$n = 15$	7.40D−03	6.41D+00
$n = 20$	4.12D−02	3.57D+01
$n = 25$	1.57D−01	1.36D+02

The second column indicates the absolute fraction whereas the third column indicates the relative importance with respect to the hydrogen ground state 1*s*. The plasma parameters are $kT_e = 3$ eV, $n_e = 10^{13}$ cm^{-3}. The populations n_H are normalized according to $\Sigma\, n_H = 1$. Note that the neutral fraction depends strongly on temperature but also on the flow of neutrals from the wall to the plasma center

$\sigma^{Cx} \propto n^4$ (classical scaling). In fact, under typical conditions of ITER, the hydrogen excited states populations increase rapidly due to the increasing statistical weights. Combined with the charge exchange scaling, this results finally in an effective divergence $\propto n^6$. This charge exchange-driven divergence is therefore much more pronounced than the well-known divergence of the partition sum (quadratic divergence).

Table 10.3 shows the importance of the excited state-driven charge exchange processes. The neutral fraction depends strongly on electron temperature but also on the neutral flow from the walls to the plasma center (to be discussed in detail below, Sect. 10.3.2). For about $n > 15$, excited state contributions become even more important than the ground state contribution. At $n = 20$, all charge exchange flow is driven by excited states rather than the ground state. Therefore, any level cutoff (see also Chap. 8) is highly critical and numerical simulations are rather instable. In this respect, also the effective rate coefficients proposed in (Abramov et al. 1985) have to be employed with caution.

In the framework of the self-consistent excited states coupling approach (Rosmej et al. 2006a), no critical level cutoff is present (or necessary) because charge exchange and collisions are treated on a unique footing: A large charge exchange flow into highly excited states is directly redistributed by collisions between even higher excited/next ionization states before radiative decay can populate the ground states. Figures 10.10a,b visualize schematically the relevant mechanisms in the self-consistent model. Figure 10.10a shows the thermal limit $n_{thermal}$ corresponding to usual collisional–radiative processes. Above this limit, Partial-Local-Thermodynamic-Equilibrium (PLTE) holds true, i.e., a Boltzmann-level population starting from a certain principal quantum number n. This corresponds to the condition that collisional de-excitation is much more important than radiative decay rates (indicated as "$C \gg A$" in Fig. 10.10a). As radiative decay rates decrease

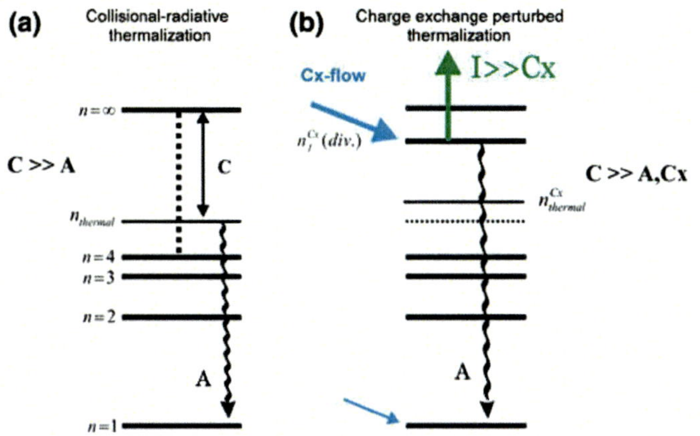

Fig. 10.10 Principle mechanisms of the self-consistent excited states coupling of charge exchange and thermalization by collisions, **a** standard collisional–radiative thermalization, **b** collisional–radiative thermalization perturbed by charge exchange flow

strongly with principal quantum number (approximately $A \propto n^{-3}$ in the hydrogenic approximation) while collisional rates are strongly increasing (approximately $C \propto n^4$ between the states $n \rightarrow n+1$ neglecting Gaunt-factor variations) PLTE starts from high lying levels. In the hydrogenic approximation, this condition can be formulated for a plasma consisting of electrons, ions, and atoms as follows:

$$n_{\mathrm{e,crit}} \geq 6 \times 10^{19} Z^7 \frac{(n_{\mathrm{thermal}}-1)^{2n_{\mathrm{thermal}}-2}}{n_{\mathrm{thermal}}^3 (n_{\mathrm{thermal}}+1)^{2n_{\mathrm{thermal}}+2}} \left(\frac{kT_{\mathrm{e}}(eV)}{Z^2\,Ry}\right)^{1/2} \left[\mathrm{cm}^{-3}\right]. \quad (10.39)$$

$n_{\mathrm{e,crit}}$ is the critical electrons density in [cm^{-3}] above which a Boltzmann population of levels, i.e.,

$$\frac{n_{\mathrm{j}}}{n_{\mathrm{i}}} = \frac{g_{\mathrm{j}}}{g_{\mathrm{i}}} \exp\left\{-\left(E_{\mathrm{i}}^{Z} - E_{\mathrm{j}}^{Z}\right)/kT_{\mathrm{e}}\right\} \quad (10.40)$$

holds true for all levels with principal quantum number larger than n_{thermal}, kT is the electron temperature in [eV], Z is the ionic charge, Ry = 13.6 eV, g_{i} and g_{j} are the statistical weights of the lower and upper levels, E_{i}^{Z} and E_{j}^{Z} are the respective state energies (note, that $E_{\mathrm{i}}^{Z} - E_{\mathrm{j}}^{Z} > 0$). For hydrogen ($Z$ = 1), n_{thermal} = 1 (corresponding that all levels are distributed according to a Boltzmann population) and kT_{e} = 1 eV from which it follows $n_{\mathrm{e,crit}} \approx 1 \times 10^{18}$ cm^{-3}. Note that, e.g., for H-like molybdenum and kT_{e} = 2 keV the critical density is very high: $n_{\mathrm{e,crit}} \approx 2 \times 10^{29}$ cm^{-3} showing that it is not the absolute density, which is of importance to obtain thermodynamic equilibrium conditions but rather the relation between the collisional and radiative decay rates. Equation (10.39) has a well-defined asymptote for large quantum numbers n_{thermal}:

$$\lim_{n_{\text{thermal}}\to\infty}\left\{\frac{(n_{\text{thermal}}-1)^{2n_{\text{thermal}}-2}}{n_{\text{thermal}}^3(n_{\text{thermal}}+1)^{2n_{\text{thermal}}+2}}\right\}$$
$$= \lim_{n_{\text{thermal}}\to\infty}\left\{\frac{1}{n_{\text{thermal}}^3(n_{\text{thermal}}+1)^4}\left(\frac{n_{\text{thermal}}-1}{n_{\text{thermal}}+1}\right)^{2n_{\text{thermal}}-2}\right\}\approx\frac{0.0183}{n_{\text{thermal}}^7} \tag{10.41}$$

because

$$\lim_{n_{\text{thermal}}\to\infty}\left\{\left(\frac{n_{\text{thermal}}-1}{n_{\text{thermal}}+1}\right)^{2n_{\text{crit}}-2}\right\}\approx\frac{1}{54.6}. \tag{10.42}$$

Therefore, we can write

$$n_{\text{e,crit}}\approx 6\times 10^{19}Z_{\text{eff}}^7\frac{1}{n_{\text{thermal}}^3}\frac{0.0183}{n_{\text{thermal}}^4}\left(\frac{kT_{\text{e}}(eV)}{Z_{\text{eff}}^2Ry}\right)^{1/2}$$
$$\approx 10^{18}\frac{Z_{\text{eff}}^7}{n_{\text{thermal}}^7}\left(\frac{kT_{\text{e}}(eV)}{Z_{\text{eff}}^2Ry}\right)^{1/2}\left[\text{cm}^{-3}\right]. \tag{10.43}$$

Equation (10.43) shows that the critical electron density scales with the 7th power of the principal quantum number and with the 7th power of the effective charge.

Figure 10.10b shows the case, when charge exchange flow (indicated by the blue arrows) populates the levels: The thermal limit n_{thermal} is changed to $n_{\text{thermal}}^{\text{Cx}}$ because collisional rates have to be compared now not only to radiative decay but also to charge transfer rates (indicated by $C \gg A$, Cx in Fig. 10.10b). As can be seen from Table 10.3, charge exchange from excited states strongly competes with the charge exchange from the ground state and at, e.g., $n = 15$, the contribution of excited states is already more than six times greater than the ground state, while, e.g., for $n = 25$, the contribution of excited states is more than 100 times greater than from the ground state. The contribution of excited states is therefore diverging (indicated schematically with $n_{\text{f}}^{\text{Cx}}(div.)$ in Fig. 10.10b). Whether the diverging charge exchange contribution strongly perturbs the standard collisional–radiative model depends, whether the radiative decay rates from the states $n_{\text{f}}^{\text{Cx}}(div.)$ transfer this diverging channel to the ground state or not. A diverging charge exchange contribution to the ground state would result into a strong perturbation of all collisional excitation–ionization processes, e.g., the ionization equilibrium and radiation loss. Therefore, a direct coupling of the excited states charge exchange contributions, via, e.g., effective charge exchange rates (Abramov et al. 1985) would be a highly critical and instable situation.

Let us therefore consider the situation more closely in the framework of the self-consistent model, where charge exchange from excited states is coupled to the excited states while all excited states (including the donor and target particles) are explicitly included in the collisional–radiative model. The final quantum number n_f for the charge transfer process from the neutrals (H/D/T) to the impurity ions with effective charge Z_{eff} can be estimated from the classical over barrier model as follows:

$$n_f \approx n_i \cdot \frac{Z_{eff}}{\sqrt{1+\sqrt{Z_{eff}}-1/\sqrt{Z_{eff}}}}, \tag{10.44}$$

where n_i is the principal quantum number of the donor projectile from which charge transfer proceeds (H/D/T in our case), Z_{eff} is the effective charge of the acceptor ion before charge transfer. For example, charge transfer from the hydrogen ground state into H-like argon: $n_i = 1$, $Z_{eff} \approx 17$ resulting in $n_i \approx 8$ (note that different models provide slightly different principal quantum numbers, e.g., according to (Ostrovsky 1995; Cornelius et al. 2000) $n_i \approx 10$). From (10.43), it follows that for a certain electron density, PLTE is achieved for principal quantum numbers larger than

$$n_{thermal} \approx \frac{373}{n_e^{1/7}(cm^{-3})} \cdot \left(\frac{kT_e(eV)}{Z_{eff}^2 Ry}\right)^{1/14}. \tag{10.45}$$

This means that all charge exchange flow into principal quantum numbers n_f that are larger than the thermal limit from (10.45) (for a certain fixed electron density and temperature) is rapidly thermalized and does not contribute to the ground state population, i.e., if the condition

$$n_f > n_{thermal} \tag{10.46}$$

holds true. In order to estimate whether condition (10.46) covers a parameter interval of practical interest for magnetically confined plasmas, let us assume an electron temperature $kT_e = 0.5 \cdot Z_{eff}^2 Ry$ and the asymptotic scaling $n_f \approx n_i \cdot Z_{eff}^{3/4}$ of (10.44). We then obtain from relation (10.46) and (10.45)

$$Z_{eff} > \frac{2500}{n_i^{4/3} \cdot n_e^{4/21}}. \tag{10.47}$$

As Table 10.3 demonstrates, excited states contributions start to rise with increasing quantum number $n_i^{(H/D/T)} \geq 4$. This increase is physically connected with the transition to PLTE for a certain high-n-quantum number. Let us therefore estimate the thermal limit (10.45) for $n_e = 10^{13}\,cm^{-3}$ and $kT_e = 3\,eV$, i.e., the parameters of Table 10.3: $n_{thermal}^{(H/D/T)} \approx 4.4$. The thermal limit therefore corresponds approximately to the quantum number from which on excited states contributions start to diverge (see Fig. 10.10), i.e., $n_{thermal}^{(H/D/T)} \approx n_f^{Cx}(\text{div.})$. We therefore can approximate n_i in (10.47) by $n_i \approx 355 \cdot n_e^{1/7}$ resulting into

$$Z_{\text{eff}}^{(\text{thermalized}-\text{Cx})} > 1, \tag{10.48}$$

where the upper index thermalized—*Cx* indicates that the divergent charge exchange flow is essentially thermalized rather than decaying to the ground state. From the kinetic point of view, the charge exchange flow decreases the impurity charge state from $Z + H \to (Z-1) + p$ while the thermalization due to collisions (which is a thermalization with the continuum) increases the charge state from $(Z-1) + e \to Z + 2e$. Therefore, the impurity charge state is essentially unchanged. As relation (10.48) demonstrates, for almost all impurities of interest thermalization takes place and is also approximately independent from the electron density.

Detailed numerical self-consistent calculations carried out with the SOPHIA code (Rosmej et al. 2006a; Rosmej 2012a) demonstrate that the thermal limit (10.39) is slightly increased if charge exchange is consistently coupled to excited states. This is indicated in Fig. 10.10b with the new thermal limit $n_{\text{thermal}}^{\text{Cx}}$. The increase, however, is rather moderate, and the general mechanism of thermalization according to (10.48) is not changed (indicated with "$I \gg Cx$" for $n_{\text{f}}^{\text{Cx}}(div.)$ in Fig. 10.10b). Therefore, the strong charge exchange flow into the excited state coupled system is naturally stabilized for almost all systems of practical interest. In consequence, this flow does not lead to a divergent population of the atomic levels. This means, that on the one hand, no artificial (and therefore uncertain) level cutoff is needed to stabilize the system and, on the other hand, the number of levels included in the simulations is not very critical (if a few principal quantum numbers are included that are larger than $n_{\text{thermal}}^{\text{Cx}}$). The last point is a very advantageous additional feature despite of the continuous controversial discussion of the ionization potential depression (see Chap. 8).

It is important to emphasize that (10.48) does *NOT* mean that excited states charge exchange contributions can effectively been neglected in fusion relevant plasmas. On the contrary, particle transport studies have to consider simultaneously charge exchange effects as both phenomena enter in a very similar manner in the general system of population equations (see 10.37). The drawback in standard methods that employ free parameters for particle transport and charge exchange is that these two parameters are very difficult to separate from each other because charge exchange effects and particle transport effects overlap (in other words: at fixed spatial position r_1 for a certain charge state Z_1 a change from Z_1 to $(Z_1 - 1)$ can be induced by charge exchange with a neutral particle, however, the charge state $(Z_1 - 1)$ can also be obtained at position r_1 if an ion with charge $(Z_1 - 1)$ diffuses from a position r_2 to the position r_1). Therefore, both cases result into the same charge state $(Z_1 - 1)$ at r_1, but their physical origin and interpretation is quite different (Rosmej et al. 2006a; Shurygin 2008). In the self-consistent model, charge exchange is not a free parameter but consistently calculated from the populations of the acceptor and donor particles and the "overlap of free parameters" does not exist. The calculations itself are stabilized including explicitly excited states for the impurity particles and also the neutral particles that are then coupled to each other via charge exchange (which is selective in *n*-quantum numbers). Therefore, the free parameter for charge exchange is removed from the system of equations (because it is calculated consistently along with all populations; see, e.g., (10.38)) and the only free parameter that remains in the system is related to the particle transport as desired for diagnostics.

10.3.2 Natural Neutral Background and Neutral Beam Injection: Perturbation of X-ray Impurity Emission

The particle transport discussion related to (10.35) was based on the ionic charge state distribution. X-ray spectroscopy, however, can provide a much more rich information via the high-resolution X-ray spectral distribution. In particular, it enables to distinguish with the help of particular selected atomic systems to extract detailed information of charge exchange and impurity transport. In this context, a dedicated experimental and theoretical analysis of the He-like lines W, X, Y, Z, the He-beta resonance line ($W3$: = $1s3p\ ^1P_1{-}1s^2\ ^1S_0$), intercombination line ($Y3$: = $1s3p\ ^3P_1{-}1s^2\ ^1S_0$) as well as the Li-like satellites $1s2l2l'{-}1s^22l''$ of highly charged impurity ions have been undertaken (Rosmej 1998; Rosmej et al. 1999a, 2006a, b; Rice et al. 2018; Rosmej and Lisitsa 1998).

Figure 10.11 shows the time-resolved soft X-ray impurity spectrum (Rosmej et al. 1999a) from the TEXTOR tokamak (solid black curve) of gas puff injected

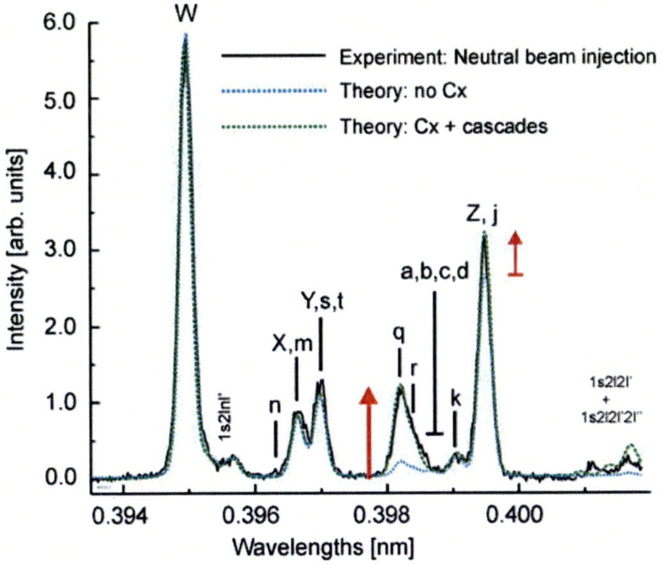

Fig. 10.11 Time-resolved X-ray impurity spectrum (t = 3.5–3.6 s) of gas puff injected argon during neutral beam injection. The red flash indicates a strong rise of Li-like satellite emission when charge exchange is included in the theory bringing the MARIA simulations in very close agreement to the data

argon during neutral beam injection with 1.2 MW. The high spectral resolution enables the distinct observation of the He-like lines $W = 1s^2\text{–}1s2p\ ^1P_1$, $X = 1s^2\text{–}1s2p\ ^3P_2$, $Y = 1s^2\text{–}1s2p\ ^3P_1$, $Z = 1s^2\text{–}1s2s\ ^3S_1$ and also to separate numerous Li-like satellites from the $1s2l2l'$-configuration (indicated in Fig. 10.11 as $m = 1s[2p^2\ ^1S]\ ^2S_{1/2}\text{–}1s^22p\ ^2P_{3/2}$, $n = 1s[2p^2\ ^1S]\ ^2S_{1/2}\text{–}1s^22p\ ^2P_{1/2}$, $s = 1s[2s2p\ ^1P]\ ^2P_{3/2}\text{–}1s^22s\ ^2S_{1/2}$, $t = 1s[2s2p\ ^1P]\ ^2P_{1/2}\text{–}1s^22s\ ^2S_{1/2}$, $q = 1s[2s2p\ ^3P]\ ^2P_{3/2}\text{–}1s^22s\ ^2S_{1/2}$, $r = 1s[2s2p\ ^3P]\ ^2P_{1/2}\text{–}1s^22s\ ^2S_{1/2}$, $a = 1s[2p^2\ ^3P]\ ^2P_{3/2}\text{–}1s^22p\ ^2P_{3/2}$, $b = 1s[2p^2\ ^3P]\ ^2P_{3/2}\text{–}1s^22p\ ^2P_{1/2}$, $c = 1s[2p^2\ ^3P]\ ^2P_{1/2}\text{–}1s^22p\ ^2P_{3/2}$, $d = 1s[2p^2\ ^3P]\ ^2P_{1/2}\text{–}1s^22p\ ^2P_{1/2}$, $k = 1s[2p^2\ ^1D]\ ^2D_{3/2}\text{–}1s^22p\ ^2P_{1/2}$, $j = 1s[2p^2\ ^1D]\ ^2D_{5/2}\text{–}1s^22p\ ^2P_{3/2}$, and $1s2lnl'\text{–}1s^2nl'$). The dotted blue curve shows the spectral collisional–radiative MARIA simulations when charge exchange is not included in the simulations. The resonance line W and the higher-order $1s2lnl'$-satellites are very well described indicating that the electron temperature is about $kT_e = 1700$ eV.

However, important discrepancies between theory and experiment are likewise observed: The qr-satellite emissions are much to low (indicated by the left red flash) and also the (Z, j)-intensity is too low (see right red flash). The MARIA simulations including line-of-sight integration effects (Rosmej 1998; Rosmej et al. 1999a) and charge exchange coupling to the neutral background result in an almost perfect agreement: The qr-satellite intensities are very well described and also the (Z, j)-intensity is in excellent agreement. Atomic structure calculations indicate that the qr-satellites have high radiative decay rates ($A(q) = 1.01 \times 10^{14}\ \text{s}^{-1}$, $A(r) = 8.73 \times 10^{13}\ \text{s}^{-1}$) while their autoionizing rates are rather moderate ($\Gamma(q) = 1.86 \times 10^{12}\ \text{s}^{-1}$, $\Gamma(r) = 1.28 \times 10^{13}\ \text{s}^{-1}$) compared to the strongest ones ($\Gamma(j) = 1.42 \times 10^{14}\ \text{s}^{-1}$). Therefore, these satellite transitions have strong contributions from electron collisional inner-shell excitation and small dielectronic recombination contribution. As charge exchange processes shift the ionic charge state distribution to lower values, Li-like population increases thereby increasing the qr-satellite intensities via inner-shell excitation. As charge exchange in the MARIA code (Rosmej 1997, 1998, 2001, 2006, 2012a, b) is not only coupled to the ground states but to excited states too, charge exchange from the H-like ground state to the $1snl$-states drives additional cascading flow (Rosmej and Lisitsa 1998) that terminates in the triplet system essentially with the states $1s2l\ ^3L$ (from which the forbidden lines X, Y, and Z originate; see also Chap. 1). This effect is strongest for the Z-line as the comparison with the blue- and green-dotted lines demonstrate (see also right red flash indicating the relative intensity difference).

Let us outline below the framework of the self-consistent simulation of X-ray impurity spectra where the impurity ions are coupled to the neutral background by charge transfer processes. The line-of-sight integrated spectral distribution $I(\omega)$ of the impurity ions is calculated according to

$$I(\omega) = \int_{-\infty}^{\infty} \left\{ \sum_{\text{ji}} \int_{r=0}^{a} I_{\text{ji}}(r,x)\mathrm{d}r \right\} \Phi(x-\omega)\mathrm{d}x. \tag{10.49}$$

The summation is performed over the various line transitions from "i" to "j"; the convolution integral takes into account the apparatus profile $\Phi(\omega)$ which can be assumed for almost all practical purposes to be a Voigt profile with user specified Gaussian and Lorentzian widths. The integration in space is carried out over the central line of sight along the minor radius a. The local spectral distribution for a single transition is given by

$$I_{\mathrm{ji}}(r,\omega) = n_{\mathrm{j}}(r)A_{\mathrm{ji}}\varphi_{\mathrm{ji}}(r,\omega). \tag{10.50}$$

n_{j} is the upper-level density, A_{ij} is the spontaneous transition probability, and $\varphi_{\mathrm{ij}}(\omega)$ is the local emission profile. The upper-level population density is obtained from the solution of the system of rate equations taking into account the temperature and density profile along the minor radius, $T_{\mathrm{e}}(r)$ and $n_{\mathrm{e}}(r)$:

$$\frac{\mathrm{d}n_{\mathrm{j}}(r)}{\mathrm{d}t} = \sum_{i=1}^{N} n_{\mathrm{i}}(r)W_{\mathrm{ji}}(r) - n_{\mathrm{j}}(r)\sum_{k=1}^{N} W_{\mathrm{jk}}(r) \tag{10.51}$$

with

$$W_{\mathrm{ji}} = C_{\mathrm{ji}} + A_{\mathrm{ji}} + I_{\mathrm{ji}} + T_{\mathrm{ji}} + D_{\mathrm{ji}} + \Gamma_{\mathrm{ji}} + R_{\mathrm{ji}} + Cx_{\mathrm{ji}}. \tag{10.52}$$

The matrix C describes the collisional excitation/de-excitation, A the spontaneous radiative decay, I the ionization, T the three-body recombination, D the dielectronic capture, Γ the autoionization, R the radiative recombination, and Cx the charge exchange process. The rates Cx_{ij} themselves depend not only on the cross sections and corresponding rate coefficients but also on the level populations of the neutral particles. If a matrix element does not exist physically, its value is zero. The sum extends over all ground and excited states (that are explicitly taken into account in the simulations). Therefore, the spectral emission is calculated simultaneously with the proper ionization balance. The convective derivative d/dt on the left-hand side of (10.51) contains the partial derivative $\partial/\partial t$ and the impurity transport that is consistently applied to all ground, single, and double excited states.

Charge exchange processes are incorporated in the system of rate equations for the impurity ions through the matrix elements Cx_{ij} (10.52). These elements are proportional to the population densities of a particular state of the neutral species. Because only relative changes in the experimental spectrum are analyzed here (relative to the electron density), these processes can be conveniently described with an effective charge exchange parameter:

$$Cx_{\alpha\beta}^{\mathrm{eff}} = \frac{\sum_{j=1}^{n_{\max}^{N}} n_{\mathrm{j}}^{N}\langle Cx_{\alpha\beta}, j\rangle}{n_{\mathrm{e}}} = \frac{Cx_{\alpha\beta}}{n_{\mathrm{e}}}. \tag{10.53}$$

n_{e} is the electron density, n_{j}^{N} are the population densities of the neutrals, $n_{\max}^{N}$ is the maximum number of high n-states present in a real plasma (typically $n_{\max}^{N}$ = 20–25), $\langle Cx, j\rangle$ are the charge exchange rate coefficients from the neutral state n_{j}^{H} (j = 1 ground state, $j > 1$ excited states). The last expression in (10.53) relates the effective charge exchange parameter $Cx^{\text{eff}}_{\alpha\beta}$ to the rate matrix $Cx_{\alpha\beta}$ in (10.52). It is further convenient to define the dimensionless relative effective fraction f_{eff} of neutrals through the relationship

$$\left\langle Cx_{\alpha\beta}\right\rangle_{\text{eff}} = f_{\text{eff}}\left\langle Cx_{\alpha\beta}, 1\right\rangle \tag{10.54}$$

with

$$\begin{aligned} f_{\text{eff}} &= \frac{n_1^N\left(1+\sum_{i=2}^{n_{\max}^N}\frac{n_i^N}{n_1^N}\frac{\left\langle Cx_{\alpha\beta}, i\right\rangle}{\left\langle Cx_{\alpha\beta}, 1\right\rangle}\right)}{n_e} \\ &= \frac{n_1^N}{Z_{\text{mean}}\, n_0^N}\left(1+\sum_{i=2}^{n_{\max}^N}\frac{n_i^N}{n_1^N}\frac{\left\langle Cx_{\alpha\beta}, i\right\rangle}{\left\langle Cx_{\alpha\beta}, 1\right\rangle}\right) \end{aligned} \tag{10.55}$$

and

$$Z_{\text{mean}} = \frac{n_e}{n_0^N}. \tag{10.56}$$

Z_{mean} is the average charge with respect to all types of impurity ions present in the plasma:

$$n_e = \sum_i Z_i n_i = n_0^N Z_{\text{mean}}. \tag{10.57}$$

The last expression in (10.57) expresses the number of free electrons per neutral particle density n_0^N and Z_{mean}. With this definition, the last expression of (10.55) has the advantage that it depends only on relative populations of the neutrals (total number of neutrals and neutrals in the ground state) and is, therefore, independent of the normalization condition. Note, that if all excited states are neglected, f_{eff} is the relative fraction of the neutrals compared to the electrons. The brackets $\langle\,\rangle$ denote the averaging over the ion distribution function. In the case of H/D/T (hydrogen/deuterium/tritium), n_0^N is the population of the neutral H/D/T.

The sum inside the brackets of (10.55) describes the influence of the charge exchange from excited states of the neutrals. The factor f_{eff} determines the contribution of charge exchange processes to the impurity kinetic system according to (10.51).

Figure 10.12 shows the sensitivity of the X-ray impurity spectra to the charge exchange parameter f_{eff}. Due to the large cross sections for the charge exchange processes, the sensitivity starts already with about $f_{\text{eff}} = 10^{-6}$. As the charge exchange cross sections scale with the 4th power of the principal quantum number of the neutral species, excited state contributions can significantly contribute to the effective cross section (Abramov et al. 1985; Rosmej and Lisitsa 1998) as demonstrated in Table 10.3. The parameter f_{eff} is, therefore, not strictly equal to the neutral fraction. The fraction of neutrals

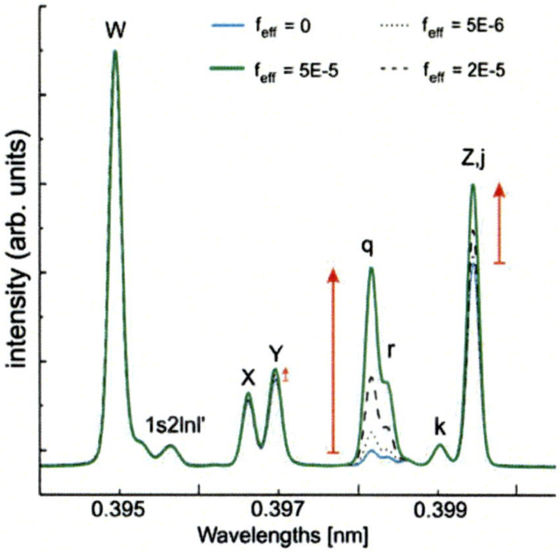

Fig. 10.12 MARIA spectral charge exchange—cascading modeling of X-ray argon impurity spectra in dependence of the effective neutral fraction f_{eff}. The electron temperature is $kT_{\mathrm{e}} = 1.5$ keV, electron density $n_{\mathrm{e}} = 2 \times 10^{13}$ cm^{-3}, heavy particle temperature $kT_{\mathrm{i}} = 0.9$ keV. The spectral resolution is $\lambda/\delta\lambda = 5000$

$$f_{\mathrm{neutral}} = \frac{\sum_{j=1}^{n_{\max}^{N}} n_{\mathrm{j}}^{N}}{n_{\mathrm{e}}} \tag{10.58}$$

can be calculated from the collisional–radiative rate equations for the neutrals because the relevant local temperature and density are usually known from various diagnostics. This, however, is not so simple as the penetration of neutrals from the wall to the center significantly increases the fraction of neutrals in the plasma center (typically many orders of magnitude). Let us therefore investigate the modeling of the neutral systems in more detail. The population densities n_{j}^{N} of the neutrals have to be determined from a system of rate equations including the penetration of neutrals from the wall to the center:

$$\frac{\mathrm{d}n_{\mathrm{j}}^{N}}{\mathrm{d}t} = \sum_{\mathrm{j}}^{n_{\max}^{N}} n_{\mathrm{i}}^{N} W_{\mathrm{ji}}^{N} - n_{\mathrm{j}}^{N} \sum_{k}^{n_{\max}^{N}} W_{\mathrm{jk}}^{N}, \tag{10.59}$$

$$W_{\mathrm{ji}}^{N} = A_{\mathrm{ji}}^{N} + C_{\mathrm{ji}}^{N} + R_{\mathrm{ji}}^{N} + I_{\mathrm{ji}}^{N} + I_{\mathrm{p,ij}}^{N} + T_{\mathrm{ji}}^{N} + D_{\mathrm{ji}}^{N} + Cx_{\mathrm{ji}}^{\mathrm{N,imp}}, \tag{10.60}$$

$$D_{\mathrm{ji}}^{N} = n_{\mathrm{e}} C_{\mathrm{ij,\,eff}}. \tag{10.61}$$

The matrix D specifies the neutral particle penetration in a global sense. A more detailed transport analysis is given below. Level populations are normalized according to

$$\sum_{i=0}^{n_{\max}^N} n_i^N = 1. \tag{10.62}$$

Normalization according to (10.62) means that the probability p_j for the population of the level j is $n_j^N = p_j$. In this case, the matrix A describes the spontaneous radiative decay rates, C the collisional excitation/de-excitation rates, R the radiative recombination rates, I the ionization rates, T the three-body recombination rates, I_p the proton ionization rates, Cx the charge exchange rates between the neutral species and the impurity ions. Introducing an "effective impurity ion" with density n_{imp} and charge Z_{imp} according to

$$1 + \frac{n_{imp}}{n_0^N} Z_{imp} = Z_{mean} \tag{10.63}$$

we can derive an explicit expression for the Cx-rate in the frame of the classical picture for charge exchange cross sections (Rosmej and Lisitsa 1998):

$$Cx^{N,imp} = n_{imp}\sigma_{Cx}V_{imp} = 8\pi a_0^2 n^4 V_{imp}(Z_{mean} - 1)n_0^N. \tag{10.64}$$

where V_{imp} is the relative velocity between the neutrals and the impurity ions. The charge exchange rates according to (10.64) have the advantage that they do not explicitly contain the impurity density n_{imp} (which is rather difficult to determine experimentally) and are instead proportional to $(Z_{mean} - 1)$. Note that in the numerical calculations, any charge exchange cross sections and semi-empirical formulas (Nakai et al. 1989) can be employed. It is important to note that proton collisions (in particular proton-induced ionization I_p of highly excited states) have an important impact on the collisional–radiative modeling of the neutral system and have therefore been included for all states in the quasi-classical approach (Garcia et al. 1969; Gryzinski 1965). Numerical studies of the neutral system show (Rosmej and Lisitsa 1998) that one can choose $C_{01,eff} > 0$ and all other $C_{ij,eff} = 0$ while $d/dt = 0$.

Then C_{eff}: = $C_{01,\,eff}$ can be determined in a self-consistent manner together with f_{eff} and in turn permits to extract the fraction of neutral atoms $f_{neutral}$ (see also (10.58)) and the characterization of the penetration of the neutral species from the wall to the center through C_{eff} (see also (10.61)). Physically, C_{eff}: = $C_{01,\,eff}$ can be interpreted as follows. We have an inflow of neutral particles in the ground state ($1s\ ^2S_{1/2}$) from the wall to the center into a volume element containing protons and neutrals with given n_e and T_e. The continuous inflow causes an effective increase of the neutral density and changes therefore the effective ionization balance for given n_e and T_e in that volume element. The normalization condition (10.62) puts into proper weight the effect of the inflow for all populations n_j.

The inflow of neutrals from the wall to the center leads to an accelerated convergence of the self-consistent model due to the effective reduction of excited state populations that originate from the three-body recombination $p + e + e \rightarrow \mathrm{H}$

$(n) + e$. In practice, the proton population decreases considerably as the fraction of neutrals may rise many orders of magnitude (e.g., from about 10^{-8} to 10^{-5} for ohmic discharges at the TEXTOR tokamak (Rosmej et al. 1999a)). Writing

$$\tau_p = \frac{1}{n_e C_{eff}} \tag{10.65}$$

the physical meaning of τ_p is an effective proton lifetime. Due to the relation (10.56), we can formulate also an expression for the electron lifetime τ_{n_e}:

$$\tau_{n_e} = \frac{1}{n_e C_{eff} Z_{mean}}. \tag{10.66}$$

The neutral fraction is thus self-consistently obtained by coupling the population kinetics of the neutrals and the impurities via charge exchange:

$$f_{neutral} = \frac{1}{Z_{mean} n_0^N} \sum_{j=1}^{n_{max}^N} n_j^N (n_e, T_e, C_{eff}, f_{eff}). \tag{10.67}$$

As mentioned in relation with (10.58) in a self-consistent simulation, the fraction according to

$$f_H = \frac{n_H(1s)}{n_e} \tag{10.68}$$

is not strictly equivalent to the relative number of neutral particles (Rosmej and Lisitsa 1998). In fact, in a self-consistent simulation, the neutral system is simultaneously also calculated in a collisional–radiative modeling in order to calculate charge exchange from excited states (see (10.38) and the discussion in relation to Table 10.3). These neutral particle simulations may likewise include also particle transport, e.g., a neutral flux from the wall to the center (Rosmej et al. 1999a; Rosmej and Lisitsa 1998). As the total number of neutrals is different from the number of neutrals in the ground state $1s$, (10.68) is not identical to the relative number of neutral particles. Simulations show (Rosmej et al. 1999a, 2006a; Rosmej and Lisitsa 1998) that (10.68) is, however, a very convenient simulation parameter: In fact, X-ray spectra can be first fitted with the help of the parameter (10.68) and then, this parameter is recalculated in the self-consistent simulation to obtain excited states contributions from the donor atoms and the neutral density according to (10.58) that may include neutral flow from the wall to the center. The best fit to the data presented in Fig. 10.11 has been obtained for $f_H = 1.7 \times 10^{-5}$.

We note that radiative cascading contributions to the q-satellite intensity from the $1s2lnl'$-states in the simple standard Corona approximation have been studied (Bernshtam et al. 2009), and it was found that it contributes considerably if compared with the dielectronic recombination channel. This is not surprising because the q-satellite has a rather ineffective dielectronic recombination channel (low satellite intensity Q-factor) and the radiative cascading contribution for the q-satellite (and also for other satellite transitions that have low Q-factors). Therefore, the cascading has to be compared rather with the inner-shell excitation channel (which is the dominating one for the q-satellite) than with the dielectronic capture channel. In this case, one can see that for typical parameters of the spectra of Fig. 10.11 cascading contributions are negligible. Moreover, the cascading contributions for the r-satellite are entirely negligible (Bernshtam et al. 2009) even if compared to the dielectronic recombination channel only. However, as the spectrum of Fig. 10.11 demonstrates, the r-satellite shows a very similar behavior as the q-satellite, and therefore, cascading properties are not at the cause of their intensity rise (left red flash in Fig. 10.11) discussed here.

We note that the general case of cascading contributions to satellite transitions including charge exchange has been investigated theoretically and experimentally (Rosmej et al. 2006c) and it was found that it gives rise to a considerable enhancement of the intercombination satellite transitions originating from the $1s2p^2\ {}^4P$ states, i.e., the transitions $h = 1s[2p^2\ {}^3P]\ {}^4P_{1/2}$–$1s^22p\ {}^2P_{3/2}$, $i = 1s[2p^2\ {}^3P]\ {}^4P_{1/2}$–$1s^22p\ {}^2P_{1/2}$, $f = 1s[2p^2\ {}^3P]\ {}^4P_{3/2}$–$1s^22p\ {}^2P_{3/2}$, $g = 1s[2p^2\ {}^3P]\ {}^4P_{3/2}$–$1s^22p\ {}^2P_{1/2}$, $e = 1s[2p^2\ {}^3P]\ {}^4P_{5/2}$–$1s^22p\ {}^2P_{3/2}$. The physical origin of the strong cascading contribution in the triplet system of the autoionizing states is that they accumulate in the lowest configuration $1s[2p^2\ {}^3P]\ {}^4P$.

Figure 10.13 shows a sensitivity study of the charge exchange enhanced q-satellite intensity. For these purposes, the intensity line ratios between the q- and k-satellites are depicted because the k-satellite is well separated in the experimental spectra from Fig. 10.11 and because it has low charge exchange sensitivity due to small inner-shell excitation channel (high autoionizing rate $\Gamma(k) = 1.34 \times 10^{14}\ \mathrm{s}^{-1}$). As Fig. 10.13 demonstrates charge exchange sensitivity starts already from fractions as low as 10^{-7} (see also Fig. 10.12). This strong sensitivity is due to the very large value of charge

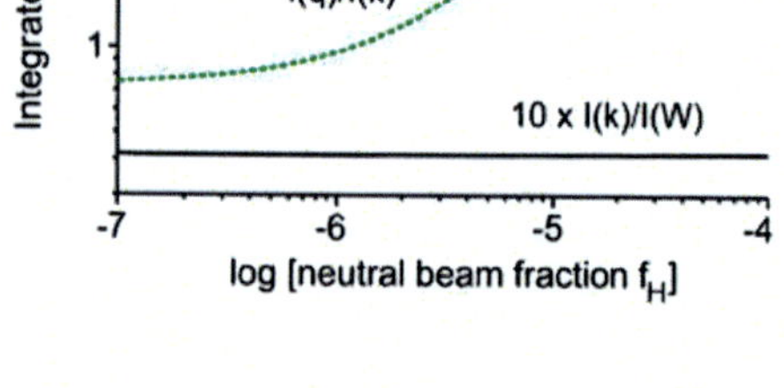

Fig. 10.13 MARIA simulations of satellite intensities in dependence of the neutral beam fraction, $kT_e = 1700$ eV, $n_e = 2 \times 10^{13}\ \mathrm{cm}^{-3}$

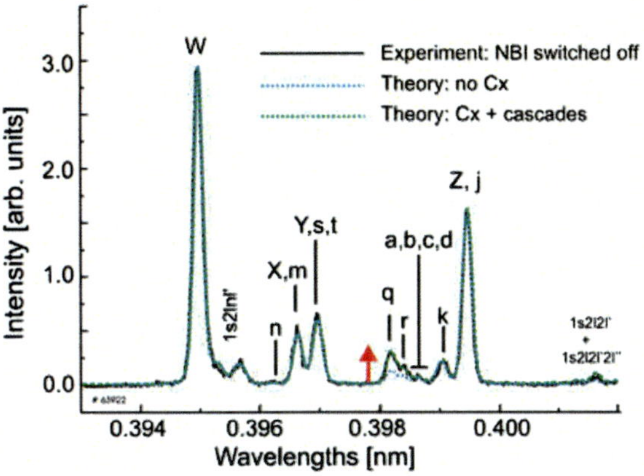

Fig. 10.14 Time-resolved X-ray argon impurity spectrum (t = 3.9–4.0 s) after switching off the neutral beam injection at t = 3.7 s. The red flash indicates that even after neutral beam injection an enhanced neutral background remains

exchange cross sections in the classical over barrier regime (see also Annex A.1 that presents a summary of numerous elementary processes).

Also included in Fig. 10.13 is a curve that shows the intensity ratio of the k-satellite and with the He-like resonance line W. The W-line intensity is essentially driven by electron collisional excitation from the He-like ground state $1s^2\ ^1S_0$ while the k-satellite intensity is essentially due to dielectronic capture from the He-like ground state. Therefore, charge exchange-driven shifts of the ionic distribution have almost no effect on this line ratio as confirmed by the simulations.

Figure 10.14 shows the time-resolved argon X-ray emission after the neutral beam injection NBI was switched off. As can be seen, the dashed blue curve (where charge exchange is not included in the simulations) results already in a rather good agreement with the data for an electron temperature of kT_e = 1300 eV. As indicated by the red flash, the q-satellite intensity, however, is again not in good agreement, albeit less pronounced as compared to Fig. 10.11. This indicates that although the

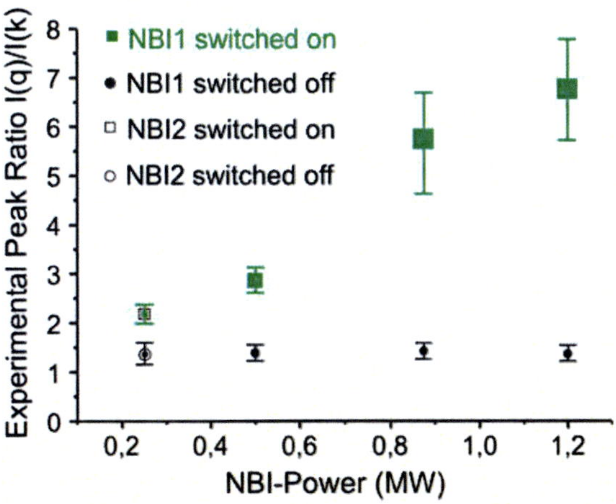

Fig. 10.15 Experimental satellite intensities in dependence of different intensities of neutral beam injection and different line of sights (NBI1 and NBI2)

neutral beam injection is switched off still charge exchange from neutrals impact on the X-ray emission, i.e., there exists an important residual neutral background. Quantification of the residual neutral background is demonstrated with the green curve that shows the MARIA simulations including charge exchange: Excellent agreement is obtained for $f_H = 6 \times 10^{-6}$.

Figure 10.15 shows the experimental peak intensity ratio of the *q*- and *k*-satellites in dependence of the neutral beam injection power. This demonstrates that with increasing NBI-power, the ratio continuously rises (solid and open squares) and relaxes to a common level (solid and open circles) after switching off the NBI injection.

For the NBI 1 (solid symbols) the line of sight for the X-ray emission crosses the injection direction, while for NBI 2, no geometrical crossing occurs (open symbols). As can be seen, open and sold circles coincide within the error bars. These results suggest that the neutral beam is rapidly thermalized creating an enhanced neutral background. Due to the large sensitivity of this method, even the neutral background in purely ohmic discharges could successfully be determined (Rosmej et al. 1999a).

As Figs. 10.11 and 10.14 demonstrate, the self-consistent charge exchange MARIA simulations provide excellent agreement between theory and experiment (Rosmej 1998, 2012a; Rosmej et al. 1999a). The impact of charge exchange on impurity spectra has later also been found to be of importance in other experiments of magnetically confined plasmas (Beiersdorfer et al. 2005). However, the statements of (Beiersdorfer et al. 2005) that large enhancement factors of 6 and more for the *Z*-line were found in (Rosmej et al. 1999a) that disagree with their measurements are incorrect. In addition, the analysis (Beiersdorfer et al. 2005) did not include any self-consistent consideration and an incomplete discussion of the *Z*-line intensity. In fact, it should be remembered that opposite to the *W*-line, the *Z*-line intensity is strongly plasma parameter dependent due to its cascade sensitivity and inner-shell ionization population channel (e.g., see above discussion of charge exchange). Therefore, the surprises announced in (Beiersdorfer et al. 2005) that their observations show quite different *Z*-line intensities as compared to (Rosmej et al. 1999a) are also irrelevant as the plasma temperatures in (Beiersdorfer et al. 2005) and (Rosmej et al. 1999a) are quite different: Much higher electron temperatures result in an entirely different proportion of recombination and inner-shell ionization contributions, and the same holds true for the different intensities of the *qr*-satellites (Rosmej and Lisitsa 1998).

Further serious discrepancies between simulations and experimental data for the *W*3 and *Y*3 argon line emission in a well-diagnosed tokamak have also been stated by (Beiersdorfer et al. 1995). However, these statements and related discussions turned out also to be in error and consistently performed multilevel multi-ion stage simulations carried out with the MARIA code (Rosmej 1997, 1998, 2001, 2006, 2012a) that included cascading, line-of-sight integration, spectral simulations including overlapping $1s2l3l'$- and $1s3lnl'$-satellites, advanced intermediate coupling, and configuration interaction demonstrated excellent agreement with the data (Rosmej 1998).

The *W*3 and *Y*3 lines have attracted renewed attention in recent very-high-resolution X-ray spectroscopic measurements for advanced confinement mode studies at the C-mod tokamak at MIT (Rice et al. 2018). The key issues concerning the X-ray diagnostics has been the simultaneous observation of the *W*3 and *Y*3 lines and the two types of satellites transitions $1s3l3l'$–$1s^2 3l'$ and $1s2l3l'$–$1s^2 2l$. The important point in this simultaneous observation of satellite transitions is that temperature can be fixed with rather high precision while studying the impact of impurity transport, charge exchange, and cascading. As has been demonstrated, the MARIA simulations provide very good agreement with the data (Rice et al. 2018). Moreover, it should be emphasized that the MARIA code analysis was based on the complete simulation of the spectral distribution that permits increased stability in the analysis. On the other hand, line ratios provide only limited information as in almost all practical applications, simultaneously several effects have to be studied: temperature, density, particle impurity transport, charge exchange, flow of neutral from the wall to the center, …. These effects are very difficult to take into account simultaneously via line ratios. For example, as demonstrated in Figs. 10.11 and 10.14, line overlap from different type of transitions can be very important (the overlap of higher-order satellites $1s2lnl'$ with the *W*-line, the overlap of the $1s3lnl'$-satellites with the *W*3 line, the overlap of the *j*-satellite with the *Z*-line, etc.). Due to this line-overlap problematic, corrected line ratios (corrected for overlapping satellite transitions) have emerged in the literature. However, these line ratios are not of great practical use because the primary line ratio becomes multiparameter dependent. If line overlap is important, only total spectral simulations (Rosmej 1998; Rosmej et al. 1999a, b, 2000, 2001c; Rosmej and Lisitsa 1998) as demonstrated with the MARIA code provide an efficient analysis.

Finally we note that charge exchange processes turned out to be also important in dense hot plasmas, e.g., in laser-produced plasmas (Rosmej et al. 1999b, 2002a, 2006c, Monot et al. 2001), Z-pinches (Rosmej et al. 2001b, 2015). Also hollow ion X-ray emission has been identified with charge exchange between highly charged ions and low-charged ions (Rosmej et al. 2015).

10.3.3 Transient Phenomena in the Start-up Phase

Figure 10.16 shows the time-resolved X-ray impurity spectra for an inductively heated tokamak discharge with residual argon (argon that remained in the machine from the gas puff injected argon of the previous discharge). Figure 10.16a shows the start-up phase at $t = 0.1$–0.2 s. Strong enhancements (indicated by red flashes) of the *qr*-satellites and the (*Z*, *j*)-lines are observed. The qr- and (*Z*, *j*)-intensities

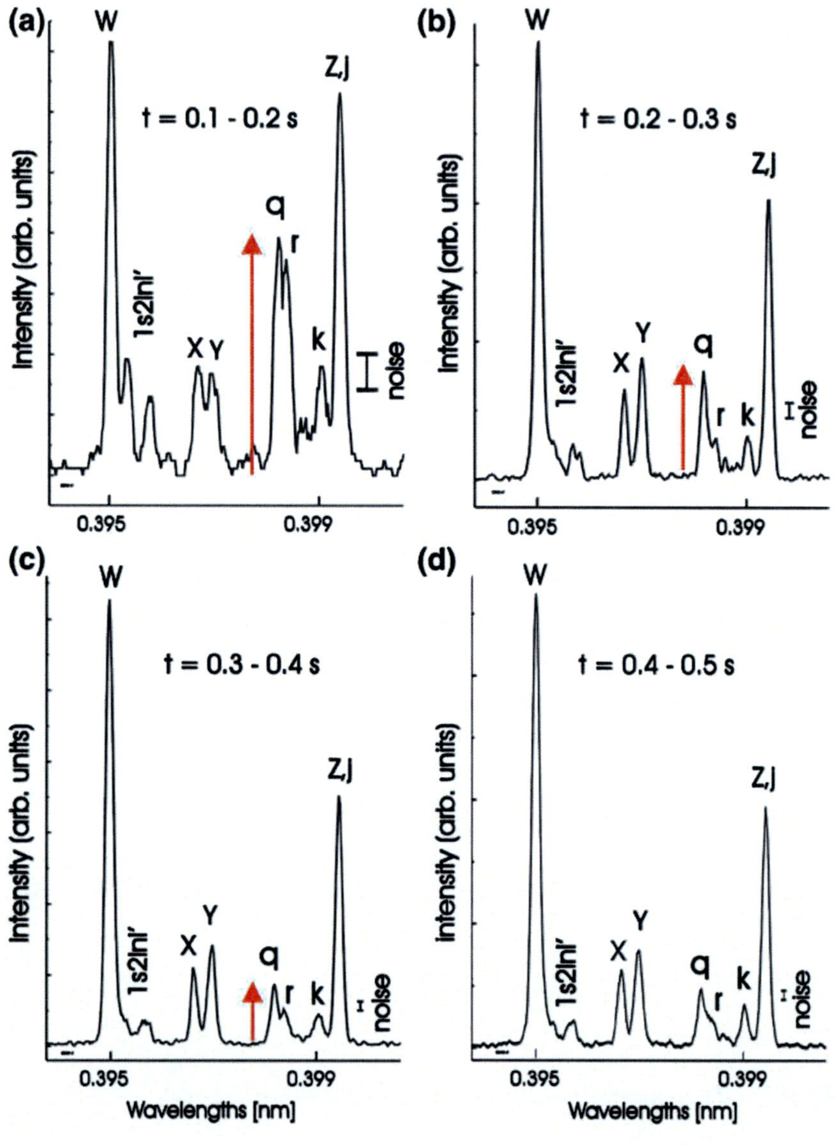

Fig. 10.16 Time-resolved X-ray impurity spectra from residual argon gas in the heating phase of an inductively driven tokamak discharge. **a** At t = 0.1–0.2 s, the Li-like satellites qr show up with large intensity that is continuously decreasing with time, **b** t = 0.2–0.3 s, and **c** t = 0.3–0.4 s. At t = 0.4–0.5 s **d** almost stationary conditions are reached

decrease continuously with time (Fig. 10.16b,c) and approach an almost stationary case for $t > 0.4$ s (see Fig. 10.16d).

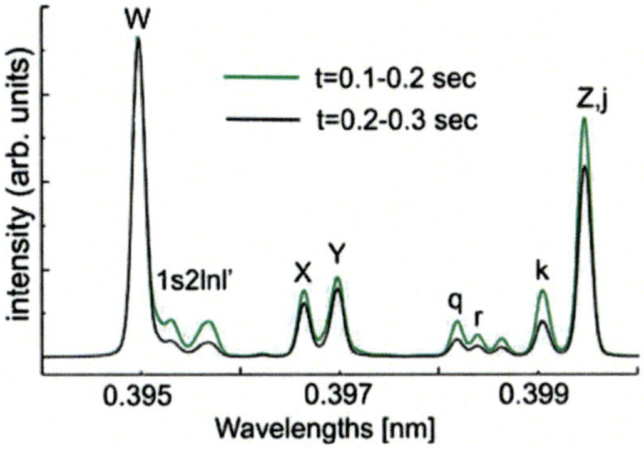

Fig. 10.17 Time-dependent MARIA simulations of argon X-ray impurity spectra from residual argon gas in the heating phase of an inductively driven tokamak discharge. The simulations take into account the experimentally measured temperature and density evolution. Time-dependent spectra have been summed up in the time intervals from 0.1 to 0.2 s and 0.2 to 0.3 s and normalized to the *W*-line peak intensity for better demonstration of transient effects

In the start-up phase, increased intensities of the qr- and (Z, *j*)-lines are due to charge exchange processes and also due to transient effects. Figure 10.17 displays therefore the transient calculation ($\partial n_j/\partial t \neq 0$; see (10.51)) of the Argon impurity spectra carried out with the MARIA code. The simulations employ the measured values of electron temperature $T_e(r = 0, t)$ and electron density $n_e(r = 0, t)$. Time-resolved spectral emission has been summed in the intervals from $t = 0.1$–0.2 s and $t = 0.2$–0.3 s in order to be compared with the time-resolved measurements of Fig. 10.16. For better demonstration of the transient effects, the spectra for both time intervals have been normalized to the peak intensity of the *W*-line. It should be emphasized that the time-dependent simulations include not only the ground states, but also all excited (single and double excited) states. Therefore, the simulations keep also track of the photon relaxation effects discussed in Sect. 6.2 along with the relaxation effects of the ionization balance.

Due to the fast rising electron temperature (e.g., $kT_e = 380$ eV at $t = 0.1$ eV, $kT_e = 1100$ eV at $t = 0.3$ s), the ionization balance lags behind the corresponding electron temperature resulting in an increased fraction of Li-like ions compared to stationary calculations. This creates the condition for a rise of the relative intensity of the *qr*-satellites (inner-shell excitation) and the Z-line (inner-shell ionization). In addition, other satellites, that have strong dielectronic recombination channels, are relatively enhanced because the electron temperature is on the average lower in the interval $t = 0.1$–0.2 s compared to the interval $t = 0.2$–0.3 s. As can be seen from the comparison of the different simulations, the satellite enhancement effects for the time interval starting from $t = 0.1$–0.2 s are rather small (see Fig. 10.17) and for $t = 0.2$–0.3 s and for later times (until the end of the discharge), these effects are found to be negligible (Fig. 10.17).

As can be seen from the comparison of the experimentally measured X-ray spectra (Fig. 10.16a) and the simulations (green curve in Fig. 10.17), the transient relaxation effects do not allow to fully explain the strong intensity increase of the *qr*-satellites (red flash in Fig. 10.16a). Therefore, most of the intensity rise can be attributed to charge exchange with the neutral background; see Fig. 10.12 green curve. The determined neutral fraction for the time interval $t = 0.1$–0.2 s is about $(2 \pm 1) \cdot 10^{-5}$, and the determined electron lifetime is about (0.1 ± 0.05) s. These results are also in good agreement with Monte Carlo simulations of the neutrals (Rosmej et al. 1999a).

If the intensity rise of the *qr*-satellites in Fig. 10.16 is essentially attributed to charge exchange, the continuous decrease of the *qr*-satellites after the start-up phase until the stationary case (Fig. 10.16d) indicates that the neutral fraction in the start-up phase is much larger than in the stationary phase of the discharge. This is understandable because in the start-up phase, the electron temperature is much lower than in the stationary phase.

10.3.4 Impurity Diffusion and τ-Approximation

Also impurity diffusion impacts on the spectral distribution, in particular on the relative intensities of the *qr*-satellites and the *X*-, *Y*-, *Z*-lines discussed above. As has been demonstrated with detailed numerical calculations of the exact radial diffusion equation (Rosmej et al. 1999a), the intensities of the *qr*-satellites and even the complex interplay of the *X*-, *Y*-, *Z*-line intensities can be reasonably described by the so-called τ-approximation. In this approximation, the diffusion term is replaced by

$$\nabla\left(n_{\mathrm{j}} V\right) \rightarrow \frac{n_{\mathrm{j}}}{\tau_{\mathrm{D}}} \tag{10.69}$$

in the population kinetic system (10.51). From the comparison of the numerical calculations of the exact diffusion equation and the τ-approximation, one can deduce the approximate relation (Rosmej et al. 1999a)

$$D \approx \alpha_{\mathrm{D}} \cdot \frac{\Delta x^2}{\tau_{\mathrm{D}}}, \tag{10.70}$$

where D is the impurity diffusion coefficient (Hulse 1983), Δx is the characteristic width of the particular charge state, α_{D} is a constant that depends on the geometrical parameters of the magnetically confined plasma and τ_{D} is the characteristic time scale for the diffusion in tau-approximation (10.69) to be used in (10.51).

Figure 10.18a shows the impact on the X-ray spectral distribution for different parameters τ_{D} in a self-consistent charge exchange simulation described above. For the simulations of the spectra of Fig. 10.18, $\alpha_{\mathrm{D}} \approx 0.16$ and $\Delta x \approx 35\,\mathrm{cm}$ have been

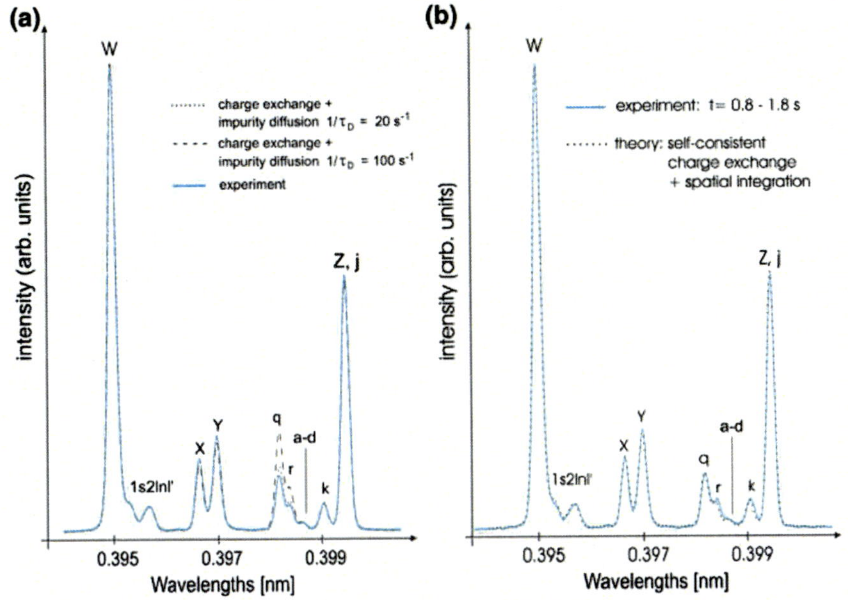

Fig. 10.18 Self-consistent MARIA simulations of charge exchange and impurity diffusion. **a** Influence of impurity diffusion on the self-consistent charge exchange simulations. Noticeable diffusion effects start only for τ-parameters $1/\tau_D > 20$ s, **b** self-consistent simulations of an inductively heated discharge. Excellent agreement with the time-resolved data in the interval $t = 0.8$–1.8 s is obtained for $f_{\mathrm{eff}} = 5.7 \times 10^{-6}$

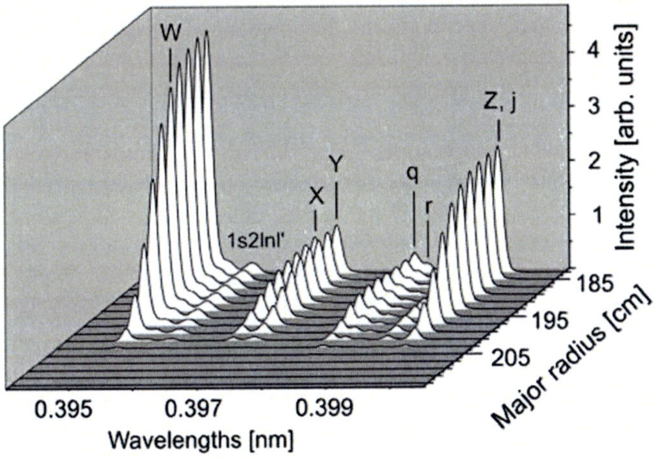

Fig. 10.19 Space-resolved self-consistent charge exchange simulation of an inductively heated discharge. Noticeable diffusion effects start only for τ-parameters $1/\tau_D > 20$ s

deduced from numerical calculations (Rosmej et al. 1999a); therefore, $D(1/\tau_{\mathrm{D}} = 20/\mathrm{s}) \approx 4 \times 10^3\,\mathrm{cm}^2/\mathrm{s}$ while $D(1/\tau_{\mathrm{D}} = 100/\mathrm{s}) \approx 2 \times 10^4\,\mathrm{cm}^2/\mathrm{s}$. Taking into account the experimental errors, the value $D(1/\tau_{\mathrm{D}} = 20/\mathrm{s}) \approx 4 \times 10^3\,\mathrm{cm}^2/\mathrm{s}$ represents an upper limit (compare the blue solid curve with the dotted black curve). As a noticeable influence on the X-ray spectra starts only with this value but this value is much larger than previously measured diffusion values $D \approx 100 - 1000\,\mathrm{cm}^2/\mathrm{s}$ (Rapp et al. 1997; Tokar 1995; Ongena et al. 1995), the essential rise of the intensity is due to charge exchange as demonstrated with Fig. 10.18b that shows excellent agreement with the data for an effective charge exchange parameter $f_{\mathrm{eff}} = 5.7 \times 10^{-6}$. Note, that the simulations include the spatial variation of the plasma parameters. Figure 10.19 visualizes the local emission (note that the local emission is correlated to the dependence on major radius, the emission presented starts at the minor radius $r = 0$ that corresponds to a major radius of $R = 183$ cm) from the simulations that have been summed up for the final fit of the line-of-sight integrated data of Fig. 10.18b. It is important to note that all spectral details are very well reproduced: (a) higher-order satellite intensities and *k*-satellite intensity with respect to the *W*-line (indicating a correct description of the electron temperature), (b) perfect agreement with the intensities of the *X*-, *Y*-, and *Z*-lines indicating that cascading driven by charge exchange and other processes are well described, (c) perfect agreement with the *qr*-satellites indicating a correct description of the charge exchange-induced shift of the ionization balance.

It is important to note that the detailed description of the *X*-, *Y*-, and *Z*-line intensities and the satellites identify and distinguish charge exchange and impurity diffusion effects. This important diagnostic property of the high-resolution X-ray diagnostic is demonstrated in Table 10.4.

The *Y*-line continuously decreases with increasing impurity diffusion parameters *D*, while the *Z*-line firstly decreases and then increases. The decrease is related with the fact that with rising diffusion parameters *D*, the recombination source (from the H-like ground state) is reduced thereby reducing the cascading contributions to the line intensity. In this respect, the *Y*- and *X*-lines behave very similar because of the similar upper state configurations $1s2p\ {}^3P_J$. The same holds true for the *Z*-line, however, for larger *D*-values inner-shell ionization contribution from the Li-like ground state $1s^22s\ {}^2S_{1/2}$ for the *Z*-line comes into play that finally enhances the intensity. This contribution is negligible for the *X*- and *Y*-lines because here inner-shell ionization proceeds from the excited states $1s^22p\ {}^2P_{3/2,1/2}$ that have very low population; see (10.33). The *qr*-satellite intensity rises continuously with increasing *D*-values because of increasing population of the Li-like ground state that enhances the inner-shell excitation channel $1s^22s + e\text{–}1s2s2p + e$.

Table 10.5 shows the corresponding simulations in the τ-approximation. It is impressive to observe that even the subtle details of the line intensities (continuous decrease of the *Y*-line, decrease and increase of the *Z*-line, continuous increase of the *qr*-satellites) are well described in the τ-approximation. One therefore observes that the τ-approximation has spectroscopic/diagnostic precision.

Table 10.4 Line-of-sight integrated intensity ratios between the *Z*-line and the *W*-line ($R_Z = I(Z)/I(W)$), the intercombination line *Y* and the *W* line ($R_Y = I(Y)/I(Y)$) and the Li-like *q*-satellite and the *W*-line ($R_q = I(q)/I\,(W)$) for various diffusion coefficients *D* (normalized to the case $D = 0$, index "$D = 0$") taking into account the spatial variations of the electron density and temperature

$D(cm^2/s)$	10^2	3×10^2	6×10^2	10^3	3×10^3	6×10^3	10^4	2×10^4	3×10^4	6×10^4	10^5
$(R_Z)/(R_Z)_{D=0}$	0.999	0.996	0.992	0.988	0.981	0.986	1.01	1.08	1.16	1.40	1.72
$(R_Y)/(R_Y)_{D=0}$	0.999	0.995	0.990	0.984	0.966	0.951	0.939	0.923	0.915	0.905	0.900
$(R_q)/(R_q)_{D=0}$	1.02	1.03	1.06	1.09	1.29	1.66	2.25	3.89	5.55	10.4	16.6

The central electron temperature is $T_e(r = 0) = 1600$ eV and central electron density $n_e(r = 0) = 2 \times 10^{13}$ cm^{-3}. The convective velocity is $C_V = 0$

Table 10.5 Same like Table 10.4, however, calculations are performed in the framework of the τ-approximation, $T_e(r = 0) = 1600$ eV and $n_e(r = 0) = 2 \times 10^{13}$ cm^{-3}

$1/\tau$ (s^{-1})	0.3	1	3	10	20	50	100	160	200	350	600
$(R_Z)/(R_Z)_{\tau=\infty}$	0.999	0.998	0.994	0.985	0.979	0.981	1.01	1.06	1.10	1.24	1.49
$(R_Y)/(R_Y)_{\tau=\infty}$	0.999	0.998	0.994	0.982	0.970	0.950	0.934	0.925	0.921	0.913	0.907
$(R_q)/(R_q)_{\tau=\infty}$	1.01	1.03	1.08	1.27	1.55	2.35	3.67	5.23	6.25	10.0	16.2

Values are normalized to the case $\tau = \infty$

Therefore, the excellent agreement in all spectral features of the *X*-, *Y*-, and *Z*-lines as demonstrated in Fig. 10.18b points on the importance of charge exchange in the data as *X*-, *Y*-, and *Z*-lines are increased and not decreased.

Finally we note that the relative interplay between cascading, inner-shell ionization, and inner-shell excitation depends on the electron temperature: For large temperatures, dielectronic satellite intensities are decreased relative to the *W*-line and the reduced population of the Li-like ionization stage reduces inner-shell ionization and inner-shell excitation channels, while recombination channels might be increased due to increased H-like ionic population. In the opposite case, i.e., small electron temperatures, dielectronic satellite contributions are enhanced relative to

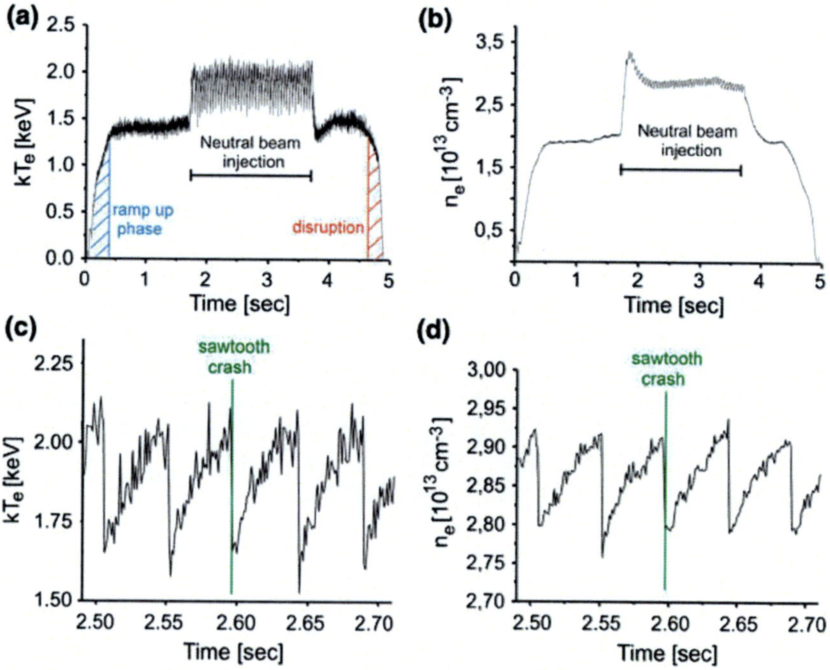

Fig. 10.20 Time-resolved temperature and density measurements for neutral beam-heated TEXTOR tokamak discharges. **a** and **b** show the total time interval of the discharge, **c** and **d** show the saw tooth oscillations in temperature and density with high temporal resolution

the X-line, inner-shell ionization and inner-shell excitation are enhanced due to high populations in the Li-like ionization stage while recombination channels are small due to small populations of the H-like ions.

10.3.5 Non-equilibrium Radiative Properties During Sawtooth Oscillations

The detailed description of the radiative properties of non-stationary and non-equilibrium plasmas plays a key role in modern tokamak research. The complexity arises due to the simultaneous presence of external sources for plasma heating and magnetic hydrodynamic plasma activity that results in sawtooth oscillations. Figure 10.20 shows an example of time-resolved measurements of the electron temperature (Fig. 10.20a, measured with the electron cyclotron method) and density (Fig. 10.20b, measured with HCN-interferometry) at a mid-size tokamak for the total period of the discharge including the ramp up phase, neutral beam heating, and the plasma disruption. During neutral beam injection, important sawtooth oscillations in temperature and density are observed that are depicted in Fig. 10.20c, d with very high temporal resolution. For the inductively driven regime, no such regular oscillations are observed. The sawtooth amplitude during neutral beam injection is very large, $\delta(kT_e) \approx 0.5\,\mathrm{keV}$ at a mean temperature of about $kT_e \approx 1.75\,\mathrm{keV}$ while the amplitude for the electron density is about $\delta(n_e) \approx 1.5 \times 10^{12}\mathrm{cm}^{-3}$ at a mean electron density of about $n_e \approx 2.85 \times 10^{13}\,\mathrm{cm}^{-3}$. The oscillation period is about $T_{\mathrm{sawtooth}} \approx 50\,\mathrm{ms}$. The rising phase of the sawtooth is well resolved (see Fig. 10.20c, d), while the so-called sawtooth crash is essentially unresolved.

The simulation of the associated non-equilibrium radiative properties is important for spectroscopic diagnostics and also for the correct prediction of the maximum radiation heat load on the inner walls. The correct description of atomic kinetics and radiative properties for these conditions requests a self-consistent solution of the kinetic equations for the electron distribution function as well as for atomic energy state populations. The fundamental quantities are the time-dependent-level populations, and we therefore start with the consideration of fluctuations in the atomic and ionic levels (Rosmej and Lisitsa 2011).

10.3.5.1 Fluctuations and Atomic Level Populations

Let us start from the Boltzmann-type kinetic equation for the seven-dimensional single-particle distribution function $f_1\left(\vec{V}, \vec{r}, t\right)$ ($\vec{r}$ is the particle position vector and $\vec{V}$ the particle velocity vector):

$$\frac{\mathrm{d}f_1\left(\vec{V},\vec{r},t\right)}{\mathrm{d}t}=\frac{\partial f_1\left(\vec{V},\vec{r},t\right)}{\partial t}+\vec{V}\frac{\partial f_1\left(\vec{V},\vec{r},t\right)}{\partial\vec{r}}+\frac{\vec{F}}{m}\frac{\partial f_1\left(\vec{V},\vec{r},t\right)}{\partial\vec{V}}=C_{\mathrm{f}}. \tag{10.71}$$

C_f indicates the collisional term which cuts the hierarchy of the many-particle distribution function. The single-particle distribution function is normalized to the total number of particles

$$n(\vec{r},t)=\int\mathrm{d}^3r f_1\left(\vec{V},\vec{r},t\right). \tag{10.72}$$

The particle distribution function according to (10.71) leads to a direct link to the atomic level populations n_{j} via the rate coefficient matrix W_{ij} (see also (10.51), (10.52)):

$$\frac{\mathrm{d}n_{\mathrm{j}}}{\mathrm{d}t}=-n_{\mathrm{j}}\sum_{i=1}^{N}W_{\mathrm{ji}}+\sum_{k=1}^{N}n_{\mathrm{k}}W_{\mathrm{kj}}, \tag{10.73}$$

$$W_{\mathrm{ji}}=C_{\mathrm{ji}}+A_{\mathrm{ji}}+I_{\mathrm{ji}}+T_{\mathrm{ji}}+D_{\mathrm{ji}}+\Gamma_{\mathrm{ji}}+R_{\mathrm{ji}}. \tag{10.74}$$

The collisional rate coefficients are linked via the particle distribution function (in particular the electron energy distribution function) according to

$$(C,I,R)_{\mathrm{ji}}=\int_{\Delta E_{\mathrm{ji}}}^{\infty}\mathrm{d}E\,\sigma_{\mathrm{ji}}^{(C,I,R)}(E)V(E)F(E) \tag{10.75}$$

with

$$E=\frac{1}{2}m\vec{V}^2, \tag{10.76}$$

$$F(E)=\frac{1}{n(\vec{r},t)}f_1\left(\vec{V},\vec{r},t\right)\frac{\partial V}{\partial E}. \tag{10.77}$$

For the dielectronic capture process and the three-body recombination, these expressions differ because the dielectronic capture is a resonance process and the three-body recombination involves the energy distribution of two particles (with energies E_1 and E_2 after ionization, energy E before ionization and double differential cross section σ_{ji}^{I}, to be discussed in more detail in Sect. 10.4):

$$D_{\mathrm{ji}}=\frac{\pi^2\hbar^3}{\sqrt{2}m_{\mathrm{e}}^{3/2}}\frac{g_{\mathrm{j}}}{g_{\mathrm{i}}}\Gamma_{\mathrm{ji}}\int_0^{\infty}\mathrm{d}E\,\delta(E_{\mathrm{S}},E)\frac{F(E)}{\sqrt{E_{\mathrm{S}}}}, \tag{10.78}$$

$$T_{\mathrm{ji}} = \frac{\pi^2\hbar^3}{m_{\mathrm{e}}^{3/2}}\frac{g_{\mathrm{i}}}{g_{\mathrm{j}}}\int_0^\infty \mathrm{d}E_1 \int_0^\infty \mathrm{d}E_2 \frac{E}{\sqrt{E_1E_2}}\sigma_{\mathrm{ji}}^{I}(E,E_1)F(E_1)F(E_2). \tag{10.79}$$

Equations (10.71)–(10.79) provide a complete link of the atomic level populations to the kinetic description of any time- and space-dependent phenomena. In optically thin plasmas, the local radiative emission of the atomic system is then given by

$$I(\omega,\vec{r},t) = \sum_{i=1}^{N}\sum_{j=1}^{N} \hbar\omega_{\mathrm{ji}} n_{\mathrm{j}} A_{\mathrm{ji}} \varphi_{\mathrm{ji}}(\omega), \tag{10.80}$$

where ω_{ji} is the transition frequency, $\varphi_{\mathrm{ji}}(\omega)$ the local emission line profile, and N the total number of levels. The observed emission is given by the integration over the line of sight and the convolution of the spectral distribution with the apparatus profile $\Phi(\omega)$:

$$I(\omega,t) = \int_{-\infty}^{+\infty} \mathrm{d}\omega' \int_{r=0}^{R} \mathrm{d}r\, I(\omega',r,t)\Phi(\omega'-\omega). \tag{10.81}$$

The spatial and temporal dependences of temperature $T(\vec{r},t)$ and density $n(\vec{r},t)$ can be directly obtained from the particle distribution function $f_1\left(\vec{V},\vec{r},t\right)$

$$T(\vec{r},t) = \frac{m}{3k}\int \vec{V}^2 f_1\left(\vec{V},\vec{r},t\right)\mathrm{d}^3r. \tag{10.82}$$

10.3.5.2 Histogram Technique

The temporal distribution of the density is given by (10.72). If the variation of temperature and density in time are independent of each other, a histogram technique can be applied to $T(\vec{r},t)$ and $n(\vec{r},t)$ to obtain the distribution functions of temperature and density, $G_{\mathrm{T}}(T,\vec{r})$ and $G_{\mathrm{n}}(n,\vec{r})$, respectively. The distribution functions G_{T} and G_{n} can then be measured in experiments, e.g., by means of probe measurements, cyclotron emission, interferometry or spectroscopic measurements of the atomic radiation emission.

For observations sufficiently long compared to the fluctuation time scale, the emission of a single bound–bound transition is given by

$$I_{\mathrm{ji}}(\omega,t)=\int_{-\infty}^{+\infty}\mathrm{d}\omega'\int_{r=0}^{R}\mathrm{d}r\int_{T=0}^{\infty}\mathrm{d}T\int_{n=0}^{\infty}\mathrm{d}n\,G_{\mathrm{T}}(T,r)G_{\mathrm{n}}(n,r)\times B_{\mathrm{ji}}\left(r,T,n,\omega_{\mathrm{ji}},\omega'\right)\Phi(\omega'-\omega),\tag{10.83}$$

$$B_{\mathrm{ji}}\left(r,T,n,\omega_{\mathrm{ji}},\omega'\right)=\hbar\omega_{\mathrm{ji}}A_{\mathrm{ji}}n_{\mathrm{j}}(r,T,n)\varphi_{\mathrm{ji}}\left(\omega_{\mathrm{ji}},\omega',r,T,n\right).\tag{10.84}$$

In order to provide practical use of the functions G_{T} and G_{n} and the expressions according to (10.83), (10.84), density and temperature fluctuations have to be independent and the time constant of the response function of the "observation system" has to be small compared to the time constant of the fluctuations. For spectroscopic measurements, the relevant relaxation constants are those of the system of differential equations (10.73).

Two classes of relaxation constants turn out to play an important role for the response function of the radiation to fluctuations: (a) the relaxation of the photon emission itself (see Sect. 6.2.2) and (b) the relaxation of the ion charge state distribution (see Sect. 6.2.1). The relaxation of the photon field is given by (see also (6.62)):

$$\tau_{\mathrm{ji}}=\frac{1}{A_{\mathrm{ji}}+C_{\mathrm{ji}}+C_{\mathrm{ji}}}.\tag{10.85}$$

A_{ji} is the radiative decay rate from level j to level i and C_{ji} at C_{ij} are the corresponding collisional processes. This means that even at extremely low-density plasmas, the relaxation time is rather fast due to the usually high radiative decay rate. For the hydrogen Balmer-alpha transitions (n = 3, n' = 2) (10.85) gives $\tau < 7.6\times10^{-9}\ \mathrm{s}^{-1}$. This time is usually much shorter than the fluctuation time. Therefore, photon relaxation does usually not play a role in turbulent plasmas. However, collisional processes from metastable levels can considerably enhance (orders of magnitude) the relaxation constant as this couples a "slow" time constant of forbidden transitions to the atomic level from which the resonance line origins. A characteristic example is the magnetic quadrupole transition $Z = 1s2s\ {}^3S_1\text{–}1s^2\ {}^1S_0$. For example, for neon (used for radiative cooling in mid-size tokamaks), the radiative decay rate is $\tau(Z) = 1.3\times10^{-4}$ s, whereas for the resonance transition $W = 1s2p\ {}^1P_1\text{–}1s^2\ {}^1S_0$ the time constant is $\tau(W) = 1.1\times10^{-11}$ s. Therefore, the photon emission of the Z-line is not relaxed on the usual time scale of turbulence, whereas the photon emission of the W-line represents almost instantaneously compared to the fluctuation time (if collisional coupling (see Sect. 6.2.3) between the singlet and triplet levels is negligible). The relaxation behavior of the ion charge state distribution is quite different (see also (6.48)):

$$\tau_{\mathrm{Z,Z+1}}=\frac{1}{I_{\mathrm{Z,Z+1}}+T_{\mathrm{Z+1,Z}}+R_{\mathrm{Z+1,Z}}+D_{\mathrm{Z+1,Z}}}.\tag{10.86}$$

I is the ionization rate, T the three-body recombination rate, R the radiative recombination rate, and D the dielectronic recombination rate. In low-density plasmas, these rates are very small and the corresponding time constant is very large.

The relaxation of the ionization balance is of importance if the intensity I of a particular line is intended to be used for turbulence analysis. The reasons are manifold: The intensity is not only given by the collisional excitation rate coefficient from the ground state, but also by the ionic fraction of the ground state itself, determined in turn by ionization and recombination processes. For highly charged ions, numerical calculations show (see (6.50)) that the K-shell emission has reached quasi-stationary conditions if

$$n_e\tau \geq 3 \times 10^{11}\,\mathrm{cm}^{-3}\,\mathrm{s}, \tag{10.87}$$

where n_e is the electron density in [cm^{-3}] and τ is the time after which the ionic fractions are in equilibrium. For magnetically confined fusion plasmas, the quasi-stationary condition (10.87) is usually a rather stringent condition for the radiation emission of highly charged impurity ions, because the density is of the order of 10^{12}–10^{15} cm^{-3} providing a relaxation time of the ionic fraction of about $\tau = 3 \times 10^{-1}$ to 3×10^{-4} s. This time constant is usually much larger than the fluctuation time scale.

The distribution functions G_T and G_n can be directly linked to the energy distribution function $F(E)$ via the single rate coefficients of the W-matrix:

$$\begin{aligned}&\int\limits_{T=0}^{\infty} \mathrm{d}T \int\limits_{n=0}^{\infty} \mathrm{d}n \int\limits_{\Delta E}^{\infty} \mathrm{d}E\,\sigma^X(E,n)V(E)F_M(E,T)G_T(T)G_n(n)\\ &\quad = n_{tot}\int\limits_{\Delta E}^{\infty} \mathrm{d}E\,\sigma^X(E,n)V(E)F(E).\end{aligned} \tag{10.88}$$

The integral equation relates the distribution functions G_T and G_n to the particle energy distribution function $F(E)$. A solution to (10.88) can be found, expanding $F(E)$ to multiple Maxwellians:

$$F(E) = \sum_{i=1}^{N} f_i F_M(E, T_i). \tag{10.89}$$

Inserting (10.89) into (10.88), we obtain

$$G_T(T)G_n(n) = \sum_{i=1}^{N} f_i \delta(T - T_i)\delta(n - n_i). \tag{10.90}$$

We note that an arbitrary distribution function might not be expanded in terms of Maxwellian functions. However, with respect to the radiative properties of the atomic system it turns out to be very useful to approximate the energy distribution function by multiple Maxwellians. In particular, the approximation by three

Maxwellian distribution functions provides an effective approximation and also a clear physical interpretation:

$$F(E) = f_1 F^M(E,T_1) + f_2 F^M(E,T_2) + f_3 F^M(E,T_3), \tag{10.91}$$

where F_M is the Maxwellian energy distribution function according to

$$F_M(E) = \frac{2\sqrt{E}}{\sqrt{\pi}} \frac{\exp(-E/kT)}{(kT)^{3/2}}. \tag{10.92}$$

The first term in (10.91) is the "bulk electron temperature $T_1 = T_{bulk}$", the second the "hot electron temperature $T_2 = T_{hot}$", and the third the "recombination temperature $T_3 = T_{rec}$". The bulk electron fraction is given by

$$f_1 = f_{bulk} = \frac{n_e(bulk)}{n_e(bulk) + n_e(hot) + n_e(rec)}. \tag{10.93}$$

The hot electrons fraction is defined by

$$f_2 = f_{hot} = \frac{n_e(hot)}{n_e(bulk) + n_e(hot) + n_e(rec)}, \tag{10.94}$$

whereas the recombination fraction is defined by

$$f_3 = f_{rec} = \frac{n_e(rec)}{n_e(bulk) + n_e(hot) + n_e(rec)}. \tag{10.95}$$

In order to ensure the normalization of the total distribution function $F(E)$, (10.91), namely

$$\int_0^\infty F(E)\mathrm{d}E = 1, \tag{10.96}$$

the fraction f_1 is determined by the relation

$$f_1 = 1 - f_2 - f_3. \tag{10.97}$$

We note that the temperatures T_1, T_2, and T_3 are not temperatures in a thermodynamic sense but just the parameters of the distribution function according to (10.92).

If $f_2 = f_3 = 0$, $F(E)$ according to (10.92) describes a Maxwellian energy distribution function with the temperature T_{bulk}. Hot electrons do have a considerable effect on ionization, excitation, and inner-shell processes and are described by the hot electron temperature T_{hot}. An excess of low energy electron will lead to enhanced recombination processes, and this effect is described by T_{rec}. The respective rate coefficients are given by

$$X_{ji} = (1 - f_2 - f_3)X_{ji}^{M}(T_1) + f_2 X_{ji}^{M}(T_2) + f_3 X_{ji}^{M}(T_3). \tag{10.98}$$

For the three-body recombination, this expression differs due to the double integration over the double differential ionization cross section:

$$T_{ji} = (1 - f_2 - f_3)^2 T_{ji}(T_1) + f_2^2 T_{ji}(T_2) + f_3^2 T_{ji}(T_3) + \xi_{ji}, \tag{10.99}$$

$$\xi_{ji} = 2f_2(1 - f_2 - f_3)\left\langle V_1 V_2 \sigma_{ji}^{T} \right\rangle + 2f_3(1 - f_2 - f_3)\left\langle V_1 V_3 \sigma_{ji}^{T} \right\rangle + 2f_2 f_3 \left\langle V_2 V_3 \sigma_{ji}^{T} \right\rangle. \tag{10.100}$$

Even in the case of multiple Maxwellians, the "mixed term" $\left\langle V_a V_b \sigma_{ji}^{T} \right\rangle$ cannot be reduced analytically and has to be evaluated numerically. This numerical integration is extremely time consuming. Numerical calculations carried out with non-Maxwellian simulations of the MARIA code show that the "mixed term" can be reasonably approximated by

$$\left\langle V_a V_b \sigma_{ji}^{T} \right\rangle \approx 0.95\sqrt{T_{ji}(T_a) \cdot T_{ji}(T_b)} \cdot \left(\frac{T_{ji}(T_a)}{T_{ji}(T_b)}\right)^{0.1}, \tag{10.101}$$

where $T_a < T_b$.

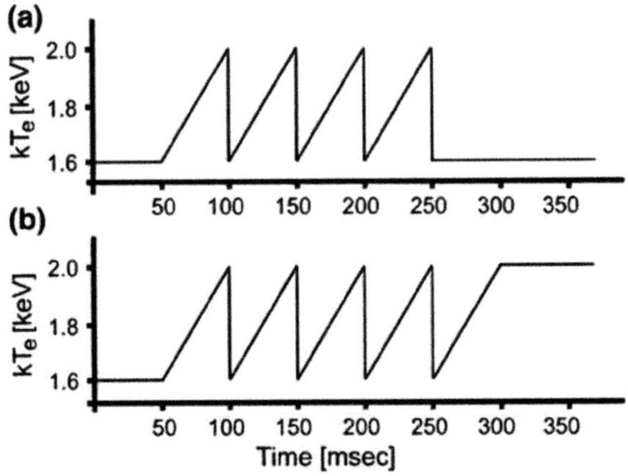

Fig. 10.21 Model sawtooth of the electron temperature, **a** initial and final temperature are identical and **b** the final temperature ends at the maximum temperature of the oscillation

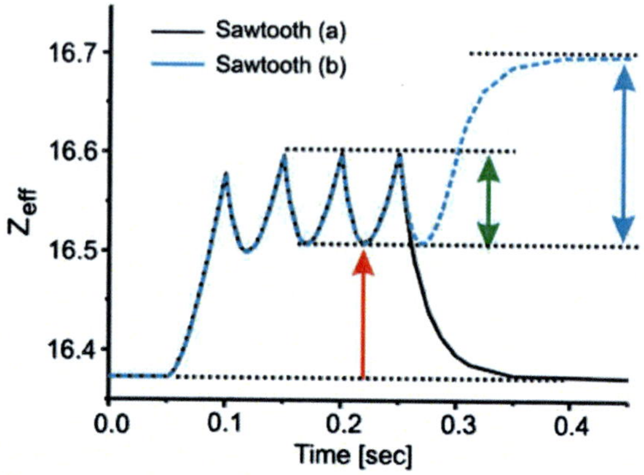

Fig. 10.22 Evolution of the average charge Z_{eff} of argon under sawtooth oscillation, **a** initial and final temperature are identical and **b** the final temperature ends at the maximum temperature of the oscillation

10.3.5.3 Time-Dependent Charge State Evolution

Figures 10.20 have shown that the neutral beam injection induces a rather periodic part of the oscillation while a small stochastic part is onset on these oscillations. In order to investigate the effect of the periodic oscillations on the radiative properties, we employ two different types of "model sawtooth" that are presented in Fig. 10.21. Figure 10.21a shows a model sawtooth, where initial and final temperatures are identical while for the model sawtooth shown in Fig. 10.21b, the final temperature is the maximum temperature of the oscillation. Similar model sawtooths are applied for the density oscillations. The time-dependent parameters $n_e(t)$ and $T_e(t)$ are then employed in the MARIA simulations to calculate time-dependent atomic populations, (10.73)–(10.80).

The two types of model sawtooths according to Fig. 10.21a, b allow explaining the basic principles of the transient radiative properties, the evolution of the charge state distribution, and spectroscopic diagnostics. Figure 10.22 shows the oscillations of the average charge Z_{eff} of argon ions under sawtooth oscillations from Fig. 10.21 taking into account temperature and also density oscillations (see Fig. 10.20). It can be seen that about 2 oscillations after the onset of the sawtooth activity are needed to obtain regular charge state oscillations (indicated by the red flash in Fig. 10.22). During the oscillation, the average charge is highly out of equilibrium and oscillates (green flash in Fig. 10.22) about between $Z_{eff} = 16.5$–16.6. As can be seen from the simulation for the model sawtooth (b), the average charge never reaches the value that corresponds to the highest temperature in the oscillation (indicated with blue flash in Fig. 10.22) and also never reaches the average charge values that correspond to the lowest temperature in the oscillation (indicated by the lower horizontal dotted line). This demonstrates that at any instant of the oscillation, the charge state is highly out of equilibrium. Moreover, as can be seen from the rise and fall of the oscillations in Fig. 10.22, the charge state evolution is not able to follow the sawtooth crash, instead a rather smooth decrease of Z_{eff} is observed.

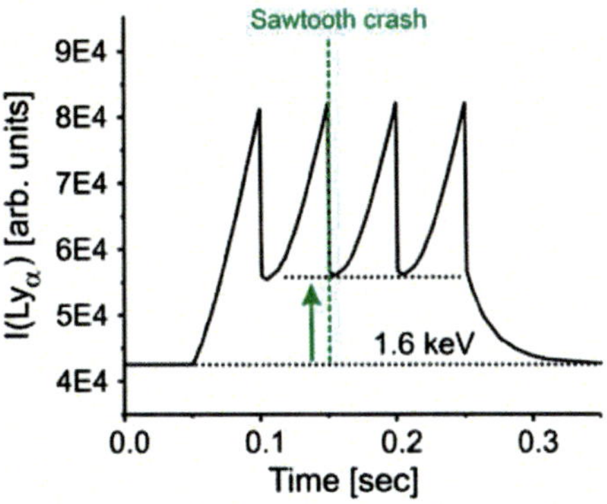

Fig. 10.23 Evolution of the intensity of Lyman-alpha of argon under sawtooth oscillation where initial and final temperatures are identical (Fig. 10.21a). The relaxation of the Lyman-alpha photons permits to follow the sawtooth crash

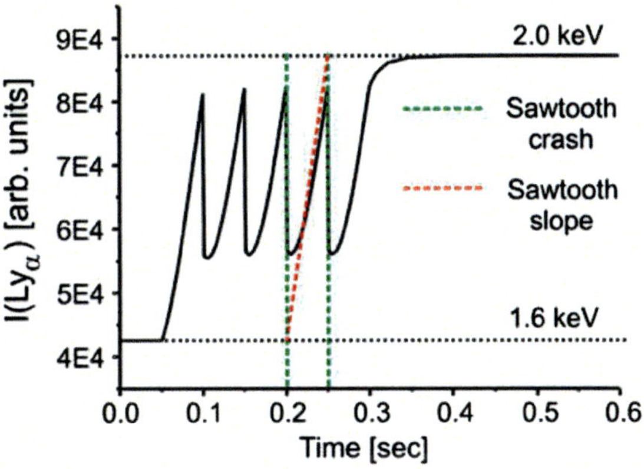

Fig. 10.24 Evolution of the intensity of Lyman-alpha of argon under sawtooth oscillation where the final temperature is identical to the maximum temperature of the oscillation (Fig. 10.21b). The sawtooth crash can be followed; however, the sawtooth emission slope is perturbed and not identical to the original one

The "out of equilibrium features" are a consequence of the characteristic ionization time scale, (10.86). It should be emphasized that these non-equilibrium effects have nothing to do with a limited time resolution, but rather with the "internal" atomic time scale that does not allow to respond quickly enough to external forces. In other words, these relaxation effects are still observed even if the time resolution is infinitely large.

10.3.5.4 Time-Dependent Evolution of Line Intensities

The temporal emission of impurity lines is of great interest for diagnostic purposes, e.g., for impurity transport investigations via temporal decay of line intensity studies, certain temporal behavior of line emissions signaling the development of a plasma disruption. Figures 10.23 and 10.24 show the temporal evolution of the H-like Lyman-alpha emission of argon under sawtooth oscillations type (a) and type (b) (Fig. 10.21). Figure 10.23 shows an almost instantaneous fall off of the intensity when the sawtooth crash appears because the photon relaxation time is

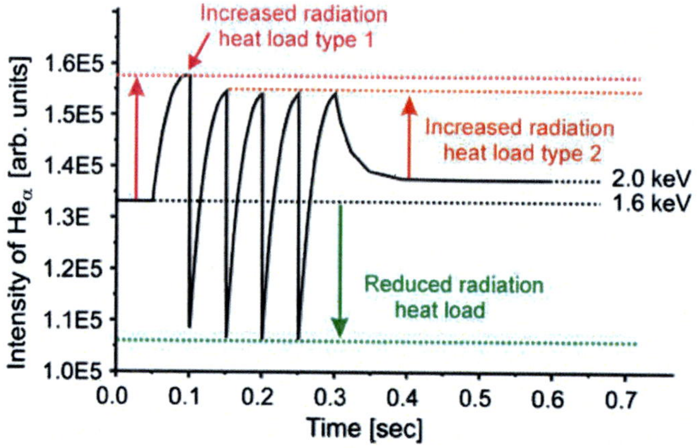

Fig. 10.25 MARIA simulations of the temporal evolution of the radiation heat load. The start of the sawtooth oscillation leads to a considerable increase of the radiation heat load "heat load 1" compared to the stationary case at 2 keV while during the oscillatory phase, periodic radiation heat load increases "heat load 2" are observed

very small due to the strong Z-scaling of the photon transition probability $\left(A \propto Z_{\text{eff}}^4\right)$. The oscillations in the fall-off phase do not completely return due to non-relaxed ion charge state distribution as discussed in relation with Fig. 10.22. In order to position the time-dependent results with respect to the stationary limits of the extreme parameters of the oscillation, Fig. 10.24 shows the Lyman-alpha emission for the model sawtooth of Fig. 10.21b. It can clearly be seen that the Lyman-alpha intensity oscillates between the stationary limits corresponding to $kT_e = 1.6$ and 2.0 keV. Also indicated in Fig. 10.24 is the analysis about the capacity to resolve the sawtooth oscillation. The two dashed vertical green lines indicate that the sawtooth crash can be well resolved; however, the slope of the Lyman-alpha intensity (solid black line) is not fully corresponding to the original slope indicated by the red dashed line. Therefore, the evolution of the absolute intensity of the H-like resonance line is not able to fully "resolve" the sawtooth oscillations.

10.3.5.5 Enhanced Radiation Heat Load

Depending on the plasma parameters with respect to the radiating atom/ion, the oscillatory behavior of electron temperature and density may also induce a considerable increase of the radiation heat load (Rosmej and Lisitsa 2011). This is demonstrated in Fig. 10.25 with the help of the He-like resonance line W of argon. It can be seen that with the onset of the oscillation, an increased radiation heat load is observed, indicated as "heat load type 1" in Fig. 10.25. After stabilization of the oscillatory response, an oscillatory increase of the radiation heat load is observed,

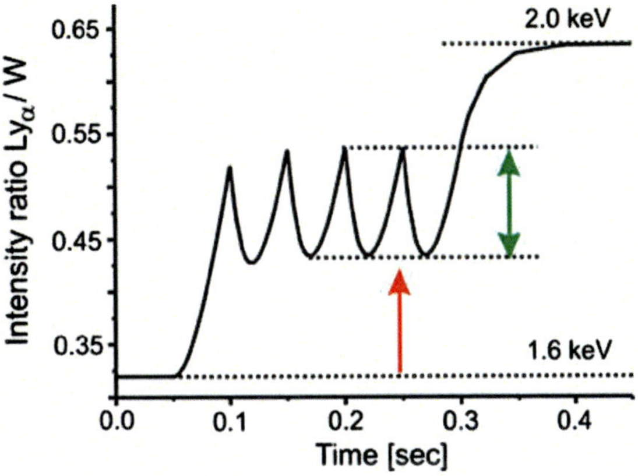

Fig. 10.26 Evolution of the line intensity ratio of Lyman-alpha and Helium-alpha of argon under sawtooth oscillation where the final temperature is identical to the maximum temperature of the oscillation (Fig. 10.21b). Neither the sawtooth crash nor the slope can correctly be followed via the line ratio

indicated as "heat load type 2" in Fig. 10.25. During the oscillatory phase, also reduced radiation heat load is observed. The increased radiation heat load (means increased compared to the stationary case indicated with dotted lines in Fig. 10.25) is due to the characteristic time scale of the ionic distribution (see (10.86)): The temperature rises more rapidly than the characteristic time scale of the ionic distribution; therefore, instantaneous electron collisional excitation at high temperatures takes place from non-relaxed (enhanced) ground states (in the current example the He-like ground state $1s^2\ ^1S_0$).

The increased radiation heat load of type 1 and type 2 may have dramatic consequences for the plasma confinement as material damages can be induced. It is therefore of great interest for the future experimental reactor ITER to have temporally resolved line emissions available that are calibrated in the sense of a "precursor diagnostic" to shut off the machine before destructive radiation heat load develops.

We note that Fig. 10.25 demonstrates the principle mechanisms of the increased radiation heat load via a transparent example. In practice, the increased radiation heat load can develop for any other ionization stage, line emission, and impurity

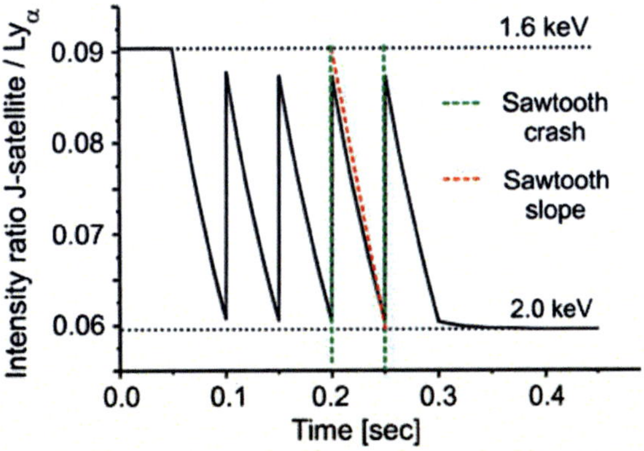

Fig. 10.27 Evolution of the line intensity ratio of the He-like dielectronic satellite $J = 2p^2\ ^1D_2$–$1s2p\ ^1P_1$ and Lyman-alpha of argon under sawtooth oscillation where the final temperature is identical to the maximum temperature of the oscillation (Fig. 10.21b). Sawtooth crash and slope can approximately be followed via the satellite resonance line ratio

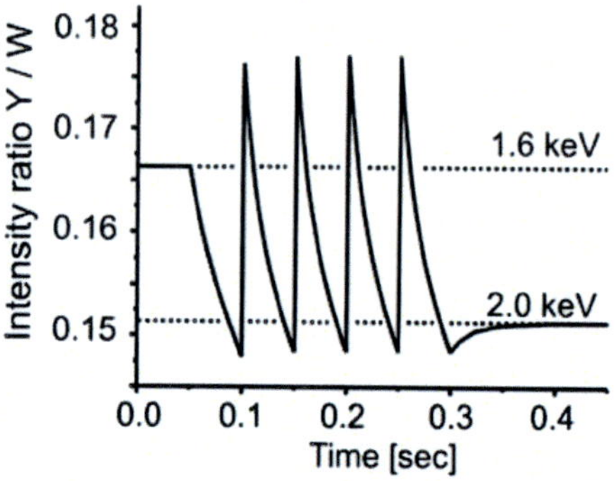

Fig. 10.28 Evolution of the line intensity ratio of the He-like intercombination line Y and resonance line W of argon under sawtooth oscillation where the final temperature is identical to the maximum temperature of the oscillation (Fig. 10.21b). Sawtooth crash and slope are strongly perturbed and differ strongly from the original model sawtooth (Fig. 10.21b) while the oscillatory amplitudes are strongly outside the interval corresponding to the stationary values for $kT_e = 1.6$ keV and $kT_e = 2.0$ keV

atom/ion in dependence of the electron temperature. For heavy element impurities (such as tungsten), increased radiation heat load can therefore develop in many different spectral ranges.

10.3.5.6 Time-Dependent Line Intensity Ratios

For spectroscopic and diagnostic applications, it is of great interest to study diagnostic line ratios during sawtooth oscillations. Figure 10.26 shows the temporal evolution of the intensity ratio of the H-like Lyman-alpha line and He-like resonance line W of argon that is frequently employed for temperature diagnostics. It can be seen that during sawtooth oscillation of type (b) (Fig. 10.21b), the line intensity ratio oscillates between the stationary values corresponding to $kT_e = 1.6$ keV and $kT_e = 2.0$ keV (dotted lines indicated with "1.6 keV" and "2.0 keV" in Fig. 10.26) while these limits are never reached during the oscillation. Therefore, this line ratio has not the capacity to serve as a time-resolved temperature diagnostic during sawtooth oscillation.

As can also be seen from Fig. 10.26, the line ratio displays only a limited capacity to resolve the slope of the sawtooth and entirely fails to describe the sawtooth crash.

Figure 10.27 shows the temporal evolution of the satellite-to-resonance line ratio, i.e., the intensity ratio of the transition $J = 2p^2\ ^1D_2{-}1s2p\ ^1P_1$ and the corresponding H-like Lyman-alpha line $2p\ ^2P_{1/2,3/2}{-}1s\ ^2S_{1/2}$. This line ratio is a very convenient temperature diagnostic as discussed above (see Sect. 10.2.1.1). It can be seen from Fig. 10.27 that this line ratio has a great capacity to resolve even the oscillatory phase of the sawtooth: rising slope and sawtooth crash are well described and close to the original form (Fig. 10.21b) of the sawtooth. Note that the sawtooth crash leads to a rise of the ratio because the satellite-to-resonance line ratio increases with decreasing temperature.

Figure 10.28 demonstrates the temporal evolution of the He-like intercombination Y and resonance line W ratio during sawtooth oscillation. It can clearly be seen that this ratio is highly out of equilibrium: The ratio may be considerably smaller and larger than the corresponding stationary values (indicated with dashed lines "1.6 keV" and "2.0 keV" in Fig. 10.28). Also slope and crash of the line ratio do not well correspond to the original sawtooth.

10.4 Suprathermal Electrons

10.4.1 Non-Maxwellian Elementary Atomic Physics Processes

According to (10.73), (10.74), the atomic populations n_j are related to the elementary atomic physics processes via the transition matrix W_{ij}. Each matrix element W_{ij} is the sum over the rates of all relevant elementary atomic physics processes. If particle correlations can be neglected in the calculation of the elementary atomic physics processes, the rates (units of [s^{-1}]) can be composed into a product of particle densities and rate coefficients, i.e.,

$$\frac{\mathrm{d}n_j}{\mathrm{d}t} = -n_j \sum_{i=1}^{N} W_{ji} + \sum_{k=1}^{N} n_k W_{kj} \tag{10.102}$$

with

$$W_{ji} = n_e \cdot C_{ji} + A_{ji} + n_e \cdot I_{ji} + n_e^2 \cdot T_{ji} + n_e \cdot D_{ji} + \Gamma_{ji} + n_e \cdot R_{ji} + \cdots . \tag{10.103}$$

Equation (10.103) explicitly lists the most important elementary processes in a plasma, i.e., electron collisional excitation/de-excitation rate coefficients C_{ij} and C_{ji} (in units of [cm^3 s^{-1}]), if the electron density is in units of [cm^{-3}]), radiative decay rates A_{ij} (units of [s^{-1}]), electron collisional ionization I_{ij} (units of [cm^3 s^{-1}]), electron-induced three-body recombination rate coefficient (units of [cm^6 s^{-1}]), dielectronic capture D_{ij} (in units of [cm^3 s^{-1}]), autoionization Γ_{ij} (in units of [s^{-1}]), and radiative recombination rate coefficient R_{ij} (in units of [cm^3 s^{-1}]). The dots on the right-hand side of (10.103) indicate further processes that might be important for particular applications (e.g., charge exchange as discussed in Sect. 10.3, heavy particle processes, radiation field terms). The rate coefficients $X = C_{ij}, I_{ij}, D_{ij}, R_{ij}, T_{ij}$ have to be determined from integrals over the respective cross sections and the electron energy distribution functions $F(E)$. For $X \neq T$, we have

$$X_{ji} = \int_{E_0}^{\infty} \mathrm{d}E\, \sigma_{ji}^{X}(E)V(E)F(E). \tag{10.104}$$

σ_{ji}^{X} is the cross section for the process "X" for the transition $i \to j$, V is the relative velocity of the colliding particles and E_0 is the threshold energy (if no threshold exist $E_0 = 0$). The energy distribution function is normalized according to

$$\int_{0}^{\infty} F(E) \cdot \mathrm{d}E = 1. \tag{10.105}$$

In the nonrelativistic approximation

$$V(E) = \sqrt{\frac{2E}{m_e}}. \tag{10.106}$$

For the three-body recombination coefficient T_{ij}, one has to take care of the fact that the energy distribution function of simultaneously 2 particles ("1" and "2") has to be taken into account:

$$T_{ji} = \frac{\pi^2\hbar^3}{m_e^2}\frac{g_i}{g_j}\int_{0}^{\infty} \mathrm{d}E_1 \int_{0}^{\infty} \mathrm{d}E_2 \frac{E}{\sqrt{E_1 E_2}}\sigma_{ji}^{I}(E, E_1)F(E_1)F(E_2) \tag{10.107}$$

or, in convenient units

$$T_{ji} = 1.3949 \times 10^{-26} \cdot \frac{g_i}{g_j} \cdot \int_{0}^{\infty} \mathrm{d}E_1 \int_{0}^{\infty} \mathrm{d}E_2 \frac{E}{\sqrt{E_1 E_2}}\sigma_{ji}^{I}(E, E_1)F(E_1)F(E_2)\, [\mathrm{cm}^6\, \mathrm{s}^{-1}] \tag{10.108}$$

with E, E_1, and E_2 in [eV] and σ in [cm^2]. $\sigma_{ji}^{I}(E, E_1)$ is the double differential ionization cross section for the transition $i \to j$, g_i and g_j are the statistical weights of the level i and j, respectively, and

$$E = \Delta E_{ji} + E_1 + E_2, \tag{10.109}$$

where ΔE_{ji} is the ionization energy. Numerical calculations have shown (Green and Sawada 1972; Clark et al. 1991; Faucher et al. 2000) that the double differential cross section can be cast into a product of a single ionization cross section $\sigma_{ji}^{I}(E)$ and a probability $\Omega(E, E_1)$:

$$\sigma_{ji}^{I}(E, E_1) = \sigma_{ji}^{I}(E) \cdot \Omega(E, E_1) \tag{10.110}$$

with

$$\int_{0}^{(E-E_1)/2} \Omega(E, E_1) \mathrm{d}E_1 = 1. \tag{10.111}$$

The advantage in the decomposition of the double differential cross sections according to (10.110) lies in the fact that the single ionization cross section $\sigma_{ji}^{I}(E)$ is readily provided by atomic physics codes while the probability function $\Omega(E, E_1)$ can be approximated by an analytical function:

$$\Omega(E, E_1) = \frac{1}{(E - E_i) \cdot \left(E^2 + aE_i^2\right)} \cdot \left\{2 \cdot (a+1) \cdot E_i^2 + \frac{b \cdot (E + E_i) \cdot (E_1 - 0.5 \cdot (E - E_i))^4}{(E - E_i)^3}\right\}, \tag{10.112}$$

where E_i is the ionization energy. From the analysis of H-like Coulomb–Born exchange ionization cross sections from 1*s* until 6 h, the fitting parameters "*a*" and "*b*" can be approximated with $a \approx 14.4$ and $b \approx 160$ (Clark et al. 1991) and also be applied for non-hydrogenic ions. For highly charged ions, the fitting parameters describe in general well (within 20%) the double differential cross sections; however, for neutral or near neutral atoms/ions and multiple-filled shells, the agreement is less accurate in particular at small parameters $E_1/(E - E_i)$.

For practical application of (10.104), it is necessary to establish the link between the direct and inverse cross sections because in non-Maxwellian plasmas, the principle of detailed balance cannot be employed to extract the inverse rate coefficient from the direct rate coefficient. For the cross section of the rate coefficient C_{ij}, we need to relate the collisional excitation cross section to the collisional de-excitation cross section (see also Sect. 7.7.2):

$$\sigma_{ji}(E') = \frac{g_i}{g_j} \sigma_{ji}\left(E' + \Delta E_{ji}\right) \frac{E' + \Delta E_{ji}}{E'}, \tag{10.113}$$

where E and E' are the electron energies before and after scattering, respectively. These two energies are related to each other via the excitation energy ΔE_{ji}:

$$E = E' + \Delta E_{ji}. \tag{10.114}$$

Relation (10.113) is known as "Klein–Rosseland equation." For a Maxwellian electron energy distribution, direct and inverse rate coefficients are directly related by

$$C_{ji} = C_{ji} \cdot \frac{g_i}{g_j} \cdot \exp(\Delta E_{ji}). \tag{10.115}$$

Applying the same method outlined in Sect. 7.6.2, we can relate the ionization cross section for the ionization rate coefficient I_{ij} to the three-body recombination equivalent cross section (note that the index characterizes a state "*i*" before ionization and the index "*j*" a state after ionization):

$$\sigma_{ji}^{T}(E_1, E_2) = \frac{\pi^2 \hbar^3}{m_e^2} \cdot \frac{g_i}{g_j} \cdot \frac{E}{\sqrt{E_1 \cdot E_2}} \cdot \sigma_{ji}^{I}(E, E_1), \tag{10.116}$$

where E is the electron energy before ionization and E_1 and E_2 are the electron energies after ionization. These three energies are related to each other via the ionization energy E_i:

$$E = E_i + E_1 + E_2. \tag{10.117}$$

Comparing (10.107) and (10.116), the three-body recombination rate coefficient can be represented as a double integral over the three-body recombination equivalent cross section, i.e.,

$$T_{ji} = \int_0^\infty dE_1 \int_0^\infty dE_2 \, \sigma_{ji}^{T}(E_1, E_2) F(E_1) F(E_2). \tag{10.118}$$

For a Maxwellian energy distribution function, the three-body recombination coefficient can be directly expressed via the ionization rate coefficient, i.e.,

$$\begin{aligned} T_{ji} &= I_{ji} \cdot \frac{g_i}{2g_j} \cdot \left(\frac{2\pi\hbar^2}{m_e k T_e}\right)^{3/2} \cdot \exp(E_i/kT_e) \\ &= 1.6564 \times 10^{-22} \cdot I_{ji} \cdot \frac{g_i}{g_j} \cdot \frac{\exp(E_i/kT_e)}{(kT_e)^{3/2}} \; [\mathrm{cm}^6\,\mathrm{s}^{-1}]. \end{aligned} \tag{10.119}$$

For the last relation in (10.119), I_{ij} in [$\mathrm{cm}^3\,\mathrm{s}^{-1}$], kT_e and E_i in [eV] (n_e in [cm^{-3}] in (10.103)).

Concerning the radiative recombination rate coefficient R_{ij}, it is conveniently expressed in terms of the photoionization cross section that is related to the radiative recombination cross section

$$g_i \cdot \sigma_{ji}^{iz}(\hbar\omega) = \frac{2 m_e c^2 E}{\hbar^2 \omega^2} \cdot g_j \cdot \sigma_{ji}^{r}(E), \tag{10.120}$$

where E is the energy of the photoionized electron and $\hbar\omega$ is the photon energy. These energies are related to each other via the ionization energy E_i:

$$\hbar\omega = E_i + E. \tag{10.121}$$

Relation (10.120) is known as the "Milne equation." The radiative recombination rate coefficient is therefore given by

$$R_{ji} = \int_0^\infty \sigma_{ji}^r(E) \cdot V(E) \cdot F(E) \cdot dE \tag{10.122}$$

or, expressed in terms of the photoionization cross section:

$$R_{ji} = \frac{g_i}{g_j} \cdot \frac{1}{\sqrt{2} \cdot m_e^{3/2} \cdot c^2} \cdot \int_0^\infty \sigma_{ji}^{iz}(E_i + E) \cdot \frac{(E_i + E)^2}{\sqrt{E}} \cdot F(E) \cdot dE \tag{10.123}$$

or, in convenient units

$$R_{ji} = 2.8616 \times 10^{19} \cdot \frac{g_i}{g_j} \cdot \int_0^\infty \sigma_{ji}^{iz}(E_i + E) \cdot \frac{(E_i + E)^2}{\sqrt{E}} \cdot F(E) \cdot dE\,(\mathrm{cm^3\,s^{-1}}) \tag{10.124}$$

with E and E_i in [eV] and σ in [cm^2] (n_e in [cm^{-3}] in (10.103)).

Because the dielectronic capture is the inverse process of autoionization, we can apply a similar method than in Sect. 7.6.2 to determine the dielectronic capture rate coefficient:

$$D_{ji} = \frac{\pi^2\hbar^3}{\sqrt{2}m_e^{3/2}} \frac{g_j}{g_i} \Gamma_{ji} \int_0^\infty \delta(E_S, E) \cdot \frac{F(E)}{\sqrt{E_S}} \cdot dE. \tag{10.125}$$

E_s is the dielectronic capture energy. The δ-function in (10.125) appears because the dielectronic capture is a resonance processes: Only continuum electrons that meet exactly the atomic resonance energy can take part in a capture process. Therefore, (10.125) takes the form

$$D_{ji} = \frac{\pi^2\hbar^3}{\sqrt{2}m_e^{3/2}} \cdot \frac{g_j}{g_i} \cdot \Gamma_{ji} \cdot \frac{F(E_S)}{\sqrt{E_S}} \tag{10.126}$$

or, in convenient units (Γ_{ji} in [s^{-1}], E_S in [eV], F in [1/eV])

$$D_{ji} = 2.9360 \times 10^{-40} \cdot \frac{g_j}{g_i} \cdot \Gamma_{ji} \cdot \frac{F(E_S)}{\sqrt{E_S}} [\mathrm{cm^3\,s^{-1}}]. \tag{10.127}$$

For a Maxwellian energy distribution function, (10.126) takes the form

$$D_{ji} = 1.6564 \times 10^{-22} \cdot \frac{g_j}{g_i} \cdot \Gamma_{ji} \cdot \frac{\exp(-E_S/kT_e)}{(kT_e)^{3/2}} \ (\mathrm{cm^3\,s^{-1}}) \tag{10.128}$$

with kT_e and E_S in [eV] (n_e in [cm^{-3}] in (10.103)).

10.4.2 Pathological Line Ratios

Due to the increased parameter space for non-Maxwellian electrons, it is difficult to derive general conclusions like for Maxwellian plasma. An important insight into non-Maxwellian atomic kinetics, however, can be obtained in the framework of the "Hot Electron Approximation". In this approximation, only f_{bulk} and f_{hot} (see (10.93), (10.94)) are retained to approximate the non-Maxwellian energy distribution function $F(E)$, i.e.,

$$F(E, T_{bulk}, T_{hot}) = (1 - f_{hot})F_M(E, T_{bulk}) + f_{hot}F_M(E, T_{hot}). \tag{10.129}$$

$F_M(E,T_{bulk})$ and $F_M(E,T_{hot})$ are Maxwellian energy distribution functions with the parameters T_{bulk} and T_{hot} (note that T_{bulk} and T_{hot} are not temperatures in a thermodynamic sense but just convenient parameters), f_{hot} is the fraction of the hot electrons which are described by the energy distribution function $F_M(E,T_{hot})$. Many experiments with hot dense plasmas have shown (e.g., dense laser-produced plasmas, dense pinch plasmas) that (10.129) is a reasonable approximation to the measured distribution function (e.g., obtained by means of the bremsstrahlung). Moreover, T_{hot} is often much larger than T_{bulk}. In this case, it is convenient to speak of a "bulk" electron temperature T_{bulk} and to interpret T_{hot} as a hot electron temperature T_{hot}, while f_{hot} is the hot electron fraction that is defined by

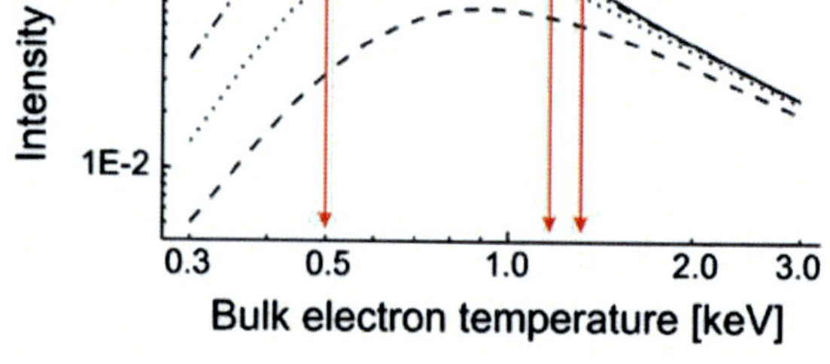

Fig. 10.29 MARIA simulations of the intensity ratio of the *J*-satellite and H-like Ly$_\alpha$ of Ar for different hot electron fractions f_{hot} in dependence of the bulk electron temperature, $n_e = 10^{22}\ \mathrm{cm^{-3}}$, $kT_{hot} = 20$ keV

$$f_{\text{hot}} = \frac{n_e(hot)}{n_e(bulk) + n_e(hot)}. \tag{10.130}$$

$n_e(bulk)$ is the density of the "bulk" electrons and $n_e(hot)$ those of the hot or suprathermal electrons.

The introduction of the "bulk" and "hot" electron temperature permits to understand the basic effects of suprathermal electrons on the radiation emission, the spectral distribution, and line intensity ratios. Figure 10.29 shows the intensity line ratios of the He-like J-satellite and the H-like Ly_α of argon for different fractions of hot electrons, $kT_{\text{hot}} = 20$ keV, $n_e(\text{hot}) + n_e(\text{bulk}) = 10^{22}$ cm^{-3}. The case $f_{\text{hot}} = 0$ corresponds to a Maxwellian plasma, and the numerical simulations are close to the analytical model discussed above (10.18)–(10.28). Deviations from the analytical model for very low temperatures are due to collisional–radiative effects. The monotonic dependence on the electron temperature indicates a strong sensitivity for electron temperature measurements. Hot electron fractions, however, lead to a non-monotonic behavior. This is connected with the different asymptotic behavior of the cross sections for the collisional excitation and the dielectronic capture: For the collisional excitation, all electrons whose energy is larger than the excitation energy contribute to the excitation; the dielectronic capture, however, is a resonance processes and only those electrons in the continuum contribute to the cross section which match the resonance energy (see (10.125)). Therefore, hot electrons contribute strongly to the collisional excitation of the resonance line but only little to the dielectronic capture of the satellite lines. As for low bulk electron temperatures, the hot electron-induced collisional excitation is large compared to those of the bulk electrons; the intensity ratio is much lower than for a Maxwellian plasma. For high bulk electron temperatures, the hot electrons do not contribute much compared to the bulk electrons and the curves for $f > 0$ approach those for $f = 0$.

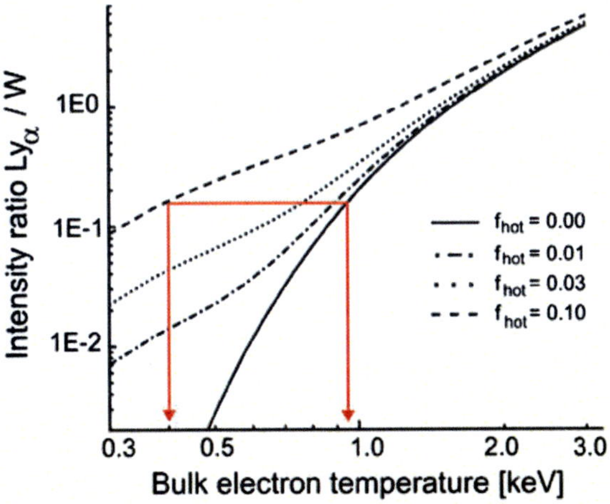

Fig. 10.30 MARIA simulations of the intensity ratio of the H-like Ly_α-line and the He-like resonance line W of Ar for different hot electron fractions f_{hot} in dependence of the bulk electron temperature, $n_e = 10^{22}$ cm^{-3}, $kT_{\text{hot}} = 20$ keV

The red arrows in Fig. 10.29 show the principle difficult for diagnostics: The same line ratio can be obtained for three different sets of parameters: (1) $kT_{\text{bulk}} = 0.5$ keV, $f = 3\%$, (2) $kT_{\text{bulk}} = 1.17$ keV, $f = 3\%$, (3) $kT_{\text{bulk}} = 1.24$ keV, $f = 0$. Of particular importance is the difference between the solutions (1) and (3): low bulk temperature and hot electrons versus a high electron temperature without hot electrons. This example indicates the general difficult to interpret the measurements: This difficulty is not connected with the particular selection of the resonance line and the satellite transitions but is based on the general asymptotic dependence of the cross sections (resonance process and threshold process). Therefore, all line ratios of any resonance line and its satellite transitions are affected in a similar manner.

Figure 10.30 shows the line ratios of the H-like Ly_α and He-like He_α of argon, $n_e = 10^{22}$ cm^{-3}, $kT_{\text{hot}} = 20$ keV. In a Maxwellian plasma ($f_{\text{hot}} = 0$), the strong monotonic dependence is very convenient for temperature diagnostics. The presence of hot electrons, however, rises considerably the intensity ratio for lower bulk electron temperatures due to increased ionization induced by hot electrons: $1s^2 + e\,(hot) \rightarrow 1s + 2e$.

The red arrows in Fig. 10.30 indicate an example of the principle difficult for diagnostics due to multiple solutions for the same line intensity ratio: (1) $kT_{\text{bulk}} = 400$ eV, $f_{\text{hot}} = 0.1$, (2) $kT_{\text{bulk}} = 930$ eV, $f_{\text{hot}} = 0$. Also this example shows that the neglect of hot electrons leads to a considerable overestimation of the bulk electron temperature. It should be noted that the intensity ratio of Fig. 10.30 poses other difficulties in transient plasmas: In ionizing plasmas, the ionic populations are lagging behind the electron temperature and therefore the Ly_α emission is lower than it would correspond to the given electron temperature. In other words, using the stationary line intensity ratio (i.e., neglecting the ionizing nature of the plasma) underestimates the electron temperature. In recombining plasmas, the electron temperature is overestimated because the Ly_α intensity is too large for the given electron temperature. The line intensity ratio of Ly_α and He_α is therefore more

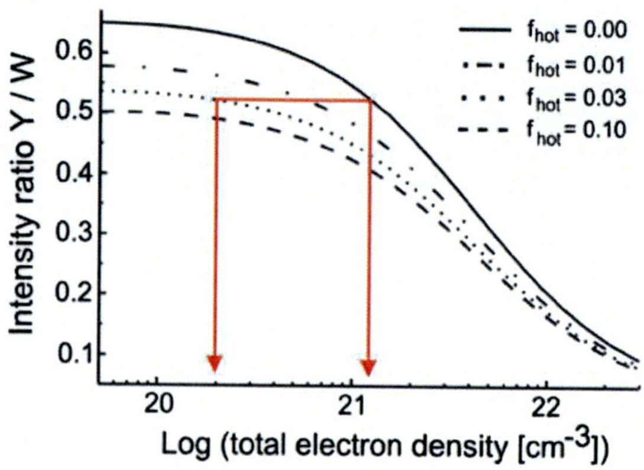

Fig. 10.31 MARIA simulations of the intensity ratio of the He-like intercombination line *Y* and the He-like resonance line *W* of Ar for different hot electron fractions f_{hot} in dependence of the total electron density, $kT_{\text{bulk}} = 600$ eV, $kT_{\text{hot}} = 20$ keV

indicative for the ionization temperature and not for the electron temperature as often erroneously stated.

Figure 10.31 shows the line intensity ratio of the He-like intercombination line $Y = 1s2p\ ^3P_1 \rightarrow 1s^2\ ^1S_0 + h\nu$ and the He-like resonance line $W = 1s2p\ ^1P_1 \rightarrow 1s^2\ ^1S_0 + h\nu$, $kT_{\text{bulk}} = 600$ eV. The continuous decrease of the line intensity ratio with increasing density is due to an effective collisional population transfer from the triplet system to the singlet system. As the intercombination transition has a rather low transition probability compared to the resonance line (spin forbidden transition in the LS-coupling scheme), the population of the triplet system is much larger than those of the singlet system. Therefore, the transfer of population from the singlet to the triplet system is smaller than opposite resulting in an effective population transfer from the triplet to the singlet levels. This in turn results in a decrease of the line intensity ratio. As Fig. 10.31 indicates, the density sensitivity of this line intensity ratio is strong and can therefore be used as an electron density diagnostic.

Hot electrons, however, result in an overall decrease of the intensity ratio for all electron densities. This effect is due to the different asymptotic dependence of direct and exchange excitation cross sections (see also discussion in Sect. 5.5.2):

$$\sigma_{\text{direct}} \propto \frac{\ln E}{E}, \tag{10.131}$$

$$\sigma_{\text{exchange}} \propto \frac{1}{E^3}. \tag{10.132}$$

In a pure LS-coupling scheme, the collisional excitation from the ground state to the triplet levels is carried only by the exchange part of the cross section. Due to the strong decrease of the exchange cross sections with increasing impact energy (compared to the direct cross section), the hot electron-induced excitation for the singlet levels is much larger than for the triplet levels. This results in a strong decrease of the line ratio and in turn to a large overestimation of the electron density when hot electrons are neglected. We note that this effect is not connected with the particular line ratio of the *Y*- and *W*-lines: It is a consequence of the general asymptotic dependence of the cross sections according to (10.131), (10.132). Therefore, all line ratios which are based on resonance and intercombination line transitions are perturbed in a similar manner. The red arrows in Fig. 10.31 (calculated for $kT_{\text{bulk}} = 600$ eV, $kT_{\text{hot}} = 20$ keV) indicate an example of the principle difficult of density diagnostics due to multiple solutions for the same line intensity ratio: (1) $n_e = 2 \times 10^{20}$ cm^{-3}, $f_{\text{hot}} = 3\%$, 2) $n_e = 1.2 \times 10^{21}$ cm^{-3}, $f_{\text{hot}} = 0.0$. This example shows that the neglect of hot electrons leads to a considerable overestimation (order of magnitude) of the electron density.

Let us consider the influence of intermediate coupling effects. For the transitions $1s^2 + e \rightarrow 1s2l\ ^{1,3}L + e$, the excitation cross sections in intermediate coupling σ_{IC} can be written (see Sect. 5.5.2) as follows:

$$\sigma_{IC} = Q_d \sigma_{direct} + Q_e \sigma_{exchange}. \tag{10.133}$$

For example, for the excitation of the resonance line W (cross section $1s^2\ ^1S_0 + e \rightarrow 1s2p\ ^1P_1 + e$) and the intercombination line Y (cross section $1s^2\ ^1S_0 + e \rightarrow 1s2p\ ^3P_1 + e$), we have in the pure LS-coupling scheme: $Q_d^{LS}(W) = 2$, $Q_e^{LS}(W) = 0.5$, $Q_d^{LS}(Y) = 0$, $Q_e^{LS}(Y) = 0.5$. For argon, the angular factors are only slightly different in the intermediate coupling scheme: $Q_d^{IC}(W) = 1.9684$, $Q_e^{LS}(W) = 0.5$, $Q_d^{LS}(Y) = 0.0316$, $Q_e^{LS}(Y) = 0.5$. The intermediate coupling angular factors Q indicates that the triplet levels have a small admixture of the direct cross section. This admixture is, however, of importance for rather heavy elements, e.g., for molybdenum we have $Q_d^{IC}(W) = 1.515$, $Q_e^{LS}(W) = 0.5$, $Q_d^{LS}(Y) = 0.485$, $Q_e^{LS}(Y) = 0.5$. In conclusion, even in the intermediate coupling scheme (note that the above simulations presented in Fig. 10.31 include intermediate coupling effects), the discussion concerning the asymptotic behavior of the excitation cross sections (10.131), (10.132) remains valid.

What is the general conclusion from Figs. 10.29, 10.30 and 10.31? The standard line ratios are excellent methods for density and temperature diagnostics in stationary Maxwellian plasmas. However, for plasmas containing hot electrons, the development of other methods is mandatory. Of primary importance are the stable determination of the bulk electron temperature and the hot electron fraction.

10.4.3 Bulk Electron Temperature

Let us consider a plasma whose electron energy distribution function is given by (10.129), (10.130). The rate coefficients for the processes "X" are then given by

$$\langle X \rangle = (1 - f_{hot}) \langle X, T_{bulk} \rangle + f_{hot} \langle X, T_{hot} \rangle. \tag{10.134}$$

For the three-body recombination rate coefficient $\langle T \rangle$, the expression is more complicated due to the need for simultaneously two energy distribution functions of the continuum electrons:

$$\begin{aligned} \langle TR \rangle = (1 - f_{hot})^2 \langle TR, T_{bulk} \rangle + f_{hot}^2 \langle TR, T_{hot} \rangle \\ + 2 f_{hot} (1 - f_{hot}) \langle TR, T_{bulk}, T_{hot} \rangle. \end{aligned} \tag{10.135}$$

The first term describes the usual three-body recombination at a temperature T_{bulk} ($\langle TR, T_{bulk} \rangle$ being the three-body recombination rate coefficient at temperature T_{bulk}) the second one those at a temperature T_{hot} ($\langle TR, T_{hot} \rangle$ being the three-body recombination rate coefficient at temperature T_{hot}). The last term is a mixed term which requests the integration over the double differential cross section (see (10.118)). This term cannot be expressed by simple combinations of usual three-body coefficients with T_{bulk} and/or T_{hot} like the first and second terms of

(10.135). As discussed in relation with (10.101), this is very inconvenient for numerical simulations because the double integration in a multilevel multi-ion stage atomic system requests considerable computational resources. The "mixed" term of (10.135) can be roughly approximated by

$$\langle TR, T_{\text{bulk}}, T_{\text{hot}}\rangle \approx 0.95\sqrt{\langle TR, T_{\text{bulk}}\rangle\langle TR, T_{\text{hot}}\rangle}\left(\frac{T_{\text{bulk}}}{T_{\text{hot}}}\right)^{0.1}, \tag{10.136}$$

where $T_{\text{bulk}} < T_{\text{hot}}$.

Let us now consider an innovative method to determine the electron bulk temperature in dense plasmas containing hot electrons (Rosmej 1995b) that is based on the analysis of the X-ray line emission of a He-like satellite and the He-like Rydberg series $1snp\ ^1P_1 \rightarrow 1s^2\ ^1S_0 + h\nu$ of highly charged ions. The intensity of a He-like satellite with high radiative and high autoionizing rate is given by (see also Fig. 5.1)

$$I_{\text{k,ji}}^{\text{sat}} = \hbar\omega_{\text{ji}} n_{\text{e}} n_{\text{k}}(k = 1s)\{(1 - f_{\text{hot}})\langle DR, T_{\text{bulk}}\rangle + f\langle DR, T_{\text{hot}}\rangle\}\,. \tag{10.137}$$

The rate coefficient of the dielectronic recombination is given by (see also (10.20)–(10.23))

$$\langle \text{DR}, T\rangle = \alpha\frac{Q_{\text{k,ji}}}{g_{\text{k}}}\frac{\exp(-E_{\text{kj}}/kT)}{(kT)^{3/2}}, \tag{10.138}$$

where $Q_{\text{k,ji}}$ is the satellite intensity factor defined in (10.23), E_{kj} is the capture energy.

If the electron density is sufficiently high to ensure a balance between the levels "$1s$" and "$1snp\ ^1P_1$" via collisional ionization and three-body recombination, we can determine the population density of the $1snp\ ^1P_1$-level analytically:

$$\begin{aligned} &n_{\text{e}} n(1snp^1P_1)\{(1 - f_{\text{hot}})\langle I, T_{\text{bulk}}\rangle + f\langle I, T_{\text{hot}}\rangle\} \\ &\approx n_{\text{e}}^2 n(1s)\left\{(1 - f_{\text{hot}})^2\langle TR, T_{\text{bulk}}\rangle + f_{\text{hot}}^2\langle TR, T_{\text{hot}}\rangle + 2f_{\text{hot}}(1 - f_{\text{hot}})\langle TR, T_{\text{bulk}}, T_{\text{hot}}\rangle\right\}. \end{aligned} \tag{10.139}$$

From (10.137) and (10.139), the intensity ratio of the dielectronically captured satellite and the He-like Rydberg series is given by [note that the term $\langle TR,\ T_{\text{hot}}\rangle$ is rather small because $\langle TR,\ T\rangle$ decreases strongly with increasing temperature; see (5.50):

$$\frac{I_{\text{k,ji}}^{\text{sat}}}{I_{\text{n}}} = \frac{\omega_{\text{ji}}}{\omega_{\text{n}}}\frac{n_{\text{e}} n(1s)\{(1 - f_{\text{hot}})\langle DR, T_{\text{bulk}}\rangle + f_{\text{hot}}\langle DR, T_{\text{hot}}\rangle\}\{(1 - f_{\text{hot}})\langle I, T_{\text{bulk}}\rangle + f\langle I, T_{\text{hot}}\rangle\}}{A_{\text{n}} n_{\text{e}} n(1s)\left\{(1 - f_{\text{hot}})^2\langle TR, T_{\text{bulk}}\rangle + f_{\text{hot}}^2\langle TR, T_{\text{hot}}\rangle + 2f_{\text{hot}}(1 - f_{\text{hot}})\langle TR, T_{\text{bulk}}, T_{\text{hot}}\rangle\right\}}. \tag{10.140}$$

A_n is the spontaneous transition probability of the Rydberg series $1snp\ ^1P_1 \rightarrow 1s^2\ ^1S_0$; ω_J and ω_n are the transition frequencies of the satellite and Rydberg transitions. Developing (10.140) in the series of f_{hot} gives

$$\frac{I_{k,ji}^{sat}}{I_n} = \frac{\omega_{ji}}{\omega_n} \frac{\langle DR, T_{bulk}\rangle \cdot \langle I, T_{bulk}\rangle}{A_n \cdot \langle TR, T_{bulk}\rangle} \cdot \{1 + f_{hot} \cdot G_1 + \cdots\} \tag{10.141}$$

with

$$G_1 = \frac{\langle I, T_{hot}\rangle}{\langle I, T_{bulk}\rangle} + \frac{\langle DR, T_{hot}\rangle}{\langle DR, T_{bulk}\rangle} - 2\frac{\langle TR, T_{bulk}, T_{hot}\rangle}{\langle TR, T_{bulk}\rangle}. \tag{10.142}$$

With

$$\langle DR, T_{bulk}\rangle = \frac{1}{g(1s)} \cdot \frac{(2\pi)^{3/2}\hbar^3}{2(m_e kT_{bulk})^{3/2}} \cdot Q_{k,ji} \cdot \exp(-E_{kj}/kT_{bulk}) \tag{10.143}$$

and

$$\langle TR, T_{bulk}\rangle = \langle I, T_{bulk}\rangle \cdot \frac{g_n}{2g(1s)} \cdot \frac{(2\pi)^{3/2}\hbar^3}{(m_e kT_{bulk})^{3/2}} \cdot \exp(E_n/kT_{bulk}) \tag{10.144}$$

the intensity ratio of (10.141) can be written as

$$\frac{I_{k,ji}^{sat}}{I_n} = \frac{\omega_{ji}}{\omega_n} \frac{Q_{k,ji}}{A_n g_n} \exp\left(-\frac{E_{kj} + E_n}{kT_{bulk}}\right) \cdot \{1 + f \cdot G_1 + \cdots\}. \tag{10.145}$$

Calculations show that for almost all practical cases, $G_1 < 1$. This indicates that the zero-order approximation of (10.145), namely

$$\left.\frac{I_{k,ji}^{sat}}{I_n}\right|_{0-order} = \frac{\omega_{ji}}{\omega_n} \cdot \frac{Q_{k,ji}}{A_n \cdot g_n} \cdot \exp\left(-\frac{E_{kj} + E_n}{kT_{bulk}}\right) \tag{10.146}$$

is of extraordinary importance: (10.146) represents the ideal case of a bulk electron temperature diagnostic in non-Maxwellian plasmas, i.e., (see also (10.12), (10.13))

$$\frac{I_{ji}}{I_{j'i'}} = G_{jij'i'}(T_{bulk}). \tag{10.147}$$

With respect to (10.147), it is important to note that the zero-order approximation does not depend on the hot electron fraction but still describes the line ratios very accurately: That is why this approximation is of great interest to determine the bulk electron temperature in non-Maxwellian plasmas. An extremely useful He-like satellite transition is the J-satellite ($2p^2\ ^1D_2 \rightarrow 1s2p\ ^1P_1 + h\nu$) because of its large autoionizing rate, high radiative decay rate, and its simplicity in experimental registration (see, e.g., Fig. 10.1).

A further advantage of (10.146) is that it contains only a few atomic data. These data can be expressed in an analytical manner with high precision. For A_n, we propose the following expression:

$$A_n = \left(A_0 - \alpha Z_{\mathrm{eff}}^2\right) \cdot \frac{n(n-1)^{2n-2}}{(n+1)^{2n+2}} Z_{\mathrm{eff}}^4 \left[s^{-1}\right], \tag{10.148}$$

$$Z_{\mathrm{eff}} = Z_n - \sigma. \tag{10.149}$$

Note that the first factor in (10.148) takes into account intermediate coupling effects which strongly depend on the nuclear charge Z_n. With $A_0 = 4.826 \times 10^{11}$, $\alpha = 7.873.10^7$, $\sigma = 0.8469$, a precision better than 6% is reached for all n for elements in the interval $Z_n = 6$–32 (practically all elements of practical interest for diagnostics). The dielectronic satellite intensity factor for the J-satellite has already been described by (10.24). The energies in (10.146) can be approximated by

$$E_{1s,J} + E_n = 13.6\,\mathrm{eV}\left[\frac{3}{4}Z_n^2 + \frac{3}{4}(Z_n - 0.4)^2 - (Z_n - 0.5)^2 \cdot \left(1 - \frac{1}{n^2}\right)\right] \tag{10.150}$$

with a precision better than 2%. For example, for $Z_n = 18$ and $n = 4$, the exact data are: $A_4 = 1.21 \times 10^{13}\ \mathrm{s}^{-1}$, $Q_J = 4.30 \times 10^{14}\ \mathrm{s}^{-1}$, $E_{1s,J} + E_4 = 2.55$ keV, whereas the analytical formulas (10.148)–(10.150) provide $A_4 = 1.19 \times 10^{13}\ \mathrm{s}^{-1}$, $Q_J = 4.36 \times 10^{14}\ \mathrm{s}^{-1}$, $E_{1s,J} + E_{i,4} = 2.56$ keV. This indicates a sufficiently high precision of the simple analytical expressions for diagnostic applications.

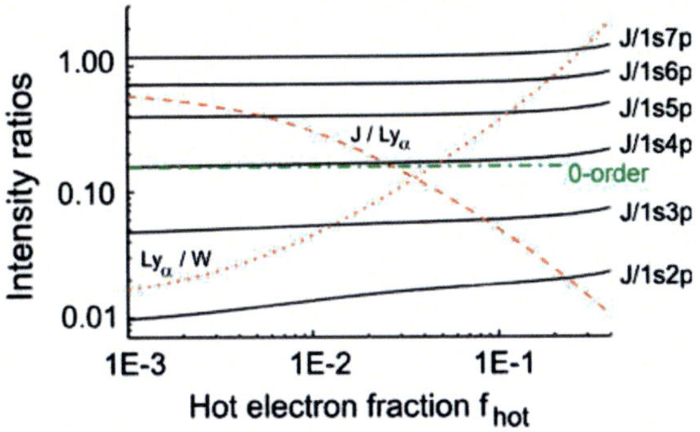

Fig. 10.32 MARIA simulations of the intensity ratios Ly$_\alpha$/W, J/Ly$_\alpha$ and J/1snp $^1P_1$ of Ar in a dense plasma containing hot electrons, $n_e = 10^{22}$ cm^{-3}, $kT_{\mathrm{bulk}} = 600$ eV, $kT_{\mathrm{hot}} = 20$ keV. Also indicated the analytical zero-order approximation for the ratio J/1$s4p$ $^1P_1$ of (10.146) to determine the bulk electron temperature. Red dashed and red dotted curves are the standard line ratios for the determination of the electron temperature that shows a strong dependence with respect to the hot electron fraction

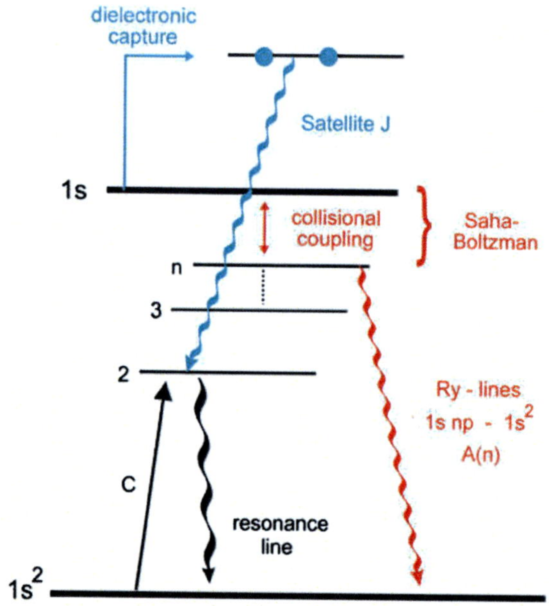

Fig. 10.33 Energy-level diagram showing the collisional coupling of Rydberg states 1*snl* with the continuum (see (10.139)), the dielectronic capture to the 2*l*2*l*′-states from the H-like ground state 1*s* $^2S_{1/2}$ (see (10.137)) and the relevant radiative decay rates

Figure 10.32 shows the numerical simulations of the intensity ratio of the He-like *J*-satellite line and the Rydberg series for $n = 2$–7 (solid black lines) in dependence of the hot electron fraction f_{hot}, the bulk electron temperature is $kT_{bulk} = 600$ eV, the total electron density is $n_e = 10^{22}$ cm^{-3}, and the hot electron temperature is $kT_{hot} = 20$ keV. Also shown in Fig. 10.32 are the intensity ratios of Ly$_\alpha$/*W* and *J*/Ly$_\alpha$ which has been discussed in connection with Figs. 10.29 and 10.30. Figure 10.32 demonstrates an impressive stability of the intensity ratios *J*/ 1*snp* $^1P_1$ even for very large hot electron fractions up to 10%. For such large hot electron fractions, the standard line intensity ratios are already off by an order of

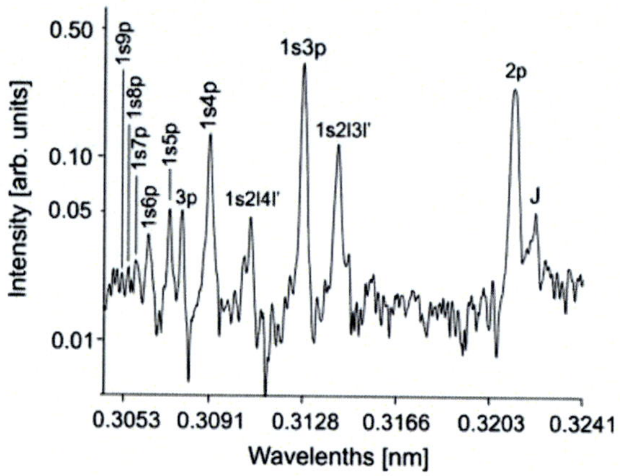

Fig. 10.34 Experimental X-ray spectrum from a dense Mega-Ampère Z-pinch driven with argon showing the H-like Lyman-alpha line (2*p*), Lyman-beta line (3*p*), the He-like *J*-satellite, and the He-like series 1*snp* $^1P_1$–$1s^2$ $^1S_0$

Table 10.6 Bulk electron temperatures deduced from the experimental spectrum of Fig. 10.34 with the help of the zero-order approximation of (10.146)

$I(J)/I(1snp)$	kT_{bulk} (keV)	$n_{e,crit}$ (cm^{-3})
$n = 4$	0.7 ± 0.1	1.7×10^{22}
$n = 5$	0.9 ± 0.1	3.4×10^{21}
$n = 6$	0.9 ± 0.1	9.4×10^{20}
$n = 7$	1.2 ± 0.2	3.2×10^{20}
$n = 8$	1.3 ± 0.2	1.2×10^{20}
$n = 9$	1.3 ± 0.2	5.3×10^{19}

Right column indicates the critical density obtained from (10.39) for $kT_{bulk} = 1.3$ keV and $Z = 17$

magnitude which means that they are meaningless for a temperature diagnostics. Figure 10.32 also indicates that for higher Rydberg series transitions, the intensity ratio is only very weakly dependent on the hot electron fraction because the thermalization threshold to ensure PLTE has already been reached. This is demonstrated schematically in Fig. 10.33 which also explains the basic characteristics of the bulk electron temperature diagnostic with the help of an energy level diagram.

Figure 10.32 presents also the results of the zero-order approximation (10.146) for the $1s4p\ ^1P_1$-line. Excellent agreement is seen between the numerical non-Maxwellian simulations and the zero-order analytical approximation. This confirms the great importance of the zero-order approximation to determine the bulk electron temperature and the practical realization of (10.147).

Let us apply the bulk electron temperature measurement to the X-ray emission spectra of a dense Mega-Ampere Pinch driven with argon (Rosmej et al. 1993). Figure 10.34 shows the soft X-ray spectrum in the spectral range from the H-like Lyman-alpha line with corresponding satellites until the He-like series limit of $1snp\ ^1P_1{-}1s^2\ ^1S_0$. Table 10.6 shows the electron temperatures deduced from the line ratios of (10.146). It can be seen that the bulk electron temperature stabilizes at about $kT_{bulk} = 1.3$ keV. This indicates that the collisional coupling of the $1snl$-states with the continuum (see Fig. 10.33) is effective starting with principal quantum number of about $n = 7$. The last column of Table 10.6 presents the critical densities obtained from (10.39) for $Z = 17$ and $kT_{bulk} = 1.3$ keV. As the line ratios stabilize at about $n = 7$, the right column can be used to estimate the electron density that are in agreement with density measurements of different methods (Rosmej et al. 1993).

We note that the satellite to Rydberg transition method for the determination of the bulk electron temperature can be transferred to different satellite transitions in order to optimize the application under various experimental constraints, e.g., the use of the Lyman-beta satellites (Rosmej et al. 2009).

10.4.4 Hot Electron Fraction

10.4.4.1 Hot Electron Perturbed Satellite and Resonance Line Intensities

Having once determined the bulk electron temperature, the intensity ratios of the He-like $2l2l'$-satellites with the Ly_α line can be used for the determination of the hot electron fraction if resonance line intensity and satellite intensity are well approximated by

$$I^{\mathrm{res}}_{\mathrm{k',j'i'}} \approx n_{\mathrm{e}} n_{\mathrm{k'}} \frac{A_{\mathrm{j'i'}}}{\sum_{\mathrm{l'}} A_{\mathrm{j'l'}}} \langle C_{\mathrm{k'j'}} \rangle, \tag{10.151}$$

$$I^{\mathrm{sat}}_{\mathrm{k,ji}} \approx \alpha n_{\mathrm{e}} n_{\mathrm{k}} \frac{Q_{\mathrm{k,ji}}}{g_{\mathrm{k}}} \frac{\exp(-E_{\mathrm{kj}}/kT_{\mathrm{e}})}{(kT_{\mathrm{e}})^{3/2}}. \tag{10.152}$$

If the energy distribution function is described by (10.129), (10.130) and the rate coefficients by (10.134) the fraction of hot electrons is then given by

$$f_{\mathrm{hot}} \approx \frac{1}{1 + \dfrac{R\langle CR, T_{\mathrm{hot}} \rangle - \langle DR, T_{\mathrm{hot}} \rangle}{\langle DR, T_{\mathrm{bulk}} \rangle - R\langle CR, T_{\mathrm{bulk}} \rangle}}, \tag{10.153}$$

where the line intensity ratio R is

$$R = \frac{I^{\mathrm{sat}}_{\mathrm{k,ji}}}{I^{\mathrm{res}}_{\mathrm{k',j'i'}}}. \tag{10.154}$$

The relevant rate coefficients of (10.153) are given by

$$\langle CR, T \rangle = \frac{A_{\mathrm{j'i'}}}{\sum_{\mathrm{l'}} A_{\mathrm{j'l'}}} \langle C_{\mathrm{k'j'}}, T \rangle, \tag{10.155}$$

$$\langle DR, T \rangle = \alpha \frac{Q_{\mathrm{k,ji}}}{g_{\mathrm{k}}} \frac{\exp(-E_{\mathrm{kj}}/kT)}{(kT)^{3/2}}. \tag{10.156}$$

In order to best fulfill the parameter range of validity of (10.156), the use of the He-like J-satellite (see also discussion above) is recommended. Equations (10.24)–(10.28) provide the necessary data for the $2l2l'$- and $1s2l2l'$-satellites and their corresponding resonance lines. In high-density plasmas, the range of validity of (10.151)–(10.156) might be limited (in fact, (10.151), (10.152) are strictly valid only in the framework of the Corona model and negligible inner-shell contribution of the dielectronic satellite emission) and other methods have to be developed.

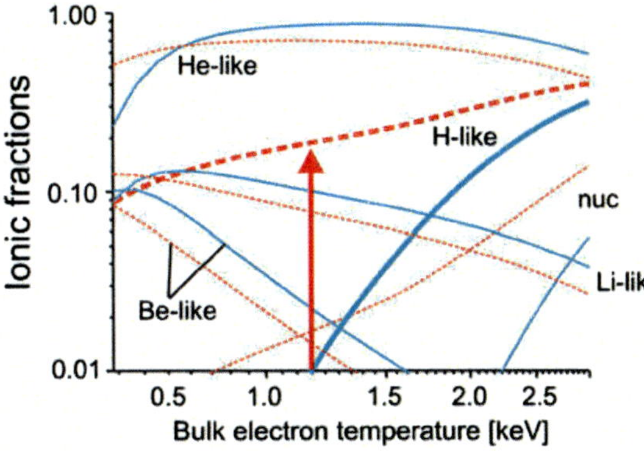

Fig. 10.35 MARIA simulations of the ionic fractions of titanium in dense optically thick plasmas containing hot electrons, $n_e = 10^{21}$ cm^{-3}, $L_{eff} = 300$ μm for different hot electron fractions, $f_{hot} = 0.0$ (solid curves), $f_{hot} = 0.09$, $kT_{hot} = 20$ keV (dashed curves)

10.4.4.2 Qualitative Distortion of the Ionic Charge State Distribution

Suprathermal electrons are routinely excited in high-intensity laser-produced plasmas via instabilities driven by nonlinear laser–plasma interaction. Their accurate characterization is crucial for the performance of inertial confinement fusion as well as astrophysical and high-energy-density experiments.

In view of the pathological line ratios in non-Maxwellian plasmas discussed above, it is therefore mandatory to develop alternative spectroscopic methods for the determination of the hot electron fraction. Let us therefore consider the ionic fractions in dense plasmas with and without fractions of hot electrons. Fig. 10.35 shows the MARIA simulations for $n_e = 10^{21}$ cm^{-3}, $L_{eff} = 300$ μm and $f_{hot} = 0.0$ (solid curves) and $f_{hot} = 0.09$ (dashed curves), $kT_{hot} = 20$ keV. The comparison between the solid and the dashed curves shows that a qualitative deformation of the ionic fractions has taken place. The arrow indicates a particular strong rise of the H-like abundance for lower bulk electron temperatures, whereas other ionic fractions (He-, Li-, Be-like) are much less influenced. This qualitative deformation can in turn be used for the determination of the hot electron fraction by visualizing the various fractions via the X-ray line emissions from the H-, He-, Li-, Be-, B-, C-like ions (Rosmej 1997).

Experimentally, it is difficult to observe simultaneously the line emission from H- and He-like ions (*K*-shell emission) and those of Li-, Be-, B-, C-like ions (*L*-shell emission) due to the strongly different spectral ranges (requesting a) different types of spectrometers and b) their relative intensity calibration). However, by means of inner-shell satellite transitions $1s2s^n2p^m \to 1s^22s^n2p^{m-1} + h\nu$ this drawback can be circumvented (Rosmej 1997): The wavelength interval of all line transitions is located in a similar spectral range (*K*-shell), and all transitions are therefore simultaneously observable with one type of spectrometer.

Figure 10.36 show the MARIA simulations of the soft X-ray emission spectra of titanium in a dense optically thick plasma, $n_e = 10^{21}$ cm^{-3}, $L_{eff} = 300$ μm. Figure 10.36a shows the simulation for $f_{hot} = 0.0$, $kT_e = 2.3$ keV, Fig. 10.36b shows the simulation for $f_{hot} = 0.0$, $kT_e = 1.0$ keV, and Fig. 10.36c shows the

Fig. 10.36 MARIA simulations of the spectral distribution (linear scale, normalized) of titanium in dense optically thick non-Maxwellian plasmas, $n_e = 10^{21}$ cm^{-3}, $L_{eff} = 300$ μm, **a** $f_{hot} = 0.0$, $kT_e = 2.3$ keV, **b** $f_{hot} = 0.0$, $kT_e = 1.0$ keV, **c** $f_{hot} = 0.09$, $kT_{bulk} = 800$ eV, $kT_{hot} = 20$ keV

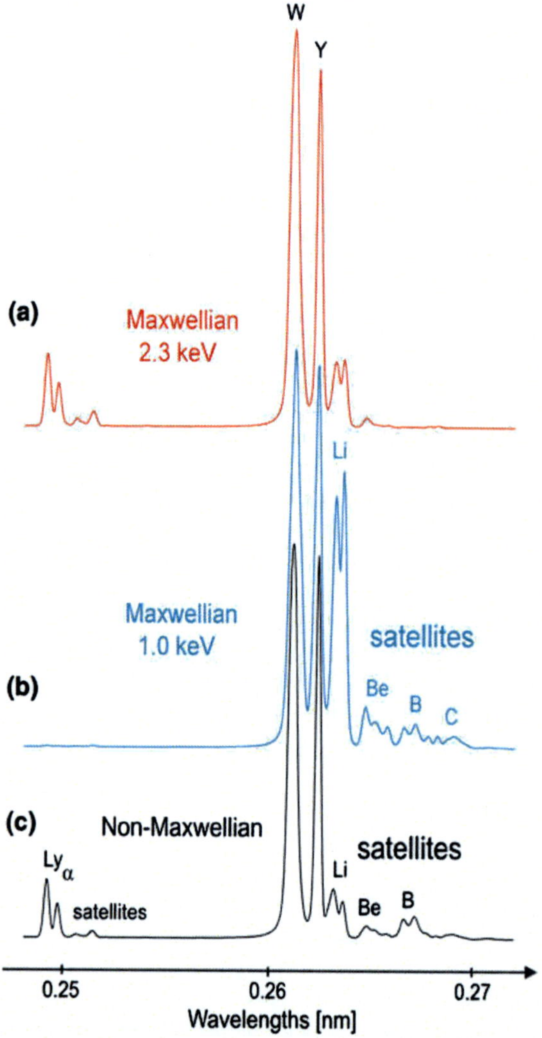

MARIA simulation for $f_{hot} = 0.09$, $kT_{bulk} = 800$ eV, and $kT_{hot} = 20$ keV. As Fig. 10.36b demonstrates the Ly$_\alpha$ emission is practically absent for low electron temperatures, whereas the K$_\alpha$-satellite series is strongly pronounced. At high electron temperatures (Fig. 10.36a), the Ly$_\alpha$-emission is strong; however, the K$_\alpha$-satellites series of Be- B- and C-like ions is practically absent. This reflects the general behavior of the ionic charge state distribution depicted in Fig. 10.35 (see also Fig. 6.5): The ionic populations of highly charged ions (nuc, H-like) are never at the same time as large as those for low-charged ions (Be-, B-, C-like). In non-Maxwellian plasmas, however, large fractions of highly and low-charged ions (see arrow in Fig. 10.35) can exist simultaneously. This circumstance is related to

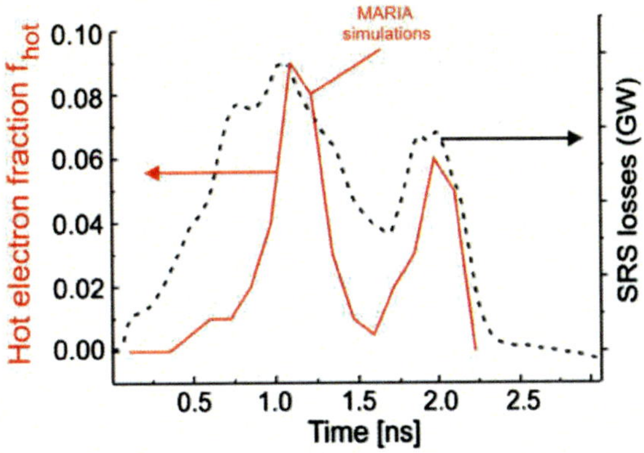

Fig. 10.37 Hot electron fraction f_{hot} as inferred from the time-dependent MARIA simulations of titanium X-ray spectra. The hot electron fractions correlate with the SRS losses and reaches fractions up to about 10%

the fact that for hot electrons, the exponential factor in the expression for the rate coefficients is close to 1 because kT_{hot} is larger than the threshold energies (excitation, ionization). Therefore, the shell structure is not anymore strongly reflected in the distribution of ionic populations. The "admixture" of a considerable fraction of H-like ions simultaneously with Li-, Be-like ions is then connected with the following: the ionization rate for the bulk electrons is in strong competition with the hot electron-induced rate. However, the ionization of the He- and H-like ions is essentially driven by the hot electrons as the rate coefficients for the bulk electrons are exponentially small. Therefore, hot electrons lead only to a minor decrease of the low-charged ions (Be-, B-, C-like) but to a strong increase of the H-like ions with corresponding simultaneously strong Ly_{α} emission and K_{α}-satellite series emission.

The characteristic distortion of the ion charge stage distribution can in turn be used for the determination of the hot electron fraction (Rosmej 1997). This method has successfully been applied in laser-driven inertial fusion experiments to determine the time- and space-resolved hot electron fraction in the NOVA-hohlraums (Glenzer et al. 1998). For these purposes, tracer elements of titanium have been mounted into the hohlraum and time-resolved X-ray spectra (spectral interval as shown in Fig. 10.36) have been recorded. Figure 10.37 shows the time-dependent hot electron fraction as inferred from the non-Maxwellian time-dependent MARIA simulations. The maxima correspond to the rise of the pulse-shaped laser irradiation of the hohlraum and reach hot electron fractions up to about 10% (Fig. 10.37, left scale). Simultaneous measurements of the stimulated Raman scattering (SRS), Fig. 10.37 (right scale), indicate a clear correlation with the spectroscopic inferred hot electron fraction. This allowed identifying the parametric SRS instability as the main source of hot electron production.

10.4.4.3 Temporal Shifts of the Hot Electron Fraction

Suprathermal electron production driven by instabilities in laser–plasma interaction (Kruer 1988) is of paramount interest for the inertial confinement fusion

(ICF) science and high-energy-density physics (HEDP). In the direct drive scheme, hot electrons can cause degradation in the performance of ICF capsules by fuel preheat and reduced compressibility of the capsule (Glenzer et al. 1998; Lindl 1995; Lindl et al. 2004, 2014). In the fast ignition scheme (Tabak et al. 2005), laser coupling to fast electrons determines the efficiency of the energy delivery to the ignition region. In the shock ignition scheme (Betti et al. 2007), the fuel is ignited from a central hot spot heated by a strong spherically convergent shock. The laser intensities required to launch this shock exceed the threshold of parametric instabilities (such as stimulated Raman scattering or two-plasmon decay), which couple a significant fraction of the laser energy to hot electrons.

Hydrodynamic simulations of laser–plasma interactions for pulse durations of the order of 0.1–10 ns and intensities $I \cdot \lambda^2 = 10^{13}-10^{16}$ W/cm^2 are highly challenging as nonlinear processes play an important role. These kinetic processes cannot be directly incorporated into large-scale hydrodynamic models because of a large disparity of temporal and spatial scales. A detailed characterization of hot electrons as well as the plasma evolution via independent methods is therefore mandatory to validate large-scale hydrodynamic approaches that are at its infancy. In this context, X-ray spectroscopy and non-thermal atomic physics are of particular interest due to their potential for a unique characterization of hot electrons inside the relevant plasma.

As demonstrated in relation with Fig. 10.37, the time evolution of the hot electron generation is also of great interest. These timing issues receive a renewed interest for the shock ignition scheme as recent experiments have demonstrated that hot electron onset has a strong influence on the shock strength (Theobald et al. 2015).

Due to the typical ns-time scale in high-energy-density research and inertial fusion applications, time-resolved information can also be extracted from the spatial evolution of the X-ray emission as the radiating ions propagate in space of the order of $V_{\text{ion}} \cdot \tau_{\text{laser}} \approx 10^7\,\text{cm/s} \cdot 1\,\text{ns} = 100\,\mu\text{m}$. Such displacements can be easily resolved with space-resolved X-ray spectroscopy employing curved X-ray Bragg crystals (Renner and Rosmej 2019). Figure 10.38 shows an example of the

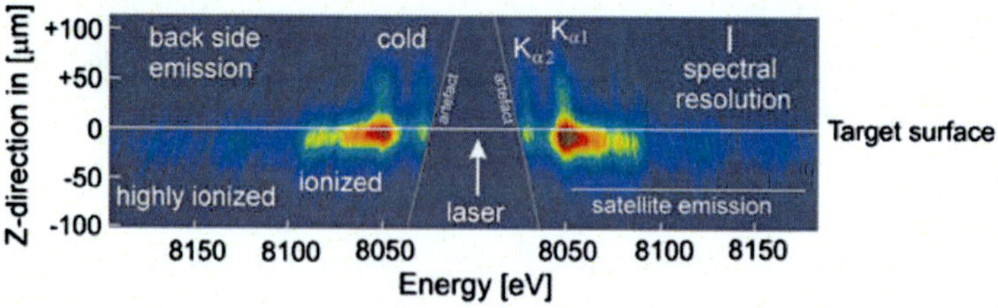

Fig. 10.38 Space-resolved X-ray spectrum of copper irradiating a solid copper foil with a ns kilojoule laser. *K*-alpha emission is excited by hot electrons and extends far behind the original target surface due to the fast-propagating ions. The double image is a particular advantageous feature of the vertical Johann geometry

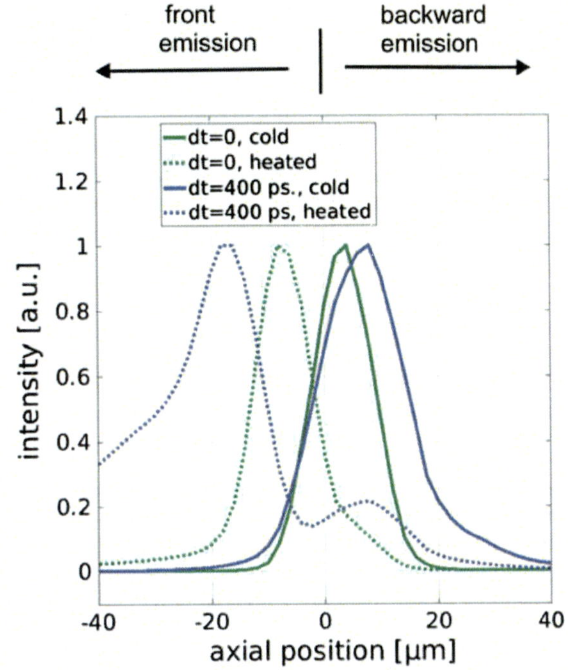

Fig. 10.39 Space-resolved simulation of the X-ray spectrum of copper irradiated with a 0.35 ns, 700 J laser pulse with $\lambda = 1.315$ μm. Cold *K*-alpha emission extends far backwards (positive axial position) with respect to the "heated emission" when the onset of the hot electrons generation is displaced relative the maximum of the laser pulse by 400 ps

space-resolved *K*-alpha emission of copper produced by the irradiation of copper foils with a ns kilojoule laser (Smid et al. 2019). Due to the vertical Johann geometry, registration of a double-sided image is possible (Renner et al. 1997) that greatly increases the experimental precision (see also Chap. 8). Due to kilojoule energy, intensities up to 10^{16} W/cm^2 are achieved that generate hot electrons. The hot electrons propagate into the cold copper foil and drive *K*-shell ionization of rather low-charged ions which results into the typical *K*-alpha radiation (indicated with $K_{\alpha 1}$ and $K_{\alpha 2}$ in Fig. 10.38). Due to the backward ion acceleration, the *K*-alpha emission extends far behind the original target position of the order of 50 μm; see Fig. 10.38. In front of the target, the plasma is strongly heated resulting in higher charge states and corresponding satellite emission from higher charge states (indicated as "satellite emission" in Fig. 10.38). Due to the high spectral resolution, the satellite emission originating from open *M*-shell configurations (indicated as "ionized" in Fig. 10.38) can be separated from open *L*-shell emission (indicated as "highly ionized" in Fig. 10.38).

The simulations demonstrate that the spatial evolution of these satellites and $K_{\alpha 1}$ and $K_{\alpha 2}$ emission is sensitive to the hot electron evolution due to the qualitative distortion of the ionic populations as discussed above (Sect. 10.4.4.2). Moreover, due to the backward acceleration of the ions, timing issues of the hot electrons can be addressed: Ions propagate backwards but X-ray emission is only excited if the hot electrons propagate into the backward moving ions. The extension of the *K*-

alpha emission behind the original target position and its relative displacement to the "heated emission" is therefore characteristic for the onset of the hot electrons as demonstrated with the simulations in Fig. 10.39. This method includes the spatial grid variation obtained from two-dimensional hydrosimulations which have been employed to deduce relative time shifts for the onset of the hot electron generation with respect to the laser pulse maximum (Smid et al. 2019). The temporal information deduced from high-resolution X-ray spectroscopy is of great interest to test and develop large-scale hydrodynamic simulations that are currently at its infancy (Smid et al. 2019).

10.5 Space-Resolved Measurements of Fast Ions

10.5.1 Spatial Resolution of Plasma Jets

Large efforts are made to create homogenous dense plasmas under extreme conditions to provide samples and emission properties, which can be directly compared with theory. Unfortunately, dense hot laboratory plasmas show almost always large variations of the plasma parameters over space, and, in consequence a large variation of the spectral emission (see, e.g., Fig. 10.38). As spatial parameter grid reconstructions are difficult to implement for grid averaged data, X-ray spectroscopy with spatial resolution is frequently applied to obtain supplementary

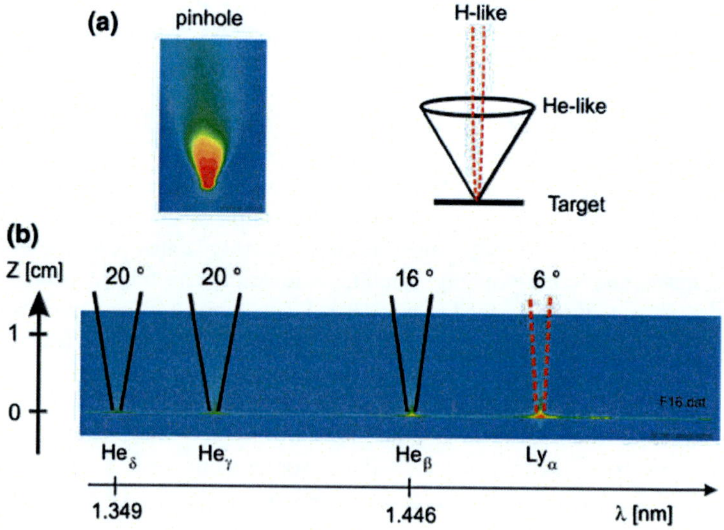

Fig. 10.40 Space-resolved plasma jets consisting of highly charged radiating aluminum ions. **a** X-ray pinhole image (left) and schematic geometry of the jet emission (right), **b** two-dimensional X-ray imaging realized with a spherically curved Bragg crystal

information. Spatial resolution can be realized with a slit mounted at a suitable distance between the source and the X-ray crystal; however, luminosity is drastically reduced. Space resolution can also be obtained without slits employing X-ray-focusing optics realized with curved X-ray crystals. The most commonly used curved crystal arrangements are the Johann geometry, the vertical Johann geometry, the Johannson geometry, the Chauchois geometry that employ cylindrically curved crystals while two-dimensional curved crystals (including the spherical ones) allow to achieve at the same time high spectral and spatial resolution while maintaining high luminosity (Renner and Rosmej 2019). Two particular methods turned out to be extremely useful for dense plasma research:

(1) the vertical Johann geometry (Renner et al. 1997) which is extremely suitable for line profile investigations: Spatial resolution of some μm can be achieved (therefore, image plates with about 50 μm resolution seriously limit high-resolution spectroscopy and line shape analysis becomes a very critical

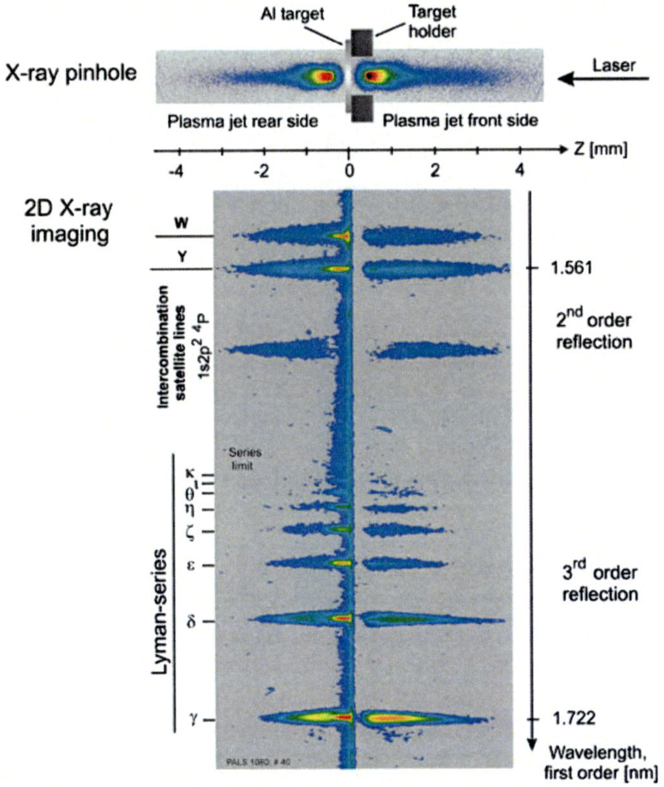

Fig. 10.41 X-ray pinhole emission and two-dimensional X-ray imaging of highly charged aluminum plasma jet interaction with the residual gas. Rear-side and front-side emission show characteristic differences that depend on the particular X-ray transition

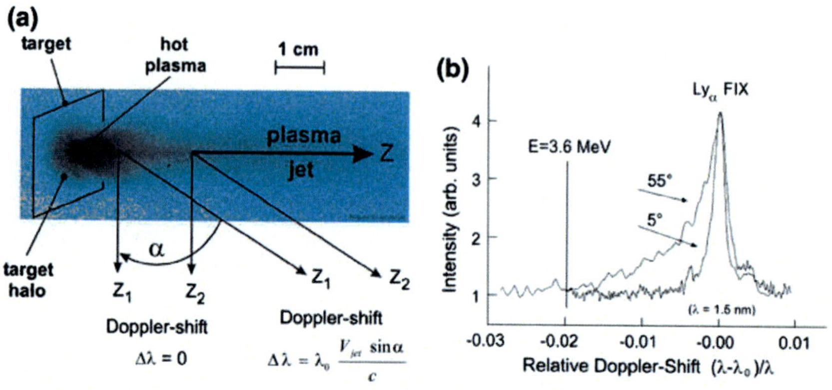

Fig. 10.42 **a** Schematic geometrical arrangement of fast ion Doppler shift measurements of X-ray spectral lines. The first spectrometer is positioned at very low angle $\alpha \approx 0$ while the second spectrometer is located typically at angles $\alpha > 30°$. **b** The comparison of these two spectra allows deducing shifts originating from the fast ion velocities

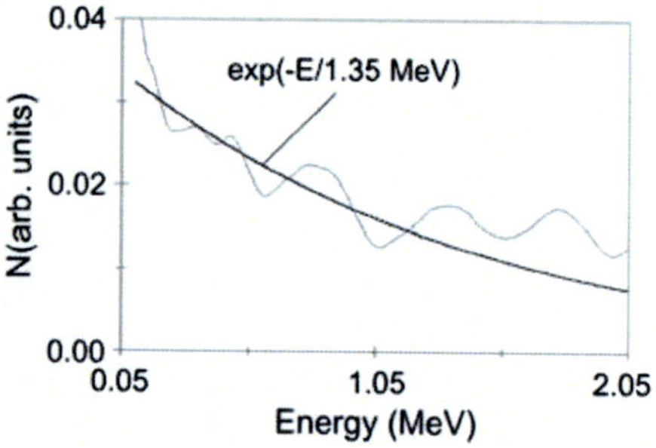

Fig. 10.43 Experimental determination of the fast ion distribution function from the optically thin part of the Doppler shifted emission. The distribution function is reasonably well approximated with a Maxwellian function with a temperature parameter kT_{ion} = 1.35 MeV

issue; note that even in case of Charged Coupled Devices CCD the 13 μm pixel size sets a serious limitation for high-resolution spectroscopy), while simultaneous extremely high spectral resolution of about $\lambda/\delta\lambda \approx 6000$ can be realized. The spectral range, however, is rather limited, permitting only to observe, e.g., the H-like aluminum Lyman$_\alpha$ line and corresponding satellite transitions. Due to the appearance of double-sided spectra, the geometry can provide line shift measurements without reference lines,

(2) the spherical X-ray crystals (Faenov et al. 1994; Skobelev et al. 1995) which do provide simultaneously high spectral ($\lambda/\delta\lambda \approx$ 1000–6000, dependent on the large geometrical variations) and spatial resolution (about 10–30 μm), large spectral windows (permitting, e.g., to observe all the K_α-satellite series until the He-like resonance line W for aluminum (Rosmej et al. 2001c)) and large spatial window (up to cm with 10 μm resolution) and the possibility of X-ray microscopic applications (two-dimensional X-ray imaging). For example, the two-dimensional

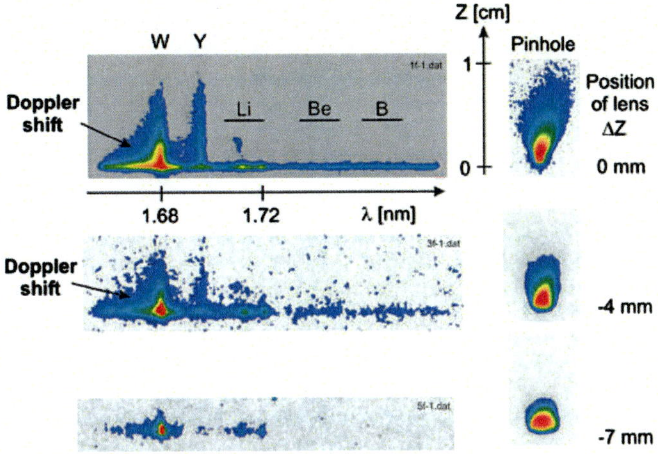

Fig. 10.44 Space-resolved X-ray emission (left) and X-ray pinhole emission (right) of highly charged fluorine plasma jet in dependence of different focal lens positions. At highest intensities, the Doppler shift of the He-like resonance line *W* is very pronounced

imaging permits to determine the angle of laser-produced plasma jet diffusion in space in dependence of the ionic charge state; see Fig. 10.40.

The two-dimensional X-ray imaging realizes a two-dimensional spatial resolution with one axis in the Z-direction of the laser propagation while the other direction is along the wavelengths scale. Therefore, the spectral resolution is limited. The key point in the experimental realization is to slightly defocus the spectrometer so that spatial resolution along the wavelengths axis is obtained but to tune the defocusing in a manner to maintain still a spectral resolution that allows identifying the line transitions of interest.

Figure 10.41 shows the spatial variation of the plasma jet emission in colliding plasmas when a highly ionized plasma jet interacts with the residual gas in the target chamber (Rosmej et al. 2006c). The upper part of Fig. 10.41 shows the geometry and the X-ray pinhole emission of the rear-side and front-side emissions. The lower part of Fig. 10.41 shows the two-dimensional X-ray imaging that allows resolving particular line emission (spectral resolution) along with spatial resolution. The spatial variation of the resonance and intercombination lines is drastically different. Moreover, usually very weak intercombination satellite transitions originating from the $1s[2p^2\ ^3P]\ ^4P$-states show very large intensities. These changes are induced by charge transfer process from the low-charged residual gas to the highly ionized plasma jets (Rosmej et al. 2006c). Therefore, space-resolved X-ray spectroscopy has identified charge exchange coupling between the colliding plasmas (jet and residual gas).

The detailed spatial information of the radiating plasma jets provides a unique characterization of (a) the interaction between the jet and the residual gas, and (b) their mutual coupling via charge transfer. Therefore, charge exchange is also of considerable interest in high-density plasmas to shed more light into the complex theory of colliding plasmas.

10.5.2 Energy Distribution of Fast Ions

A further important application of space-resolved spectroscopy is the experimental determination of the energy distribution function of fast ions in dense laser-produced plasmas measured via line shifts of spectrally highly resolved resonance lines. These shifts are induced by the directionally Doppler shift. Figure 10.42a shows schematically the principal geometrical arrangement of the measurement. The measurements request two identical spectrometers positioned at very small and at large angles. The comparison of these two spectra then allows attributing the observed relative shift to the fast ion velocity. Figure 10.42b shows the line profile of the H-like fluorine Ly_α and the strongly pronounced "blue" line wing. The Doppler shifted position which corresponds to an ion energy of $E = 3.6$ MeV is indicated.

Fast ion velocity distribution functions can then be deduced from the optically thin part of the Doppler shifted emission. Figure 10.43 shows that the ion velocity distribution function is reasonably approximated by a unidirectional Maxwellian function with a temperature parameter of $kT_{\text{ion}} = 1.35$ MeV (Rosmej et al. 1999c, 2002b). These types of measurements provide critical information to test kinetic plasma simulations.

Figure 10.44 shows the space-resolved Doppler line shift measurements for different laser irradiation conditions employing a laser pulse with duration 15 ns, laser wavelength 1.06 μm, and laser energy of 10–60 J. The different focus conditions are realized via different positions of the focusing lens. Also seen from Fig. 10.44 is a correlation of the spatial extension of the spectrally resolved plasma jets (Z-axis, direction of the expanding plasma) and the X-ray pinhole measurements (Fig. 10.44).

Fast ion velocities in dense plasmas lead to differential shifts of emission and absorption coefficients which may lead to considerable modification of optically thick lines shapes as discussed in Sect. 1.1.4 (Fig. 1.4). For this reason, the analysis of the energy distribution function as discussed in Fig. 10.43 has been limited to the line wings, which are optically thin.

Fast ion velocities may also contribute directly to excitation and ionization processes via the *W*-matrix in the atomic population kinetics. Important examples are the redistribution of population in the fine structure of H-like ions ($2s\ ^2S_{1/2}$, $2p\ ^2P_{1/2}$, $2p\ ^2P_{3/2}$) due to a collisional coupling of the $2s\ ^2S_{1/2}$ and $2p\ ^2P_{1/2}$ levels. Their energy difference is very small (Lamb shift), and heavy particle collisions are therefore effective. Other examples are the collisional proton ionization of hydrogen levels with large principal quantum numbers, which turned out to be an important effect in magnetic fusion research (Rosmej and Lisitsa 1998) to analyze charge exchange processes (see Sects. 10.3.1 and 10.3.2).

A very rough estimate of the importance of collisional cross sections induced by fast ions can be made (in some cases) via the classical approach: If the ion velocity V_{ion} is much smaller than the effective Bohr velocity V_n for an electron with principal quantum number n, the ion-induced cross section might be negligible:

$$\sigma \approx 0 : V_{ion} \ll V_n = V_0 \frac{Z_{eff}}{n}. \tag{10.157}$$

V_0 is the Bohr velocity (2.19×10^8 cm/s) and Z_{eff} is the effective charge of the target ion where the atomic transitions are induced. More detailed calculations of heavy particle collisional cross sections are proposed (Sobelman and Vainshtein 2006; Gryzinsky 1965).

10.6 Atomic Physics in Dense Plasmas with X-ray Free Electron Lasers

XUV and X-ray Free Electron Lasers (XFELs) have provided the high-energy-density physics community with outstanding tools to investigate and to create matter under extreme conditions never achieved in laboratories so far. The key parameters of existing XFEL installations (LCLS in USA, EU-XFEL in Germany, SACLA in Japan) are micro- and even sub-microfocusing to achieve intensities in access to 10^{16} W/cm^2, short pulse lengths (10–100 fs), tunable photon energy (1–30 keV), small bandwidth (some 10 eV at about 10 keV in SASE mode and about 1 eV in self-seeded mode), and high repetition frequency (some 10 Hz up to 100 Hz), allowing to accumulate thousands of shots to improve signal-to-noise ratios.

This makes XFEL installations distinct different from well-known synchrotron radiation facilities. The brilliance of XFEL's is more than ten orders of magnitude higher than modern synchrotrons, and this allows to photoionize inner-shells of almost every atom in a solid crystal in a single pulse. As the pulse duration is of the order of the Auger time scale an exotic state of matter, a "Hollow Crystal" can be created. The decay of crystalline order can be initiated by a burst of Auger electrons with energies in the X-ray range that heat up the hollow crystal as identified with high-resolution spectroscopy (Galtier et al. 2011) in the first high-energy-density experiment at the XUV–FEL FLASH (Riley et al. 2009). This is distinct different to synchrotrons: Auger electron production is rare compared to the total number of atoms, and Auger electrons do not allow changing the physical properties of the crystal.

Next, the tunable photon energy combined with the small bandwidth permits to pump selected atomic transitions in the X-ray range. Compared to the well-known pumping of low energy transitions by optical lasers, X-ray pumping will allow outstanding steps forward. The first experiment of X-ray pumping of dense plasmas that have been produced by a powerful auxiliary optical laser has been performed at LCLS at SXR (Seely et al. 2011). In this experiment, it has successfully been demonstrated that efficient changes in the atomic populations of highly charged ions induced by XFEL can be achieved (Moinard et al. 2013; Rosmej et al. 2016). As it has been the case for laser-induced fluorescence (LIF) with standard optical lasers, a revolutionary impact is expected via the photopumping of X-ray transitions. In addition to standard LIF, X-LIF will allow to study isoelectronic sequences due to the large range of tunability of the XFEL photons. As synchrotrons might not allow selective efficient X-ray pumping (drastic change of atomic populations), XFEL facilities have opened a new world for scientific activity.

An important further technical process has been made at XFEL's by reducing the width of the energy distribution function: In the so-called seeded-mode, a bandwidth down to some eV is reached at some keV photon energies. This has led to the first experiment to scan X-ray line shapes in dense plasmas (created with a powerful auxiliary optical laser) at LCLS at MEC (Rosmej et al. 2018).

10.6.1 *Scaling Laws to Move Atomic Populations with XFEL*

10.6.1.1 Description of Time- and Energy-Dependent XFEL Radiation

Let us assume that time and energy dependence of the XFEL radiation are independent from each other and can therefore be described by the functions $f_{\mathrm{FEL}}(t)$ and $\tilde{N}_{\mathrm{FEL}}(E)$. The number of photons per volume/time/energy is then given by:

$$\tilde{N}_{\mathrm{FEL}}(E,t) = \tilde{N}_{\mathrm{FEL}}(E) f_{\mathrm{FEL}}(t), \qquad (10.158)$$

$$\int_{-\infty}^{+\infty} f_{\mathrm{FEL}}(t)\mathrm{d}t = 1. \qquad (10.159)$$

We assume a Gaussian energy dependence to simulate the narrow bandwidth of the XFEL:

$$\tilde{N}_{\mathrm{FEL}}(E) = \tilde{N}_0 \frac{1}{\sqrt{\pi}\Gamma_{\mathrm{FEL}}} \exp\left(-\frac{(E - E_{\mathrm{FEL}})^2}{\Gamma_{\mathrm{FEL}}^2}\right), \qquad (10.160)$$

$$\Gamma_{\mathrm{FEL}} = \delta E / 2\sqrt{\ln 2}. \qquad (10.161)$$

E_{FEL} is the central energy of the radiation field, $\tilde{N}_{\mathrm{FEL}}(E)$ is the number of photons/volume/energy, $\tilde{N}_0$ is the peak number of photons/volume, δE is the bandwidth. Assuming a Gaussian time dependence, the number of photons $N_{\mathrm{tot},\tau}$ per pulse length τ is given by

$$N_{\text{tot},\tau} = \int_0^{\infty} \mathrm{d}E \int_{\text{volume}} \mathrm{d}V \int_{-\tau/2}^{+\tau/2} \mathrm{d}t\, \tilde{N}_{\text{FEL}}(E,t) \approx 2Ac\tau\tilde{N}_0\, erf\left(\sqrt{\ln 2}\right)$$
$$\approx 0.761 \cdot Ac\tau\tilde{N}_0. \tag{10.162}$$

A is the focal spot area, τ is the XFEL pulse width (FWHM). For estimations, the error function can be approximated by

$$erf(x) = \frac{1}{\sqrt{\pi}} \int_0^x e^{-t^2} \cdot \mathrm{d}t$$
$$\approx \frac{1}{2} - \frac{e^{-x^2}}{2} \cdot \left[\frac{0.3480242}{1+0.47047 \cdot x} - \frac{0.0958798}{(1+0.47047 \cdot x)^2} + \frac{0.7478556}{(1+0.47047 \cdot x)^3}\right]. \tag{10.163}$$

The laser intensity $\tilde{I}_{\text{FEL}}(E,t)$ per bandwidth energy and time interval is related to the photon density $\tilde{N}_{\text{FEL}}(E,t)$ via

$$\tilde{I}_{\text{FEL}}(E,t)\mathrm{d}E\mathrm{d}A\mathrm{d}t = \tilde{N}_{\text{FEL}}(E,t)E \cdot \mathrm{d}E\mathrm{d}V\mathrm{d}t. \tag{10.164}$$

Integrating the XFEL beam over a full width at half maximum with respect to energy and time, $\bar{I}_{\text{FEL},\delta E,\tau}$ (energy/time/surface) is given by (assuming a Gaussian time dependence):

$$\bar{I}_{\text{FEL},\delta E,\tau} = \int_{-\delta E/2}^{\delta E/2} \mathrm{d}E \int_{-\tau/2}^{\tau/2} c \cdot \mathrm{d}tE \cdot \tilde{N}(E,t) \approx 4E_{\text{FEL}}c\tilde{N}_0 erf^2\left(\sqrt{\ln 2}\right)$$
$$\approx 0.579 \cdot c \cdot E_{\text{FEL}} \cdot \tilde{N}_0 \tag{10.165}$$

or, in convenient units

$$\bar{I}_{\text{FEL},\delta E,\tau} \approx 2.8 \times 10^{-9} \left(\frac{\tilde{N}_0}{cm^3}\right)\left(\frac{E_{\text{FEL}}}{eV}\right)\left[\frac{\text{W}}{\text{cm}^2}\right]. \tag{10.166}$$

The number of photons $N_{\text{tot},\tau}$ is related to the intensity $\bar{I}_{\text{FEL},\tau}$ via (d is the focal spot diameter)

$$\bar{I}_{\text{FEL},\tau} = 2 \cdot erf\left(\sqrt{\ln 2}\right) \cdot \frac{N_{\text{tot},\tau} \cdot E_{\text{FEL}}}{\pi\tau \cdot d^2/4} \approx \frac{N_{\text{tot},\tau} \cdot E_{\text{FEL}}}{\tau \cdot d^2}. \tag{10.167}$$

10.6.1.2 Photoionization

In order to change atomic populations by irradiation of matter with XFEL, photoionization rates need to be larger than corresponding electron ionization rates and, in case of photopumping, photoexcitation rates need to be larger than corresponding spontaneous radiative decay rates. In order to obtain analytical formulas, we consider the hydrogen-like approximation for an arbitrary atom with effective charge Z_{eff} and an atomic level with principal quantum number n and energy

$$E_{\text{n}} = \frac{Z_{\text{eff}}^2 \cdot Ry}{n^2}, \tag{10.168}$$

where $Ry = 13.6057\,\text{eV}$. For the case of photoionization this leads to the following relation:

$$\int_{-\tau/2}^{+\tau/2} \mathrm{d}t \int_{E_{\text{n}}}^{\infty} \mathrm{d}E\sigma_{\text{n}}^{\text{iz}}(E)c\tilde{N}_{\text{FEL}}(E,t) > n_{\text{e}} I_{\text{n}}. \tag{10.169}$$

$\sigma_{\text{n}}^{\text{iz}}(E_{\text{FEL}})$ is the photoionization cross section from level n, n_{e} is the electron density, I_{n} is the electron collisional ionization rate, c the speed of light. Employing the Kramers classical cross section for the photoionization (see also Sects. 3.3 and 3.7 and review [Rosmej et al. 2020b]),

$$E > E_{\text{n}} : \sigma_{\text{n}}^{\text{iz}}(E) = 2.9 \times 10^{-17} \frac{E_{\text{n}}^{5/2}}{Z \cdot E^3}\left[\text{cm}^2\right] \tag{10.170}$$

and the Lotz-formula for the electron collisional ionization (see Sect. 5.3.1), i.e.,

$$I_{\text{n}} = 6 \times 10^{-8} \cdot \left(\frac{Ry}{E_{\text{n}}}\right)^{3/2} \cdot \sqrt{\beta_{\text{n}}} \cdot e^{-\beta_{\text{n}}} \cdot \ln\left[1 + \frac{0.562 + 1.4\beta_{\text{n}}}{\beta_{\text{n}}(1 + 1.4\beta_{\text{n}})}\right]\left[\text{cm}^3\text{s}^{-1}\right], \tag{10.171}$$

$$\beta_{\text{n}} = \frac{E_{\text{n}}}{kT_{\text{e}}}. \tag{10.172}$$

Equations (10.168)–(10.172) provide the following estimate (peak intensity $I_{\text{FEL}} = cE_{\text{FEL}}\tilde{N}_0$):

$$I_{\text{FEL}} > 3 \times 10^{-8} \cdot n_{\text{e}}\left(\text{cm}^{-3}\right) \cdot Z \cdot \sqrt{\beta_{\text{n}}} \cdot e^{-\beta_{\text{n}}} \cdot \ln\left[1 + \frac{0.562 + 1.4\beta_{\text{n}}}{\beta_{\text{n}}(1 + 1.4\beta_{\text{n}})}\right]\left[\frac{\text{W}}{\text{cm}^2}\right] \tag{10.173}$$

assuming $E_{\text{FEL}} = E_{\text{n}} + 3\delta E (\delta E \ll E_{\text{FEL}})$ for effective photoionization. For $n_{\text{e}} = 10^{21}\,\text{cm}^{-3}$, $Z_{\text{eff}} = 13$, $\beta_{\text{n}} = 2$ (10.173) delivers $I_{\text{FEL}} > 3 \times 10^{13}\,\text{W/cm}^2$.

10.6.1.3 Photoexcitation

Let us now consider the relations for the X-ray photopumping of line transitions (i.e., resonant photopumping). In order to influence strongly via photoexcitation on the atomic populations, photoexcitation rates need to be larger than corresponding spontaneous radiative decay rates:

$$\int_{-\tau/2}^{+\tau/2} \mathrm{d}t \int_{\Delta E_{\mathrm{nm}}-\delta\tilde{E}/2}^{\Delta E_{\mathrm{nm}}+\delta\tilde{E}/2} \mathrm{d}E\, \sigma_{\mathrm{nm}}^{\mathrm{abs}}(E) c \tilde{N}_{\mathrm{FEL}}(E,t) \geq A_{\mathrm{mn}}. \tag{10.174}$$

$\sigma_{\mathrm{nm}}^{\mathrm{abs}}(E)$ is the photoabsorption cross section for the transition from level n to level m and A_{mn} is the spontaneous radiative decay rate from level m to level $n, \delta\tilde{E}$ is an effective width for the XFEL interaction with the atomic transition. The photoabsorption cross section is given by

$$\sigma_{\mathrm{nm}}^{\mathrm{abs}}(E) = \frac{E}{4\pi} B_{\mathrm{nm}} \varphi_{\mathrm{nm}}(E). \tag{10.175}$$

B_{nm} is the Einstein coefficient of stimulated absorption that is related to the Einstein coefficient of spontaneous radiative decay according to

$$B_{\mathrm{nm}} = \frac{4\pi^3 \hbar^3 c^2}{E^3} \frac{g_{\mathrm{m}}}{g_{\mathrm{n}}} A_{\mathrm{mn}}, \tag{10.176}$$

$\varphi_{\mathrm{nm}}(E)$ is the normalized local absorption line profile:

$$\int_{-\infty}^{+\infty} \varphi_{\mathrm{nm}}(E) \mathrm{d}E = 1. \tag{10.177}$$

We assume a Gaussian line profile with full width at half maximum, i.e.,

$$FWHM = 2\sqrt{\ln 2}\, \Gamma_{\mathrm{G}} \tag{10.178}$$

and a width parameter

$$\Gamma_{\mathrm{G}} = \Delta E_{\mathrm{nm}} \sqrt{2kT_{\mathrm{i}}/Mc^2} \tag{10.179}$$

corresponding to a Doppler profile to obtain analytical estimates:

$$\varphi_{\text{nm}}(E) = \frac{1}{\sqrt{\pi}\,\Gamma_{\text{G}}} \exp\left[-\left(\frac{E - \Delta E_{\text{nm}}}{\Gamma_{\text{G}}}\right)^2\right] \tag{10.180}$$

and ($\delta\tilde{E}$ is the bandwidth)

$$\frac{4\ln 2}{\delta\tilde{E}^2} = \frac{1}{\Gamma_{\text{G}}^2} + \frac{1}{\Gamma_{\text{FEL}}^2}. \tag{10.181}$$

If the XFEL photon energy is exactly tuned to the transition energy, e.g., $E_{\text{XFEL}} = \Delta E_{\text{nm}}$ $E_{\text{XFEL}} = \Delta E_{\text{nm}}$, (10.160), (10.161), (10.174)–(10.181) provide the following estimate:

$$I_{\text{FEL}} > 2 \times 10^5 \Delta E_{\text{nm}}^3 \frac{g_{\text{n}}}{g_{\text{m}}} \sqrt{FWHM_{\text{FEL}}^2 + FWHM_{\text{G}}^2} \left[\frac{\text{W}}{\text{cm}^2}\right] \tag{10.182}$$

with ΔE_{nm} and *FWHM* in [eV]. For H-like Al Ly$_\alpha$, $\Delta E_{\text{nm}} = 1728\,\text{eV}$, $g_{\text{n}} = 2$, $g_{\text{m}} = 6$, we obtain $I_{\text{XFEL}} = 4 \times 10^{15}\,\text{W/cm}^2$ (assuming $\sqrt{FWHM_{\text{FEL}}^2 + FWHM_G^2} = 10\,\text{eV}$). The relation (10.182) indicates an important scaling law:

$$I_{\text{FEL}} \propto Z^6. \tag{10.183}$$

Therefore, extremely high brilliance of XFEL's is needed to pump X-ray transitions. Assuming a spot diameter of $d = 2\,\mu\text{m}$, pulse length $\tau = 100\,\text{fs}$, photon energy $E_{\text{XFEL}} = 1.7\,\text{keV}$ and a laser intensity of $\bar{I}_{\text{FEL},\tau} = 10^{16}\,\text{W/cm}^2$, a minimum of about $N_{\text{tot},\tau} = 2 \times 10^{11}$ photons in the XFEL pulse is requested according (10.167) to effectively move atomic populations in the X-ray energy range. Currently operating/planed Free Electron Laser facilities fulfill these requirements. We note that relation (10.183) does not depend on the electron density and the intensity estimate holds therefore true for low- and high-density plasmas.

10.6.2 Atomic Kinetics Driven by Intense Short Pulse Radiation

Radiation field quantum mechanics in second quantization (Heitler 1954; Pike and Sarkar 1995) and quantum kinetics (Rautian and Shalagin 1991) is the most general approach to study the interaction of radiation fields with atoms while the quantum kinetic approach via the density matrix theory (see Chap. 7) is the most general approach to determine atomic populations (that are at the heart of almost all properties of matter). On a unique footing, these approaches allow describing atomic population and coherences and provide all necessary quantum matrix elements to take into account the relevant elementary atomic processes (cross sections) and field-atom interactions.

Under the assumption of broadband illumination and/or large collisional broadening, the non-diagonal density matrix elements are negligible compared to the diagonal ones (atomic populations) and the so-called rate equation atomic population kinetic approach becomes valid (Loudon 2000). In order to keep the presentation transparent, we employ for the below presentation the rate equation kinetics in its most general form as outlined in Chap. 6. This approach will be supplemented below to take into account an external intense short pulse radiation field to describe the XFEL interaction with matter. We therefore start from (6.22)–(6.28); however, the transition matrix elements have to be supplemented with a radiation field term driven by the XFEL, i.e.,

$$W_{\mathrm{ji}} = W_{\mathrm{ji}}^{\mathrm{col}} + W_{\mathrm{ji}}^{\mathrm{rad}} + W_{\mathrm{ji}}^{\mathrm{FEL}}. \quad (10.184)$$

The radiation field matrix elements $W_{\mathrm{ji}}^{\mathrm{FEL}}$ for the external XFEL radiation are given by

$$W_{\mathrm{ji}}^{\mathrm{FEL}} = W_{\mathrm{ji}}^{\mathrm{FEL,PI}} + W_{\mathrm{ji}}^{\mathrm{FEL,SR}} + W_{\mathrm{ji}}^{\mathrm{FEL,SA}} + W_{\mathrm{ji}}^{\mathrm{FEL,SE}}, \quad (10.185)$$

$$W_{\mathrm{ji}}^{\mathrm{FEL,PI}} = \int_{\hbar\omega_{\mathrm{ji}}}^{\infty} \mathrm{d}(\hbar\omega)\sigma_{\mathrm{ji}}^{\mathrm{PI}}(\hbar\omega)c\tilde{N}(\hbar\omega), \quad (10.186)$$

$$W_{\mathrm{ji}}^{\mathrm{FEL,SR}} = n_{\mathrm{e}}\frac{\pi^2 c\hbar^3}{\sqrt{2}m_{\mathrm{e}}^{3/2}}\frac{g_{\mathrm{i}}}{g_{\mathrm{j}}}\int_0^{\infty} \mathrm{d}E\,\frac{F(E)}{\sqrt{E}}\,\sigma_{\mathrm{ji}}^{\mathrm{PI}}(\hbar\omega_{\mathrm{ji}}+E)\tilde{N}(\hbar\omega_{\mathrm{ji}}+E), \quad (10.187)$$

$$W_{\mathrm{ji}}^{\mathrm{FEL,SA}} = \pi^2 c^3\hbar^3 A_{\mathrm{ji}}\frac{g_{\mathrm{j}}}{g_{\mathrm{i}}}\int_0^{\infty} \mathrm{d}(\hbar\omega)\frac{\varphi_{\mathrm{ji}}(\omega)}{\hbar}\frac{\tilde{N}(\hbar\omega)}{(\hbar\omega)^2}, \quad (10.188)$$

$$W_{\mathrm{ji}}^{\mathrm{FEL,SE}} = \pi^2 c^3\hbar^3 A_{\mathrm{ji}}\int_0^{\infty} \mathrm{d}(\hbar\omega)\frac{\varphi_{\mathrm{ji}}(\omega)}{\hbar}\frac{\tilde{N}(\hbar\omega)}{(\hbar\omega)^2}. \quad (10.189)$$

$W_{\mathrm{ji}}^{\mathrm{FEL,PI}}$ describes photoionization, $W^{\mathrm{FEL,SR}}$ stimulated radiative recombination, $W_{\mathrm{ji}}^{\mathrm{FEL,SA}}$ stimulated photoabsorption, $W_{\mathrm{ji}}^{\mathrm{FEL,SE}}$ stimulated photoemission. $\sigma_{\mathrm{ji}}^{\mathrm{PI}}$ is the photoionization cross section, $F(E)$ the energy distribution function of the continuum electrons, A_{ji} the Einstein coefficient of spontaneous emission, φ_{ji} the line profile, c the speed of light, $\hbar$ is the Planck constant, g_{i} the statistical weight of a bound state, ω the angular frequency of the external radiation field, ω_{ji} the atomic transition frequency, and $\tilde{N}$ is the number of external photons (those of the Free Electron Laser) per unit volume and energy. We note, that (10.186–10.189) assume that the concept of standard ionization and excitation probability per unit time in

Fermi's golden rule and in Einstein's theory of stimulated emission and absorption is still valid. This assumption breaks down for ultra-short pulses [Rosmej et al. 2021]: numerical calculations carried out in terms of a generalized probability for the total duration of pulses in the near-threshold regime demonstrate essentially non-linear behavior, while absolute values may change by orders of magnitude for typical ultra-short XFEL and High-Harmonic Generated HHG pulses [Rosmej et al. 2021].

In optically thick plasmas, the spectral intensity distribution of an atomic transition $i \rightarrow j$ with frequency ω_{ji} is given by (see also Sect. 1.1.4):

$$I_{j_Z i_Z}(\omega) = \frac{\hbar\omega_{j_Z i_Z}}{4\pi} \cdot n_{j_Z} \cdot A_{j_Z i_Z} \cdot \Lambda_{j_Z i_Z} \cdot \Phi_{j_Z i_Z}\left(\omega, \omega_{j_Z i_Z}\right). \tag{10.190}$$

n_{j_Z} is the population of the upper-level j, $A_{j_Z i_Z}$ is the spontaneous transition probability for the transition $j \rightarrow i$, $\Lambda_{j_Z i_Z}$ is the lambda-operator to describe radiation transport effects [Sect. 1.1.4, e.g., (1.33)] and $\Phi_{j_Z i_Z}\left(\omega, \omega_{j_Z i_Z}\right)$ is the associated optically thick emission line profile (see Sect. 1.1.4, (1.34)). Equation (10.190) indicates a strong interplay between the atomic structure (means transition probabilities $A_{j_Z i_Z}$) and atomic populations kinetics (population densities n_{j_Z}). The total spectral local distribution is then given by

$$I(\omega) = \sum_{Z=0}^{Z_n} \sum_{j_Z=1}^{N_Z} \sum_{i_Z=1}^{N_Z} I_{j_Z i_Z}(\omega). \tag{10.191}$$

Equation (10.191) is of great practical interest: It is the spectral distribution that is accessible via X-ray spectroscopic measurements. In plasmas where opacity in line transitions is important $\Lambda_{j_Z i_Z} < 1$ and a generalized optically thick line profile has to be invoked (see (1.34) of Sect. 1.1.4). If also radiation transport in the continuum is important, the generalized radiation transport equation has to be solved (Mihalas 1978). Such experimental situations, however, should be avoided: The photon absorption in the continuum of a line transition is not at all redistributed in the line itself, but redistributed in a large spectral interval due to the large "width" of the continuum (in other words the "line photon" is lost). Therefore, under such circumstances diagnostic ratios would rather be meaningless than useful. In order to properly design experiments that avoid radiation transport in the continuum, the continuum opacities τ_{ff} (free–free opacity) and τ_{fb} (free–bound opacity) should be $\tau_{ff} < 0.01$ and $\tau_{fb} < 0.01$ (the opacities can be estimated from (1.54) and (1.55). It is also desirable to avoid as much as possible the bound–bound opacity and ensure $\tau_{ji} < 0.1$ (see, e.g., (1.56), (1.57)). This strong estimate is related to the fact that in most cases of XFEL research, X-ray transitions from autoionizing states are recorded: Radiation transport related to autoionizing levels is very harmful for spectroscopic diagnostic studies, as the reemission after absorption is almost suppressed due to the large autoionizing rates (Kienle et al. 1995), i.e., a small branching factor

$$p_{ji} = \left(\frac{A_{ji}}{\sum_l A_{jl} + \sum_k \Gamma_{jk} + \sum_m C_{jm}} \right)^{N_{ji}}. \tag{10.192}$$

p_{ji} is the probability that a photon from a line transition $j \rightarrow i$ is reemitted, A_{ji} is the radiative decay, $\sum_l A_{jl}, \sum_k \Gamma_{jk}$ and $\sum_m C_{jm}$ are the total radiative decay rates, autoionizing rates and collisional rates leading to a depopulation of the upper-level j, and N_{ji} is the number of photon scatterings. As an example let us consider aluminum and the j-dielectronic satellite (discussed in Table 10.1): $A_{ji} = 1.31 \times 10^{13}\,\mathrm{s}^{-1}, \Gamma_{jk} = 1.29 \times 10^{14}\,\mathrm{s}^{-1}$. Even if we set $\sum_m C_{jm} = 0$ the remission probability is very low, $p_{ji} = 9.18 \times 10^{-2}$.

Finally we note that bound–bound opacity of a transition that is emitted, absorbed, and reemitted in a dense plasma should not be confused with the opacity of XFEL-induced resonant pumping.

10.6.3 Interaction of XFEL with Dense Plasmas

10.6.3.1 General Features of XFEL Interaction with Dense Plasmas: Simulations

Figure 10.45 shows an experimental scheme for a typical pump probe experiment. A powerful optical laser is irradiating a solid target to create a dense plasma plume. The target is mounted on a rotating cylinder in order to accumulate spectra to improve the signal-to-noise ratio. The XFEL is then triggered with respect to the optical laser and focused into the dense plasma plume to pump X-ray transitions. A high-resolution (high spectral and spatial) X-ray spectrometer is employed to record the spectral distribution of the pumped X-ray transitions in dependence of the target distance (along Y-axis in Fig. 10.45).

Let us follow the principle steps of XFEL interaction with a dense plasma with the MARIA simulations described above taking also into account all radiation field terms of (10.184)–(10.189) and a detailed LSJ-split atomic atomic/ionic level system including ground, single, and multiple excited and hollow ion states from various charge states (Rosmej and Lee 2007).

Figure 10.46 shows the evolution of the average charge (solid curve) when an intense pulsed radiation field (dashed curve) is interacting with dense magnesium plasma:

$$\langle Z \rangle = \sum_{Z=0}^{Z_n} n_Z Z, \tag{10.193}$$

where n_Z is the ionic population of charge Z (see (6.29)). The plasma density is $n_e = 10^{21}\,\mathrm{cm}^{-3}$, the temperature $kT_e = 40\,\mathrm{eV}$. Opacity effects of the internal

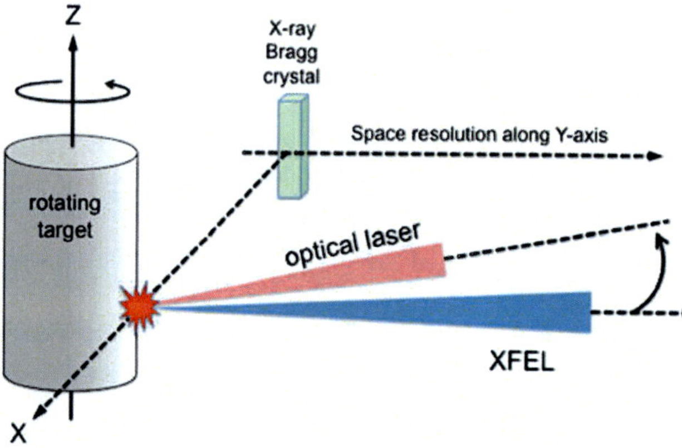

Fig. 10.45 Principle scheme of a pump probe experiment when intense XFEL radiation interacts with dense plasma produced by a powerful auxiliary optical laser

Fig. 10.46 MARIA simulations of the temporal evolution of the XFEL pulse and the average charge state of a dense Mg plasma, $E_{FEL} = 1850$ eV, $\tau = 100$ fs, $I_{max} = 2.2 \times 10^{17}$ W/cm^2, $n_e = 10^{21}$ cm^{-3}, $kT_e = 40$ eV, $L_{eff} = 10$ μm

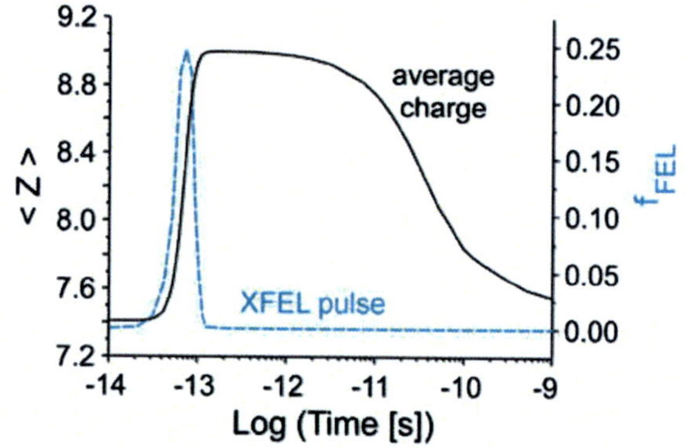

atomic/ionic radiation are included via an effective photon path length of $L_{eff} = 10$ μm (see Sect. 1.1.4). The XFEL pulse duration is $\tau = 100\,\text{fs}$, photon energy $E_{XFEL} = 1850\,\text{eV}$ and the photon density is $\tilde{N}_0 = 10^{23}\,\text{cm}^{-3}$. The maximum laser intensity is then

$$I_{max} = cEf_{FEL,max} \cdot \tilde{N}_0 = 4.8 \times 10^{-9} f_{FEL,max} \left(\frac{\tilde{N}_0}{\text{cm}^3}\right)\left(\frac{E}{\text{eV}}\right)\left[\frac{\text{W}}{\text{cm}^2}\right], \tag{10.194}$$

where $f_{XFEL,\,max}$ is the maximum value ($f_{XFEL,\,max} = 0.246$ in Fig. 10.46) of the normalized time-dependent function of the laser intensity; see (10.158), (10.159), $I_{max} = 2.2 \times 10^{17}\,\text{W/cm}^2$. Before the XFEL pulse interacts with the Mg plasma plume, the average charge state is about $\langle Z \rangle \approx 7.4$ that rises dramatically during the interaction with the XFEL pulse. The system shows shock characteristics: After laser pulse maximum, the average charge state is still increasing (at about $t = 10^{-13}$ s), then stays almost constant for a few ps, then decreases on a 100 ps time scale followed by a very slow final equilibration phase (10–100 ns). We note, that general characteristics of shocked atomic systems have been explored in [Deschaud et al. 2020].

Let us follow the shock characteristics in more detail via the ionic populations. Figure 10.47 shows the charge state evolutions of the bare nucleus (nuc), H-like ions (H), He-like ions (He), and Li-like ions (Li). Figure 10.47a shows the case when the photon energy $E_{XFEL} = 1850\,\text{eV}$ is larger than the ionization potential of the He-like Mg ground state $(E_i(1s^2\,{}^1S_0) = 1762\,\text{eV})$. Before the XFEL pulse, the ionic fractions nuc, H, and He are negligibly small due to the low electron temperature of the plasma plume. With the onset of the XFEL pulse, He-like and H-like ionic fractions rise rapidly. The population of the fully stripped Mg ("nuc" in Fig. 10.47a) stays very small, because the XFEL photon energy does not allow direct photoionization of the H-like ground state. The slight rise at the beginning of the XFEL pulse is connected with collisional excitation–ionization processes in the dense plasma.

Figure 10.47b shows a simulation when the photon energy is larger than the ionization potential of the H-like ground state. Like in Fig. 10.47a, before the XFEL pulse, the ionic fractions of the bare nucleus, H- and He-like ions are negligibly small due to the low electron temperature of the plasma plume. With the onset of the XFEL pulse, Li-like, He-like, and H-like ionic fractions rise rapidly. At about laser pulse maximum, the fraction of H-, He-, and Li-like ions drop again because the XFEL photons are photoionizing the H-like ground state $1s\ ^2S_{1/2}$ because the photon energy of $E_{XFEL} = 3100\,\text{eV}$ keV is larger than the ionization potential of H-like Mg ground state $\left(E_i\left(1s\,^2S_{1/2}\right) = 1963\text{ eV}\right)$. The depletion of

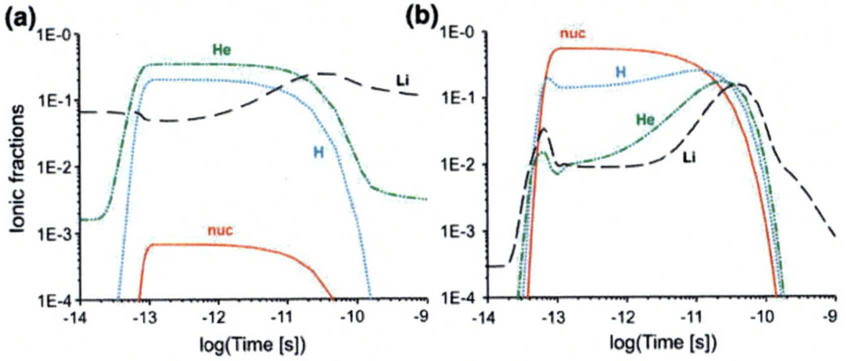

Fig. 10.47 MARIA simulations of the temporal evolution of the ionic fractions after interaction of the XFEL pulse with a dense Mg plasma plume, **a** $E_{FEL} = 1850$ eV, $\tau = 100$ fs, $\tilde{N}_0 = 10^{23}\text{cm}^{-3}$, $I_{max} = 2.2 \times 10^{17}$ W/cm^2, $n_e = 10^{21}$ cm^{-3}, $kT_e = 40$ eV, $L_{eff} = 30$ µm, **b** $E_{FEL} = 3100$ eV, $\tau = 100$ fs, $\tilde{N}_0 = 10^{23}\text{cm}^{-3}$, $I_{max} = 3.7 \times 10^{17}$ W/cm^2, $n_e = 10^{21}$ cm^{-3}, $kT_e = 30$ eV, $L_{eff} = 30$ µm

almost all electrons from the atomic system makes the plume transparent to the XFEL radiation as no more absorption is possible: The absorption is saturated. When the pulse is off, H-like, He-like, and Li-like ionic fractions increase as recombination starts from the bare nucleus. At even later times (about $t = 10^{-10}$ s), all ionic fractions (nuc, H, He, Li) decrease due to the overall cooling of the plume (rise of ionic fractions of low-charged ions not shown in Fig. 10.47b).

Figures 10.47 demonstrate that in the photoionization regime, the tuning of the XFEL beam allows selection of different charge states (i.e., cutoff of charge states that have ground state ionization energies above the photon pumping energy).

10.6.3.2 X-ray Pumping of Dense Plasmas

In the resonant pumping scheme, the tuning of the XFEL energy allows to selectively pump bound–bound transitions and induce corresponding fluorescence, so-called X-ray laser-induced fluorescence, so-called X-LIF (Rosmej et al. 2016, 2022).

Figure 10.48 presents the first X-LIF spectra of dense aluminum plasma observed at LCLS for different X-ray pumping energies and for X-ray pulses delayed by 100 ps (Moinard et al. 2013; Rosmej et al. 2016, 2022). For reference, the upper spectrum (red curve) shows the spectral emission of the Al plasma created by the optical YAG laser only (i.e., X-ray pump was absent). Dominating transitions are the He-like resonance line $1s^2\ ^1S_0$–$1s2p\ ^1P_1$ and the intercombination line $1s^2\ ^1S_0$–$1s2p\ ^3P_1$ (W and Y in Fig. 10.48) of 11 times ionized aluminum. XFEL pumping at 1597 eV (curve labeled as "He-like" for pumping of helium-like atomic states) matches the atomic transition W that in turn results in a strong intensity increase (factor 4) of the W-line (note that all spectra have been normalized to the peak intensity).

The pumping at lower energies introduces abrupt changes in the spectral distribution. In particular, the pumping at 1587 eV corresponds to the transitions $K^2L^1 + h\nu_{XFEL} \rightarrow K^1L^2$. The pumped states are therefore the Li-like (10 times ionized aluminum) autoionizing levels $1s2l2l'$. Although the transitions that originate from these states are of very low intensity for the upper red curve (no pump), the X-ray pumping results in a strong intensity increase. X-ray pumping at 1570, 1551, and 1531 eV corresponds to X-LIF of the autoionizing Be-like (nine times ionized aluminum) states K^1L^3, the B-like (eight times ionized aluminum) states K^1L^4, and C-like (seven times ionized aluminum) states K^1L^5. Here, K and L denote the population of the inner atomic shells.

It is important to emphasize that emission lines, being weak or entirely absent in the dense optical laser-produced plasma (e.g., the Be-, B-, and C-like transitions) have been pumped to intensities that exceed even those of the resonance line W (that is not pumped and its occurrence is only due to thermal excitation inside the dense optical laser-produced plasma, upper red curve in Fig. 10.48).

The possibility of strong signal rise in the X-ray range from non-observable intensities to excellent signal-to-noise ratios is further demonstrated in Fig. 10.49. The figure demonstrates a rise in signal levels for the Be-like satellite X-ray

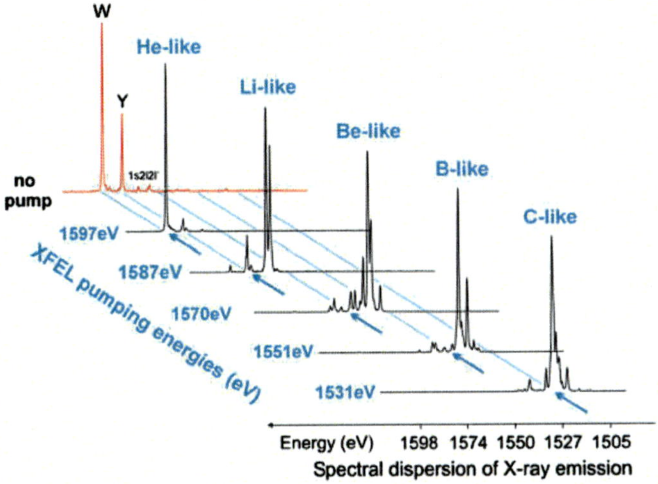

Fig. 10.48 Experimental spectra of resonant XFEL pumping of dense plasmas produced with an auxiliary optical laser. The upper spectrum (red, linear scale) shows the X-ray emission from the optical laser-produced plasma only that is dominated by the He-like resonance line *W* and intercombination line *Y*. The pumping energies (blue numbers, spectral positions indicted with flashes) fall into the spectral window of the X-ray spectral distribution of the $K^2L^N + E_{\mathrm{XFEL}} \rightarrow K^1L^{N+1}$ transitions ($N = 1$–4). The induced emission (solid black curves, linear scale, normalized to peak emission) from the He-like to C-like charge states corresponds to emission from ionic core hole states K^1L^{N+1}. When the X-ray pump is on, strong intensity rises of X-ray transitions (X-ray laser-induced fluorescence X-LIF) from Li-like until C-like ions are observed. The X-ray pumping moves the intensities by many orders of magnitude

transitions by a factor of 300 (the lower red spectrum has been shifted down by a factor of 10 for better comparison) implying that the charge state distribution can be probed in low-temperature dense plasmas (strongly coupled) that would not be observable otherwise.

Figure 10.49 demonstrates likewise that X-LIF can move non-observable bound–bound transitions to excellent signal levels. The two-electron transitions of the Be-like satellites, i.e., $1s2s^22p^1\ {}^1P_1{-}1s^22p^2\ {}^1D_2$ are not observable in the spectrum from the dense optical laser-produced plasma (red dotted curve in Fig. 10.49) but moves to a good signal-to-noise ratio due to X-ray pumping.

Figure 10.50 demonstrates transient resonant pumping by setting different delay times between the maximum of the optical laser pulse and the XFEL pulse in the range of $\Delta t = -50$ ps to +300 ps. Figure 10.50a shows the spectral distribution in the high-quality X-ray emission spectra of aluminum for $\Delta t = 0$ ps. The blue curve shows the best fit from a transient non-LTE simulation carried out with the MARIA collisional–radiative kinetics code described above. The simulations include the exact overlap integrals of line profiles of the atomic transitions and the energy distribution function of the XFEL for resonant and non-resonant pumping. In addition, the XFEL pulse was resolved in time and brought to interaction with the plasma at a given

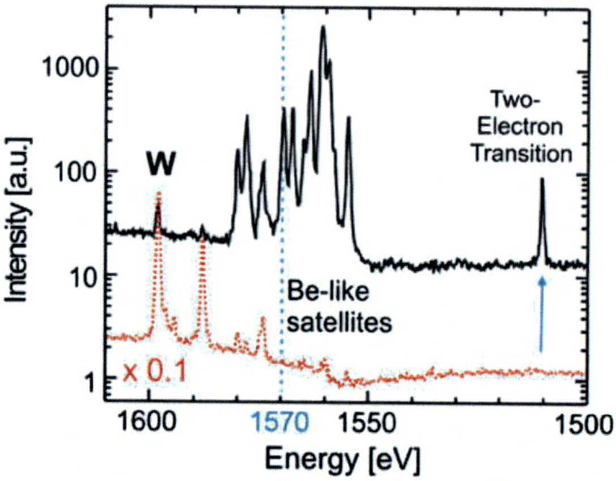

Fig. 10.49 Experimental spectrum of XFEL pumped dense aluminum plasma at E_{XFEL} = 1570 eV (indicated with dashed blue line). X-LIF increases the Be-like satellite intensity by about two–three orders of magnitude (compared to the emission from optical laser only, red dotted curve) moving even non-detectable transitions (e.g., the two-electron transitions indicated by blue flash) to excellent signal-to-noise ratios

temperature and density. A spectral distribution was then calculated at every time step and then summed up until convergence was reached to compare simulations with data. The simulations presented in Fig. 10.50a (blue curve) are obtained for an electron temperature of 100 eV and electron density of 1×10^{21} cm^{-3}. Despite a simple zero-dimensional model, the blue curve provides a rather good fit to the

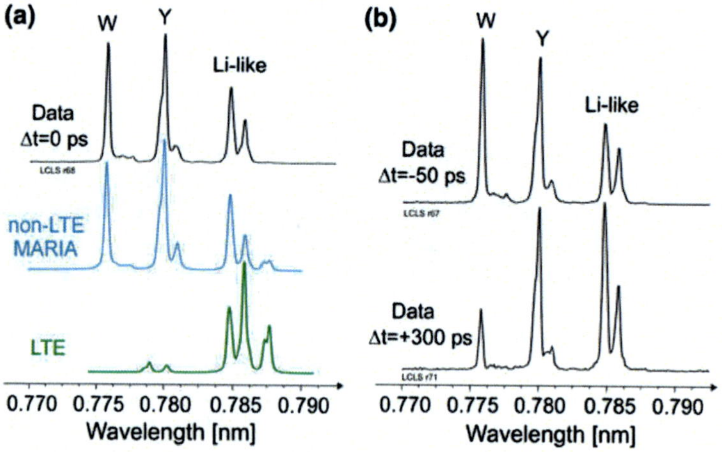

Fig. 10.50 Experimental spectra of transient X-ray pumping at different delay times Δt with respect to the optical laser pulse. **a** Time-dependent non-LTE simulations for Δt = 0 ps carried out with the MARIA code, **b** experimental X-LIF spectra for Δt = −50 ps and +300 ps

measured data. We note that the plasma density is distinctly higher (order of magnitude) than that obtained from the non-LTE fit of the optical laser-produced plasma (upper red curve in Fig. 10.48). This is consistent because XFEL absorption and subsequent X-LIF take place at higher densities for the geometry shown in Fig. 10.45. The spectral distribution is far from LTE, as demonstrated by the comparison of the blue and green curves, cf. Figure 10.50a. Non-LTE effects are therefore important even for high densities because high autoionizing rates and fs time scales do not permit effective collisional redistribution.

In conclusion, the first transient X-ray pumping of dense optical laser-produced plasmas presented in Figs. 10.48 and 10.49 demonstrates the great potential of X-ray laser-induced fluorescence (X-LIF) in dense plasma atomic physics and high-energy-density physics (Rosmej et al. 2016, 2022).

10.6.4 Beating the Auger Clock

10.6.4.1 Photoionization Versus Autoionization

Photoionization of inner atomic shells creates multiple excited states that can decay via non-radiative transitions. Let us consider the photoionization from the *K*-shell:

$$K^2L^XM^YN^Z + h\nu_{\mathrm{XFEL}} \to K^1L^XM^YN^Z + e_{\mathrm{photo}} \tag{10.195}$$

(e.g., iron is described by the configuration $K^2L^8M^{14}N^2$). The photoionized state is multiple excited and can decay via radiative and non-radiative (autoionization, known as Auger effect in solid-state physics) transitions. Let us consider a simple example ($Y = 0$, $Z = 0$):

$$K^1L^X \to \left\{ \begin{array}{ll} \textit{radiative decay}: & K^2L^{X-1} + h\nu_{K_\alpha} \\ \textit{non-radiative decay}: & K^2L^{X-2} + e_{\mathrm{Auger}} \end{array} \right\}. \tag{10.196}$$

Radiative and non-radiative decay processes in the X-ray energy range have extensively been studied in the very past (Flügge 1957). Particularly synchrotrons have been employed for advanced studies of X-ray interaction with solid matter. Synchrotron radiation, however, is not very intense, allowing occurrence of photoionization of inner-shells only as a rare process (means a negligible fraction of the atoms in the crystal are photoionized thereby leaving the solid system almost unperturbed). This situation is quite different for XFEL's outlined above: Their brilliance is more than ten orders of magnitude higher than those of most advanced synchrotrons. Photoionization of inner-shells may therefore concern almost every atom in the crystal structure leading to essential perturbations and corresponding dramatic changes in the physical properties of matter.

In terms of elementary processes, XFEL-driven photoionization rates allow to compete even with the Auger rates (autoionizing rates Γ are very large, order of

10^{12}–10^{16} s^{-1}). The necessary XFEL intensities to "compete" with the Auger effect can be estimated according to (see also (10.169)):

$$\int_{-\tau/2}^{+\tau/2} \mathrm{d}t \int_{E_n}^{\infty} \mathrm{d}E\sigma_n^{iz}(E)c\tilde{N}_{FEL}(E,t) > \Gamma. \qquad (10.197)$$

Assuming a photon energy E_{FEL} of the XFEL which is sufficient to proceed toward effective photoionization, namely $E_{FEL} = E_n + 3\delta E$, ($\delta E \ll E_n$, E_n is the ionization energy of the inner-shell with principal quantum number "n"), we obtain the following estimate:

$$I_{FEL} > 4 \times 10^{-1} \cdot \Gamma \cdot \frac{Z^4}{n^3}\left[\frac{\mathrm{W}}{\mathrm{cm}^2}\right]. \qquad (10.198)$$

As autoionizing rates scale approximately like $\Gamma \propto Z^0$ (means almost independent of Z in the hydrogenic approximation) the Z-scaling of (10.198) is approximately given by

$$I_{FEL} \propto Z^4. \qquad (10.199)$$

Let us consider the photoionization of the K-shell of Al I as an example: $Z \approx 10.8$, $n = 1$, $\Gamma \approx 10^{14}$ s^{-1}, $I_{FEL} > 5 \times 10^{17}$ W/cm^2. As microfocusing is a standard setup at the XFEL installations (usually realized with Beryllium lenses), intensities of the order of 10^{17} W/cm^2 can be achieved allowing photoionization of inner-shells to compete with the Auger rate. Also sub-microfocusing was demonstrated employing a four-mirror focusing system in Kirkpatrick–Baez geometry. This allowed to reach focal spot sizes down to 0.05 μm and intensities up to 10^{20} W/cm^2 (achieved at the SACLA XFEL (Mimura et al. 2014)). We note that the competition between the photoionization of inner-shells and the autoionization means that the change in atomic populations due to photoionization is essential compared to the Auger rate that destroys the inner-shell excited autoionizing state.

10.6.4.2 Hollow Ion Formation

Apart the threshold intensity of (10.198), the characteristic Auger time scale is another important issue. Before XFEL's became available for dense plasma physics experiments, it was proposed (Rosmej 2007, Rosmej and Lee 2007) on the basis of simulations carried out with the MARIA code that "beating the Auger clock" will allow massive creation of hollow ions and permit their observation via the characteristic X-ray emission. Let us consider the relevant physics via an example: creation of hollow ion K^0L^X configurations and corresponding characteristic inner-shell X-ray emission. We start from the K^2L^X configurations. Photoionization of the K-shell creates the state

$$K^2L^N + \hbar\omega_{\mathrm{XFEL}} \to K^1L^N + e_{\mathrm{photo},1}. \tag{10.200}$$

In order to proceed with interesting processes from the XFEL-produced single hole state K^1L^X, the duration of the XFEL pulse (being responsible for the first photoionization) must be of the order of the characteristic Auger time scale. As the operating VUV/X-ray FEL facilities propose the requested pulse durations (order of 10–100 fs), photoionization may further proceed from the single K-hole state to produce a second K-hole (hollow ion):

$$K^1L^N + \hbar\omega_{\mathrm{XFEL}} \to K^0L^N + e_{\mathrm{photo},2}. \tag{10.201}$$

The existence of the double K-hole configuration K^0L^X can easily be identified via the characteristic hollow ion X-ray transitions that are located approximately between Ly_α and He_α of highly charged ions (Faenov et al. 1999; Rosmej et al. 2015):

$$K^0L^N \to K^1L^{N-1} + \hbar\omega_{\mathrm{hollow}}. \tag{10.202}$$

Ab initio calculations with the MARIA code that include radiation field physics (see Sect. 10.6.2) demonstrate that hollow ion production is effective and observable levels of characteristic X-ray emission are achieved. These simulations have lead to a proposal for hollow ion research in dense plasmas at planed XFEL installations (Rosmej 2007, Rosmej and Lee 2007).

The central wavelengths of hollow ion emission groups can be approximated by (Rosmej et al. 2015):

$$\lambda(K^0L^N \to K^1L^{N-1}) \approx \frac{1.215 \times 10^{-7}\, m}{\left(Z_n - \sigma_{Ly_\alpha} - \sigma_L \times (N-1)\right)^2}, \tag{10.203}$$

$$\sigma_{Ly_\alpha} = -a_1(Z_n - 3) - a_3(Z_n - 3)^3 \tag{10.204}$$

with $a_1 = 6.17094 \times 10^{-4}$ and $a_3 = 9.15902 \times 10^{-6}$ resulting in an approximation with spectroscopic precision for $Z_n = 3$–56 (Z_n is the nuclear charge). Equation (10.204) includes the negative screening that is important for higher Z-elements, e.g., for $Z_n = 56$, $\sigma_{\mathrm{Ly}\alpha} = -1.397$. From Hartree–Fock calculations, we obtain a screening constant for the L-shell electrons, $\sigma_L \approx 0.07$ for He-like until O-like ions for each supplementary electron in the L-shell.

In order to estimate which hollow ion transitions are located between the H-like Ly_α and He-like He_α transitions, we can estimate the transition energies in the optical electron model:

$$\Delta E = \frac{3}{4} Z_{\text{eff}}^2 Ry, \tag{10.205}$$

where $Ry = 13.6058$ eV and the effective charge is given by

$$Z_{\text{eff}} = Z_n - \sigma. \tag{10.206}$$

Z_n is the nuclear charge and σ is a screening constant. For the He-like resonance line $\text{He}_\alpha = 1s2p\ ^1P_1 - 1s^2\ ^1S_0$, the screening constant can be approximated by

$$\sigma_{He_\alpha} = \sigma_0 - a_1(Z_n - 8) - a_3(Z_n - 8)^3, \tag{10.207}$$

where $\sigma_0 = 0.50417$, $a_1 = 3.4565 \times 10^{-3}$, $a_3 = 1.16632 \times 10^{-5}$. The application of this formula results in wavelength calculations with spectroscopic precision for $Z_n = 6$–56 (note that the approximation according to (10.207) includes also negative screening effects for high Z-elements).

Figure 10.51 shows the time evolution of the characteristic X-ray emission of H-like Ly_α ($2p$–$1s$), He-like resonance line ($1s2p\ ^1P_1 - 1s^2\ ^1S_0$) as well as the X-ray emission originating from hollow ions: K^0L^2–K^1L^1 and K^0L^3–K^1L^2. The MARIA simulations have been carried out for an intense XFEL beam that is interacting with a dense Mg plasma (see Fig. 10.45) with electron density $n_e = 10^{21}$ cm^{-3} and electron temperature $kT_e = 30$ eV. The photon energy is $E_{\text{FEL}} = 3100$ eV, pulse duration $\tau = 100$ fs, and a photon density $\tilde{N}_0 = 10^{23}\,\text{cm}^{-3}$ (corresponding to an intensity of $I_{\max} = 3.7 \times 10^{17}$ W/cm^2). As can be seen from Fig. 10.51, the intensity of the hollow ion X-ray emission is of the order of the H-like and He-like resonance line emissions that are known to be observable.

Let us clearly identify the real importance of the successive photoionization for the hollow ion X-ray emission (10.200)–(10.202). Figure 10.52 shows the temporal evolution when all photoionization channels are included in the simulations (solid curves) and when photoionization from and to the states that involve a K^1-electron

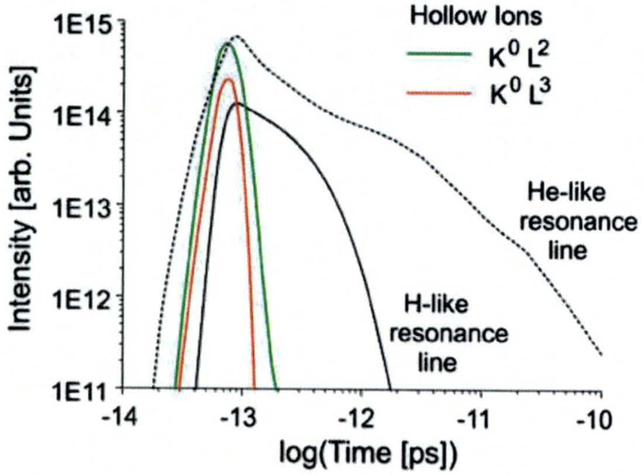

Fig. 10.51 MARIA simulations of the temporal evolution of various line intensities after interaction of the XFEL pulse with a dense Mg plasma plume, $E_{\text{FEL}} = 3100$ eV, $\tau = 100$ fs, $\tilde{N}_0 = 10^{23}\,\text{cm}^{-3}$, $I_{\max} = 3.7 \times 10^{17}$ W/cm^2, $n_e = 10^{21}$ cm^{-3}, $kT_e = 30$ eV

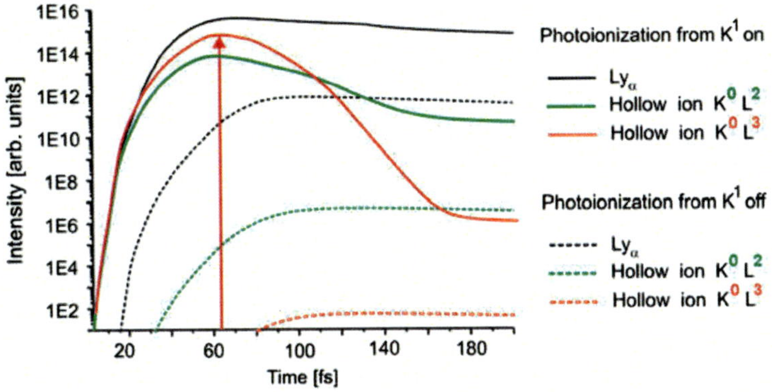

Fig. 10.52 MARIA simulations of the temporal evolution of various line intensities after interaction of the XFEL pulse with a dense Mg plasma plume, $E_{FEL} = 3100$ eV, $\tau = 100$ fs, $\tilde{N}_0 = 10^{23}\ \mathrm{cm}^{-3}$, $I_{max} = 3.7 \times 10^{17}$ W/cm^2, $n_e = 10^{21}\ \mathrm{cm}^{-3}$, $kT_e = 30$ eV

are artificially switched off (dashed curves in Fig. 10.52). It can clearly be seen that the hollow ion X-ray emission is practically absent when photoionization from K^1 is off: The remaining intensities are due to collisional effects. This means that in a proof of principal simulation with the MARIA code, hollow ion production and corresponding X-ray emission have been identified as driven by successive photoionization from K^2 and K^1-electron states (see flash in Fig. 10.52). This is equivalent to say that the XFEL allows beating the Auger clock to proceed toward successive K-shell ionization before the autoionization/Auger effect disintegrates the upper states.

The above-predicted double K-hole states produced via double inner-shell photoionization (see relations (10.200)–(10.202)) (Rosmej 2007; Rosmej and Lee 2007) had later been observed in experiments at the XFEL LCLS (Cryan et al. 2010).

Hollow ions can also be excited in a regime where photoionization is followed by resonant photoexcitation (Rosmej et al. 2015):

$$K^2L^N + \hbar\omega_{\mathrm{XFEL}} \to K^1L^N + e_{\mathrm{photo},1}, \tag{10.208}$$

$$K^1L^N + \hbar\omega_{\mathrm{XFEL}} \to K^0L^{N+1}, \tag{10.209}$$

$$K^0L^{N+1} \to K^1L^N + \hbar\omega_{\mathrm{hollow}}. \tag{10.210}$$

Contrary to the regime of double photoionization (10.200)–(10.202), photoionization followed by resonant photoexcitation depends strongly on the photon energy because the photon energy has to be large enough for single K-shell photoionization and, at the same time, need to match the resonance energy. In this respect, ionization potential depression may therefore have an important influence on the spectral distribution as it may permit to excite emission from higher charge states. This has been extensively explored in [Deschaud et al. 2020] where a general

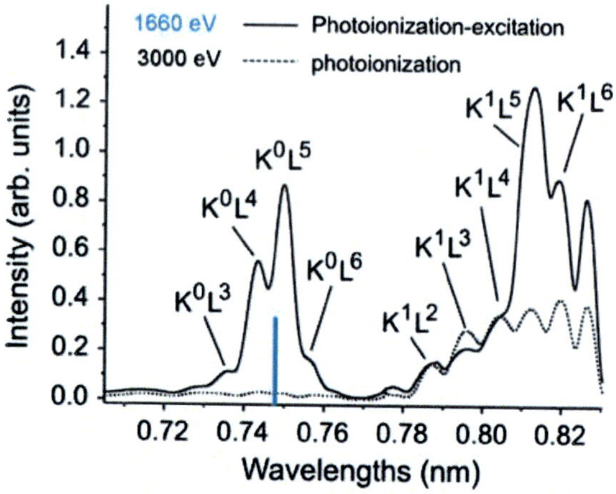

Fig. 10.53 Hollow ion X-ray emission excited by a combination of photoionization–resonance photoexcitation, $E_{XFEL} = 1660$ eV, $\tau_{XFEL} = 80$ fs, $I_{XFEL} = 10^{17}$ W/cm^2 including the temporal evolution of the plasma

method of ionization potential depression analysis has been proposed that was based on a 2-dimensional map (frequency of target emission versus XFEL photon frequency) of highly resolved hollow ion X-ray emission.

Figure 10.53 shows a simulation of the hollow ion X-ray emission of aluminum coupling the time-dependent atomic population kinetics (see (10.184)–(10.189) and (6.22)–(6.28)) of the energy balance equation (Deschaud et al. 2014; Rosmej et al. 2015)

$$\frac{dE}{dt} = P_{abs} - P_{rad}, \tag{10.211}$$

where P_{abs} is the power absorbed from the radiation field and P_{rad} is the power emitted via radiative recombination, spontaneous and stimulated emission. A photon energy of $E_{XFEL} = 1660$ eV was chosen permitting strong emission around the inner-shell transitions originating from the K^0L^5 configuration (see blue bar in Fig. 10.53). In dense plasmas, collisional ionization may further populate higher charge states and emission up to K^0L^3 is therefore observed. Also shown in Fig. 10.53 is a simulation for $E_{XFEL} = 3000$ eV (dashed line). The comparison between the solid black and dashed line in Fig. 10.53 demonstrates that the regime of photoionization-excitation (10.208)–(10.210) can strongly increase the population of a certain hollow ion state.

10.6.4.3 X-ray Emission Switches for Ultrafast Dense Matter Investigations

The X-ray emission of hollow ions discussed in the forgoing paragraph provides outstanding possibilities to investigate dense exotic states of matter that just exist during the interaction of the XFEL pulse with matter (Rosmej et al. 2007). Figure 10.54 shows the temporal evolution of the hollow ion X-ray emission

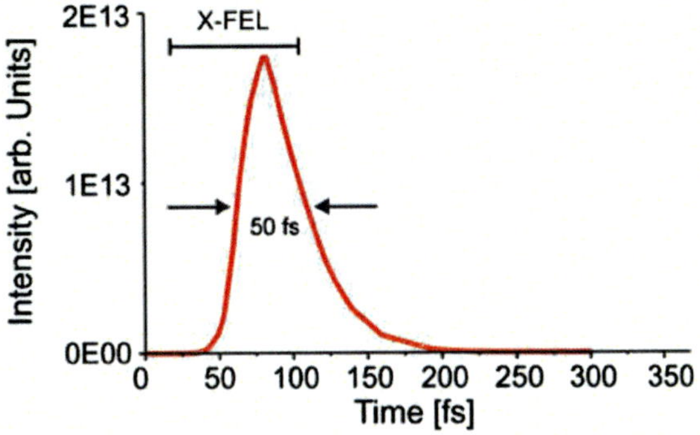

Fig. 10.54 MARIA simulations of the temporal evolution of the hollow ion X-ray emission induced by the interaction of a XFEL pulse with a dense Mg plasma plume, $E_{FEL} = 3100$ eV, $\tau = 100$ fs, $\tilde{N}_0 = 10^{23}\,\text{cm}^{-3}$, $I_{max} = 3.7 \times 10^{17}$ W/cm^2, $n_e = 10^{21}$ cm^{-3}, $kT_e = 30$ eV

$K^0L^3 \rightarrow K^1L^2 + h\nu_{\text{Hollow ion}}$ of Fig. 10.51 on a linear intensity scale (Rosmej and Lee 2007). The simulations demonstrate that the FWHM of the X-ray emission is only 50 fs and temporally located very close to the XFEL pulse. Therefore, dense matter properties that are just produced during the XFEL interaction can be studied via the hollow ion X-ray emission emitted from the matter under study.

Let us discuss in more detail the outstanding properties of the characteristic hollow ion X-ray emission for the K^0L^X-configurations for dense matter investigations:

(a) *Small opacity of hollow ion X-ray transitions*
Spectroscopic investigations of dense matter suffer usually from opacity effects because even in the X-ray range, a line photon can be absorbed and reemitted many times. In this respect, X-ray hollow ion transitions have the advantageous feature of almost negligible opacity. This can be understood as follows. The hollow ion opacity effects can be estimated considering a simple balance equation for the populations K^1L^N (absorbing lower states) and K^2L^{N-2} (atomic ground state) that is driven by dielectronic capture:

$$n(K^1L^N)\{A+\Gamma\} \approx n(K^2L^{N-2})n_e\langle DC\rangle, \tag{10.212}$$

where $n_e\langle DC\rangle$ (where n_e is the electron density) is the dielectronic capture rate into the state K^2L^{N-2} that is balanced by radiative decay A and autoionization Γ of state K^1L^N. This provides

$$\frac{n(K^1L^N)}{n(K^2L^{N-2})} \approx n_e \cdot \frac{g(K^1L^N)}{g(K^2L^{N-2})} \cdot \alpha \cdot \frac{\Gamma}{A+\Gamma} \cdot \frac{\exp(-E_C/kT_e)}{(kT_e)^{3/2}}, \tag{10.213}$$

where α is the constant for dielectronic capture ($\alpha = 1.66 \times 10^{-22}\,\text{cm}^3\,\text{s}^{-1}$), n_e the electron density in [cm^{-3}], the electron temperature kT_e in [eV], E_C is the capture energy in [eV], autoionizing rate Γ in [s^{-1}] and radiative decay rate A in [s^{-1}]. Let us estimate the right-hand side of (10.213):

$$\frac{n(K^1L^N)}{n(K^2L^{N-2})} \approx \frac{n_e}{\text{cm}^{-3}} \cdot 10 \cdot 2 \times 10^{-22} \cdot 1 \cdot \left(10^{-5} \ldots 10^{-9}\right). \tag{10.214}$$

Therefore even for an electron density of 10^{23} cm^{-3}, the population of K^1L^N is about three–seven orders of magnitude smaller than those of the states K^2L^{N-2}. The line center bound–bound opacity for a Doppler broadened line can then be estimated by (see also Sect. 1.1.4)

$$\tau_{0,\text{ij}} \approx 1.08 \times 10^{-10} \frac{\lambda_{\text{ji}}}{(\text{m})} \sqrt{\frac{M(amu)}{T_{\text{i}}(eV)}} f_{\text{ji}} \frac{n(K^1L^N)}{(\text{cm}^{-3})} \frac{L_{\text{eff}}}{(\mu\text{m})}, \tag{10.215}$$

where T_{i} is the ion temperature, M the number of nucleons, λ the wavelength, f the oscillator strength, $n(K^1L^N)$ the absorbing lower-level density, and L_{eff} the effective photon path length. Let us assume a dense laser-produced plasma with an electron density of $n_e = 10^{23}$ cm^{-3}, ion temperature of $T_{\text{i}} = 1$ keV, and the plasma dimension of $L_{\text{eff}} = 30$ μm. Assuming an aluminum plasma, $M = 27$, $\lambda = 7 \times 10^{-10}$ m, $f = 0.2$, average charge state of 10, a fraction of 0.1 in the charge states K^2L^{N-2} (resulting in a density $n(K^2L^{N-2}) \approx 0.1 \times 10^{23}$ cm^{-3}/10 = 10^{21} cm^{-3} and, according to (10.124), in an absorbing ground state density of $n(K^1L^N) \approx 10^{14}$–$10^{18}$ cm^{-3}). The line center opacity is therefore about $\tau_0(\text{hollow}) \approx 10^{-6} \ldots 10^{-1} \ll 1$. Note that at the same time resonance line opacity is many orders of magnitude larger, $\tau_0(\text{resonance}) \approx 10^3 \gg 1$. Hollow ion X-ray emission is therefore expected to escape even from a super-dense plasma, whereas resonance line emission is strongly absorbed.

(b) *Suppression of low-density recombining plasmas*
A further advantageous property concerns the transient evolution (Rosmej et al. 2007). Resonance line emission is not only sensitive to collisional excitation at high densities but to the low-density recombination regime too (due to radiative cascades). Therefore, interesting high-density features that usually appear at the beginning of the laser pulse may be seriously masked by high intensity but low-density emission.

K^0L^N-Hollow ion formation ($N = 3$–8), however, is rather insensitive to the radiative recombination regime because this population process has to proceed from the hollow ion states K^0L^{N-1}: $X(K^0L^{N-1}) + e \to X(K^0L^N) + \hbar\omega_{\text{rad.recom}}$. The states K^0L^{N-1} are barely populated as dielectronic capture $X(K^1L^{N-2}) + e \to X(K^0L^N)$ and collisional inner-shell excitation $X(K^1L^{N-1}) + e \to X(K^0L^N) + e$ have to proceed from single K-hole states (rather than from highly populated ground states). In addition, dielectronic capture rate coefficients $\langle DC \rangle$ are very small at low temperatures (radiative recombination regime) because in the X-ray spectral range the capture energy E_C is very large (order of keV). This results practically in an almost exponential cutoff of the capture rate

$$\langle DC \rangle \propto (kT_e)^{-3/2} \cdot \exp(-E_C/kT_e). \tag{10.216}$$

Therefore, single K-hole configurations are barely populated and in turn double K-hole configurations too. The same holds true for inner-shell excitation from K- to L-shell. Therefore, the population of hollow ion configurations is not expected to be sensitive to the long lasting radiative recombination regime (see also Fig. 10.51).

In conclusion due to small opacity, negligible influence of overlapping low-density recombining regimes and the short time scale (some 10 fs, see Fig. 10.54) hollow ion X-ray emission is a very suitable diagnostic for the study of short living dense matter samples because interesting properties of dense matter samples exist essentially only during the period of laser-matter interaction itself (Rosmej and Lee 2007; Rosmej et al. 2007, 2009; Deschaud et al. 2020). We note that X-ray streak cameras may help to suppress emission from the recombination regime, see Fig. 10.51; however, they will hardly be able to streak down to 50 fs (current limits are about 0.5 ps).

10.6.4.4 Transparent Materials and Saturated Absorption

A material is transparent to photons at certain energies, if neither photoabsorption nor photopumping is effective at these photon wavelengths. As the photon absorption is related to the various photon opacities (see (1.54)–(1.56)), this effect is strongly related to the density of the atomic populations: In the case of photoionization, it is the population density of the state that is photoionized, in the case of photopumping it is the lower state of the atomic transition.

As has been shown in the forgoing paragraphs, XFEL radiation allows to effectively change atomic populations in the X-ray energy range. This permits also to selectively deplete atomic populations. If these populations are related to photoionization/photopumping transparency to the XFEL radiation itself is induced and a so-called saturated absorption regime is achieved.

Saturated absorption has been observed at the XUV–FEL FLASH (Nagler et al. 2009) in the framework of the spectroscopic beam time proposal at FLASH (Riley et al. 2009) irradiating solid Al foils with a 92 eV FEL beam in the photoionization regime:

$$1s^2 2s^2 2p^6 3s^2 3p^1 + \hbar\omega_{\mathrm{XFEL}} \rightarrow 1s^2 2s^2 2p^5 3s^2 3p^1 + e_{\mathrm{photo}}. \tag{10.217}$$

As photoionization of a $2p$-electron from the $2p^6$-configuration is the most effective and a second photoionization (means the creation of a $2p^4$-configuration) seems energetically not probable the ionization of almost all $2p^6$-configurations will induce transparence to the 92 eV XUV-laser radiation. Solid aluminum has therefore turned transparent for 92 eV photons. We note that transparency is limited by the principle of detailed balance: Stimulated photoemission (10.189) and stimulated radiative recombination (10.187) sets a definite limit to that what can actually be observed. Also three-body recombination in dense matter will destroy the hole

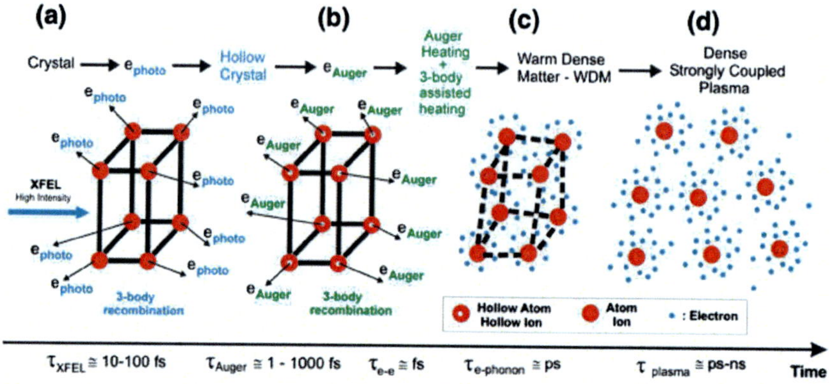

Fig. 10.55 Schematic evolution of XFEL interaction with matter. **a** Interaction with cold solid, **b** interaction with heated solid, **c–d** evolution of dense matter; see text

states, thereby driving the saturation regime to higher intensities (see also Sect. 10.6.4.6).

Saturated absorption implies enhanced homogeneity of the irradiated material, as no more geometrical energy deposition peaks exist. This effect is well known from the stopping of relativistic heavy ion beams in matter: If the Bragg peak is placed outside the target, almost homogenous parameter conditions are meet (Kozyreva et al. 2003; Tauschwitz et al. 2007).

The term "transparent aluminum" is also known to the non-scientific society from the science fiction series "Stark Trek" (Wiki 2021): The chief engineer M. Scott has invented transparent aluminum to fabricate windows that have the strength and density of solid aluminum for its use to transport whales in an aquarium. This has moved XFEL research to the frontiers of science fiction (Larousserie 2009).

10.6.4.5 Exotic States of Dense Matter: Hollow Crystals

According to (10.165), a typical XFEL intensity at some keV photon energy implies a photon density $\tilde{N}_0$ of the order of solid density:

$$\tilde{N}_0 = \frac{I_{FEL,\delta E,\tau}}{4 \cdot c \cdot erf^2\left(\sqrt{\ln 2}\right) \cdot E_{FEL}} \approx 3.6 \times 10^8 \cdot \frac{I_{FEL,\delta E,\tau}\left(\mathrm{W/cm^2}\right)}{E_{FEL}(eV)} \left[1/\mathrm{cm^3}\right]. \tag{10.218}$$

E.g., for $I_{FEL,\delta E,\tau} = 10^{17}\,\mathrm{W/cm^2}$ and $E_{FEL} = 2000\,\mathrm{eV}$ we obtain $\tilde{N}_0 = 1.8 \times 10^{22}\,\mathrm{cm^{-3}}$. Therefore, inside the XFEL light pencil there exists a photon for almost every atom in the lattice structure. Due to X-ray pulse durations

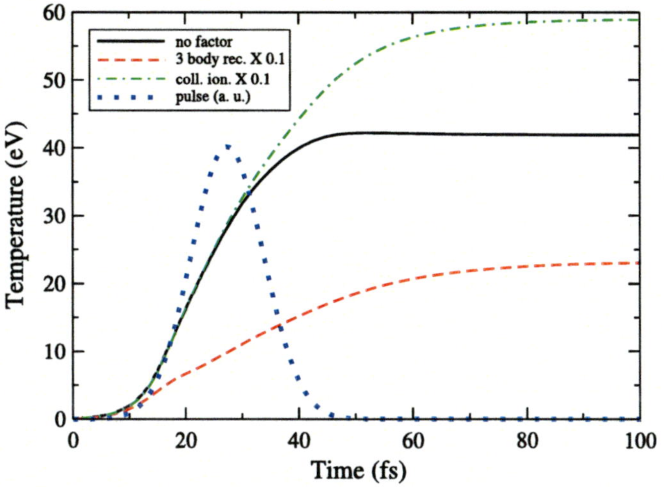

Fig. 10.56 Simulation of the time evolution of the electron temperature in XUV–FEL interaction with solid aluminum (solid black curve). Dashed red curve shows a simulation suppressing the three-body recombination rates by a factor of 10. The dramatic difference between the red dashed and solid black curves identifies three-body recombination as an important mechanism in the material heating. Simulation parameters: $I_{FEL} = 10^{16}$ W/cm^2, $E_{FEL} = 92$ eV, $\tau_{FEL} = 15$ fs

that are of the order of (or even smaller) than the characteristic Auger time scale, the irradiation of solids with XFEL allows a sudden photoionization of inner-shells for almost all atoms in the lattice. Consequently, almost every atom is transformed to an autoionizing state followed by a massive burst of Auger electrons on the time scale of some 10 fs.

Figure 10.55 schematically illustrates these processes and the subsequent evolution. In Fig. 10.55a, a high-intensity XFEL beam interacts with the solid and creates for almost every atom a photoelectron. Therefore, almost every atom has lost one electron in the inner-shell leading to the creation of a very exotic state, a dense hollow crystal (Fig. 10.55b). Due to the core hole vacancies, the hole states are autoionizing and decay on the Auger time scale (some 10 fs). Therefore, the hollow crystal is a very short living exotic state of matter with subsequent transitions to Warm Dense Matter (Fig. 10.55c) and strongly coupled plasmas (Fig. 10.55d).

10.6.4.6 New Role of Elementary Processes: Auger Electron and Three-Body Recombination Heating

As demonstrated by (10.218), almost every atom in the lattice structure is concerned by the creation of photo- and Auger electrons. Therefore, Auger and photoelectrons have also near solid density. The photoelectrons have kinetic energy

E_{photo} given by the difference of the XFEL photon energy E_{XFEL} and the ionization energy E_{core} of the core electron, i.e.

$$E_{photo} = E_{XFEL} - E_{core}, \tag{10.219}$$

while the Auger electrons have characteristic energies that correspond to the dielectronic capture energy (see also Fig. 5.1). The photoelectron kinetic energy can therefore be close to zero if the XFEL is tuned exactly to the ionization energy. This is distinct different for the Auger electrons that have kinetic energies corresponding to the capture energy that is independent of the photon energy (if the photon energy exceeds the ionization energy to create a core hole state). As the capture energy of core hole states is of the order of $0.5 \cdot Z_{eff}^2 Ry$, the Auger electron kinetic energy is very large because it is of the order of the core hole ionization energy. Therefore, Auger electron kinetic energy constitutes an important contribution to the material heating in XFEL solid matter interaction.

Due to the high density, Auger and photoelectrons equilibrate rapidly with the "cold" conduction band electrons (fs…10 fs time scale) thereby creating a common "bath" of dense electrons with a temperature much less than the original kinetic energies. Therefore, efficient three-body recombination (due to high density and low temperature) into the original hole states can take place from this bath of dense electrons. As can be seen from (10.117), three-body recombination between the atom/ion and the two continuum electrons with energies E_1 and E_2 transfers back the ionization energy E_i to the continuum electron with energy E, i.e.,

$$E - E_1 - E_2 = E_i. \tag{10.220}$$

Three-body recombination influences therefore on the heating of dense matter (Deschaud et al. 2015) as indicated schematically in Fig. 10.55b, c. Therefore, the well-known elementary processes photoionization, autoionization, and three-body recombination contribute also directly to the material heating and attribute a new role to these elementary atomic physics cross sections in the XFEL-induced material heating.

The importance of three-body recombination for the material heating is demonstrated in Fig. 10.56. via a zero-dimensional simulation of the XUV–FEL interaction with solid aluminum (Deschaud et al. 2015) solving (10.184)–(10.189), (10.211) for an intensity of $I_{FEL} = 10^{16}$ W/cm^2, photon energy of $E_{XFEL} = 92$ eV, and pulse duration of $\tau = 15$ fs. The solid black curve shows the time evolution of the electron temperature, while the dashed red curve shows the electron temperature when the three-body recombination rates are reduced by a factor of 10. The green dashed-dot curve shows the electron temperature when the collisional ionization rates are reduced by a factor of 10. The comparison of the red dashed curve and the solid black curves identifies the important role of three-body recombination as a heating mechanism: Suppressing the three-body recombination by a factor of 10 reduces the maximum temperature considerably (note that changing one elementary processes in a consistent simulation may induce also changes for other processes;

therefore, the change in the electron temperature as shown in Fig. 10.56 is only presented to qualitatively indicate the influence of three-body recombination). Therefore, turning back the ionization energy to the continuum according to (10.220) is important for the material heating. At the same time, the absorbing ground states are recreated from which further photoabsorption in the light pencil can take place. This phenomenon strongly influences on the absorption of the material and at high intensities, one atom can even absorb more than one photon during the FEL pulse duration.

A corresponding effect is seen for collisional ionization. If it is reduced by a factor of 10 (see green dashed-dotted line in Fig. 10.56), the electron temperature increases. This indicates that collisional ionization is an important energy loss mechanism for the energy balance (10.211).

In order to study and validate the various heating mechanisms with independent methods, high-resolution spectroscopic studies have been undertaken (Galtier et al. 2011) in the framework of the first high-energy-density experiment at the XUV–FEL FLASH (Riley et al. 2009). Figure 10.57 shows the experimental spectrum of aluminum in the spectral range from 12 to 20 nm. The blue curve in Fig. 10.57a shows a simulation of the Ne-like transitions $1s^2 2s^2 2p^5 3l \rightarrow 1s^2 2s^2 2p^6 + \hbar\omega_{\mathrm{Al\,VI}}$ of Al IV. The comparison of the data (back curve) with the simulations (blue curve) identifies the principal transitions near 16 nm that are originating from the resonance transitions $1s^2 2s^2 2p^5 3s \rightarrow 1s^2 2s^2 2p^6 + \hbar\omega$ but shows strong discrepancies near 13 nm where potentially the resonance transitions $1s^2 2s^2 2p^5 3d \rightarrow 1s^2 2s^2 2p^6 + \hbar\omega$ are located. Also systematic discrepancies are observed on the red wings of both types of resonance transitions (indicated with red flashes in Fig. 10.57a). Atomic structure calculations show that the experimentally observed intensities on the red wing of the $3s$- and $3d$-resonance transitions near 13 and 16 nm can be attributed to Na-like (red dashed curve in Fig. 10.57b) and Mg-like (red solid curve in Fig. 10.57b) dielectronic satellite transitions of the type $K^2L^7M^2 \rightarrow K^2L^8M^1 + \hbar\omega_{\mathrm{sat}}$ and $K^2L^7M^3 \rightarrow K^2L^8M^2 + \hbar\omega_{\mathrm{sat}}$, respectively. Taking into account these satellite transitions results in a perfect agreement with the data as demonstrated in Fig. 10.57b (green solid curve).

The simulations show that the spectral distributions of these Na- and Mg-like satellite transitions are very sensitive to the electron temperature. Figure 10.58 demonstrates this temperature sensitivity for the Na-like satellites. The best fit to the data (Fig. 10.57b) is obtained for $kT_{\mathrm{e}} = 25\,\mathrm{eV} \pm 10\,\mathrm{eV}$. The intriguing point of the use of the Na-like and Mg-like dielectronic satellite transitions is that their corresponding upper states are autoionizing with very large Auger rates to ground and excited states (Petitdemange and Rosmej 2013). Therefore, the characteristic time of photon emission τ_{ji} is very small as it is dominated by the Auger rates:

$$\tau_{\mathrm{ji}} = \frac{1}{A_{\mathrm{ji}} + C_{\mathrm{ji}} + C_{\mathrm{ji}} + \sum_k \Gamma_{\mathrm{jk}}} \approx \frac{1}{\sum_k \Gamma_{\mathrm{jk}}} \approx 1 \ldots 10\,\mathrm{fs}. \tag{10.221}$$

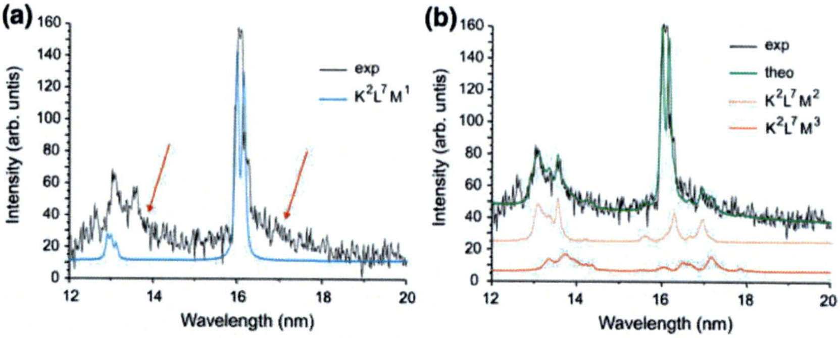

Fig. 10.57 MARIA simulations of the XUV spectra induced by XUV–FEL interaction with solid aluminum. a Simulations taking into account only the Ne-like transitions $K^2L^7M^1$–K^2L^8 (blue solid curve), **b** simulations taking into account the Ne-like transitions $K^2L^7M^1$–K^2L^8 as well as transitions from dielectronic satellites of Na-like (red dashed curve) and Mg-like (red solid curve) aluminum, $K^2L^7M^2$–$K^2L^8M^1$ and $K^2L^7M^3$–$K^2L^8M^2$ respectively

Fig. 10.58 MARIA simulation of the spectral distribution of the Na-like dielectronic satellites $K^2L^7M^2$–$K^2L^8M^1$ of aluminum in dependence of the electron temperature

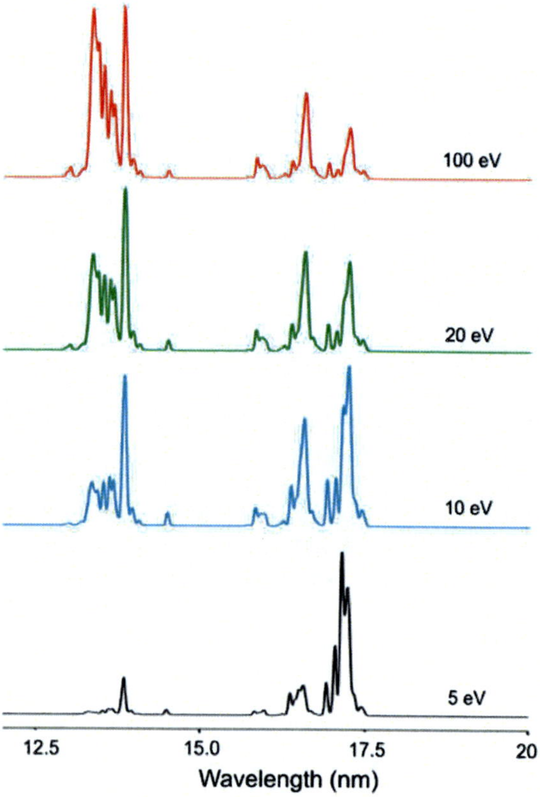

Consequently, the corresponding emission is closely related to the instants of XUV–FEL interaction with the near solid matter (see also discussion in relation with Figs. 10.51 and 10.54). It is important to note that for the temperature determination, only the spectral distribution of the Na-like and Mg-like dielectronic satellites has been used while the intensity contribution from the Ne-like resonance lines (blue curve in Fig. 10.57a) is only taking into account to obtain the correct contribution of the Na-like and Mg-like satellites. In fact, the simulations demonstrate that the Ne-like resonance contributions show emission at about 8 eV, which is significantly lower than the temperature obtained from the satellite contribution (Galtier et al. 2011). This is understandable because the resonance line intensities have also contributions from the low-density, low-temperature recombining regime (see also discussion in relation with Fig. 10.51) because their radiative decay rates are of the order of 10^9–10^{11} s indicating self emission about 0.1 ns after interaction, i.e., much after the XFEL-pulse interaction with matter.

In order to develop a consistent picture for the spectral interpretation, we can estimate the electron density from the Saha–Boltzmann equation (see (6.11)) applied to the two different charge states corresponding to the upper states of the Na- and Mg-like dielectronic satellite transitions. This can be realized because the best fit of the data presented in Fig. 10.57b delivers likewise the integral intensity ratio of the Na- and Mg-like satellite emission. This results in a density estimate of $n_e \approx 5 \times 10^{22}\,\text{cm}^{-3}$ (Galtier et al. 2011) and indicates that the emission spectra have been recorded while the material density is still at a significant fraction of the solid density (but still some expansion of the heated solid aluminum has already taken place). Assuming a simple adiabatic model for the plasma expansion, electron temperature decreases about a factor of 0.42 (assuming a density decrease from solid density to $n_e \approx 5 \times 10^{22}\,\text{cm}^{-3}$) compared to the case where immobile matter (zero-dimensional model) is heated. Therefore, the maximum temperature of the simulations presented in Fig. 10.56 has to be decreased by about a factor of 0.42 to be compared with the data. With these corrections, we obtain from the simulations (Fig. 10.56) of the experiment (Fig. 10.57) an electron temperature of about $kT_e \approx 0.42 \times (40-50)\,\text{eV} = 17-21\,\text{eV}$ (note that the experimentally measured intensity is uncertain by about a factor 2 resulting in the given temperature interval of about 40–50 eV for the simulations (Deschaud et al. 2014). This value is in reasonable agreement with the spectroscopic measurement of $kT_e \approx (25 \pm 10)\,\text{eV}$.

The particular remarkable point in this agreement is that the spectroscopically determined temperature is significantly higher than expected from a model that invokes photoelectron heating only. Assuming an ionization potential of the $2p^6$ electron of $E_{core} = 73$ eV (L_{II}- and L_{III}-edges of solid cold aluminum), the photoelectrons have kinetic energies of $E_{photo} = E_{XFEL} - E_{core} = 92\text{ eV} - 73\text{ eV} = 19$ eV. Assuming photoionization to the bottom of the valence band (due to heating the Fermi distribution creates free places below the Fermi energy) results in a kinetic energy of $E_{photo} = E_{XFEL} - E_{core} + \varepsilon_F = 92\text{ eV} - 73\text{ eV} + 12\text{ eV} = 31$ eV ($\varepsilon_F = 12$ eV is the Fermi energy of solid cold aluminum). Thus, a kinetic energy of 19–31 eV is rapidly distributed among the four electrons in the valence band (because cold solid aluminum has three valence band electrons). Assuming an equal distribution of the kinetic energy among the four electrons results in an electron temperature of about 8 eV. This value is significantly lower than the spectroscopic measurements.

Therefore, for the first time spectroscopic analysis has identified a new heating mechanisms, Auger electron and three-body recombination heating, as important processes in the material heating in FEL interaction with solids (Galtier et al. 2011) while reasonable agreement with emerging simulations was obtained.

It is interesting to see from the simulations of Fig. 10.56 that an electron temperature of about 28 eV is obtained at maximum pulse (see blue dashed line in Fig. 10.56). This value is close to a temperature, obtained from photoelectron and Auger heating only. This can be seen as follows. The Auger kinetic energy is about 70 eV resulting into an additional heating of the valence band electrons that have already been heated to about 8 eV from the photoelectron kinetic access energy. Assuming equal energy partition between 3 electrons at 8 eV and one electron at 70 eV results into an average electron temperature of about 24 eV.

With respect to Fig. 10.55, we arrive therefore at the following approximate qualitative picture of solid material heating in XUV- and X-ray FEL interaction. After intense XFEL irradiation of a solid, almost every atom has a core hole from photoionization, Fig. 10.55a. Photoelectron kinetic energy preheats the valence band electrons. As the core hole states are autoionizing, Auger electron kinetic energy rapidly contributes to the material heating, Fig. 10.55b. As almost every atom is concerned, photoelectrons and Auger electrons have near solid density and three-body recombination from the valence band heated electrons refills the core holes resulting into a further material heating, Fig. 10.55c. The three-body recombination results also into a recreation of absorbing states thereby increasing further absorption and heating. After the laser pulse is off, a warm dense matter sample and ion displacements develop (Fig. 10.55c) resulting finally in a strongly coupled plasma, Fig. 10.55d. The various steps shown in Fig. 10.55a–d are selected in order to roughly guide the origin of some physical mechanism. Really, there is considerably overlap between these regimes and impact of other phenomena (e.g., non-Maxwellian energy distributions).

We finally note that in principle synchrotron radiation may also produce Auger electrons via photoionization of inner-shells; however, the low intensity makes Auger emission a rare process compared to an important number of atoms that are not concerned (note that this is not a contradiction to the fact that Auger electron spectra can be well measured). Therefore, no heating of the crystal is induced via Auger electron heating. Moreover, synchrotron radiation does not allow photoionization on the Auger time scale. Therefore, creation of exotic states of matter such as "hollow crystals" and "transparent solids" are almost impossible.

10.6.5 Generalized Atomic Physics Processes

Numerical atomic structure calculations show that transitions in solids involving hole states are reasonably approximated by the free atom approach. For example, for aluminum, that has been discussed above Hartree–Fock calculations (Cowan 1981) that contain only the Al II configurations $1s^2 2s^2 2p^6 3s^2$, $1s^2 2s^2 2p^5 3s^2 3p^1$, $1s^2 2s^1 2p^6 3s^2 3p^1$, and $1s^1 2s^2 2p^6 3s^2 3p^1$ provide for the K_α-transitions a wavelength of about 0.833 nm whereas the measured values are about 0.834 nm. Even for the K_β-transition, the calculated value of 0.793 nm is in reasonable agreement with the measurements of 0.796 nm. This indicates that the framework of the standard atomic population kinetics as outlined in Chap. 6 might be useful (including some corrections) for the populations of inner-shells that play a key role for the energy deposition in the XFEL interaction with matter (and also for spectroscopic diagnostic).

10.6.5.1 Generalized Three-Body Recombination and Autoionization

In order to maintain the standard rate equation approach for the core hole population kinetics, it looks therefore reasonable for the above-discussed example of aluminum to designate the atomic structure of the solid according to $1s^2 2s^2 2p^6 (VB)^3$ where $(VB)^3$ indicates the valence band that is occupied with three electrons (i.e., the conduction band for the present case of aluminum). In order to make this spectroscopic designation meaningful, we need to establish a population kinetic link between the core $1s^2 2s^2 2p^6$ and the valence band $(VB)^3$. This link has to be established from the cold solid to the heated solid. This is equivalent to the definition of transition matrix elements $W_{k_{Z'} j_Z}$ that create this link via generalized atomic physics processes (Deschaud et al. 2014) in order to establish the core hole kinetics via the set of equations

$$\frac{dn_{j_Z}}{dt} = -n_{j_Z} \sum_{Z'=0}^{Z_n} \sum_{i_{Z'}=1}^{N_{Z'}} W_{j_Z i_{Z'}} + \sum_{Z'=0}^{Z_n} \sum_{k_{Z'}=1}^{N_{Z'}} n_{k_{Z'}} W_{k_{Z'} j_Z}, \tag{10.222}$$

where the levels j_Z are generalized levels that contain the core hole states and the valence band while the corresponding matrix elements $W_{k_{Z'} j_Z}$ are generalized matrix elements. For transparence of the discussion, let us first consider the Auger effect (directly related to core hole states) that has been identified above as an direct important heating mechanism:

$$\left(1s^2 2s^2 2p^5, 1s^2 2s^1 2p^6, 1s^1 2s^2 2p^6\right)(VB)^3 \rightarrow 1s^2 2s^2 2p^6 (VB)^1 + e_{\text{Auger}}. \tag{10.223}$$

The processes according to relation (10.223) are important characteristic transitions in the cold solid state. These processes need to be matched to the free atom and extended up to high temperatures. For these purpose, let us consider for a moment the valence band electrons $(VB)^3$ as "free" electrons that recombine via three-body recombination with the cores $1s^22s^22p^5$, $1s^22s^12p^6$, and $1s^12s^22p^6$. For better transparency, we split the valence band $(VB)^3$ into $(VB)^2(VB)^1$ in order to depict two electrons that recombine with the cores:

$$(1s^22s^22p^5, 1s^22s^12p^6, 1s^12s^22p^6)(VB)^2(VB)^1 \rightarrow 1s^22s^22p^6(VB)^1 + e. \quad (10.224)$$

Comparing relation (10.223) with (10.224), we realize that left-hand sides and right-hand sides are formally equivalent for the same core hole states. This hints to the idea that the characteristic Auger process in solids might be described by a "generalized three-body recombination":

$$(1s^22s^22p^5, 1s^22s^12p^6, 1s^12s^22p^6)(VB)^2(VB)^1 \rightarrow 1s^22s^22p^6(VB)^1(VB)^1. \quad (10.225)$$

The generalization consists of two parts: First, the three-body recombination turns into the Auger rate for cold solid matter, and, second, turns into the standard three-body recombination if the temperature is high. Is it possible to generalize also other atomic physics processes in order to be consistent for the whole population kinetics?

10.6.5.2 Generalized Collisional Excitation, Ionization, and Dielectronic Capture

Let us continue with collisional excitation. Excitation from the core to the valence band (*VB*) has to be considered as ionization:

$$1s^22s^22p^6(VB)^1(VB)^1 \rightarrow (1s^22s^22p^5, 1s^22s^12p^6, 1s^12s^22p^6)(VB)^1(VB)^2. \quad (10.226)$$

Within the same philosophy resonant capture to the *VB* according to

$$1s^22s^22p^6(VB)^1(VB)^1 \rightarrow (1s^22s^22p^5, 1s^22s^12p^6, 1s^12s^22p^6)(VB)^2(VB)^1 \quad (10.227)$$

is therefore a generalized ionization process for the core $1s^22s^22p^6$. This is consistent with the principle of detailed balance: Resonant capture is the inverse process of autoionization; autoionization is considered as a generalized three-body recombination.

10.6.5.3 Generalized Fluorescence and Radiative Recombination

What remains is fluorescence of an electron from the *VB* to a core level:

$$\left(1s^2 2s^2 2p^5, 1s^2 2s^1 2p^6, 1s^1 2s^2 2p^6\right)(VB)^1 (VB)^2 \rightarrow 1s^2 2s^2 2p^6 (VB)^2 + \hbar\omega. \tag{10.228}$$

This corresponds to a generalized radiative recombination process. This generalized process represents therefore fluorescence at low temperature and turns into standard radiative recombination for high temperatures.

Relations (10.223)–(10.228) indicate that the link from the core to the valence band is realized by generalized processes while they are readily modified for different ionization stages. We note that transitions within the core, such as radiative decay (e.g., *K*-alpha transitions), are incorporated in the standard manner as motivated above.

10.6.5.4 The Heated Solid and Generalized Atomic Fermi–Dirac Rate Coefficients

We now need to treat the case of a heated solid. As the Pauli principle drives the behavior of the *VB* electrons in the cold solid, the Fermi–Dirac statistics for the electron energy distribution function and also for the calculation of the rate matrix elements $W_{j_Z i_{Z'}}$ of (10.222) has been proposed for the calculation of the *W*-matrix elements (Deschaud et al. 2014, 2020). The use of the Fermi–Dirac distribution function leads to important differences compared to the Maxwell–Boltzmann statistics that is usually employed in atomic population kinetics. The Fermi–Dirac distribution function of electrons in the valence band is given by

$$F_{\mathrm{FD}}(\varepsilon, T_e) = \frac{1}{2\pi^2 n_e}\left(\frac{2m_e}{\hbar^2}\right)^{3/2} \frac{\sqrt{\varepsilon}}{e^{[\varepsilon-\mu(T_e)]/kT_e} + 1}, \tag{10.229}$$

where $\mu(T)$ is the chemical potential that is determined from the normalization condition

$$\int_0^{\infty} F_{\mathrm{FD}} \cdot \mathrm{d}\varepsilon = 1. \tag{10.230}$$

At high temperatures, the Fermi–Dirac electron energy distribution function naturally tends to the Maxwell–Boltzmann distribution function ensuring the connection between the solid state and the plasma state. Thus, the notation $1s^2 2s^2 2p^6 (VB)^3$ (see relation (10.223)) corresponding to the cold solid state will designate, in the plasma picture, the configuration $1s^2 2s^2 2p^6$ with three free electrons per atom in the electron energy distribution function.

Particular attention must be paid to the Pauli exclusion principle. For each process of the transition matrix element $W_{k_{Z'}j_Z}$ of (10.222), the vacancy of the free states has to be accounted for. For example, in the cold solid, all the free states are occupied with energy located below the Fermi energy ε_F. Therefore, any transition into these states is forbidden. The solution to the Pauli principle in atomic kinetics is the use Pauli-blocking factors according to

$$P(E, T_e) = \left\{1 - \frac{1}{1 + \exp([E - \mu(T_e)]/kT_e)}\right\}. \tag{10.231}$$

At high temperature, (10.231), it is equal to one (note that the high temperature limit of the chemical potential is given by $\mu(T_e) \approx -kT_e \cdot \ln\left[(m_e kT_e)/\left(2\pi\hbar^2 n_e^{2/3}\right)\right]^{3/2}$) and the model reproduces the classical one but is smaller than one for low temperatures where the occupation of the free states below the Fermi edge decreases. At $kT_e = 0$, there is no space in the distribution below ε_F and thus, the factor is equal to 0 and blocks every rate transferring electrons below ε_F.

Let us illustrate the introduction of this factor for the process of collisional excitation included in (10.222). Its rate coefficient between the state configurations $i_{Z'}$ and j_Z is calculated from

$$C_{i_{Z'}j_Z} = \int\limits_{\Delta E_{i_{Z'}j_Z}}^{\infty} \sigma_{i_{Z'}j_Z}^{ex}(\varepsilon) \cdot \sqrt{\frac{2\varepsilon}{m_e}} \cdot F_{FD}(\varepsilon) \cdot P\left(\varepsilon - \Delta E_{i_{Z'}j_Z}\right) \cdot d\varepsilon, \tag{10.232}$$

where $\sigma_{j_Z i_{Z'}}^{ex}(\varepsilon)$ is the collisional excitation cross section and $\Delta E_{i_{Z'}j_Z}$ is the difference of energy between the two configurations. A free electron with energy ε collides with a bound electron and loses a part of its energy. In this process, the electron moves in the electron energy distribution function from the energy ε to the energy $\varepsilon' = \varepsilon - \Delta E_{i_{Z'}j_Z}$. The Pauli-blocking factor accounts here for the space available in the EEDF at ε'.

Let us investigate the principle of detailed balance if Fermi–Dirac distribution function and Pauli-blocking factor are involved. The number of collisional excitations $N_{i_{Z'}j_Z}^{ex}$ per unit time and unit volume between the states $i_{Z'}$ and j_Z driven by electron collisions with electrons of energy ε in the energy interval $d\varepsilon$ and secondary electrons of energy ε' in the interval $d\varepsilon'$ is given by (n_e is the electron density)

$$N_{i_{Z'}j_Z}^{\mathrm{ex}} = n_\mathrm{e} \cdot n_{i_{Z'}} \cdot \sigma_{i_{Z'}j_Z}^{\mathrm{ex}}(\varepsilon) \cdot \sqrt{\frac{2\varepsilon}{m_\mathrm{e}}} \cdot F_{\mathrm{FD}}(\varepsilon) \cdot P(\varepsilon') \cdot \mathrm{d}\varepsilon. \tag{10.233}$$

In the same manner, the number of collisional de-excitations per unit time from j_Z to $i_{Z'}$ is given by

$$N_{j_Z i_{Z'}}^{\mathrm{de-ex}} = n_\mathrm{e} \cdot n_{j_Z} \cdot \sigma_{j_Z i_{Z'}}^{\mathrm{de-ex}}(\varepsilon') \cdot \sqrt{\frac{2\varepsilon'}{m_\mathrm{e}}} \cdot F_{\mathrm{FD}}(\varepsilon') \cdot P(\varepsilon) \cdot \mathrm{d}\varepsilon'. \tag{10.234}$$

In local thermodynamic equilibrium (LTE), these two quantities (10.233) and (10.234) are equal. Using the relation $\varepsilon = \varepsilon' + \Delta E_{i_{Z'}j_Z}$ with gives $\mathrm{d}\varepsilon = \mathrm{d}\varepsilon'$ we obtain

$$\frac{N_{i_{Z'}}}{N_{j_Z}} \cdot \sigma_{i_{Z'}j_Z}^{\mathrm{ex}}(\varepsilon) \cdot \sqrt{\frac{\varepsilon}{\varepsilon'}} \cdot \frac{F_{\mathrm{FD}}(\varepsilon)}{F_{\mathrm{FD}}(\varepsilon')} \cdot \frac{P(\varepsilon')}{P(\varepsilon)} = \sigma_{j_Z i_{Z'}}^{\mathrm{de-ex}}(\varepsilon'). \tag{10.235}$$

If we substitute the population ratio $N_{i_{Z'}}/N_{j_Z}$ by the Boltzmann population (that holds true in thermodynamic equilibrium), we obtain from (10.235)

$$\sigma_{j_Z i_{Z'}}^{\mathrm{de-ex}}(\varepsilon') = \sigma_{i_{Z'}j_Z}^{\mathrm{ex}}(\varepsilon) \cdot \frac{g_{i_{Z'}}}{g_{j_Z}} \cdot \frac{\varepsilon}{\varepsilon'}. \tag{10.236}$$

Relation (10.236) is equivalent to the Klein–Rosseland formula of (10.113) and does not anymore depend on the electron temperature (see also discussion in Sect. 7.7.2). Therefore, the Fermi–Dirac electron energy distribution function combined with the Pauli-blocking factor from (10.231) is consistent with the principle of microreversibility. In the classical case, i.e., using a Maxwell–Boltzmann electron energy distribution function without Pauli-blocking factors, we obtain exactly the same relation, namely (10.113). In the degenerate case, the Pauli-blocking factors are necessary to assure the principle of microreversibility.

We can also verify that a consistent introduction of the Pauli-blocking factors, for all the other processes of the transition matrix $W_{i_{Z'}j_Z}$ involving the *VB*, allows one to maintain detailed balance from low to high temperatures. In particular, two Pauli-blocking factors have to be invoked for the collisional ionization process as two electrons are ejected. The inverse processes are then calculated using the principle of detailed balance, as in the Maxwell–Boltzmann classical case.

The number of collisional ionization events per second and unit volume $N_{i_{Z'}j_Z}^{\mathrm{iz}}$ from the configurations $i_{Z'}$ to j_Z driven by impacting electrons in the energy interval $\mathrm{d}\varepsilon$ with energy ε and secondary electrons in the interval $\mathrm{d}\varepsilon'$ with energy ε' is given by:

$$N_{i_{Z'}j_Z}^{\mathrm{iz}} = n_\mathrm{e} n_{i_{Z'}} \cdot \frac{\mathrm{d}\sigma_{i_{Z'}j_Z}^{\mathrm{iz}}}{\mathrm{d}\varepsilon'} \cdot \sqrt{\frac{2\varepsilon}{m_\mathrm{e}}} \cdot F_{\mathrm{FD}}(\varepsilon) \cdot P(\varepsilon') \cdot P(\varepsilon'') \cdot \mathrm{d}\varepsilon \cdot \mathrm{d}\varepsilon'. \tag{10.237}$$

$\mathrm{d}\sigma^{iz}_{i_{Z'}j_Z}/\mathrm{d}\varepsilon'$ is the differential collisional ionization cross section. The two Pauli-blocking factors in expression (10.237) account for the available space at the energies of the secondary electrons. In the same manner, the number of three-body recombination events $N^{3b}_{j_Z i_{Z'}}$ from configuration j_Z to $i_{Z'}$ is given by

$$N^{3b}_{j_Z i_{Z'}} = n_e^2 \cdot n_{j_Z} \cdot \frac{\mathrm{d}\sigma^{3b}_{j_Z i_{Z'}}}{\mathrm{d}\varepsilon'} \cdot P(\varepsilon) \cdot \sqrt{\frac{2\varepsilon'}{m_e}} \cdot \sqrt{\frac{2\varepsilon''}{m_e}} \cdot F_{\mathrm{FD}}(\varepsilon') \cdot F_{\mathrm{FD}}(\varepsilon'') \cdot \mathrm{d}\varepsilon' \cdot \mathrm{d}\varepsilon'', \tag{10.238}$$

where $\mathrm{d}\sigma^{3b}_{j_Z i_{Z'}}/\mathrm{d}\varepsilon'$ is the three-body differential cross section, ε' and ε'' the energies of two incoming electrons, and ε the energy of the secondary electron. For a system in thermodynamic equilibrium, the two quantities from (10.237) and (10.238) are equal. Employing the relation of energy conservation $\varepsilon = \Delta E_{i_{Z'}j_Z} + \varepsilon' + \varepsilon''$ (see also (10.117)) where for a fixed value of $\mathrm{d}\varepsilon'$, one has $\mathrm{d}\varepsilon = \mathrm{d}\varepsilon''$, we obtain

$$\frac{1}{n_e} \cdot \frac{N_{i_{Z'}}}{N_{j_Z}} \cdot \frac{F_{\mathrm{FD}}(\varepsilon)}{F_{\mathrm{FD}}(\varepsilon') \cdot F_{\mathrm{FD}}(\varepsilon'')} \cdot \frac{P(\varepsilon') \cdot P(\varepsilon'')}{P(\varepsilon)} \cdot \frac{\mathrm{d}\sigma^{iz}_{i_{Z'}j_Z}}{\mathrm{d}\varepsilon'} \cdot \sqrt{\frac{m_e \cdot \varepsilon}{2\varepsilon' \cdot \varepsilon''}} = \frac{\mathrm{d}\sigma^{3b}_{j_Z i_{Z'}}}{\mathrm{d}\varepsilon'}. \tag{10.239}$$

In thermodynamic equilibrium, the relation between the populations of two configurations of different ionization stages is given by the generalized Saha–Boltzmann law, i.e.,

$$\frac{n_{i_{Z'}}}{n_{j_Z}} = \frac{g_{i_{Z'}}}{g_{j_Z}} \cdot \exp\left[\left(\Delta E_{i_{Z'}j_Z} + \mu(T_e)\right)/kT_e\right]. \tag{10.240}$$

Using (10.231), (10.237)–(10.240), we obtain the following relation between the differential cross sections of ionization and three-body recombination:

$$\frac{g_{i_{Z'}}}{g_{j_Z}} \cdot \frac{\varepsilon}{\varepsilon' \cdot \varepsilon''} \cdot \frac{\pi^2 \hbar^3}{2m_e} \cdot \frac{\mathrm{d}\sigma^{iz}_{i_{Z'}j_Z}}{\mathrm{d}\varepsilon'} = \frac{\mathrm{d}\sigma^{3b}_{j_Z i_{Z'}}}{\mathrm{d}\varepsilon'}. \tag{10.241}$$

This expression is equivalent to the microreversibility relation and therefore provides the proof of a consistent implementation of the Fermi–Dirac distribution function and the Pauli principle. Indeed, the microreversibility relation does not depend on the plasma parameters. One can verify that the same microreversibility relation is obtained with a classical treatment, i.e., a Maxwell–Boltzmann distribution, the classical Saha–Boltzmann law and without Pauli-blocking factors (see also (10.107), (10.116), note that expression (10.116) employs the single electron ionization cross section rather than the differential one as in (10.241)). In the framework of the Fermi–Dirac statistics, it can only be obtained with the inclusion of the Pauli-blocking factors. The total rate for the collisional ionization is therefore given by

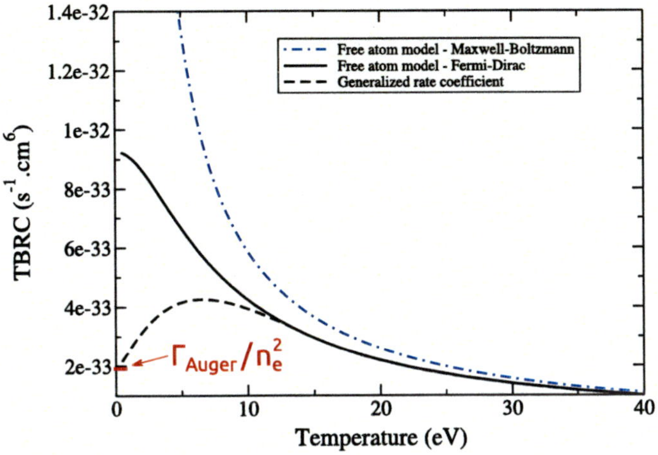

Fig. 10.59 Generalized three-body recombination rate coefficient "TBRC" of Al. The Fermi–Dirac and Maxwell–Boltzmann rates are also displayed

$$I_{i_{Z'}j_Z} = n_e \cdot \int_0^{\infty} \int_{\Delta E_{i_{Z'}j_Z}}^{\infty} \frac{d\sigma^{iz}_{i_{Z'}j_Z}}{d\varepsilon'} \cdot \sqrt{\frac{2\varepsilon}{m_e}} \cdot F_{FD}(\varepsilon) \cdot P(\varepsilon') \cdot P(\varepsilon'') \cdot d\varepsilon' \cdot d\varepsilon, \qquad (10.242)$$

whereas the three-body recombination rate is given by

$$T_{j_Z i_{Z'}} = n_e^2 \cdot \int_0^{\infty} \int_0^{\infty} \frac{d\sigma^{3b}_{j_Z i_{Z'}}}{d\varepsilon'} \cdot \sqrt{\frac{4\varepsilon'\varepsilon''}{m_e^2}} \cdot F_{FD}(\varepsilon') \cdot F_{FD}(\varepsilon'') \cdot P(\varepsilon) \cdot d\varepsilon' \cdot d\varepsilon''. \qquad (10.243)$$

In the classical case (high temperature), the Pauli-blocking factors are $P = 1$ and the Fermi–Dirac distribution function turns into the Maxwell–Boltzmann distribution. In order to establish a link between the cold and heated solid based on relations (10.223)–(10.228), we employ the probability formalism discussed in Sect. 6.3.2. The generalized three-body recombination rate can therefore be written as

$$T^{(G)}_{j_Z i_{Z'}} = (1 - f(T)) \cdot \Gamma_{j_Z i_{Z'}} + f(T) \cdot T_{j_Z i_{Z'}}. \qquad (10.244)$$

$\Gamma_{j_{Z'} i_{Z'}}$ is the Auger rate for the cold solid from state j_Z to state $i_{Z'}$ and $T_{j_Z i_{Z'}}$ is the free atom three-body recombination rate. The driving term for the probability is the "free space" determined from the Fermi–Dirac statistics:

$$f(T) = \int_{\mu(T)}^{\infty} F_{FD}(\varepsilon, T) d\varepsilon. \qquad (10.245)$$

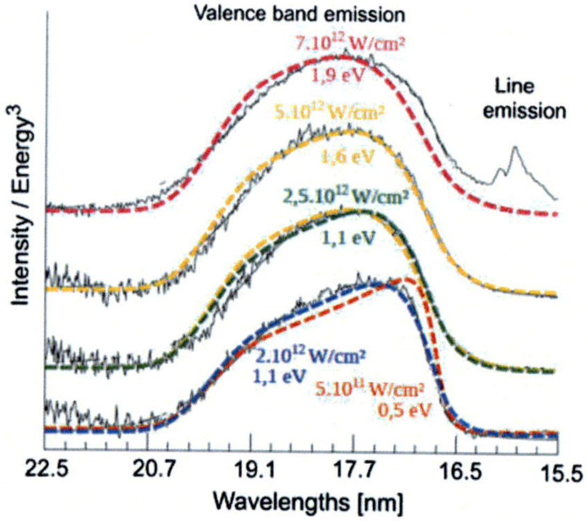

Fig. 10.60 Experimental XUV fluorescence spectra of aluminum irradiated with XUV–FEL photons at 92 eV and pulse duration of 15 fs in dependence of the irradiation intensity. Ab initio simulations with the generalized atomic physics approach that only varies the XUV–FEL intensity results in a good agreement with the data

At zero temperature $f(T = 0) = 0$, the rate (10.244) is equal to $\Gamma_{j_Z i_{Z'}}$ while at high temperatures $f(T)$ approaches 1 and the rate tends to $T_{j_Z i_{Z'}}$ as it should be.

Figure 10.59 shows the generalized three-body recombination rate coefficient from (10.244) using a solid-state Auger rate. For comparison, the three-body coefficient TBRC integrated over a Fermi–Dirac distribution (10.243) is also shown. It can clearly be seen that the probability formalism provides the transition from the cold solid to the heated solid. With this new approach, we generalize the role of the Auger decay presented above: From a state with an inner-shell hole, the generalized three-body recombination recombines an electron and provides a new target for a further photoionization. At high temperature, the collisional ionization can compete with the three-body recombination and take out this electron. The equilibrium between the three-body recombination, the photoionization, and the collisional ionization is very important to properly calculate the energy deposition.

We now consider radiative recombination and apply likewise the probability method for the generalized processes depicted in relation (10.228):

$$R^{(G)}_{j_Z i_{Z'}} = (1 - f(T)) \cdot A_{j_Z i_{Z'}} + f(T) \cdot R_{j_Z i_{Z'}}, \tag{10.246}$$

where $A_{j_Z i_{Z'}}$ is the fluorescence probability in the cold solid from state j_Z to state $i_{Z'}$ and $R_{j_Z i_{Z'}}$ is the radiative recombination rate given by (see also (10.122)):

$$R_{j_Z i_{Z'}} = \int_0^\infty \sigma^r_{j_Z i_{Z'}}(\varepsilon) \cdot \sqrt{\frac{2\varepsilon}{m_e}} \cdot F_{\mathrm{FD}}(\varepsilon) \cdot \mathrm{d}\varepsilon. \tag{10.247}$$

The radiative recombination cross section is related to the photoionization cross section via the Milne relation of (10.120).

Finally, we note that Fermi–Dirac rate coefficients have also been considered by (Aslanyan and Tallents 2015), however, all essential calculations concerning the Fermi–Dirac rates and the Pauli-blocking factors as well as their impact on the XFEL interaction with matter is essentially a repetition of the work of (Deschaud et al. 2014).

10.6.5.5 Fluorescence Emission of Warm Dense Matter

Figure 10.60 shows the simulations (Deschaud et al. 2015) of the XUV fluorescence spectra and the comparison with the measurements. Despite of the simplicity of the generalized approach that employs a simple Fermi–Dirac distribution function (instead of the much more complex molecular dynamics (MD) or density functional theory (DFT) simulations of the solid electronic structure and density), the simulation of the spectral distribution of the fluorescence spectra (using (10.245)) demonstrates very good agreement with the data [Vinko et al. 2010] obtained during the FLASH experimental campaign [Riley et al. 2009]. Note, that the calculations based on the density functional theory DFT [Vinko et al. 2010] have been questioned [Iglesias 2011] and the critics to the DFT calculations of [Vinko et al. 2014] have been renewed by [Rosmej 2018, Karasiev and Hu 2021]).

The simulations include the full time history for a certain irradiation intensity from which then the final electron temperature (see, e.g., Fig. 10.56) has been deduced. Note that in the experiments, the variation of the irradiation intensity has been realized by defocusing the focusing optics and experimental measurements of the irradiation intensity are somewhat uncertain (within a factor of about two). The simulations for different intensities shown in Fig. 10.60 provide a quite good match for the various fluorescence spectra (spectra have different offset for better presentation). For the two lowest intensities (first two spectra from bottom), simulations for different intensities have been presented to demonstrate the sensitivity of the spectral distribution with respect to the data. At the highest intensities shown in Fig. 10.60 (spectrum at the top), line intensity develops due to the strong heating of the solid. The line emission near 16 nm corresponds to those observed for very high irradiation intensities as demonstrated in Fig. (10.57) and is due to 3*s*–2*p* transitions.

Figure 10.60 demonstrates that the electron excitation due to the XUV–FEL irradiation of a solid changes considerably the electron band structure. At temperatures much above 1 eV, line emission of ionized aluminum develops indicating that the heated valence band structure starts to disappear while for temperatures below some eV, a warm dense matter sample that emits fluorescence radiation exists.

References

V.A. Abramov, V.S. Lisitsa, AYu. Pigarov, Changes in effective charge-exchange cross sections in a plasma. JETP Lett. **42**, 356 (1985)

V.A. Aleseyev, I.I. Sobelman, Influence of collisions on stimulated random scattering in gases. JETP **28**, 991 (1969)

M. Alonso, E.J. Finn, *Quantum and Statistical Physics* (Addison-Wesly Publishing Company, 1968). http://books.pakchem.net/fundamental-university-physics—quantum-and-statistical-physics-by-alonso-finn.html

V. Aslanyan, G.J. Tallents, Ionization rate coefficients in warm dense plasmas. Phys. Rev. E **91**, 063106 (2015)

S. Atzeni, Laser driven inertial fusion: the physical basis of current and recently proposed ignition experiments. Plasma Phys. Control. Fusion **51**, 124029 (2009)

A. Bar-Shalom, J. Oreg, W.H. Goldstein, D. Shvarts, A. Zigler, Super-transition-arrays: a model for the spectral analysis of hot, dense plasma. Phys. Rev. A **40**, 3183 (1989)

P. Beiersdorfer, A.L. Osterheld, T.W. Phillips, M. Bitter, K.W. Hill, S. von Goeler, High-resolution measurement, line identification, and spectral modeling of the Kβ spectrum of heliumlike $Ar16^{+}$. Phys. Rev. E **52**, 1980 (1995)

P. Beiersdorfer, M. Bitter, M. Marion, R.E. Olson, Charge-exchange-produced K-shell X-ray emission from Ar^{16+} in a tokamak plasma with neutral-beam injection. Phys. Rev. A **72**, 032725 (2005)

F.N. Beg, A.R. Bell, A.E. Dangor, C.N. Danson, A.P. Fews, M.E. Glinsky, B.A. Hammel, P. Lee P, P.A. Norreys, M. Tatarakis, A study of picosecond laser-solid interactions up to 10^{19} W cm^{-2}. Phys. Plasmas **4**, 447 (1997)

V.A. Bernshtam, Yu. Zarnitsky, Yu. Ralchenko, L.A. Vainshtein, L. Weingarten, Y. Maron, Effect of radiative cascades on intensitites of dielectronic satellites to Heα. Phys. Scr. **79**, 035303 (2009)

R. Betti, C.D. Zhou, K.S. Anderson, L.J. Perkins, W. Theobald, A.A. Solodov, Shock ignition of thermonuclear fuel with high areal density. Phys. Rev. Lett. **98**, 155001 (2007)

V.A. Boiko, V.A. Vinogradov, S.A. Pikuz, IYu. Skobelev, A.Ya. Faenov, X-ray spectroscopy of laser produced plasmas. J. Sov. Laser Res. **6**, 82 (1985)

R.E.H. Clark, J. Abdallah Jr., J.B. Mann, Integral and differential cross section for electron impact ionization. Astrophys. J. **381**, 597 (1991)

K.R. Cornelius, K. Wojtkowski, R.E. Olson, State-selective cross section scalings for electron capture collisions. J. Phys. B **33**, 2017 (2000)

R.B. Cowan, *The Theory of Atomic Structure and Spectra* (University of California Press, 1981)

J.P. Cryan, J.M. Glownia, J. Andreasson, A. Belkacem, N. Berrah, C.I. Blaga, C. Bostedt, J. Bozek, C. Buth, L.F. DiMauro, L. Fang, O. Gessner, M. Guehr, J. Hajdu, M.P. Hertlein, M. Hoener, O. Kornilov, J.P. Marangos, A.M. March, B.K. McFarland, H. Merdji, V.S. Petrovic, C.S. Raman, D. Ray, D. Reis, F. Tarantelli, M. Trigo, J. White, W. White, L. Young, P.H. Bucksbaum, R.N. Coffee, Auger electron angular distribution of double core-hole states in the molecular reference frame. Phys. Rev. Lett. **105**, 083004 (2010)

B. Deschaud, O. Peyrusse, F.B. Rosmej, Generalized atomic physics processes when intense femtosecond XUV- and X-ray radiation is interacting with solids. Europhys. Lett. **108**, 53001 (2014)

B. Deschaud, O. Peyrusse, F.B. Rosmej, Atomic kinetics for isochoric heating of solid aluminum under short intense XUV free electron laser irradiation. HEDP **15**, 22 (2015)

B. Deschaud, O. Peyrusse, F.B. Rosmej, *Simulations of XFEL induced fluorescence spectra of hollow ions and studies of dense plasma effects*, Physics of Plasmas **27**, 063303 (2020)

R.C. Elton, J.A. Cobble, H.R. Griem, D.S. Montgomery, R.C. Mancini, V.L. Jacobs, E. Behar, Anomalous satellite-line intensities form a TRIDENT laser-produced plasma. JQSRT **65**, 185 (2000)

W. Engelhardt, in *Course on diagnostics for fusion reactor conditions*, vol. 1, p. 1, Varenna, Italy, 1982, EUR 8351-IEN

A.Y. Faenov, S.A. Pikuz, A.I. Erko, B.A. Bryunetkin, V.M. Dyakin, G.V. Ivanenkov, A.R. Mingaleev, T.A. Pikuz, V.M. Romanova, T.A. Shelkovenko, High-performance X-ray spectroscopic devices for plasma microsources investigations. Phys. Scr. **50**, 333 (1994)

A.Y. Faenov, A.I. Magunov, T.A. Pikuz, I. Yu Skobelev, S.A. Pikuz, A.M. Urnov, J. Abdallah, R. E.H. Clark, J. Cohen, R.P. Johnson, G.A. Kyrala, M.D. Wilke, A. Maksimchuk, D. Umstadter, N. Nantel, R. Doron, E. Behar, P. Mandelbaum, J.J. Schwob, J. Dubau, F.B. Rosmej, A. Osterheld, High-resolved X-ray spectra of hollow atoms in a femtosecond laser-produced solid plasma. Phys. Scr. **T80**, 536 (1999)

P. Faucher, N. Peyraud-Cuenca, F.B. Rosmej, Effect of a highly energetic electron beam on the electron distribution function in a hot dense plasma. Application to an argon plasma. J. Plasma Physics **63**, 255 (2000)

E. Flügge, *Encyclopedia of Physics*, vol. XXX (X-rays) (Springer, 1957)

T. Fujimoto, *Plasma Spectroscopy* (Clarendon Press, Oxford, 2004)

A.H. Gabriel, Dielectronic satellite spectra for highly-charge helium-like ion lines. Mon. Not. R. Astron. Soc. **160**, 99 (1972)

E. Galtier, F.B. Rosmej, D. Riley et al., Decay of crystalline order and equilibration during solid-to-plasma transition induced by 20-fs microfocused 92 eV free electron laser pulses. Phys. Rev. Lett. **106**, 164801 (2011)

D. Garcia, E. Gerjuoy, J.E. Welker, Classical approximation for ionization by proton impact. Phys. Rev. **165**, 66 (1969) and Erratum: Phys. Rev. A **3**, 2150 (1971)

S.J. Gitomer, R.D. Jones, F. Begay et al., Fast ions and hot-electrons in laser plasma interactions. Phys. Phys. Fluids **29**, 2679 (1986)

S.H. Glenzer, F.B. Rosmej, R.W. Lee, C.A. Back, K.G. Estabrook, B.J. MacCowan, T.D. Shepard, R.E. Turner, Measurements of suprathermal electrons in hohlraum plasmas with X-ray spectroscopy. Phys. Rev. Lett. **81**, 365 (1998). http://journals.aps.org/prl/abstract/10.1103/PhysRevLett.81.365

A.E.S. Green, T. Sawada, Ionisation cross sections and secondary electron distributions. J. Atomos. Terr. Phys. **34**, 1719 (1972)

H.R. Griem, *Plasma Spectroscopy* (McGraw-Hill, New York, 1964)

H.R. Griem, *Spectral Line Broadening by Plasmas* (Academic Press, New York, 1974)

H.R. Griem, *Principles of Plasma Spectroscopy* (Cambridge University Press, New York, 1997)

M. Gryzinski, Classical theory of atomic collisions. I. Theory of inelastic collisions. Phy. Rev. **138**, A336 (1965)

W. Heitler, *The Quantum Theory of Radiation*, 3rd edn. (Oxford University Press, 1954)

K. Huang, *Statistical Mechnanics* (Wiley, New York, 1963)

R.A. Hulse, Numerical studies of impurities in fusion plasmas. Nucl. Technol./Fusion **3**, 259 (1983)

C.A. Iglesias, Comment on "Free-free opacity in warm aluminum". HEDP **7**, 38 (2011)

ITER, https://www.iter.org/construction/construction (2021)

V.J. Jacobs, M. Blaha, Effects of angular-momentum-changing collisions on dielectronic satellite spectra. Phys. Rev. A **21**, 525 (1980)

V.V. Karasiev, S.X. Hu, *Unraveling the intrinsic atomic physics behind x-ray absorption line shifts in warm dense silicon plasmas*, Phys. Rev. E **103**, 033202 (2021)

S. Kienle, F.B. Rosmej, H. Schmidt, Investigation of photoabsorption effects of Li-like satellites in inhomogenous hot dense pinching plasmas. J. Phys. B: At. Mol. Opt. Phys. **28**, 3675 (1995)

A. Kozyreva, M. Basko, F.B. Rosmej, T. Schlegel, A. Tauschwitz, D.H.H. Hoffmann, Dynamic confinement of targets heated quasi-isochorically with heavy ion beam. Phys. Rev. E **68**, 056406 (2003)

W.L. Kruer, *The Physics of Laser Plasma Interactions* (Redwood City, CA, Addison-Wesley, 1988)

H.-J. Kunze, *Introduction to Plasma Spectroscopy* (Springer, Berlin, 2009)

D. Larousserie, L'aluminium devient transparent aux rayons X. Science et Avenir **09**, 20 (2009)

J.D. Lindl, Development of the indirect drive approach to inertial confinement fusion and the target physics basis for ignition and gain. Phys. Plasmas **2**, 3933 (1995)

J.D. Lindl, P. Amendt, R.L. Berger, S.G. Glendinning, S.H. Glenzer, S.W. Haan, R.L. Kauffman, O.L. Landen, L.J. Suter, The physics basis for ignition using indirect-drive targets on the national ignition facility. Phys. Plasmas **11**, 339 (2004)

J. Lindl, O. Landen, J. Edwards, E. Moses, N.I.C. Team, Review of the National ignition campaign 2009-2012. Phys. Plasmas **21**, 020501 (2014)

V.S. Lisitsa, *Atoms in Plasmas* (Springer, 1994)

W. Lochte-Holtgreven, *Plasma Diagnostics* (North-Holland Publishing Company, Amsterdam, 1968)

R. Loudon, *The Quantum Theory of Light* (Oxford University Press, New York, 2000). ISBN 0-19-8501767-3

R.W.P. McWhirter, Spectral intensities, in *Plasma Diagnostic Techniques*, ed. by R.H. Huddelstone, S.L. Leonard (Academic Press, New York, 1965)

C. de Michelis, M. Mattioli, Soft X-ray spectroscopic diagnostics of laboratory plasmas. Nucl. Fusion **21**, 677 (1981)

D. Mihalas, *Stellar Atmospheres*, 2nd edn. (W.H. Freeman, San Francisco, 1978)

H. Mimura, H. Yumoto, S. Matsuyama, T. Koyama, K. Tono, Y. Inubushi, T. Togashi, T. Sato, J. Kim, R. Fukui, Y. Sano, M. Yabashi, H. Ohashi, T. Ishikawa, K. Yamauchi, Generation of 10^{20} W/cm^{-2} hard X-ray laser pulses with two-stage reflective focusing system. Nat. Commun. **5**, 3539 (2014)

A. Moinard, F.B. Rosmej, O. Renner, E. Galtier, H. J. Lee, B. Nageller, P. A. Heimann, W. Schlotter, J. J. Turner, R.W. Lee, M. Makita, D. Riley, J. Seely, XFEL resonant photo-pumping of dense plasmas and dynamic evolution of autoionizing core hole states, in *12th International Conference on Fusion Science and Applications—IFSA*, Nara, Japan (2013)

P. Monot, P. D' Oliveira, S. Hulin, A. Ya. Faenov, S. Dobosz, T. Auguste, T. A. Pikuz, A. I. Magunov, I. Yu. Skobelev, F. B. Rosmej, N. E. Andreev, E. Lefebvre, Study of the interaction of a 10 TW femtosecond laser with a high-density long-scale pulsed gas jet. Phys. Plasmas **8**, 3766 (2001)

B. Nagler, U. Zastrau, R. Fäustlin, S.M. Vinko, T. Whitcher, A.J. Nelson, R. Sobierajski, J. Krzywinski, J. Chalupsky, E. Abreu, S. Bajt, T. Bornath, T. Burian, H. Chapman, J. Cihelka, T. Döppner, S. Düsterer, T. Dzelzainis, M. Fajardo, E. Förster, C. Fortmann, E. Galtier, S.H. Glenzer, S. Göde, G. Gregori, V. Hajkova, P. Heimann, L. Juha, M. Jurek, F.Y. Khattak, A.R. Khorsand, D. Klinger, M. Kozlova, T. Laarmann, H.J. Lee, R.W. Lee, K.-H. Meiwes-Broer, P. Mercere, W.J. Murphy, A. Przystawik, R. Redmer, H. Reinholz, D. Riley, G. Röpke, F.B. Rosmej, K. Saksl, R. Schott, R. Thiele, J. Tiggesbäumker, S. Toleikis, T. Tschentscher, I. Uschmann, H.J. Vollmer, J. Wark, Transparency induced in solid density aluminum by ultra-intense XUV radiation. Nat. Phys. **5**, 693 (2009)

Y. Nakai, T. Shirai, T. Tabata, R. Ito, A semiempirical formula for single-electron-capture cross sections of multiply charged ions colliding with H, H_2, He. Phys. Scr. **T28**, 77 (1989)

V.N. Ostrovsky, Rydberg atom-ion collisions: classical overbarrier model for charge exchange. J. Phys. B **28**, 3901 (1995)

J. Ongena, A.M. Messiaen, M. Tokar, U. Samm, B. Unterberg, N. Schoon, P. Dumortier, H.G. Esser, F. Durodie, H. Euringer, G. Fuchs, E. Hintz, F. Hoenen, R. Koch, L. Könen, A. Krämer-Flecken, A. Pospieszczyk, B. Schweer, H. Soltwisch, G. Telesca, P.E. Vandenplas, R. Van Nieuwenhove, G. Van Ost, G. Van Wassenhove, R.R Weynants, G. Waidmann, J. Winter, G.H. Wolf, Results and modelling of high power edge radiation cooling in Textor. Phys. Scr. **52**, 449 (1995)

D. Pasini, M. Mattioli, A.W. Edwards, R. Gianella, R.D. Gill, N.C. Hawkes, G. Magyar, B. Saoutic, Z. Wang, D. Zasche, Impurity transport in JET using laser injected impurities in ohmic and radiofrequency heated plasmas. Nucl. Fusion **30**, 2049 (1990)

F. Petitdemange, F.B. Rosmej, Dielectronic satellites and Auger electron heating: irradiation of solids by intense XUV-free electron laser radiation, in *New Trends in Atomic & Molecular*

Physics - Advanced Technological Applications, vol. 76, ed. by M. Mohan (Springer, 2013), p. 91–114. ISBN 978-3-642-38166-9. http://www.springer.com/de/book/9783642381669
E.R. Pike, S. Sarkar, *The Quantum Theory of Radiation* (Oxford Science Publications, Clarendon Press, Oxford, 1995)
S.G. Rautian, A.M. Shalagin, *Kinetic Problems of Non-linear Spectroscopy* (North-Holland, Amsterdam, 1991)
J. Rapp, M.Z. Tokar, L. Könen, H.R. Koslowski, G. Bertschinger, M. Brix, H. Claassen, R. Jaspers, A. Krämer-Flecken, K. Ohya, V. Philipps, A. Pospieszczyk, U. Samm, T. Tanabe, G. Telesca, B. Unterberg, G. Van Ost, Transport studies of high-Z elements in neon edge radiation cooled discharges in TEXTOR-94. Plasma Phys. Controlled Fusion **39**, 1615 (1997)
F. Reif, *Fundamentals of Statistical and Thermal Physics* (McGraw Hill, 1965). https://ourphysics.org/wiki/index.php/File:Fundamentals_of_Statistical_And_Thermal_Physics-F_Reif.pdf
O. Renner, E. Krouský, F.B. Rosmej, P. Sondhauss, M.P. Kalachnikov, P.V. Nickles, I. Uschmann, E. Förster, Observation of H-like Al Lyα disappearance in dense cold laser produced plasmas. Appl. Phys. Lett. **79**, 177 (2001)
O. Renner, T. Missalla, P. Sondhauss, E. Krousky, E. Förster, C. Chenais-Popovics, O. Rancu, High luminosity, high resolution X-ray spectroscopy of laser produced plasma by vertical geometry Johann spectrometer. Rev. Sci. Instrum. **68**, 2393 (1997)
O. Renner, F.B. Rosmej, Challenges of X-ray spectroscopy in investigations of matter under extreme conditions. Matter Radiat. Extremes, Review **4**, 024201 (2019)
J.E. Rice, F.B. Rosmej, N. Cao, M. Chilenski, N.T. Howard, A.E. Hubbard, J.W. Hughes, J.H. Irby, Y. Lin, P. Rodriguez-Fernandez, S.M. Wolfe, S.J. Wukitch, M. Bitter, L. Delgado-Aparicio, K. Hill, M.L. Reinke, X-ray observations of K-beta emission from medium Z He-like Ions in C-Mod Tokamak plasmas. J. Phys. B. **51**, 035702 (2018)
D. Riley, F.B. Rosmej, R.W. Lee, M. Farjado, J. Wark, L. Juha, S. Toleikis, *K-shell X-ray spectroscopy of FEL irradiated solids*, approved beam time proposal at FLASH at BL3 n° F-20080259, experiment performed in 2009
F.B. Rosmej, Diagnostic properties of Be-like and Li-like satellites in dense transient plasmas under the action of highly energetic electrons. JQSRT **51**, 319 (1994)
F.B. Rosmej, Spectra simulations of Be-like $1s2x2ynz$-$1s^22xnz$ satellite transitions of highly ionized ions. Nuc. Instr. Meth. Phys. Res. B **98**, 33 (1995a)
F.B. Rosmej, A spectroscopic method for the determination of the bulk-electron temperature in highly ionized plasmas containing Non-Maxwellian electrons. J. Phys. B Lett.: At. Mol. Opt. Phys. **28**, L747 (1995b)
F.B. Rosmej, Hot electron X-ray diagnostics. J. Phys. B. Lett.: At. Mol. Opt. Phys. **30**, L819 (1997). http://iopscience.iop.org/0953-4075/30/22/007/pdf/b72217.pdf
F.B. Rosmej, K_β-line emission in fusion plasmas. Rapid Commun. to Phys. Rev. E **58**, R32 (1998)
F.B. Rosmej, A new type of analytical model for complex radiation emission of hollow ions in fusion and laser produced plasmas. Europhys. Lett. **55**, 472 (2001)
F.B. Rosmej, An alternative method to determine atomic radiation. Europhys. Lett. **76**, 1081 (2006)
F.B. Rosmej, Hollow ion emission, in *The European X-Ray Free-Electron Laser Technical Design Report*, Chap. 6.4.2 (DESY 2007), pp. 290–291. ISBN: 978-3-935702-17-1. http://xfel.desy.de/technical_information/tdr/tdr/
F.B. Rosmej, X-ray emission spectroscopy and diagnostics of non-equilibrium fusion and laser produced plasmas, in *Highly Charged Ion Spectroscopic Research*, ed by Y. Zou, R. Hutton (Taylor and Francis, 2012a), pp. 267–341. ISBN: 9781420079043. http://www.crcnetbase.com/isbn/9781420079050
F.B. Rosmej, Exotic states of high density matter driven by intense XUV/X-ray free electron lasers, in *Free Electron Laser*, ed. by S. Varró (InTech, 2012b). ISBN: 978-953-51-0279-3. The download from the webside is free of charge: https://www.intechopen.com/books/free-electron-lasers/exotic-states-of-high-density-matter-driven-by-intense-xuv-x-ray-free-electron-lasers

F.B. Rosmej, *Ionization potential depression in an atomic-solid-plasma picture*, J. Phys. B. Lett.: At. Mol. Opt. Phys. **51**, 09LT01 (2018)
F.B. Rosmej, O.N. Rosmej, S.A. Komarov, V.O. Mishensky, J.G. Utjugov, Soft X-ray spectra analysis in a high current Z-Pinch. AIP Conf. Proc. **299**, 552 (1993)
F.B. Rosmej, A.Y. Faenov, T.A. Pikuz, I.Yu. Skobelev, A.E. Stepanov, A.N. Starostin, B. S. Rerich, B.A. Makhrov, F. Flora, S. Bollanti, P. Di Lazzaro, T. Letardi, K. Wigli-Papadaki, A. Nottola, A. Grilli, L. Palladino, A. Reale, A. Scafati, L. Reale, Dominant role of dielectronic satellites in the radiation spectra of a laser plasma near the target surface. JETP Lett. **65**, 708 (1997)
F.B. Rosmej, J. Abdallah Jr, Blue satellite structure near He_{α} and He_{β} and redistribution of level populations. Phys. Lett. A **245**, 548 (1998)
F.B. Rosmej, V.S. Lisitsa, A self-consistent method for the determination of neutral density from X-ray impurity spectra. Phys. Lett. A **244**, 401 (1998)
F.B. Rosmej, A.Y. Faenov, T.A. Pikuz, F. Flora, P. Di Lazzaro, S. Bollanti, N. Lizi, T. Letardi, A. Reale, L. Palladino, O. Batani, S. Bossi, A. Bornardinello, A. Scafati, L. Reale, Line formation of high intensity He_{β}-Rydberg dielectronic satellites 1s3lnl' in laser produced plasmas. J. Phys. B Lett.: At. Mol. Opt. Phys. **31**, L921 (1998)
F.B. Rosmej, V.S. Lisitsa, D. Reiter, M. Bitter, O. Herzog, G. Bertschinger, H.-J. Kunze, Influence of charge exchange processes on X-ray spectra in tokamak plasmas: experimental and theoretical investigation. Plasma Phys. Control. Fusion **41**, 191 (1999a)
F.B. Rosmej, A.Ya. Faenov, T.A. Pikuz, A.I. Magunov, I.Yu. Skobelev, T. Auguste, P.D'Oliveira, S. Hulin, P. Monot, N.E. Andreev, M.V. Chegotov, M.E. Veisman, Charge exchange induced formation of hollow atoms in high intensity laser produced plasmas. J. Phys. B. Lett. : At. Mol. Opt. Phys. **32**, L107 (1999b)
F.B. Rosmej, D.H.H. Hoffmann, W. Süß, M. Geißel, P. Pirzadeh, M. Roth, W. Seelig, A.Ya. Faenov, I.Yu. Skobelev, A.I. Magunov, T.A. Pikuz, R. Bock, U.N. Funk, U. Neuner, S. Udrea, A. Tauschwitz, N.A. Tahir, B.Yu. Sharkov, N.E. Andreev, Observation of MeV-ions in long pulse large-scale laser produced plasmas. JETP Lett. **70**, 270 (1999c)
F.B. Rosmej, U.N. Funk, M. Geißel, D.H.H. Hoffmann, A. Tauschwitz, A.Ya. Faenov, T.A. Pikuz, I.Y. Skobelev, F. Flora, S. Bollanti, P.D. Lazzaro, T. Letardi, A. Grilli, L. Palladino, A. Reale, A. Scafati, L. Reale, T. Auguste, P. D'Oliveira, S. Hulin, P. Monot, A. Maksimchuk, S.A. Pikuz, D. Umstadter, M. Nantel, R. Bock, M. Dornik, M. Stetter, S. Stöwe, V. Yakushev, M. Kulisch, N. Shilkin, X-ray radiation from ions with K-shell vacancies. JQSRT **65**, 477 (2000)
F.B. Rosmej, D.H.H. Hoffmann, M. Geißel, M. Roth, P. Pirzadeh, A.Ya. Faenov, T.A. Pikuz, I.Yu. Skobelev, A.I. Magunov, Space resolved observation of Si Li-like high energy Rydberg-transitions from autoionizing levels in dense laser produced plasmas. Phys. Rev. A **63**, 063409 (2001a)
F.B. Rosmej, D.H.H. Hoffmann, W. Süß, M. Geißel, O.N. Rosmej, A.Ya. Faenov, T.A. Pikuz, T. Auguste, P. D'Oliveira, S. Hulin, P. Monot, J.E. Hansen, G. Verbookhaven, High resolution X-ray imaging spectroscopy diagnostic of hollow ions in dense plasmas. Nuc. Instrum. Methods A **464**, 257 (2001b)
F.B. Rosmej, D.H.H. Hoffmann, W. Süß, M. Geißel, A.Ya. Faenov, T.A. Pikuz, Direct observation of forbidden X-ray transitions from autoionizing levels in dense laser produced plasmas. Phys. Rev. A **63**, 032716 (2001c)
F.B. Rosmej, H.R. Griem, R.C. Elton, V.L. Jacobs, J.A. Cobble, A.Y. Faenov, T.A. Pikuz, M. Geißel, D.H.H. Hoffmann, W. Süß, D.B. Uskov, V.P. Shevelko, R.C. Mancini, Investigation of charge exchange induced formation of two electron satellite transitions in dense laser produced plasmas. Phys. Rev. E **66**, 056402 (2002a)
F.B. Rosmej, D.H.H. Hoffmann, W. Süß, A.E. Stepanov, Y.A. Satov, Y.B. Smakovskii, V.K. Roerich, S.V. Khomenko, K.N. Makarov, A.N. Starostin, A.Y. Faenov, I.Y. Skobelev, A.I. Magunov, M. Geissel, P. Pirzadeh, W. Seelig, T.A. Pikuz, R. Bock, T. Letardi, F. Flora, S. Bollanti, P.D. Lazarro, A. Reale, A. Scafati, G. Tomassetti, T. Auguste, P. D'Oliveira, S.Hulin, P. Monot, B.Y. Sharkov: Study of electron and ion energy distribution in nanosecond

laser-produced plasma by combination of X-ray and time-of-flight diagnostics. JETP **94**, 60 (2002b)

F.B. Rosmej, A. Calisti, R. Stamm, B. Talin, C. Mossé, S. Ferri, M. Geißel, D.H.H. Hoffmann, A. Ya. Faenov, T.A. Pikuz, Strongly coupled laser produced plasmas: investigation of hollow ion formation and line shape analysis. JQSRT **81**, 395 (2003)

F.B. Rosmej, R. Stamm, V.S. Lisitsa, Convergent coupling of Helium to the H/D background in magnetically confined plasmas. Europhys. Lett. **73**, 342 (2006a)

F.B. Rosmej, R. Stamm, V.S. Lisitsa, Convergent coupling of helium to the H/D background in magnetically confined plasmas, in *Symposium on the 18th International Conference on Spectral Line Shapes ICSLS*, Auburn, USA, AIP **874**, 276 (2006b)

F.B. Rosmej, V.S. Lisitsa, R. Schott, E. Dalimier, D. Riley, A. Delserieys, O. Renner, E. Krousky, Charge exchange driven X-ray emission from highly ionized plasma jets. Europhys. Lett. **76**, 815 (2006c)

F.B. Rosmej, R.W. Lee, Hollow ion emission driven by pulsed X-ray radiation fields. Europhys. Lett. **77**, 24001 (2007)

F.B. Rosmej, R.W. Lee, D.H.G. Schneider, Fast X-ray emission switches driven by intense X-ray free electron laser radiation. High Energy Density Phys. **3**, 218 (2007)

F.B. Rosmej, R. Schott, E. Galtier, P. Angelo, O. Renner, F.Y. Khattak, V.S. Lisitsa, D. Riley, Lyman-β satellite emission in dense non-Maxwellian laser produced plasmas. High Energy Density Phys. **5**, 191 (2009)

F.B. Rosmej, V.S. Lisitsa, Non-equilibrium radiative properties in fluctuating plasmas. Plasma Phys. Rep. **37**, 521 (2011)

F.B. Rosmej, R. Dachicourt, B. Deschaud, D. Khaghani, M. Dozières, M. Smîd, O. Renner, Exotic X-ray emission from dense plasmas. J. Phys. B: Rev. Spec. Top. **48**, 224005 (2015). Available online: http://iopscience.iop.org/article/10.1088/0953-4075/48/22/224005

F.B. Rosmej, A. Moinard, O. Renner, E. Galtier, H.J. Lee, B. Nagler, P.A. Heimann, W. Schlotter, J.J. Turner, R.W. Lee, M. Makita, D. Riley, J. Seely, XFEL resonant photo-pumping of dense plasmas and dynamic evolution of autoionizing core hole states. J. Phys: Conf. Ser. **688**, 012093 (2016)

F.B. Rosmej (PI), S. Glenzer, F. Condamine, D. Khaghani, E. Galtier, O. Renner, in *Solving solar opacity problems*, approved beam time proposal at LCLS (MEC), LR21, experiment performed in 2018

F.B. Rosmej, V.A. Astapenko, V.S. Lisitsa, L.A. Vainshtein, *Dielectronic recombination in non-LTE plasmas*, Review on Atomic and Molecular Physics for Controlled Fusion and Astrophysics, Matter and Radiation at Extremes (Review) **5**, 064201 (2020a)

F.B. Rosmej, L.A. Vainshtein, V.A. Astapenko, V.S. Lisitsa, *Statistical and quantum photoionization cross sections in plasmas: analytical approaches for any configurations including inner shells*, Matter and Radiation at Extremes (Review) **5**, 064202 (2020b)

F.B. Rosmej, V.A. Astapenko, E. Khramov, *XFEL and HHG interaction with matter: effects of ultrashort pulses and random spikes*, Letter to Matter and Radiation at Extremes **6**, 034001 (2021).

F.B. Rosmej, O. Renner, E. Galtier, H.J. Lee, B. Nagler, W.F. Schlotter, M.P. Minitti, J.J. Turner, R.W. Lee, S.H. Glenzer, J.F. Seely, X-ray pumping of dense transient matter. Submitted (2022)

J. Seely, F.B. Rosmej, R. Shepherd, D. Riley, R.W. Lee, in *Proposal to Perform the 1st High Energy Density Plasma Spectroscopic Pump/Probe Experiment*, approved beam time proposal at LCLS (SXR), L332, performed in (2011)

V.A. Shurygin, Kinetics of impurity charge-state distributions in tokamak plasmas. Plasma Phys. Rep. **30**, 443 (2004)

V.A. Shurygin, Analytical impurity transport model: Coupling between particle and charge state transports in tokamak plasmas. Phys. Plasmas **15**, 012506 (2008)

I.Yu. Skobelev, A.Ya. Faenov, B.A. Bryunetkin, V.M. Dyakin, Investigating the emission properties of plasma structures with x-ray imaging spectroscopy. JETP **81**, 692 (1995)

I.Yu. Skobelev, A.Ya. Faenov, T.A. Pikuz, A.I. Magunov, F. Flora, S. Bollanti, P. DiLazzaro, D. Murra, A. Reale, L. Reale, G. Tomassetti, A. Ritucci, G. Petrocelli, S. Martellucci, N. Lisi, F.B.

Rosmej, Spectral Transitions from the Rydberg autoionization states of a Li-like Mg X ion. JETP **95**, 421 (2002)

M. Smid, O. Renner, A. Colaitis, V.T. Tikhonchuk, T. Schlegel, F.B. Rosmej, *Characterization of suprathermal electrons inside a laser accelerated near solid-density-matter via highly-resolved K-alpha emission*. Nature Communications **10**, 4212 (2019)

I.I. Sobelman, L.A. Vainshtein, *Excitation of Atomic Spectra* (Alpha Science, 2006). ISBN 978-1842652336

M. Tabak, D.S. Clark, S.P. Hatchett, M.H. Key, B.F. Lasinski, R.A. Snavely, S.C. Wilks, R.P. J. Town, Review of progress in fast ignition. Phys. Plasmas **12**, 057305 (2005)

B. Talin, A. Calisti, L. Godbert, R. Stamm, R.W. Lee, L. Klein, Frequency-fluctuation model for line-shape calculations in plasma spectroscopy. Phys. Rev. **51**, 1918 (1995)

B. Talin, A. Calisti, S. Ferri, M. Koubiti, T. Meftah, C. Mossé, L. Mouret, R. Stamm, S. Alexiou, R.W. Lee, L. Klein, Ground work supporting the codes based upon the frequency fluctuation model. JQSRT **58**, 953 (1997)

A. Tauschwitz, J.A. Maruhn, D. Riley, G. Shabbir Naz, F.B. Rosmej, S. Borneis, A. Tauschwitz, Quasi-isochoric ion beam heating using dynamic confinement in spherical geometry for X-ray scattering experiments in WDM regime. High Energy Density Phys. **3**, 371 (2007)

W. Theobald, R. Nora, W. Seka, M. Lafon, K.S. Anderson, M. Hohenberger, F.J. Marshall, D.T. Michel, A.A. Solodov, C. Stöckl, D.H. Edgell, B. Yaakovi, A. Casner, C. Reverdin, X. Ribeyre, A. Shvydky, A. Vallet, J. Peebles, F.N. Beeg, M.S. Wei, R. Betti, Spherical strong-shock generation for shock-ignition inertial fusion. Plasma Phys. **22**, 056310 (2015)

M.Z. Tokar, Non-linear phenomena in textor plasmas caused by impurity radiation. Phys. Scr. **51**, 665 (1995)

L.A. Vainshtein, V.P. Shevelko, The structure and characteristics of ions in hot plasmas. Physika i Technika Spektroscopii (1986) (in Russian)

S.M. Vinko, G. Gregori, B. Nagler, T.J. Whitcher, J.S. Wark, U. Zastrau, E. Förster, S. Mazevet, J. Andreasson, S. Bajt, R.R. Fäustlin, S. Toleikis, T. Tschentscher, J. Chalupsky, J. Cihelka, V. Hajkova, L. Juha, H. Chapman, T. Dzelzainis, D. Riley, E. Galtier, F.B. Rosmej, P.A. Heimann, M. Jurek, J. Krzywinski, R.W. Lee, A.J. Nelson, R. Sobierajski, Electronic structure of an XUV photo-generated solid-density aluminum plasma. Phys. Rev. Lett. **104**, 225001 (2010)

S.M. Vinko, O. Ciricosta, J.S. Wark, *Density functional theory calculations of continuum lowering in strongly coupled plasmas*, Nature Communications **5**, 3533 (2014)

V.A. Vinogradov, I.Yu. Skobelev, E.A. Yukov, Effect of collisions on the intensities of the dielectronic satellites of resonance lines of hydrogenlike ions. Sov. Phys. JETP **45**, 925 (1977)

J. Wesson, *Tokamaks*, 4th edn. (Oxford University Press, 2004). ISBN 978-0-19-959223-4

Wiki, https://en.wikipedia.org/wiki/Star_Trek, https://en.wikipedia.org/wiki/Star_Trek:_The_Original_Series (2021)

L.A. Woltz, V.L. Jacobs, C.F. Hooper, R.C. Mancini, Effects of electric microfields on argon dielectronic satellite spectra in laser-produced plasmas. Phys. Rev. A **44**, 1281 (1991)

N. Yamamoto, T. Kato, F.B. Rosmej, X ray spectral diagnostics for satellite lines of H-like Mg ions measured by high resolution spectrometer. JQSRT **96**, 343 (2005)

Annexes

A.1 Summary of Simple General Formulae of Some Elementary Atomic Physics Processes

The hydrogenic approximations combined with the concept of an optical electron which moves into an effective Coulomb potential with effective charge Z_{eff} as well as the classical model allow reasonable estimates for quite a large number of important elementary processes. Below, there are given useful analytical expressions to estimate various rate coefficients (averaged over a Maxwellian energy distribution function) and cross sections.

Electron temperatures and energies are in units of [eV], $Ry = 13.6$ eV, n and m are principal quantum numbers unless otherwise stated.

A.1.1 Transition Energies and Radiative Decay Rates

Transition energies (n and m are principal quantum numbers of the lower and upper states, respectively):

$$\Delta E = Z_{\text{eff}}^2 Ry\left(\frac{1}{n^2} - \frac{1}{m^2}\right).$$

Effective charge (P_{i} is the number of electrons in the atomic shell with principal quantum number i):

$$Z_{\text{eff}}(n) = Z_{\text{nuc}} - \sum_{i=1}^{Z_{\text{nuc}}-1} \sigma_{\text{n,i}} P_{\text{i}}.$$

Screening constants:

F. B. Rosmej et al., *Plasma Atomic Physics*, Springer Series on Atomic, Optical, and Plasma Physics 104, https://doi.org/10.1007/978-3-030-05968-2

	$n = 1$	$n = 2$	$n = 3$	$n = 4$	$n = 5$	$n = 6$	$n = 7$
$i = 1$	0.3125	0.9380	0.9840	0.9954	0.9970	0.9980	0.9990
$i = 2$	0.2345	0.6038	0.9040	0.9722	0.9979	0.9880	0.9900
$i = 3$	0.1093	0.4018	0.6800	0.9155	0.9796	0.9820	0.9860
$i = 4$	0.0622	0.2430	0.5150	0.7100	0.9200	0.9600	0.9750
$i = 5$	0.0399	0.1597	0.3527	0.5888	0.7320	0.8300	0.9000
$i = 6$	0.0277	0.1098	0.2455	0.4267	0.5764	0.7248	0.8300
$i = 7$	0.0204	0.0808	0.1811	0.3184	0.4592	0.6098	0.7374

Rate of spontaneous radiative transitions:

$$A_{\mathrm{mn}} = A(m \rightarrow n) \approx \frac{1.57 \times 10^{10} Z_{\mathrm{eff}}^4}{nm^3(m^2 - n^2)} \quad [\mathrm{s}^{-1}], \quad \mathrm{m} > \mathrm{n}.$$

Oscillator strength:

$$f_{\mathrm{nm}} \approx \frac{1.96}{n^5 m^3 \left(\frac{1}{n^2} - \frac{1}{m^2}\right)^3}, \quad \mathrm{m} > \mathrm{n}.$$

A.1.2 Electron Collisional Excitation and De-excitation

If $n < m$, the electron collisional excitation rate coefficient can be approximated by

$$C_{\mathrm{ex}}(n \rightarrow m) = 3.15 \times 10^{-7} f_{\mathrm{nm}} \left(\frac{Ry}{\Delta E}\right)^{3/2} \sqrt{\beta}\, e^{-\beta} p(\beta) \quad [\mathrm{cm}^3\mathrm{s}^{-1}].$$

For $Z_{\mathrm{eff}} > 0$:

$$p(\beta) = 0.2757 e^{-1.3\beta}\left(\beta - \frac{\beta^2}{4} - \ln(\beta) - 0.5772\right) + 0.2\left(1 - e^{-4.5\beta}\right).$$

For $Z_{\mathrm{eff}} = 0$:

$$p(\beta) = \left\{ \begin{array}{ll} \mathit{if}\ \beta \leq 0.4: & 0.27566 \cdot \left(\beta - \frac{\beta^2}{4} + \frac{\beta^3}{12} - \ln(\beta) - 0.57722\right) \\ \mathit{else} & 0.066 \frac{\sqrt{\beta + 2}}{\beta + 0.127} \end{array} \right\},$$

$$\Delta E = E_{\mathrm{m}} - E_{\mathrm{n}},$$

$$\beta = \frac{\Delta E}{kT_{\mathrm{e}}}.$$

The de-excitation rate coefficient is given by

$$C_{\mathrm{dex}}(m \to n) = C_{\mathrm{ex}}(n \to m)\frac{g_{\mathrm{n}}}{g_{\mathrm{m}}}e^{\beta}.$$

We remember the relation between the oscillator strength and radiative decay rate:

$$A_{\mathrm{ji}}(s^{-1}) = \frac{0.667027}{\lambda^2(\mathrm{cm})}\frac{g_{\mathrm{i}}}{g_{\mathrm{j}}}f_{\mathrm{ij}}.$$

A.1.3 Electron Collisional Ionization and Three-Body Recombination

If P_{n} is the number of electrons in the quantum shell, the ionization rate coefficient can be approximated by

$$I_{\mathrm{n}} = 6 \times 10^{-8} P_{\mathrm{n}}\left(\frac{Ry}{E_{\mathrm{n}}}\right)^{3/2} \sqrt{\beta_{\mathrm{n}}}e^{-\beta_{\mathrm{n}}}\alpha(\beta_{\mathrm{n}}) \quad [\mathrm{cm}^3\mathrm{s}^{-1}],$$

$$\alpha(\beta) = -e^{\beta}\, Ei\,(-\beta) \approx \ln\left[1 + \frac{0.562 + 1.4\beta}{\beta(1 + 1.4\beta)}\right],$$

Ei(x) is the exponential integral:

$$Ei(x) = -\int_{-x}^{\infty}\frac{e^{-t}}{t}dt, \quad x > 0,$$

$$\beta_{\mathrm{n}} = \frac{E_{\mathrm{n}}}{kT_{\mathrm{e}}}.$$

The three-body recombination rate coefficient can be estimated according to

$$T_{\mathrm{ji}} = 2 \times 10^{-31}\left(\frac{Ry}{E_{\mathrm{n}}}\right)^3 \frac{P_{\mathrm{n}}g_{\mathrm{n}}}{g_{\mathrm{j}}}\beta_{\mathrm{n}}^2\,\alpha(\beta_{\mathrm{n}}) \quad [\mathrm{cm}^6\mathrm{s}^{-1}].$$

A.1.4 Radiative Recombination

If Q_{n} is an angular factor which takes into account the Pauli principle the recombination rate is given by

$$R(n) = 5.2 \times 10^{-14}\, Q_{\mathrm{n}}\, Z_{\mathrm{eff}}\, \beta_{\mathrm{n}}^{3/2}\, \gamma(\beta_{\mathrm{n}}) \quad [\mathrm{cm^3 s^{-1}}],$$

$$\gamma(\beta) = -e^{\beta} Ei(-\beta) \approx \ln\left[1 + \frac{0.562 + 1.4\beta}{\beta(1 + 1.4\beta)}\right],$$

$$\beta_{\mathrm{n}} = \frac{E_{\mathrm{n}}}{kT_{\mathrm{e}}},$$

$$Q_{\mathrm{n}} \approx 1 - \frac{N}{2n^2}.$$

Also the sum (for $n \geq n_1$) of the radiative recombination rates can be expressed in analytical form:

$$R^{\mathrm{tot}}(n \geq n_1) = 2.6 \times 10^{-14}\, Z_{\mathrm{eff}} n_1\, \beta_{\mathrm{n}}^{1/2}$$
$$\times \{\ln(1.78\beta_1) + g(\beta_1)(1 + \beta_1/n_1)\}[\mathrm{cm^3 s^{-1}}],$$

$$\beta_1 = \frac{Z_{\mathrm{eff}}^2 Ry}{n_1^2 kT_{\mathrm{e}}},$$

$$g(\beta) = -e^{\beta} Ei(-\beta),$$

$$g(\beta) \approx \ln\left[1 + \frac{0.562 + 1.4\beta}{\beta(1 + 1.4\beta)}\right].$$

n_1 is the principal quantum number from which the sum is taken over all higher-lying excited states. Note that the summation formula should *not* be applied to ground states as it does not include the Pauli principle.

A.1.5 Dielectronic Recombination

The effective dielectronic recombination rate coefficient can be calculated with the help of the Burgess formula:

$$D_{Z+1,Z}(\alpha_0\alpha) = 4.8 \times 10^{-11} f_{\alpha_0\alpha} B_{\mathrm{d}}\, \beta^{3/2}\, e^{-\beta\chi_{\mathrm{d}}} \quad [\mathrm{cm^3 s^{-1}}],$$

$$\beta = \frac{(z+1)^2 Ry}{kT_{\mathrm{e}}},$$

$$\chi_{\mathrm{d}} = \frac{\chi}{1 + 0.015\dfrac{z^3}{(z+1)^2}},$$

$$\chi = \frac{\Delta E}{(z+1)^2 Ry}.$$

z is the spectroscopic symbol ($z = Z_{\mathrm{n}} - N_{\mathrm{bound}}$, where N_{bound} is the number of bound electrons) and $f_{\alpha_0\alpha}$ is the oscillator strength for the resonance transitions (transitions to the ground state): dipole transition $\alpha_0 \to \alpha$ with transition energy ΔE (in eV). It is usually sufficient to consider only the first, second, and third transition to the ground state as the oscillator strength is rapidly decreasing with increasing principal quantum number.

If the first resonance transition is a $\Delta n = 0$ transition, the branching factor B_d is given by

$$B_{\mathrm{d}} = \left(\frac{z\chi}{z^2 + 13.4}\right)^{1/2} \frac{1}{1 + 0.105(z+1)\chi + 0.015(z+1)^2\chi^2}.$$

For $\Delta n \neq 0$:

$$B_{\mathrm{d}} = \left(\frac{z\chi}{z^2 + 13.4}\right)^{1/2} \frac{0.5}{1 + 0.210(z+1)\chi + 0.030(z+1)^2\chi^2}.$$

Note, that the Burgess formula should be used with care, as it might overestimate the dielectronic recombination by orders of magnitude (see Sect. 5.6).

A.1.6 *Charge Exchange*

A.1.6.1 Single-electron Charge Exchange

If charge exchange is classically allowed (over-barrier transitions), the cross section can be estimated according to (Z is the charge of the receptor ion)

$$\sigma^{(\mathrm{Cx})} \approx 4\pi a_0^2 p^4 \left(1 + \sqrt{Z}\right)^2.$$

p is the effective principal quantum number of the donor atom (I is the ionization potential of quantum shell p)

$$p = \sqrt{\frac{I}{Ry}}.$$

Charge exchange is selective in principle quantum number q of the receptor ion:

$$q_{Z-1} \approx p \frac{Z}{\sqrt{(1 + \sqrt{Z} - 1/\sqrt{Z})}}.$$

The orbital distribution is given by

$$W_{\mathrm{nl}} = \frac{\sigma_{\mathrm{nl}}}{\sigma_{\mathrm{n}}}.$$

The regime of weak mixing (at low-collision velocities, the Stark mixing of states is low) can be estimated according to

$$W_{\mathrm{nl}} = \frac{(2l+1)\cdot[(n-1)!]^2}{(n+l)!\cdot(n-1-l)!}.$$

If $n, l \gg 1$, the W_{nl} distribution takes the form

$$W_{\mathrm{nl}} \approx \frac{2l+1}{n}\cdot\exp\left[-\frac{l\cdot(l+1)}{n}\right].$$

In the case of strong mixing, a statistical distribution is obtained:

$$W_{\mathrm{nl}} = \frac{2l+1}{n^2}.$$

A.1.6.2 Multiple-electron Charge Exchange

Multiple-electron charge exchange in the over-barrier regime at low velocities can be estimated within the framework of the so-called classical absorbing sphere model (Janev et al. 1985). Single-, double-, and triple-charge exchange cross sections are given by

$$\sigma_1^{(Cx)} \approx \pi a_0^2\left(R_1^2 - R_2^2 - R_3^2\right),$$
$$\sigma_2^{(Cx)} \approx \pi a_0^2\left(R_2^2 - R_3^2\right),$$
$$\sigma_3^{(Cx)} \approx \pi a_0^2 R_3^2,$$
$$R_1 \approx 2Ry\frac{2\sqrt{Z}+1}{I_1},$$
$$R_2 \approx 2Ry\frac{2\sqrt{Z-1}+1}{I_2},$$
$$R_3 \approx 2Ry\frac{2\sqrt{Z-2}+1}{I_3},$$

where σ_1^{Cx}, σ_2^{Cx} and σ_3^{Cx} are the single-, double-, and triple-charge exchange cross sections, respectively. R_1,R_2, and R_3 are the corresponding classical radii, and I_1, I_2, and I_3 are the corresponding ionization potentials in [eV], Ry = 13.6 eV. As before, Z is the charge of the receptor ion.

A.2 Simple General Formulae for Collisional–Radiative Processes in Hydrogen

A.2.1 Energies

For hydrogen, the energy levels are specified by the principal quantum number n and are given by

$$E_{\mathrm{H}}(n) = \frac{Ry}{1+\frac{m_{\mathrm{e}}}{m_{\mathrm{p}}}} \frac{Z_{\mathrm{eff,H}}(n)}{n^2} = 13.5983\,\mathrm{eV}\,\frac{Z_{\mathrm{eff,H}}(n)}{n^2},$$

$$Z_{\mathrm{eff,H}}(n) = 1.0000.$$

For the energy difference between the $2s$ and $2p$ levels, we take

$$\Delta E = \{E(n=2) - E(n=1)\}\frac{\mathrm{d}\lambda}{\lambda} = 10.2\,\mathrm{eV}\frac{4.45 \times 10^{-13}}{1215 \times 10^{-10}} \approx 3.7 \times 10^{-5}\,\mathrm{eV}.$$

A.2.2 Spontaneous Transition Probabilities

For transitions $np \rightarrow 1s$, $np \rightarrow 2s$, $ns \rightarrow 2p$, $nd \rightarrow 2p$, the following exact analytical formulas have been adopted for the so-called Lyman- and Paschen-series:

$$A(np \rightarrow 1s) = \frac{8 \times 10^9 \times 2^8}{9} \frac{n(n-1)^{2n-2}}{(n+1)^{2n+2}} \quad \left[\mathrm{s}^{-1}\right],$$

$$A(np \rightarrow 2s) = \frac{8 \times 10^9 \times 2^{11}}{9} \frac{n(n^2-1)(n-2)^{2n-3}}{(n+2)^{2n+3}} \quad \left[\mathrm{s}^{-1}\right],$$

$$A(ns \rightarrow 2p) = \frac{8 \times 10^9 \times 2^9}{9} \frac{n^3(n-2)^{2n-3}}{(n+2)^{2n+3}} \quad \left[\mathrm{s}^{-1}\right],$$

$$A(nd \rightarrow 2p) = \frac{8 \times 10^9 \times 2^{14}}{9 \cdot 5} \frac{n^3(n^2-1)(n-2)^{2n-4}}{(n+2)^{2n+4}} \quad \left[\mathrm{s}^{-1}\right].$$

These transition probabilities have to be averaged over the respective statistical weights $g_{\mathrm{n}} = 2n^2$ for the quantum level n:

$$A(n \to 1) = \frac{8 \times 10^9 \cdot 2^8}{3} \frac{(n-1)^{2n-2}}{n(n+1)^{2n+2}} \quad [\mathrm{s}^{-1}],$$

$$A(n \to 2) = \frac{8 \times 10^9 \cdot 2^9}{9} \frac{n(n-2)^{2n-3}}{(n+2)^{2n+3}} \left\{1 + \frac{12(n^2-1)}{n^2} + \frac{2^5(n^2-1)}{(n+2)(n-2)}\right\} \quad [\mathrm{s}^{-1}],$$

$$A(n \to 2s) = \frac{6}{2n^2} A(np \to 2s) = \frac{8 \times 10^9 \cdot 2^{11}}{3} \frac{(n^2-1)(n-2)^{2n-3}}{n(n+2)^{2n+3}} \quad [\mathrm{s}^{-1}],$$

$$A(n \to 2p) = \frac{8 \times 10^9 \cdot 2^9}{9} \frac{n(n-2)^{2n-3}}{(n+2)^{2n+3}} \left\{1 + \frac{2^5(n^2-1)}{(n+2)(n-2)}\right\} \quad [\mathrm{s}^{-1}].$$

For larger principal quantum numbers, the following analytical asymptotic formula is proposed that is accurate within a few %:

$$A(n' \to n) \approx \frac{1.57 \times 10^{10}}{nn'^3(n'^2 - n^2)} \cdot \left\{1 + \frac{2}{5\sqrt{n+n'}} - \frac{1}{\sqrt{n}+3} - \frac{1}{\sqrt{n'}+3}\right\} \quad [\mathrm{s}^{-1}].$$

The hydrogen 2s-level decays via two-photon decay and its transition probability is given by

$$A(2s\,^2S_{1/2} - 1s\,^2S_{1/2}) = 8.2291\,\mathrm{s}^{-1}.$$

A.2.3 Radiative Recombination

The radiative recombination rate for the processes

$$p + e \to H(n) + \hbar\omega$$

has been fitted into the following formula (for $n < 10$):

$$\langle V\sigma_r \rangle = 10^{-8} A \sqrt{\beta} \frac{\beta + D}{\beta + \chi} \quad [\mathrm{cm}^{-3}\mathrm{s}^{-1}],$$

$$\beta = \frac{Ry}{T_e}.$$

For $\frac{1}{16} \le \beta \le 8$, (i.e., $1.7\,\mathrm{eV} \le \beta \le 220\,\mathrm{eV}$), the fitting formula provides an accuracy better than about 10% for all values of β and n. The following table contains the fitting parameters.

nl	A	χ	D
1	3.98×10^{-6}	0.419	0.02
2	1.89×10^{-6}	1.07	0.02
2*s*	5.57×10^{-7}	0.306	0.0
2*p*	1.37×10^{-6}	1.76	0.0
3	1.16×10^{-6}	2.00	0.03
4	7.76×10^{-7}	2.78	0.03
5	5.54×10^{-7}	3.54	0.03
6	4.00×10^{-7}	3.83	0.02
7	3.09×10^{-7}	4.44	0.02
8	2.45×10^{-7}	5.01	0.02
9	1.98×10^{-7}	5.54	0.02

For principal quantum numbers $n > 9$, the following expression is adopted:

$$\langle V\sigma_r \rangle = 5.20 \times 10^{-14}\, \beta_n^{3/2}\, e^{\beta_n}\, Ei(\beta_n),$$
$$\beta_n = \frac{E_H(n)}{T_e}.$$

The exponential integral $Ei(\beta_n)$ can be approximated within a few percent by

$$Ei(\beta_n) \approx e^{-\beta_n} \ln\left[1 + \frac{0.562 + 1.4\beta_n}{\beta_n(1 + 1.4\beta_n)}\right].$$

A.2.4 Electron Collisional Excitation and De-excitation

Quantum mechanical calculations of cross sections for the transitions $nl \to n'l'$ have been summed over all angular momenta l, l' in order to obtain cross sections and rate coefficients for the processes

$$H(n) + e \to H(n') + e.$$

Rate coefficients for the transitions from the ground state, i.e., $n = 1 \to n'$ have been fitted into the following formula:

$$\langle V\sigma_{ex}(1 \to n)\rangle = \frac{10^{-8}}{n^2} f(n) C_1 \sqrt{\beta_n} \frac{\beta_n + 1 + C_2}{\beta_n + C_3} \ln\left(C_4 + \frac{C_5}{\beta_n}\right) e^{-\Delta E/T_e},$$

$$f(n) = n^{\alpha}\left(\frac{n^2}{n^2-1}\right)^{3/2} \frac{n(n-1)^{2n-2}}{(n+1)^{2n+2}},$$

$$\beta_n = \frac{E_H(n)}{T_e},$$

$$\Delta E = E_H(1) - E_H(n).$$

For $\frac{1}{16} \leq \beta_n n^2 \leq 32$ (i.e., $0.43\,\mathrm{eV} \leq T_e \leq 220\,\mathrm{eV}$), the fitting formula provides an accuracy better than about 4% for all values of β and n. The following table contains the fitting parameters. α has been fixed to $\alpha = 0.75$.

n	C_1	C_2	C_3	C_4	C_5	$\Delta_{max}/\%$
2	1.5485×10^2	1.0123×10^2	1.1168×10^1	1.6272×10^0	2.4689×10^{-1}	0.6
3	1.5464×10^0	7.9704×10^3	7.8003×10^0	2.0652×10^0	7.1971×10^{-2}	2.4
4	1.0358×10^0	8.0008×10^3	4.5476×10^0	2.1340×10^0	2.9540×10^{-2}	3.4
5	7.7698×10^{-1}	7.9794×10^3	3.0132×10^0	2.1026×10^0	1.5099×10^{-2}	3.6
6	5.8235×10^{-1}	7.9708×10^3	2.1232×10^0	2.1147×10^0	9.7984×10^{-3}	3.8
7	4.5254×10^{-1}	7.9900×10^3	1.5852×10^0	2.1244×10^0	6.9830×10^{-3}	3.9
8	3.6339×10^{-1}	8.0109×10^3	1.1923×10^0	2.0889×10^0	4.8699×10^{-3}	4.1
9	2.4012×10^{-1}	9.9983×10^3	9.4891×10^{-1}	2.0875×10^0	3.7632×10^{-3}	4.1

For large principal quantum numbers, we derived a n-scaled formula based on the numerical calculations for n = 4, 5, 6, 7, 8, and 9:

$$\langle V\sigma_{ex}(1 \to n)\rangle = \frac{10^{-8}}{n^2} f(n) C_1 \sqrt{\beta_n} \frac{\sqrt{\beta_n + 1 + C_2}}{\beta_n + C_3} \ln\left(n^{0.6} C_4 + \frac{C_5}{\beta_n}\right) e^{-\Delta E/T_e},$$

$$f(n) = n^{\alpha}\left(\frac{n^2}{n^2-1}\right)^{3/2} \frac{n(n-1)^{2n-2}}{(n+1)^{2n+2}},$$

$$\beta_n = \frac{E_H(n)}{T_e},$$

$$\Delta E = E_H(1) - E_H(n).$$

Taking the above formula with $\alpha = 2.3304 \times 10^{-1}$, $C_1 = 3.0877 \times 10^3$, $C_2 = 1.4249 \times 10^1$, $C_3 = 3.5330 \times 10^0$, $C_4 = 1.0098 \times 10^0$, $C_5 = 3.5575 \times 10^{-2}$ an accuracy better than 15% is achieved for all values of n and β.

Electron collisional process from the $1s$ level into the $2s$- and $2p$-level as well as from the $2s$-level into the $2p$-level are described by the following fitting formula:

$$\langle V\sigma_{\mathrm{ex}}(ns \to 2l)\rangle = 10^{-8}A\left(\frac{E(2l)}{E(ns)}\right)^{3/2} \cdot \frac{\sqrt{\beta}}{\sqrt{1+\frac{\Delta E}{T_e}}} \cdot \frac{\beta+1+D}{\beta+\chi} \cdot \exp\left(-\frac{\Delta E}{T_e}\right) \quad [\mathrm{cm^3 s^{-1}}],$$

$$\beta = \frac{Ry}{T_e},$$
$$\Delta E = E(ns) - E(2l).$$

n	A	χ	D
1*s*-2*s*	1.83×10^1	8.96	2.20
1*s*-2*p*	3.89×10^1	6.64×10^{-2}	−0.70
2*s*-2*p*	1.30×10^3	2.53	3.00

For $\frac{1}{16} \le \beta \le 32$ (i.e., $0.43\,\mathrm{eV} \le T_e \le 220\,\mathrm{eV}$), the fitting formula provides an accuracy better than about 23% for all values of β.

For transitions between excited states, an accuracy as provided by a Regemorter-type expression is sufficient:

$$\langle V\sigma_{\mathrm{ex}}(n \to n')\rangle = 3.15 \times 10^{-7} f(n \to n')\left(\frac{Ry}{\Delta E}\right)^{3/2} \sqrt{\beta}\, e^{-\beta_n} p(\beta_n),$$
$$\Delta E = E_{\mathrm{H}}(n) - E_{\mathrm{H}}(n'),$$
$$\beta_n = \frac{\Delta E}{T_e},$$

f is the oscillator strength and p is an effective Gaunt factor. The required oscillator strengths are obtained from the transition probabilities determined in the above paragraphs and the general relation between the oscillator strength and the transition probability, namely

$$f(n \to n') = \frac{1}{4.339 \times 10^7 (E_n - E_{n'})^2} \frac{g(n')}{n} A(n' \to n)$$

with E_n and $E_{n'}$ in [eV]. For the Gaunt factor the following expression is proposed

$$p(\beta) = \left\{ \begin{array}{ll} \text{if } \beta \le 0.4: & 0.27566 \cdot \left(\beta - \frac{\beta^2}{4} + \frac{\beta^3}{12} - \ln(\beta) - 0.57722\right) \\ \text{else} & 0.066 \frac{\sqrt{\beta} + 2}{\beta + 0.127} \end{array} \right\}.$$

These formulas provide the correct asymptotic behavior for low and high energies and an accuracy better than 3% for all values of β.

De-excitation rates are obtained from the principle of detailed balance (note that ΔE_{ij} is a positive quantity):

$$\langle V\sigma_{\rm ex}(i \to j)\rangle = \langle V\sigma_{\rm dex}(j \to i)\rangle \frac{g_{\rm i}}{g_{\rm j}} \exp\left(\Delta E_{\rm ij}/T_{\rm e}\right).$$

A.2.5 Ionization and Three-Body Recombination

Ionization rates for the processes

$$H(n) + e \to p + 2e$$

are fitted to the following formula:

$$\langle V\sigma_{\rm iz}\rangle = 10^{-8}\frac{1}{n^2}\sqrt{\beta}\frac{A}{\beta+\chi}\frac{\beta+1+D}{\beta+1}\left(\frac{1}{1+\beta\frac{E_{\rm H}(n)}{Ry}}\right)^{1/2} e^{-E_{\rm H}(n)/T_{\rm e}} \quad \left[{\rm cm^3 s^{-1}}\right],$$

$$\beta = \frac{Ry}{T_{\rm e}}.$$

The following table contains the corresponding fitting parameters. For all values of n = 1–9 and $\frac{1}{16} \le \beta \le 32$ (i.e., $0.5\,{\rm eV} \le \beta \le 220\,{\rm eV}$), the fitting is accurate within 10%.

n	A	χ	D
1	8.72	0.276	−0.59
2	6.13×10^3	47.3	0.72
3	2.93×10^6	3.28×10^3	0.48
4	8.98×10^6	2.62×10^3	0.35
5	1.18×10^7	1.33×10^3	0.43
6	3.99×10^6	1.91×10^2	0.36
7	3.15×10^6	1.34×10^2	0.180
8	3.56×10^6	8.88×10^1	0.110
9	7.74×10^6	7.65×10^1	0.110

Ionization out of the levels 2*s* and 2*p* is described by the following fitting formula:

$$\langle V\sigma_{\rm iz}\rangle = \frac{10^{-8}}{2l_0+1}\sqrt{\beta}\frac{A}{\beta+\chi}\frac{\beta+1+D}{\beta+1}\left(\frac{1}{1+\beta\frac{E_{\rm H}(n)}{Ry}}\right)^{1/2} e^{-E_{\rm H}(n)/T_{\rm e}} \quad \left[{\rm cm^3 s^{-1}}\right],$$

$$\beta = \frac{Ry}{T_{\rm e}}.$$

nl	A	χ	D	l_0
$2s$	2.72×10^4	7.68×10^2	0.60	0
$2p$	4.35×10^3	5.15×10^1	0.90	1

For the $2s$ level, the accuracy is better than 17%, for the $2p$ level better than 6% for all values of β in the interval $\frac{1}{16} \leq \beta \leq 32$ (i.e., $0.5\,\text{eV} \leq \beta \leq 220\,\text{eV}$).

In order to provide ionization rates for any values of $n \geq 7$ and β the following fitting formula has been developed:

$$\langle V\sigma_{\text{iz}}\rangle = \frac{10^{-8}}{n^2}\sqrt{\beta}\frac{An^6}{\beta+\chi}\frac{\beta+D}{\beta+f(n)}e^{-\beta} \quad \left[\text{cm}^3\text{s}^{-1}\right],$$

$$f(n) = 10 + \frac{\gamma}{n^4},$$

$$\beta = \frac{E_{\text{H}}(n)}{T_{\text{e}}}.$$

With A = 5.7669, χ = 0.2401, D = 0.6719 and $\gamma = -4.252 \times 10^3$, the fitting formula provides an accuracy better than 30% for all values of $n \geq 7$ and β $(\frac{1}{16} \leq Ry/T_{\text{e}} \leq 32)$.

The three-body recombination rate coefficient $\langle V\sigma_{\text{tr}}(j \rightarrow i)\rangle$ is calculated from the principle of detailed balance:

$$\langle V\sigma_{\text{tr}}(j \rightarrow i)\rangle = \langle V\sigma_{\text{iz}}(i \rightarrow j)\rangle \frac{g_{\text{i}}}{2g_{\text{j}}}\left(\frac{2\pi\hbar^2}{m_{\text{e}}kT_{\text{e}}}\right)^{3/2} \exp\left[\Delta E_{\text{ij}}/kT_{\text{e}}\right]$$

or, written in convenient units (energies and temperatures in eV):

$$\langle V\sigma_{\text{tr}}(j \rightarrow i)\rangle = \langle V\sigma_{\text{iz}}(i \rightarrow j)\rangle \cdot 1.656 \times 10^{-22}\frac{g_{\text{i}}}{g_{\text{j}}}\frac{\exp\left[\Delta E_{\text{ij}}/T_{\text{e}}\right]}{(T_{\text{e}})^{3/2}} \quad \left[\frac{\text{cm}^6}{\text{s}}\right].$$

g_{i} and g_{j} are the statistical weights of the states before and after ionization, respectively.

A.2.6 Matrix Elements Including Phase Sign, Oscillator Strengths, and Energies of nlj-Split Levels

Matrix elements including phase sign, oscillator strengths, and energies for an LSJ-split level structure have been calculated for the states nlj from n = 1–6, l = 0–5 and all j. Designations of levels, principal quantum numbers, orbital quantum numbers, statistical weights, and energies are given in the following table:

N°	n	l	g	E (eV)
1	1	0	2	0.0000000000D+00
2	2	1	2	1.0198832906D+01
3	2	0	2	1.0198834783D+01
4	2	1	4	1.0198878221D+01
5	3	1	2	1.2087521013D+01
6	3	0	2	1.2087521649D+01
7	3	2	4	1.2087534174D+01
8	3	1	4	1.2087534440D+01
9	3	2	6	1.2087538645D+01
10	4	1	2	1.2748559996D+01
11	4	0	2	1.2748560297D+01
12	4	2	4	1.2748565493D+01
13	4	1	4	1.2748565660D+01
14	4	3	6	1.2748567261D+01
15	4	2	6	1.2748567379D+01
16	4	3	8	1.2748568202D+01
17	5	1	2	1.3054525858D+01
18	5	0	2	1.3054526031D+01
19	5	2	4	1.3054528639D+01
20	5	1	4	1.3054528758D+01
21	5	3	6	1.3054529499D+01
22	5	2	6	1.3054529605D+01
23	5	4	8	1.3054529897D+01
24	5	3	8	1.3054529981D+01
25	5	4	10	1.3054530186D+01
26	6	0	2	1.3220729319D+01
27	6	1	2	1.3220729379D+01
28	6	2	4	1.3220730958D+01
29	6	1	4	1.3220731054D+01
30	6	3	6	1.3220731411D+01
31	6	2	6	1.3220731517D+01
32	6	4	8	1.3220731580D+01
33	6	5	10	1.3220731636D+01
34	6	3	8	1.3220731690D+01
35	6	5	12	1.3220731748D+01
36	6	4	10	1.3220731748D+01

Note that the total quantum number j can be calculated from the relation $g = 2j+1$. Matrix elements [in atomic units], oscillator strengths and wavelengths [in Angstroem] are given in the following table:

N° of level i	N° of level j	$\langle i\|r\|j\rangle$ [a_0]	f_{ij}	λ_{ji} [10^{-10} m]
1	2	1.0543386372D+00	1.3887965000D−01	1.2156706897D+03
1	4	1.4910583918D+00	2.7776000000D−01	1.2156647299D+03
1	5	4.2249347333D−01	2.6430490000D−02	1.0257206358D+03
1	8	5.9750643218D−01	5.2862900000D−02	1.0257197873D+03
1	10	2.4932974013D−01	9.7081700000D−03	9.7253483531D+02
1	13	3.5261371639D−01	1.9417245000D−02	9.7253407245D+02
1	17	−1.7117838949D−01	4.6858355000D−03	9.4974071836D+02
1	20	−2.4208910812D−01	9.3721600000D−03	9.4974071836D+02
1	27	−1.2790755570D−01	2.6495730000D−03	9.3780136952D+02
1	29	−1.8089361240D−01	5.2994400000D−03	9.3780136952D+02
2	3	4.2443122876D+00	4.1420190000D−07	6.6053632268D+09
3	4	6.0024434964D+00	1.9171535000D−05	2.8542596510D+08
3	5	−2.5045849469D+00	1.4513075000D−01	6.5645738360D+03
2	6	−7.6737985611D−01	1.3624130000D−02	6.5645634088D+03
4	6	1.0853346629D+00	1.3626207500D−02	6.5647232960D+03
2	7	−5.4859663395D+00	6.9630150000D−01	6.5645217004D+03
4	7	−2.4534483987D+00	6.9631325000D−02	6.5646781097D+03
3	8	−3.5419042309D+00	2.9024485000D−01	6.5645251760D+03
4	9	−7.3603593382D+00	6.2668550000D−01	6.5646642064D+03
3	10	−1.0486923895D+00	3.4349250000D−02	4.8626493838D+03
2	11	3.1332304525D−01	3.0662410000D−03	4.8626455695D+03
4	11	−4.4313865789D−01	3.0666425000D−03	4.8627313916D+03
2	12	1.9763026422D+00	1.2199145000D−01	4.8626341268D+03
4	12	8.8382902926D−01	1.2198922500D−02	4.8627218557D+03
3	13	−1.4830525288D+00	6.8696650000D−02	4.8626379410D+03
4	15	2.6515146649D+00	1.0979262500D−01	4.8627180413D+03
3	17	6.3392490629D−01	1.4057715000D−02	4.3416527558D+03
2	18	1.8769032278D−01	1.2323195000D−03	4.3416497151D+03
4	18	−2.6545276469D−01	1.2324720000D−03	4.3417181319D+03
2	19	−1.1282249318D+00	4.4527800000D−02	4.3416451541D+03
4	19	−5.0455371026D−01	4.4526425000D−03	4.3417150911D+03
3	20	8.9649755435D−01	2.8114965000D−02	4.3416481948D+03
4	22	−1.5136815662D+00	4.0074875000D−02	4.3417135707D+03
2	26	1.3175152958D−01	6.4256800000D−04	4.1028608198D+03
4	26	−1.8633727395D−01	6.4264400000D−04	4.1029219177D+03
3	27	4.4446123041D−01	7.3126550000D−03	4.1028621775D+03
2	28	7.6792644965D−01	2.1829640000D−02	4.1028581044D+03
4	28	3.4342316671D−01	2.1828827500D−03	4.1029192022D+03
3	29	6.2855952145D−01	1.4625150000D−02	4.1028608198D+03
4	31	1.0302849858D+00	1.9646542500D−02	4.1029192022D+03
5	6	1.0396520293D+01	8.4204300000D−07	1.9495576770D+10
5	7	1.1624128376D+01	2.1783980000D−05	9.4205821143D+08
6	8	1.4703026865D+01	3.3873555000D−05	9.6927466337D+08
8	9	1.5595423907D+01	6.2628600000D−06	2.9490757393D+09
6	10	−4.4699726888D+00	1.6179470000D−01	1.8755976318D+04

(continued)

(continued)

N° of level i	N° of level j	$\langle i\|r\|j\rangle$ [a_0]	f_{ij}	λ_{ji} [10^{-10} m]
7	10	1.5063615968D+00	9.1870275000D−03	1.8756330993D+04
5	11	1.9980383976D+00	3.2326810000D−02	1.8755947944D+04
8	11	−2.8258780048D+00	3.2331300000D−02	1.8756328156D+04
5	12	8.7422227653D+00	6.1887400000D−01	1.8755800404D+04
8	12	3.9097694628D+00	6.1890225000D−02	1.8756180609D+04
6	13	−6.3212383461D+00	3.2356580000D−01	1.8755814590D+04
7	13	−6.7360646606D−01	1.8370990000D−03	1.8756169260D+04
9	13	2.0208821893D+00	1.1023203333D−02	1.8756296944D+04
7	14	−1.5858488843D+01	1.0182272500D+00	1.8756123861D+04
9	14	−4.2383754190D+00	4.8487066667D−02	1.8756251545D+04
8	15	1.1729225657D+01	5.5700575000D−01	1.8756129536D+04
9	16	−1.8954619524D+01	9.6974500000D−01	1.8756223171D+04
6	17	1.8488134109D+00	4.0489515000D−02	1.2821473474D+04
7	17	−5.6051251458D−01	1.8607637500D−03	1.2821639214D+04
5	18	7.9483539821D−01	7.4836100000D−03	1.2821462867D+04
8	18	−1.1241406831D+00	7.4844875000D−03	1.2821640540D+04
5	19	−3.4326859317D+00	1.3958090000D−01	1.2821428394D+04
8	19	−1.5351641773D+00	1.3958262500D−02	1.2821606065D+04
6	20	2.6145563148D+00	8.0975500000D−02	1.2821435024D+04
7	20	2.5064955494D−01	3.7209675000D−04	1.2821600762D+04
9	20	−7.5196846381D−01	2.2326883333D−03	1.2821660429D+04
7	21	5.1467926724D+00	1.5689012500D−01	1.2821591480D+04
9	21	1.3755319930D+00	7.4708550000D−03	1.2821649821D+04
8	22	−4.6054939461D+00	1.2562457500D−01	1.2821592806D+04
9	24	6.1516048435D+00	1.4941910000D−01	1.2821644517D+04
5	26	4.7273192521D−01	3.1021840000D−03	1.0940991151D+04
8	26	−6.6858438156D−01	3.1025225000D−03	1.0941116666D+04
6	27	1.1168369771D+00	1.7314775000D−02	1.0940991151D+04
7	27	−3.2345454192D−01	7.2615475000D−04	1.0941116666D+04
5	28	2.0185098460D+00	5.6558750000D−02	1.0940971841D+04
8	28	9.0271019453D−01	5.6558750000D−03	1.0941097356D+04
6	29	1.5794184552D+00	3.4628405000D−02	1.0940981496D+04
7	29	1.4464272260D−01	1.4520992500D−04	1.0941097356D+04
9	29	−4.3393884511D−01	8.7129916667D−04	1.0941145631D+04
7	30	2.7937873777D+00	5.4173900000D−02	1.0941097356D+04
9	30	7.4666516295D−01	2.5796583333D−03	1.0941135976D+04
8	31	2.7081390689D+00	5.0903225000D−02	1.0941097356D+04
9	34	3.3392163147D+00	5.1594066667D−02	1.0941135976D+04
10	11	−1.8980909754D+01	1.3298330000D−06	4.1146319042D+10
10	12	−2.4009761258D+01	3.8819515000D−05	2.2553807150D+09
11	13	−2.6843222983D+01	4.7335870000D−05	2.3119223326D+09
12	14	−2.4603540035D+01	6.5539475000D−06	7.0138505576D+09
13	15	−3.2212503975D+01	1.0924952500D−05	7.2126308692D+09
15	16	−2.9406866902D+01	2.9084850000D−06	1.5052324060D+10

(continued)

(continued)

N° of level i	N° of level j	$\langle i\|\|r\|\|j\rangle$ [a_0]	f_{ij}	λ_{ji} [10^{-10} m]
11	17	−6.9631817018D+00	1.8172555000D−01	4.0522263614D+04
12	17	3.5240377639D+00	2.3272565000D−02	4.0522952317D+04
10	18	3.7627294360D+00	5.3064900000D−02	4.0522210638D+04
13	18	−5.3217004845D+00	5.3071825000D−02	4.0522952317D+04
10	19	−1.2758886267D+01	6.1014100000D−01	4.0521866296D+04
13	19	−5.7061678764D+00	6.1017725000D−02	4.0522607962D+04
14	19	2.5798359576D+00	8.3149066667D−03	4.0522819872D+04
11	20	−9.8469851870D+00	3.6342190000D−01	4.0521879540D+04
12	20	−1.5758701934D+00	4.6538050000D−03	4.0522568230D+04
15	20	4.7277463446D+00	2.7924250000D−02	4.0522819872D+04
12	21	−2.1807194345D+01	8.9118500000D−01	4.0522475520D+04
15	21	−5.8282661256D+00	4.2437816667D−02	4.0522727161D+04
13	22	−1.7118326853D+01	5.4914975000D−01	4.0522488764D+04
14	22	−6.8946878035D−01	5.9388700000D−04	4.0522700673D+04
16	22	3.0834473067D+00	8.9085625000D−03	4.0522819872D+04
14	23	−3.2833273019D+01	1.3468010000D+00	4.0522660940D+04
16	23	−6.3187733718D+00	3.7411112500D−02	4.0522780139D+04
15	24	−2.6064733772D+01	8.4875366667D−01	4.0522660940D+04
16	25	−3.7382402069D+01	1.3093925000D+00	4.0522740406D+04
10	26	1.4703917819D+00	1.2505235000D−02	2.6258417690D+04
13	26	−2.0795734663D+00	1.2506592500D−02	2.6258729123D+04
11	27	−2.8319393241D+00	4.6386690000D−02	2.6258428813D+04
12	27	1.3090618343D+00	4.9557700000D−03	2.6258718000D+04
10	28	5.0806138416D+00	1.4929995000D−01	2.6258323150D+04
13	28	2.2721646048D+00	1.4930402500D−02	2.6258640141D+04
14	28	−8.8653017717D−01	1.5152575000D−03	2.6258729123D+04
11	29	−4.0048569401D+00	9.2768600000D−02	2.6258334272D+04
12	29	−5.8539029891D−01	9.9102175000D−04	2.6258623458D+04
15	29	1.7562128282D+00	5.9463933333D−03	2.6258729123D+04
12	30	−8.0396697518D+00	1.8692555000D−01	2.6258606774D+04
15	30	−2.1486931542D+00	8.9011933333D−03	2.6258712439D+04
13	31	6.8164683584D+00	1.3437280000D−01	2.6258606774D+04
14	31	2.3692871354D−01	1.0822686667D−04	2.6258695755D+04
16	31	−1.0595917334D+00	1.6234437500D−03	2.6258751368D+04
14	32	−9.7392645400D+00	1.8287400000D−01	2.6258695755D+04
16	32	−1.8743152170D+00	5.0797862500D−03	2.6258745807D+04
15	34	−9.6092629583D+00	1.7802466667D−01	2.6258695755D+04
16	36	−1.1088643482D+01	1.7779412500D−01	2.6258740246D+04
17	18	3.0008063230D+01	1.9066210000D−06	7.1730805657D+10
17	19	−3.9697519168D+01	5.3693400000D−05	4.4575875479D+09
18	20	4.2438024636D+01	6.0169200000D−05	4.5460080592D+09
19	21	−4.6489972766D+01	1.1378732500D−05	1.4424107130D+10
20	22	−5.3259848444D+01	1.4712262500D−05	1.4641468202D+10
21	22	−1.2424948019D+01	6.6811433333D−08	1.1697989859D+11

(continued)

(continued)

N° of level i	N° of level j	$\langle i\|r\|j\rangle$ [a_0]	f_{ij}	λ_{ji} [10^{-10} m]
21	23	−4.1676378708D+01	2.8221466667D−06	3.1158280125D+10
22	24	−5.5566154993D+01	4.7434316667D−06	3.2953466163D+10
23	24	−8.0206132240D+00	1.6609812500D−08	1.4705672547D+11
24	25	−4.7450655110D+01	1.4137125000D−06	6.0472480537D+10
17	26	−6.0705011099D+00	7.5026900000D−02	7.4597819541D+04
20	26	8.5855986133D+00	7.5036325000D−02	7.4599121182D+04
18	27	−9.9963119971D+00	2.0344515000D−01	7.4597909307D+04
19	27	−6.2944488986D+00	4.0331675000D−02	7.4599076298D+04
17	28	−1.7545320353D+01	6.2675050000D−01	7.4597101413D+04
20	28	−7.8468417326D+00	6.2679300000D−02	7.4598403030D+04
21	28	5.6873588360D+00	2.1951433333D−02	7.4598717220D+04
18	29	−1.4136252495D+01	4.0685580000D−01	7.4597146295D+04
19	29	2.8147481441D+00	8.0651875000D−03	7.4598313261D+04
22	29	−8.4444742419D+00	4.8393450000D−02	7.4598762104D+04
19	30	2.8837469167D+01	8.4654575000D−01	7.4598133726D+04
22	30	7.7072164276D+00	4.0312333333D−02	7.4598582566D+04
23	30	−3.7627216579D+00	7.2062137500D−03	7.4598717220D+04
20	31	−2.3540221142D+01	5.6410175000D−01	7.4598133726D+04
21	31	−1.5199670384D+00	1.5678746667D−03	7.4598492798D+04
24	31	6.7976006282D+00	2.3518750000D−02	7.4598717220D+04
21	32	4.1834311008D+01	1.1877046667D+00	7.4598447914D+04
24	32	8.0510400288D+00	3.2991912500D−02	7.4598672335D+04
23	33	5.7424650817D+01	1.6784187500D+00	7.4598627451D+04
25	33	8.6570941799D+00	3.0516660000D−02	7.4598717220D+04
22	34	3.4467587709D+01	8.0624116667D−01	7.4598447914D+04
23	34	7.2412472009D−01	2.6688875000D−04	7.4598582566D+04
25	34	−4.2840205760D+00	7.4730180000D−03	7.4598717220D+04
25	35	6.3616431347D+01	1.6479030000D+00	7.4598672335D+04
24	36	4.7630564288D+01	1.1547157500D+00	7.4598582566D+04
26	27	4.3450707612D+01	1.3939010000D−06	2.0571022716D+11
27	28	5.8754619678D+01	6.6773250000D−05	7.8519159535D+09
26	29	6.1448830864D+01	8.0238100000D−05	7.1472820147D+09
28	30	7.2404340281D+01	1.4539105000D−05	2.7381406714D+10
29	31	7.8827627571D+01	1.7611017500D−05	2.6793951358D+10
30	31	1.9350853880D+01	1.6104865000D−07	1.1771079042D+11
30	32	−7.4476870257D+01	3.8314950000D−06	7.3290524041D+10
32	33	−6.2881464558D+01	6.7429212500D−07	2.2265447502D+11
31	34	8.6539758390D+01	5.3128683333D−06	7.1363350246D+10
32	34	−1.4333068597D+01	6.9141012500D−08	1.1281761253D+11
33	36	−9.4797280670D+00	2.4604620000D−08	1.1094285446D+11
34	36	−8.4795679514D+01	1.2670612500D−06	2.1546859883D+11

A.3 Simple General Formulae for Collisional–Radiative Processes in He^{0+} and He^{1+}

A.3.1 Energies

Helium II (He^{1+})

Energy levels specified by the main quantum number n are given by

$$E_{He^{1+}}(n) = \frac{Ry}{1+\dfrac{m_e}{2m_p+2m_n}} \frac{Z^2_{eff,He^{1+}}(n)}{n^2} = 13.6038\,\text{eV}\frac{Z^2_{eff,He^{1+}}(n)}{n^2}.$$

With

$$E_{He^{1+}}(n) = 54.4182\,\text{eV}\frac{1}{n^2}$$

we obtain

$$Z^2_{eff,He^{1+}}(n) = 4.00019.$$

Statistical weights:

$$g_{He^{1+}}(n) = 2n^2.$$

For the energy difference between the $2s$ and $2p$ levels, we take

$$\Delta E = \{E(n=2) - E(n=1)\}\frac{d\lambda}{\lambda} = 40.8\,\text{eV}\frac{5.26\times 10^{-13}}{303.7\times 10^{-10}} \approx 7.1\times 10^{-4}\,\text{eV}.$$

Helium I (He^{0+})

$$E_{He^{0+}}(n,S) = \frac{Ry}{1+\dfrac{m_e}{2m_p+2m_n}} \frac{Z^2_{eff,He^{0+}}(n,S)}{n^2} = 13.6038\,\text{eV}\frac{Z^2_{eff,He^{0+}}(n,S)}{n^2}$$

with

$$Z^2_{eff,He^{0+}}(n=1,S=0) = 1.80739$$

for $n = 1$. For $n > 1$, we develop the following formulas:

$$Z^2_{\text{eff,He}^{0+}}(n, S=0) = 1 - \frac{c_1 + c_2 n^2}{c_3 + n^{c_4}}$$

with $c_1 = -4.1589 \times 10^{-2}$, $c_2 = 2.6773 \times 10^{-2}$, $c_3 = -1.1905$ and $c_4 = 3.0429$ and

$$Z^2_{\text{eff,He}^{0+}}(n, S=1) = 1 + \frac{c_1 + c_2 n^2}{c_3 + n^{c_4}}$$

with $c_1 = -3.1897 \times 10^{-1}$, $c_2 = 1.6946 \times 10^{-1}$, $c_3 = -2.9979$ and $c_4 = 3.0850$. For all n- and S-quantum numbers, an accuracy better than 10^{-4} is achieved.

For the neutral helium, we consider the metastable-resolved energy levels $1s2s\ ^1S_0$, $1s2s\ ^3S_1$, $1s2p\ ^1\text{P}_1$, and $1s2p\ ^3P_{012}$ (averaged over *J*-quantum number):

Level	Energy (eV)
$1s2s\ ^1S_0$	3.97162
$1s2s\ ^3S_1$	4.76404
$1s2p\ ^1P_1$	3.36575
$1s2p\ ^3P_{012}$	3.61938

Statistical weights:

$$g_{\text{He}^{0+}}(n, S=0) = n^2,$$
$$g_{\text{He}^{0+}}(n, S=1) = 3n^2.$$

A.3.2 Spontaneous Transition Probabilities

Helium II (He^{1+})

$$A_{\text{He}^{1+}}(n \to 1) = \frac{8 \times 10^9 \cdot 2^8}{3} \frac{(n-1)^{2n-2}}{n(n+1)^{2n+2}} Z^4_{\text{eff,He}^{1+}}(n)\ \left[\text{s}^{-1}\right],$$

$$A_{\text{He}^{1+}}(n \to 2) = \frac{8 \times 10^9 \cdot 2^9}{9} \frac{n(n-2)^{2n-3}}{(n+2)^{2n+3}} \times \left\{1 + \frac{12(n^2-1)}{n^2} + \frac{2^5(n^2-1)}{(n+2)(n-2)}\right\} Z^4_{\text{eff,He}^{1+}}(n)\ \left[\text{s}^{-1}\right].$$

For larger principal quantum numbers, we employ the formula

$$A_{\text{He}^{1+}}(n' \to n) \approx \frac{1.57 \times 10^{10}\ Z^4_{\text{eff,He}^{1+}}(n)}{nn'^3(n'^2 - n^2)} \cdot \left\{1 + \frac{2}{5\sqrt{n'+n}} - \frac{1}{\sqrt{n'}+3} - \frac{1}{\sqrt{n}+3}\right\}\ \left[\text{s}^{-1}\right].$$

The H-like 2*s*-level decays via two-photon decay and its transition probability is given by

$$A(2s\,^2S_{1/2} - 1s\,^2S_{1/2}) = 5.2661 \times 10^2\,\mathrm{s}^{-1}.$$

Transitions from high *n*-levels to the 2*s* and 2*p* states are obtained from the following formulas:

$$A(n \to 2s) = \frac{2}{2n^2}A(np \to 2s) = \frac{8 \times 10^9 \cdot 2^{11}}{9}\frac{(n^2-1)(n-2)^{2n-3}}{n(n+2)^{2n+3}} Z^4_{\mathrm{eff,He^{1+}}}(n)\,\left[\mathrm{s}^{-1}\right],$$

$$A(n \to 2p) = \frac{8 \times 10^9 \cdot 2^9}{9}\frac{n(n-2)^{2n-3}}{(n+2)^{2n+3}}\left\{1 + \frac{2^5(n^2-1)}{(n+2)(n-2)}\right\} Z^4_{\mathrm{eff,He^{1+}}}(n)\,\left[\mathrm{s}^{-1}\right].$$

Helium I (He^{0+})

Transition probabilities $A(1s^2\,^1S_0 \to 1snp\,^1P_1)$, $A(1s^2\,^1S_0 \to 1s2p\,^3P_1)$, and $A(1s^2\,^1S_0 \to 1s3p\,^3P_1)$ are given in the following table:

n	$A(1s^2\,^1S_0 \to 1snp\,^1P_1)$/s	$A(1s^2\,^1S_0 \to 1snp\,^3P_1)$/s
2	1.80×10^9	1.76×10^2
3	5.66×10^8	5.61×10^1
4	2.46×10^8	–
5	1.28×10^8	–
6	7.19×10^7	–
7	5.07×10^7	–
8	3.43×10^7	–
9	2.37×10^7	–
10	1.81×10^7	–

The averaged transition probabilities $A(n \to 1,\ S = 0) = \frac{3}{n^2} \cdot A(1s^2\,^1S_0 \to 1snp\,^1P_1)$ are fitted into the following formula:

$$A_{\mathrm{He^{0+}}}(n \to 1, S = 0) = 3A\left(2 - \frac{\sigma}{n}\right)^4 \frac{(n-1)^{2n-2}}{n(n+1)^{2n+2}}\,\left[\mathrm{s}^{-1}\right]$$

with $A = 6.469 \times 10^{10}$ and $\sigma = 0.4306$, the fitting formula provides an accuracy better than 6% even for all values n = 2–10. For the transition probabilities from the triplet system to the ground state, only two values are of importance:

$$A_{\mathrm{He^{0+}}}(2 \to 1, S = 1) = 4.40 \times 10^1\,\mathrm{s}^{-1},$$
$$A_{\mathrm{He^{0+}}}(3 \to 1, S = 1) = 6.23 \times 10^0\,\mathrm{s}^{-1}.$$

For larger main quantum numbers, the intercombination transitions are obtained from a $1/n^3$-scaling law:

$$A_{\mathrm{He}^{0+}}(n \rightarrow 1, S = 1) = \frac{1.38 \times 10^3}{n^5}\ \mathrm{s}^{-1}$$

(note that the n^5-scaling law arises from the statistical weight average of the triplet levels). For all other transitions in the singlet system into $n = 2$, we use the following formula:

$$A_{\mathrm{He}^{0+},S}(n \rightarrow 2, S) = 0.95 \frac{8 \times 10^9 \cdot 2^9}{9} \frac{n(n-2)^{2n-3}}{(n+2)^{2n+3}} Z^4_{\mathrm{eff,He}^{0+}}(2, S) \times \left\{1 + \frac{12(n^2-1)}{n^2} + \frac{2^5(n^2-1)}{(n+2)(n-2)}\right\}\left[\mathrm{s}^{-1}\right]$$

with $S = 0$ and 1.

For larger principal quantum numbers, the averaged transition probabilities are dominated by transitions with high angular momentum, i.e., $l' = n' - 1 \rightarrow l = l' - 1$. Z_{eff}-values for these high angular momenta energy levels are very close to 1 and singlet and triplet values practically do not differ, see table:

n	l	$E(n, S = 0)$/eV	$Z^2_{\mathrm{eff}}(S = 0)$	$E(n, S = 1)$/eV	$Z^2_{\mathrm{eff}}(S = 1)$
3	d	1.51332	1.00118	1.51374	1.00146
4	f	0.85038	1.00017	0.85039	1.00018
5	g	0.54418	1.00005	0.54418	1.00005
6	h	0.37789	1.00002	0.37789	1.00002
7	i	0.27763	1.00001	0.27763	1.00001

For n', $n > 2$, we propose the following formula:

$$A_{\mathrm{He}^{0+},S}(n' \rightarrow n, S) \approx \frac{1.57 \times 10^{10} Z^4_{\mathrm{eff,He}^{0+},S}(n, S)}{n n'^3 (n'^2 - n^2)} \left\{1 + \frac{2}{5\sqrt{n'+n}} - \frac{1}{\sqrt{n}+3} - \frac{1}{\sqrt{n'}+3}\right\}\left[\mathrm{s}^{-1}\right]$$

with

$$Z^2_{\mathrm{eff,He}^{0+}}(n \geq 3, S) \approx 1.00.$$

with $S = 0$ and 1. Note that the above formula contains only the Z_{eff}-value of the lower energy.

The above formulas permit to determine radiative decay rates for H-like ions within about 5% for all values of n' and n and also increases the accuracy for neutral helium to about 5% for $n', n > 2$.

Transition probabilities involving metastable and ground states are given in the following table:

Transition	A/s^{-1}
$1s2s\ ^1S_0 \rightarrow 1s^2\ ^1S_0$	5.13×10^1 s^{-1}
$1s2s\ ^3S_1 \rightarrow 1s^2\ ^1S_0$	1.27×10^{-4} s^{-1}
$1s2p\ ^1P_1 \rightarrow 1s^2\ ^1S_0$	1.71×10^9 s^{-1}
$1s2p\ ^3P_{012} \rightarrow 1s^2\ ^1S_0$	5.95×10^1 s^{-1}
$1s2p\ ^1P_1 \rightarrow 1s2s\ ^1S_0$	1.93×10^6 s^{-1}
$1s2p\ ^3P_{012} \rightarrow 1s2s\ ^3S_1$	1.15×10^7 s^{-1}
$1s2p\ ^1P_1 \rightarrow 1s2s\ ^3S_1$	1.55×10^0 s^{-1}
$1s2s\ ^1S_0 \rightarrow 1s2s\ ^3S_1$	1.51×10^{-7} s^{-1}
$1s2p\ ^3P_{012} \rightarrow 1s2s\ ^1S_0$	0.0

Finally, we note very convenient general formulas for the determination of spin-allowed and spin-forbidden transitions (Rosmej et al. 2007) for the cases $n, n' > 2$:

$$A_{\mathrm{He}^{0+}}(n'S' \rightarrow nS) = \frac{1.57 \times 10^{10}}{nn'^3(n'^2 - n^2)} \cdot g(n'S' \rightarrow nS)\left[\mathrm{s}^{-1}\right].$$

For spin-allowed transitions in the singlet system ($S' = 0 \rightarrow S = 0$, i.e., ss-transition) and in the triplet system ($S' = 1 \rightarrow S = 1$, i.e., tt-transition), the following g-function is proposed:

$$g_{\mathrm{ss,tt}} = C_1 \cdot \left\{1 + \frac{C_2}{n' + n} + \frac{C_3 \cdot n'^2}{(n' + n)^2 \cdot (n'^2 - n^2)}\right\}.$$

For the intercombination transitions ($S' = 0 \rightarrow S = 1$ and $S' = 1 \rightarrow S = 0$ i.e., *st*- and *ts*-transitions, respectively) one has to take into account intermediate coupling effects that are very important for high-*n* and high-*l*-values. Due to the wave function mixing, intercombination transitions can be of the same order of magnitude as allowed transitions [Rosmej et al. 2007] and general formulas are very rare. The following *g*-function is proposed:

$$g_{\mathrm{st,ts}} = C_1 \cdot (n' - n)^{C_2} \cdot n'^{C_3}.$$

The table presents the fitting parameters for spin-allowed and intercombination transitions:

Transition type	C_1	C_2	C_3	Accuracy (%)
ss	0.9510	−1.6234	−3.3480	3.9
tt	0.8598	0.3796	−5.3992	3.7
st	2.8856	0.8862	−2.2736	31
ts	1.2884	0.9304	−2.4436	28

Due to the large wave function mixing, collisional processes for intercombination transitions can be estimated with the help of a Regemorter-type formula (in particular for suprathermal electrons) employing the intercombination transition probabilities given above.

A.3.3 Radiative Recombination

Helium II (He^{1+})

The radiative recombination rate for the processes

$$He^{2+} + e \rightarrow He^{1+}(n) + \hbar\omega$$

has been fitted into the following formula (for $n < 10$):

$$\langle V\sigma_r \rangle = 10^{-8} A\sqrt{\beta}\,\frac{\beta + D}{\beta + \chi} \quad \left[\mathrm{cm^{-3}s^{-1}}\right],$$

$$\beta = \frac{Ry}{T_e}$$

(note that above formula contains the ratio Ry/T_e and *not* 4 Ry/T_e). For $\frac{1}{16} \le \beta \le 8$ (i.e., $1.7\,\mathrm{eV} \le \beta \le 220\,\mathrm{eV}$), the fitting formula provides an accuracy for all values of n better than 10%. The following table contains the fitting parameters:

n	A	χ	D
1	1.64×10^{-5}	0.133	0.01
2	8.53×10^{-6}	0.418	0.02
2*s*	2.39×10^{-6}	0.108	0.0
2*p*	6.05×10^{-6}	0.529	0.0
3	5.46×10^{-6}	0.725	0.02
4	3.87×10^{-6}	1.06	0.02
5	3.00×10^{-6}	1.59	0.03
6	2.35×10^{-6}	1.99	0.02
7	1.90×10^{-6}	2.38	0.03
8	1.56×10^{-6}	2.77	0.03
9	1.31×10^{-6}	3.15	0.03

For principal quantum numbers $n > 9$, the following expression is adopted:

$$\langle V\sigma_r \rangle = 5.20 \times 10^{-14} \beta_n^{3/2} e^{\beta_n}\, Ei(\beta_n),$$
$$\beta_n = \frac{E_{He^{1+}}(n)}{T_e}.$$

Helium I (He^{0+})

The radiative recombination rate for the processes

$$He^{1+}\left(1s\,^2S_{1/2}\right) + e \rightarrow He^{0+}\left(n,^{2S+1}L\right) + \hbar\omega$$

has been fitted into the following formula (for $n < 10$):

$$\langle V\sigma_r \rangle = 10^{-8}\, QA\sqrt{\beta}\,\frac{\beta + D}{\beta + \chi}\left[\mathrm{cm^{-3}s^{-1}}\right],$$
$$\beta = \frac{Ry}{T_e},$$
$$Q = \begin{Bmatrix} 1/4 & if\ S = 0 \\ 3/4 & if\ S = 1 \end{Bmatrix}$$

for excited states while $Q = 1$ if $n = 1$ (note, that the Pauli principle is included in the fitting parameter A). For $\frac{1}{16} \leq \beta \leq 8$ (i.e., $1.7\,\mathrm{eV} \leq \beta \leq 220\,\mathrm{eV}$), the fitting formula provides an accuracy better than about 10% for all values of β and n. The following table contains the fitting parameters:

nl	A	χ	*D*
1	4.66×10^{-6}	0.0736	0.00
2	1.91×10^{-6}	0.51	0.04
$1s2s\ ^1S_0$	6.19×10^{-7}	8.99×10^{-2}	0.0
$1s2s\ ^3S_1$	7.21×10^{-7}	5.69×10^{-2}	0.0
$1s2p\ ^3P_{012}$	1.39×10^{-6}	1.53	0.0
$1s2p\ ^1P_1$	1.79×10^{-6}	0.939	0.0
3	1.10×10^{-6}	1.37	0.08
4	7.19×10^{-7}	2.07	0.08
5	5.02×10^{-7}	2.60	0.07
6	3.73×10^{-7}	3.24	0.07
7	2.88×10^{-7}	3.85	0.07
8	2.23×10^{-7}	4.14	0.06
9	1.81×10^{-7}	4.66	0.06

For principal quantum numbers $n > 10$, the following expression is adopted:

$$\langle V\sigma_r\rangle = 5.20\times10^{-14}\,Q\,\beta_n^{3/2}\,e^{\beta_n}\,Ei(\beta_n),$$
$$\beta_n = \frac{E_{He^{0+}}(n)}{T_e},$$
$$Q = \begin{Bmatrix} 1/4 & if\,S=0 \\ 3/4 & if\,S=1 \end{Bmatrix}.$$

A.3.4 Dielectronic Recombination

Helium II (He^{1+})

For the dielectronic recombination that exist only for He^{1+}, two types of processes are considered:

$$He^{1+}(1s)+e\rightarrow 2l\,nl\rightarrow \begin{Bmatrix} 1s\,nl+\hbar\omega_2 \\ 1s\,2l+\hbar\omega_n \end{Bmatrix}\rightarrow 1s^2+\begin{Bmatrix} \hbar\omega_n' \\ \hbar\omega_2' \end{Bmatrix}$$

and

$$He^{1+}(1s)+e\rightarrow 3l\,nl\rightarrow \begin{Bmatrix} 1s\,nl+\hbar\omega_3 \\ 1s\,3l+\hbar\omega_n \end{Bmatrix}\rightarrow 1s^2+\begin{Bmatrix} \hbar\omega_n' \\ \hbar\omega_3' \end{Bmatrix}.$$

The dielectronic recombination rate coefficient is approximated by the following formula:

$$\langle V\sigma_d\rangle = 10^{-8}A\,\beta^{3/2}\,e^{-\beta\chi}\quad [\mathrm{cm^3s^{-1}}],$$
$$\beta = \frac{4Ry}{T_e}.$$

Autoionizing states	A	χ
2lnl'	3.10×10^{-4}	0.744
3lnl'	5.48×10^{-6}	0.888

The dielectronic recombination rates are dominated by the configurations *2lnl'*. Contributions from higher orders, like the *nln'l'*-configurations (*n, n'* > 2), are negligible. The above formulas can be applied coupling the H-like ground state to the He-like like ground state via the given dielectronic recombination rate coefficient.

More detailed considerations of the dielectronic recombination coupled to excited states can be found in Guedda et al. (2006, 2007).

A.3.5 Electron Collisional Excitation and De-excitation

Helium II (He^{1+})

Quantum mechanical calculations of cross sections for the transition $nl \to n'l'$ have been summed over all angular momenta l, l' in order to obtain cross sections and rate coefficients for the processes

$$He^{1+}(n) + e \to He^{1+}(n') + e.$$

Rate coefficients for the transitions from the ground state, i.e., $n = 1 \to n'$ have been fitted into the following formula:

$$\langle V\sigma_{ex}(1 \to n')\rangle = 10^{-8}\left(\frac{E_{He^{1+}}(n')}{E_{He^{1+}}(1)}\right)^{3/2} A\sqrt{\beta}\,\frac{\beta + 1 + D}{\beta + \chi}\,e^{-\Delta E/T_e} \quad \left[cm^3 s^{-1}\right],$$

$$\Delta E = E_{He^{1+}}(1) - E_{He^{1+}}(n'),$$

$$\beta = \frac{Ry}{T_e}$$

(note that the expression for β does not contain the factor 2^2, because a scaling factor of 0.25 has been introduced for the calculations in order to obtain the same energy interval like for hydrogen). For $\frac{1}{16} \leq \beta \leq 32$ (i.e., $0.43\,eV \leq \beta \leq 220\,eV$), the fitting formula provides an accuracy better than about 2% for all values of β and n. The following table contains the fitting parameters.

n'	A	χ	D
2	5.87	4.77×10^{-2}	−0.72
3	4.99	3.30×10^{-2}	−0.84
4	4.79	2.35×10^{-2}	−0.88
5	4.71	2.28×10^{-2}	−0.89
6	4.66	2.51×10^{-2}	−0.89
7	4.63	2.63×10^{-2}	−0.89
8	4.61	2.70×10^{-2}	−0.89
9	4.60	2.78×10^{-2}	−0.89

For large principal quantum numbers, we derived a n-scaled formula based on the numerical calculations for $n' = 4$, 5, 6, 7, 8, and 9:

$$\langle V\sigma_{ex}(1 \to n)\rangle = \frac{10^{-8}}{n^2} f(n) C_1 \sqrt{\beta_n}\, \frac{\sqrt{\beta_n + 1 + C_2}}{\beta_n + C_3} \ln\left(n \cdot C_4 + \frac{C_5}{\beta_n}\right) e^{-\Delta E/T_e},$$

$$f(n) = n^{\alpha}\left(\frac{n^2}{n^2 - 1}\right)^{3/2} \frac{n(n-1)^{2n-2}}{(n+1)^{2n+2}},$$

$$\beta_n = \frac{E_H(n)}{T_e},$$

$$\Delta E = E_{He^{1+}}(1) - E_{He^{1+}}(n).$$

It is remarkable that with $\alpha = 0.38753$, $C_1 = 4.2011 \times 10^4$, $C_2 = 1.2997 \times 10^2$, $C_3 = 1.3126 \times 10^2$, $C_4 = 7.8875 \times 10^{-1}$, $C_5 = 2.7627 \times 10^{-2}$ an accuracy better than 5% is achieved for all values of n and β. $E_H(n)$ is given in A.2.1.

For transitions between excited states with large principal quantum numbers, we employ a Regemorter-type expression:

$$\langle V\sigma(n \to n')\rangle = 3.15 \times 10^{-7} f(n \to n')\left(\frac{Ry}{\Delta E}\right)^{3/2} \sqrt{\beta}\, e^{-\beta_n} p(\beta_n),$$

$$\Delta E = E_{He^{1+}}(n) - E_{He^{1+}}(n'),$$

$$\beta_n = \frac{\Delta E}{T_e},$$

where f is the oscillator strength and p the effective Gaunt factor. The required oscillator strengths are obtained from the transition probabilities determined in the above paragraphs and the general relation between the oscillator strength and the transition probability. We employ the following Gaunt factor:

$$p(\beta) = 0.2757 e^{-1.3\beta}\left(\beta - \frac{\beta^2}{4} - \ln(\beta) - 0.5772\right) + 0.2\left(1 - e^{-4.5\beta}\right).$$

This formula provides the correct asymptotic behavior for low and high energies and an accuracy better than 5% for all values of β.

For the electron collisional process from the 2s-level into the 2p-level, we employ the following fitting formula:

$$\langle V\sigma_{ex}(2s \to 2p)\rangle = 10^{-8} A \sqrt{\beta}\, \frac{\beta + 1 + D}{\beta + \chi} \quad \left[\mathrm{cm^3 s^{-1}}\right],$$

$$\beta = \frac{Ry}{T_e}.$$

n	A	χ	D
2s-2p	2.14×10^2	2.25	3.35

For $\frac{1}{16} \leq \beta \leq 32$ (i.e., $0.43\,\text{eV} \leq T_e \leq 220\,\text{eV}$), the fitting formula provides an accuracy better than about 5% for all values of β (note that Ry/T_e is employed and not $4Ry/T_e$).

Collisional de-exciation is calculated from the excitation rates as described in Annex A.3.1.

Helium I (He^{0+})

Due to the singlet and triplet structures, the calculations of the cross sections in neutral Helium are much more complicated: also cross sections from singlet to triplet states are requested which cannot be approximated by any Regemorter-type expression. We distinguish therefore four different types of transitions.

I. $He^{0+}(n = 1, S = 0) + e \rightarrow He^{0+}(n', S = 0) + e,$

these transitions lead to a strong population of the excited singlet levels through collisions from the ground state.

Rate coefficients for the transitions from the ground state, i.e., $n = 1 \rightarrow n'$ have been fitted into the following formula:

$$\langle V\sigma_{ex}(1 \rightarrow n)\rangle = \frac{10^{-8}}{n^2} f(n) C_1 \sqrt{\beta_n} \frac{\sqrt{\beta_n + 1 + C_2}}{\beta_n + C_3} \ln\left(C_4 + \frac{C_5}{\beta_n}\right) e^{-\Delta E/T_e},$$

$$f(n) = n^{\alpha} \left(\frac{n^2}{n^2 - 1}\right)^{3/2} \frac{n(n-1)^{2n-2}}{(n+1)^{2n+2}},$$

$$\beta_n = \frac{E_H(n)}{T_e},$$

$$\Delta E = E_{He^{0+}}(1) - E_{He^{0+}}(n).$$

For $\frac{1}{16} \leq \beta_n n^2 \leq 32$ (i.e., $0.43\,\text{eV} \leq T_e \leq 220\,\text{eV}$), the fitting formula provides an accuracy better than about 4% for all values of β and n. The following table contains the fitting parameters. α has been fixed to $\alpha = 0.3$.

n'	C_1	C_2	C_3	C_4	C_5	$\Delta_{max}/\%$
2	1.8128×10^3	4.4878×10^0	5.4878×10^0	1.5134×10^0	1.0478×10^{-1}	1.3
3	4.5137×10^2	2.1152×10^2	5.0594×10^0	1.5360×10^0	3.7746×10^{-2}	1.6
4	1.3811×10^3	4.4394×10^0	1.7340×10^0	1.5156×10^0	1.6433×10^{-2}	1.9
5	1.9392×10^2	2.8594×10^2	1.5315×10^0	1.5255×10^0	1.1232×10^{-2}	1.7
6	2.0475×10^2	1.5987×10^2	1.0326×10^0	1.5171×10^0	7.4141×10^{-3}	1.8
7	1.0388×10^3	1.8963×10^0	5.9016×10^{-1}	1.4962×10^0	4.7582×10^{-3}	2.0
8	4.6740×10^2	1.3220×10^1	5.4899×10^{-1}	1.5017×10^0	3.8854×10^{-3}	2.0
9	1.3774×10^2	1.2765×10^2	4.4985×10^{-1}	1.5036×10^0	3.0905×10^{-3}	2.1

For large principal quantum numbers, we derived a n-scaled formula based on the numerical calculations for n = 4, 5, 6, 7, 8, and 9:

$$\langle V\sigma_{ex}(1 \rightarrow n)\rangle = \frac{10^{-8}}{n^2} f(n) C_1 \sqrt{\beta_n} \frac{\sqrt{\beta_n + 1 + C_2}}{\beta_n + C_3} \ln\left(n^{0.1} C_4 + \frac{C_5}{\beta_n}\right) e^{-\Delta E/T_e},$$

$$f(n) = n^{\alpha} \left(\frac{n^2}{n^2 - 1}\right)^{3/2} \frac{n(n-1)^{2n-2}}{(n+1)^{2n+2}},$$

$$\beta_n = \frac{E_H(n)}{T_e},$$

$$\Delta E = E_{He^{0+}}(1) - E_{He^{0+}}(n).$$

Taking the above formula with $\alpha = 2.8828 \times 10^{-1}$, $C_1 = 1.6571 \times 10^3$, $C_2 = -3.4064 \times 10^{-1}$, $C_3 = 6.5951 \times 10^{-1}$, $C_4 = 1.3230 \times 10^0$, $C_5 = 1.3372 \times 10^{-2}$, an accuracy better than 15% is achieved for all values of n and β.

II. $He^{0+}(n = 1, S = 0) + e \rightarrow He^{0+}(n', S = 1) + e,$

these transitions lead to a strong population of the excited triplet levels through exchange collisions from the ground state. Rate coefficients for the transitions from the ground state, i.e., $n = 1 \rightarrow n'$ have been fitted into the following formula:

$$\langle V\sigma_{ex}(1 \rightarrow n)\rangle = \frac{10^{-8}}{n^2} f(n) C_1 \sqrt{\beta} \frac{\beta^{C_2}}{\beta + C_3} e^{-\Delta E/T_e},$$

$$f(n) = \left(\frac{n^2}{n^2 - 1}\right)^{3/2} \frac{n(n-1)^{2n-2}}{(n+1)^{2n+2}},$$

$$\beta = \frac{Ry}{T_e}.$$

For $\frac{1}{16} \leq \beta \leq 32$ (i.e., $0.43\,\text{eV} \leq T_e \leq 220\,\text{eV}$), the fitting formula provides an accuracy better than about 9%. We note that the introduction of the fitting exponent C_2 provides a better description of the high energy fall than the standard formulas being proportional to $\beta_n^{3/2}$. The following table contains the fitting parameters.

n'	C_1	C_2	C_3	Δ_{max}/%
2	1.4455×10^2	6.9433×10^{-1}	3.0607×10^{-1}	6.8
3	2.9826×10^2	6.6088×10^{-1}	1.7122×10^{-1}	9.6
4	3.7799×10^2	6.1672×10^{-1}	1.1785×10^{-1}	8.5
5	4.1927×10^2	5.9992×10^{-1}	8.1190×10^{-2}	7.9
6	4.4504×10^2	5.8875×10^{-1}	5.9438×10^{-2}	7.8
7	4.6308×10^2	5.9198×10^{-1}	4.2183×10^{-2}	8.5
8	4.7360×10^2	5.7756×10^{-1}	3.4819×10^{-2}	9.8
9	4.8923×10^2	5.8679×10^{-1}	2.6089×10^{-2}	9.1

For large principal quantum numbers, we derived a n-scaled formula based on the numerical calculations for n = 4, 5, 6, 7, 8, and 9:

$$\langle V\sigma_{ex}(1 \to n)\rangle = \frac{10^{-8}}{n^2} f(n) C_1 \sqrt{\beta} \frac{\beta^{C_2}}{\beta + C_3} e^{-\Delta E/T_e},$$

$$f(n) = n^{\alpha} \left(\frac{n^2}{n^2 - 1}\right)^{3/2} \frac{n(n-1)^{2n-2}}{(n+1)^{2n+2}},$$

$$\beta = \frac{Ry}{T_e}.$$

Taking the above formula with $\alpha = 1.7440 \times 10^{-1}$, $C_1 = 2.1931 \times 10^2$, $C_2 = 6.1083 \times 10^{-1}$, $C_3 = 1.9344 \times 10^0$, an accuracy better than 9% is achieved for all values of n and β.

III. $He^{0+}(n, S = 0) + e \to He^{0+}(n' = n, S = 1) + e,$

these transitions are very important to obtain a decreasing importance of metastables of the triplet system with increasing density (note that these transitions are very rare in the literature). Correct population of the triplet excited states of helium is very important to incorporate, e.g., charge exchange processes. The numerical calculation of these cross section data include all combinations of angular momenta (monopole, dipole, quadruple, octupole….) and all multiplicities κ (e.g., for $n = n' = 5$: l = 0–4 and l' = 0–4, κ = 0–8, see also Sect. 5.5.2):

$$\sigma_{ex}(n \to n') = \frac{1}{n^2} \sum_{\kappa=|l-l'|}^{|l+l'|} \sum_{l=0}^{n-1} \sum_{l'=0}^{n'-1} (2l+1)\sigma_{\kappa}(nl \to n'l').$$

Summed rates are fitted into the following expression:

$$\langle V\sigma_{ex}(n, S = 0 \to n' = n, S = 1)\rangle = 10^{-8} \left(\frac{E_{He^{1+}}(n', S = 1)}{E_{He^{1+}}(n, S = 0)}\right)^{3/2}$$

$$\times A\sqrt{\beta} \frac{\beta + D}{\beta + \chi} e^{-\Delta E/T_e} \left[\mathrm{cm^3 s^{-1}}\right],$$

$$\Delta E = E_{He0+}(n, S = 0) - E_{He0+}(n' = n, S = 1),$$

$$\beta = \frac{Ry}{T_e}.$$

(note that ΔE is negative as line strengths have been calculated from singlet to triplet system). For $\frac{1}{16} \le \beta \le 32$ (i.e., $0.43\,\mathrm{eV} \le \beta \le 220\,\mathrm{eV}$), the fitting formula provides an accuracy better than about 2% for all values of β and n. The following table contains the fitting parameters.

$n = n'$	A	χ	D
2	6.39×10^1	2.17	0.01
3	2.78×10^2	4.53	0.00
4	4.91×10^2	8.17	0.00
5	1.23×10^3	1.33×10^1	0.00
6	2.04×10^3	2.04×10^1	0.00
7	2.65×10^3	2.78×10^1	0.00
8	4.69×10^3	4.61×10^1	0.00
9	4.58×10^3	5.40×10^1	0.00

For large principal quantum numbers, we derived a n-scaled formula based on the numerical calculations for n = 4, 5, 6, 7, 8, and 9:

$$\langle V\sigma(n \to n')\rangle = \frac{10^{-8}}{n^2} C_1 n^{\alpha} \sqrt{\beta} \frac{\beta^{C_2}}{\beta + C_3} e^{-\Delta E/T_e},$$
$$\Delta E = E_{\mathrm{He0+}}(n, S = 0) - E_{\mathrm{He0+}}(n' = n, S = 1),$$
$$\beta = \frac{Ry}{T_e}.$$

Taking the above formula with $\alpha = 1.3339 \times 10^0$, $C_1 = 2.6840 \times 10^2$, $C_2 = 9.4216 \times 10^{-1}$, $C_3 = 4.2964 \times 10^1$, an accuracy better than 65% is achieved for all values of n and β (note that more accurate fits may not provide necessarily improved accuracy as the accuracy of calculation for $\Delta n = 0$ transitions is about a factor of 2).

IV. $$He^{0+}(n \geq 2, S = 0) + e \to He^{0+}(n' \geq 2, S = 0) + e,$$
$$He^{0+}(n \geq 2, S = 1) + e \to He^{0+}(n' \geq 2, S = 1) + e,$$

these transitions provide the necessary exchange of populations between the excited states. For all transitions $\Delta S = 0$ and Regemorter-type formulas provide a good approximation:

$$\langle V\sigma_{\mathrm{ex}}(n \to n')\rangle = 3.15 \times 10^{-7} f(n \to n') \left(\frac{Ry}{\Delta E}\right)^{3/2} \sqrt{\beta_{\mathrm{n}}}\, e^{-\beta} p(\beta_{\mathrm{n}}),$$
$$\Delta E = E_{\mathrm{He}^{01+}}(n) - E_{\mathrm{He}^{0+}}(n'),$$
$$\beta_{\mathrm{n}} = \frac{\Delta E}{T_e}.$$

f is the oscillator strength and $p(\beta)$ the effective Gaunt factor. The required oscillator strengths are obtained from the transition probabilities determined above. For the Gaunt factor, we employ the expression

$$p(\beta) = \left\{ \begin{array}{ll} \text{if } \beta \le 0.4: & 0.27566 \cdot \left(\beta - \frac{\beta^2}{4} + \frac{\beta^3}{12} - \ln(\beta) - 0.57722 \right) \\ \text{else} & 0.066 \frac{\sqrt{\beta + 2}}{\beta + 0.127} \end{array} \right\}.$$

Concerning the metastable level, we consider the electron collisional process from the 2*s*-level into the 2*p*-level employing the following fitting formula:

$$\langle V\sigma(1s2lLS \rightarrow 1s2l'L'S')\rangle = 10^{-8} \left(\frac{E_{up}}{E_{low}} \right)^{3/2} \frac{A}{2l_0 + 1} \frac{\sqrt{\beta}}{\beta + \chi} F(\Delta S) e^{-\Delta E/T_e} \left[\mathrm{cm^3 s^{-1}} \right],$$

$$F(\Delta S) = \left\{ \begin{array}{ll} \beta + 1 + D & \text{if } \Delta S = 0 \\ \beta + D & \text{if } \Delta S = 1 \end{array} \right\},$$

$$\Delta E = E_{low} - E_{up},$$

$$\beta = \frac{Ry}{T_e}.$$

Transition	A	χ	D	ΔS	l_0
$1s2s\ ^3S_1 \rightarrow 1s2s\ ^1S_0$	2.79	1.66	0.0	1	0
$1s2s\ ^3S_1 \rightarrow 1s2p\ ^3P_{012}$	49.5	0.95	8.55	0	0
$1s2s\ ^3S_1 \rightarrow 1s2p\ ^1P_1$	3.64	1.52	0.0	1	0
$1s2s\ ^1S_0 \rightarrow 1s2p\ ^3P_{012}$	5.30	1.08	0.0	1	0
$1s2s\ ^1S_0 \rightarrow 1s2p\ ^1P_1$	53.5	1.22	13.3	0	0
$1s2p\ ^3P_{012} \rightarrow 1s2p\ ^1P_1$	8.76	1.59	0.0	1	1

For $\frac{1}{16} \le \beta \le 32$ (i.e., $0.43\,\mathrm{eV} \le T_e \le 220\,\mathrm{eV}$), the fitting formula provides an accuracy better than about 10% for all values of β (note that even in the case $\Delta S = 0$, there is an exchange part in the cross section and the total sum of direct and exchange part has been fitted in that case into the formula for $\Delta S = 0$).

Collisional de-excitation is calculated from the excitation rates as described in Annex A.2.4.

A.3.6 Ionization and Three-Body Recombination

Helium II (He^{1+})

Ionization rates for the processes

$$He^{1+}(n) + e \rightarrow He^{2+} + 2e$$

are fitted to the following formula:

$$\langle V\sigma_{\mathrm{iz}}\rangle = \frac{10^{-8}}{8n^2}\sqrt{\beta}\frac{A}{\beta+\chi}\frac{\beta+1+D}{\beta+1} e^{-E_{\mathrm{HeI}^+}(n)/T_\mathrm{e}} \quad \left[\mathrm{cm^3 s^{-1}}\right],$$
$$\beta = \frac{4Ry}{T_\mathrm{e}}.$$

The following table contains the corresponding fitting parameters. For all values of n = 1–9 and $\frac{1}{4} \le \beta \le 128$ (i.e., $0.5\,\mathrm{eV} \le \beta \le 220\,\mathrm{eV}$), the fitting is accurate within 10%.

n	A	χ	D
1	6.22	0.216	−0.590
2	9.08×10^2	1.10×10^1	1.44
3	1.21×10^4	2.04×10^1	1.14
4	6.61×10^4	3.00×10^1	1.23
5	2.00×10^5	2.76×10^1	0.780
6	4.93×10^5	2.61×10^1	0.530
7	7.47×10^5	3.30×10^1	0.230
8	2.09×10^6	3.41×10^1	0.180
9	4.44×10^6	3.61×10^1	0.120

In order to provide ionization rates for any values of $n \geq 7$ and β, the following fitting formula has been developed:

$$\langle V\sigma_{\mathrm{iz}}\rangle = \frac{10^{-8}}{8n^2}\sqrt{\beta}\frac{An^5}{\beta+\chi}\frac{\beta+D}{\beta+f(n)}e^{-\beta} \quad \left[\mathrm{cm^3 s^{-1}}\right],$$
$$f(n) = 3 + \frac{\gamma}{n^4},$$
$$\beta = \frac{4 \cdot E_\mathrm{H}(n)}{T_\mathrm{e}}.$$

With A = 1.1426×10^1, $\chi = 3.5913 \times 10^{-1}$, $D = 2.4579 \times 10^0$ and $\gamma = 6.8156 \times 10^3$, the above fitting formula provides an accuracy better than 20% for all values of $n \geq 7$ and β ($\frac{1}{4} \le 4Ry/T_\mathrm{e} \le 128$).

Concerning metastable levels, we consider here the ionization out of the levels 2s and 2p and fit the rates to the following fitting formula:

$$\langle V\sigma_{\mathrm{iz}}\rangle = \frac{10^{-8}}{8(2l_0+1)}\sqrt{\beta}\frac{A}{\beta+\chi}\frac{\beta+1+D}{\beta+1} e^{-E_{\mathrm{HeI}^+}(n)/T_\mathrm{e}} \quad \left[\mathrm{cm^3 s^{-1}}\right],$$
$$\beta = \frac{4Ry}{T_\mathrm{e}}.$$

n	A	χ	D	I_0
2s	2.46×10^2	2.52×10^1	3.90	0
2p	6.84×10^2	9.10	0.95	1

For the 2s level, the accuracy is better than 8%, for the 2p level better than 5% for all values of β in the interval $\frac{1}{4} \le \beta \le 128$ (i.e., $0.5\,\text{eV} \le T_e \le 220\,\text{eV}$).

Three-body recombination is calculated from the ionization rates as described in Annex A.2.5.

Helium I (H^{0+})

Ionization rates for the processes

$$He(n,S) + e \rightarrow He^{1+} + 2e$$

are fitted to the following formula:

$$\langle V\sigma_{iz}\rangle = 10^{-8}\frac{1}{n^2}\sqrt{\beta}\frac{mA}{\beta+\chi}\frac{\beta+1+D}{\beta+1} \times \left(\frac{1}{1+\beta\frac{E_{He^{0+}}(n,S)}{Ry}}\right)^{1/2} e^{-E_{He_{0+}}(n,S)/T_e} \quad \left[\text{cm}^3\text{s}^{-1}\right],$$

$$\beta = \frac{Ry}{T_e},$$

$$m = \begin{Bmatrix} 2 & \text{if } n = 1 \\ 1 & \text{if } n \ge 2 \end{Bmatrix}.$$

The following table contains the corresponding fitting parameters. For all values of n = 1–9 and $\frac{1}{16} \le \beta \le 32$ (i.e., $0.5\,\text{eV} \le T_e \le 220\,\text{eV}$), the fitting is accurate within 10%.

n	A	χ	D
1	1.79	0.100	−0.510
2	4.04×10^3	3.93×10^1	0.950
3	1.65×10^6	1.91×10^3	0.460
4	6.30×10^6	1.86×10^3	0.330
5	1.28×10^7	1.45×10^3	0.430
6	3.86×10^6	1.85×10^2	0.350
7	4.84×10^6	1.13×10^2	0.310
8	6.23×10^6	7.61×10^1	0.230
9	9.30×10^6	6.49×10^1	0.170

In order to provide ionization rates for any values of $n \geq 6$ and β, the following fitting formula has been developed:

$$\langle V\sigma_{\mathrm{iz}}\rangle = \frac{10^{-8}}{n^2}\sqrt{\beta}\frac{An^5}{\beta+\chi\beta+f(n)}\frac{\beta+D}{}e^{-\beta}\quad[\mathrm{cm^3 s^{-1}}],$$

$$f(n) = 10+\frac{\gamma}{n^4},$$

$$\beta = \frac{E_{\mathrm{H}}(n)}{T_{\mathrm{e}}}.$$

With A = 8.4667 × 10^1, χ = 1.8865 × 10^{-1}, D = 5.3907 × 10^{-1}, and γ = 8.3805 × 10^2, the fitting formula provides an accuracy better than 10% for all values of $n \geq 6$ and β ($\frac{1}{16} \leq Ry/T_{\mathrm{e}} \leq 32$).

Concerning metastable levels, we consider here the ionization out of the levels $1s2s\ ^1S_0$, $1s2s\ ^3S_1$, $1s2p\ ^1P_1$ and $1s2p\ ^3P_{012}$. Data are fitted to the following formula:

$$\langle V\sigma_{\mathrm{iz}}\rangle = 10^{-8}\frac{1}{2l_0+1}\sqrt{\beta}\frac{A}{\beta+\chi}\frac{\beta+1+D}{\beta+1}$$

$$\times\left(\frac{1}{1+\beta\frac{E_{\mathrm{He^{0+}}}(n,S)}{Ry}}\right)^{1/2} e^{-E_{He_{0+}}(n,S)/T_{\mathrm{e}}}\ [\mathrm{cm^3 s^{-1}}],$$

$$\beta = \frac{Ry}{T_{\mathrm{e}}}.$$

Level	A	χ	D	l_0
$1s2s\ ^1S_0$	7.76 × 10^2	120.6	0.1	0
$1s2s\ ^3S_1$	4.17 × 10^2	15.8	0.05	0
$1s2p\ ^1P_1$	4.14 × 10^3	50.5	1.0	1
$1s2p\ ^3P_{012}$	3.28 × 10^3	46.6	1.15	1

For the $1s2s$-levels, the accuracy is better than 22%, for the $1s2p$-levels better than 6% for all values of β in the interval $\frac{1}{16} \leq \beta \leq 32$ (i.e., $0.5\,\mathrm{eV} \leq T_{\mathrm{e}} \leq 220\,\mathrm{eV}$).

Three-body recombination is calculated from the ionization rates as described in Annex A.2.5.

A.3.7 Matrix Elements Including Phase Sign, Oscillator Strengths, and Energies of nlj-Split Levels

Matrix elements including phase sign, oscillator strengths, and energies for an LSJ-split level structure have been calculated for the H-like He II (He^{1+}) states *nlj* from n = 1–6, l = 0–5 and all j. Designations of levels, principal quantum numbers, orbital quantum numbers, statistical weights, and energies are given in the following table:

N°	*n*	*l*	*g*	E (eV)
1	1	0	2	0.0000000000D+00
2	2	1	2	4.0813272140D+01
3	2	0	2	4.0813305651D+01
4	2	1	4	4.0813998142D+01
5	3	1	2	4.8371547614D+01
6	3	0	2	4.8371557179D+01
7	3	2	4	4.8371762682D+01
8	3	1	4	4.8371762735D+01
9	3	2	6	4.8371834300D+01
10	4	1	2	5.1016914617D+01
11	4	0	2	5.1016918535D+01
12	4	1	4	5.1017005364D+01
13	4	2	4	5.1017005503D+01
14	4	2	6	5.1017035715D+01
15	4	3	6	5.1017035836D+01
16	4	3	8	5.1017050924D+01
17	5	1	2	5.2241329093D+01
18	5	0	2	5.2241331054D+01
19	5	1	4	5.2241375557D+01
20	5	2	4	5.2241375708D+01
21	5	2	6	5.2241391176D+01
22	5	3	6	5.2241391315D+01
23	5	3	8	5.2241399040D+01
24	5	4	8	5.2241399083D+01
25	5	4	10	5.2241403718D+01
26	6	0	2	5.2906436533D+01
27	6	1	2	5.2906437979D+01
28	6	1	4	5.2906464806D+01
29	6	2	4	5.2906464971D+01
30	6	2	6	5.2906473913D+01
31	6	3	6	5.2906474035D+01
32	6	3	8	5.2906478506D+01

(continued)

(continued)

N°	n	l	g	E (eV)
33	6	4	8	5.2906478564D+01
34	6	4	10	5.2906481246D+01
35	6	5	10	5.2906481250D+01
36	6	5	12	5.2906483038D+01

Note that the total quantum number j can be calculated from the relation $g = 2j+1$. Matrix elements [in atomic units], oscillator strengths, and wavelengths [in Angstroem] are given in the following table:

N° of level i	N° of level j	$\langle i\|r\|j\rangle$ [a_0]	f_{ij}	λ_{ji} [10^{-10} m]
1	2	5.2682085636D−01	1.3875710000D−01	3.0378400702D+02
1	4	7.4499002699D−01	2.7748380000D−01	3.0377857353D+02
1	5	2.1099507736D−01	2.6379260000D−02	2.5631634091D+02
1	8	2.9839170775D−01	5.2758600000D−02	2.5631522814D+02
1	10	1.2441232688D−01	9.6731700000D−03	2.4302566933D+02
1	12	1.7594874397D−01	1.9347070000D−02	2.4302519297D+02
1	17	−8.5289593601D−02	4.6551585000D−03	2.3732969088D+02
1	19	−1.2062103245D−01	9.3108350000D−03	2.3732946373D+02
1	27	−6.3478326321D−02	2.6114840000D−03	2.3434611552D+02
1	28	−8.9774842625D−02	5.2233150000D−03	2.3434602693D+02
2	3	2.1213549918D+00	1.8473335000D−06	3.6997769406D+08
3	4	3.0002088177D+00	7.6356300000D−05	1.7904088992D+07
3	5	−1.2514149539D+00	1.4499455000D−01	1.6403839279D+03
2	6	−3.8323688474D−01	1.3598340000D−02	1.6403745956D+03
4	6	5.4217667006D−01	1.3606960000D−02	1.6405321745D+03
2	7	−2.7413928866D+00	6.9583400000D−01	1.6403298886D+03
4	7	−1.2261012383D+00	6.9589550000D−02	1.6404874589D+03
3	8	−1.7695205778D+00	2.8991610000D−01	1.6403372672D+03
4	9	−3.6782605813D+00	6.2629700000D−01	1.6404720478D+03
3	10	−5.2367309981D−01	3.4276955000D−02	1.2151011946D+03
2	11	1.5621586858D−01	3.0502365000D−03	1.2150964312D+03
4	11	−2.2099194090D−01	3.0519350000D−03	1.2151833691D+03
3	12	−7.4053101163D−01	6.8544350000D−02	1.2150904770D+03
2	13	9.8731127826D−01	1.2184155000D−01	1.2150869045D+03
4	13	4.4154526347D−01	1.2183627500D−02	1.2151726500D+03
4	14	1.3246570743D+00	1.0965652500D−01	1.2151690771D+03
3	17	3.1619517158D−01	1.3996175000D−02	1.0849139833D+03
2	18	9.3277283933D−02	1.2180140000D−03	1.0849101860D+03
4	18	−1.3195340857D−01	1.2186640000D−03	1.0849794921D+03
3	19	4.4714413496D−01	2.7989550000D−02	1.0849092366D+03
2	20	−5.6323515853D−01	4.4410080000D−02	1.0849063886D+03
4	20	−2.5188303201D−01	4.4406075000D−03	1.0849747449D+03

(continued)

(continued)

N° of level i	N° of level j	$\langle i\|r\|j\rangle$ [a_0]	f_{ij}	λ_{ji} [10^{-10} m]
4	21	−7.5566875360D−01	3.9967625000D−02	1.0849737954D+03
2	26	6.4905812278D−02	6.2407250000D−04	1.0252422609D+03
4	26	−9.1817320562D−02	6.2439775000D−04	1.0253033052D+03
3	27	2.2099568857D−01	7.2349300000D−03	1.0252448043D+03
3	28	3.1252217611D−01	1.4468695000D−02	1.0252422609D+03
2	29	3.8255318319D−01	2.1679660000D−02	1.0252397176D+03
4	29	1.7107847530D−01	2.1677210000D−03	1.0253007615D+03
4	30	5.1325133580D−01	1.9510722500D−02	1.0252999137D+03
5	6	5.1964061126D+00	3.1638560000D−06	1.2962348253D+09
5	7	5.8100687326D+00	8.8933300000D−05	5.7649045043D+07
6	8	7.3490809850D+00	1.3599535000D−04	6.0316471756D+07
8	9	7.7950603030D+00	2.6633950000D−05	1.7324773478D+08
6	10	−2.2332723176D+00	1.6162020000D−01	4.6868602990D+03
7	10	7.5242596190D−01	9.1722550000D−03	4.6872235320D+03
5	11	9.9789242413D−01	3.2268740000D−02	4.6868354949D+03
8	11	−1.4116906054D+00	3.2287050000D−02	4.6872164439D+03
6	12	−3.1577918066D+00	3.2314250000D−01	4.6866990771D+03
7	12	−3.3637515058D−01	1.8332060000D−03	4.6870622850D+03
9	12	1.0092598727D+00	1.1001866667D−02	4.6871898641D+03
5	13	4.3680651666D+00	6.1831100000D−01	4.6866813610D+03
8	13	1.9537261369D+00	6.1843025000D−02	4.6870622850D+03
8	14	5.8609613290D+00	5.5655250000D−01	4.6870091291D+03
7	15	−7.9253093721D+00	1.0176557500D+00	4.6870091291D+03
9	15	−2.1181800823D+00	4.8460866667D−02	4.6871349334D+03
9	16	−9.4727135678D+00	9.6920783333D−01	4.6871083544D+03
6	17	9.2291223527D−01	4.0377025000D−02	3.2039145200D+03
7	17	−2.7920337529D−01	1.8475725000D−03	3.2040850834D+03
5	18	3.9622424428D−01	7.4421200000D−03	3.2039054128D+03
8	18	−5.6049632184D−01	7.4457075000D−03	3.2040834274D+03
6	19	1.3050652667D+00	8.0738950000D−02	3.2038764355D+03
7	19	1.2482435800D−01	3.6928700000D−04	3.2040461669D+03
9	19	−3.7451634894D−01	2.2161933333D−03	3.2041057841D+03
5	20	−1.7142785676D+00	1.3931000000D−01	3.2038681564D+03
8	20	−7.6669325542D−01	1.3931855000D−02	3.2040461669D+03
8	21	−2.3000620886D+00	1.2538520000D−01	3.2040337469D+03
7	22	2.5712976433D+00	1.5670107500D−01	3.2040329189D+03
9	22	6.8720222027D−01	7.4616866667D−03	3.2040925357D+03
9	23	3.0732996173D+00	1.4923771667D−01	3.2040859115D+03
5	26	2.3430506454D−01	3.0497050000D−03	2.7340070948D+03
8	26	−3.3144081804D−01	3.0510975000D−03	2.7341367207D+03
6	27	5.5605781257D−01	1.7176460000D−02	2.7340119178D+03
7	27	−1.5967977257D−01	7.0818050000D−04	2.7341361178D+03
6	28	7.8632466099D−01	3.4347865000D−02	2.7339956400D+03
7	28	7.1389288497D−02	1.4155095000D−04	2.7341198385D+03

(continued)

(continued)

N° of level *i*	N° of level *j*	⟨*i*‖*r*‖*j*⟩ [a_0]	f_{ij}	λ_{ji} [10^{-10} m]
9	28	−2.1419176570D−01	8.4948200000D−04	2.7341626474D+03
5	29	1.0061669877D+00	5.6239000000D−02	2.7339902142D+03
8	29	4.4998536072D−01	5.6239775000D−03	2.7341198385D+03
8	30	1.3499587692D+00	5.0616050000D−02	2.7341144121D+03
7	31	1.3938573724D+00	5.3961550000D−02	2.7341144121D+03
9	31	3.7251643883D−01	2.5694516667D−03	2.7341572209D+03
9	32	1.6659753963D+00	5.1391016667D−02	2.7341548091D+03
10	11	−9.4874850284D+00	4.3197240000D−06	3.1647591541D+09
11	12	−1.3417669371D+01	1.9149105000D−04	1.4279069213D+08
10	13	−1.2001284991D+01	1.6035470000D−04	1.3641682040D+08
12	14	−1.6101464875D+01	4.8195025000D−05	4.0850062140D+08
13	15	−1.2297803148D+01	2.8096875000D−05	4.0875367637D+08
14	16	−1.4698713513D+01	1.3417165000D−05	8.1521240307D+08
11	17	−3.4781790607D+00	1.8145075000D−01	1.0126026881D+04
13	17	1.7595814849D+00	2.3217370000D−02	1.0126746433D+04
10	18	1.8787374628D+00	5.2940650000D−02	1.0125985531D+04
12	18	−2.6577456899D+00	5.2968975000D−02	1.0126729891D+04
11	19	−4.9179764535D+00	3.6278080000D−01	1.0125646470D+04
13	19	−7.8664993011D−01	4.6405950000D−03	1.0126365968D+04
14	19	2.3602425325D+00	2.7849783333D−02	1.0126614094D+04
10	20	−6.3736618286D+00	6.0932750000D−01	1.0125613392D+04
12	20	−2.8508487737D+00	6.0947950000D−02	1.0126365968D+04
15	20	1.2876782181D+00	8.2894016667D−03	1.0126614094D+04
12	21	−8.5521100361D+00	5.4848250000D−01	1.0126233639D+04
15	21	−3.4410064572D−01	5.9195016667D−04	1.0126490029D+04
16	21	1.5389825539D+00	8.8805025000D−03	1.0126614094D+04
13	22	−1.0896416503D+01	8.9039675000D−01	1.0126233639D+04
14	22	−2.9123126164D+00	4.2402383333D−02	1.0126481759D+04
14	23	−1.3023966457D+01	8.4801516667D−01	1.0126423863D+04
15	24	−1.6408257294D+01	1.3459900000D+00	1.0126423863D+04
16	24	−3.1578243335D+00	3.7389425000D−02	1.0126547926D+04
16	25	−1.8681697519D+01	1.3086012500D+00	1.0126506571D+04
10	26	7.3130397524D−01	1.2378725000D−02	6.5616694063D+03
12	26	−1.0344767148D+00	1.2384267500D−02	6.5619854337D+03
11	27	−1.4116474719D+00	4.6124505000D−02	6.5616798243D+03
13	27	6.5001137343D−01	4.8895750000D−03	6.5619819607D+03
11	28	−1.9961384017D+00	9.2228850000D−02	6.5615860635D+03
13	28	−2.9061190681D−01	9.7737750000D−04	6.5618881913D+03
14	28	8.7192842213D−01	5.8654183333D−03	6.5619923797D+03
10	29	2.5342204992D+00	1.4865350000D−01	6.5615721733D+03
12	29	1.1334334804D+00	1.4867155000D−02	6.5618847184D+03
15	29	−4.3903461013D−01	1.4870816667D−03	6.5619923797D+03
12	30	3.4002205381D+00	1.3379865000D−01	6.5618534625D+03
15	30	1.1732287429D−01	1.0619518333D−04	6.5619611228D+03

(continued)

(continued)

N° of level i	N° of level j	$\langle i\|\|r\|\|j\rangle$ [a_0]	f_{ij}	λ_{ji} [10^{-10} m]
16	30	−5.2471954713D−01	1.5931312500D−03	6.5620132178D+03
13	31	−4.0127029969D+00	1.8634237500D−01	6.5618534625D+03
14	31	−1.0724525916D+00	8.8735166667D−03	6.5619611228D+03
14	32	−4.7961306414D+00	1.7746900000D−01	6.5619437580D+03
15	33	−4.8634295293D+00	1.8248433333D−01	6.5619437580D+03
16	33	−9.3596542645D−01	5.0689487500D−03	6.5619958527D+03
16	34	−5.5372513801D+00	1.7741375000D−01	6.5619889067D+03
17	18	1.5001143084D+01	5.4071150000D−06	6.3208735236D+09
18	19	2.1215239903D+01	2.4536205000D−04	2.7860093707D+08
17	20	−1.9845199301D+01	2.2488710000D−04	2.6597487289D+08
19	21	−2.6625200749D+01	6.7817925000D−05	7.9378989543D+08
20	22	−2.3240464435D+01	5.1629475000D−05	7.9443075941D+08
21	23	−2.7777699949D+01	2.4775700000D−05	1.5766659664D+09
22	24	−2.0833402241D+01	1.3766800000D−05	1.5961010389D+09
23	25	−2.3719889904D+01	8.0600025000D−06	2.6504784510D+09
17	26	−3.0267465352D+00	7.4640050000D−02	1.8641228018D+04
19	26	4.2817182363D+00	7.4678475000D−02	1.8642528580D+04
18	27	−4.9885775301D+00	2.0275560000D−01	1.8641242032D+04
20	27	−3.1371520667D+00	4.0089550000D−02	1.8642492139D+04
18	28	−7.0535358063D+00	4.0536915000D−01	1.8640488124D+04
20	28	1.4025392294D+00	8.0132350000D−03	1.8641740934D+04
21	28	−4.2081126630D+00	4.8089600000D−02	1.8642175392D+04
17	29	−8.7577720434D+00	6.2492150000D−01	1.8640429272D+04
19	29	−3.9172845936D+00	6.2509825000D−02	1.8641732525D+04
22	29	2.8318636637D+00	2.1778166667D−02	1.8642172589D+04
19	30	−1.1751167180D+01	5.6253000000D−01	1.8641480268D+04
22	30	−7.5675345540D−01	1.5552163333D−03	1.8641923124D+04
23	30	3.3845496376D+00	2.3331375000D−02	1.8642138953D+04
20	31	1.4399946756D+01	8.4470600000D−01	1.8641483071D+04
21	31	3.8487468547D+00	4.0227316667D−02	1.8641914715D+04
24	31	−1.8715341660D+00	7.1340150000D−03	1.8642136150D+04
21	32	1.7211615450D+01	8.0450483333D−01	1.8641791386D+04
24	32	3.6014935336D−01	2.6418412500D−04	1.8642012818D+04
25	32	−2.1307899176D+00	7.3979220000D−03	1.8642141756D+04
22	33	2.0896249892D+01	1.1858288333D+00	1.8641794189D+04
23	33	4.0215877936D+00	3.2940987500D−02	1.8642010015D+04
23	34	2.3791555176D+01	1.1528925000D+00	1.8641934335D+04
24	35	2.8691833454D+01	1.6767175000D+00	1.8641934335D+04
25	35	4.3255334009D+00	3.0486630000D−02	1.8642066075D+04
25	36	3.1785489261D+01	1.6462250000D+00	1.8642015621D+04
26	27	2.1735529073D+01	8.3705800000D−06	8.5719452155D+09
26	28	3.0739190428D+01	3.2725270000D−04	4.3852644457D+08
27	29	2.9392173223D+01	2.8564190000D−04	4.5934155734D+08
28	30	3.9433824907D+01	8.6734850000D−05	1.3614712155D+09

(continued)

(continued)

N° of level i	N° of level j	$\langle i\|\|r\|\|j\rangle$ [a_0]	f_{ij}	λ_{ji} [10^{-10} m]
29	31	3.6221721643D+01	7.2840575000D−05	1.3678204895D+09
30	32	4.3293277359D+01	3.5155733333D−05	2.6990924233D+09
31	33	−3.7260079667D+01	2.5670983333D−05	2.7379030420D+09
32	34	−4.2422489500D+01	1.5102075000D−05	4.5247011583D+09
33	35	−3.1460467629D+01	8.1423575000D−06	4.6154534293D+09
34	36	−3.4852649778D+01	5.3334240000D−06	6.9181466042D+09

A.4 Ionization Potential Depression: Level Delocalization and Line Shifts

In a low-density environment, where atoms and ions are essentially free, atomic population kinetics of gases and plasmas has been very successful in many different scientific and technical disciplines. As density increases, the free atom model breaks down resulting in a perturbation of the atomic energy levels. The perturbation of atomic levels manifests itself essentially in a broadening and a shift. These perturbations can be observed in high-resolution spectroscopic experiments via the analysis of the line broadening, the line shift and the disappearance of the line emission (corresponding to the ionization potential depression IPD of the upper level).

The IPD is of great interest for applications in thermodynamics and also for the understanding of the various radiative properties (emission, absorption, scattering). Since decades, the working horse in dense plasma atomic physics has been a plasma screening potential that acts as a perturber for the free atom Hamiltonian. Of particular interest is the self-consistent finite temperature ion sphere SCFTIS model: it allows to combine dense plasma effects in a very general manner with the numerical solution of the Schrödinger or Dirac equation and allows reaching spectroscopic precision. Recent measurements have, however questioned our understanding in the precision of the Multi-Configuration-Dirac-Fock Self-Consistent Finite Temperature Ion Sphere model MCDF-SCFTIS: it was claimed [Beiersdorfer et al. 2019] that measured line shift values are significantly smaller than predictions from the self-consistent-field ion-sphere model. However, a more profound analysis of the situation has discovered [Li and Rosmej 2020] that the self-consistent ion sphere model is in excellent agreement with the data and that the claimed discrepancies of [Beiersdorfer et al. 2019] are no conceptual ones of the ion sphere model but are just due to a limited precision of the employed 2nd-order approximation of the arbitrary perturbation potential method APPM while the original APPM [Rosmej et al. 2011] provides excellent agreement with the measurements.

Due to the complexity of the self-consistent numerical calculations of the MCDF and the Poisson equations applications are not very convenient neither for dense plasma atomic physics studies nor for integrated simulations in high-energy density physics. Moreover, analytical approximations for precise quantum states are rather rare and had been developed mostly for rather specific cases. Application of the SCFTIS model for non specialists was therefore very difficult. In the following, we close this gap proposing general and easy to use analytical formulas that do not request specific knowledge in atomic physics nor numerical resources. The developed formulas are based on the analytical b-potential method proposed recently [Li and Rosmej 2020].

A.4.1 The Analytical b-potential Method

The energy shift of an N-electron system is approximated by (in atomic units, i.e. ΔE in units of $2Ry = 2 \times 13.6057\,\mathrm{eV}$, R in units of the Bohr radius $a_0 = 0.529177 \cdot 10^{-8}\,cm$)

$$\Delta E_\mathrm{i} \approx \sum_{i=1}^{N} \frac{N_f}{R} \cdot \left[1 + \frac{1}{x-1} \cdot \left(1 - \frac{\langle r_\mathrm{i}^{x-1} \rangle}{R^{x-1}}\right)\right], \tag{A4.1}$$

where

$$x = 3 - \frac{b}{\pi} \cdot \sqrt{\frac{N_f}{R \cdot kT_\mathrm{e}}}. \tag{A4.2}$$

The ion sphere radius R is given by

$$\frac{4}{3}\pi R^3 \cdot n_\mathrm{e} = Z_\mathrm{n} - N_\mathrm{b}, \tag{A4.3}$$

n_e the electron density in units of $1/a_0^3$, kT_e the electron temperature in units of $2Ry$, Z_n is the nuclear charge, N_b is the number of bound electrons and $\mathrm{N_f}$ is the number of free electrons in the ion sphere that is determined from the quasi-neutrality condition:

$$N_f = Z_\mathrm{n} - N_\mathrm{b}. \tag{A4.4}$$

$\langle r_\mathrm{i}^{x-1} \rangle$ is the mean value of r_i^{x-1} of the ith-bound electron (in units of a_0) that can be approximated analytically according

$$\langle r_{\mathrm{i}}^{x-1}\rangle \approx \frac{(-1)^{n-l-1}}{2n}\cdot\left(\frac{n}{2\cdot Z_{\mathrm{i}}^{\mathrm{eff}}}\right)^{x-1}$$
$$\times\sum_{i=0}^{n-l-1}\frac{(-1)^{i}}{\Gamma(i+1)}\cdot\frac{\Gamma(2l+3+i+\beta)\cdot\Gamma(2+i+\beta)}{\Gamma(2l+2+i)\cdot\Gamma(n-l-i)\cdot\Gamma(l+3-n+i+\beta)},\tag{A4.5}$$

$$\beta = x-1,\tag{A4.6}$$

where n and l are the principal and orbital quantum number of the ith-electron, respectively. The effective nuclear charge can be calculated according

$$Z_{\mathrm{i}}^{\mathrm{eff}} = n\sqrt{2|E_{\mathrm{i}}|},\tag{A4.7}$$

where E_{i} is the energy of the ith-electron in units of $2Ry$ (means E_{i} is the ionization energy of the ith-electron). $\Gamma(z)$ is the Gamma function:

$$\Gamma(z) = \int_{0}^{\infty} t^{z-1}\cdot e^{-t}\mathrm{d}t.\tag{A4.8}$$

If the argument of the Gamma function is integer, we have simply

$$\Gamma(n+1) = n!.\tag{A4.9}$$

Due to possible non-integer values of the x-parameter (A4.2), the Gamma function for non-integer arguments is in generally requested. In that case, the integral can be calculated by an analytical approximation for all relevant values $z > 0$:

$$\Gamma(z+1)\approx\sqrt{2\pi}\cdot(z+\gamma+0.5)^{z+0.5}\cdot e^{-z-\gamma-0.5}$$
$$\times\left(c_0+\frac{c_1}{z+1}+\frac{c_2}{z+2}+\frac{c_3}{z+3}+\frac{c_4}{z+4}+\frac{c_5}{z+5}+\frac{c_6}{z+6}+\varepsilon\right).\tag{A4.10}$$

With $\gamma = 5.0$, $c_0 = 1.0$, $c_1 = 76.18009173$, $c_2 = -86.50532033$, $c_3 = 24.01409822$, $c_4 = -1.231739516$, $c_5 = 0.120858003\cdot 10^{-2}$, $c_6 = -0.536382\cdot 10^{-5}$ the error $|\varepsilon| < 2\cdot 10^{-10}$ which is more than sufficient for the present purpose.

Physically, the b-parameter describes the slope of the free electron distribution function inside the ion sphere and can be approximated by an analytical representation according

$$P_{\mathrm{f}}(r) = N_{\mathrm{f}} \cdot \left(\frac{r}{R}\right)^{x}. \tag{A4.11}$$

$P_{\mathrm{f}}(r)$ has the correct limit of the uniform electron gas model for which $x = 3$ (infinite temperature). b is a parameter that characterizes the self-consistent electron distribution in the ion sphere, e.g. for a Maxwellian electron distribution the MCDF-SCFTIS simulations are well matched for $b \approx 2$ while IPD values of near solid density matter could be described with $b \approx 4$, the so-called b4-potential (remaining lattice structure) [Li et al. 2019, Li and Rosmej 2020].

A.4.2 Simple Analytical Formulas for Line Shifts

If only one electron is involved in the atomic transition (corresponding to the overwhelming number of cases), the corresponding spectral shift $\Delta E_{\mathrm{shift}}$ can be easily calculated by the difference of the energy shift between the two orbits i and j involved, i.e. for the transition $j \to i$:

$$\Delta E_{\mathrm{shift}}(j \to i) = N_{\mathrm{f}} \cdot \frac{1}{x-1} \cdot \frac{1}{R^{x}} \cdot \left(\left\langle r_{\mathrm{j}}^{x-1}\right\rangle - \left\langle r_{\mathrm{i}}^{x-1}\right\rangle\right). \tag{A4.12}$$

Let us apply the analytical b-potential method for demonstration to the line shift in dense plasmas. We consider $He_{\beta} = 1s3p\,^{1}P_{1} \to 1s^{2}\,^{1}S_{0}$ of Cl^{15+} for $kT_{\mathrm{e}} = 600eV$, $n_{\mathrm{e}} = 5 \cdot 10^{23} cm^{-3}$, $b = 2$, $i = 1s^{2}\,^{1}S_{0}$, $E_{\mathrm{i}} = 3658.34eV$, $j = 1s3p\,^{1}P_{1}$, $E_{\mathrm{j}} = 386.69eV$ from which it follows (in atomic units): $N_{\mathrm{f}} = Z_{\mathrm{n}} - N_{\mathrm{b}} = 17 - 2 = 15$, $n_{\mathrm{e}} = 0.07409$, $kT_{\mathrm{e}} = 22.05$, $E_{\mathrm{i}} = 134.44$, $Z_{\mathrm{i}}^{\mathrm{eff}} = 16.40$, $E_{\mathrm{j}} = 14.211$, $Z_{\mathrm{j}}^{\mathrm{eff}} = 15.99$, $R = 3.643$, $x = 2.725$, $\beta = 1.725$, $\left\langle r_{\mathrm{i}}^{x-1}\right\rangle = 1.942 \cdot 10^{-2}$, $\Delta E_{\mathrm{i}} = 6.500$, $\left\langle r_{\mathrm{j}}^{x-1}\right\rangle = 0.7171$, $\Delta E_{\mathrm{j}} = 6.321$ and finally $\Delta E_{\mathrm{shift}}(j \to i) = 6.321 - 6.500 = -0.179$, i.e. $\Delta E_{\mathrm{shift}}(j \to i) = -4.87eV$ (red shift). This value (shown with black solid square in Fig. A.4.1) is in excellent agreement with the numerical calculations in the framework of the MCDF-SCFTIS (red crosses in Fig. A.4.1) [Li and Rosmej 2020]. The figure shows also the measurements [Beiersdorfer et al. 2019] of the line shift of He_{β} of Cl^{15+} in dependence of electron density that demonstrates also excellent agreement for all other densities. The curve with solid black squares of Figure A.4.1 shows the analytical result of the b-potential method according (A4.1–A4.12): it demonstrates very good agreement with the numerical MCDF-SCFTIS simulations and the data for all densities. The analytical b-potential method has therefore spectroscopic precision.

Figure A.4.2 demonstrates the application of the b-potential method for near solid density plasma, where lattice effects may remain (for more detailed discussion see [Li et al. 2019; Li and Rosmej 2020]. The figure shows the experimental measurements of $He_{\alpha} = 1s2p\,^{1}P_{1} \to 1s^{2}\,^{1}S_{0}$ of Al^{11+} [Stillman et al. 2017] and the numerical MCDF-SCFTIS simulations (red crosses connected with solid red line). It can be seen, that the MCDF-SCFTIS simulations are very far from the data. The solid blue circles show MCDF-simulations that include the b-potential for b = 4 (so-called b4-potential method [Li and Rosmej 2020]) while the dashed blue curve includes to these simulations the apparent shifts due to line overlap from higher order satellites and other effects (for details see [Stillman et al. 2017]). Excellent agreement with the measurements is observed. Figure A.4.2 includes also the results form the analytical b-potential method (i.e. (A4.1–A4.12)) for $b = 4$, black solid curve. Excellent agreement with the much more complex MCDF simulations (solid blue circles connected with solid blue curve) is observed. Figure A.4.2 demonstrates therefore the capacity of the b-potential method to imitate lattice effects in ionization potential depression and also line shifts [Li et al. 2019, Li and Rosmej 2020]. However, to what extend the b-potential method could imitate lattice effects for $b > 2$ at temperatures much larger than the Fermi temperature remains an open question and is subject to active research. We note, that for near solid density plasmas with temperatures near the Fermi temperature, the Atomic-Solid-Plasma Model ASP (see Chapter 8) has provided excellent agreement with the data.

A.4.3 Quantum Number Dependent Line Shift and Level Delocalization: High Precision 4th-order Analytical Formulas

A more simplified representation of the arbitrary perturbation potential method APPM [Rosmej et al. 2011] has been developed in [Li and Rosmej 2012, 2020] via 2nd and 4th order expansions in terms of non-fractional matrix elements (i.e. with integer β-values of (A4.6) that have then analytical solutions).

Below, we recommend the 4th-order quantum number dependent analytical approximation of the IPD in the optical electron approximation for dense hot plasmas that is given by [Rosmej 2018, Li and Rosmej 2020] (in atomic units):

$$\Delta E_{\text{IPD}} = \frac{Z}{R_0} \cdot \frac{F(<r>, <r^2>, <r^3>, <r^4>)}{F\left(R_0, R_0^2, R_0^3, R_0^4\right)}, \tag{A4.13}$$

$$F(x_1, x_2, x_3, x_4) \approx \frac{R_0^2}{2} - \frac{x_2^2}{6} + \frac{4}{3\sqrt{\pi}} \cdot \sqrt{\frac{Z}{k_B T_e}} \cdot \left\{ R_0^{3/2} - \frac{2}{5} \cdot \left[a_1 x_1 + a_2 x_2^2 + a_3 x_3^3 + a_4 x_4^4 \right] \right\} \tag{A4.14}$$

$$<r> = \frac{1}{2Z_{\text{eff}}} \left[3n^2 - l(l+1)\right], \tag{A4.15}$$

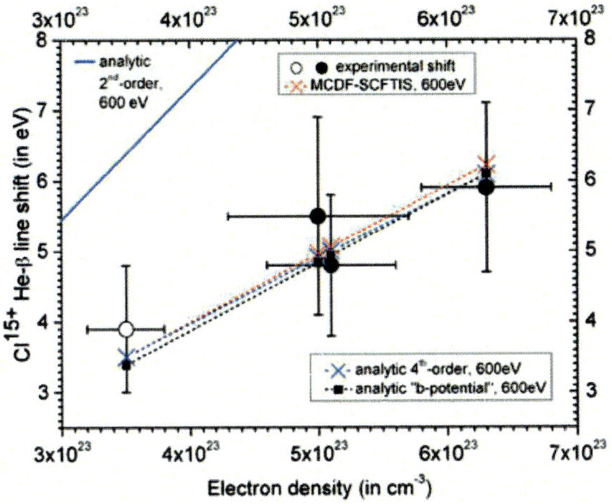

Figure. A.4.1 Comparison of line shift measurements of He_β of Cl^{15+} in dense laser produced plasma with different models for an electron temperature of $kT_e = 600$ eV: analytic 2nd-order expansion (blue solid line), analytic 4th-order expansion (blue crosses connected with blue dashed line, (A4.13–A4.20)), multi-configuration Dirac-Fock self-consistent finite temperature ion sphere model (red crosses connected with red dashed lines) [Li and Rosmej 2020] and the analytic b-potential method (solid black squares connected with dashed black line, (A4.1–A4.12))

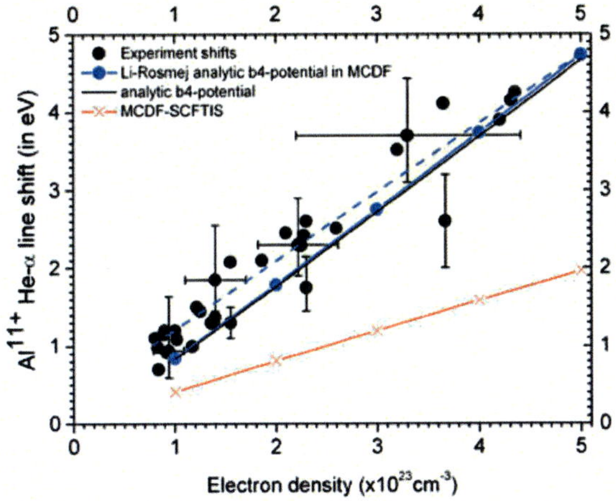

Figure. A.4.2 Comparison of line shift measurements of He_α of Al^{11+} in dense laser produced plasma with different models for $kT_e = 250$ eV: b4-potential in MCDF calculations (blue solid circles connected with blue solid line), analytic b4-potential line shift model (solid black line, (A4.1–A4.12)), MCDF-SCFTIS model (red crosses connected with solid red line). The dashed blue line includes apparent line shift corrections due to line overlap and other effects, see text

$$<r^2> = \frac{n^2}{2Z_{\mathrm{eff}}^2}\left[5n^2+1-3l(l+1)\right], \tag{A4.16}$$

$$<r^3> = \frac{n^2}{8Z_{\mathrm{eff}}^3}\left[35n^2(n^2-1)-30n^2(l+2)(l-1)+3(l+2)(l+1)l(l-1)\right], \tag{A4.17}$$

$$<r^4> = \frac{n^4}{8Z_{\mathrm{eff}}^4}\left[63n^4-35n^2(2l^2+2l-3)+5l(l+1)(3l^2+3l-10)+12\right], \tag{A4.18}$$

$$Z_{\mathrm{eff}} = n\sqrt{2\left|E_{nl}^{LSJ}\right|}, \tag{A4.19}$$

$$R_0 \approx 0.62035\cdot(Z/n_{\mathrm{i}}<Z>)^{1/3}. \tag{A4.20}$$

ΔE_{IPD} is the ionization potential depression energy in atomic units [units of $2Ry$], $1Ry = 13.6057eV$, $E_{\mathrm{nl}}^{\mathrm{LSJ}}$ is the ionization potential (in [eV]) of the optical electron with principal/orbital quantum numbers nl that moves in an effective Coulomb potential with charge Z_{eff} in the free atom/ion picture, z is the spectroscopic symbol, i.e. $z = Z + 1$ where Z is the ion charge $Z = Z_n - N_b$, n_i is the ion density in atomic units (i.e. in units of $1/a_0^3$, R_0 is the ion-sphere radius in atomic units (i.e. in units of $a_0 = 0.529177\cdot10^{-8}cm$) (vanishing wavefunction at $r = R_0$ has been assumed), $<Z>$ is the average charge of the plasma, $k_{\mathrm{B}}T_{\mathrm{e}}$ is the electron temperature in atomic units (i.e. in units of $2Ry$) and the fit parameters are $a_1 = 0.77383$, $a_2 = 0.36112$, $a_3 = -0.01605$, $a_4 = 0.000385182$.

From its conception the above representation is in form of an analytical Optical Electron Finite Temperature Ion Sphere (OEFTIS) model [Rosmej 2018] (see also Chapter 8): an effective potential is calculated from an effective charge Z_{eff} seen by the optical electron. The ionization potential depression formula is cast into an entirely analytical description that only needs energies and nl-quantum numbers of the free atom/ion configuration as it employs only matrix elements with integer exponents (i.e. $<r^1>$, $<r^2>$, $<r^3>$, $<r^4>$). The overall necessary data for application of the 4th-order formula are therefore readily available even for non-specialists in atomic physics.

As the above ionization potential depression formula is nl-quantum number dependent and takes into account the effective potential via Z^{eff} for each state (i.e. the exact ionization potential $E_{\mathrm{nl}}^{\mathrm{LSJ}}$), the formulas can readily be employed to calculate line shifts for the transition $j \rightarrow i$:

$$\Delta E_{\mathrm{shift}}(j \rightarrow i) = \Delta E_{\mathrm{IPD}}(j) - \Delta E_{\mathrm{IPD}}(i). \tag{A4.21}$$

Note, that the well-known and widely applied formulas for the ionization potential depression like the Stewart-Pyatt formula [Stewart and Pyatt 1996] and the Ecker-Kröll formula [Ecker and Kröll 1963] (see also Chapter 8) do not have the capacity to determine line shifts as upper and lower states do not depend on quantum numbers.

Let us apply the analytical 4th-order approximation for demonstration to the line shift in dense plasmas. We consider $He_\beta = 1s3p\,^1P_1 \rightarrow 1s^2\,^1S_0$ of Cl^{15+} for $kT_e = 600eV$, $<Z> = 15$, $n_e = n_i <Z>$, $n_e = 5 \cdot 10^{23} cm^{-3}$, $i = 1s^2\,^1S_0$, $E_i = 3658.34eV$, $j = 1s3p\,^1P_1$, $E_j = 386.69eV$ from which it follows (in atomic units): $Z = Z_n - N_b = 17 - 2 = 15$, $n_e = 0.07409$, $n_i = 6.734 \cdot 10^{-3}$, $R = 3.643$, $kT_e = 22.05$, $E_i = 134.44$, $Z_i^{eff} = 16.40$, $E_j = 14.211$, $Z_j^{eff} = 15.99$, $\langle r_i^1 \rangle = 9.1477 \cdot 10^{-2}$, $\langle r_i^2 \rangle = 1.1157 \cdot 10^{-2}$, $\langle r_i^3 \rangle = 1.7010 \cdot 10^{-3}$, $\langle r_i^4 \rangle = 3.1121 \cdot 10^{-4}$, $E_j = 14.211$, $Z_j^{eff} = 15.99$, $\left\langle r_j^1 \right\rangle = 7.8156 \cdot 10^{-1}$, $\left\langle r_j^2 \right\rangle = 7.0368 \cdot 10^{-1}$, $\left\langle r_j^3 \right\rangle = 6.9296 \cdot 10^{-1}$, $\left\langle r_j^4 \right\rangle = 7.3656 \cdot 10^{-1}$, $\Delta E_i = 6.407$, $\Delta E_j = 6.227$ and finally $\Delta E_{shift}(j \rightarrow i) = 6.227 - 6.407 = -0.180$, i.e. $\Delta E_{shift}(j \rightarrow i) = -4.90eV$ (red shift). This value (shown with blue cross in Fig. A.4.1) is in excellent agreement with the numerical calculations in the framework of the MCDF-SCFTIS (red crosses in Fig. A.4.1) [Li and Rosmej 2020]. For completeness, we mention the results from the 2nd-order expansion (see also Chapter 8): $\Delta E_{shift}^{2nd-order}(j \rightarrow i) = -9.14eV$.

The curve with blue crosses (connected with dashed blue line) in Figure A.4.1 shows the application of the 4th-order formulas ((A.4.13–A.4.21) to the line shift of $He_\beta = 1s3p\,^1P_1 \rightarrow 1s^2\,^1S_0$ of Cl^{15+} for $kT_e = 600eV$. Also very good agreement with the numerical calculations and the data is observed. The Figure depicts also the 2nd-order quantum number dependent approximation developed in [Li and Rosmej 2012]: it provides reasonable estimates of the line shift within about a factor of two. Although the 2nd-order approximation has limited precision, it has the advantage to allow more easily and partially analytic studies of scaling relations with respect to quantum numbers, effective charges and other parameters.

A.5 Atomic Units and Constants

	cgs	MKSA
Planck constant:	$\hbar = 1.05457 \cdot 10^{-27} erg\, s$	$1.05457 \cdot 10^{-34}$ Js
Boltzmann constant:	$k = 1.38066 \cdot 10^{-16} erg/K$	$1.38066 \cdot 10^{-23}$ J/K
Electron mass:	$m_e = 9.10939 \cdot 10^{-28} g$	$9.10939 \cdot 10^{-31}$ kg
Speed of light:	$c = 2.99792 \cdot 10^{10} cm/s$	$2.99792 \cdot 10^{8}$ m/s
Electron charge:	$e = 4.80321 \cdot 10^{-10} stat\, coulomb$	$1.60218 \cdot 10^{-19}$ C
Rydberg:	$Ry = \frac{m_e e^4}{2\hbar^2} = 2.17987 \cdot 10^{-11} erg$	$2.17987 \cdot 10^{-18}$ J
Bohr radius:	$a_0 = \frac{\hbar^2}{m_e e^2} = 5.29177 \cdot 10^{-9} cm$	$5.29177 \cdot 10^{-11}$ m
Fine structure constant:	$\alpha = \frac{e^2}{\hbar c} = 1/137.036$	1/137.036
Proton mass:	$m_p = 1.67262 \cdot 10^{-24} g$	$1.67262 \cdot 10^{-27}$ kg
Neutron mass:	$m_n = 1.67493 \cdot 10^{-24} g$	$1.67493 \cdot 10^{-27}$ kg
Atomic mass unit:	$u = 1.66054 \cdot 10^{-24} g$	$1.66054 \cdot 10^{-27}$ kg
Loschmidt number:	$L = 6.02274 \cdot 10^{23}$	$6.02274 \cdot 10^{23}$
Gas constant:	$R = 8.31451 \cdot 10^{7} erg\, mol^{-1} K^{-1}$	8.31451 J mol^{-1} K^{-1}
Electrical permittivity:	$\varepsilon_0 = 1$	$8.85419 \cdot 10^{-12}$ C^2/m^2N
Magnetic permittivity:	$\mu_0 = 1$	$4\pi \cdot 10^{-7}$ Vs/Am

Some useful relations:

$1J = 10^7 erg$, $1eV \hat{=} 11604.4\, K$, $1Ry = 13.60569253\, eV$, $1Ry \hat{=} 109.737315685\, cm^{-1}$, $Ry = 2.179870 \cdot 10^{-11} erg$, $1eV = 1.602176 \cdot 10^{-12} erg$, $1 stat\, Volt = 299.792\, Volt$ $hc/e = 1.239841930 \cdot 10^{-6}\, JmC^{-1}$, $e/hc = 8.06554429 \cdot 10^5\, J^{-1} m^{-1} C$, $E(1000\, cm^{-1}) \hat{=} E(eV)/8.065545$, $E(1000\, cm^{-1}) \hat{=} E(Ry)/109.7373$.

Atomic units:

Length:	$a_0 = \frac{\hbar^2 \varepsilon_0}{\pi m_e e^2} = 5.29177 \cdot 10^{-11}$ m	Energy:	$2Ry = 2 \cdot \frac{m_e e^4}{8\varepsilon_0^2 h^2} = 27.2114$ eV
Velocity:	$V_0 = \frac{e^2}{2h\varepsilon_0} = 2.18770 \cdot 10^6$ m/s	Time:	$T_0 = \frac{a_0}{V_0} = 2.41887 \cdot 10^{-17}$ s
E-Field:	$E_0 = \frac{e}{4\pi\varepsilon_0 a_0^2} = 5.14222 \cdot 10^{11}$ V/m	Mass:	$m_e = 9.10939 \cdot 10^{-31}$ kg
Angular momentum:	$L_0 = \hbar = m_e a_0 V_0 = 1.05457 \cdot 10^{-34}$ Js		
Speed of light:	$c_{a.u} = \frac{c_{MKSA}}{V_0} = \frac{1}{V_0} \cdot \frac{1}{\sqrt{\varepsilon_0 \mu_0}} = \frac{1}{\alpha} = 137.036$		

For more details on atomic constants see [Mohr et al. 2012].

References

P. Beiersdorfer, G. Brown, A. McKelvey, R. Shepherd, D. Hoarty, C. Brown, M. Hill, L. Hobbs, S. James, J. Morton, L. Wilson, *High resolution measurements of Cl^{+15} line shifts in hot, solid-density plasmas*, Phys. Rev. A **100**, 012511 (2019).

G. Ecker, W. Kröll, *Lowering of the ionization energy for a plasma in thermodynamic equilibrium*, Physics Fluids **6**, 62 (1963).

E.H. Guedda, M. Koubiti, F.B. Rosmej, R. Stamm, V.S. Lisitsa, *Dense Plasma Effects on atomic data and line emission of HeI for divertor plasma conditions*, Journal of Nuclear Materials **363–365**, 1420 (2007).

E.H. Guedda, F.B. Rosmej, P. Genesio, R. Stamm, L.A. Vainshtein, M.F. Gu, E. Lindroth, F. Khelfaoui, *Double excited high-n spin dependent atomic structure scaling laws for HeI: Application to radiative properties for edge plasma conditions*, Contrib. Plasma Physics **46**, 655 (2006).

R.K. Janev, L.P. Presnyakov, V.P. Shevelko, *Physics of Highly Charged Ions*, Springer, Berlin 1985.

X. Li, F.B. Rosmej, *Quantum number dependent energy level shifts of ions in dense plasmas: a generalized analytical approach*, Europhysics Letters **99**, 33001 (2012).

X. Li, F.B. Rosmej, V.A. Astapenko, V.S. Lisitsa, *An analytical plasma screening potential based on the Self-Consistent-Field Ion-Sphere model*, Phys. Plasmas **26**, 03301 (2019).

X. Li, F. B. Rosmej, Analytical approach to level delocalization and line shifts in finite temperature dense plasmas, Phys. Lett A **384**, 126478 (2020).

P.J. Mohr, B.N. Taylor, D.B. Newell, *CODATA Recommended Values of the Fundamental Physical Constants: 2010*, Rev. Mod. Phys. **84**, 1527 (2012).

F. B. Rosmej, *Ionization potential depression in an atomic-solid-plasma picture*, Letter J. Phys. B: Atomic molecular and optical physics **51**, 09LT01 (2018).

F.B. Rosmej, E.H. Guedda, R. Stamm, *Universal high-n scaling law for intercombination transitions in helium*, J. Phys. B: At. Mol. Opt. Phys. **40**, 49 (2007).

F. B. Rosmej, K. Bennadji, V. Lisitsa, *Effect of dense plasmas on exchange-energy shifts in highly charged ions: an alternative approach for arbitrary perturbation potentials*, Phys. Rev. A **84**, 032512 (2011).

J. Stewart, K. Pyatt, *Lowering of Ionization Potentials in Plasmas*, Astrophys. J. **144**, 1203 (1996).

C. Stillman, P. Nilson, S. Ivancic, I. Golovkin, C. Mileham, I. Begishev, D. Froula, *Picosecond time-resolved measurements of dense plasma line shifts*, Phys. Rev. E **95** (2017) 063204.

Index

F. B. Rosmej et al., *Plasma Atomic Physics*, Springer Series on Atomic, Optical, and Plasma Physics 104, https://doi.org/10.1007/978-3-030-05968-2